DATE DUE FOR RETURN

The loan period may be shortened if the item is requested.

EVOLUTION

THIRD EDITION

MARK RIDLEY

Blackwell Publishing

© 2004 by Blackwell Science Ltd
a Blackwell Publishing company

350 Main Street, Malden, MA 02148-5020, USA
108 Cowley Road, Oxford OX4 1JF, UK
550 Swanston Street, Carlton, Victoria 3053, Australia

First published 1993
Second edition 1996
Third edition 2004

Library of Congress Cataloging-in-Publication Data
Ridley, Mark.
        Evolution / Mark Ridley. – 3rd ed.
                    p.   cm.
Includes bibliographical references (p.   ).
        ISBN 1-4051-0345-0 (pbk.: alk. paper)
    1. Evolution (Biology) I. Title.
QH366.2.R524 2003
576.8–dc21   2003000140

A catalogue record for this title is available from the British Library.

Set in 9.5/12pt Minion
by Graphicraft Limited, Hong Kong
Printed and bound in Italy
by G. Canale & C. SpA, Turin

1003361399

For further information on
Blackwell Publishing, visit our website:
http://www.blackwellpublishing.com

Brief Contents

PART 5.   MACROEVOLUTION   521

Color plate section between pp. 70 and 71

Full Contents

PART 5.   MACROEVOLUTION   521

18.   The History of Life   523

19.   Evolutionary Genomics   556

20. Evolutionary Developmental Biology 572

21. Rates of Evolution 590

22.   Coevolution   613

23.   Extinction and Radiation   643

Color plate section between pp. 70 and 71

Preface

The theory of evolution is outstandingly the most important theory in biology, and it is always a pleasure to be a member, facing in either direction, of the class that is fortunate enough to be studying it. No other idea in biology is so powerful scientifically, or so stimulating intellectually. Evolution can add an extra dimension of interest to the most appealing sides of natural history — we shall see, for example, how modern evolutionary biologists tend to argue that the existence of sex is the profoundest puzzle of all, and quite possibly a mistake that half the living creatures of this planet would be better off without. Evolution gives meaning to the drier facts of life too, and it is one of the delights of the subject to see how there are ideas as well as facts within the disorienting technicalities of the genetics laboratory, and how deep theories about the history of life can hinge on measurements of the width of a region called "prodissoconch II" in the larval shell of a snail, or the number of ribs in a trilobite's tail. So great is the depth and range of evolutionary biology that every other classroom on campus must feel (as you can sort of tell) locked in with more superficial and ephemeral materials.

The theory of evolution, as I have arranged it here, has four main components. Population genetics provides the fundamental theory of the subject. If we know how any property of life is controlled genetically, population genetics can be applied to it directly. We have that knowledge particularly for molecules (together with some, mainly morphological, properties of whole organisms), and molecular evolution and population genetics are therefore well integrated. Part 2 of the book considers them together. The second component is the theory of adaptation, and it is the subject of Part 3. Evolution is also the key to understanding the diversity of life, and in Part 4 we consider such topics as what a species is, how new species originate, and how to classify and reconstruct the history of life. Finally, Part 5 is about evolution on the grand scale — on a timescale of tens or hundreds of millions of years. We look at the history of life, both genetically and paleontologically, at rates of evolution, and at mass extinctions.

Controversy is always tricky to deal with in an introductory text, and evolutionary biology has more than its fair share of it. When I have come to a controversial topic, my first aim has been to explain the competing ideas in such a way that they can be understood on their own terms. In some cases (such as cladistic classification) I think the controversy is almost settled and I have taken sides. In others (such as the relative empirical importance of gradual and punctuated change in fossils) I have not. I am well aware that not everyone will agree with the positions I have taken, or indeed with my decisions in some cases not to take a position; but in a way these are secondary matters. The book's success mainly depends on how well it enables a reader who has not studied the subject much before to understand the various ideas and come to a sensible viewpoint about them.

The great (or at any rate, one of the great) events in evolutionary biology as I have been writing the third edition is the way genetics is becoming a macroevolutionary, as well as a microevolutionary, subject. Historically, there has been a good working distinction between evolutionary research on short and long timescales — between micro- and macroevolutionary research. The distinction was one not simply of timescales but of research methods and even institutionalized academic disciplines. Genetics, and experimental methods generally, were used to study evolution on the timescale of research projects — of a few years, at most. That work was done mainly in departments of biology. Long-term evolution, over approximately 10–1,000 million years, was studied by comparative morphology in living and fossil life forms. That work was done more in museums and departments of geology or earth sciences, than in biology departments.

I see the distinction between micro- and macroevolutionary research as breaking down, in perhaps three ways. The first is through the use of molecular phylogenetics. A phylogeny is a family tree for a group of species, and they were classically inferred from morphological evidence. Molecular evidence started to be used in the 1960s, but it somehow trapped itself (I caricature a little) in about 20 years of obsessive behavior, as a small number of case studies — particularly human evolution — were endlessly rehashed. Molecular phylogenetics broke out into life as a whole during the 1980s, and the result has been a huge increase in the number of species for which we know, or have evidence concerning, their phylogenetic relations.

The research program of molecular phylogenetics may have been established for almost an academic generation, and it is certainly flourishing, but it has still only just begun. A recent estimate is that only about 50,000 of the 1.75 million or so described species have been put in any kind of "minitree" — that is, a phylogenetic tree with their close relations. Sydney Brenner has remarked that the next generation of biologists has the prospect of finding the tree of life, something that all previous generations of post-Darwinian biologists could only dream about. In Chapter 15, we look at how the work is being done. The new phylogenetic knowledge is not only interesting in itself, but is also enabling many other kinds of work that were formerly impossible. We shall see how phylogenies are being exploited in studies of coevolution and biogeography, among other topics.

The other two ways in which molecular genetics is being used in macroevolutionary research are more recent. I have added chapters on evolutionary genomics (Chapter 19) and "evo-devo" (Chapter 20). The addition of these two chapters in Part 5 of the book is a small symbol of the way macroevolution has become genetic as well as paleobiological: in my first two editions, Part 5 was almost exclusively paleontological. The introduction of new techniques into the study of macroevolution creates an excitement of its own. It has also resulted in a number of controversies, where the two methods (molecular genetic and paleontological) seem to point to conflicting conclusions. We shall look at several of those controversies, including the nature of the Cambrian explosion and the significance of the Cretaceous–Tertiary mass extinction.

This book is about evolution as a "pure" science, but that science has practical applications — in social affairs, in business, in medicine. Stephen Palumbi has recently estimated that evolutionary change induced by human action costs the US economy about $33–50 billion a year (Palumbi 2001a). The costs come from the way microbes evolve resistance to drugs, pests evolve resistance to pesticides, and fish evolve back against our fishing procedures. Palumbi's estimate is approximate and preliminary, and probably an underestimate. But whatever the exact number is, the economic consequences of evolution must be huge. The economic benefits of understanding evolution could be proportionally huge. In this edition I have added a number of special boxes within chapters, on "Evolution and human affairs." The examples I discuss are only a sample, which happen to fit in with themes in the text. Bull & Wichman (2001) discuss many further examples of "applied evolution," from directed evolution of enzymes to evolutionary computation.

The book is intended as an introductory text, and I have subordinated all other aims to that end. I have aimed to explain concepts, wherever possible by example, and with a minimum of professional clutter. The principal interest, I believe, of the theory of evolution is as a set of ideas to think about, and I have therefore tried in every case to move on to the ideas as soon as possible. The book is not a factual encyclopedia, nor (primarily) a reference work for research biologists. I do not provide many references in the main text, though in this edition I have referred to the sources for the examples in formal "scientific" reference format. For readers who are unfamiliar with this format, I should say that references are given in the way I wrote "Palumbi (2001a)" and "Bull & Wichman (2001)" in the previous paragraph. The reference has the author's (or authors') name and a date. In the reference list at the end you will find the full bibliographic details, listing the authors alphabetically. There is also a convention for papers with multiple authors. When a paper has more than two (or three, with some publishers), it is referred to by the name of the first author with an "*et al.*" and the date: Losos *et al.* (1998), for example. The "*et al.*" means "and others." It is a space-saving device, and is abbreviated to avoid problems with Latin declension. For instance, if the reference is the subject of the sentence the full version would be "Losos *et alii* (1998) studied lizards. . . ." But other phases require other full versions: "the work of Losos *et aliorum* (1998) . . ." or "the work by Losos *et aliis* (1998)." In all, "*al.*" could stand for 12–18 full versions. Anyhow, all the authors are usually listed in the main reference list — I say "usually" because some authorial teams have grown so huge that they are not all given. Blackwell house style is that for papers with more than seven authors, the reference list has the first three and then an "*et al.*"

Although I have referred to the specific papers under discussion in the text, I do not give general references there. The reason is that I do not want to spoil the most powerful textual positions, such as the end of a paragraph or a section, with a list of further reading. The way I have things, those textual positions can be occupied by summary sentences and other more useful matter. The "further reading" section at the end of each chapter is the main vehicle for general references, and for references to other studies like those in the text. I have referred to recent reviews when they exist, and the historic bibliography of each topic can be traced through them.

In summary, this new edition contains:

- **two types of box** — one featuring practical applications and the other related information, which supply added depth without interrupting the flow of the text
- **margin comments** that paraphrase and highlight key concepts
- **study and review questions** to help students review their understanding at the end of each chapter, while new challenge questions prompt students to synthesize the chapter concepts to reinforce the learning at a deeper level

- **two new chapters** — one on evolutionary genomics and one on evolution and development bring state-of-the-art information to the coverage of evolutionary study

There is also **a dedicated website** at **www.blackwellpublishing.com/ridley** which provides an interactive experience of the book, with illustrations downloadable to PowerPoint, and a full supplemental package complementing the book. Scattered **margin icons** indicate where there is relevant information included on the dedicated website.

Finally, my thanks to the many people who have helped me with queries and reviews as I have prepared the new edition — Theodore Garland, University of Wisconsin; Michael Whiting, Brigham Young University; William Brown, SUNY Fredonia; Geoff Oxford, University of York; C.P. Kyriacou, Leicester University; Chris Austin, University of North Dakota; David King, University of Illinois; Paul Spruell, University of Montana; Daniel J. O'Connell, University of Texas, Arlington; Susan J. Mazer, University of California, Santa Barbara; Greg C. Nelson, University of Oregon — and to those students (now "Evil Syst" at Oxford rather than ANT 362 or BIO 462 at Emory) who, perhaps not on purpose, inspire much of the writing.

Mark Ridley

Part one

Introduction

When Darwin put forward his theory of evolution by natural selection, he lacked a satisfactory theory of inheritance, and the importance of natural selection was widely doubted until it was shown in the 1920s and 1930s how natural selection could operate with Mendelian inheritance. The two key events in the history of evolutionary thought are therefore Darwin's discovery of evolution by natural selection and the synthesis of Darwin's and Mendel's theories — a synthesis variously called the modern synthesis, the synthetic theory of evolution, and neo-Darwinism. Chapter 1 discusses the rise of evolutionary theory historically, and introduces some of its main figures. During the twentieth century, the sciences of evolutionary biology and genetics have developed together and some knowledge of genetics is essential for understanding the modern theory of evolution. Chapter 2 provides an elementary review of the main genetic mechanisms. In Chapter 3, we move on to consider the evidence for evolution — the evidence that species have evolved from other, ancestral species rather than having separate origins and remaining forever fixed in form. The classic case for evolution was made in Darwin's *On the Origin of Species* and his general arguments still apply; but it is now possible to use more recent molecular and genetic evidence to illustrate them. Chapter 4 introduces the concept of natural selection. It considers the conditions for natural selection to operate, and the main kinds of natural selection. One crucial condition is that the population should be variable, that is, individuals should differ from one another; the chapter shows that variation is common in nature. New variants originate in mutation. Chapter 2 reviews the main kinds of mutation, and how mutation rates are measured. Chapter 4 looks at how mutations contribute to variation, and discusses why mutation can be expected to be adaptively undirected.

1

The Rise of Evolutionary Biology

The chapter first defines biological evolution, and contrasts it with some related but different concepts. It then discusses, historically, the rise of modern evolutionary biology: we consider Darwin's main precursors; Darwin's own contribution; how Darwin's ideas were received; and the development of the modern "synthetic theory" of evolution.

1.1 Evolution means change in living things by descent with modification

Evolutionary biology is a large science, and is growing larger. A list of its various subject areas could sound rather daunting. Evolutionary biologists now carry out research in some sciences, like molecular genetics, that are young and move rapidly, and in others like morphology and embryology, that have accumulated their discoveries at a more stately speed over a much longer period. Evolutionary biologists work with materials as diverse as naked chemicals in test tubes, animal behavior in the jungle, and fossils collected from barren and inhospitable rocks.

Evolution is a big theory in biology

However, a beautifully simple and easily understood idea — evolution by natural selection — can be scientifically tested in all these fields. It is one of the most powerful ideas in all areas of science, and is the only theory that can seriously claim to unify biology. It can give meaning to facts from the invisible world in a drop of rain water, or from the many colored delights of a botanic garden, to thundering herds of big game. The theory is also used to understand such topics as the geochemistry of life's origins and the gaseous proportions of the modern atmosphere. As Theodosius Dobzhansky, one of the twentieth century's most eminent evolutionary biologists, remarked in an often quoted but scarcely exaggerated phrase, "nothing in biology makes sense except in the light of evolution" (Dobzhansky 1973).

Evolution means change, change in the form and behavior of organisms between generations. The forms of organisms, at all levels from DNA sequences to macroscopic morphology and social behavior, can be modified from those of their ancestors during evolution. However, not all kinds of biological change are included in the definition (Figure 1.1). Developmental change within the life of an organism is not evolution in the strict sense, and the definition referred to evolution as a "change between generations" in order to exclude developmental change. A change in the composition of an ecosystem, which is made up of a number of species, would also not normally be counted as evolution. Imagine, for example, an ecosystem containing 10 species. At time 1, the individuals of all 10 species are, on average, small in body size; the average member of the ecosystem is therefore "small." Several generations later, the ecosystem may still contain 10 species, but only five of the original small species remain; the other five have gone extinct and have been replaced by five species with large-sized individuals, that have immigrated from elsewhere. The average size of an individual (or species) in the ecosystem has changed, even though there has been no evolutionary change within any one species.

Evolution can be defined . . .

Most of the processes described in this book concern change between generations within a population of a species, and it is this kind of change we shall call evolution. When the members of a population breed and produce the next generation, we can imagine a *lineage* of populations, made up of a series of populations through time. Each population is ancestral to the descendant population in the next generation: a lineage is an "ancestor–descendant" series of populations. Evolution is then change between generations within a population lineage. Darwin defined evolution as "descent with modification," and the word "descent" refers to the way evolutionary modification takes place in a series of populations that are descended from one

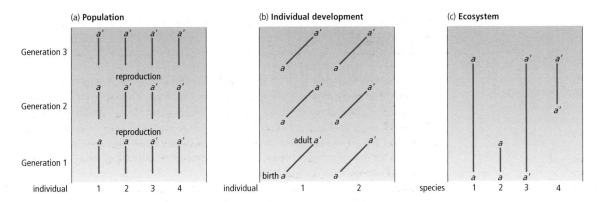

Figure 1.1

Evolution refers to change within a lineage of populations between generations. (a) Evolution in the strict sense of the word. Each line represents one individual organism, and the organisms in one generation are reproduced from the organisms in the previous generation. The composition of the population has changed, evolutionarily, through time. The letter a' represents a different form of the organism from a. For instance, a organisms might be smaller in size than a' organisms. Evolution has then been in the direction of increased body size. (b) Individual developmental change is not evolution in the strict sense. The composition of the population has not changed between generations and the developmental changes (from a to a') of each organism are not evolutionary. (c) Change in an ecosystem is not evolution in the strict sense. Each line represents one species. The average composition of the ecosystem changes through time: from $2a : 1a'$ at generation 1 to $1a : 2a'$ at generation 3. But within each species there is no evolution.

another. Recently, Harrison (2001) defined evolution as "change over time via descent with modification."

. . . and has distinct properties

Evolutionary modification in living things has some further distinctive properties. Evolution does not proceed along some grand, predictable course. Instead, the details of evolution depend on the environment that a population happens to live in and the genetic variants that happen to arise (by almost random processes) in that population. Moreover, the evolution of life has proceeded in a branching, tree-like pattern. The modern variety of species has been generated by the repeated splitting of lineages since the single common ancestor of all life.

Changes that take place in human politics, economics, history, technology, and even scientific theories, are sometimes loosely described as "evolutionary." In this sense, evolutionary means mainly that there has been change through time, and perhaps not in a preordained direction. Human ideas and institutions can sometimes split during their history, but their history does not have such a clear-cut, branching, tree-like structure as does the history of life. Change, and splitting, provide two of the main themes of evolutionary theory.

1.2 Living things show adaptations

Adaptation is another of evolutionary theory's crucial concepts. Indeed, it is one of the main aims of modern evolutionary biology to explain the forms of adaptation that we find in the living world. *Adaptation* refers to "design" in life — to those properties of living things that enable them to survive and reproduce in nature. The concept is easiest to understand by example. Many of the attributes of a living organism could be used to illustrate the concept of adaptation, because many details of the structure, metabolism, and behavior of an organism are well designed for life.

Examples exist of adaptation

The woodpecker provided Darwin's favorite examples of adaptation. The woodpecker's most obvious adaptation is its powerful, characteristically shaped beak. It enables the woodpecker to excavate holes in trees. They can thus feed on the year-round food supply of insects that live under bark, insects that bore into the wood, and the sap of the tree itself. Tree holes also make safe sites to build a nest. Woodpeckers have many other design features as well as their beaks. Within the beak is a long, probing tongue, which is well adapted to extract insects from inside a tree hole. They have a stiff tail that is used as a brace, short legs, and their feet have long curved toes for gripping on to the bark; they even have a special type of molting in which the strong central pair of feathers (that are crucial in bracing) are saved and molted last. The beak and body design of the woodpecker is adaptive. The woodpecker is more likely to survive, in its natural habitat, by possessing them.

Camouflage is another, particularly clear, example of adaptation. Camouflaged species have color patterns and details of shape and behavior that make them less visible in their natural environment. Camouflage assists the organism to survive by making it less visible to its natural enemies. Camouflage is adaptive. Adaptation, however, is not an isolated concept referring to only a few special properties of living things — it applies to almost any part of the body. In humans, hands are adapted for grasping, eyes for seeing, the alimentary canal for digesting food, legs for movement: all these functions assist us to survive. Although most of the obvious things we notice are adaptive, not every detail of an organism's form and behavior is necessarily adaptive (Chapter 10).

Adaptation has to be explained, . . .

Adaptations are, however, so common that they have to be explained. Darwin regarded adaptation as the key problem that any theory of evolution had to solve. In Darwin's theory — as in modern evolutionary biology — the problem is solved by natural selection.

. . . and is, by natural selection

Natural selection means that some kinds of individual in a population tend to contribute more offspring to the next generation than do others. Provided that the offspring resemble their parents, any attribute of an organism causing it to leave more offspring than average will increase in frequency in the population over time. The composition of the population will then change automatically. Such is the simple, but immensely powerful, idea whose ramifying consequences we shall be exploring in this book.

1.3 A short history of evolutionary biology

We shall begin with a brief sketch of the historic rise of evolutionary biology, in four main stages:

1. Evolutionary and non-evolutionary ideas before Darwin.
2. Darwin's theory (1859).
3. The eclipse of Darwin (*c.* 1880–1920).
4. The modern synthesis (1920s to 1950s).

1.3.1 *Evolution before Darwin*

The history of evolutionary biology really begins in 1859, with the publication of Charles Darwin's *On the Origin of Species*. However, many of Darwin's ideas have an older pedigree. The most immediately controversial claim in Darwin's theory is that species are not permanently fixed in form, but that one species evolves into another. ("Fixed" here means unchanging.) Human ancestry, for instance, passes through a continuous series of forms leading back to a unicellular stage. Species fixity was the orthodox belief in Darwin's time, though that does not mean that no one then or before had questioned it. Naturalists and philosophers a century or two before Darwin had often speculated about the transformation of species. The French scientist Maupertuis discussed evolution, as did *encyclopédistes* such as Diderot. Charles Darwin's grandfather, Erasmus Darwin, is another example. However, none of these thinkers put forward anything we would now recognize as a satisfactory theory to explain why species change. They were mainly interested in the factual possibility that one species might change into another.

Evolutionary thinkers existed before Darwin, but either lacked, . . .

The question was brought to an issue by the French naturalist Jean-Baptiste Lamarck (1744–1829). The crucial work was his *Philosophie Zoologique* (1809), in which he argued that species change over time into new species. The way in which he thought species changed was importantly different from Darwin's and our modern idea of evolution. Historians prefer the contemporary word "transformism" to describe Lamarck's idea.[1]

Figure 1.2 illustrates Lamarck's conception of evolution, and how it differs from Darwin's and our modern concept. Lamarck supposed that lineages of species persisted indefinitely, changing from one form into another; lineages in his system did not branch and did not go extinct. Lamarck had a two-part explanation of why species change. The principal mechanism was an "internal force" — some sort of unknown mechanism within an organism causing it to produce offspring slightly different from itself, such that when the changes had accumulated over many generations the lineage would be visibly transformed, perhaps enough to be a new species.

. . . or proposed, unsatisfactory mechanisms to drive evolution

Lamarck's second (and possibly to him less important) mechanism is the one he is now remembered for: the inheritance of acquired characters. Biologists use the word "character" as a short-hand for "characteristic." A character is any distinguishable property of an organism; it does not here refer to character in the sense of personality. As an organism develops, it acquires many individual characters, in this biological sense, due to its particular history of accidents, diseases, and muscular exercises. Lamarck suggested that a species could be transformed if these individually acquired modifications were inherited by the individual's offspring. In his famous discussion of the giraffe's neck, he argued that ancestral giraffes had stretched to reach leaves higher

---

[1]   The historic change in the meaning of the term "evolution" is a fascinating story in itself. Initially, it meant something more like what we mean by development (as in growing up from an egg to an adult) than by evolution: an unfolding of predictable forms in a preprogramed order. The course of evolution, in the modern sense, is not preprogramed; it is unpredictable in much the same way that human history is unpredictable. The change of meaning occurred around the time of Darwin; he did not use the word in *The Origin of Species* (1859), except in the form "evolved," which he used once as the last word in the book. However, he did use it in *The Expression of the Emotions* (1872). It took a long time for the new meaning to become widespread.

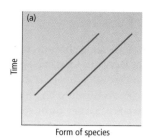

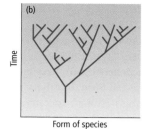

Figure 1.2

(a) Lamarckian "transformism," which differs in two crucial respects from evolution as Darwin imagined it. (b) Darwinian evolution is tree-like, as lineages split, and allows for extinction.

up trees. The exertion caused their necks to grow slightly longer. Their longer necks were inherited by their offspring, who thus started life with a propensity to grow even longer necks than their parents. After many generations of neck stretching, the result was what we can now see. Lamarck described the process as being driven by the "striving" of the giraffe, and he often described animals as "wishing" or "willing" to change themselves. His theory has, therefore, sometimes been caricatured as suggesting that evolution happens by the will of the organism. However, the theory does not require any conscious striving on the part of the organism — only some flexibility in individual development and the inheritance of acquired characters.

Lamarck did not invent the idea of the inheritance of acquired characters. The idea is ancient — it was discussed in ancient Greece by Plato, for example. However, most modern thinking about the role of the process in evolution has been inspired by Lamarck, and the inheritance of acquired characters is now conventionally, if unhistorically, called Lamarckian inheritance.

Lamarck, as a person, lacked the genius for making friends, and his main rival, the anatomist Georges Cuvier (1769–1832), knew how to conduct a controversy. Lamarck had broad interests, in chemistry and meteorology as well as biology, but his contributions did not always receive the attention he felt they deserved. By 1809, Lamarck had already persuaded himself that there was a conspiracy of silence against his ideas. The meteorologists ignored his weather forecasting system, the chemists ignored his chemical system, and when the *Philosophie Zoologique* (Lamarck 1809) was finally published, Cuvier saw to it that this, too, was greeted with silence. However, in reality it was an influential book. It was at least partly in reaction to Lamarck that Cuvier and his school made a belief in the fixity of species a virtual orthodoxy among professional biologists. Cuvier's school studied the anatomy of animals to discover the various fundamental plans according to which the different types of organism were designed. Cuvier in this way established that the animal kingdom had four main branches (called *embranchements* in French): vertebrates, articulates, mollusks, and radiates. A slightly different set of main groups is recognized in modern biology, but the modern groupings do not radically contradict Cuvier's four-part system. Cuvier also established, contrary to Lamarck's belief, that species had gone extinct (Section 23.2, p. 646).

Lamarck's ideas mainly became known in Britain through a critical discussion by the British geologist Charles Lyell (1797–1875). Lyell's book *Principles of Geology* (1830–33) had a wide influence, and incidentally criticized Lamarck (though Lamarckism was not the main theme of the book). Cuvier's influence came more through Richard Owen (1804–1892), who had studied with Cuvier in Paris before returning to England. Owen

Most biologists in the years just before Darwin, accepted that species do not evolve

Figure 1.3
Charles Robert Darwin (1809–82), in 1840.

became generally thought of as Britain's leading anatomist. By the first half of the nineteenth century, most biologists and geologists had come to accept Cuvier's view that each species had a separate origin, and then remained constant in form until it went extinct.

1.3.2 *Charles Darwin*

Meanwhile, Charles Darwin (Figure 1.3) was forming his own ideas. Darwin, after graduating from Cambridge, had traveled the world as a naturalist on board the *Beagle* (1832–37). He then lived briefly in London before settling permanently in the country. His father was a successful doctor, and his father-in-law controlled the Wedgwood china business; Charles Darwin was a gentleman of independent means. The crucial period of his life, for our purposes, was the year or so after the *Beagle* voyage (1837–38). As he worked over his collection of birds from the Galápagos Islands, he realized that he should have recorded which island each specimen came from, because they varied from island to island. He had initially supposed that the Galápagos finches were all one species, but it now became clear that each island had its own distinct species. How easy to imagine that they had evolved from a common ancestral finch! He was similarly struck by the way the ostrich-like birds called rheas differed between one region and another in South America. These observations of geographic variation probably first led Darwin to accept that species can change.

The next important step was to invent a theory to explain why species change. The notebooks Darwin kept at the time still survive. They reveal how he struggled with several ideas, including Lamarckism, but rejected them all because they failed to explain

Darwin developed evolutionary views . . .

a crucial fact — adaptation. His theory would have to explain not only why species change, but also why they are well designed for life. In Darwin's own words (in his autobiography):

. . . looked for a mechanism . . .

> It was equally evident that neither the action of the surrounding conditions, nor the will of the organisms [an allusion to Lamarck], could account for the innumerable cases in which organisms of every kind are beautifully adapted to their habits of life — for instance a woodpecker or tree-frog to climb trees, or a seed for dispersal by hooks or plumes. I had always been much struck by such adaptations, and until these could be explained it seemed to me almost useless to endeavour to prove by indirect evidence that species have been modified.

Darwin came upon the explanation while reading Malthus's *Essay on Population*. He continued:

> In October 1838, that is fifteen months after I had begun my systematic enquiry, I happened to read for amusement 'Malthus on population', and being well prepared to appreciate the struggle for existence which everywhere goes on from long-continued observation of the habits of animals and plants, it at once struck me that under these circumstances favourable variations would tend to be preserved and unfavourable ones to be destroyed. The result of this would be the formation of a new species.

. . . and discovered natural selection

Because of the struggle for existence, forms that are better adapted to survive will leave more offspring and automatically increase in frequency from one generation to the next. As the environment changes through time (for example, from humid to arid), different forms of a species will be better adapted to it than were the forms in the past. The better adapted forms will increase in frequency, and the now poorly adapted forms will decrease in frequency. As the process continues, eventually (in Darwin's words) "the result of this would be the formation of a new species." This process provided Darwin with what he called "a theory by which to work." And he started to work. He was still at work, fitting facts into his theoretical scheme, 20 years later when he received a letter from another traveling British naturalist, Alfred Russel Wallace (Figure 1.4). Wallace had independently arrived at a very similar idea to Darwin's natural selection. Darwin's friends, Charles Lyell and Joseph Hooker (Figure 1.5a), arranged for a simultaneous announcement of Darwin and Wallace's idea at the Linnean Society in London in 1858. By then Darwin was already writing an abstract of his full findings: that abstract is the scientific classic *On the Origin of Species*.

1.3.3 *Darwin's reception*

The reactions to Darwin's two connected theories — evolution and natural selection — differed. The idea of evolution itself become controversial mainly in the popular sphere only, rather than among biologists. Evolution seemed to contradict the Bible, in which the various kinds of living things are said to have been created separately. In Britain, Thomas Henry Huxley (Figure 1.5b) particularly defended the new evolutionary view against religious attack.

Figure 1.4
Alfred Russel Wallace (1823–1913), photographed in 1848.

Evolution was less controversial among professional scientists. Many biologists came almost immediately to accept evolution. The new theory in some cases made remarkably little difference to day-to-day biological research. The kind of comparative anatomy practiced by the followers of Cuvier, including Owen, lent itself equally well to a post-Darwinian search for pedigrees as to the pre-Darwinian search for "plans" of nature. The leading anatomists were by now mainly German. Carl Gegenbauer (1826–1903), one of the major figures, had soon reorientated his work to the tracing

Figure 1.5
Darwin's British supporters: (a) Joseph Dalton Hooker (1817–1911) on a botanical expedition in Sikkim in 1849 (after a sketch by William Tayler), and (b) Thomas Henry Huxley (1825–95). Darwin called Huxley "my general agent."

(a)   (b)

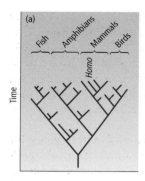
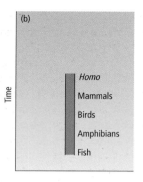

Figure 1.6

(a) Darwin's theory suggests that evolution has proceeded as a branching tree; note that it is arbitrary where *Homo* is positioned across the top of the diagram. *Homo* is often placed at the extreme right, but does not have to be. The tree should be contrasted with the popular idea (b) that evolution is a one-dimensional progressive ascent of life. Darwinian evolution is more like a tree than a ladder (cf. Figure 1.2).

of evolutionary relationships between animal groups. The famous German biologist Ernst Haeckel (1834–1919) vigorously investigated the same problem, as he applied his "biogenetic law" — the theory of recapitulation (which we shall meet in Section 20.2, p. 573) — to reveal phylogenetic pedigrees.

Although some kind of evolution was widely accepted among biologists, probably few of those biologists shared Darwin's own idea of it. In Darwin's theory, evolution is not inherently or automatically progressive. The local conditions at each stage mainly determine how a species evolves. The species does not have an inherent tendency to rise to a higher form. If Darwinian evolution does proceed in a progressive way, in some sense, then that is just how things turned out. Most evolutionists of the late nineteenth and early twentieth centuries had a different conception of evolution from this. They imagined evolution instead as one-dimensional and progressive. They often concerned themselves with thinking up mechanisms to explain why evolution should have an unfolding, predictable, progressive pattern (Figure 1.6).

While evolution — of a sort — was being accepted, natural selection was just as surely being rejected. People disliked the theory of natural selection for many reasons. This first chapter is not going to explain the arguments in any depth. What follows here is only an introduction to the history of the ideas that we shall consider in more detail in later chapters.

One of the more sophisticated objections to Darwin's theory was that it lacked a satisfactory theory of heredity. There were various theories of inheritance at that time, and all of them are now known to be wrong. Darwin preferred a "blending" theory of inheritance, in which the offspring blend their parental attributes; for example, if a red male mated with a white female, and inheritance "blended," the offspring would be pink. One of the deepest hitting criticisms of the theory of natural selection pointed out that it could hardly operate at all if heredity blended (Section 2.9, p. 37).

At a more popular level, many objections were raised against natural selection. One was that natural selection explains evolution by chance. This was (and still is) a misunderstanding of natural selection, which is a non-random process. Almost every chapter in this book after Chapter 4 illustrates how natural selection is non-random, but the topic is particularly discussed in Chapters 4 and 10. Chapters 6–7 discuss an evolutionary process called random drift. Random drift is random, but it is a completely different process from natural selection.

Evolution was accepted, but often confused with progressive change

Natural selection was widely rejected . . .

A second objection was that gaps exist between forms in nature — gaps that could not be crossed if evolution was powered by natural selection alone. The anatomist St George Jackson Mivart (1827–1900), for instance, in his book *The Genesis of Species* (1871), listed a number of organs that would not (he thought) be advantageous in their initial stages. In Darwin's theory, organs evolve gradually, and each successive stage has to be advantageous in order that it can be favored by natural selection. Mivart retorted that although for a bird a fully formed wing, for example, is advantageous, the first evolutionary stage — of a tiny protowing — might not be.

. . . leading to the development of theories of directed variation

Biologists who accepted the criticism sought to get round the difficulty by imagining processes other than selection that could work in the early stages of a new organ's evolution. Most of these processes belong to the class of theories of "directed mutation," or directed variation. These theories suggest that the offspring, for some unspecified reason to do with the hereditary mechanism, consistently tend to differ from their parents in a certain direction. In the case of wings, the explanation by directed variation would say that the wingless ancestors of birds somehow tended to produce offspring with protowings, even though there was no advantage to it. (Chapter 10 deals with this general question, and Chapter 4 discusses variation.)

Lamarckian inheritance was the most popular theory of directed variation. Variation is "directed" in this theory because the offspring tend to differ from their parents in the direction of characteristics acquired by their parents. If the parental giraffes all have short necks and acquire longer necks by stretching, their offspring have longer necks to begin with, before any elongation by stretching. Darwin accepted that acquired characters can be inherited. He even produced a theory of heredity ("my much abused hypothesis of pangenesis," as he called it) that incorporated the idea. In Darwin's time, the debate was about the relative importance of natural selection and the inheritance of acquired characteristics; but by the 1880s the debate moved into a new stage. The German biologist August Weismann (1833–1914) then produced strong evidence and theoretical arguments that acquired characteristics are not inherited. After Weismann, the question became whether Lamarckian inheritance had any influence in evolution at all. Weismann initially suggested that practically all evolution was driven by natural selection, but he later retreated from this position.

Weismann was a rare early supporter of the theory of natural selection

Around the turn of the century, Weismann was a highly influential figure, but few biologists shared his belief in natural selection. Some, such as the British entomologist Edward Bagnall Poulton, were studying natural selection. However, the majority view was that natural selection needed to be supplemented by other processes. An influential history of biology written by Erik Nordenskiöld in 1929 could even take it for granted that Darwin's theory was wrong. About natural selection, he concluded "that it does not operate in the form imagined by Darwin must certainly be taken as proved;" the only remaining question, for Nordenskiöld, was "does it exist at all?"

Mendel's ideas were rediscovered around 1900

By this time, Mendel's theory of heredity had been rediscovered. *Mendelism* (Chapter 2) has been the generally accepted theory of heredity since the 1920s, and is the basis of all modern genetics. Mendelism eventually allowed a revival of Darwin's theory, but its initial effect (around 1900–20) was the exact opposite. The early Mendelians, such as Hugo de Vries and William Bateson, all opposed Darwin's theory of natural selection. They mainly did research on the inheritance of large differences between organisms, and generalized their findings to evolution as a whole. They

Figure 1.7

Early Mendelians and biometricians. (a) Early Mendelians studied large differences between organisms, and thought that evolution happened when a new species evolved from a "macromutation" in its ancestor. (b) Biometricians studied small interindividual differences, and explained evolutionary change by the transition of whole populations. Mendelians were less interested in the reasons for small interindividual variations. The figure is a simplification — no historic debate between two groups of scientists lasting for three decades can be fully represented in a single diagrammatic contrast.

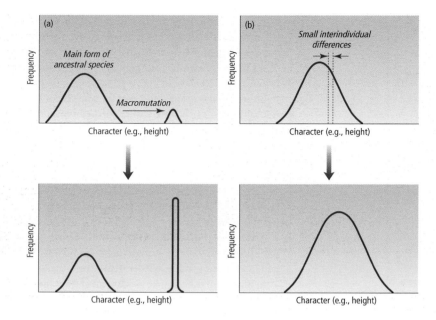

Biometricians rejected Mendelian theory

suggested that evolution proceeded in big jumps, by macromutations. A *macromutation* is a large and genetically inherited change between parent and offspring (Figure 1.7a). (Chapters 10 and 20 discuss various perspectives on the question of whether evolution proceeds in small or large steps.)

Mendelism was not universally accepted in the early twentieth century, however. Members of the other principal school, which rejected Mendelism, called themselves biometricians; Karl Pearson was one of the leading figures. *Biometricians* studied small, rather than large, differences between individuals and developed statistical techniques to describe how frequency distributions of measurable characters (such as height) passed from parent to offspring population. They saw evolution more in terms of the steady shift of a whole population rather than the production of a new type from a macromutation (Figure 1.7b). Some biometricians were more sympathetic to Darwin's theory than were the Mendelians. W.F. Weldon, for instance, was a biometrician, and he attempted to measure the amount of selection in crab populations on the seashore.

1.3.4 *The modern synthesis*

By the second decade of the twentieth century, research on Mendelian genetics had already become a major enterprise. It was concerned with many problems, most of which are more to do with genetics than evolutionary biology. But within the theory of evolution, the main problem was to reconcile the atomistic Mendelian theory of genetics with the biometrician's description of continuous variation in real populations. This reconciliation was achieved by several authors in many stages, but a 1918 paper by

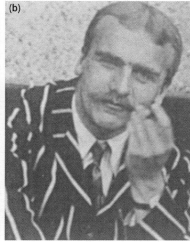

Figure 1.8
(a) Ronald Aylmer Fisher (1890–1962) in 1912, as a Steward at the First International Eugenics Conference.
(b) J.B.S. Haldane (1892–1964) in Oxford, UK in 1914.
(c) Sewall Wright (1889–1988) in 1928 at the University of Chicago.

Fisher, Haldane, and Wright created a synthesis of Darwinism and Mendelism. The synthesis began with population genetics . . .

R.A. Fisher is particularly important. Fisher demonstrated there that all the results known to the biometricians could be derived from Mendelian principles.

The next step was to show that natural selection could operate with Mendelian genetics. The theoretical work was mainly done, independently, by R.A. Fisher, J.B.S. Haldane, and Sewall Wright (Figure 1.8). Their synthesis of Darwin's theory of natural selection with the Mendelian theory of heredity established what is known as *neo-Darwinism*, or the *synthetic theory of evolution*, or the *modern synthesis*, after the title of a book by Julian Huxley, *Evolution: the Modern Synthesis* (1942). The old dispute between Mendelians and Darwinians was ended. Darwin's theory now possessed what it had lacked for half a century: a firm foundation in a well tested theory of heredity.

The ideas of Fisher, Haldane, and Wright are known mainly from their great summary works all written around 1930. Fisher published his book *The Genetical Theory of Natural Selection* in 1930. Haldane published a more popular book, *The Causes of Evolution*, in 1932; it contained a long appendix under the title "A mathematical theory of artificial and natural selection," summarizing a series of papers published from 1918 onwards. Wright published a long paper on "Evolution in Mendelian populations" in 1931; unlike Fisher and Haldane, Wright lived to publish a four-volume treatise (1968–78) at the end of his career. These classic works of theoretical population genetics demonstrated that natural selection could work with the kinds of variation observable in natural populations and the laws of Mendelian inheritance. No other processes are needed. The inheritance of acquired characters is not needed. Directed variation is not needed. Macromutations are not needed. This insight has been incorporated into all later evolutionary thinking, and the work of Fisher, Haldane, and Wright is the basis for much of the material in Chapters 5–9.

Figure 1.9
Theodosius Dobzhansky
(1900–75) in a group photo in
Kiev in 1924; he is seated second
from the left at the front, in the
great boots.

. . . and inspired research in the
field and lab . . .

The reconciliation between Mendelism and Darwinism soon inspired new genetic research in the field and laboratory. Theodosius Dobzhansky (Figure 1.9), for example, began classic investigations of evolution in populations of fruitflies (*Drosophila*) after his move from Russia to the USA in 1927. Dobzhansky had been influenced by the leading Russian population geneticist Sergei Chetverikov (1880–1959), who had an important laboratory in Moscow until he was arrested in 1929. Dobzhansky, after he had emigrated, worked both on his own ideas and collaborated with Sewall Wright. Dobzhansky's major book, *Genetics and the Origin of Species*, was first published in 1937 and its successive editions (up to 1970 (retitled)) have been among the most influential works of the modern synthesis. We shall encounter several examples of Dobzhansky's work with fruitflies in later chapters.

E.B. Ford (1901–88) began in the 1920s a comparable program of research in the UK. He studied selection in natural populations, mainly of moths, and called his subject "ecological genetics." He published a summary of this work in a book called *Ecological Genetics*, first published in 1964 (Ford 1975). H.B.D. Kettlewell (1901–79) studied melanism in the peppered moth *Biston betularia*, and this is the most famous piece of ecological genetic research (Section 5.7, p. 108). Ford collaborated closely with Fisher. Their best known joint study was an attempt to show that the random processes emphasized by Wright could not account for observed evolutionary changes in the scarlet tiger moth *Panaxia dominula*. Julian Huxley (Figure 1.10a) exerted his influence more through his skill in synthesizing work from many fields. His book *Evolution: the Modern Synthesis* (1942) introduced the theoretical concepts of Fisher, Haldane, and Wright to many biologists, by applying them to large evolutionary questions.

From population genetics, the modern synthesis spread into other areas of evolutionary biology. The question of how one species splits into two — the event is called speciation — was an early example. Before the modern synthesis had penetrated the subject, speciation had often been explained by macromutations or the inheritance of

Figure 1.10
(a) Julian Huxley (1887–1975) in 1918. (b) Ernst Mayr (1904–), on the right, on an ornithological expedition in New Guinea in 1928, with his Malay assistant.

(a)   (b)

acquired characters. A major book, *The Variation of Animals in Nature*, by two systematists, G.C. Robson and O.W. Richards (1936), accepted neither Mendelism nor Darwinism. Robson and Richards suggested that the differences between species are non-adaptive and have nothing to do with natural selection. Richard Goldschmidt (1878–1958), most famously in his book on *The Material Basis of Evolution* (1940), argued that speciation was produced by macromutations, not the selection of small variants.

The question of how species originate is closely related to the questions of population genetics, and Fisher, Haldane, and Wright had all discussed it. Dobzhansky and Huxley emphasized the problem even more. They all reasoned that the kinds of changes studied by population geneticists, if they took place in geographically separated populations, could cause the populations to diverge and eventually evolve into distinct species (Chapter 14). The classic work, however, was by Ernst Mayr: *Systematics and the Origin of Species* (1942). Like many classic books in science, it was written as a polemic against a particular viewpoint. It was precipitated by Goldschmidt's *Material Basis* but criticized Goldschmidt from the viewpoint of a complete and differing theory — the modern synthesis — rather than narrowly refuting him and it therefore has a much broader importance. Both Goldschmidt and Mayr (Figure 1.10b) were born and educated in Germany and later emigrated to the USA. Mayr left in 1930 as a young man, but Goldschmidt was 58 and had built a distinguished career when he left Nazi Germany in 1936.

A related development is often called the "new systematics," after the title of a book edited by Julian Huxley (1940). It refers to the overthrow of what Mayr called the "typological" species concept and its replacement by a species concept better suited to modern population genetics (Chapter 13). The two concepts differ in what sense they

. . . and led to a new understanding of speciation . . .

. . . and biological classification . . .

make of variation between individuals within a species. Species, in the typological conception, had been defined as a set of more or less similar-looking organisms, where similarity was measured relative to a standard (or "type") form for the species. A species then contains some individuals of the standard type, and other individuals who deviate from that type. The type individuals are conceptually privileged, whereas the deviants show some sort of error.

However, the concept of a species as type plus deviants was inappropriate in the theory of population genetics. The changes in gene frequencies analyzed by population geneticists take place within a "gene pool" — that is, a group of interbreeding organisms, who exchange genes when they reproduce. The crucial unit is now the set of interbreeding forms, regardless of how similar looking they are to each other. The idea of a "type" for a species is meaningless in a gene pool containing many genotypes. One genotype is no more of a standard form for the species than any other genotype. A gene pool does not contain one, or a few, "type" genotypes that are the standard forms for a species, with other genotypes being deviants from that "type." No type form exists that could be used as a reference point for defining the species. Population geneticists therefore came to define the members of a species by the ability to interbreed rather than by their morphological similarity to a type form. The modern synthesis had spread to systematics.

. . . and research on fossils

A similar treatment was given to paleontology by George Gaylord Simpson (Figure 1.11) in *Tempo and Mode in Evolution* (1944). Many paleontologists in the 1930s still persisted in explaining evolution in fossils by what are called orthogenetic processes — that is, some inherent (and unexplained) tendency of a species to evolve in a certain direction. Orthogenesis is an idea related to the pre-Mendelian concept of directed mutation, and the more mystical internal forces we saw in the work of Lamarck. Simpson argued that no observations in the fossil record required these processes. All the evidence was perfectly compatible with the population genetic mechanisms discussed by Fisher, Haldane, and Wright. He also showed how such topics as rates of evolution and the origin of major new groups could be analyzed by techniques derived from the assumptions of the modern synthesis (Chapters 18–23).

The modern synthesis was established by the 1940s

By the mid-1940s, therefore, the modern synthesis had penetrated all areas of biology. The 30 members of a "committee on common problems of genetics, systematics, and paleontology" who met (with some other experts) at Princeton in 1947 represented all areas of biology. But they shared a common viewpoint, the viewpoint of Mendelism and neo-Darwinism. A similar unanimity of 30 leading figures in genetics, morphology, systematics, and paleontology would have been difficult to achieve before that date. The Princeton symposium was published as *Genetics, Paleontology, and Evolution* (Jepsen *et al.* 1949) and is now as good a symbol as any for the point at which the synthesis had spread throughout biology. Of course, there remained controversy within the synthesis, and a counterculture outside. In 1959, two eminent evolutionary biologists — the geneticist Muller and the paleontologist Simpson — could still both celebrate the centenary of *The Origin of Species* with essays bearing (almost) the same memorable title: "One hundred years without Darwinism are enough" (Muller 1959; Simpson 1961a).

In this book, we shall look in detail at the main ideas of the modern synthesis, and see how they are developing in recent research.

Figure 1.11
George Gaylord Simpson (1902–84) with a baby guanaco in central Patagonia in 1930.

Summary

1 Evolution means descent with modification, or change in the form, physiology, and behavior of organisms over many generations of time. The evolutionary changes of living things occur in a diverging, tree-like pattern of lineages.

2 Living things possess adaptations: i.e., they are well adjusted in form, physiology, and behavior, for life in their natural environment.

3 Many thinkers before Darwin had discussed the possibility that species change through time into other species. Lamarck is the best known. But in the mid-nineteenth century most biologists believed that species are fixed in form.

4 Darwin's theory of evolution by natural selection explains evolutionary change and adaptation.

5 Darwin's contemporaries mainly accepted his idea of evolution, but not his explanation of it by natural selection.

6 Darwin lacked a theory of heredity. When Mendel's ideas were rediscovered at the turn of the twentieth century, they were initially thought to count against the theory of natural selection.

7 Fisher, Haldane, and Wright demonstrated that Mendelian heredity and natural selection are compatible; the synthesis of the two ideas is called neo-Darwinism or the synthetic theory of evolution.

8 During the 1930s and 1940s, neo-Darwinism gradually spread through all areas of biology and became widely accepted. It unified genetics, systematics, paleontology, and classic comparative morphology and embryology.

Further reading

A popular essay about the adaptations of woodpeckers is by Diamond (1990). Bowler (1989) provides a general history of the idea of evolution. On Lamarck and his context, see Burkhardt (1977) and Barthélemy-Madaule (1982); and Rudwick (1997) for Cuvier. There are many biographies of Darwin; Browne (1995–2002) is as near to a "standard" modern biography as any. Darwin's autobiography is an interesting source. A pleasant (if more demanding) way to follow Darwin's life is through his correspondence: a modern scholarly edition is under way (Burkhardt & Smith 1985–). Bowler (1989) discusses and gives references about the reception and fate of Darwin's ideas. Berry (2002) is a readable anthology from Wallace's writings. On the modern synthesis, see also Provine (1971), Mayr & Provine (1980), Bowler (1996), and Gould (2002b). Numbers (1998) is about the American reception of Darwinism.

There are biographies of many of the key figures: Box (1978) for Fisher; Clark (1969) for Haldane; Provine (1986) for Wright. Huxley (1970–73) and Simpson (1978) wrote autobiographies. Laporte (2000) is an intellectual biography of Simpson. See Adams (1994) for Dobzhansky, and Powell (1997) for the contributions of the "Drosophila model" to evolution. See the papers in a dedicatory issue of *Evolution* (1994), vol. 48, pp. 1–44, for Mayr. See the special issue of *Proceedings of the National Academy of Sciences USA* (2000), vol. 97, pp. 6941–7055 for Stebbins.

Evolution is probably better covered than any other scientific theory by popular science writers. Dawkins (1986, 1989a, 1996) introduces many ideas in evolution, particularly those to do with adaptation and natural selection. Gould's popular essays, which first appeared in *Natural History* magazine from 1977 to 1990, have been anthologized in a series of books and introduce many aspects of evolutionary biology (Gould 1977b, 1980, 1983, 1985, 1991, 1993, 1996, 1998, 2000, 2002a). Jones (1999) is a popular update of Darwin's *Origin of Species*: it keeps Darwin's original structure, and uses modern examples. Mayr (2001) is an overview of the subject for a general reader, as well as containing the current views of an authoritative writer.

Pagel (2002) and the encyclopedia of the life sciences (www.els.com) are encyclopedias of evolution and of biology, respectively. The encyclopedia of life sciences is comprehensive on evolution. Evolution is covered in many web pages, and links are provided to them from the web page associated with this book (www.blackwellscience.com/evolution). Zimmer (2001) is a popular book about evolution, accompanying a PBS TV series. *Trends in Ecology and Evolution* is a good one-stop source to follow a wide range of evolutionary research.

Study and review questions

1 Review the ways in which biological evolution differs from individual development, changes in the species composition of ecosystems, and some other kinds of change that you can think of.

2 What property of nature must any theory of evolution explain, if it is not (in Darwin's words) to be "almost useless."

3 How did the main popular concept of evolution in the late nineteenth and early twentieth centuries differ from the conception of evolution in Darwin's theory?

4 What are the two theories that are combined in the synthetic theory of evolution?

2 Molecular and Mendelian Genetics

The chapter is an introduction to the genetics that we need to understand the evolutionary biology contained in this book. It begins with the molecular mechanism of inheritance, and moves on to Mendelian principles. It then considers how Darwin's theory almost required heredity to be Mendelian, because natural selection can hardly operate at all with a blending mechanism of inheritance.

2.1 Inheritance is caused by DNA molecules, which are physically passed from parent to offspring

The molecule called DNA (deoxyribose nucleic acid) provides the physical mechanism of heredity in almost all living creatures. The DNA carries the information used to build a new body, and to differentiate its various body parts. DNA molecules exist inside almost all the cells of a body, and in all the reproductive cells (or gametes). Its precise location in the cell depends on cell type.

DNA is carried in different ways in prokaryotic and in eukaryotic cells

There are two main types of cell: *eukaryotic* and *prokaryotic* (Figure 2.1). Eukaryotic cells have a complex internal structure, including internal organelles and a distinct region, surrounded by a membrane, called the nucleus. Eukaryotic DNA exists within the nucleus. Prokaryotic cells are simpler and have no nucleus. Prokaryotic DNA lies within the cell, but in no particular region. All complex multicellular organisms, including all plants and animals, are built of eukaryotic cells. Fungi are also eukaryotic; some fungi are multicellular (such as mushrooms) others are unicellular (such as baker's and brewer's yeast, *Saccharomyces cerevisiae*). Protozoans, most of which (such as amebas) are unicellular, are the other main group of eukaryotes. Bacteria and Archaea are the two kinds of life in which the cells are prokaryotic.

Within a eukaryotic cell nucleus, the DNA is physically carried in structures called *chromosomes*. Chromosomes can be seen through a light microscope at certain stages in the cell cycle. Individuals of different species characteristically have different numbers of chromosomes — each individual human has 46, for example, whereas a fruitfly *Drosophila melanogaster* has eight, and other species have other numbers. The finer structure of the DNA is too small to be seen directly, but it can be inferred by the method of X-ray diffraction. The molecular structure of DNA was worked out by Watson and Crick in 1953.

Figure 2.1

The cells of a body have a fine structure (or "ultrastructure") made up of a number of organelles. Not all the organelles illustrated here are found in all cells. Animal and fungal cells, for example, lack plastids; but all photosynthesizing organisms have them. Eukaryotes (i.e., all plants and animals) have complex cells with a separate nucleus. Within the nucleus the DNA is here illustrated in the diffuse form called chromatin; when the cell divides, the chromatin coalesces into structures called chromosomes. Prokaryotes are simpler organisms, particularly bacteria, and they lack a distinct nucleus; their DNA lies naked within the cell.

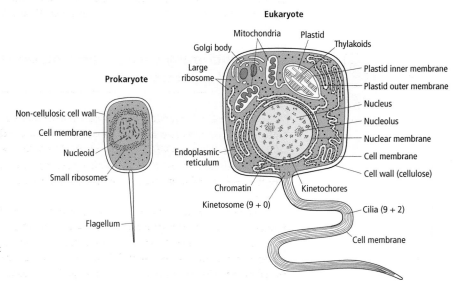

(a) **Structure of single strand**

5' end of chain

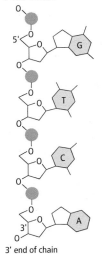

3' end of chain

(b) **Structure of double strand**

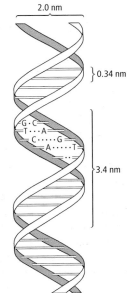

Figure 2.2

The structure of DNA. (a) Each strand of DNA is made up of a sequence of nucleotide units. Each nucleotide consists of a phosphate (P), a sugar, and a base (of which there are four types, here called G, C, T, and A). (b) The full DNA molecule has two complementary strands, arranged in a double helix.

The DNA molecule consists of a sequence of units; each unit, called a *nucleotide*, consists of a phosphate and a sugar group with a *base* attached. The alternating sugar and phosphate groups of successive nucleotides form the backbone of the DNA molecule. The full DNA molecule consists of two paired, complementary strands, each made up of sequences of nucleotides. The nucleotides of opposite strands are chemically bonded together. The two strands exist as a double helix (Figure 2.2).

2.2 DNA structurally encodes information used to build the body's proteins

How does the DNA encode the information to build a body? The DNA in an individual human cell contains about 3×10^9 nucleotide units. This total length can be divided into *genes* and various kinds of *non-coding* DNA. We will consider genes first. Some genes lie immediately next to neighboring genes; others are separated by more or less lengthy regions of non-coding DNA. Genes contain the information that codes for proteins.

A crude but workable way to describe protein biology is to say that bodies are built from proteins and are regulated, maintained, and defended by proteins. Different parts of the body have their distinct characteristics because of the kinds of proteins they are made of. Skin, for example, is mainly made of a protein called keratin; oxygen is carried in red blood cells by a protein called hemoglobin; eyes are sensitive to light as a result of pigment proteins such as rhodopsin (actually, rhodopsin is made up of a protein called opsin combined with a derivative of vitamin A); and metabolic processes are catalyzed by a whole battery of proteins called enzymes (cytochrome *c*, for instance, is a respiratory enzyme and alcohol dehydrogenase is a digestive enzyme). Other proteins, such as the immunoglobulins, defend the body from parasites. The expression of genes is regulated by yet other proteins, such as the transcription factors coded for by the *Hox* genes that we shall discuss in Chapter 20.

Most genes code for proteins . . .

Proteins are made up of particular sequences of amino acids. Twenty different amino acids are found in most kinds of living things. Each amino acid behaves chemically in distinct ways, such that different sequences of amino acids result in proteins with very different properties. A protein's exact sequence of amino acids determines its nature. Hemoglobin, for example, is made up of two α-globin molecules, which are 141 amino acids long in humans, and two β-globins, which are 144 amino acids long; insulin has another sequence of 51 amino acids. Hemoglobin binds oxygen in the blood, whereas insulin stimulates cells (particularly muscle cells, and others) to absorb glucose from the blood. The different behavior of hemoglobin and insulin is caused by the chemical properties of their different amino acids, arranged in their characteristic sequence. (As Chapters 4 and 7 discuss in more detail, the particular sequence of any one protein can vary within and between species. Thus turkey hemoglobin differs on average from human hemoglobin, though it binds oxygen in both species. There are also different variants of hemoglobin within a species, a condition called *protein polymorphism*. However, the sequences of all variants of hemoglobin from the same or different species are similar enough for them to be recognizably hemoglobins.)

. . . but there are complications, such as alternative splicing

The idea that one gene encodes for one protein is a simplification. On the one hand, some proteins are assembled from more than one gene. For example, hemoglobin is assembled from four genes, in two main positions in the DNA. On the other hand, one gene can be used to produce more than one protein. For example, the process of *alternative splicing* generates a number of proteins from one gene. Alternative splicing can be illustrated by the gene *slo*, which works in the development of our acoustic sensory system. We are sensitive to a range of frequencies because we have a series of tiny hairs in our inner ears; some of the hairs are bent by high frequency sounds, others by low frequency sounds. The frequency that a hair is sensitive to depends on the chemical properties of the proteins it is made of. *Slo* is one of the key genes that code for a protein in the hair cells. It might be thought that we should contain a series of genes, coding for a series of proteins, that produce the series of hair cells with a range of sensitivities. In fact *slo* is read in many ways. The *slo* gene is made up of several units, which can be combined in a large number of ways. Exactly how many ways *slo* can be read is uncertain, but the alternative splicing of *slo* contributes part of the molecular diversity underlying our acoustic sense. Thus, it is not strictly speaking correct to say that one gene codes for one protein. Nevertheless, for many purposes it is not grievously wrong to describe DNA as made up of genes (and non-coding regions), and genes as coding for proteins.

How exactly do genes in the DNA code for proteins? The answer is that the sequence of nucleotides in a gene specifies the sequence of amino acids in the protein. There are four types of nucleotide in the DNA. They differ only in the base part of the nucleotide unit; the sugar and phosphate group are the same in all four. The four are called adenine (A), cytosine (C), guanine (G), and thymine (T). Adenine and guanine belong to the chemical group called purines; cytosine and thymine are pyrimidines. In the double helix, an A nucleotide in one strand always pairs with a T nucleotide in the other; and a C always pairs with a G (as in Figure 2.2b). If the nucleotide sequence in one strand was ...AGGCTCCTA..., then the complementary strand would be ...TCCGAGGAT.... Because the sugar and phosphate are constant, it is often more convenient to imagine a DNA strand as a sequence of bases, like the ...AGGCTCCTA... sequence above.

2.3   Information in DNA is decoded by transcription and translation

There are four types of nucleotide, but 20 different amino acids. A one-to-one code of nucleotide encoding an amino acid would therefore be impossible. In fact a triplet of bases encodes one amino acid; the nucleotide triplet for an amino acid is called a *codon*. The four nucleotides can be arranged in 64 ($4 \times 4 \times 4$) different triplets, and each one codes for a single amino acid. The relation between triplet and amino acid has been deciphered and is called the *genetic code*.

Information encoded in the DNA . . .

The mechanism by which the amino acid sequence is read off from the nucleotide sequence of the DNA is understood in molecular detail. The full detail is unnecessary for our purposes, but we should distinguish two main stages. RNA (ribonucleic acid) is a class of molecules that has a similar composition to DNA. Messenger RNA (mRNA) is one of the main forms of RNA. Messenger RNA is transcribed from the DNA and the process is called *transcription*. Messenger RNA is single-stranded and, unlike DNA, uses a base called uracil (U) instead of thymine (T). The DNA sequence AGGCTCCTA would therefore have an mRNA with the following sequence transcribed from it: UCC-GAGGAU. The genetic code is usually expressed in terms of the codons in the mRNA (Table 2.1). The mRNA sequence UCCGAGGAU, for example, codes for three amino acids: serine, glutamic acid, and aspartic acid. The beginning and end of a gene are signaled by distinct base sequences, which (in a sense) punctuate the DNA message. As Table 2.1 shows, three of the 64 triplets in the genetic code are for "stop." Only 61 of the 64 code for amino acids.

. . . is first transcribed to mRNA . . .

. . . and then translated into protein

Transcription takes place in the nucleus. After the mRNA molecule has been assembled on the gene, it then leaves the nucleus and travels to one of the structures in the cytoplasm called ribosomes (see Figure 2.1); ribosomes are made of another kind of RNA called ribosomal RNA (rRNA). The ribosome is the site of the second main stage in protein production. It is where the amino acid sequence is read off from the mRNA sequence and the protein is assembled. The process is called *translation*. The actual translation is achieved by yet another kind of RNA, called transfer RNA (tRNA).[1]

---

[1]   That completes the three main kinds of RNA: mRNA, rRNA, and tRNA. By the way, both rRNA and tRNA molecules originate by transcription from genes in the DNA. It is therefore not always true that genes code for proteins, as stated above — some genes code for RNA.

Table 2.1

The genetic code. The code is here expressed for mRNA. Each triplet encodes one amino acid, except for the three "stop" codons, which signal the end of a gene.

First base in the codon	Second base in the codon				Third base in the codon
	U	C	A	G	
U	Phenylalanine	Serine	Tyrosine	Cysteine	U
	Phenylalanine	Serine	Tyrosine	Cysteine	C
	Leucine	Serine	Stop	Stop	A
	Leucine	Serine	Stop	Tryptophan	G
C	Leucine	Proline	Histidine	Arginine	U
	Leucine	Proline	Histidine	Arginine	C
	Leucine	Proline	Glutamine	Arginine	A
	Leucine	Proline	Glutamine	Arginine	G
A	Isoleucine	Threonine	Asparagine	Serine	U
	Isoleucine	Threonine	Asparagine	Serine	C
	Isoleucine	Threonine	Lysine	Arginine	A
	Methionine	Threonine	Lysine	Arginine	G
G	Valine	Alanine	Aspartic acid	Glycine	U
	Valine	Alanine	Aspartic acid	Glycine	C
	Valine	Alanine	Glutamic acid	Glycine	A
	Valine	Alanine	Glutamic acid	Glycine	G

The tRNA molecule has a base triplet recognition site, which binds to the complementary triplet in the mRNA, and has the appropriate amino acid attached at the other end (Figure 3.7, p. 58, shows the structure of tRNA). Cells use less than the theoretical maximum of 61 different kinds of tRNA. A single tRNA can be used for more than one codon, for example, in some cases where the same amino acid is coded for by two closely related codons. The ability of a single tRNA to bind to more than one codon is called "wobble." Cells in fact use about 45 kinds of tRNA. In summary, protein assembly consists of tRNA molecules lining up on the mRNA at a ribosome. Other molecules are also needed to supply energy and attach the RNAs correctly. Figure 2.3 summarizes the transfer of information in the cell.

In addition to the DNA on the chromosomes in the nucleus, there are much smaller quantities of DNA in certain organelles in the cytoplasm (see Figure 2.1). Mitochondria — the organelles that control respiration — have some DNA, and in plants the organelles called chloroplasts that control photosynthesis also have their own DNA. Mitochondrial DNA is inherited maternally: mitochondria are passed on through eggs but not through sperms.

Figure 2.3
The transfer of information in a cell.

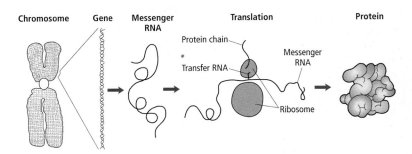

2.4 Large amounts of non-coding DNA exist in some species

Much of the DNA does not code for genes

The human genome is about 3,000 million (3×10^9) nucleotides long. The human genome project made a preliminary estimate of the number of genes in a human being as about 30,000 (3×10^4). The average length of a human gene is about 5,000 (5×10^3) nucleotides. Thus only about 5% ($1.5 \times 10^8/3 \times 10^9$) of human DNA codes for genes. Even if the preliminary figure of 30,000 genes turns out to underestimate the true figure by a factor of two, still only 10% of our DNA would code for genes. Most human DNA is not used to code for proteins, or for molecules that control the production of proteins. Most human DNA is "non-coding" DNA.

The fraction of non-coding DNA varies between species. Bacteria and viruses contain little non-coding DNA; bacterial and viral genomes are economically organized. At the other extreme, some salamanders contain 20 times as much DNA as humans do. Because it is difficult to believe that salamanders contain many more genes than we do, we can infer that more than 99% of those salamanders' DNA is non-coding.

The function of non-coding DNA is uncertain. Some biologists argue that it has no function and refer to it as "junk DNA." Others argue that it has structural or regulatory functions. Something is known about the sequence of non-coding DNA. Most non-coding DNA is repetitive. Some of it consists of side-by-side (or "tandem") stretches of repeats of a short (2–20 nucleotide) unit sequence (for example ...ACCACCACC...). Some of it consists of repeats of longer (a hundred, or a few hundred nucleotides) sequences. We can partly understand how non-coding DNA originates after we have considered our next topic: mutation.

2.5 Mutational errors may occur during DNA replication

When a cell reproduces, its DNA and genes are physically replicated. Normally an exact copy of the parental DNA is produced, but sometimes a copying error happens. The set of enzymes that replicate the DNA include proof-reading and repair enzymes. These enzymes detect and correct most of the copying errors, but some errors persist even after proof-reading and repair. These errors are called *mutations*. The new sequence of

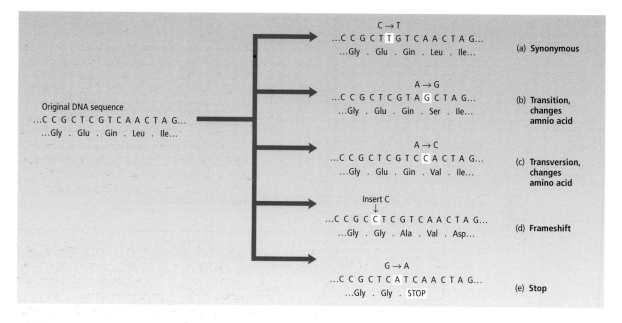

Figure 2.4

Different sorts of mutation. (a) Synonymous mutations — the base changes but the amino acid encoded does not. (b) Transition — a change between purine types or between pyrimidine types. (c) Transversion — a change from purine to pyrimidine or vice versa. (d) Frameshift mutation — a base is inserted. (e) Stop mutation — an amino acid-encoding triplet mutates to a stop codon. The terms transition and transversion can apply to synonymous or amino acid-changing mutations, but it has only been illustrated here for mutations that alter amino acids. The base sequence here is for the DNA. The genetic code is conventionally written for the mRNA sequence; thus G has to be transcribed to C, etc. when comparing the figure with Table 2.1 (the genetic code). (The figure is stereochemically unconventional because the 3′ end has been put at the left and 5′ at the right; but this detail is unimportant here.)

Different kinds of mutation can be distinguished, such as . . .

DNA that results from a mutation may code for a form of protein with different properties from the original. Mutations can happen in any cell, but the most important mutations, for the theory of evolution, are those occuring in the production of the gametes. These mutations are passed on to the offspring, who may differ from their parents because of the mutation.

. . . point mutations . . .

Various kinds of mutation can occur. One is *point mutation*, in which a base in the DNA sequence changes to another base. The effect of a point mutation depends on the kind of base change (Figure 2.4a–c). Synonymous, or silent, mutations (Figure 2.4a) are mutations between two triplets that code for the same amino acid, and have no effect on the protein sequence. Non-synonymous, or meaningful, point mutations do change the amino acid. Because of the structure of the genetic code (Table 2.1), most synonymous mutations are in the third base position of the codon. About 70% of changes in the third position are synonymous, whereas all changes in the second and most (96%) at the first position are meaningful. Another distinction for point mutations is between transitions and transversions. Transitions are changes from one

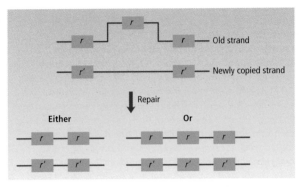

Figure 2.5

Slippage happens when a stretch of DNA is copied twice or not at all. *r* in the figure is a certain sequence of DNA. It is repeated three times in a region of the DNA molecule. This molecule is being copied, and *r'* refers to the same unit sequence in the new DNA molecule. In this case the DNA polymerase has slipped past one repeat and the new molecule has two rather than three repeats. It is also possible for the polymerase to slip back and copy a unit repeat twice; then the new molecule has four repeats. The old and new copies will have different numbers of repeats. They may be repaired to create a mutant DNA (with two or four repeats) or to restore the original number of three repeats. Sections of DNA with many repeats of a similar sequence may be particularly vulnerable to slippage.

pyrimidine to the other, or from one purine to the other: between C and T, and between A and G. Transversions replace a purine base by a pyrimidine, or vice versa: from A or G to T or C (and from C or T to A or G). The distinction is interesting because transitional changes are much commoner in evolution than transversions.

Successive amino acids are read from consecutive base triplets. If, therefore, a mutation inserts a base pair into the DNA, it can alter the meaning of every base "downstream" from the mutation (Figure 2.4d). These are called *frameshift mutations*, and will usually produce a completely nonsensical, functionless protein. Another kind of mutation is for a previously coding triplet to mutate to a "stop" codon (Figure 2.4e); the resulting protein fragments will probably again be functionless.

Some stretches of non-coding DNA consist of repeats of a short unit sequences. These sequences are particularly vulnerable to a kind of error called *slippage* (Figure 2.5). In slippage, the DNA strand that is being copied from slips relative to the new strand that is being created. A short stretch of nucleotides is then missed out or copied twice. Slippage contributes to the origin of non-coding DNA that consists of repeats of short unit sequences. However, slippage can also occur in DNA other than repetitive non-coding DNA. Slippage can cause frameshift mutations (Figure 2.4d), for instance.

The mutational mechanisms we have considered so far concern single nucleotides, or short stretches of nucleotides. Other mutational mechanisms can influence larger chunks of DNA. *Transposition* is an important example. Transposable elements — informally known as "jumping genes" — can copy themselves from one site in the DNA to another (Figure 2.6a). If a transposable element inserts itself into an existing gene, it will corrupt that gene; if it inserts itself into a region of non-coding DNA, it may do less or even no damage to the body. Transposable elements can pick up a stretch of DNA and copy it as well as itself into the new insertion site. Transposition usually alters the total length of the genome, because it creates a new duplicated stretch of DNA. This contrasts with a simple miscopying of a nucleotide, in which the total length of the genome is unchanged. Unequal crossing-over is another kind of mutation that can duplicate (or, unlike transposition, delete) a long stretch of DNA (Figure 2.6b).

Finally, mutations may influence large chunks of chromosomes, or even whole chromosomes (Figure 2.7). A length of chromosome may be translocated to another

... frameshift mutations ...

... slippage ...

... transposition ...

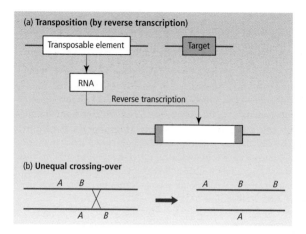

Figure 2.6

Transposition and unequal crossing-over are mutation mechanisms that affect stretches of DNA longer than one or two nucleotides. They duplicate DNA laterally through the genome. (a) Transposition can occur by more than one mechanism. Here transposition occurs via an RNA intermediate that is copied back into the DNA by reverse transcription. Transposable elements of this kind are called retroelements. (b) Unequal crossing-over happens when the sequences of the two chromosomes are misaligned at recombination (for recombination, see Figure 2.9 below). In the simple case illustrated here, chromosomes with three and with one copy of a gene could be generated from two chromosomes with two genes each. In practice, misalignment is more likely if there are a long series of copies of similar sequences.

Figure 2.7

Chromosomes can mutate by: (a) deletion; (b) duplication of a part; (c) inversion; or (d) translocation. Translocation may be either "reciprocal" (in which the two chromosomes exchange equal lengths of DNA) or "non-reciprocal" (in which one chromosome gains more than the other). In addition, whole chromosomes may fuse, and whole chromosomes (or the whole genome) may duplicate.

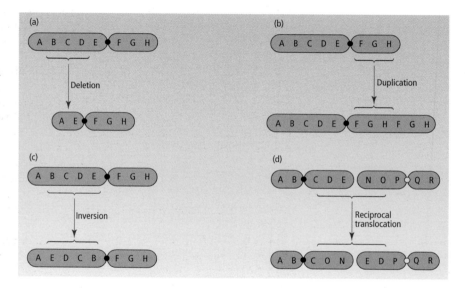

... and chromosomal mutations

chromosome, or to another place on the same chromosome, or be inverted. Whole chromosomes may fuse, as has happened in human evolution; chimps and gorillas (our closest living relatives) have 24 pairs of chromosomes whereas we have 23. Some or all of the chromosomes may be duplicated. The phenotypic effects of these chromosomal mutations are more difficult to generalize about. If the break-points of the mutation divide a protein, that protein will be lost in the mutant organism. But if the break is between two proteins, any effect will depend on whether the expression of a gene depends on its position in the genome. In theory, it might not matter whether a protein is transcribed from one chromosome or another; though in practice gene expression is probably at least partly regulated by relations between neighboring genes and a chromosomal mutation will then have phenotypic consequences.

Mutations can also delete, or duplicate, a whole chromosome. Mutations on the largest scale can duplicate all the chromosomes in the genome. The duplication of the whole genome is called *polyploidy*. For example, suppose that the members of a normal diploid species have 20 chromosomes (10 from each parent). If all 20 are duplicated in a mutation, the offspring has 40 chromosomes. Polyploidy has been an important process in evolution, particularly plant evolution (Chapters 3, 14, and 19).

That concludes our review of the main types of mutation. It is not a complete list of all known mutation mechanisms, but is enough for an understanding of the evolutionary events described later in this book.

2.6 Rates of mutation can be measured

What is the rate of mutation? The mutation rate can be estimated from the rate at which detectable new genetic variants arise in laboratory populations. Novel genetic variants used to be detectable only if they influenced the organism's visible phenotype. Now we also have rapid DNA sequencing methods, and these can be used to detect nucleotide differences between parent and offspring.

Mutation rates are problematic to measure . . .

The measuring conditions must be such as to minimize, and ideally to eliminate, the action of natural selection. The reason is as follows. Mutations may be advantageous (if they increase the survival of the mutant bearer), neutral (if they have no effect on the survival of the mutant bearer), or disadvantageous (if they decrease the survival of the mutant bearer). In natural conditions, many mutations are disadvantageous and their bearers die before the mutation can be detected. The mutation rate will then be underestimated. Thus the measuring conditions need to be such as to neutralize the damage done by disadvantageous mutations. Then all mutant bearers will survive and the underlying mutation rate can be detected. Natural selection usually cannot be completely neutralized, however, and the estimates that we have for mutation rates are only approximate.

. . . and can be expressed per nucleotide . . .

Mutation rates can be expressed per nucleotide, or per gene, or per genome. Also, they can be expressed per molecular replication, or per organismic generation, or per year. Table 2.2 gives some numbers. The mutation rate per nucleotide per molecular replication is the most basic number, and it depends on the hereditary molecule and the enzymatic machinery used by the organism. RNA viruses such as HIV use RNA as their hereditary molecule; they have a replicase enzyme (called reverse transcriptase) but lack proof-reading and repair enzymes. RNA viruses have a relatively high mutation rate, of about 10^{-4} per nucleotide. All cellular life forms, including bacteria and human beings, have a similar set of proof-reading and repair enzymes, and use DNA as their hereditary molecule. DNA is less mutable than RNA, partly because DNA is a double-stranded molecule, and the proof-reading and repair enzymes further reduce the mutation rate. Bacteria, and humans, have a mutation rate of about 10^{-9} to 10^{-10} per nucleotide per molecular replication (or per cell cycle, in these cellular life forms). The mutation rate per nucleotide per cell cycle seems to be approximately constant in cellular life forms, at least relative to the much higher figure in RNA viruses, but it may not be exactly constant. Some evidence suggests that the mutation rate is an order of

Table 2.2

Mutation rates in various life forms. Genome sizes are diploid figures. The "worm" is *Caenorhabditis elegans*. RNA viruses do not literally have cell cycles, but the number in the column refers to the number of times the RNA is replicated per generation. All numbers are approximate. In the cases of RNA viruses and bacteria, there are many species with a range of genome sizes. After various sources, see Ridley (2001).

Life form	Mutation rate per nucleotide per replication	Total genome size (nucleotides)	Cell cycles per generation	Mutation rate per genome per generation
RNA virus	10^{-4}	$\approx 10^4$	1	≈ 1
Bacteria		$\approx 2 \times 10^6$	1	$\approx 10^{-3}$
Worm	10^{-9} to 10^{-10}	2×10^8	10	≈ 2
Fruitfly		3.6×10^8	20	≈ 4
Human being		6.6×10^9	200	≈ 200

magnitude higher in bacteria than in humans (Ochman *et al.* 1999), but the measurements are uncertain. The trends, if any, in the mutation rate per nucleotide copying event within cellular life forms are unknown.

. . . per genome . . .

The mutation rate per genome varies between bacteria and human beings despite the approximate constancy of the rate per nucleotide copying event because of the effects of generation time and because we have larger genomes. In humans, for instance, the number of cell divisions per generation in a man (from his conception to his sperm cells, when he is an adult) goes up with age by about 23 divisions per year. The sperm of a 20-year-old many have about 200 cell divisions behind them, the sperm of a 30-year-old man have about 430 cell division behind them. The number of cell divisions per generation in a woman (from her conception to her egg cells) are constant independent of age, at about 33 cell divisions. The average number of cell divisions in a human generation is therefore about 100–200 or more, depending on the father's age.

. . . and per gene

Mutation rates are also sometimes expressed per gene per generation. The rate will depend on the size of the gene and the generation length of the organism. But with mutation rates per nucleotide copying event of 10^{-9} to 10^{-10}, generations ranging from 1 to 100 cell divisions (10^0–10^2), and genes ranging from 10^3 to 10^4 nucleotides, mutation rates per gene per generation are going to range around 10^{-3} to 10^{-7}. A classic memorable figure for the mutation rate per gene per generation is one in a million (10^{-6}).

Mutation rates per year can also be useful, particularly when using the molecular clock to date evolutionary events — which is a big theme in modern evolutionary biology (Chapters 7 and 15 and much of Part 5 of this book). Mutation rates per nucleotide per cell cycle can be translated into rates per year. The translation depends on the species, particularly because species differ in generation times, as discussed in Chapter 7. In later chapters we shall use figures per year for particular species rather than the more general figures such as in Table 2.2.

The numbers we have looked at here are averages. Some regions of DNA have higher, or lower, than average rates. For instance, we saw that short repeats (for example, ...ACCACCACC...) are vulnerable to slippage. These regions expand and contract by mutation, such that a parent with three repeats of the unit sequence may have offspring with two or four repeats of it. The mutation rates are high, up to 10^{-2} (Jeffreys *et al.* 1988). The high mutation rates make these regions of DNA useful in genetic fingerprinting.

2.7 Diploid organisms inherit a double set of genes

Humans, and many other creatures, have two sets of genes, one inherited from each parent

DNA is physically carried on chromosomes. Humans, as noted above, have 46 chromosomes. However, the 46 consist of two sets of 23 distinct chromosomes. (To be exact, an individual has a pair of sex chromosomes — which are similar (XX) in females, but noticeably different (XY) in males — plus a double set of 22 non-sex chromosomes, called *autosomes*.) The condition of having two sets of chromosomes (and therefore two sets of the genes carried on them) is called *diploidy*. The figure given above of 3×10^9 nucleotides in the human genome is for only one of the sets of 23 chromosomes: the total DNA library of a human cell has about 6×10^9 nucleotides (and 6.6×10^9 is a more exact figure).

Diploidy is important in reproduction. An adult individual has two sets of chromosomes. Its gametes (eggs in the female, sperm or pollen in the male) have only one set: a human egg, for example, has only 23 chromosomes before it is fertilized. Gametes are said to be *haploid*. They are formed by a special kind of cell division, called *meiosis*; in meiosis, the double set of chromosomes is reduced to result in a gamete with only one set. When male and female gametes fuse, at fertilization, the resulting *zygote* (the first cell of the new organism) has the double chromosome set restored, and it develops to produce a diploid adult. The cycle of genesis can then repeat itself. (In some species, organisms are permanently haploid; but in this book we shall mainly be concerned with diploid species. Most familiar, non-microscopic species are diploid.)

Various technical terms apply to diploid organisms

Because each individual possesses a double set of chromosomes, it also possesses a double set of each of its genes. Any one gene is located at a particular place on a chromosome, called its *genetic locus*. An individual is therefore said to have two genes at each genetic locus in its DNA. One gene comes from its father and the other from its mother. The two genes at a locus are called a *genotype*. The two copies of a gene in an individual may be the same, or slightly different (i.e., the amino acid sequences of the proteins encoded by the two copies may be identical or have one or two differences). If they are the same the genotype is a *homozygote*; if they differ it is a *heterozygote*. The different forms of the gene that can be present at a locus are called *alleles*. Genes and genotypes are usually symbolized by alphabetic letters. For instance, if there are two alleles at the genetic locus under consideration, we can call them *A* and *a*. An individual can then have one of three genotypes: it can be *AA*, or *Aa*, or *aa*.

The genotype at a locus should be distinguished from the *phenotype* it produces. If there are two alleles at a locus in a population, the two can combine into three possible genotypes: *AA*, *Aa*, and *aa*. (If there are more than two alleles, there will be more than three genotypes.) The genes will influence some property of the organism, and the property may or may not be easily visible. Suppose they influence color. The *A* gene might encode a black pigment and *AA* individuals would be black; *aa* individuals, lacking the pigment, would be (let us say) white. The coloration is then the phenotype controlled by the genotype at that locus: an individual's phenotype is its body and behavior as we observe them.

If we consider only the *AA* and *aa* genotypes and phenotypes in this example, there is a one-to-one relation between genotype and phenotype. But there does not have to be a one-to-one relation, as can be illustrated by considering two possibilities for the phenotype of the *Aa* genotype. One possibility is that the color of *Aa* individuals is intermediate between the two homozygotes — they are gray. In this case, there are three phenotypes for the three genotypes and there is still a one-to-one relation between them. The second possibility is that the *Aa* heterozygotes resemble one of the homozygotes; they might be black, for instance. The *A* allele is then called *dominant* and the *a* allele *recessive*. (An allele is dominant if the phenotype of the heterozygote looks like the homozygote of that allele; the other allele in the heterozygote is called recessive.) If there is dominance, there will be only two phenotypes for the three genotypes and there is no longer a one-to-one relation between them. If all you know is that an organism has a black phenotype, you do not know its genotype.

At different genetic loci, there can be any degree of dominance. Full dominance, in which the heterozygote resembles one of the homozygotes, and no dominance, in which it is intermediate between the homozygotes, are extreme cases. The phenotype of the heterozygote could be anywhere between the two homozygotes. Instead of being either black or gray, it could have had any degree of grayness. Dominance is only one of a number of factors that complicate the relation between genotype and phenotype. The most important such factor is the environment in which an individuals grows up (Chapter 9).

Dominance complicates the relation between genotype and phenotype

2.8 Genes are inherited in characteristic Mendelian ratios

Mendelism explains the ratios of genotypes in the offspring of particular parents

Mendelian ratios express the proportions of different genotypes in the offspring of parents of particular combinations of genotypes. The easiest case is a cross between an *AA* male and an *AA* female (Figure 2.8a). After meiosis, all the male gametes contain the *A* allele and all the female gametes also contain the *A* allele. They combine to produce *AA* offspring. The Mendelian ratio is therefore 100% *AA* offspring.

Now consider a mating between an *AA* homozygote and an *Aa* heterozygote (Figure 2.8b). Again, all the *AA* individual's gametes contain a single *A* gene. When a heterozygote reproduces, half its gametes contain an *A* gene, and half an *a*. The pair will produce *AA* : *Aa* offspring in a 50 : 50 ratio.

Finally, consider a cross between two heterozygotes (Figure 2.8c). Both male and female produces half *a* gametes and half *A* gametes. If we consider the female gametes (eggs or ovules), half of them are *a*, and half of them will be fertilized by *a* sperm, and half by *A* sperm; the other half are *A*, and half of them will be fertilized by *a* sperm and half by *A* sperm. The resulting ratio of offspring is 25% *AA* : 50% *Aa* : 25% *aa*.

The separation of an individual's two genes at a locus into its offspring is called *segregation*. The ratios of offspring types produced by different kinds of matings are

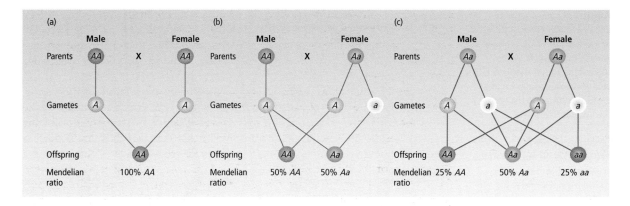

Figure 2.8
Mendelian ratios for: (a) an $AA \times AA$ cross, (b) an $AA \times Aa$ cross, and (c) an $Aa \times Aa$ cross.

examples of Mendelian ratios. They were discovered by Gregor Mendel in about 1856−63. Mendel was a monk, later Abbot, in St Thomas's Augustinian monastery in what was then Brünn in Austro-Hungary and is now Brno in the Czech Republic.

Mendelian ratios can also be for more than one genetic locus. If the alleles at one locus are A and a, and at a second B and b, then an individual will have a double genotype, such as Ab/Ab (double homozygote) or Ab/ab (single heterozygote). It has a double set of genes at each locus, one set from each parent. The segregation ratios now depend on whether the genetic loci are on the same or different chromosomes. Recall that an individual human has a haploid number of 23 chromosomes and about 30,000 genes. That means there must be on average about 1,300 genes per chromosome. Different genes on the same chromosome are described as being *linked*. Genes that are very close together are tightly linked, those further apart are loosely linked. Genes that are not on the same chromosome are unlinked.

> The Mendelian ratios for combinations of genes depend on whether the genes are linked or unlinked

The easy case is for two unlinked loci; the genes at the two loci then segregate independently. Imagine first a cross in which only one of the loci is heterozygous, such as a cross between an Ab/Ab male and an Ab/ab female. All the genes at the B locus are the same, while at the A locus the male is AA and the female is Aa. The ratio of offspring will be 50% $AAbb$ and 50% $Aabb$, a simple extension of the one-locus case.

A more complicated cross is for a male AB/Ab and a female AB/ab. Both parents are heterozygous for at least one locus. Again, the ratios of B locus genotypes associated with each A locus genotype are those predicted by applying Mendel's principles independently to each locus. A cross between two Bb heterozygotes produces a ratio of offspring of 25% BB : 50% Bb : 25% bb, and this ratio will be the same within each A locus genotype. Thus, in the cross between a male AB/Ab and a female AB/ab, there will be 50% AA and 50% Aa offspring. Of the half which are AA, 25% are AB/AB, 50% are AB/Ab, and 25% are Ab/Ab. Likewise for the 50% Aa genotypes. Add the two A genotypes and the total offspring ratios are:

AB/AB	AB/Ab	Ab/Ab	AB/aB	AB/ab	Ab/ab
1/8	1/4	1/8	1/8	1/4	1/8

Figure 2.9

Recombination seen at the level of: (a) chromosomes, (b) genes, and (c) nucleotides. At recombination, the strands of a pair of chromosomes break at the same point and the two recombine. The post-recombinational sequence of genes, or nucleotides, combine one strand from one side of the break-point with the other strand from the other side. In (b) the gene sequence in chromosome 1 changes from *ABC* to *ABc*; in (c) the nucleotide sequence of the chromosome with bases A and T (stippled nucleotides) changes to A and A. (For the nucleotide sequence only one of the strands of the double helix is shown: each of the pair of chromosomes has a full double helix with complementary base pairs, as in Figure 2.2.)

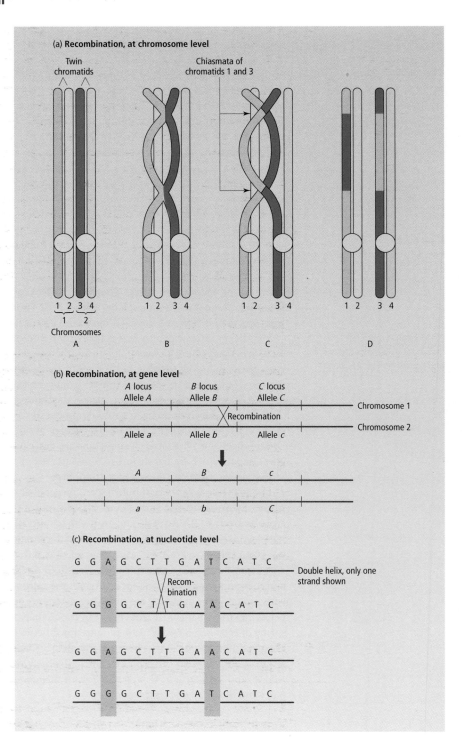

For linked genes, inheritance depends on recombination . . .

The offspring ratios for other crosses can be worked out by the same principle. The segregation of unlinked genotypes is called *independent segregation*.

When the loci are linked on the same chromosome, they do not segregate independently. At meiosis, when haploid gametes are formed from a diploid adult, an additional process called *recombination* occurs. The pairs of chromosomes physically line up and, at certain places, their strands join together and recombine (Figure 2.9). Recombination shuffles the combinations of genes. If an individual inherited *AB* from its mother and *ab* from its father, and recombination occurred between the two loci, it will produce *Ab* and *aB* gene combinations in its gametes.

Recombination is a random process; it may or may not "hit" any point in the DNA. It occurs with a given probability, usually symbolized by r, between any two points on a chromosome. r can be defined between nucleotide sites or genes. If the A locus and the B locus are linked, the chance of recombination between them in an individual is r and the chance of no recombination is $(1 - r)$. In any one individual, recombination either does or does not happen, but the chance of recombination determines the frequencies of genotypes in the gametes produced by a population. If we consider a large number of *AB/ab* individuals, they will produce gametes in the following proportions:

Gamete	*AB*	*Ab*	*aB*	*ab*
Proportion	$\frac{1}{2}(1 - r)$	$\frac{1}{2}r$	$\frac{1}{2}r$	$\frac{1}{2}(1 - r)$

. . . and Mendelian ratios can be calculated

These fractions can be used in the standard way to calculate the Mendelian ratio for a cross involving an *AB/ab* individual. The principle is logically easy to understand, but the ratios can be laborious to work out in practice. The case of independent segregation corresponds to $r = 0.5$. That is, when the A and B loci are on separate chromosomes, $r = (1 - r) = 0.5$ and the *Ab/aB* parent produces *Ab*, *aB*, *AB*, and *ab* gametes in the ratio $1 : 1 : 1 : 1$. For genes on the same chromosome, the value of r ranges from just above 0 for two sites that are next to each other up to 0.5 for two sites at opposite ends of the chromosome.

For any two genes, recombination can "hit" more than once between them in an individual (this is called "multiple hits"). If two recombinational hits occur between a pair of loci, they cancel each other. The chromosome has the same combination of genes at these two loci as if recombination had not occurred. It is more exact to say that the probability of recombination r equals the probability of an odd number of hits, and the probability $(1 - r)$ is the chance of no hits plus the chance of an even number of hits.

The Mendelian ratios, in which paired diploid genes segregate into haploid gametes and the gametes of different individuals then combine at random, is the basis of all the theory of population genetics that is discussed in Chapters 5–9.

2.9 Darwin's theory would probably not work if there was a non-Mendelian blending mechanism of heredity

As Chapter 1 described, Mendel's theory of heredity plugged a dangerous leak in Darwin's original theory, and the two theories together eventually came to form the

Figure 2.10
(a) Blending inheritance. The parental genes for dark green (*A*) and white (*a*) color blend in their offspring, who produce a new type of gene (*A'*) coding for light green color. (b) Mendelian inheritance. The parental genes are passed on unaltered by the offspring.

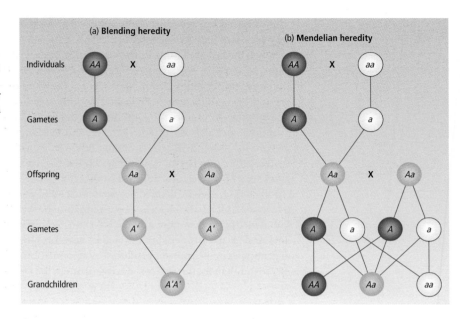

"Blending" heredity is a (theoretical) alternative to Mendelian heredity

synthetic theory of evolution, or neo-Darwinism. The problem was Darwin's lack of a sound theory of heredity, and indeed it had even been shown in Darwin's time that natural selection would not work if heredity was controlled in the way that, before Mendel, most biologists thought it was. Before Mendel, most theories of heredity were *blending* theories. We can see the distinction in much the same terms as have just been used for Mendelism (Figure 2.10). Suppose there is a gene *A* that causes its bearers to grow up dark green in color, and another gene *a* that causes its bearers to grow up white. We can imagine that, as in the real world of Mendelism, so in our imaginary world of blending heredity, individuals are diploid and have two copies of each "gene." An individual could then either inherit an *AA* genotype from its parents and have a dark green phenotype, or inherit an *aa* genotype and have a white phenotype, or an *Aa* genotype and have a light green phenotype. (Thus in the Mendelian version of the system, we should say there is no dominance between the *A* and *a* genes.)

The interesting individuals for this argument are the ones that have inherited an *Aa* genotype and have grown up to be light green. They could have been produced in a cross of a dark green and a white parent: then the offspring will be light green whether inheritance is Mendelian (with no dominance) or blending. But now consider the next generation. Under Mendelian heredity, the light green *Aa* heterozygote passes on intact to its offspring the *A* and *a* genes it had inherited from its father and mother. Under blending heredity, the same is not true. An individual does not pass on the same genes as it inherited. If an individual inherited an *A* and an *a* gene, the two would physically blend in some way to form a new sort of gene (let us call it *A'*) that causes light green coloration. And instead of producing 50% *A* gametes and 50% *a*, it would then produce all *A'* gametes. This makes a difference in the second generation. Whereas in Mendelian heredity, the dark green and white colors segregate out again in a cross between two heterozygotes, in the analogous cross with blending heredity they do not — all the grandchildren are light green (see Figure 2.10).

Mendelian heredity conserves
genetic variation

Mendelism is an atomistic theory of heredity. Not only are there discrete genes that encode discrete proteins, but the genes are also preserved during development and passed on unaltered to the next generation. In a blending mechanism, the "genes" are not preserved. The genes that an individual inherits from its parents are physically lost, as the two parental sets are blended together. In Mendelism, it is perfectly possible for the *phenotypes* of the parents to be blended in the offspring (as they are in the initial $AA \times aa$ cross in Figure 2.10), but the genes do not blend. Indeed, the phenotypes of real mothers and fathers often do blend in their offspring, and it was because they do that most students of heredity before Mendel thought that inheritance must be controlled by some blending mechanism. However, the case of heterozygotes that are intermediate between the two homozygotes (i.e., no dominance) shows that the blending of phenotypes need not mean blending of genotypes. In fact, the underlying genes are preserved.

One way of expressing the importance of Mendelism for Darwin's theory is to say that it efficiently preserves genetic variation. In blending inheritance, variation is rapidly lost as extreme types mate together and their various "genes" are blended out of existence into some general mean form. In Mendelian inheritance, variation is preserved because the extreme genetic types (even if disguised in heterozygotes) are passed down from generation to generation.

Why does this preservation of genes matter for Darwinism? Our full discussion of natural selection comes in later chapters, and some readers may prefer to return to this point after they have read about natural selection in more detail; but even with only the elementary account of natural selection in Chapter 1, it is possible to understand why Darwin, so to speak, needed Mendel. Figure 2.11 illustrates the argument.

Natural selection is more powerful
with Mendelian, than with
blending, heredity

Suppose that a population of individuals is white in color, and has the *aa* genotype, and heredity blends (in the manner of Figure 2.10). For some reason, it is advantageous for individuals in this population to be dark green in color: dark green individuals would survive better and leave more offspring. Moreover, it is better to be a bit green (i.e., light green) than to be white. Suppose now that a single new light green individual somehow crops up by mutation, and it has an *Aa* genotype. This *Aa* individual will survive better than its *aa* fellow members of the population and will produce more offspring. However, the advantageous gene cannot last long with blending. In the first generation it produces *A′* gametes; these combine with *a* gametes (because every one else in the population is white) and produce *A′a* offspring. We can suppose these individuals are a bit lighter green in color than the original *Aa* mutant; they still have an advantage, but it is lower.

The *A′a* individual's genes in turn blend, such that all its gametes will have an *A″* gene. When that unites with an *a* gamete (because still almost everyone else is white) an *A″a* offspring results, which is even lighter green in color. It is only a matter of time and the original favorable mutation will be blended almost out of existence (Figure 2.11a). The best result possible would be a population that was very slightly less white than it was to begin with. A population of dark green individuals cannot be produced from the original mutation. That original mutation, which potentially was able to produce dark green individuals, will cease to exist after one generation. This objection to the theory

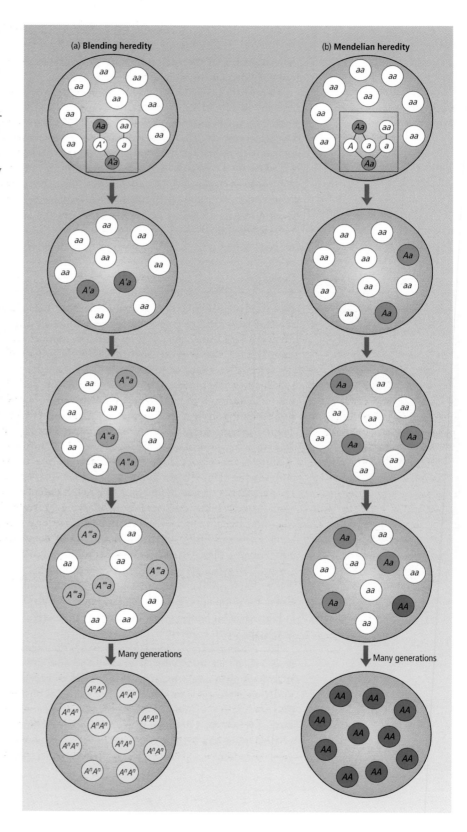

Figure 2.11

Two populations with 10 individuals each (real populations would have many more members), one with blending heredity and the other with Mendelian heredity. (a) Under blending heredity, a rare new advantageous gene is soon blended away. (b) Under Mendelian heredity, a rare new favorable gene can increase in frequency and eventually become established in the population. See text for explanation.

of natural selection was known to Darwin. Darwin was very worried by it and never did find a wholly satisfactory way round it.

Mendelism was what he needed. In the example just given, the original light green mutation will be in an *Aa* heterozygote, and fully half its offspring will be light green like it — because they are also *Aa* heterozygotes. There is ample time for natural selection to increase the proportion of light green individuals, and eventually there would be enough of them for there to be a chance that two will mate together and produce some *AA* homozygotes among their offspring. A population of dark green *AA* individuals can now theoretically be produced (Figure 2.11b). Thus natural selection is a powerful process with Mendelian heredity, because Mendelian genes are preserved over time; whereas it is at best a weak process with blending inheritance, because potentially favorable genes are diluted before they can be established.

Summary

1 Heredity is determined by a molecule called DNA. The structure and mechanisms of action of DNA are understood in detail.

2 The DNA molecule can be divided into regions called genes that encode for proteins. The code in the DNA is read off to produce a protein in two stages: transcription and translation. The genetic code has been deciphered.

3 DNA is physically carried on structures called chromosomes. Each individual has a double set of chromosomes (one inherited from its father, the other from its mother), and therefore two sets of all its genes. An individual's particular combination of genes is called its genotype.

4 New genetic variation originates by mutational changes in the DNA. Rates of mutation can be estimated by direct observation.

5 When two individuals, of given genotypes, mate together, the proportions of genotypes in their offspring appear in predictable Mendelian ratios. The exact ratios depend on the genotypes in the cross.

6 Different genes are preserved over the generations under Mendelian heredity, and this enables natural selection to operate. Before Darwin, it was generally (but wrongly) thought that the maternal and paternal hereditary materials blended in an individual rather than being preserved. If heredity did blend, natural selection would be much less powerful than under Mendelian heredity.

Further reading

Any genetics text, such as Lewin (2000), Griffiths *et al.* (2000), or Weaver & Hedrick (1997), explains the subject in detail. I include an account of mutation rate measurements in a popular book (Ridley 2001); see also the reviews referred to in Chapter 12 below. The classic statement of why Darwinism requires Mendelism, and does not work with blending heredity, is in the first chapter of Fisher (1930), which was reprinted, editorially reduced, in Ridley (1997). Graveley (2001) explains alternative splicing.

Study and review questions

1 Review your understanding of the following genetic terms: DNA, chromosome, gene, protein, genetic code, transcription, translation, mRNA, tRNA, rRNA, mutation, synonymous, non-synonymous, frameshift, inversion, genetic locus, meiosis, genotype, phenotype, homozygote, heterozygote, dominant, recessive, linked, unlinked, and recombination.

2 What are the Mendelian ratios for the following crosses: (i) $AA \times AA$, (ii) $AA \times Aa$, (iii) $Aa \times Aa$, (iv) $AB/AB \times AB/AB$, when the A and B loci are unlinked, and (v) $AB/AB \times AB/AB$, when the A and B loci are tightly linked ($r = 0$)?

3 What are the ratio, or fractions, of gametes (according to their gene combinations) produced by an individual with the two-locus genotype AB/ab if: (i) the A and B loci are unlinked, and (ii) the A and B loci are linked and the rate of recombination between the two loci is r?

3

The Evidence for Evolution

How can it be shown that species change through time, and that modern species share a common ancestor? We begin with direct observations of change on a small scale and move out to more inferential evidence of larger scale change. We then look at what is probably the most powerful general argument for evolution: the existence of certain kinds of similarity (called homologies) between species — similarities that would not be expected to exist if each species had originated independently. Homologies fall into hierarchically arranged clusters, as if they had evolved through a tree of life and not independently in each species. The order in which the main groups of animals appear in the fossil record makes sense if they arose by evolution, but would be highly improbable otherwise. Finally, the existence of adaptation in living things has no non-evolutionary explanation, though the exact way that adaptation can be used to suggest evolution depends on what alternative is being argued against.

3.1 We distinguish three possible theories of the history of life

Life could have had various kinds of history

In this chapter, we shall be asking whether, according to the scientific evidence, one species has evolved into another in the past, or whether each species had a separate origin and has remained fixed in form ever since that origin. For purposes of argument, it is useful to have some articulate alternatives to argue between. We can discuss three theories (Figure 3.1): (a) evolution; (b) "transformism," in which species do change, but there have been as many origins of species as there have been species; and (c) separate creation, in which species originated separately and remain fixed. The chapter will therefore look at evidence for two evolutionary claims. One is that species have changed in Darwin's sense of "descent with modification." The other is that all species share a common ancestor — that the change has been through a tree-like history.

Whether species have separate origins, and whether they change after their origin, are two distinct questions; some kinds of evidence, therefore, may bear upon one of question but not the other. At this stage, we need not have any particular mechanism in mind to explain either how species spring into existence so easily in the theories of transformism and separate creation (Figure 3.1b–e), or how they change in form in the theories of evolution and transformism (Figure 3.1a,b). We merely suppose it could happen by some natural mechanism, and ask which of the three patterns is supported by the evidence.

We shall consider a number of lines of biological evidence. We do so because people differ in what they see as the main objection to the idea of evolution, and different kinds of evidence, or argument, are persuasive for different people. For instance, someone who had not thought about the matter before might suppose that the world has always been much like it is now, because the plants and animals do not seem to change much from year to year in their yard — or their neighbor's yard for that matter. For them, the mere demonstration of bizarre extinct animals, like dinosaurs or the animals of the Burgess Shale, would suggest that the world has not always been the same, and might make them open to the idea of evolution.

The existence of fossil species unlike anything alive today, however, does not distinguish between the three theories of life in Figure 3.1. An extinct species could just as well have been separately created as any modern species. The theory of separate creation can easily be modified to account for extinct forms. Either there was one period in which all species separately originated and some have subsequently gone extinct (Figure 3.1d) or there were rounds of extinction followed by rounds of creation

Figure 3.1
Three theories of the history of life: (a) evolution, (b) transformism, and (c–e) creationism. (a) In evolution, all species have a common origin, and they may change through time. (b) In transformism, species have separate origins, but they may change. (c–e) In separate creation, species have separate origins and do not change; each are different versions of the theory of separate creation that might be proposed to explain extinct fossil forms, and they do not differ in their two essential features (species have separate origins and do not change). Each line represents a species in time. If the line moves up vertically the species is constant, if it deviates to the left or right the species is changing in form.

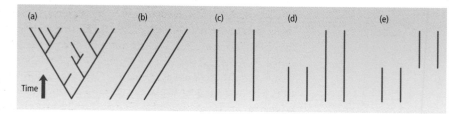

(Figure 3.1e). All three versions of separate creation (Figure 3.1c–e) share the key features that species have separate origins and do not change in form after their origin. As it happens, some early paleontologists, who worked before the theory of evolution had been accepted, were well aware how different past faunas were from the present. They suggested that the history of life looked rather like the pattern in Figure 3.1e. The history of life was thought of as a succession of rounds of extinction followed by the creation of new species.

We concentrate here on evidence that can be used to test between the three theories in Figure 3.1. We begin with straightforward observation, on the small scale. If someone doubts that species can change at all, this evidence will be useful. Other people allow that change happens on the small scale, and doubt that it can accumulate to produce large-scale change, such as a new species, or a new major group like the mammals. We work out from small-scale change to see how the case for larger scale evolutionary change can be made.

3.2 On a small scale, evolution can be observed in action

HIV illustrates evolution, on a timescale of days

The virus — human immunodeficiency virus (HIV) — that causes AIDS uses RNA as its hereditary material. It reproduces by having a DNA copy made of its RNA, inside a human cell. The normal transcription machinery of the cell will then run off multiple copies of the RNA version of the virus. Most of the reproductive process is performed by enzymes supplied by the host cell, but the virus supplies the enzyme called reverse transcriptase that makes the DNA version of the virus from the RNA version. Reverse transcriptase is not normally present in human cells, because humans do not normally convert RNA into DNA. The reverse transcriptase is a favorite target for anti-HIV drugs. If reverse transcriptase can be inactivated by a drug, the virus is stopped from reproducing without any damaging side effects on the cell.

Many drugs have been developed against reverse transcriptase. One large class of these drugs consists of nucleoside inhibitors. (A nucleoside is a nucleotide without the phosphate; it is a base plus a sugar, either ribose or deoxyribose.) The drug 3TC, for example, is a molecule similar to the nucleotide cytosine (symbolized by C), the normal constituent of DNA. The reverse transcriptase of drug-susceptible HIV will incorporate 3TC instead of C into a growing DNA chain. The 3TC then inhibits future reproduction, and thus prevents the HIV from copying itself.

A paper by Schuurman *et al.* (1995) describes what happens when human AIDS patients are treated with 3TC. Initially the HIV population in the human body decreases by a huge amount. But then, within days, 3TC-resistant strains of HIV start to be detected. The drug-resistant HIV then increases in frequency. In eight of 10 patients, drug-resistant strains had increased to 100% of the viral population in the patient's body within 3 weeks of the start of the drug treatment (it took 7 and 12 weeks in the other two patients). The change, from a viral population that was susceptible to 3TC to a viral population that was resistant to 3TC, is an example of evolution by natural selection. The evolution takes place within a single human body, and is exceptionally rapid relative to most examples of evolution. But the process observable over a few weeks in

Figure 3.2

Evolution of drug resistance in HIV. 3TC is a nucleoside inhibitor and it resembles C. (a) Drug-susceptible reverse transcriptase binds both 3TC and C. When 3TC is incorporated into a growing DNA chain, it inhibits further replication. (b) Drug resistance is achieved by the evolution of reverse transcriptase that binds only C, and not 3TC.

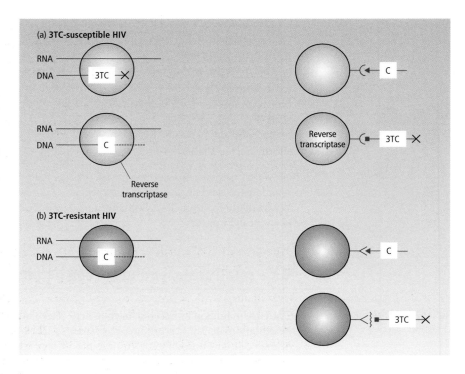

an AIDS patient is a microcosm of the process that has caused much of the diversity of life on Earth.

HIV evolves drug resistance

The evolution of drug resistance can be followed at the molecular level. The change from 3TC-susceptible HIV to 3TC-resistant HIV is achieved by a change in one codon in the gene that codes for reverse transcriptase. The amino acid methionine is changed to one of three other amino acids. The methionine is in a part of the reverse transcriptase that interacts with the nucleosides. Probably what is happening is that the normal reverse transcriptase is a relatively undiscriminating enzyme that does not distinguish between C and 3TC. The change makes the enzyme more discriminating, such that it binds C but does not bind 3TC. The virus can then reproduce in the presence of 3TC (Figure 3.2). The superior discrimination is paid for by slower reproduction, and the 3TC-resistant version of HIV is therefore at a disadvantage when the drug is not present. In the presence of the drug it is adaptive for HIV to reproduce slowly but carefully. In the absence of the drug it is adaptive to reproduce faster, and in a molecularly care-free manner.

Other examples exist too

Drug resistance in HIV is one of many examples in which evolution has been observed on a small scale. In other examples, evolutionary change has been detected in periods of years rather than days. In Section 5.7 (p. 108) we look at the famous example of evolution in the peppered moth (*Biston betularia*). In Section 9.1 (p. 223) we look at changes in the average beak size of a population of a finch species in the Galápagos islands. In Section 13.4.1 (p. 359) we look at geographic variation in the house sparrow

(*Passer domesticus*) in North America. This is another example of evolution on a human timescale. The differences between sparrows in California (where they are smaller, with a wing length averaging 2.96 in (76 mm)) and in Canada (where they are larger, with a wing length averaging 3.08 in (79 mm)) have all evolved from a colony of sparrows that was introduced to Brooklyn, New York, in 1852. The differences had evolved at least by the 1940s, which means that they evolved in less than 100 generations (Johnston & Selander 1971). Most species do not evolve as fast as North American house sparrows, British peppered moths, or HIV in countries where drug treatment is affordable, but all these examples are useful to illustrate that evolution is an observable fact.

3.3 Evolution can also be produced experimentally

Artificial selection produces evolutionary change

In a typical artificial selection experiment, a new generation is formed by allowing only a selected minority of the current generation to breed (Figure 3.3). The population in almost all cases will respond: the average in the next generation will have moved in the selected direction. The procedure is routinely used in agriculture — artificial selection has, for example, been used to alter the numbers of eggs laid by hens, the meat properties of bullocks, and the milk yield of cows. We shall meet several more examples of artificial experiments later (Section 9.7, p. 236), but we can look at a curiosity here for purposes of illustration (Figure 3.4). In an experiment, rats were selected for increased or decreased susceptibility to dental caries on a controlled diet. As the graph shows, the rats could be successfully selected to grow better or worse teeth. Evolutionary change can therefore be generated artificially.

(a) **Generation 1**

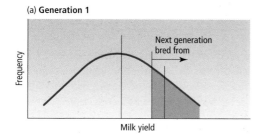

(b) **Generation 2**

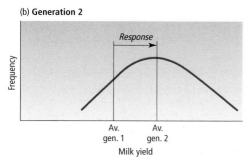

Figure 3.3

An artificial selection experiment. Generation 2 is formed by breeding from a selected minority (shaded area) of the members of generation 1. Here, for example, we imagine a population of cows and selectively breed for high milk yield. In nearly all cases, the average in the second generation changes from the first in the selected direction.

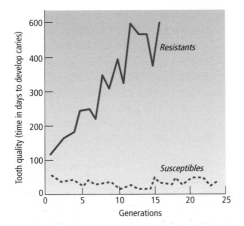

Figure 3.4

Selection for better and worse teeth in rats. Hunt *et al.* (1955) selectively bred each successive generation of rats from parental rats that developed caries later (resistants) or earlier (susceptibles) in life. The age (in days) at which their descendants developed caries was measured.

Artificial selection can produce dramatic change, if continued for long enough. A kind of artificial selection, for example, has generated almost all our agricultural crops and domestic pets. No doubt the artificial selection in these cases — begun thousands of years ago in some cases — employed less formal techniques than would a modern breeder. However, the longer timespan has led to some striking results. Darwin (1859) was impressed by the varieties of domestic pigeons, and chapter 1 of *On the Origin of Species* begins with a discussion of those birds. The point here of these, and similar, examples is to illustrate further how, on a small scale, species can be shown experimentally not to be fixed in form.

3.4 Interbreeding and phenotypic similarity provide two concepts of species

We are now close to the stage in the argument when we can consider evidence for the evolution of new species. Most of the evidence so far has been for small-scale change within a species. The amounts of artificially selected change in pigeons and other domestic animals borders on the species level, but to decide whether the species barrier has been crossed we need a concept of what a biological species is.

Living creatures are classified into species, and higher taxa

All living creatures are classified into a Linnaean hierarchy. The species is the lowest important level in the hierarchy. Species, in turn, are grouped into genera, genera into families, and so on up through a series of levels. Figure 3.5 gives a fairly complete Linnaean classification of the wolf, as an example. If all life has descended from a single common ancestor, evolution must be capable of producing new groups at all levels in the hierarchy, from species to kingdom. We shall be looking at the evidence in the rest of this chapter. Here, however, we are at the species stage. What does it mean to say a new species has evolved?

The question unfortunately lacks a simple answer that would satisfy all biologists. We shall discuss the topic fully in Chapter 13, and we shall see that there are several concepts of species. What we can do here is to take two of the most important species

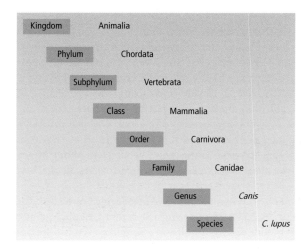

Figure 3.5
Each species in a biological classification is a member of a group at each of a succession of more inclusive hierarchical levels. The figure gives a fairly complete classification of the gray wolf *Canis lupus*. This way of classifying living things was invented by the eighteenth-century Swedish biologist who wrote under the latinized name Carolus Linnaeus.

concepts and see for each what the evidence for the evolution of new species is. In arguing for evolution, we do not have to say what a species is. If someone says, what's the evidence that evolution can produce a new species, we can reply "you tell me what you mean by species, and I'll tell you the evidence."

Species can be defined by interbreeding, . . .

One important species concept is reproductive, and defines a species as a set of organisms that interbreed among themselves but do not breed with members of other species. Humans (*Homo sapiens*) are a separate reproductive species from the common chimpanzee (*Pan troglodytes*): any human can interbreed with any other human (of appropriate sex), but not with a chimp.

. . . or by similarity of appearance

The second important concept uses phenotypic appearance: it defines a species as a set of organisms that are sufficiently similar to one another and sufficiently different from members of other species. This is a less objective definition than the reproductive definition — it is clear whether the members of two population interbreed or not, but it is less clear whether the two are sufficiently different to count as two phenotypic species. The final answer often lies with an expert who has studied the forms in question for years and has acquired a good knowledge of the difference between species; formal methods of answering the question also exist. However, for relatively familiar animals we all have an intuitive phenotypic species concept. Again, humans and common chimpanzees belong to different species, and they are clearly distinct in phenotypic appearance. Common suburban birds, such as robins, mockingbirds, and starlings are separate species, and can be seen to have distinct coloration. Thus, without attempting a general and exact answer to the question of how different two organisms must be to belong to separate species, we can see that phenotypic appearance might provide another species concept in addition to reproduction.

Because some biologists reject one or other concept, we should look at the evidence for the evolution of new species according to both concepts. As we move up the Linnaean hierarchy, to categories above the species level, the members of a group become less and less similar. Two members of the same species, such as two wolves, are more similar than are two members of the same genus but different species, such as a wolf and a silver-backed jackal (*Canis mesolemas*); and two members of the same class (Mammalia) can be as different as a bat, a dolphin, and a giraffe.

Artificial selection has produced larger differences than between natural species

What degree of difference, in these taxonomic terms, has been produced by artificial selection in domestic animals? All domestic pigeons can interbreed, and are members of the same species in a reproductive sense. The answer is different for their phenotypic appearance. Museum experts often have to classify birds from dead specimens, of unknown reproductive habits, and they make use of phenotypic characters of the bones, beak, and feathers. Darwin kept many varieties of pigeons, and in April 1856, when Lyell came for a visit, Darwin was able to show him how the 15 pigeon varieties he had at the time differed enough to make "three good genera and about fifteen species according to the received mode of species and genera-making of the best ornithologists."

The variety of dogs (*Canis familiaris*) is comparable. To most human observers, the difference between extreme forms, such as a pekinese and a St Bernard, is much greater than that between two species in nature, such as a wolf and a jackal, or even two species in different genera, such as a wolf and an African hunting dog (*Lycaon pictus*). However, most domestic dogs are interfertile and belong to the same species in a reproductive sense. The evidence from domestic animals suggests that artificial selection can produce extensive change in phenotypic appearance — enough to produce new species and even new genera — but has not produced much evidence for new reproductive species. We shall come to evidence for the evolution of new reproductive species in a later section.

3.5 Ring "species" show that variation within a species can be extensive enough to produce a new species

At any one time and place, there do appear to be an array of distinct species in nature. For example, a naturalist in southern California might have noticed two forms of the salamander *Ensatina*. One form, the species *Ensatina klauberi*, is strongly blotched in color whereas the other, the species *E. eschscholtzii*, is more uniformly and lightly pigmented. It had been suspected since the work of Stebbins in the 1940s that they were two good species in the sense that they are distinct forms that do not interbreed where they coexist. For one site, 4,600 feet (1400 m) up the Cuyamaca Mountains, San Diego County, Wake *et al.* (1986) confirmed that the two are indeed behaving as separate species. At that site, called Camp Wolahi, the two species coexist; but no hybrid forms between them were found, and the genetic differences between the two species there suggested they had not interbred in the recent past. Salamander naturalists who visited Camp Wolahi would have no doubt they were looking at two ordinary, different species.

Two Californian salamanders interbreed in some places . . .

. . . but not in others

However, if those naturalists looked further for the two salamander species in other areas of southern California, the two species do not seem to be as distinct as at Camp Wolahi. Wake *et al.* sampled the salamanders from three more sites nearby, and at all of them a small proportion (up to 8%) of individuals in the sample were hybrids between *E. eschscholtzii* and *E. klauberi*. The picture becomes clearer as we expand the geographic scale. The salamanders can be traced westward from Camp Wolahi to the coast, and northward up the mountain range (see Plate 1, opposite p. 68). However, in either direction, only one of the salamanders is present. Along the coast there is the lightly pigmented, unblotched form *E. eschscholtzii*, while inland there is the blotched *E. klauberi*. The forms can be traced up to northern California, but they vary in form toward the north; the various forms have been given a series of taxonomic names, as can be seen in Plate 1. They meet again in northern California and Oregon, but here only one form is found; the eastern and western forms have apparently merged completely.

The salamanders are an example of a ring species, . . .

The classic interpretation of the salamanders' geographic pattern is as follows. There was originally one species, living in the northern part of the present range. The population then expanded southwards, and as it did so it split down either side of the central San Joaquin Valley. The subpopulation on the Pacific side evolved the color pattern and genetic constitution characteristic of the coastal *E. eschscholtzii*, while the subpopulation inland evolved the blotches, and the genetic constitution characteristic of *E. klauberi*. At various points down California, subpopulations leaked across and met the other form. At some of these meeting areas the two forms interbreed to some extent, and hybrids can be found: there, they have not evolved apart enough to be separate reproductive species. But by the southern tip of California, the two lines of population have evolved far enough apart that when they meet, such as at Camp Wolahi, they do not interbreed: there they are two normal species. Thus the two species at Camp Wolahi are connected by a continuous set of intermediate populations, looped around the central valley.

. . . though there are complications

The detailed picture is more complicated, but recent work supports essentially the same interpretation. One of the complications can be seen in Plate 1, which shows that the set of populations may not be perfectly continuous: the map shows a gap in the southeastern part of the ring. Jackman & Wake (1994) showed that the salamander populations on either side of the gap are genetically no more different than are salamanders separated by an equivalent distance elsewhere in the ring. They suggest two interpretations. One is that salamanders lived in the gap until recently but are now extinct there; the other is that the blotched *Ensatina* are there and waiting to be found "in the rugged San Gabriel Mountains."

The salamander species *E. eschscholtzii* and *E. klauberi* in southern California are an example (not the only one) of a *ring species*. A ring species can be imagined in the abstract as follows. First imagine a species that is geographically distributed more or less in a straight line in space, say from east to west across America. It could be that the forms in the east and west are so different that they could not interbreed; but we are unlikely to know because the two forms do not meet each other. Now imagine taking the line and bending it into a circle, such that the end-points (formerly in the east and west) come to overlap in space. It will then be possible to find out whether the two extremes do interbreed. Either they do or they do not. If they do interbreed then the geographic distribution of the species will be in the shape of a ring, but it will not be a "ring species" in the technical sense.

A proper ring species is one in which the extreme forms do not interbreed in the region of overlap. A ring species has an almost continuous set of intermediates between two distinct species, and these intermediates happen to be arranged in a ring. At most points in the ring, there is only one species; but there are two where the the end-points meet. (The statement above that the extremes either do or do not interbreed is too categorical for real cases, which are typically more complicated. In the salamanders, for instance, there is hybridization at some sites but not at others in southern California where the ring closes up. The real situation is then not a simple ring, but can be understood as a ring species, with due allowance for real world complications.)

Ring species can provide important evidence for evolution, because they show that intraspecific differences can be large enough to produce an interspecies difference. The differences between species are therefore the same in kind (though not in degree) as the differences between individuals, and populations, within a species. The argument can be spelled out more.

Natural variation comes in all degrees. At the smallest level, there are slight differences between individuals. Populations of a species show rather larger differences, and species are more different still. In a normal species, whose members are perhaps distributed in something like the line we imagined above, the extreme forms may be very different from one another; but we do not know whether they are different enough to count as separate species in the reproductive sense. A supporter of the theory of separate creation might then argue that although individuals do vary within a species, nevertheless that variation is too limited ever to give rise to a new species. The origin of new species is then not a magnified extension of the kind of variation we see within a species. But in ring species the extremes meet, and we can see that they form two species. It is then almost impossible to deny that natural variation can, at least sometimes, be large enough to generate new species. At least some species, therefore, have arisen without separate creation.

There is a slippery slope from interindividual variation all the way up to the difference between two species. Small individual differences, we know, arise by the ordinary processes of reproduction and development: we can *see* that each individual is not separately created. By extension, the slightly larger differences between local populations, are easily seen to arise without separate creation. In the case of the ring species of salamanders, this process can be seen to extend far enough to produce a new species. To deny it would require an arbitrary decision about where evolution stopped and separate creation started.

Suppose, for example, someone claimed that all salamanders to the west of a point in northern California were separately created as a different species from all those to the east of it (though he or she allows that the variation within each of the species on either side of the point arose by ordinary natural evolutionary processes). The claim is clearly arbitrary and absurd. If evolution has produced the variation between salamanders in northern California and in mid-California on the coast, and between northern California and mid-California inland, it is absurd to suggest that the populations in the east and the west of northern California were separately created. The variation between any two points in the ring is of much the same kind, and the variation across the arbitrarily picked point will be just like the variation among two points to the left or right of it. Ring species show that there is a continuum from interindividual to interspecies variation. Natural variation is sufficient to break down the idea of a distinct species boundary.

The same argument, we shall see, can be applied to larger groups than species, and by extension to all life. The idea that nature comes in discrete groups, with no variation

Ring species show there is nothing special about species differences

Natural variation comes in all degrees

between, is a naive perception. If the full range of natural forms, in time and space, is studied, all the apparent boundaries become fluid.

3.6 New, reproductively distinct species can be produced experimentally

New species have been produced artificially

The species barrier can be broken by experiment too. The varieties of artificially produced domestic animals and plants can differ in appearance at least as much as natural species; but they may be able to interbreed. Dog breeds that differ greatly in size probably in practice interbreed little, but it is still interesting to know whether we can make new species that unambiguously do not interbreed. Reduced interbreeding between two forms can be directly selected for (Section 14.6.3, p. 402).

More extreme, and more abundant, examples of new, reproductively isolated species come from plants. The typical procedure is as follows. We begin with two distinct, but related species. The pollen of one is painted on the stigma of the other. If a hybrid offspring is generated, it is usually sterile: the two species are reproductively isolated. However, it may be possible to treat the hybrid in such a way as to make it fertile. The chemical colchicine can often restore hybrid fertility. It does so by causing the hybrid to double its number of chromosomes (a condition called polyploidy). Hybrids so produced may be interfertile with other hybrids like themselves, but not with the parental species. They are then a new reproductive species. They provide clear evidence that new species in the reproductive sense can be produced. If we add them to the examples of dogs and pigeons, we have now seen evidence for the evolution of new species according to both the reproductive and the phenotypic species concepts.

The Kew primrose was the first example

The first artificially created hybrid polyploid species was a primrose, *Primula kewensis*. It was formed by crossing *P. verticillata* and *P. floribunda*. *P. kewensis* is a distinct species: a *P. kewensis* individual will breed with another *P. kewensis* individual, but not with members of *P. verticillata* or *P. floribunda*. *P. verticillata* and *P. floribunda* have 18 pairs of chromosomes each, and simple hybrids between them also have 18 chromosomes. These hybrids are sterile. *P. kewensis* has 36 chromosomes and is a fertile species. The chromosome doubling in this case was not induced artificially, by colchicine treatment, but occurred spontaneously in a hybrid plant.

The same mechanism occurs in nature

Hybridization, followed by the artificial induction of polyploidy, is now a common method of producing new agricultural and horticultural varieties. Most garden varieties of irises, tulips, and dahlias, for example, are artificially created species. But their numbers are dwarfed by the huge numbers of artificial hybrid species of orchids, which it has been estimated are being formed at the rate of about 300 per month. Polyploid hybridization is also important in natural plant evolution. Section 14.7 (p. 405) discusses hybrid speciation in plants further, and we shall meet there the example of *Tragopogon* in the Washington–Idaho region. In these plants, two new species have originated in the past century by natural hybridization and polyploidy.

The most powerful method to show that a natural species originated as a hybrid is to recreate it from its ancestors, by hybridizing the conjectural parental species experimentally. This was first done for a common European herb, *Galeopsis tetrahit*, which

Müntzing in 1930 successfully created by hybridizing *G. pubescens* and *G. speciosa*. The artificially generated *G. tetrahit* can successfully interbreed with naturally occurring members of the species. This method is more time consuming than simple chromosome counts and has only been used with a small number of species. In conclusion, it is possible to make new, reproductively isolated species, using a method that has been highly important in the origin of new natural species.

3.7 Small-scale observations can be extrapolated over the long term

We have now seen that evolution can be observed directly on a small scale. The extreme forms within a species can be as different as two distinct species, and in nature and experiments, species will evolve into forms highly different from their starting point. It would be impossible, however, to observe in the same direct way the whole evolution of life from its common, single-celled ancestor a few billion years ago. Human experience is too brief. As we extend the argument from small-scale observations, like those described in HIV, dogs, and salamanders, to the history of all life we must shift from observation to inference. It is possible to imagine, by extrapolation, that if the small-scale processes we have seen were continued over a long enough period they could have produced the modern variety of life. The reasoning principle here is called *uniformitarianism*. In a modest sense, uniformitarianism means merely that processes seen by humans to operate could also have operated when humans were not watching; but it also refers to the more controversial claim that processes operating in the present can account, by extrapolation over long periods, for the evolution of Earth and life. For instance, the long-term persistence of the processes we have seen in moths and salamanders could result in the evolution of life. This principle is not peculiar to evolution. It is used in all historic geology. When the persistent action of river erosion is used to explain the excavation of deep canyons, the reasoning principle again is uniformitarianism.

Differences, it may be argued, can be of kind as well as degree. For instance, many creationists believe that evolution can operate within a species, but cannot produce a new species. Their reason is a belief that species differences are not simply a magnified version of the differences we see between individuals. As a matter of fact, this particular argument is false. For the salamanders (*Ensatina*) in California, we saw the smooth continuum of increasing difference, from the variation between individual salamanders in a region, to interregional variation, to speciation. Someone who permits uniformitarian extrapolation only up to a certain point in this continuum will inevitably be making an arbitrary decision. The differences immediately above and below the point will be just like the differences across it.

Analogous arguments to the one about species are sometimes made for higher taxonomic levels. It may be said, for example, that evolution is only possible within defined "types" (a type might be something like "dogs" or "cats," or even "birds" or "mammals"). But the evolutionist will advance the same counterargument as for species. Nature only appears to be divided up into discrete types at any one time and place.

Human observation is too short term to witness the whole history of life

But human-scale observations can be extrapolated

And many facts fit in with these extrapolations

Further study erodes the impression away. The fossil record contains a continuous set of intermediates between the mammals and reptiles, and these fossils destroy the impression that "mammals" are a discrete type (Section 18.6.2, p. 542). *Archaeopteryx* does the same for the bird type, and there are many further examples. In any case, if someone tries to argue that differences of kind arise at a certain level in the taxonomic hierarchy, they will be faced with these sorts of counterexample. If we draw on enough specimens from time and space, a strong argument can be made that organic variation is continuous, from the smallest difference between a pair of twins through to the whole history of life.

The argument for evolution does not have to rely only on small-scale observations and the principle of uniformitarianism. Other kinds of evidence also suggest that living things are descended from a common ancestor. The evidence comes from certain similarities between species, and from the fossil record.

3.8 Groups of living things have homologous similarities

If we take any two living species, they will show some similarities in appearance. Here, we need to distinguish two sorts of similarity: *homologous* and *analogous* similarity.[1] An analogous similarity, in this non-evolutionary, pre-Darwinian sense, is one that can be explained by a shared way of life. Sharks, dolphins, and whales all have a hydrodynamic shape which can be explained by their habit of swimming through water. Their similar shape is analogous; it is a functional requirement. Likewise, the wings of insects, birds, and bats are all needed for flying: they too are analogous structures.

Living creatures show similarities that would not be expected if they had independent origins

Other similarities between species are less easily explained by functional needs. The pentadactyl (five digit) limb of tetrapods is a classic example (Figure 3.6). (Tetrapods are the group of vertebrates with four legs. Amphibians, reptiles, birds, and mammals are tetrapods; fish are not.) Tetrapods occupy a wide variety of environments, and use their limbs for many differing functions. There is no clear functional or environmental reason why all of them should need a five-digit, rather than a three- or seven- or 12-digit limb. And yet they all do; or, rather, all modern tetrapods do — fossil tetrapods are known from the time in the Devonian when tetrapods were evolving from fish that have six-, seven-, and eight-digited limbs (see Figure 18.1, p. 526, for geological periods such as the Devonian). Some modern tetrapods, in the adult form, do not appear to have five-digit limbs (Figure 3.6). The wings of birds and bats are in different ways supported by less than five digits, and the limbs of horses and of some lizards also have less than five digits. However, all these limbs develop embryologically from five-digited precursor stages, showing that they are fundamentally pentadactyl. Even the boneless hind fin of the whale conceals the vestiges of the characteristic tetrapod five-digit pattern. In Darwin's (1859) words,

[1] In this chapter, these terms have a non-evolutionary meaning, which was common before Darwin. They should not be confused with the evolutionary meanings (Section 15.3, p. 427). The non-evolutionary usage is needed here in order to avoid a circular argument: evolutionary concepts cannot be used as evidence for evolution.

Figure 3.6
All modern tetrapods have a basic pentadactyl (five digit) limb structure. The forelimbs of a bird, human, whale, and bat are all constructed from the same bones even though they perform different functions. Adapted with permission from Strickberger (1990). © 1990 Boston: Jones & Bartlett Publishers.

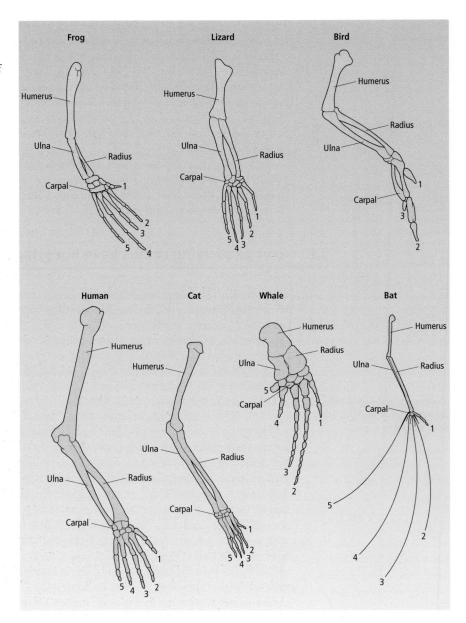

> What could be more curious than that the hand of man formed for grasping, that of a mole, for digging, the leg of a horse, the paddle of a porpoise and the wing of a bat, should all be constructed on the same pattern and should include similar bones and in the same relative positions?

The pentadactyl limb is a homology in the pre-Darwinian sense: it is a similarity between species that is not functionally necessary. Pre-Darwinian morphologists thought that homologies indicate a "plan of nature," in some more or less mystical sense. For evolutionary biologists, they are evidence of common ancestry. The evolutionary explanation of the pentadactyl limb is simply that all the tetrapods have descended from a common ancestor that had a pentadactyl limb and, during evolution, it has turned out to be easier to evolve variations on the five-digit theme, than to recompose the limb structure. If species have descended from common ancestors, homologies make sense; but if all species originated separately, it is difficult to understand why they should share homologous similarities. Without evolution, there is nothing forcing the tetrapods all to have pentadactyl limbs.

The pentadactyl limb is a morphological homology. It has a wide distribution, being found in all tetrapods; but at the molecular level there are homologies that have the widest distribution possible: they are found in all life. The genetic code is an example (Table 2.1, p. 26). The translation between base triplets in the DNA and amino acids in proteins is universal to all life, as can be confirmed, for instance, by isolating the mRNA for hemoglobin from a rabbit and injecting it into the bacterium *Escherichia coli*. *E. coli* do not normally make hemoglobin, but when injected with the mRNA they make rabbit hemoglobin. The machinery for decoding the message must therefore be common to rabbits and *E. coli*; and if it is common to them it is a reasonable inference that all living things have the same code. (Recombinant DNA technology is built on the assumption of a universal code.) Minor variants of the code, which have been found in mitochondria and in the nuclear DNA of a few species, do not affect the argument to be developed here.

Why should the code be universal? Two explanations are possible: that the universality results from a chemical constraint, or that the code is a historic accident.

In the chemical theory, each particular triplet would have some chemical affinity with its amino acid. GGC, for example, would react with glycine in some way that matched the two together. Several lines of evidence suggest this is wrong. One is that no such chemical relation has been found (and not for want of looking), and it is generally thought that one does not exist. Secondly, the triplet and the amino acid do not physically interact in the translation of the code. They are both held on a tRNA molecule, but the amino acid is attached at one end of the molecule, while the site that recognizes the codon on the mRNA is at the other end (Figure 3.7).

Finally, certain mutations can change the relation between the triplet code and amino acid (Figure 3.8). These mutations suppress the action of another class of mutants. Some of the triplets in the genetic code are "stop" codons: they act as a signal that the protein has come to an end. If a triplet within a coding region mutates to a stop codon, the protein is not made. Examples of these mutations are well known in bacterial genetics, and a mutation to the stop codon UAG, for example, is called an amber mutation. Now, once a bacterial culture with an amber mutation has been formed, it is sometimes possible to find other mutations that suppress the amber mutation: these mutants are normal, or near normal, bacteria. It turns out that the amber-suppressing mutants work by changing the coding triplet on a class of amino acid-bearing tRNA to make it bind to UAG. The UAG codon then encodes an amino acid rather than causing transcription to stop. The fact that the relation between amino acid and codon can be changed in this way shows that the same genetic code has not been forced on all species by some unalterable chemical constraint.

The genetic code is a universal homology

Mutations can alter the genetic code

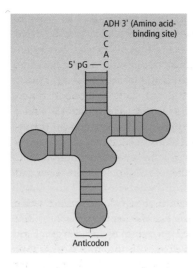

Figure 3.7

Transfer RNA molecule. The amino acid is held at the other end of the molecule from the anticodon loop where the triplet code of the mRNA molecule is read.

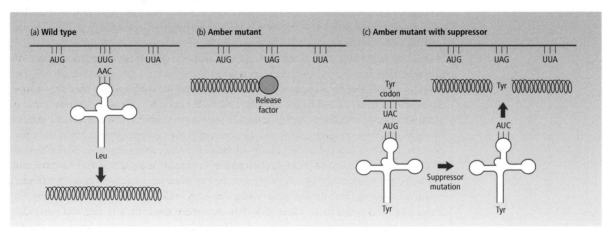

Figure 3.8

Mutations that suppress amber mutations suggest that the genetic code is chemically alterable. For example, (a) the normal codon is UUG and encodes leucine. (b) The UUG mutates to the stop codon UAG (this is called an amber mutation). (c) A tRNA for tyrosine mutates from AUG to AUC (which recognizes UAG) and suppresses the amber mutation by inserting a tyrosine.

If the genetic code is not chemically determined, why is it the same in all species? The most popular theory is as follows. The code is arbitrary, in the same sense that human language is arbitrary. In English the word for a horse is "horse," in Spanish it is "caballo," in French it is "cheval," in Ancient Rome it was "equus." There is no reason why one particular sequence of letters rather than another should signify that familiar perissodactylic mammal. Therefore, if we find more than one people using the same word, it implies they have both learned it from a common source. It implies common ancestry. When the starship *Enterprise* boldly descends on one of those extragalactic planets where the aliens speak English, the correct inference is that the locals share a common ancestry with one of the English-speaking peoples of the Earth. If they had evolved independently, they would not be using English.

All living species use a common, but equally arbitrary, language in the genetic code. The reason is thought to be that the code evolved early on in the history of life, and one early form turned out to be the common ancestor of all later species. (Notice that saying all life shares a common ancestor is not the same as saying life evolved only once.) The code is then what Crick (1968) called a "frozen accident." That is, the original coding relationships were accidental, but once the code had evolved, it would be strongly maintained. Any deviation from the code would be lethal. An individual that read GGC as phenylalanine instead of glycine, for example, would bungle all its proteins, and probably die at the egg stage.

The genetic code is an example of a frozen accident

The universality of the genetic code is important evidence that all life shares a single origin. In Darwin's time, morphological homologies like the pentadactyl limb were known; but these are shared between fairly limited groups of species (like all the tetrapods). Cuvier (Section 1.3.1, p. 8) had arranged all animals into four large groups according to their homologies. For this reason, Darwin suggested that living species may have a limited number of common ancestors, rather than just one. Molecular homologies, such as the genetic code, now provide the best evidence that all life has a single common ancestor.

Homologous similarities between species provide the most widespread class of evidence that living and fossil species have evolved from a common ancestor. The anatomy, biochemistry, and embryonic development of each species contains innumerable characters like the pentadactyl limb and the genetic code — characters that are similar between species, but would not be if the species had independent origins. Homologies, however, are usually more persuasive for an educated biologist than for someone seeking immediately intelligible evidence for evolution. The most obvious evidence for evolution is that from direct observation of change. No one will have any difficulty in seeing how the examples of evolution in action, from moths and artificial selection, suggest that species are not fixed in form. The argument from homology is inferential, and more demanding. You have to understand some functional morphology, or molecular biology, to appreciate that tetrapods would not share the pentadactyl limb, or all species the genetic code, if they originated independently.

Homologies are evidence of evolution that do not require long-term direct observation of change

But some homologies are immediately persuasive, such as vestigial organs, in which the shared form appears to be positively inefficient. If we stay with the vertebrate limb, but move in from its extremities to the junction where it joins the spine, we find another set of bones — at the pectoral and pelvic articulations — that are recognizably homologous in all tetrapods. In most species, these bones are needed in order for the limb to be able to move. But in a few species the limbs have been lost (Figure 3.9). Modern whales, for instance, do not have hindlimbs with bony supports. If we dissect a whale, we find at the appropriate place down the spine a set of bones that are clearly homologous with the pelvis of any other tetrapod. They are vestigial in the sense that they are no longer used to provide articulation for the hindlimb. Their retention suggests that whales evolved from tetrapods rather than being independently created. Modern snakes also have vestigial hindlimbs, though the bones that have been retained in vestigial form differ from those in whales.

Figure 3.9
Whales have a vestigial pelvic girdle, even though they do not have bony hindlimbs. The pelvic bones are homologous with those of other tetrapods. Snakes have vestigial hindlimb bones, homologous with those of other tetrapods — but snakes do not use them for locomotion.

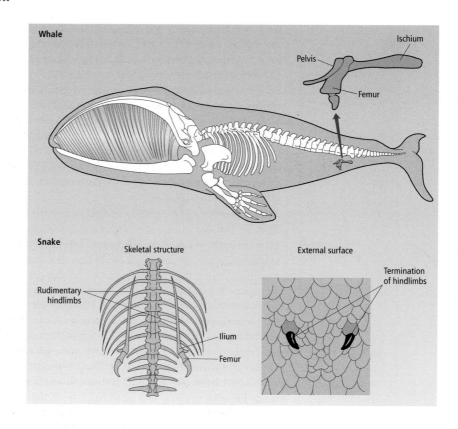

Vestigial organs are further examples of homology

An organ that is described as vestigial may not be functionless. Some vestigial organs may be truly functionless, but it is always difficult to confirm universal negative statements. Fossil whales called *Basilosaurus*, living 40 million years ago, had functional pelvic bones (Gingerich *et al.* 1990) and may have used them when copulating; and the vestigial pelvis of modern whales arguably is still needed to support the reproductive organs. However, that possibility does not count against the argument from homology: why, if whales originated independently of other tetrapods, should whales use bones that are adapted for limb articulation in order to support their reproductive organs? If they were truly independent, some other support would likely be used.

In homologies like the pentadactyl limb and the genetic code, the similarity between species is not actively disadvantageous. One form of genetic code would probably be as good as almost any other, and no species suffers for using the actual genetic code found in nature. However, some homologies do look positively disadvantageous (Section 10.7.4, p. 281). One of the cranial nerves, as we shall see, goes from the brain to the larynx via a tube near the heart (Figure 10.12, p. 282). In fish this is a direct route. But the same nerve in all species follows the same route, and in the giraffe it results in an absurd detour down and up the neck, so that the giraffe has to grow maybe 10–15 feet (3–4.5 m) more nerve than it would with a direct connection. The recurrent laryngeal nerve, as it is called, is surely inefficient. It is easy to explain such an inefficiency if giraffes have evolved in small stages from a fish-like ancestor. But it is difficult to imagine why giraffes should have such a nerve if they originated independently.

Homologies can be used to argue for evolution in several ways. Darwin was particularly impressed by a biogeographic version of the argument from homology. The species in one biogeographic area tend to be relatively similar. Species living in different areas tend to differ more, even if the species occupy a similar ecological niche. Thus, ecologically different species in one area will share similarities that are lacking between ecologically similar species in different areas. This suggests the species in any one area are descended from a common ancestor. The argument works for homologous similarities between species. In the next section we shall see a further way in which homologous similarities can be used to argue for evolution.

Homology underlies biogeographic arguments for evolution

3.9 Different homologies are correlated, and can be hierarchically classified

Different species share homologies, which suggests they are descended from a common ancestor. But the argument can be made both stronger and more revealing. Homologous similarities are the basis of biological classifications (Chapter 16): groups like "flowering plants," "primates," or "cats" are formally defined by homologies. The reason homologies are used to define groups is that they fall into a nested, or hierarchical, pattern of groups within groups; and different homologies consistently fall into the same pattern.

A molecular study by Penny *et al.* (1982) illustrates the point, and shows how it argues for evolution. Different species can be more or less similar in the amino acid sequences of their protein, just as they can be more or less similar in their morphology. The pre-Darwinian distinction between analogy and homology is more difficult to apply to proteins. Our functional understanding of protein sequences is less well advanced than for morphology, and it can be difficult to specify an amino acid's function in the way we can for the pentadactyl limb. Actually, the functions of many protein sequences are understood, but the chemistry takes a lot of explaining. For the argument here, it only needs to be accepted that *some* of the amino acid similarities between species are not functionally necessary, in the same way that all tetrapods do not have to have five-digited limbs. There are a large number of amino acids in a protein, so this need not be controversial. If we accept that some amino acids are homologous in the pre-Darwinian sense, we can see how their distribution among species suggests evolution.

Some molecular similarities between species are homologous

Penny *et al.* (1982) examined protein sequences in a group of 11 species. They used the pattern of amino acid similarities to work out the "tree" for the species. Some species have more similar protein sequences than others, and the more similar species are grouped more closely in the tree (Chapter 15). The observation that suggests evolution is as follows. We start by working out the tree for one protein. We can then work it out for another protein, and compare the trees. Penny *et al.* worked out the tree for the 11 species using each of the five proteins. The key observation was that the trees for all

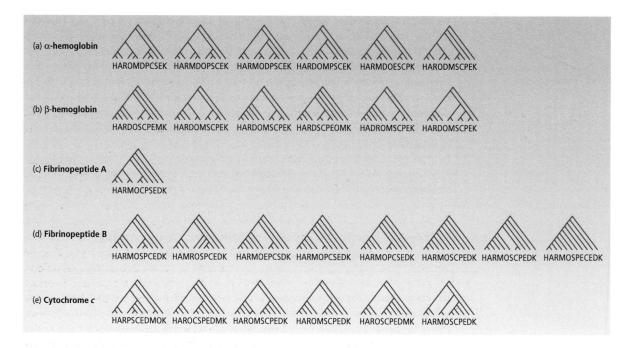

Figure 3.10

Penny *et al.* constructed the best estimate of the phylogenetic tree for 11 species using five different proteins. The "best estimate" of the phylogenetic tree is the tree that requires the smallest number of evolutionary changes in the protein. For (a) α-hemoglobin, and (b) β-hemoglobin there were six equally good estimates of the tree for the 11 species. All six trees in each case require the same number of changes. (c) For fibrinopeptide A there was one best tree; (d) for fibrinopeptide B there were eight equally good trees; and (e) for cytochrome *c* there were six equally good trees. The important point is how similar these trees are for all five proteins, given the large number of possible trees for 11 species. A, ape (*Pan troglodytes* or *Gorilla gorilla*); C, cow (*Bos primogenios*); D, dog (*Canis familiaris*); E, horse (*Equus caballus*); H, human (*Homo sapiens*); K, kangaroo (*Macropus conguru*); M, mouse (*Mus musculus*) or rat (*Rattus norvegicus*); O, rabbit (*Oryctolagus ainiculus*); P, pig (*Sus scrufa*); R, rhesus monkey (*Macaca mulatta*); S, sheep (*Ovis amnion*). Redrawn, by permission of the publisher, from Penny *et al.* (1982). © 1982 Macmillan Magazines Ltd.

Species that are more similar in one protein are also more similar in other proteins . . .

five proteins are very similar (Figure 3.10). For 11 species, there are 34,459,425 possible trees, but the five proteins suggest trees that form a small subclass from this large number of possible trees.

The similarities and differences in the amino acid sequences of the five proteins are correlated. If two species have more amino acid homologies for one of the proteins, they are also likely to for the other proteins. That is why any two species are likely to be grouped together for any of the five proteins. If the 11 species had independent origins, there is no reason why their homologies should be correlated. In a group of 11 separately created species, some would no doubt show more similarities than others for any particular protein. But why should two species that are similar for, say, cytochrome *c*, also be similar for β-hemoglobin and fibrinopeptide A? The problem is more difficult than that, because, as Figure 3.10 shows, all five proteins show a similar pattern of branching at all levels in the 11-species tree. It is easy to see how a set of independently created objects might show hierarchical patterns of similarity in any one respect. But these 11 species have been classified hierarchically for five different proteins, and the hierarchy in all five cases is similar.

. . . which suggests the species evolved from a common ancestor

If the species are descended from a common ancestor, the observed pattern is exactly what we expect. All of the five proteins have been evolving in the same pattern of evolutionary branches, and we therefore expect them to show the same pattern of similarities. The hierarchical pattern of, and correlations among, homologies are evidence for evolution.

Consider an analogy. Consider a set of 11 buildings, each of which was independently designed and built. We could classify them into groups according to their similarities; some might be built of stone, others of brick, others of wood; some might have vaults, others ceilings; some arched windows, others rectangular windows; and so on. It would be easy to classify them hierarchically with one of these properties, such as building material. This classification would be analogous, in Penny et al.'s study (1982), to making the tree of the 11 species for one protein. The same buildings could then be classified by another property, such as window shape; this is analogous to classifying the species by a second protein. There would probably be some correlations between the two classifications of the buildings, because of functional factors. Maybe buildings with arched windows would be more likely to be built of brick or stone, than of wood. However, other similarities would just be non-functional, chance associations in the particular 11 buildings in the sample. Maybe, in this 11, the white-colored buildings also happened to have garages, whereas the red buildings tended not to. The argument for evolution concentrates on these inessential, rather than functional, patterns of similarity.

An analogy with human constructions illustrates the argument

The analogy of Penny et al.'s result in the case of the buildings would be as follows. We should classify 11 buildings by five independent sets of characters. We should then look to see whether the five classifications all grouped the buildings in the same way. If the buildings were erected independently, there is no reason why they should show functionally unnecessary correlations. There would be no reason to expect that buildings that were similar for, say, window shape, would also be similar with respect to, say, number of chimney pots, or angle of roof, or the arrangement of chairs indoors.

Of course, some innocent explanation might be found for any such correlations. (Indeed if correlations were found in a real case, there would have to be some explanation.) Maybe they could all be explained by class of owner, or region, or common architects. But that is another matter; it is just to say that the buildings were not really independently created. If they were independently created, it would be very puzzling if they showed systematic, hierarchical similarity in functionally unrelated characteristics.

In the case of biological species, we do find this sort of correlation between characters. Figure 3.10 shows how similar the branching patterns are for five proteins, and the same conclusion could be drawn from any well researched classification in biology. Biological classifications, therefore, provide an argument for evolution. If species had independent origins, we should not expect that, when several different (and functionally unrelated) characters were used to classify them, all the characters would produce strikingly similar classifications.

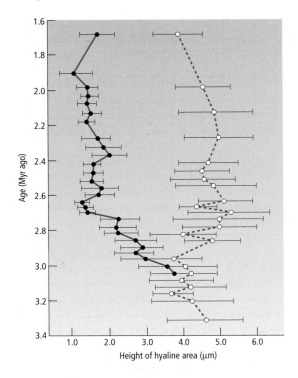

Figure 3.11

Evolution of the diatom *Rhizosolenia*. The form of the diatom is measured by the height of the hyaline (glass-like) area of the cell wall. Open circles indicate forms classified as *R. praebergonii*, closed circles indicate *R. bergonii*. Bars indicate the range of forms at each time. Redrawn, by permission of the publisher, from Cronin & Schneider (1990).

3.10 Fossil evidence exists for the transformation of species

Diatoms are single-celled, photosynthetic organisms that float in the plankton. Many species grow beautiful glass-like cell walls, and these can be preserved as fossils. Figure 3.11 illustrates the fossil record for the diatom *Rhizosolenia* between 3.3 and 1.7 million years ago. About 3 million years ago, a single ancestral species split into two; and there is a comprehensive fossil record of the change at the time of the split.

The fossil record is complete enough in some cases to illustrate continuous evolutionary transformations

The diatoms in Figure 3.11 show that the fossil record can be complete enough to reveal the origin of a new species; but examples as good as this are rare. In other cases, the fossil record is less complete and there are large gaps between successive samples (Section 21.4, p. 602). There is then only less direct evidence of smooth transitions between species. The gaps are usually long, however (maybe 25,000 years in a good case, and millions of years in less complete records). There is enough time within one of the gaps for large evolutionary changes, and no one need be surprised that fossil samples from either side of a gap in the record show large changes.

In other respects, as we saw at the beginning of the chapter (Section 3.1), the fossil record provides important evidence for evolution. Against alternatives other than separate creation and transformism, the fossil record is valuable because it shows that the living world has not always been like it is now. The existence alone of fossils shows that there has been some kind of change, though it does not have to have been change in the sense of descent with modification.

Figure 3.12

(a) Anatomic analysis of modern forms indicates that amphibians and reptiles are evolutionarily intermediate between fish and mammals. This order fits with (b) the geological succession of the major vertebrate groups. The width of each group indicates the diverity of the group at that time. Redrawn, by permission of the publisher, from Simpson (1949).

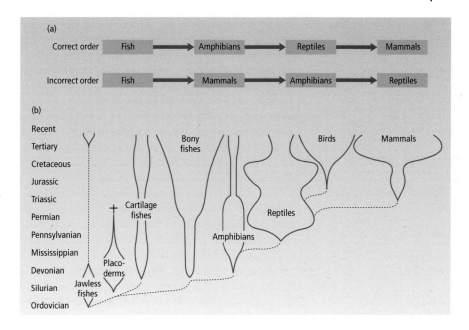

3.11 The order of the main groups in the fossil record suggests they have evolutionary relationships

Groups of animals can be arranged in a series according to their similarity

The main subgroups of vertebrates, on a conventional classification, are: fish, amphibians, reptiles, birds, and mammals. It is possible to deduce that their order of evolution must have been fish then amphibia then reptiles then mammals; and not, for example, fish then mammals then amphibia then reptiles (Figure 3.12a). The deduction follows from the observation that an amphibian, such as a frog, or a reptile, such as an alligator, is intermediate in form between a fish and a mammal. Amphibians, for instance, have gills as fish do, but have four legs, like reptiles and mammals, and not fins. If fish had evolved into mammals, and then mammals had evolved into amphibians, the gills would have been lost in the evolution of mammals and then regained in the evolution of amphibia. This is much less likely than that amphibia evolved from fish, retaining their gills, and the gills were then lost in the origin of mammals. (Chapter 15 discusses these arguments more fully.) Gills and legs are just two examples: the full list of characters putting amphibians (and reptiles, by analogous arguments) between fish and mammals would be long indeed. The forms of modern vertebrates alone, therefore, enable us to deduce the order in which they evolved.[2]

---

[2]   Strictly speaking, on the argument given here, it could also be that mammals came first and evolved into reptiles, the reptiles evolved into amphibia, and the amphibia into fish. However, we can extend the argument by including more groups of animals, back to a single-celled stage; the fish would then be revealed in turn as an intermediate stage between amphibians and simpler animals.

The groups appear in the same order in the fossil record

The inference, from the modern forms, can be tested against the fossil record. The fossil record supports it: fish, amphibians, reptiles, and mammals, appear in the fossil record in the same order as they should have evolved (Figure 3.12b). The fit is good evidence for evolution, because if fish, amphibians, reptiles, and mammals had been separately created, we should not expect them to appear in the fossil record in the exact order of their apparent evolution. Fish, frogs, lizards, and rats would probably appear as fossils in some order, if they did not appear at the same time; but there is no reason to suppose they would appear in one order rather than another. It is therefore a revealing coincidence when they turn out to be in the evolutionary order. Similar analyses have been done with other large and well fossilized groups of animals, such as the echinoderms, and have found the same result.

Haldane discussed a Precambrian rabbit

The argument can be stated another way. Haldane once said he would give up his belief in evolution if someone found a fossil rabbit in the Precambrian. The reason is that the rabbit, which is a fully formed mammal, must have evolved through reptilian, amphibian, and piscine stages and should not therefore appear in the fossil record 100 million years or so before its fossil ancestors. Creationists have appreciated the power of this argument. Various claims have been made for fossil human footprints contemporary with dinosaur tracks. Whenever one of these claims has been properly investigated, it has been exploded: some have turned out to have been carved fraudulently, others were carved as tourist exhibits, others are perfectly good dinosaur footprints. But the principle of the argument is valid. If evolution is correct, humans could not have existed before the main radiation of mammals and primates, and these took place after the dinosaurs had gone extinct. The fact that no such human fossils have been found — that the order of appearance of the main fossil groups matches their evolutionary order — is the way in which the fossil record provides good evidence for evolution.

3.12 Summary of the evidence for evolution

We have met three main classes of evidence for evolution: from direct observation on the small scale; from homology; and from the order of the main groups in the fossil record. The small-scale observations work most powerfully against the idea of species fixity; by themselves, they are almost equally good evidence for evolution and for transformism (see Figure 3.1a,b). They show, by uniformitarian extrapolation, that evolution could have, in theory, produced the whole history of life. Stronger arguments for large-scale evolution come from classification and the fossil record. The geological succession of the major groups and most classic morphological homologies strongly suggest that these large groups have a common ancestor. The more recently discovered molecular homologies, such as the universal genetic code, extend the argument to the whole of life — and favor evolution (Figure 3.1a) over both transformism and creationism (Figure 3.1b–e).

Such is the standard argument for evolution. Moreover, the theory of evolution can also be used to make sense of, and to analyze, a large array of additional facts. As we study the different areas of evolutionary biology, it is worth keeping the issue of this chapter in mind. How, for example, could we explain the molecular clock (Section 7.3, p. 164) if species have independent origins? Or the difficulties of deciding whether closely related forms are different species (Chapter 13)? Or the unique branching pattern of chromosomal inversions in the Hawaiian fruitflies (Section 15.14, p. 463)? Or the way new species of Hawaiian fruitflies tend to be most closely related to species on neighboring islands (Section 17.6, p. 503)?

3.13 Creationism offers no explanation of adaptation

Another powerful reason why evolutionary biologists reject creationism is that creationism offers no explanation for adaptation. Living things are well designed, in innumerable respects, for life in their natural environments. They have sensory systems to find their way around, feeding systems to catch and digest food, and nervous systems to coordinate their actions. The theory of evolution has a mechanical, scientific theory for adaptation: natural selection.[3]

Any theory of life has to explain adaptation

Creationism, by contrast, has no explanation for adaptation. When each species originated, it must have already been equipped with adaptations for life, because the theory holds that species are fixed in form after their origin. An unabashedly religious version of creationism would attribute the adaptiveness of living things to the genius of God. However, even this does not actually explain the origin of the adaptation; it just pushes the problem back one stage (Section 10.1, p. 256). In the scientific version of creationism (see Figure 3.1c–e) we are concerned with here, supernatural events do not take place, and we are left with no theory of adaptation at all. Without a theory of adaptation, as Darwin realized (Section 1.3.2, p. 10), any theory of the origin of living things is a non-starter.

3.14 Modern "scientific creationism" is scientifically untenable

That life has evolved is one of the great discoveries in all the history of science, and it is correspondingly interesting to know the arguments in favor of it. In modern evolutionary biology, the question of whether evolution happened is no longer a topic of research, because the question has been answered; but it is still controversial outside science. Christian fundamentalists — some of them politically influential — in the USA have supported various forms of creationism and have been trying since the 1920s,

---

[3] The modern school of "intelligent design" creationism denies that natural selection explains adaptation — opening up the possibility that some further (supernatural?) force may be operating. Intelligent design creationists are not concerned to deny evolution, or to argue that species have separate origins and are fixed in form. They are therefore not included in this chapter. In Chapter 10, we look at how well natural selection explains adaptation.

sometimes successfully, sometimes unsuccessfully, to intrude them into school biology curricula.

The scientific evidence counts against creationism

What relevance do the arguments of this chapter have for these forms of creationism? For a purely scientific form of creationism, the relevance is straightforward. The creationism of Figure 3.1c–e, which simply suggests that species have had separate origins and have been fixed since then, has been the subject of the whole chapter and we have seen that it is refuted by the evidence. The scientific creationism of Figure 3.1c–e said nothing about the mechanism by which species originated and therefore need not assert that the species were created by God. A supporter of Figure 3.1c–e might merely say that species originated by some natural mechanism, the details of which are not yet understood. However, it is unlikely that anyone would now seriously support the theory of Figure 3.1c–e unless they also believed that the species originated supernaturally. Then we are not dealing with a scientific theory.

Scientists ignore supernatural agents

This chapter has confined itself to the scientific resources of logical argument and public observation. Scientific arguments only employ observations that anybody can make, as distinct from private revelations, and consider only natural, as distinct from supernatural, causes. Indeed, two good criteria to distinguish scientific from religious arguments are whether the theory invokes only natural causes, or needs supernatural causes too, and whether the evidence is publicly observable or requires some sort of faith. Without these two conditions, there are no constraints on the argument. It is, in the end, impossible to show that species were not created by God and have remained fixed in form, because to God (as a supernatural agent) everything is permitted. It equally cannot be shown that the building (or garden) you are in, and the chair you are sitting on, were not created supernaturally by God 10 seconds ago from nothing — at the time, He would also have to have adjusted your memory and those of all other observers, but a supernatural agent can do that. That is why supernatural agents have no place in science.

Two final points are worth making. The first is that, although modern "scientific creationism" closely resembles the theory of separate creation in Figure 3.1c–e, it also possesses the added feature of specifying the time when all the species were created. Theologians working after the Reformation were able to deduce, from some plausible astronomical theory and rather less plausible Biblical scholarship, that the events described in Genesis chapter 1 happened about 6,000 years ago; and fundamentalists in our own time have retained a belief in the recent origin of the world. A statement of creationism in the 1970s (and the one legally defended in court at Arkansas in 1981) included, as a creationist tenet, that there was "a relatively recent inception of the earth and living kinds." Scientists accept a great age for the Earth because of radioactive dating and cosmological inferences from the background radiation. Cosmological and geological time are important scientific discoveries, but we have ignored them in this chapter because our subject has been the scientific case for evolution: religious fundamentalism is another matter.

Science and religion, properly understood, can coexist peacefully

Finally, it is worth stressing that there need be no conflict between the theory of evolution and religious belief. This is not an "either/or" controversy, in which accepting evolution means rejecting religion. No important religious beliefs are contradicted by the theory of evolution, and religion and evolution should be able to coexist peacefully in anyone's set of beliefs about life.

Summary

1 A number of lines of evidence suggest that species have evolved from a common ancestor, rather than being fixed in form and created separately.

2 On a small scale, evolution can be seen taking place in nature, such as in the color patterns of moths, and in artificial selection experiments, such as those used in breeding agricultural varieties.

3 Natural variation can cross the species border, for example in the ring species of salamanders, and new species can be made artificially, as in the process of hybridization and polyploidy by which many agricultural and horticultural varieties have been created.

4 Observation of evolution on the small scale, combined with the extrapolative principle of uniformitarianism, suggests that all life could could have evolved from a single common ancestor.

5 Homologous similarities between species (understood as similarities that do not have to exist for any pressing functional reason), suggest that species descended from a common ancestor. Universal homologies — such as the genetic code — found in all living things suggest that all species are descended from a single common ancestor.

6 The fossil record provides some direct evidence of the origin of new species.

7 The order of succession of major groups in the fossil record is predicted by evolution, and contradicts the separate origin of the groups.

8 The independent creation of species does not explain adaptation; evolution, by the theory of natural selection, offers a valid explanation.

Further reading

Eldredge (2000), Futuyma (1997), and Moore (2002) have written books about creationism and the case for evolution. The latest version of creationism is "intellgent design" creationism, which does not challenge evolution in the sense of this chapter: on it see Chapter 10 in this book, and Pennock (2000, 2001). Chapters 10–14 of *On the Origin of Species* (Darwin 1859) are the classic account of the evidence for evolution. Jones (1999) remakes Darwin's case, using modern examples, including drug resistance in HIV.

Palumbi (2001a, 2001b) describes many examples of evolution in response to environmental changes that humans have caused, including HIV evolution; he also does some interesting sums on the economic cost of that evolution. Reznick *et al.* (1997) describe another good example of evolution in action: changes to the life histories of guppies in Trinidad. See Ford (1975), Endler (1986), and the references in Hendry & Kinnison (1999) for further examples. Huey *et al.* (2000) discuss another example of rapid evolution of a cline within a species, like the house sparrow example in the text but with the addition that the newly formed cline in North America parallels one in Europe.

Irwin *et al.* (2001b) review ring species, including the Californian salamander. On polyploidy in plants, see the references in Chapter 14. On the genetic code, see Osawa (1995). Zimmer (1998) describes fossil whales and tetrapods. Ahlberg (2001) includes material on Devonian tetrapods with non-pentadactyl limbs. Gould (1989) describes the animals of the Burgess Shale. Wellnhofer (1990) describes *Archaeopteryx*. On adaptation, see Dawkins (1986). For the broader context, see Numbers (1992) for the history, and Antolin & Herbers (2001) on educational, and Larson (2003) on legal, business.

Study and review questions

1 The average difference between two individuals increases as they are sampled from the same local population, two separate populations, two species, two genera, and so on up to two kingdoms (such as plants and animals). Up to approximately what stage in this sequence can evolution be observed in a human lifetime?

2 In what sense is the range of forms of life on Earth (i) arranged, and (ii) not arranged, in distinct "kinds"?

3 Which of the following are homologies and which analogies, in the pre-Darwinian sense of the terms? (a) A dolphin flipper and a fish fin. (b) The five-digit skeletal structure of the dolphin flipper and of a frog foot. (c) The white underside coloration of gulls, albatrosses, and ospreys (all of which are seabirds and catch fish by air raids from above). (d) The number of vertebrae in the necks of camels, mice, and humans (they all have seven vertebrae).

4 The genetic code has been called a "frozen accident." In what sense is it an accident, and why was it frozen?

5 Imagine a number of sets of about 10 objects each: such as 10 books, 10 dishes for dinner, 10 gems, 10 vehicles, 10 politicians, . . . or whatever. For each set, devise two or three different ways of classifying them in hierarchical groups. (For example, 10 politicians might be classified first into two groups such as left of center/right of center; then those groups could be divided by such criteria as average length of sound-bites, number of scandals per year, gender, region represented, etc.) Do the different hierarchical classifications recognize the same sets of groups, or similar sets of groups, or are they unrelated? Think about why for some sets of groups and for some classificatory criteria, the different classifications are similar, whereas for others they differ.

6 Why would Haldane have given up his belief in evolution if someone discovered a fossil rabbit in the Precambrian?

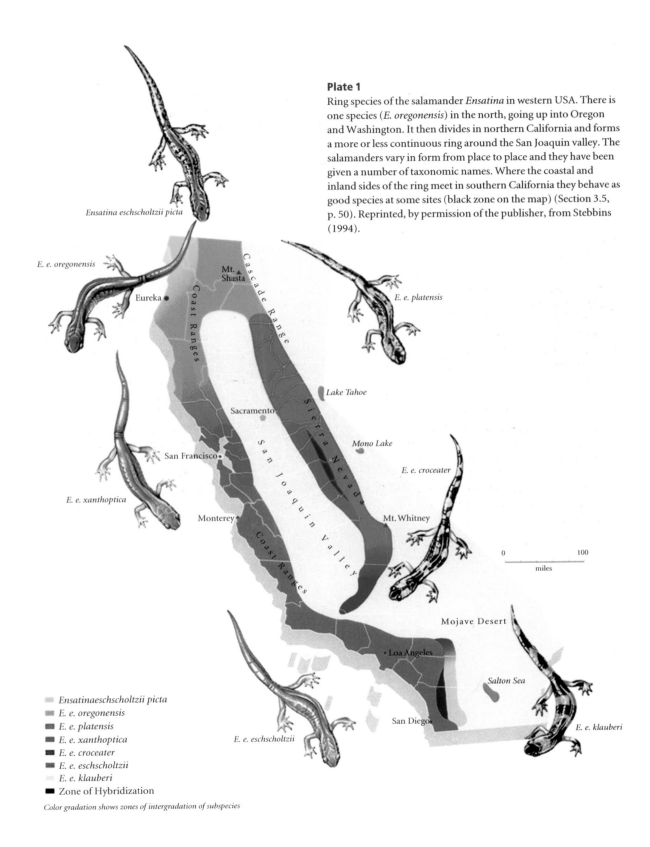

Plate 1

Ring species of the salamander *Ensatina* in western USA. There is one species (*E. oregonensis*) in the north, going up into Oregon and Washington. It then divides in northern California and forms a more or less continuous ring around the San Joaquin valley. The salamanders vary in form from place to place and they have been given a number of taxonomic names. Where the coastal and inland sides of the ring meet in southern California they behave as good species at some sites (black zone on the map) (Section 3.5, p. 50). Reprinted, by permission of the publisher, from Stebbins (1994).

Ensatina eschscholtzii picta

E. e. oregonensis

Mt. Shasta

Cascade Range

Eureka

Coast Ranges

E. e. platensis

Lake Tahoe

Sacramento

Sierra Nevada

Mono Lake

San Francisco

San Joaquin Valley

E. e. croceater

E. e. xanthoptica

Monterey

Mt. Whitney

Coast Ranges

0   100
miles

Mojave Desert

Loa Angeles

Salton Sea

San Diego

E. e. eschscholtzii

E. e. klauberi

Ensatinaeschscholtzii picta
E. e. oregonensis
E. e. platensis
E. e. xanthoptica
E. e. croceater
E. e. eschscholtzii
E. e. klauberi
Zone of Hybridization

Color gradation shows zones of intergradation of subspecies

Plate 2

Large-beaked (left) and small-beaked (right) forms of the African finch formally named *Pyrenestes ostrinus* and informally known as the black-bellied seedcracker. The polymorphism is an example of disruptive selection (Section 4.4, p. 80). (Courtesy of T.B. Smith.)

(a) (b) (c) (d) (e) (f)

(g) (h) (i) (j) (k) (l)

(m)

Plate 3

Here in the lower row are six of the many forms of *Papilio memnon*, beneath the model species that they may mimic. (a–f) Six suspected models: (a, b) two forms of the female *Losaria coon*; (c) *L. aristolochiae*; (d) *Triodes helena*; (e) *T. amphrysus*; (f) *Atrophaneura sycorax*. (g–l) Six forms of *Papilio memnon*. Three of the forms (g–i) mimic species (a–c) that have tails, and three (j–l) mimic species (d–f) that lack tails. (m) Another form of *P. memnon*, the rare probable recombinant form *anura*, from Java. It is like the normal mimetic form called *achates* (illustrated in g–i), but it lacks *achates'* tail. It may be a recombinant between *achates* and a tailless form such as in (d–f) (Section 8.1, p. 195). From Clarke *et al.* (1968) and Clarke & Sheppard (1969).

Plate 4

(a) *Geospiza magnirostris* on Daphne, Galápagos Islands. (b) *G. fortis*, also on Daphne. These two species are closely related, although *G. magnirostris* has a larger beak and is more efficient at eating larger seeds. (c) The crater on Daphne in the normal weather conditions of 1976. The birds in the foreground are boobies. (d) The same crater on Daphne in the El Niño year of 1983. The distinctive vegetation consists mainly of *Heliotropium angiospermum* and *Cacabus miersii*. (See Sections 9.1, p. 223, and 13.7.3, p. 373.) (Photos courtesy of Peter Grant (a–c) and Nicola Grant (d).)

(a)

(b)

Plate 5

These stalk-eyed flies from Malaya have an eye span that is longer than their body. (a) *Cyrtodiopsis dalmanni*. There is an allometric relation between eye span and body length, and Wilkinson has artificially selected the flies to alter the slope of the allometric relation. (b) The closely related species *C. whitei*. (Section 10.7.3, p. 280.) (Photos courtesy of Jerry Wilkinson.)

Plate 6

Scrub jays (*Aphelocoma coerulescens*) in Florida breed in cooperative groups of a parental pair and a number of "helpers." Kin selection is probably the reason why altruistic helping is favored in this species in Florida (Section 11.2.4, p. 299).

(a)

(b)

Plate 7

Prezygotic isolation by color differences in two cichlids. (a) In normal light, the two species differ in coloration. *Pundamilia nyererei* (above) has red colors and *P. pundamilia* has blue (look at the tail fins, for instance). The red females mate only with red males, and blue females only with blue males. (b) In an experiment with monochromatic orange light, the two species were indistinguishable. Now the red females mated indiscriminately with red and blue males, as did the blue females. The offspring were all viable and fertile. The experiment shows that the two species are held apart by the color-based mating preferences. It also suggests that the species have evolved very recently because there is no postzygotic isolation (Section 13.3.3, p. 358). (Photos courtesy of Ole Seehausen.)

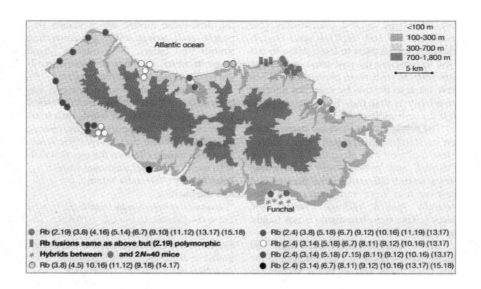

Plate 8

Chromosomal races of the house mouse (*Mus musculus*) in Maderia. Circles and squares represent samples, and the different symbols represent different chromosome forms. Rb stands for Robertsonian fusion, which is the fusion between two chromosomes that (before fusion) had centromeres at their ends. The numbers in parentheses are the two chromosomes that fused. Diploid numbers (2N) and sample sizes (*n*) are as follow: red dot, 2N = 22, *n* = 43; red rectangle, 2N = 23–24, *n* = 5; red star, 2N = 24–40, *n* = 38; yellow dot, 2N = 28–30, *n* = 5; blue dot, 2N = 25–27, *n* = 10; white dot, 2N = 24–26, *n* = 11; green dot, 2N = 24–27, *n* = 25; black dot, 2N = 24, *n* = 6. (See Section 13.4.2, p. 361.) Reprinted, by permission of the publisher, from Britton-Davidian *et al.* (2000).

(a)

(b)

Plate 9

Hybrid speciation in irises. (a) The three "parental" species: *Iris hexagona* (left), *I. fulva* (center), and *I. brevicaulis* (right). (b) These parental species have contributed to the recent origin of *I. nelsonii*, shown here in the woods of Louisiana (Section 14.7, p. 405). (Photos courtesy of Mike Arnold.)

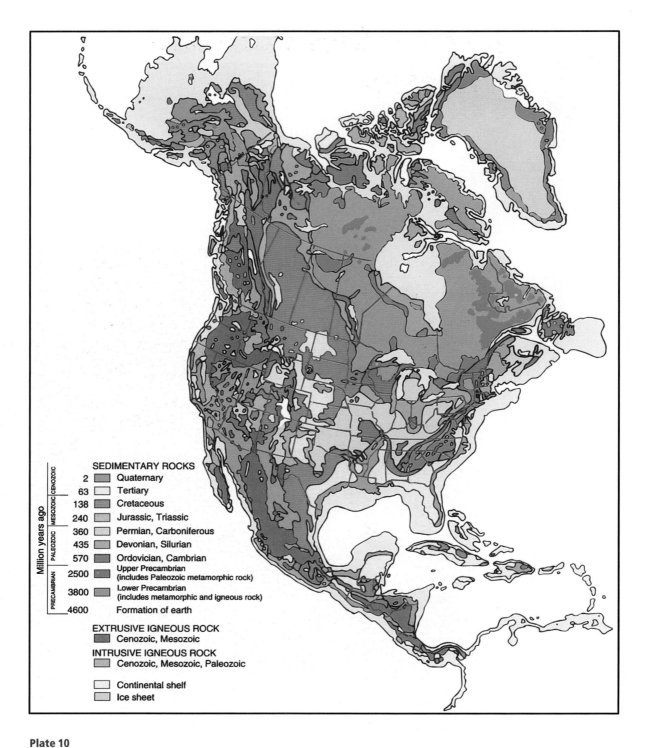

SEDIMENTARY ROCKS

Million years ago

CENOZOIC
- 2 — Quaternary
- 63 — Tertiary

MESOZOIC
- 138 — Cretaceous
- 240 — Jurassic, Triassic

PALEOZOIC
- 360 — Permian, Carboniferous
- 435 — Devonian, Silurian
- 570 — Ordovician, Cambrian

PRECAMBRIAN
- 2500 — Upper Precambrian (includes Paleozoic metamorphic rock)
- 3800 — Lower Precambrian (includes metamorphic and igneous rock)
- 4600 — Formation of earth

EXTRUSIVE IGNEOUS ROCK
- Cenozoic, Mesozoic

INTRUSIVE IGNEOUS ROCK
- Cenozoic, Mesozoic, Paleozoic

- Continental shelf
- Ice sheet

Plate 10

Geological map of North America, showing the age of the bedrock (the rock that is either at the surface of the Earth, or immediately below the top soil) (Section 18.1, p. 525).

4 Natural Selection and Variation

This chapter first establishes the conditions for natural selection to operate, and distinguishes directional, stabilizing, and disruptive forms of selection. We then consider how widely in nature the conditions are met, and review the evidence for variation within species. The review begins at the level of gross morphology and works down to molecular variation. Variation originates by recombination and mutation, and we finish by looking at the argument to show that when new variation arises it is not "directed" toward improved adaptation.

4.1   In nature, there is a struggle for existence

Cod produce far more eggs than are needed to propagate the population

The Atlantic cod (*Gadus callarias*) is a large marine fish, and an important source of human food. They also produce a lot of eggs. An average 10-year-old female cod lays about 2 million eggs in a breeding season, and large individuals may lay over 5 million (Figure 4.1a). Female cod ascend from deeper water to the surface to lay their eggs; but as soon as they are discharged, a slaughter begins. The plankton layer is a dangerous place for eggs. The billions of cod eggs released are devoured by innumerable planktonic invertebrates, by other fish, and by fish larvae. About 99% of cod eggs die in their first month of life, and another 90% or so of the survivors die before reaching an age of 1 year (Figure 4.1b). A negligible proportion of the 5 million or so eggs laid by a female cod in her lifetime will survive and reproduce — an average female cod will produce only two successful offspring.

This figure, that on average two eggs per female survive to reproduce successfully, is not the result of an observation. It comes from a logical calculation. Only two can survive, because any other number would be unsustainable over the long term. It takes a pair of individuals to reproduce. If an average pair in a population produce less than two offspring, the population will soon go extinct; if they produce more than two, on average, the population will rapidly reach infinity — which is also unsustainable. Over a small number of generations, the average female in a population may produce more or less than two successful offspring, and the population will increase or decrease accordingly. Over the long term, the average must be two. We can infer that, of the 5 million or so eggs laid by a female cod in her life, 4,999,998 die before reproducing.

A life table can be used to describe the mortality of a population (Table 4.1). A life table begins at the egg stage and traces what proportion of the original 100% of eggs die off at the successive stages of life. In some species, mortality is concentrated early in life, in others mortality has a more constant rate throughout life. But in all species there is mortality, which reduces the numbers of eggs produced to result in a lower number of adults.

As do all other life forms

The condition of "excess" fecundity — where females produce more offspring than survive — is universal in nature. In every species, more eggs are produced than can survive to the adult stage. The cod dramatizes the point in one way because its fecundity, and mortality, are so high; but Darwin dramatized the same point by considering the opposite kind of species — one that has an extremely low reproductive rate. The

Figure 4.1
(a) Fecundity of cod. Notice both the large numbers, and that they are variable between individuals. The more fecund cod lay perhaps five times as many eggs as the less fecund; much of the variation is associated with size, because larger individuals lay more eggs. (b) Mortality of cod in their first 2 years of life. Redrawn, by permission of the publisher, from May (1967) and Cushing (1975).

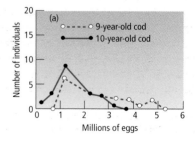

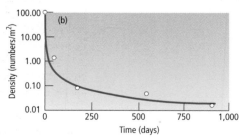

Table 4.1

A life table for the annual plant *Phlox drummondii* in Nixon, Texas. The life table gives the proportion of an original sample (cohort) that survive to various ages. A full life table may also give the fecundity of individuals at each age. Reprinted, by permission of the publisher, from Leverich & Levin (1979).

Age interval (days)	Number surviving to end of interval	Proportion of original cohort surviving	Proportion of original cohort dying during interval	Mortality rate per day
0–63	996	1.000	0.329	0.005
63–124	668	0.671	0.375	0.009
124–184	295	0.296	0.105	0.006
184–215	190	0.191	0.014	0.002
215–264	176	0.177	0.004	0.001
264–278	172	0.173	0.005	0.002
278–292	167	0.168	0.008	0.003
292–306	159	0.160	0.005	0.002
306–320	154	0.155	0.007	0.003
320–334	147	0.148	0.043	0.021
334–348	105	0.105	0.083	0.057
348–362	22	0.022	0.022	1.000
362–	0	0	—	

fecundity of elephants is low, but even they produce many more offspring than can survive. In Darwin's words:

> The elephant is reckoned the slowest breeder of all known animals, and I have taken some pains to estimate its probable minimum rate of natural increase; it will be safest to assume it begins breeding when thirty years old, and goes on breeding until ninety years old, bringing forth six young in the interval, and surviving till one hundred years old; if this be so, after a period of 740 to 750 years there would be nearly nineteen million elephants alive, descended from the first pair.[1]

In elephants, as in cod, many individuals die between egg and adult; they both have excess fecundity. This excess fecundity exists because the world does not contain enough resources to support all the eggs that are laid and all the young that are born. The world contains only limited amounts of food and space. A population may expand to some extent, but logically there will come a point beyond which the food supply must limit its further expansion. As resources are used up, the death rate in the population increases, and when the death rate equals the birth rate the population will stop growing.

Organisms, therefore, in an ecological sense compete to survive and reproduce — both directly, for example by defending territories, and indirectly, for example by eating food that could otherwise be eaten by another individual. The actual competitive

Excess fecundity results in competition, to survive and reproduce

---

[1] The numerical details are questionable, but Darwin's exact numbers can be obtained on the assumption of overlapping generations. See Ricklefs & Miller (2000, p. 300). The general point stands anyhow.

factors limiting the sizes of real populations make up a major area of ecological study. Various factors have been shown to operate. What matters here, however, is the general point that the members of a population, and members of different species, compete in order to survive. This competition follows from the conditions of limited resources and excess fecundity. Darwin referred to this ecological competition as the "struggle for existence." The expression is metaphorical: it does not imply a physical fight to survive, though fights do sometimes happen.

The struggle for existence takes place within a web of ecological relations. Above an organism in the ecological food chain there will be predators and parasites, seeking to feed off it. Below it are the food resources it must in turn consume in order to stay alive. At the same level in the chain are competitors that may be competing for the same limited resources of food, or space. An organism competes most closely with other members of its own species, because they have the most similar ecological needs to its own. Other species, in decreasing order of ecological similarity, also compete and exert a negative influence on the organism's chance of survival. In summary, organisms produce more offspring than — given the limited amounts of resources — can ever survive, and organisms therefore compete for survival. Only the successful competitors will reproduce themselves.

The struggle for existence refers to ecological competition

4.2 Natural selection operates if some conditions are met

The theory of natural selection can be understood as a logical argument

The excess fecundity, and consequent competition to survive in every species, provide the preconditions for the process Darwin called natural selection. Natural selection is easiest to understand, in the abstract, as a logical argument, leading from premises to conclusion. The argument, in its most general form, requires four conditions:

1. Reproduction. Entities must reproduce to form a new generation.
2. Heredity. The offspring must tend to resemble their parents: roughly speaking, "like must produce like."
3. Variation in individual characters among the members of the population. If we are studying natural selection on body size, then different individuals in the population must have different body sizes. (See Section 1.3.1, p. 7, on the way biologists use the word "character.")
4. Variation in the *fitness* of organisms according to the state they have for a heritable character. In evolutionary theory, fitness is a technical term, meaning the average number of offspring left by an individual relative to the number of offspring left by an average member of the population. This condition therefore means that individuals in the population with some characters must be more likely to reproduce (i.e., have higher fitness) than others. (The evolutionary meaning of the term fitness differs from its athletic meaning.)

If these conditions are met for any property of a species, natural selection automatically results. And if any are not, it does not. Thus entities, like planets, that do not reproduce, cannot evolve by natural selection. Entities that reproduce but in which parental characters are not inherited by their offspring also cannot evolve by natural selection. But when the four conditions apply, the entities with the property conferring higher fitness will leave more offspring, and the frequency of that type of entity will increase in the population.

HIV illustrates the logical argument The evolution of drug resistance in HIV illustrates the process (we looked at this example in Section 3.2, p. 45). The usual form of HIV has a reverse transcriptase that binds to drugs called nucleoside inhibitors as well as the proper constituents of DNA (A, C, G, and T). In particular, one nucleoside inhibitor called 3TC is a molecular analog of C. When reverse transcriptase places a 3TC molecule, instead of a C, in a replicating DNA chain, chain elongation is stopped and the reproduction of HIV is also stopped. In the presence of the drug 3TC, the HIV population in a human body evolves a discriminating form of reverse transcriptase — a form that does not bind 3TC but does bind C. The HIV has then evolved drug resistance. The frequency of the drug-resistant HIV increases from an undetectably low frequency at the time the drug is first given to the patient up to 100% about 3 weeks later.

The increase in the frequency of drug-resistant HIV is almost certainly driven by natural selection. The virus satisfies all four conditions for natural selection to operate. The virus reproduces; the ability to resist drugs is inherited (because the ability is due to a genetic change in the virus); the viral population within one human body shows genetic variation in drug-resistance ability; and the different forms of HIV have different fitnesses. In a human AIDS patient who is being treated with a drug such as 3TC, the HIV with the right change of amino acid in their reverse transcriptase will reproduce better, produce more offspring virus like themselves, and increase in frequency. Natural selection favors them.

4.3 Natural selection explains both evolution and adaptation

When the environment of HIV changes, such that the host cell contains nucleoside inhibitors such as 3TC as well as valuable resources such as C, the population of HIV changes over time. In other words, the HIV population evolves. Natural selection produces evolution when the environment changes; it will also produce evolutionary change in a constant environment if a new form arises that survives better than the current form of the species. The process that operates in any AIDS patient on drug treatment has been operating in all life for 4,000 million years since life originated, and has driven much larger evolutionary changes over those long periods of time.

Natural selection can not only produce evolutionary change, it can also cause a population to stay constant. If the environment is constant and no superior form arises in the population, natural selection will keep the population the way it is. Natural selection can explain both evolutionary change and the absence of change.

Natural selection also explains adaptation. The drug resistance of HIV is an example of an adaptation (Section 1.2, p. 6). The discriminatory reverse transcriptase enzyme enables HIV to reproduce in an environment containing nucleoside inhibitors. The new adaptation was needed because of the change in the environment. In the drug-treated AIDS patient, a fast but undiscriminating reverse transcriptase was no longer adaptive. The action of natural selection to increase the frequency of the gene coding

Natural selection drives evolutionary change . . .

. . . and generates adaptation

for a discriminating reverse transcriptase resulted in the HIV becoming adapted to its environment. Over time, natural selection generates adaptation. The theory of natural selection therefore passes the key test set by Darwin (Section 1.3.2, p. 8) for a satisfactory theory of evolution.

4.4 Natural selection can be directional, stabilizing, or disruptive

Many characters have continuous distributions

In HIV, natural selection adjusted the frequencies of two distinct types (drug susceptible and drug resistant). However, many characters in many species do not come in distinct types. Instead, the characters show continuous variation. Human body size, for instance, does not come in the form of two distinct types, "big" and "small." Body size is continuously distributed. A sample of humans will show a range of sizes, distributed in a "bell curve" (or normal distribution). In evolutionary biology, it is often useful to think about evolution in continuous characters such as body size slightly differently from evolution in discrete characters such as drug resistance and drug susceptibility. However, no deep difference exists between the two ways of thinking. Discrete variation blurs into continuous variation, and evolution in all cases is due to changes in the frequency of alternative genetic types.

Natural selection alters the form of continuous distributions: it can be directional . . .

Natural slection can act in three main ways on a character, such as body size, that is continuously distributed. Assume that smaller individuals have higher fitness (that is, produce more offspring) than larger individuals. Natural selection is then *directional*: it favors smaller individuals and will, if the character is inherited, produce a decrease in average body size (Figure 4.2a). Directional selection could, of course, also produce an evolutionary increase in body size if larger individuals had higher fitness.

For example, pink salmon (*Onchorhynchus gorbuscha*) in the Pacific Northwest have been decreasing in size in recent years (Figure 4.3). In 1945, fishermen started being paid by the pound, rather than per individual, for the salmon they caught and they increased the use of gill netting, which selectively takes larger fish. The selectivity of gill netting can be shown by comparing the average size of salmon taken by gill netting with those taken by an unselective fishing technique: the difference ranged from 0.3 to 0.48 lb (0.14–0.22 kg). Therefore, after gill netting was introduced, smaller salmon had a higher chance of survival. The selection favoring small size in the salmon population was intense, because fishing effort is highly efficient — about 75–80% of the adult salmon swimming up the rivers under investigation were caught in these years. The average weight of salmon duly decreased, by about one-third, in the next 25 years. (Box 4.1 describes a practical application of this kind of evolution.)

. . . or stabilizing . . .

A second (and in nature, more common) possibility is for natural selection to be *stabilizing* (Figure 4.2b). The average members of the population, with intermediate body sizes, have higher fitness than the extremes. Natural selection now acts against changes in body size, and keeps the population constant through time.

Studies of birth weight in humans have provided good examples of stabilizing selection. Figure 4.4a illustrates a classic result for a sample in London, UK, in 1935–46 and similar results have been found in New York, Italy, and Japan. Babies that are heavier or

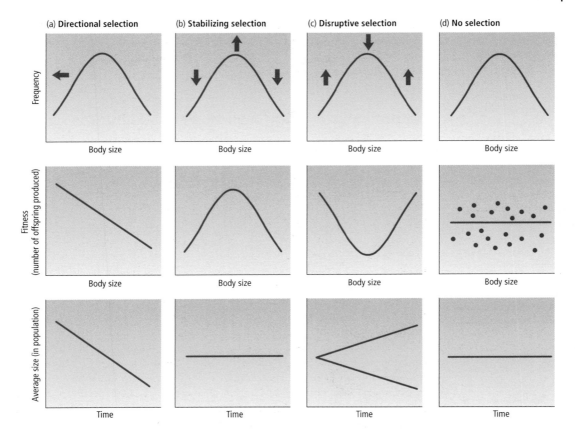

| (a) Directional selection | (b) Stabilizing selection | (c) Disruptive selection | (d) No selection |

Figure 4.2

Three kinds of selection. The top line shows the frequency distribution of the character (body size). For many characters in nature, this distribution has a peak in the middle, near the average, and is lower at the extremes. (The normal distribution, or "bell curve," is a particular example of this kind of distribution.) The second line shows the relation between body size and fitness, within one generation, and the third the expected change in the average for the character over many generations (if body size is inherited).

(a) Directional selection. Smaller individuals have higher fitness, and the species will decrease in average body size through time. Figure 4.3 is an example. (b) Stabilizing selection. Intermediate-sized individuals have higher fitness. Figure 4.4a is an example. (c) Disruptive selection. Both extremes are favored and if selection is strong enough, the population splits into two. Figure 4.5 is an example. (d) No selection. If there is no relation between the character and fitness, natural selection is not operating on it.

lighter than average did not survive as well as babies of average weight. Stabilizing selection has probably operated on birth weight in human populations from the time of the evolutionary expansion of our brains about 1–2 million years ago until the twentieth century. In most of the world it still does. However, in the 50 years since Karn and Penrose's (1951) study, the force of stabilizing selection on birth weight has relaxed in wealthy countries (Figure 4.4b), and by the late 1980s it had almost disappeared. The pattern has approached that of Figure 4.2d: percent survival has become almost the same for all birth weights. Selection has relaxed because of improved care for premature

Figure 4.3

Directional selection by fishing on pink salmon, *Onchorhynchus gorbuscha*. The graph shows the decrease in size of pink salmon caught in two rivers in British Columbia since 1950. The decrease has been driven by selective fishing for the large individuals. Two lines are drawn for each river: one for the salmon caught in odd-numbered years, the other for even years. Salmon caught in odd years are consistently heavier, which is presumably related to the 2-year life cycle of the pink salmon. (5 lb ≈ 2.2 kg.) From Ricker (1981). Redrawn with permission of the Minister of Supply and Services Canada, 1995.

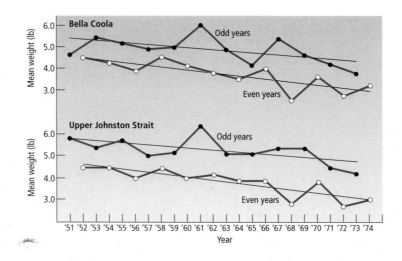

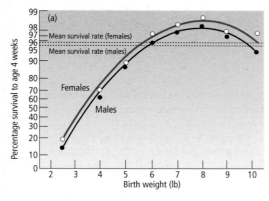

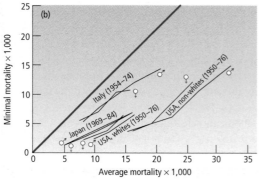

Figure 4.4

(a) The classic pattern of stabilizing selection on human birth weight. Infants weighing 8 lb (3.6 kg) at birth have a higher survival rate than heavier or lighter infants. The graph is based on 13,700 infants born in a hospital in London, UK, from 1935 to 1946. (b) Relaxation of stabilizing selection in wealthy countries in the second half of the twentieth century. The *x*-axis is the average mortality in a population; the *y*-axis is the mortality of infants that have the optimal birth weight in the population (and so the minimum mortality achieved in that population). In (a), for example, females have a minimum mortality of about 1.5% and an average mortality of about 4%. When the average equals the minimum, selection has ceased: this corresponds to the 45° line (the "no selection" case in Figure 4.2d would give a point on the 45° line.) Note the way in Italy, Japan, and the USA, the data approach the 45° line through time. By the late 1980s the Italian population had reached a point not significantly different from the absence of selection. From Karn & Penrose (1951) and Ulizzi & Manzotti (1988). Redrawn with permission of Cambridge University Press.

Box 4.1
Evolving fisheries

When large fish are selectively caught, the fish population evolves smaller size. Figure 4.3 in this chapter shows an example from the salmon of the Pacific Northwest. The evolutionary response of fished populations was the subject of a further study by Conover & Munch (2002). They looked at the long-term yield obtained from fish populations that were exploited in various ways.

Selective fishing of large individuals can set up selection in favor not only of small size but also of slow growth. The advantage (to the fish) of slow growth is easiest to see in a species in which (unlike salmon) each individual produces eggs repeatedly over a period of time. An individual that grows slowly will have a longer period of breeding before it reaches the size at which it is vulnerable to fishing. Slow growth can also be advantageous in a species in which individuals breed only once. The slower growing individuals may (depending on the details of the life

cycle) be smaller at the time of breeding, and less likely to be fished.

The evolution of slow growth has commercial consequences. The supply of fish reaching the fishable size will decrease, and the total yield for the fishery will go down. Fishery yields are highest when the fish grow fast, but selective fishing of large individuals tends to cause evolution to proceed in the opposite direction.

Conover & Munch (2002) kept several populations of Atlantic silverside (*Menidia menidia*) in the laboratory. They experimentally fished some of the populations by taking individuals larger than a certain size each generation, other populations by taking individuals smaller than a certain size each generation, and yet other populations by taking random-sized fish. They measured various properties of the fish populations over four generations.

Figure B4.1a shows the evolution of growth rate. The populations in which large individuals were fished out

evolved toward slow growth rates. This had the predicted effect on the total success of the experimental fishery. Figure B4.1b shows how the total harvest of the fish decreased. As the fish evolved slower growth, they had evolved in such a way that fewer fish were available to be fished. In populations in which small individuals were fished, evolution, and the success of the fishery, went in the other direction.

Conservationists and fishery scientists have been concerned about the maintenance of sustainable fisheries. They have often recommended regulations that result in the fishing of large individuals. What has often been overlooked is the way the fish population will evolve in relation to fishing practices. In general, exploited populations will "evolve back," depending how we exploit them. Conover and Munch's experiment illustrates this point and shows how one commonly recommended fishing practice also causes the evolution of reduced yields.

Figure B4.1
Evolution in an experimental fishery. (a) Growth rates in populations in which large (squares), small (black circles), or random-sized (open circles) fish have been experimentally removed each generation. (b) Total yield of the experimental fisheries. The total yield is the number of fish caught multiplied by the average weight of the caught fish. (1 lb ≈ 450 g.) From Conover & Munch (2002).

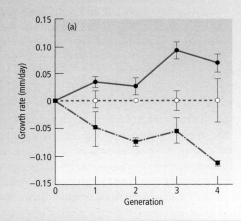

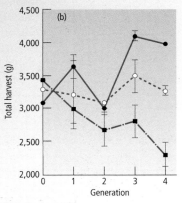

Figure 4.5

Disruptive selection in the seedcracking finch *Pyrenestes ostrinus*. (a) Beak size is not distributed in the form of a bell curve; it has large and small forms, but with some blurring between them. The bimodal distribution is only found for beak size. (b) General body size, such as measured by tail size, shows a classic normal distribution. The distributions shown are for males. (c) Fitness shows twin peaks. Notice that the peaks and valleys correspond to the peaks and valleys in the frequency distribution in (a). Fitness was measured by the survival of marked juveniles over the 1983–90 period. Performance was measured as the inverse of the time to crack seeds. (1 in ≈ 25 mm.) Modified from Smith & Girman (2000).

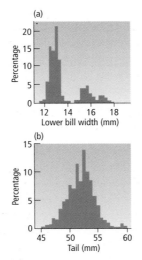

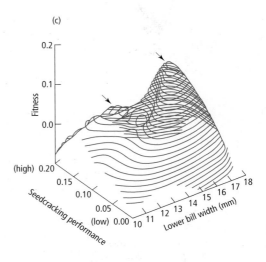

. . . or disruptive

deliveries (the main cause of lighter babies) and increased frequencies of Cesarian deliveries for babies that are large relative to the mother (the lower survival of heavier babies was mainly due to injury to the baby or the mother during birth). By the 1990s in wealthy countries, the stabilizing selection that had been operating on human birth weight for over a million years had all but disappeared.

The third type of natural selection occurs when both extremes are favored relative to the intermediate types. This is called *disruptive* selection (Figure 4.2c). T.B. Smith has described an example in the African finch *Pyrenestes ostrinus*, informally called the black-bellied seedcracker (Smith & Girman 2000) (see Plate 2, between pp. 68 and 69). The birds are found through much of Central Africa, and specialize on eating sedge seeds. Most populations contain large and small forms that are found in both males and females; this is not an example of sexual dimorphism. As Figure 4.5a illustrates, this is a case in which the character is not clearly either discretely or continuously distributed. The categories of discrete and continuous variation blur into each other, and the beaks of these finches are in the blurry zone. We shall look more at the meaning of continuous variation in Chapter 9, but here we are using the example only to illustrate disruptive selection and it does not much matter whether Figure 4.5a is called discrete or continuous variation.

Several species of sedge occupy the finch's environment, and the sedge seeds vary in how hard they are to crack open. Smith measured how long it took a finch to crack open a seed, depending on the finch's beak size. He also measured fitness, depending on beak size, over a 7-year period. Figure 4.5c summarizes the results and shows two fitness peaks. The twin peaks primarily exist because there are two main species of sedge. One sedge species produces hard seeds, and large finches specialize on it; the other sedge species produces soft seeds and the smaller finches specialize on it. In an evironment with a bimodal resource distribution, natural selection drives the finch population to have a bimodal distribution of beak sizes. Natural selection is then disruptive. Disruptive selection is of particular theoretical interest, both because it can increase the genetic diversity of a population (by frequency-dependent selection — Section 5.13, p. 127) and because it can promote speciation (Chapter 14).

A final theoretical possibility is for there to be no relation between fitness and the character in question: then there is *no natural selection* (Figure 4.2d; Figure 4.4b provides an example, or a near example).

4.5 Variation in natural populations is widespread

Natural selection will operate whenever the four conditions in Section 4.2 are satisfied. The first two conditions need little more to be said about them. It is well known that organisms reproduce themselves: this is often given as one of the defining properties of living things. It is also well known that organisms show inheritance. Inheritance is produced by the Mendelian process, which is understood down to a molecular level. Not all the characters of organisms are inherited; and natural selection will not adjust the frequencies of non-inherited characters. But many are inherited, and natural selection can potentially work on them. The third and fourth conditions do need further comment.

The extent of variation, particularly in fitness, matters for understanding evolution

How much, and with respect to what characters, do natural populations show variation and, in particular, variation in fitness? Let us consider biological variation through a series of levels of organization, beginning with the organism's morphology, and working down to more microscopic levels. The purpose of this section is to give examples of variation, to show how variation can be seen in almost all the properties of living things, and to introduce some of the methods (particularly molecular methods) that we shall meet again and that are used to study variation.

Morphological level

At the morphological level, the individuals of a natural population will be found to vary for almost any character we may measure. In some characters, like body size, every individual differs from every other individual; this is called continuous variation. Other morphological characters show discrete variation — they fall into a limited number of categories. Sex, or gender, is an obvious example, with some individuals of a population being female, others male. This kind of categorical variation is found in other characters too.

Variation exists in morphological, . . .

A population that contains more than one recognizable form is *polymorphic* (the condition is called polymorphism). There can be any number of forms in real cases, and they can have any set of relative frequencies. With sex, there are usually two forms. In the peppered moth (*Biston betularia*), two main color forms are often distinguished, though real populations may contain three or more (Section 5.7, p. 108). As the number of forms in the population increases, the polymorphic, categorical kind of variation blurs into the continuous kind of variation (as we saw in the seedcracker finch, Figure 4.5).

Cellular level

Variation is not confined to morphological characters. If we descend to a cellular character, such as the number and structure of the chromosomes, we again find variation.

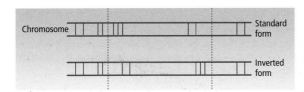

Figure 4.6
Chromosomes can exist in standard and inverted forms. It is arbitrary which is called "standard" and which "inverted." The inversion can be detected by comparing the fine structure of bands, as is diagrammatically illustrated here, or by the behavior of the chromosomes at meiosis.

Figure 4.7
The Australian grasshopper *Keyacris scurra* is polymorphic for inversions for two chromosomes. The two chromosomes are called the CD and the EF chromosomes. The standard and inverted forms of the CD chromosome are called *St* and *Bl*; the standard and inverted forms of the EF chromosome are called *St′* and *Td*. *v* is the relative viability at a site at Wombat, New South Wales, expressed relative to the viability of the *St/Bl St′/St′* form, which is arbitrarily set as 1. *n* is the sample size, *x* is the mean live weight, and the pictures illustrate the relative sizes of the grasshoppers. From White (1973).

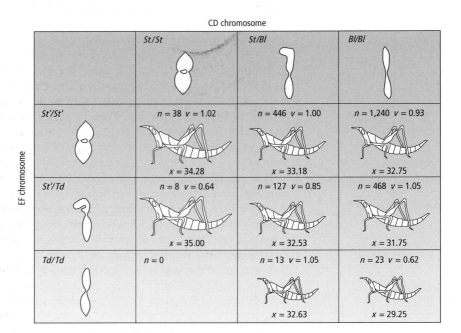

... cellular, such as chromosomal, ...

In the fruitfly *Drosophila melanogaster*, the chromosomes exist in giant forms in the larval salivary glands and they can be studied with a light microscope. They turn out to have characteristic banding patterns, and chromosomes from different individuals in a population have subtly varying banding patterns. One type of variant is called an *inversion* (Figure 4.6), in which the banding pattern — and therefore the order of genes — of a region of the chromosome is inverted. A population of fruitflies may be polymorphic for a number of different inversions.

Chromosomal variation is less easy to study in species that lack giant chromosomal forms, but it is still known to exist. Populations of the Australian grasshopper *Keyacris scurra*, for example, may contain two (normal and inverted) forms for each of two chromosomes; that makes nine kinds of grasshopper in all because an individual may be homozygous or heterozygous for any of the four chromosomal types. The nine differ in size and viability (Figure 4.7).

Chromosomes can vary in other respects too. Individuals may vary in their number of chromosomes, for example. In many species, some individuals have one or more extra chromosomes, in addition to the normal number for the species. These "super-numerary" chromosomes, which are often called B chromosomes, have been particularly studied in maize and in grasshoppers. In the grasshopper *Atractomorpha australis*, normal individuals have 18 autosomes, but individuals have been found with from one to six supernumary chromosomes. The population is polymorphic with respect to chromosome number. Inversions and B chromosomes are just two kinds of chromosomal variation. There are other kinds too; but these are enough to make the point that individuals vary at the subcellular, as well as the morphological level.

Biochemical level

The story is the same at the biochemical level, such as for proteins. Proteins are molecules made up of sequences of amino acid units. A particular protein, like human hemoglobin, has a particular characteristic sequence, which in turn determines the molecule's shape and properties. But do all humans have exactly the same sequence for hemoglobin, or any other protein? In theory, we could find out by taking the protein from several individuals and then working out the sequence in each of them; but it would be excessively laborious to do so. *Gel electrophoresis* is a much faster method. Gel electrophoresis works because different amino acids carry different electric charges. Different proteins — and different variants of the same protein — have different net electric charges, because they have different amino acid compositions. If we place a sample of proteins (with the same molecular weight) in an electric field, those with the largest electric charges will move fastest. For the student of biological variation, the importance of the method is that it can reveal different variants of a particular type of protein. A good example is provided by a less well known protein than hemoglobin — the enzyme called alcohol dehydrogenase, in the fruitfly.

Fruitflies, as their name suggests, lay their eggs in, and feed on, decaying fruit. They are attracted to rotting fruit because of the yeast it contains. Fruitflies can be collected almost anywhere in the world by leaving out rotting fruit as a lure; and drowned fruitflies are usually found in a glass of wine left out overnight after a garden party in the late summer. As fruit rots, it forms a number of chemicals, including alcohol, which is both a poison and a potential energy source. Fruitflies cope with alcohol by means of an enzyme called alcohol dehydrogenase. The enzyme is crucial. If the alcohol dehydrogenase gene is deleted from fruitflies, and those flies are then fed on mere 5% alcohol, "they have difficulty flying and walking, and finally, cannot stay on their feet" (quoted in Ashburner 1998).

Gel electrophoresis reveals that, in most populations of the fruitfly *Drosophila melanogaster*, alcohol dehydrogenase comes in two main forms. The two forms show up as different bands on the gel after the sample has been put on it, an electric current put across it for a few hours, and the position of the enzyme has been exposed by a specific stain. The two variants are called slow (*Adh-s*) or fast (*Adh-f*) according to how far they have moved in the time. The multiple bands show that the protein is polymorphic. The enzyme called alcohol dehydrogenase is actually a class of two polypeptides with slightly different amino acid sequences. Gel electrophoresis has been applied to a large number of proteins in a large number of species and different proteins show different degrees of variability (Chapter 7). But the point for now is that many of these

. . . biochemical, such as in enzymes, . . .

proteins have been found to be variable — extensive variation exists in proteins in natural populations.

DNA level

. . . and genetic characters

If variation is found in every organ, at every level, among the individuals of a population, variation will almost inevitably also be found at the DNA level too. The inversion polymorphisms of chromosomes that we met above, for example, are due to inversions of the DNA sequence. However, the most direct method of studying DNA variation is to sequence the DNA itself. Let us stay with alcohol dehydrogenase in the fruitfly. Kreitman (1983) isolated the DNA encoding alcohol dehydrogenase from 11 independent lines of *D. melanogaster* and individually sequenced them all. Some of the 11 had *Adh-f*, others *Adh-s*, and the difference between *Adh-f* and *Adh-s* was always due to a single amino acid difference (Thr or Lys at codon 192).

The amino acid difference appears as a base difference in the DNA, but this was not the only source of variation at the DNA level. The DNA is even more variable than the protein study suggests. At the protein level, only the two main variants were found in the sample of 11 genes, but at the DNA level there were 11 different sequences with 43 different variable sites. The amount of variation that we find is therefore highest at the DNA level. At the level of gross morphology, a *Drosophila* with two *Adh-f* genes is indistinguishable from one with two *Adh-s* genes; gel electrophoresis resolves two classes of fly; but at the DNA level, the two classes decompose into innumerable individual variants.

Restriction enzymes provide another method of studying DNA variation. Restriction enzymes exist naturally in bacteria, and a large number — over 2,300 — of restriction enzymes are known. Any one restriction enzyme cuts a DNA strand wherever it has a particular sequence, usually of about 4–8 base pairs. The restriction enzyme called *EcoR1*, for instance, which is found in the bacterium *Escherichia coli*, recognizes the base sequence ...GAATTC... and cuts it between the initial G and the first A. In the bacterium, the enzymes help to protect against viral invasion by cleaving foreign DNA, but the enzymes can be isolated in the laboratory and used to investigate DNA sequences. Suppose the DNA of two individuals differs, and that one has the sequence GAATTC at a certain site whereas the other individual has another sequence such as GTATT. If the DNA of each individual is put with *EcoR1*, only that of the first individual will be cleaved. The difference can be detected in the length of the DNA fragments: the pattern of fragment lengths will differ for the two individuals. The variation is called *restriction fragment length polymorphism* and has been found in all populations that have been studied.

Conclusion

In summary, natural populations show variation at all levels, from gross morphology to DNA sequences. When we move on to look at natural selection in more detail, we can assume that in natural populations the requirement of variation, as well as of reproduction and heredity, is met.

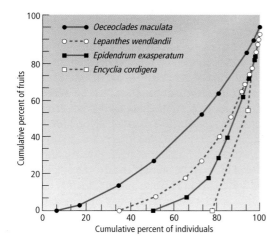

Figure 4.8
Variation in reproductive success within populations, illustrated by four species of orchids. The graphs plot the cumulative percentage of offspring produced by the plants, with the individual plants ranked from the least to the most successful. For instance, in *Epidendrum exasperatum*, the least successful 50% of individuals produce none of the offspring: they fail to reproduce. The next 17% or so of individuals, moving up in the ranking of success, produce about 5% of the fruit in the population; and the next 10% produce about 13%; and so on. If every individual produced the same number of offspring the cumulative percentage graph would be the 45° line. Graphs of this kind can be used to express inequality in a population generally; they were first used to express inequality in human wealth and are sometimes called Lorenz curves. Redrawn, by permission of the publisher, from Calvo (1990).

4.6 Organisms in a population vary in reproductive success

If natural selection is to operate, it is not enough that characters vary. The different forms of the character must also be associated with reproductive success (or fitness) — in the degree to which individuals contribute offspring to the next generation. Reproductive success is more difficult to measure than a phenotypic character like body size, and there are far fewer observations of variation in reproduction than in phenotype. However, there are still a good number of examples. We have met some already this chapter (Section 4.4) and we shall meet more later in the book. Here we can look at an even more abundant sort of evidence, and at an abstract argument.

Individuals differ in reproductive success in all populations

Whenever reproductive success in a biological population has been measured, it has been found that some individuals produce many more offspring than others. Figure 4.8 illustrates this variation in four species of orchids in the form of a cumulative percentage graph. If every individual produced the same number of fruit (that is, the same number of offspring), the points would fall along the 45° line. In fact the points usually start some way along the *x*-axis and fall below the 45° line. The reason is that some individuals fail to reproduce and a successful minority contribute a disproportionate number of offspring.

Four orchid species provide an example

The differences between the four orchid species in Figure 4.8 can be understood in terms of their relationships with insect pollinators. The reproductively egalitarian species *Oeceoclades maculata* reproduces by self-fertilization, and has no use for pollinators. The two intermediate species *Lepanthes wendlandii* and *Epidendrum exasperatum* are each capable of self-fertilization but can also be pollinated by insects. The highly inegalitarian *Encyclia cordigera*, in which 80% of the individuals fail to reproduce, requires insect pollination. However, this species is unattractive to pollinating insects. It is one of the orchids that have evolved "deceptive" flowers that produce and receive pollen but do not supply nectar. The orchids "cheat" the insect and insects tend to avoid them (though not completely) in consequence. The amount of reproductive failure in orchids with these deceptive flowers can be remarkably high — even higher than the 80% in *Encyclia cordigera*.

More extreme examples exist. Gill (1989) measured reproduction in a population of almost 900 individuals of the pink lady's slipper orchid *Cypripedium acaule* in Rockingham County, Virginia, from 1977 to 1986. In that 10-year period only 2% of the individuals managed to produce fruit: the rest had been avoided by pollinators and failed to breed. In four of the years none of the orchids bred at all. Thus the ecological factor determining variation in reproductive success in orchids is the availability of, and need for, pollinating insects. If pollinating insects are unnecessary, all the orchids in a population produce a similar number of fruit. But if pollinating insects are necessary and scarce, because of the way the orchid "cheats" the pollinators, only a small minority of individuals may succeed in reproducing. Pollinators happen to be a key factor in orchids; but in other species other factors will operate and ecological study can reveal why some individuals are more reproductively successful than others.

The results in Figure 4.8 show the amount of reproductive variation among the adults that exist in a population, but this variation is only for the final component of the life cycle. Before it, individuals differ in survival, and a life table like Table 4.1 at the beginning of the chapter quantifies that variation. A full description of the variation in lifetime success of a population would combine variation in survival from conception to adulthood and variation in adult reproductive success.

The conditions for natural selection to operate are often met

Examples such as HIV, or the pink salmon, show that natural selection can operate; but that leaves open the question of how often natural selection operates in natural populations, and in what proportion of species. We could theoretically find out how widespread natural selection is by counting how frequently all four conditions apply in nature. That, however, would at the least be hard work. The evidence of variation in phenotypic characters and of ecological competition suggests that the preconditions required for natural selection to operate are widespread, indeed probably universal. Whenever anyone has looked they have found variation in the phenotypic characters of populations, and ecological competition within them.

Natural selection is likely at work in natural populations all the time

Indeed, you do not need to be a professional biologist to know about variation and the struggle for existence. They are almost obvious facts of nature. It is logically possible that individual reproductive success varies in all populations in the manner of Figure 4.8, but that natural selection does not operate in any of them, because the variation in reproductive success is not associated with any inherited characters. However, though it is logically possible, it is not ecologically probable. In almost every species, a high proportion of individuals are doomed to die. Any attribute that increases the chance of survival, in a way that might appear trivial to us, is likely to result in a higher than average fitness. Any tendency of individuals to make mistakes, slightly increasing their risk of death, will result in lowered fitness. Likewise, once an individual has survived to adulthood, there will be many ways in which its phenotypic attributes can influence its chance of reproductive success. The struggle for existence, and phenotypic variation, are both universal conditions in nature. Variation in fitness associated with some of those phenotypic characters is therefore also likely to be very common. The argument is one of plausibility, rather than certainty: it is not logically inevitable that in a population showing (inherited) variation in a phenotypic character there will also be an association between the varying character and fitness. But if there is, natural selection will operate.

4.7 New variation is generated by mutation and recombination

Long-term evolutionary change requires an input of new variation

The variation that exists in a population is the resource on which natural selection works. Imagine a population evolving increased body size. To begin with there is variation and average size can increase. However, the population could only evolve a limited amount if the initial variation were all there was to work with; it would soon reach the edge of available variation (Figure 4.9a). In existing human populations, for instance, height does not range much beyond about 8 feet (2.4 m). The evolution of humans more than 8 feet high would be impossible if natural selection only had the currently existing variation to work on. Evolution from the origin of life to the level of modern diversity must have required more variation than existed in the original population. Where did the extra variation come from?

It comes from recombination . . .

Recombination (in sexual populations) and mutation are the two main answers. As a population evolves toward individuals of larger body size, the genotypes encoding larger body size increase in frequency. At the initial stage, large body size was rare and there might have been only one or two individuals possessing genotypes for large body size. The chances are that they would interbreed with other individuals closer to the average size for the population and produce offspring of less extreme size. But as the

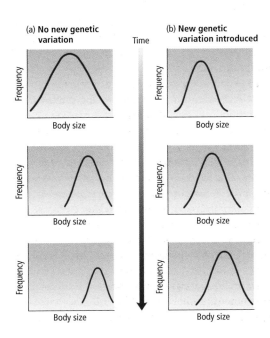

Figure 4.9
Natural selection produces evolution by working on the variation in a population. (a) In the absence of new variation, evolution soon reaches the limit of existing variation and comes to a stop. (b) However, recombination generates new variation as the frequencies of the genotypes change during evolution. Evolution can then proceed further than the initial range of variation.

genotypes for large body size itself become the average, they are more likely to inter-breed and produce new genotypes encoding even larger body size. As evolution proceeds, recombination among the existing genotypes generates a new range of variation (Figure 4.9b).

. . . and mutation

Mutation also introduces new variation. Chapter 2 (Table 2.2, p. 32) gave some figures for typical mutation rates. The exceptionally rapid evolution of drug resistance in HIV occurs not only because of the huge selective force imposed by the drug itself (which effectively sterilizes the virus), but it also has huge population sizes, even within one human body, rapid reproduction, and a relatively high mutation rate. Consider some figures. In an average AIDS patient, at least 10^{12} new individual HIV are generated per day. The virus is about 10^4 nucleotides long and has a mutation rate of about one mutation per 10^4 nucleotides. Each new virus contains an average of about one mutation. With an input of 10^{12} new viruses per day, we can be sure that every nucleotide position down the 10^4 nucleotide length of the virus will be mutated every day within one AIDS patient. Indeed, every possible single nucleotide mutation will occur several times over, along with most possible combinations of two-nucleotide mutation. Given that resistance to 3TC requires a change in only one amino acid, we can see that natural selection is an overwhelmingly powerful counterforce against human medicine operating with single-drug treatments. A combination of several drugs is needed to overpower an evolving HIV population. Mutation introduces less variation in other life forms that have lower population sizes, lower reproductive rates, and lower mutation rates. But in all species, mutation is an abundant source of new variation, providing raw material for evolutionary change.

4.8 Variation created by recombination and mutation is random with respect to the direction of adaptation

A basic property of Darwinism is that the direction of evolution, particularly of adaptive evolution, is uncoupled from the direction of variation. When a new recombinant or mutant genotype arises, there is no tendency for it to arise in the direction of improved adaptation. Natural selection imposes direction on evolution, using undirected variation. In this section, we define the alternative viewpoint (the theory of directed variation) and consider why it is not accepted.

Consider HIV again. When the environment changed, a new form of HIV was favored. According to Darwin's theory, that environmental change does not itself cause mutations of the right form to appear. New mutations of all sorts are constantly arising but independently of what is required for adaptation to the current environment. The alternative would be some kind of *directed mutation*. For mutation to be directed would mean that when the environment changed to favor a drug-resistant virus, the mutational process itself selectively tended to produce drug-resistant mutations.

Directional mutation is a theoretical alternative to natural selection

The strongest reason to doubt that mutations are adaptively directed is theoretical. The drug treatment imposed an environment on the virus that it had never encountered before. The environment (probably) was completely new. A particular genetic change was needed for the virus to continue to reproduce. Could it arise by directed mutation? At the genetic level, the mutation consisted of a set of particular changes in the base sequence of a gene. No mechanism has been discovered that could direct the right base changes to happen.

Adaptively directed mutation is unlikely, for theoretical reasons

If we reflect on the kind of mechanism that would be needed, it becomes clear that an adaptively directed mutation would be practically impossible. The virus would have to recognize that the environment had changed, work out what change was needed to adapt to the new conditions, and then cause the correct base changes. It would have to do so for an environment it had never previously experienced. As an analogy, this ability would be like humans describing subject matter they had never encountered before in a language they did not understand; like a seventeenth century American using Egyptian hieroglyphics to describe how to change a computer program. (Hieroglyphics were not deciphered until the discovery of the Rosetta Stone in 1799.) Even if it is just possible to imagine, as an extreme theoretical possibility, directed mutations in the case of viral drug resistance, the changes in the evolution of a more complex organ (like the brain, or circulatory system, or eye) would require a near miracle. Mutations are therefore thought not to be directed toward adaptation.

Although mutation is random and undirected with respect to the direction of improved adaptation, that does not exclude the possibility that mutations are non-random at the molecular level. For example, the two-nucleotide sequence CG tends to mutate, when it has been methylated, to TG. (The DNA in a cell is sometimes methylated, for reasons that do not matter here.) After replication a complementary pair of CG on the one strand and GC on the other will then have produced TG and AC. Species with high amounts of DNA methylation have (perhaps for this reason) low amounts of CG in their DNA.

But some non-adaptive mutational processes are directed

Molecular mutational biases are not the same as changes toward improved adaptation, however. You cannot change a drug-susceptible HIV into a drug-resistant HIV just by converting some of its CG dinucleotides into TG. Some critics of Darwinism have read that Darwinian theory describes mutation as "random," and have then trotted out these sorts of molecular mutational biases as if they contradicted it. But mutation can be non-random at the molecular level without contradicting Darwinian theory. What Darwinism rules out is mutation directed toward new adaptation. Because of this confusion about the word random, it is often better to describe mutation not as random, but as "undirected" or "accidental" (which was the word Darwin used).

Summary

1 Organisms produce many more offspring than can survive, which results in a "struggle for existence," or competition to survive.

2 Natural selection will operate among any entities that reproduce, show inheritance of their characteristics from one generation to the next, and vary in "fitness" (i.e., the relative number of offspring they produce) according to the characteristic they possess.

3 The increase in the frequency of drug-resistant, relative to drug-susceptible, HIV illustrates how natural selection causes both evolutionary change and the evolution of adaptation.

4 Selection may be directional, stabilizing, or disruptive.

5 The members of natural populations vary with respect to characteristics at all levels. They differ in their morphology, their microscopic structure, their chromosomes, the amino acid sequences of their proteins, and in their DNA sequences.

6 The members of natural populations vary in their reproductive success: some individuals leave no offspring, others leave many more than average.

7 In Darwin's theory, the direction of evolution, particularly of adaptive evolution, is uncoupled from the direction of variation. The new variation that is created by recombination and mutation is accidental, and adaptively random in direction.

8 Two reasons suggest that neither recombination nor mutation can alone change a population in the direction of improved adaptation: there is no evidence that mutations occur particularly in the direction of novel adaptive requirements, and it is theoretically difficult to see how any genetic mechanism could have the foresight to direct mutations in this way.

Further reading

An ecology text, such as Ricklefs & Miller (2000), will introduce life tables. For the theory of natural selection, see Darwin's original account (1859, chapters 3 and 4), Endler (1986), and Bell (1997a, 1997b). Law (1991) describes the selective effects of fishing. Travis (1989) reviews stabilizing selection. Ulizzi et al. (1998) update the human birthweight story. Greene et al. (2000) describe another possible example of disruptive selection. Chapter 3 in this text gave references for HIV.

Genetic variation is described in all the larger population genetics texts, such as Hartl (2000), Hartl & Clark (1997), and Hedrick (2000). White (1973) and Dobzhansky (1970) describe chromosomal variation. Variation in proteins and DNA will be discussed further in Chapter 7, which gives references. The authors in Clutton-Brock (1988) discuss natural variation in reproductive sucess.

I have concentrated on the theoretical argument against directed mutation, but experiments have also been done. The classic one was by Luria & Delbruck (1943). It was challenged by Cairns et al. (1988) but modern interpretations of results such as Cairns et al. rule out directed mutation: see Andersson et al. (1998) and Foster (2000). Two other themes are the evolution of mutation rates (see Sniegowski et al. 2000), and the possibility that the high mutation rates of HIV could be used against them by triggering a mutational meltdown. The underlying theory is discussed in Chapter 12 later in this text. See Holmes (2000a) for the HIV possibilities. Biases at the molecular level in the mutation process are set to be revealed by genomic data (see, for instance, Silva & Kondrashov 2002), and Mukai-style mutation–accumulation experiments, discussed in Chapter 12 of this text.

Study and review questions

1 Use Figure 4.1b to construct a life table, like Table 4.1, for cod. (Use the densities per meter squared as numbers; and you may prefer to ignore the right-hand column, for daily mortality rates, which require logarithms.)

2 (a) Review the four conditions needed for natural selection to operate. (b) What would happen in a population in which only conditions 1, 2, and 3 were satisfied? (c) And in one in which only 1, 3, and 4 were satisfied?

3 Variation in reproductive success has been found in all populations in which it has been measured. Why is this observation alone insufficient to show that natural selection operates in all populations?

4 It is occasionally suggested that mutation is adaptively directed rather than random. Think through what a genetic mechanism of adaptively directed mutation would have to do. For each component of the mechanism, how plausible is it that it could really exist?

5 What sort of selection is taking place in populations a, b, and c in the graph?

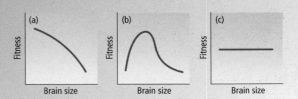

6 [These question are more for further thought than review of chapter content.] (a) On average only two offspring survive per parental pair: why therefore does every pair in the population not produce exactly two offspring (rather than the more variable reproductive success we see in nature)? This would lead to the same end consequence. (b) Why in some species is the "excess" far greater than in others?

Evolutionary Genetics

The theory of population genetics is the most important, most fundamental body of theory in evolutionary biology. It is the proving ground for almost all ideas in evolutionary biology. The coherence of an evolutionary hypothesis usually remains in doubt until the hypothesis is expressed in the form of a population genetic model. We start with the simplest, and move on to the more complex, cases. The simplest case is when the population is large, large enough that we can ignore random effects; models of this kind are called deterministic. In Chapter 5, we look at a simple deterministic model of natural selection. The model has only one genetic locus, and one allele of higher fitness is being substituted for an inferior allele. We also look at how natural selection can maintain variation at a single locus, in three circumstances, and look at examples of each.

Chapter 6 considers random effects in population genetics. The transfer of genes from one generation to the next is not a perfectly exact process, because random sampling may change the frequency of a gene. The effects of random sampling are most powerful when the different genotypes all have the same fitness, and when population sizes are small. The theory of random drift has been most important for thinking about molecular evolution. Chapter 7 looks at the relative contributions of random drift and natural selection to molecular evolution. The question of their relative contributions has stimulated one of the richest research programs in evolutionary biology. We shall concentrate on modern research, but look at its conceptual roots too.

In Chapter 8, we move on to consider natural selection working simultaneously on more than one locus. Linkage between loci complicates the one-locus model. With more than one locus, the genes at different loci may interact and influence each other's fitness. Evolution at one locus can be influenced by genes at other loci. It is a matter of controversy how important higher level interactions between gene loci are, and how far the one-locus model is an adequate description of the real world. As we move from two-locus evolution to multiple-locus evolution we abandon Mendelian exactitude and use a quite different method: quantitative genetics. In quantitative genetics (Chapter 9), the relations between individuals and between successive generations are described approximately and abstractly. Quantitative genetics is concerned with "continuous" characters, at the morphological level. As we saw in Chapter 4, morphological characters show variation in natural populations, and we shall consider how to account for the level of variation that is observed.

5

The Theory of Natural Selection

This chapter introduces formal population genetic models. We first establish what the variables are that the models are concerned with, and the general structure of population genetic models. We look at the Hardy–Weinberg equilibrium, and see how to calculate whether a real population fits it. We then move on to models of natural selection, concentrating on the specific case of selection against a recessive homozygote. We apply the model to two examples: the peppered moth and resistance to pesticides. The second half of the chapter is mainly about how natural selection can maintain genetic polymorphism. We look at selection–mutation balance, heterozygous advantage, and frequency-dependent selection; and we finish by looking at models that include migration in a geographically subdivided population. The theory in this chapter all assumes that the population size is large enough for random effects to be ignored. Chapters 6 and 7 consider how random effects can interact with selection in small populations.

5.1 Population genetics is concerned with genotype and gene frequencies

The human genome, on current estimates, contains something like 30,000 gene loci. Let us focus on just one of them — on a locus at which there is more than one allele, because no evolutionary change can happen at a locus for which every individual in the population has two copies of the same allele. We shall be concerned in this chapter with models of evolution at a single genetic locus; these are the simplest models in population genetics. Chapters 8 and 9 discuss more complex models in which evolutionary change occurs simultaneously at more than one locus.

We define genotype frequency . . .

The theory of population genetics at one locus is mainly concerned to understand two closely connected variables: *gene frequency* and *genotype frequency*. They are easy to measure. The simplest case is one genetic locus with two alleles (*A* and *a*) and three genotypes (*AA*, *Aa*, and *aa*). Each individual has a genotype made up of two genes at the locus and a population can be symbolized like this:

Aa   AA   aa   aa   AA   Aa   AA   Aa

This is an imaginary population with only eight individuals. To find the genotype frequencies we simply count the numbers of individual with each genotype. Thus:

Frequency of *AA* = 3/8 = 0.375
Frequency of *Aa* = 3/8 = 0.375
Frequency of *aa* = 2/8 = 0.25

In general we can symbolize genotype frequencies algebraically, as follows.

Genotype	*AA*	*Aa*	*aa*
Frequency	*P*	*Q*	*R*

P, *Q*, and *R* are expressed as percentages or proportions, so in our population, *P* = 0.375, *Q* = 0.375, and *R* = 0.25 (they have to add up to 1, or to 100%). They are measured simply by observing and counting the numbers of each type of organism in the population, and dividing by the total number of organisms in the population (the population size).

. . . and gene frequency

The gene frequency is likewise measured by counting the frequencies of each gene in the population. Each genotype contains two genes, and there are a total of 16 genes per locus in a population of eight individuals. In the population above,

Frequency of *A* = 9/16 = 0.5625
Frequency of *a* = 7/16 = 0.4375

Algebraically, we can define *p* as the frequency of *A*, and *q* as the frequency of *a*. *p* and *q* are usually called "gene" frequencies, but in a strict sense they are allele frequencies: they are the frequencies of the different alleles at one genetic locus. The gene frequencies can be calculated from the genotype frequencies:

$$p = P + \tfrac{1}{2}Q \qquad\qquad (5.1)$$
$$q = R + \tfrac{1}{2}Q$$

(and $p + q = 1$). The calculation of the gene frequencies from genotype frequencies is highly important. We shall make recurrent use of these two simple equations in the chapter. Although the gene frequencies can be calculated from the genotype frequencies (P, Q, R), the opposite is not true: the genotype frequencies cannot be calculated from the gene frequencies (p, q).

Now that we have defined the key variables, we can see how population geneticists analyze changes in those variables through time.

5.2   An elementary population genetic model has four main steps

Population geneticists try to answer the following question: if we know the genotype (or gene) frequencies in one generation, what will they be in the next generation? It is worth looking at the general procedure before going into particular models. The procedure is to break down the time from one generation to the next into a series of stages. We then work out how genotype frequencies are affected at each stage. We can begin at any arbitrarily chosen starting point in generation n and then follow the genotype frequencies through to the same point in generation $n + 1$. Figure 5.1 shows the general outline of a population genetics model.

We start with the frequencies of genotypes among the adults in generation n. The first step is to specify how these genotypes combine to breed (called a mating rule); the

Figure 5.1
The general model of
population genetics.

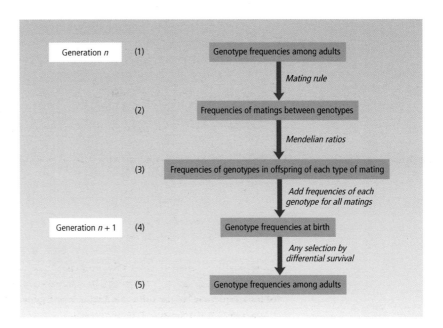

second step is to apply the Mendelian ratios (Chapter 2) for each type of mating; we then add the frequencies of each genotype generated from each type of mating to find the total frequency of the genotypes among the offspring, at birth, in the next generation. If the genotypes have different chances of survival from birth to adulthood, we multiply the frequency of each genotype at birth by its chance of survival to find the frequency among the adults. When the calculation at each stage has been completed, the population geneticist's question has been answered.

Population genetic models track gene frequencies over time

Natural selection can operate in two ways: by differences in survival among genotypes or by differences in fertility. There are two theoretical extremes. At one, the surviving individuals of all genotypes produce the same number of offspring, and selection operates only on survival; at the other, individuals of all genotypes have the same survival, but differ in the number of offspring they produce (that is, their fertility). Both kinds of selection probably operate in many real cases, but the models we shall consider in this chapter all express selection in terms of differences in chance of survival. This is not to suggest that selection always operates only on survival; it is to keep the models simple and consistent.

The model, in the general form of Figure 5.1, may look rather complicated. However, we can cut it down to size by making some simplifying assumptions. The first two simplifying assumptions to consider are random mating and no selection (no differences in survival between genotypes from stages 4 to 5).

5.3 Genotype frequencies in the absence of selection go to the Hardy–Weinberg equilibrium

We can stay with the case of one genetic locus with two alleles (A and a). The frequencies of genotypes AA, Aa, and aa are P, Q, and R. Our question is, if there is random mating and no selective difference among the genotypes, and we know the genotype frequencies in one generation, what will the genotype frequencies be in the next generation? The answer is called the *Hardy–Weinberg equilibrium*. Let us see what that means.

We deduce the frequencies of pairings, with random pairing . . .

Table 5.1 gives the calculation. The mating frequencies follow from the fact that mating is random. To form a pair, we pick out at random two individuals from the population. What is the chance of an $AA \times AA$ pair? Well, to produce this pair, the first individual we pick has to be an AA and the second one also has to be an AA. The chance that the first is an AA is simply P, the genotype's frequency in the population. In a large population, the chance that the second one is AA is also P.[1] The chance of drawing out two AA individuals in a row is therefore P^2. (The frequency of $Aa \times Aa$ and $aa \times aa$ matings are likewise Q^2 and R^2, respectively.) Similar reasoning applies for the frequencies of matings in which the two individuals have different genotypes. The chance of

---

[1] "Large" populations are not a separate category from "small" ones; populations come in all sizes. The random effects we consider in Chapter 6 become increasingly important as a population becomes smaller. However, one rough definition of a large population is one in which the sampling of one individual to form a mating pair does not affect the genotype frequencies in the population: if one AA is taken out, the frequency of AA in the population, and the chance of picking another AA, remains effectively P.

Table 5.1
Calculations needed to derive the Hardy–Weinberg ratio for one locus and two alleles, A and a. (Frequency of $AA = P$, of $Aa = Q$, and of $aa = R$.) The table shows the frequencies of different matings if the genotypes mate randomly, and the genotype proportions among the progeny of the different matings.

Mating type	Frequency of mating	Offspring genotype proportions
$AA \times AA$	P^2	1 AA
$AA \times Aa$	PQ	$\frac{1}{2} AA : \frac{1}{2} Aa$
$AA \times aa$	PR	1 Aa
$Aa \times AA$	QP	$\frac{1}{2} AA : \frac{1}{2} Aa$
$Aa \times Aa$	Q^2	$\frac{1}{4} AA : \frac{1}{2} Aa : \frac{1}{4} aa$
$Aa \times aa$	QR	$\frac{1}{2} Aa : \frac{1}{2} aa$
$aa \times AA$	RP	1 Aa
$aa \times Aa$	RQ	$\frac{1}{2} Aa : \frac{1}{2} aa$
$aa \times aa$	R^2	1 aa

picking an AA and then an Aa (to produce an $AA \times Aa$ pair), for example, is PQ; the chance of picking an AA and then an aa is PR; and so on.

. . . and use Mendel's rules to deduce the genotype frequencies in the offspring

The genotypic proportions in the offspring of each type of mating are given by the Mendelian ratios for that cross. We can work out the frequency of a genotype in the next generation by addition. We look at which matings generate the genotype, and add the frequencies generated by all the matings. Let us work it out for the genotype AA. AA individuals, Table 5.1 shows, come from $AA \times AA$, $AA \times Aa$ (and $Aa \times AA$), and $Aa \times Aa$ matings. We can ignore all the other types of mating. $AA \times AA$ matings have frequency P^2 and produce all AA offspring, $AA \times Aa$ and $Aa \times AA$ matings each have frequency PQ and produce 50% AA offspring, and $Aa \times Aa$ matings have frequency Q^2 and produce 25% AA offspring. The frequency of AA in the next generation,[2] P', is then:

$$P' = P^2 + \tfrac{1}{2}PQ + \tfrac{1}{2}PQ + \tfrac{1}{4}Q^2 \tag{5.2}$$

This can be rearranged to:

$$P' = (P + \tfrac{1}{2}Q)(P + \tfrac{1}{2}Q)$$

We have seen that $(P + \tfrac{1}{2}Q)$ is simply the frequency of the gene A, p. Therefore:

$$P' = p^2$$

---

[2] Population geneticists conventionally symbolize the frequency of variables one generation on by writing a prime. If P is the frequency of genotype AA in one generation, P' is its frequency in the next; if p is the frequency of an allele in one generation, p' is its frequency in the next generation. We shall follow this convention repeatedly in this book.

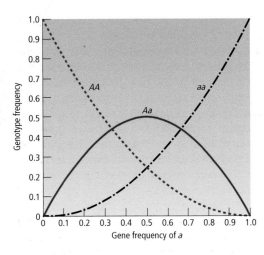

Figure 5.2

Hardy–Weinberg frequencies of genotypes AA, Aa, and aa in relation to the frequency of the gene a (q).

The result is the Hardy–Weinberg equilibrium

The frequency of genotype AA after one generation of random mating is equal to the square of the frequency of the A gene. Analogous arguments show that the frequencies of Aa and aa are $2pq$ and q^2. The Hardy–Weinberg frequencies are then:

Genotype $AA : Aa : aa$
Frequency $p^2 : 2pq : q^2$

Figure 5.2 shows the proportions of the three different genotypes at different frequencies of the gene a; heterozygotes are most frequent when the gene frequency is 0.5.

The Hardy–Weinberg genotype frequencies are reached after a single generation of random mating from any initial genotype frequencies. Imagine, for example, two populations with the same gene frequency but different genotype frequencies. One population has 750 AA, 0 Aa, and 250 aa; the other has 500 AA, 500 Aa, and 0 aa. $p = 0.75$ and $q = 0.25$ in both. After one generation of random mating, the genotype frequencies in both will become 558 AA, 375 Aa, and 67 aa if the population size remains 1,000. (Fractions of an individual have been rounded to make the numbers add to 1,000. The proportions are 9/16, 6/16, and 1/16.) After reaching those frequencies immediately, in one generation, the population stays at the Hardy–Weinberg equilibrium for as long as the population size is large, there is no selection, and mating is random.

As we saw in Section 5.1, it is not in general possible to calculate the genotype frequencies in a generation if you only know the gene frequencies. We can now see that it is possible to calculate, from gene frequencies alone, what the genotype frequencies will be in the next generation, provided that mating is random, there is no selection, and the population is large. If the gene frequencies in this generation are p and q, in the next generation the genotype will have Hardy–Weinberg frequencies.

The proof of the Hardy–Weinberg theorem we have worked through was long-winded. We worked though it all in order to illustrate the general model of population genetics in its simplest case. However, for the particular case of the Hardy–Weinberg equilibrium, a more elegant proof can be given in terms of gametes.

A simpler proof of the Hardy–Weinberg equilibrium

Diploid organisms produce haploid gametes. We could imagine that the haploid gametes are all released into the sea, where they combine at random to form the next generation. This is called random union of gametes. In the "gamete pool" A gametes will have frequency p and a gametes frequency q. Because they are combining at random, an a gamete will meet an A gamete with chance p and an a gamete with chance q. From the a gametes, Aa zygotes will therefore be produced with frequency pq and aa gametes with frequency q^2. A similar argument applies for the A gametes (which have frequency p): they combine with a gametes with chance q, to produce Aa zygotes (frequency pq) and A gametes with chance p to form AA zygotes (frequency p^2). If we now add up the frequencies of the genotypes from the two types of gamete, the Hardy–Weinberg genotype frequencies emerge. We have now derived the Hardy–Weinberg theorem for the case of two alleles; the same argument easily extends to three or more alleles (Box 5.1).

(Some people may be puzzled by the 2 in the frequency of the heterozygotes. It is a simple combinatorial probability. Imagine flipping two coins and asking what the chances are of flipping two heads, or two tails, or one head and one tail. The chance of two heads is $(1/2)^2$ and of two tails $(1/2)^2$; the chance of a head and a tail is $2 \times (1/2)^2$, because a tail then a head, and a head then a tail, both give one head and one tail. The head is analogous to allele A, the tail to a; two heads to producing an AA genotype, and one head and one tail to a heterozygote Aa. The coin produces heads with probability $1/2$, and is analogous to a gene frequency of $p = 1/2$. The frequency $2pq$ for heterozygotes is analogous to the chance of one head and one tail, $2 \times (1/2)^2$. The 2 arises because there are two ways of obtaining one head and one tail. Likewise there are two ways of producing an Aa heterozygote: either the A gene can come from the father and the a from the mother, or the a gene from the father and the A from the mother. The offspring is Aa either way.)

Box 5.1
The Hardy–Weinberg Theorem for Three Alleles

We can call the three alleles A_1, A_2, and A_3, and define their gene frequencies as p, q, and r, respectively. We form new zygotes by sampling two successive gametes from a large pool of gametes. The first gamete we pick could be A_1, A_2, or A_3. If we first pick (with chance p) an A_1 allele from the gamete pool, the chance that the second allele is another A_1 allele is p, the chance that it is an A_2 allele is q, and the chance that it is an A_3 allele is r: from these three, the frequencies of A_1A_1, A_1A_2, and A_1A_3 zygotes are p^2, pq, and pr.

Now suppose that the first allele we picked out had been an A_2 (which would happen with chance q). The chances that the second allele would again be A_1, A_2, or A_3 would be p, q, and r, respectively, giving A_1A_2, A_2A_2, and A_2A_3 zygotes in frequency pq, q^2, and qr.

Finally, if we had picked (with chance r) an A_3 allele, we produce A_1A_3, A_2A_3, and A_3A_3 zygotes in frequency pr, qr, and r^2.

The only way to form the homozygotes A_1A_1, A_2A_2, and A_3A_3 is by picking two of the same kind of gamete and the frequencies are p^2, q^2, and r^2. The heterozygotes can be formed from more than one kind of first gamete and their frequencies are obtained by addition. The total chance of forming an A_1A_3 zygote is $pr + rp = 2pr$; of forming an A_1A_2 zygote is $pq + qp = 2pq$; and of an A_2A_3 zygote is $2qr$. The complete Hardy–Weinberg proportions are:

$$A_1A_1 : A_1A_2 : A_1A_3 : A_2A_2 : A_2A_3 : A_3A_3$$
$$p^2 \quad 2pq \quad 2pr \quad q^2 \quad 2qr \quad r^2$$

5.4 We can test, by simple observation, whether genotypes in a population are at the Hardy–Weinberg equilibrium

Natural populations may or may not fit the Hardy–Weinberg equilibrium

The Hardy–Weinberg theorem depends on three main assumptions: no selection, random mating, and large population size. In a natural population, any of these could be false; we cannot assume that natural populations will be at the Hardy–Weinberg equilibrium. In practice, we can find out whether a population is at the Hardy–Weinberg equilibrium for a locus simply by counting the genotype frequencies. From those frequencies, we first calculate the gene frequencies; then, if the observed homozygote frequencies equal the square of their gene frequencies, the population is in Hardy–Weinberg equilibrium. If they do not, it is not.

The MN blood group system in humans is a good example, because the three genotypes are distinct and the genes have reasonably high frequencies in human populations. Three phenotypes, M, MN, and N are produced by three genotypes (MM, MN, NN) and two alleles at one locus. The phenotypes of the MN group, like the better known ABO group, are recognized by reactions with antisera. The antisera are made by injecting blood into a rabbit, which then makes an antiserum to the type of blood that was injected. If the rabbit has been injected with M-type human blood, it produces

Table 5.2
The frequencies of the MM, MN, and NN blood groups in three American populations. The figures for expected proportions and numbers have been rounded.

Population		MM	MN	NN	Total	Frequency M	Frequency N
African Americans	Observed number	79	138	61	278		
	Expected proportion	0.283	0.499	0.219		0.532	0.468
	Expected number	78.8	138.7	60.8			
European Americans	Observed number	1,787	3,039	1,303	6,129		
	Expected proportion	0.292	0.497	0.211		0.54	0.46
	Expected number	1,787.2	3,044.9	1,296.9			
Native Americans	Observed number	123	72	10	205		
	Expected proportion	0.602	0.348	0.05		0.776	0.224
	Expected number	123.3	71.4	10.3			

Specimen calculation for African Americans:

Frequency of M allele $= 79 + (^1\!/_2 \times 138) = 0.532 = p$
Frequency of N allele $= 61 + (^1\!/_2 \times 138) = 0.468 = q$

Expected proportion of MM $= p^2 = (0.532)^2 = 0.283$
Expected proportion of MN $= 2pq = 2(0.532)(0.468) = 0.499$
Expected proportion of NN $= q^2 = (0.468)^2 = 0.219$

Expected numbers $=$ expected proportion $\times$ total number (n)

Expected number of MM $= p^2 n = 0.283 \times 278 = 78.8$
Expected number of MN $= 2pqn = 0.499 \times 278 = 138.7$
Expected number of NN $= q^2 n = 0.219 \times 278 = 60.8$

anti-M serum. Anti-M serum agglutinates blood from humans with one or two M alleles in their genotypes; likewise anti-N blood agglutinates the blood of humans with one or two N alleles. Therefore MM individuals are recognized as those whose blood reacts only with anti-M, NN individuals react only with anti-N, and MN individuals react with both.

The MN human blood group system is close to Hardy–Weinberg equilibrium

Table 5.2 gives some measurements of the frequencies of the MN blood group genotypes for three human populations. Are they at Hardy–Weinberg equilibrium? In European Americans, the frequency of the M gene (calculated from the usual $p = P + \frac{1}{2}Q$ relation) is 0.54. If the population is at the Hardy–Weinberg equilibrium, the frequency of MM homozygotes (p^2) will be $2 \times 0.54 = 0.2916$ (1,787 in a sample of 6,129 individuals); and the frequency of MN heterozygotes ($2pq$) will be $2 \times 0.54 \times 0.46 = 0.497$ (3,045 in a sample of 6,129). As the table shows, these are close to the observed frequencies. In fact all three populations are at Hardy–Weinberg equilibrium. We shall see in Section 5.6 that the same calculations do not correctly predict the genotype frequencies after selection has operated.

5.5 The Hardy–Weinberg theorem is important conceptually, historically, in practical research, and in the workings of theoretical models

We have just seen how to find out whether a real population is in Hardy–Weinberg equilibrium. The importance of the Hardy–Weinberg theorem, however, is not mainly as an empirical prediction. We have no good reason to think that genotypes in natural populations will generally have Hardy–Weinberg frequencies, because it would require both no selection and random mating, which are rarely found. The interest of the theorem lies elsewhere, in three other areas.

The Hardy–Weinberg theorem matters conceptually, . . .

One is historical and conceptual. We saw in Section 2.9 (p. 37) how with blending inheritance the genetic variation in a population is rapidly blended out of existence and the population becomes genetically uniform. With Mendelian genetics, variation is preserved and the Hardy–Weinberg theorem gives quantitative demonstration of that fact. The theorem was published in the first decade of the twentieth century, as Mendelism was becoming accepted, and it was historically influential in proving to people that Mendelian inheritance did allow variation to be preserved.

. . . in research . . .

A second interest of the theorem is as a kind of springboard, that launches us toward interesting empirical problems. If we compare genotype frequencies in a real population with Hardy–Weinberg ratios, then if they deviate it suggests something interesting (such as selection or non-random mating) may be going on, which would merit further research.

. . . and in theory

A third interest is theoretical. In the general model of population genetics (Section 5.2) there were five stages, joined by four calculations. The Hardy–Weinberg theorem simplifies the model wonderfully. If we assume random mating, we can go directly from the adult frequencies in generation n to the genotype frequencies at birth in generation $n + 1$, collapsing three calculations into one (Figure 5.3). If we know the adult genotype frequencies in generation n (stage 1), we only need to calculate the gene

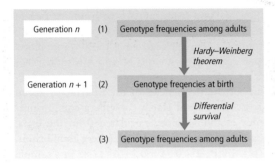

Figure 5.3
The general model of population genetics simplified by the Hardy–Weinberg theorem.

frequencies: the genotype frequencies at birth in the next generation (stage 2) must then have Hardy–Weinberg frequencies, because the gene frequencies do not change between the adults of one generation and the newborn members of the next generation. A simple model of selection can concentrate on how the genotype frequencies are modified between birth and the adult reproductive stage (from stage 2 to stage 3 of Figure 5.3).

5.6 The simplest model of selection is for one favored allele at one locus

We shall start with the simplest case. It is the case of natural selection operating on only one genetic locus, at which there are two alleles, one dominant to the other. Suppose that individuals with the three genotypes have the following relative chances of survival from birth to the adult stage:

Genotype	Chance of survival
AA, Aa	1
aa	$1 - s$

Population genetic models specify the fitness of all genotypes

s is a number between 0 and 1, and is called the *selection coefficient*. Selection coefficients are expressed as reductions in fitness relative to the best genotype. If s is 0.1 then *aa* individuals have a 90% chance of survival, relative to 100% for *AA* and *Aa* individuals. These are relative values: in a real case the chance of survival from birth to reproduction of an individual with the best genotype might be 50%, much less than 100%. If it was 50%, then an s of 0.1 would mean that *aa* individuals really had a 45% chance of survival. (The convention of giving the best genotype a relative 100% chance of survival simplifies the algebra. If you are suspicious, check whether it makes any difference in what follows if the chances of survival are 50%, 50%, and 45% for *AA*, *Aa*, and *aa*, respectively, rather than 100%, 100%, and 90%.) The chance of survival is the *fitness* of the genotype (we are assuming that all surviving individuals produce the same number of offspring). Fitnesses are, like the chances of survival, expressed relative to a figure of 1 for the best genotype. This can be spelled out more by referring to fitnesses as

Table 5.3

(a) Algebraic calculation of genotype frequences after selection, with selection against a recessive genotype. (b) A numerical illustration. See text for further explanation.

(a)

	Genotype		
	AA	*Aa*	*aa*
Birth			
Frequency	p^2	$2pq$	q^2
Fitness	1	1	$1-s$
Adult			
Relative frequency	p^2	$2pq$	$q^2(1-s)$
Frequency	$p^2/(1-sq^2)$	$2pq/(1-sq^2)$	$q^2(1-s)/(1-sq^2)$

(b)

	Genotype			
	AA	*Aa*	*aa*	**Total**
Birth				
Number	1	18	81	100
Frequency	0.01	0.18	0.81	
Fitness	1	1	0.9	
Adult				
Number	1	18	73	92
Frequency	1/92	18/92	73/92	

"relative fitnesses." However, biologists usually just say "fitness." With the fitnesses given above, selection will act to eliminate the *a* allele and *fix* the *A* allele. (To "fix" a gene is genetic jargon for carry its frequency up to 1. When there is only one gene at a locus, it is said to be "fixed" or in a state of "fixation.") If *s* were 0, the model would lapse back to the Hardy–Weinberg case and the gene fequencies would be stable.

Notice that alleles do not have any tendency to increase in frequency just because they are dominant, or to decrease because they are recessive. Dominance and recessivity only describe how the alleles at a locus interact to produce a phenotype. Changes in gene frequency are set by the fitnesses. If the recessive homozygote has higher fitness, the recessive allele will increase in frequency. If, as here, the recessive homozygote has lower fitness, the recessive allele decreases in frequency.

We construct a model for the change in gene frequency per generation

How rapidly will the population change through time? To find out, we seek an expression for the gene frequency of A (p') in one generation in terms of its frequency in the previous generation (p). The difference between the two, $\Delta p = p' - p$, is the change in gene frequency between two successive generations. The model has the form of Figure 5.3, and we shall work through both the general algebraic version and a numerical example (Table 5.3).

To begin with, at birth the three genotypes have Hardy–Weinberg frequencies as they are produced by random mating among adults of the previous generation. Selection then operates; *aa* individuals have a lower chance of survival and their frequency among the adults is reduced. As the numerical example shows (Table 5.3b), the total number of adults is less than the number at birth and we have to divide the adult numbers of each genotype by the total population size to express the adult numbers as frequencies comparable to the frequencies at birth. In the algebraic case, the relative frequencies after selection do not add up to 1, and we correct them by dividing by the *mean fitness*.

$$\text{Mean fitness} = p^2 + 2pq + q^2(1 - s) = 1 - sq^2 \tag{5.3}$$

Dividing by mean fitness in the algebraic case is the same as dividing by the population size after selection in the numerical example. Notice that now the adult genotype frequencies are not in Hardy–Weinberg ratios. If we tried to predict the proportion of *aa* from q^2, as in the MN blood group (Section 5.4), we should fail. The frequency of *aa* is $q^2(1 - s)/1 - sq^2$, not q^2.

What is the relation between p' and p? Remember that the frequency of the gene *A* at any time is equal to the frequency of *AA* plus half the frequency of *Aa*. We have just listed those frequencies in the adults after selection:

$$p' = \frac{p^2 + pq}{1 - sq^2} = \frac{p}{1 - sq^2} \tag{5.4}$$

(remember $p + q = 1$, and therefore $p^2 + pq = p(p + q) = p$.) The denominator $1 - sq^2$ is less than 1, because *s* is positive, so p' is greater than p: selection is increasing the frequency of the *A* gene. We can now derive a result for Δp, the change in gene frequency in one generation. The algebra looks like this.

$$\Delta p = p' - p = \frac{p}{1 - sq^2} - p$$

$$= \frac{p - p + spq^2}{1 - sq^2}$$

$$= \frac{spq^2}{1 - sq^2} \tag{5.5}$$

For example, if $p = q = 0.5$ and *aa* individuals have fitness 0.9 compared with *AA* and *Aa* individuals ($s = 0.1$) then the change in gene frequency to the next generation will be $(0.1 \times 0.5 \times (0.5)^2)/(1 - 0.1 \times (0.5)^2) = 0.0128$; the frequency of *A* will therefore increase to 0.5128.

The model predicts the rate of change in gene frequency as the superior gene is fixed

We can use this result to calculate the change in gene frequency between successive generations for any selection coefficient (*s*) and any gene frequency. The result in this simple case is that the *A* gene will increase in frequency until it is eventually fixed (that is, has a frequency of 1). Table 5.4 illustrates how gene frequencies change when selection acts against a recessive allele, for each of two selection coefficients. There are two points to notice in the table. One is the obvious one that with a higher selection coefficient against the *aa* genotype, the *A* gene increases in frequency more rapidly. The other is the more interesting observation that the increase in the frequency of *A* slows down when it becomes common, and it would take a long time finally to eliminate the *a* gene. This is because the *a* gene is recessive. When *a* is rare it is almost always found in *Aa* individuals, who are selectively equivalent to *AA* individuals: selection can no longer "see" the *a* gene, and it becomes more and more difficult to eliminate them. Logically, selection cannot eliminate the one final *a* gene from the population, because if there is only one copy of the gene it must be in a heterozygote.

Table 5.4
A simulation of changes in gene frequency for selection against the recessive gene *a*, using two selection coefficients: *s* = 0.05 (i.e., *aa* individuals have a relative chance of survival of 95%, against 100% for *AA* and *Aa*) and *s* = 0.01 (i.e., *aa* individuals have a relative chance of survival of 99%, against 100% for *AA* and *Aa*). The change between generation 0 and 100 is found by applying the equation in the text 100 times successively.

Generation	Gene frequency, *s* = 0.05		Gene frequency, *s* = 0.01	
	A	*a*	*A*	*a*
0	0.01	0.99	0.01	0.99
100	0.44	0.56	0.026	0.974
200	0.81	0.19	0.067	0.933
300	0.89	0.11	0.15	0.85
400	0.93	0.07	0.28	0.72
500	0.95	0.05	0.43	0.57
600	0.96	0.04	0.55	0.45
700	0.96	0.04	0.65	0.35
800	0.97	0.03	0.72	0.28
900	0.97	0.03	0.77	0.23
1,000	0.98	0.02	0.80	0.20

Just as equation 5.4 can be used to calculate a gene frequency change given the fitnesses, so it can be used to calculate the fitnesses given the frequency changes. If we know the gene frequency in two successive generations then equations 5.4 and 5.5 can be rearranged to:

$$s = \frac{\Delta p}{p'q^2} \tag{5.6}$$

to find *s*.

Haldane (1924) first produced this particular model of selection. One important feature of the model is that it shows how rapidly, in evolutionary time, natural selection can produce change. When we look at the complex organs and behavior patterns of living creatures, including ourselves, it is easy to wonder whether there has really been enough time for them to have evolved in the manner suggested by Darwin's theory. To find out, for any particular organ, such as the heart, liver, or brain, we need answers to two questions: (i) how many genetic changes did its evolution require; and (ii) how long did each change take.

A model like the one in this section gives us an idea of the answer to the second question. (We shall look more at the first question in Section 10.5, p. 266.) The fitness differences of 1–5% in Table 5.4 are small, relative to many of the risks we take though our lives; but they are enough to carry a gene up from being negligibly rare to being the

We need to know more to understand completely the rate of evolution of whole organs

majority form in the population in 1,000 to 10,000 generations. On the evolutionary timescale, 10,000 generations are an eye-blink: too short a period to be resolved in the fossil record. A quantitative model such as Haldane's was needed to answer the quantitative question of how rapidly selection can drive evolution.

The model can be extended

The model can be extended in various ways. The modifications for different degrees of dominance, and separate selection on heterozygotes and homozygotes, are conceptually straightforward, though they make the algebra more complex. Other modifications can be made to analyze the other stages in the general picture of Figure 5.1: to analyze non-random mating, non-Mendelian inheritance, or fitnesses that vary according to fertility rather than survival. However, for our purposes it is mainly important to see how an exact model of selection can be built and exact predictions made from it. The model is simplified, but it can help us to understand a number of real cases — as we shall now see.

5.7 The model of selection can be applied to the peppered moth

5.7.1 *Industrial melanism in moths evolved by natural selection*

The peppered moth *Biston betularia* provides one of the best known stories in evolutionary biology (Figure 5.4). In collections made in Britain in the eighteenth century, the form of the moth was always a light, peppered color. A dark (melanic) form was first recorded in 1848 near Manchester. That melanic form then increased in frequency until it made up more than 90% of the populations in polluted areas in the mid-twentieth century. In unpolluted areas, the light form remained common. Clean air laws were passed in the mid-twentieth century, and the frequency of the melanic form decreased in formerly polluted areas.

We can estimate the fitness differences during peppered moth evolution

The peppered moth can be used to illustrate the simple model of the previous section. A controversy has grown up about the peppered moth concerning the reason why the melanic and light-colored moths differed in fitness, although this does not matter while we are simply estimating fitnesses. The increase in frequency of the melanic form in polluted areas has classically been explained by bird predation. Some doubts have been raised about the evidence for this view. Section 5.7.4 looks at the controversy, but we begin by looking at estimates of fitness. All we need to know for these estimates is that natural selection is acting — just how it is acting, whether by bird predators or other factors, is another question.

Before we can apply the theory of population genetics to a character, we need to know its genetics. Breeding experiments initially suggested that the difference in color was controlled by one main locus. The original, peppered form was one homozygote (cc) and the melanic form was another homozygote (CC), and the C allele is dominant. However, in other experiments the melanic allele was less dominant and the heterozygotes were intermediate; there seem to be a number of different melanic alleles. It may be that selection initially favored a melanic allele with no or weak dominance, and subsequently some other melanic alleles with stronger dominance. In any case, the degree of dominance of the melanic allele that was originally favored in the nineteenth century

Figure 5.4
Peppered moths naturally settle on the undersides of twigs in higher branches of trees (and not on tree trunks, as is sometimes said). Melanic forms are better camouflaged in polluted areas: compare (a) the peppered form and (b) the melanic form, both photographed in a polluted area. (c) and (d) show that peppered forms are well camouflaged in unpolluted areas. Reprinted, by permission of the publisher, from Brakefield (1987).

is uncertain, and it may have differed from the dominance shown by the melanic alleles that exist in modern populations.

The first estimates of fitnesses were made by Haldane (1924), and he dealt with the problem of varying degrees of dominance by making two estimates of fitness, one assuming that the C allele is dominant and the other assuming that the heterozygote is intermediate. The real average degree of dominance was probably between the two. Here we shall look only at the estimate for a dominant C gene.

5.7.2 *One estimate of the fitnesses is made using the rate of change in gene frequencies*

What were the relative fitnesses of the genes controlling the melanic and light coloration during the phase from the early nineteenth to the mid-twentieth centuries, while the melanic form increased in frequency in polluted areas? For the first method we need measurements of the frequencies of the different color forms for at least two times. We can then estimate the gene frequencies from the genotype frequencies, and substitute them in equation 5.6 to solve s, the selection coefficient.

The melanic form was first seen in 1848; but it was probably not a new mutation then. It probably existed at a low frequency in the population, in what is called "mutation–selection balance." Mutation–selection balance means that the gene is disadvantageous, and exists at a low frequency determined by a balance between being formed by mutation and being lost by selection (Section 5.11). We shall see that the frequency of a gene can be calculated from its mutation rate m and its selective disadvantage s. The values of m and s are unknown for the gene in the early nineteenth century. However, typical mutation rates for genes are about 10^{-6} and a selective disadvantage of about 10% for the melanic mutants in preindustrial times may be approximately correct. With these figures, and using equation 5.9 below, the melanic C gene would have had a frequency of 10^{-5} up to the year 1848. By 1898, the frequency of the light-colored genotype was 1–10% in polluted areas (it was not more than 5% near the industrial city of Manchester, for example, implying a gene frequency of about 0.2). There would have been about 50 generations between 1848 and 1898.

We now know all we need. What selective coefficient would generate an increase in its frequency from 10^{-5} to 0.8 in 50 generations? Equation 5.6 gives the selection coefficient in terms of gene frequencies in two successive generations, but between 1848 and 1898 there would have been 50 generations. The formula therefore has to be applied 50 times over, which is most easily done by computer. A change from 10^{-5} to 0.8 in 50 generations, it turns out, requires $s \approx 0.33$: the peppered moths had two-thirds the survival rate of melanic moths (Table 5.5). The calculations are rough, but they show how fitness can be inferred from the observed rate of change in gene frequency.

The observed gene frequency changes suggest $s \approx 0.33$

Table 5.5

Theoretical changes in gene frequencies in the evolution of melanism in the peppered moth, starting with an initial frequency of C of 0.00001 (rounded to 0 in the table). C is dominant and c is recessive: genotypes CC and Cc are melanic and cc is peppered in color. 1848 is generation zero in the simulation. Selection coefficient $s = 0.33$.

Generation date	Gene frequency	
	C	c
1848	0.00	1.00
1858	0.00	1.00
1868	0.03	0.97
1878	0.45	0.55
1888	0.76	0.24
1898	0.86	0.14
1908	0.90	0.10
1918	0.92	0.08
1928	0.94	0.06
1938	0.96	0.04
1948	0.96	0.04

Table 5.6

Frequencies of melanic and light peppered moths in samples recaptured at two sites in the UK: Birmingham (polluted) and Deanend Wood, Dorset (unpolluted). The observed numbers are the actual numbers recaught; the expected numbers are the numbers that would have been recaught if all morphs survived equally (equals proportion in released moths times the number of moths recaptured). The recaptured moths at Birmingham were taken over a period of about 1 week, at Deanend Wood over about 3 weeks. Data from Kettlewell (1973).

	Light moths	Melanic moths
Birmingham (polluted)		
Numbers recaptured		
Observed	18	140
Expected	36	122
Relative survival rate	0.5	1.15
Relative fitness	5/1.15 = 0.43	1.15/1.15 = 1
Deanend wood (unpolluted)		
Numbers recaptured		
Observed	67	32
Expected	53	46
Relative survival rate	1.26	0.69
Relative fitness	1.26/1.26 = 1	0.69/1.26 = 0.55

5.7.3 A second estimate of the fitnesses is made from the survivorship of the different genotypes in mark–recapture experiments

The estimate of fitness can be checked against other estimates. The gene frequency change was (and still is) thought to be produced by survival differences between the two forms of moth in nature, rather than differential fertility. We can measure the rate of survival of the two forms in nature, and see how they differ. Kettlewell (1973) measured survival rates by mark–recapture experiments in the field. He released melanic and light-colored peppered moths in known proportions in polluted and unpolluted regions, and then later recaught some of the moths (which are attracted to mercury vapor lamps). He counted the proportions of melanic and light-colored moths in the moths recaptured from the two areas.

Table 5.6 gives some results for two sites, Birmingham (polluted) and Deanend Wood, an unpolluted forest in Dorset, UK. The proportions in the recaptured moths are as we would expect: more light-colored moths in the Deanend Wood samples and more melanic moths in the Birmingham samples. In Birmingham, melanic moths were recaptured at about twice the rate of light-colored ones, implying $s = 0.57$. This is a higher fitness difference than the $s = 0.33$ implied by the change in gene frequency.

Mark–recapture experiments suggest $s \approx 0.57$

The discrepancy is unsurprising because both estimates are uncertain; it could have a number of causes. Possible causes include sampling error in the mark–recapture experiments (the numbers in Table 5.6 are small) and errors in the assumptions of the estimate from gene frequency changes. For instance, the initial gene frequency may have been less than 10^{-5}. Also, the relative fitness of the two moth forms probably changed over time and moths may have migrated between polluted and unpolluted areas. Whatever the cause of the discrepancy, the two calculations do illustrate two important methods of estimating fitness.

5.7.4 *The selective factor at work is controversial, but bird predation was probably influential*

So far we have concentrated on estimating fitnesses, and have ignored the factors that cause the fitness difference between the melanic and light-colored forms of the moth. The material thus far is uncontroversial. The gene frequency changes have undoubtedly occurred, and provide an excellent example of evolution by natural selection. Now we can move on to ask what the agent, or agents, of natural selection were in this example.

Peppered moth evolution is classically explained by bird predation

The classic answer, due to the research of Kettlewell (1973), has been bird predation. The light-colored form is better camouflaged in unpolluted woods and therefore less likely to be eaten by visually hunting birds. But smoke pollution killed the lichens that covered the trees, after which the melanic form was better camouflaged (Figure 5.4). Several lines of evidence support Kettlewell's explanation. Birds do eat the moths, and have been photographed in the act. Birds also have been shown to take more of the poorly camouflaged form, in various experimental set-ups. Also, the gene frequency changes closely match the rise and fall of air pollution. The melanic form increased in frequency following the industrial revolution, and then decreased in frequency after air pollution decreased in the late twentieth century. Indeed, the case for Kettlewell's explanation is arguably now stronger than when he worked. The decrease in frequency of the melanic form has become particularly clear from 1970 to 2000, adding a new line of evidence that was unavailable to Kettlewell (whose main work was in the 1950s).

Some of the classic experiments have been criticized

However, not everyone accepts that bird predation is the selective agent. Some of Kettlewell's research has itself been criticized. We looked above at fitness estimates from gene frequency changes and from mark–recapture experiments. Kettlewell and others also estimated fitnesses by pinning out dead moths of the two forms on tree trunks in polluted and unpolluted areas. He then measured how many moths of each form disappeared over time. These experiments were particularly criticized after it was discovered in the 1980s that peppered moths do not naturally settle on tree trunks, but on the higher branches and twigs of trees (Figure 5.4). Other criticisms were also made. However, Kettlewell's case does not depend on these pin-out experiments. As we saw, he also did mark–recapture experiments in which he released live moths. Those moths presumably settled, and behaved, in a natural manner. The results of all the experiments — pin-outs and mark–recapture — were similar, so the fact that the moths were pinned out in the wrong place did not bias the fitness estimates.

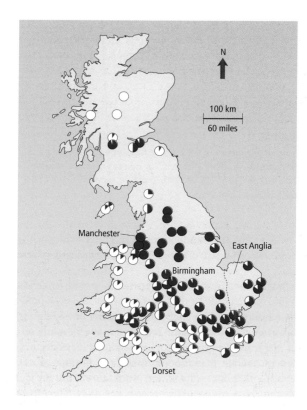

Manchester

East Anglia

Birmingham

Dorset

Figure 5.5

Frequency of melanic and light-colored forms of the peppered moth in different parts of Britain when the frequency of the melanic form was near its peak. The green part of each pie diagram is the frequency of the melanic form in that area. Melanic moths are generally higher in industrial areas, such as central England; but note the high proportion in East Anglia. Melanic frequencies have subsequently decreased (see Figure 5.6, for instance). Redrawn, by permission of the publisher, from Lees (1971).

The fitness estimates have been repeated many times

Cook (2000) reviewed about 30 experimental fitness estimates, done by several teams of biologists,[3] and they all gave similar results. The fitness estimates for the two forms of the peppered moth are about the most repeated result in evolutionary biology, and do not depend on the details of any particular experiment. The repeated results amount to an almost overwhelming case that the rise and fall of the melanic form of the peppered moth depended on air pollution. The evidence that air pollution exerted its effect via bird predation is also strong, if not overwhelming.

Other factors have been suggested

Evidence has also been put forward for other factors, in addition to bird predation. Migration is one extra factor. The geographic distribution of the two forms does not exactly fit Kettlewell's theory. The melanic form, for example, had a frequency of up to 80% in East Anglia, where pollution is low (Figure 5.5). And in some polluted areas, the dark form did not seem to have a high enough frequency. It never exceeded about 95% even though it was clearly better camouflaged and ought for that reason to have had a frequency of 100%. However, male moths can fly long distances to find females, and a male peppered moth mates on average 1.5 miles (2.5 km) away from where it is born.

---

[3] It has even been suggested that Kettlewell faked his results. The charge has only been supported by indirect evidence that is open to innocent interpretations. But however that may be, Kettlewell's explanation for evolution in the peppered moth — bird predation — does not depend on Kettlewell's own research. His results have been independently repeated.

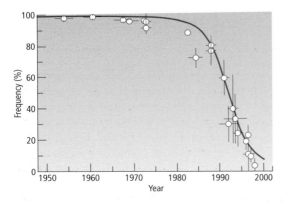

Figure 5.6

Decrease in frequency of the melanic form of the peppered moth in the region around Manchester. The decrease did not become really noticeable until about 1990. Redrawn, by permission of the publisher, from Cook *et al.* (1999).

Migration may explain why melanic moths are found in some unpolluted areas such as East Anglia and why light-colored moths persisted in polluted areas where they were less well camouflaged.

A second additional factor is that the two forms may differ in fitness independently of bird predation. Creed *et al.* (1980) collected all the measurements that had been made on survival to adulthood in the laboratory. They analyzed the results of 83 broods, containing 12,569 offspring; the original measurements had been made by many different geneticists in the previous 115 years. The viability of light-colored homozygotes, it turned out, was about 30% less on average than that of the melanic homozygote in the laboratory, where there is no bird predation — the reason is not known, but the fact alone implies there is some "inherent" advantage to the melanic genotype. The fitness advantage detected in the lab implies that melanic moths would replace light ones even without bird predation in polluted areas. In unpolluted areas, light-colored moths may remain only because birds eat more of the conspicuous melanic moths.

Some biologists have suggested that three factors — bird predation, inherent advantage to melanic genotypes, and migration — are needed to explain peppered moth evolution. The importance of migration in addition to bird predation is generally accepted, but the inherent advantage to the melanic form is controversial. Since the measurements compiled in Creed *et al.* (1980) were made, the decrease in the melanic form's frequency has been more and more widely documented. The decrease did not happen around the formerly industrial Manchester region until the 1990s (Figure 5.6). The decrease makes sense if the advantage to the melanic form depends on air pollution, but not if it has an inherent advantage. Therefore, other biologists explain the observations in terms of bird predation (supplemented by migration) alone, and rule out the inherent advantage.

In conclusion, the industrial melanism of the peppered moth is a classic example of natural selection. It can be used to illustrate the one-locus, two-allele model of selection. The model can be used to make a rough estimate of the difference in fitness between the two forms of moth using their frequencies at different times; the fitnesses can also be estimated from mark–recapture experiments. Good evidence exists that bird predation is at least partly the agent of selection, but some biologists suggest other factors are at work too.

The melanic form may have an "inherent" advantage

But the decrease in melanic frequency since the air became cleaner supports the classic explanation

5.8 Pesticide resistance in insects is an example of natural selection

Malaria is caused by a protozoan blood parasite (Section 5.12.2), and humans are infected with it by mosquitoes (family Culicidae — genera include *Aedes*, *Anopheles*, *Culex*). It can therefore be prevented by killing the local mosquito population, and health workers have recurrently responded to malarial outbreaks by spraying insecticides such as DDT in affected areas. DDT, sprayed on a normal insect, is a lethal nerve poison. When it is first sprayed on a local mosquito population, the population goes into abrupt decline. What happens then depends on whether DDT has been sprayed before.

Pests, such as mosquitoes, evolve resistance to pesticides, such as DDT

On its first use, DDT is effective for several years; in India, for example, it remained effective for 10–11 years after its first widespread use in the late 1940s. DDT, on a global scale, was one reason why the number of cases of malaria reduced to 75 million or so per year by the early 1960s. But by then, DDT-resistant mosquitoes had already begun to appear. DDT-resistant mosquitoes were first detected in India in 1959, and they have increased so rapidly that when a local spray program is begun now, most mosquitoes become resistant in a matter of months rather than years (Figure 5.7). The malarial statistics reveal the consequence. The global incidence of the disease almost exploded, up to somewhere between 300 and 500 million people at present. Malaria currently kills over 1 million people per year, mainly children aged 1–4 years. Pesticide resistance was not the only reason for the increase, but it was important.

DDT becomes ineffective so quickly now because DDT-resistant mosquitoes exist at a low frequency in the global mosquito population and, when a local population is sprayed, a strong force of selection in favor of the resistant mosquitoes is immediately created. It is only a matter of time before the resistant mosquitoes take over. A graph such as Figure 5.7 allows a rough estimate of the strength of selection. As for the peppered moth, we need to understand the genetics of the character, and to measure the genotype frequencies at two or more times. We can then use the formula for gene frequency change to estimate the fitness.

The fitnesses can be estimated, . . .

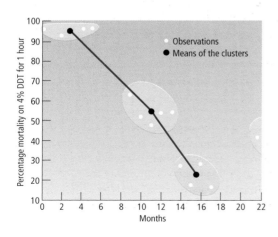

Figure 5.7

Increase in frequency of pesticide resistance in mosquitoes (*Anopheles culicifacies*) after spraying with DDT. A sample of mosquitoes was captured at each time indicated and the number that were killed by a standard dose of DDT (4% DDT for 1 hour) in the laboratory was measured. Redrawn, by permission of the publisher, from Curtis *et al.* (1978).

. . . given certain assumptions

We have to make a number of assumptions. One is that resistance is controlled by a single allele (we shall return to this below). Another concerns the degree of dominance: the allele conferring resistance might be dominant, recessive, or intermediate, relative to the natural susceptibility allele. The case of dominant resistance is easiest to understand. (If resistance is recessive we follow the same general method, but the exact result differs.) Let us call the resistance allele R and the susceptibility allele r. All the mosquitoes that die, in the mortality tests used in Figure 5.7, would then have been homozygous (rr) for susceptibility. Assuming (for simplicity rather than exact accuracy) Hardy–Weinberg ratios, we can estimate the frequency of the susceptibility gene as the square root of the proportion of mosquitoes that die in the tests. The selection coefficients are defined as follows, where fitness is measured as the chance of survival in the presence of DDT:

Genotype	RR	Rr	rr
Fitness	1	1	$1 - s$

If we define p as the frequency of R and q as the frequency of r, equation 5.5 again gives the change in gene frequency: selection is working against a recessive gene. Figure 5.7 shows the decline in frequency of the susceptible mosquitoes, which are the recessive homozygotes. We therefore need a formula for the change in q in one generation (Δq), rather than Δp (as on p. 106). The decrease in q is the mirror image of the increase in p, and we just need to put a minus sign in front of equation 5.5:

$$\Delta q = \frac{-spq^2}{1 - sq^2} \tag{5.7}$$

The selection coefficient $s \approx 0.5$

The generation time is about 1 month. (The generations of mosquitoes overlap, rather than being discrete as the model assumes; but the exact procedure is similar in either case, and we can ignore the detailed correction for overlapping generations.) Table 5.7 shows how the genotype frequencies were read off Figure 5.7 in two stages, giving two estimates of fitness. Again, the formula for one generation has to be applied

---

Table 5.7
Estimated selection coefficients against DDT-susceptible *Anopheles culicifacies*, from Figure 5.7, where the relative fitness of the susceptible type is $(1 - s)$. The estimate assumes the resistance allele is dominant. Simplified from Curtis *et al.* (1978).

Frequency of susceptible type			
Before	After	Time (months)	Selection coefficient
0.96	0.56	8.25	0.4
0.56	0.24	4.5	0.55

Figure 5.8

The mortality of mosquitos (*Culex quinquifasciatus*) of three genotypes at a locus when exposed to various concentrations of permethrin. The susceptible homozygote (*SS*) dies at lower concentrations of the poison than the resistant homozygote (*RR*). The heterozygote (*RS*) has intermediate resistance. Redrawn, by permission of the publisher, from Taylor (1986).

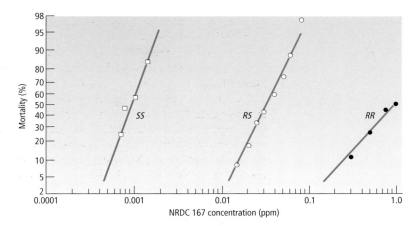

The real genetics of resistance is known in some cases

recurrently, for 8.25 and 4.5 generations in this case, to give an average fitness for the genotypes through the period. It appears that in Figure 5.7 the resistant mosquitoes had about twice the fitness of the susceptible ones — which is very strong selection.

The genetics of resistance in this case are not known, and the one-locus, two-allele model is an assumption only; but they are understood in some other cases. Resistance is often controlled by a single resistance allele. For example, Figure 5.8 shows that the resistance of the mosquito *Culex quinquifasciatus* to permethrin is due to a resistance (*R*) allele, which acts in a semidominant way, with heterozygotes intermediate between the two homozygotes. In houseflies, resistance to DDT is due to an allele called *kdr*. *kdr* flies are resistant because they have fewer binding sites for DDT on their neurons. In other cases, resistance may be due not to a new point mutation, but to gene amplification. *Culex pipiens*, for instance, in one experiment became resistant to an organophosphate insecticide called temephos because individuals arose with increased numbers of copies of a gene for an esterase enzyme that detoxified the poison. In the absence of temephos, the resistance disappeared, which suggests that the amplified genotype has to be maintained by selection. A number of mechanisms of resistance are known, and Table 5.8 summarizes the main ones that have been identified.

When an insect pest has become resistant to one insecticide, the authorities often respond by spraying it with another insecticide. The evolutionary pattern we have seen here then usually repeats itself, and on a shorter timescale. On Long Island, New York, for example, the Colorado potato beetle (*Leptinotarsa septemlineata*) was first attacked with DDT. It evolved resistance to it in 7 years. The beetles were then sprayed with azinphosmethyl, and evolved resistance in 5 years; next came carbofuran (2 years), pyrethroids (another 2 years), and finally pyrethroids with synergist (1 year). The decreasing time to evolve resistance is probably partly due to detoxification mechanisms that work against more than one pesticide. Pesticides cost money to develop, and the evolution of resistance reduces the economic lifetime of a pesticide. Box 5.2 looks at how the lifetime of a pesticide may be lengthened by slowing the evolution of resistance.

The theory has practical applications

Insecticide resistance matters not only in the prevention of disease, but also in farming. Insect pests at present destroy about 20% of world crop production, and it has been estimated that in the absence of pesticides as much as 50% would be lost. Insect pests

Table 5.8
The main mechanisms of resistance to insecticides. Reprinted, by permission of the publisher, from Taylor (1986).

Mechanism	Insecticides affected
Behavioral	
Increased sensitivity to insecticide	DDT
Avoid treated microhabitats	Many
Increased detoxification	
Dehydrochlorinase	DDT
Microsome oxidase	Carbamates
	Pyrethroids
	Phosphorothioates
Glutathione transferase	Organophosphates (*O*-dimethyl)
Hydrolases, esterases	Organophosphates
Descreased sensitivity of target site	
Acetylcholinesterase	Organophosphates
	Carbamates
Nerve sensitivity	DDT
	Pyrethroids
Cyclodiene-resistance genes	Cyclodienes (organochlorines)
Decreased cuticular penetration	Most insecticides

are a major economic and health problem. The evolution of resistance to pesticides causes misery to millions of people, whether through disease or reduced food supply. The fact that insects can rapidly evolve resistance is not the only problem with using pesticides against pests — the pesticides themselves (as is well known) can cause ecological side effects that range from the irritating to the dangerous. But however that may be, pesticides did not exist during the hundreds of millions of years that insects lived for before they were introduced in the 1940s, and the rapid evolution since then of resistance to pesticides provides a marvellously clear example of evolution by natural selection (Section 10.7.3, p. 276, extends the story, and Box 8.1, p. 213, looks at drug resistance in the malaria organism itself).

5.9 Fitnesses are important numbers in evolutionary theory and can be estimated by three main methods

Fitness can be measured . . .

The fitness of a genotype, in the theory and examples we have met, is its relative probability of survival from birth to adulthood. The fitness also determines the change in gene frequencies between generations. These two properties of fitness allow two methods of measuring it.

Box 5.2
Resistance Management

The evolution of resistance to each new pesticide, and antibiotic, is probably ultimately inevitable. However, we may be able to prolong the economically useful lives of these defensive chemicals by slowing down the evolution of resistance. The time it takes for resistance to evolve will be influenced by several factors. Two such factors can be seen in the simple models of selection we have been considering.

1. The degree of genetic dominance. The frequency of an advantageous dominant gene increases much more rapidly by natural selection than does the frequency of an advantageous recessive gene. An advantageous gene, such as one producing resistance to a pesticide, will initially be present only in one copy, in a heterozygote. If the gene is recessive, it is not expressed in that heterozygote. Natural selection cannot "see" the gene until it is found in a homozygote. If the gene is dominant, it is immediately expressed and natural selection immediately favors it. A recessive resistance gene will increases in frequency much more slowly than a dominant resistance gene.

2. The relative fitness of the resistant and non-resistant genotypes. A genotype with a large fitness advantage will increase in frequency more rapidly than one with a low fitness advantage. For instance, in Table 5.4 we can see that a genotype with a 1% advantage takes five times as long to reach a frequency of 80% than does a genotype with a 5% advantage.

Thus the evolution of resistance could be slowed down if we could make the resistance gene more recessive (or less dominant), and if we could reduce its fitness advantage relative to the non-resistant types.

One way to make the resistance gene recessive might be to apply the pesticide in large doses. The resistance gene may code for a protein that somehow neutralizes the pesticide. If there are small quantities of pesticide, a single copy of the resistance gene (in a heterozygote) may produce enough of the protein to cope with the pesticide. The gene is then effectively dominant, because it produces resistance in heterozygotes. The gene will spread fast. But if large amounts of the pesticide are used, the single gene may be overwhelmed. Two resistance genes (in a

homozygote) may be needed to cope. The large amount of pesticide makes the resistance gene effectively recessive.

The relative fitnesses of the resistant and non-resistant genotypes may be influenced by the way the pesticide is applied in space. If pesticides are applied in some places but not others, the non-resistant genotypes will have a selective advantage in the localities where there is no pesticide. The average fitness of the resistant genotype will then not be so high, relative to the non-resistant genotype, as it would be if the pesticide were applied indiscriminately in the whole region.

Rausher (2001) refers to the combination of these two policies as the "high dose/refuge strategy." However, the strategy requires certain conditions to succeed in slowing the evolution of resistance, even in theory, and very little practical work has been done to test it. Currently, it is a research problem for the future. However, the idea does illustrate how the evolutionary models of this chapter can have practical applications. The economic value of these models could even turn out to be huge.

Further reading: Rausher (2001).

. . . by relative survival within a generation . . .

The first method is to measure the relative survival of the genotypes within a generation. Kettlewell's mark–recapture experiment with the peppered moth is an example. If we assume that the relative rate of recapture of the genotypes is equal to their relative chance of survival from egg to adulthood, we have an estimate of fitness. The assumption may be invalid. The genotypes may, for instance, differ in their chances of survival at some stage of life other than the time of the mark–recapture experiment. If the survival of adult moths is measured by mark–recapture, any differences among genotypes in survival at the egg and caterpillar stages will not be detected. Also, the genotypes may differ in fertility: fitnesses estimated by differences in survival are only accurate if all the genotypes have the same fertility. These assumptions can all be tested by further work. For instance, survival can be measured at the other life stages too, and fertility can also be assessed. In a few cases, lifetime fitnesses have been measured comprehensively, by tracing survival and reproduction from birth to death.

... or rate of gene frequency change between generations ...

The second method is to measure changes in gene frequencies between generations. We then substitute the measurements into the formula that expresses fitness in terms of gene frequencies in successive generations (equation 5.6). Both methods have been used in many cases; the main problems are the obvious difficulties of accurately measuring survival and gene frequencies, respectively. Apart from them, in the examples we considered there were also difficulties in understanding the genetics of the characters: we need to know which phenotypes correspond to which genotypes in order to estimate genotype fitnesses.

... or other methods

We shall meet a third method of estimating fitness below, in the case of sickle cell anemia (see Table 5.9, p. 126). It uses deviations from the Hardy–Weinberg ratios. It can be used only when the gene frequencies in the population are constant between the stages of birth and adulthood, but the genotypes have different survival. It therefore cannot be used in the examples of directional selection against a disadvantageous gene that we have been concerned with so far, because in them the gene frequency in the population changes between birth and adult stages.

We have discussed the inference of fitness in detail because the fitnesses of different genotypes are among the most important variables — perhaps the most important variables — in the theory of evolution. They determine, to a large extent, which genotypes we can expect to see in the world today. The examples we have looked at, however, illustrate that fitnesses are not easy to measure. We require long time series and large sample sizes, and even then the estimates may be subject to "other things being equal" assumptions. Therefore, despite their importance, they have been measured in only a small number of the systems that biologists are interested in. (That does not mean that the absolute number of such studies is small. A review of research on natural selection in the wild by Endler in 1986 contains a table (24 pages long) listing all the work he had located. Fitnesses have only been measured in a minority — an unknown minority — of those 24 pages' worth of studies of natural selection, but the number could still be non-trivial.) Many unsolved controversies in evolutionary biology implicitly concern values of fitnesses, but in systems in which it has not been possible to measure fitnesses directly with sufficient accuracy or in a sufficiently large number of cases. The controversy about the causes of molecular evolution in Chapter 7 is an example. When we come to discuss controversies of this sort it is worth bearing in mind what would have to be done to solve them by direct measurements of fitness.

5.10 Natural selection operating on a favored allele at a single locus is not meant to be a general model of evolution

Evolutionary change in which natural selection favors a rare mutation at a single locus, and carries it up to fixation, is one of the simplest forms of evolution. Sometimes evolution may happen that way. But things can be more complicated in nature. We have considered selection in terms of different chances of survival from birth to adulthood; but selection can also take place by differences in fertility, if individuals of different genotypes — after they have survived to adulthood — produce different numbers of offspring. The model had random mating among the genotypes: but mating may be non-random. Moreover, the fitness of a genotype may vary in time and space, and depend on what genotypes are present at other loci (a subject we shall deal with in Chapter 8). Much of evolutionary change probably consists of adjustments in the frequencies of alleles at polymorphic loci, as fitnesses fluctuate through time, rather than the fixation of new favorable mutations.

Other factors will be at work in real examples

These complexities in the real world are important, but they do not invalidate — or trivialize — the one-locus model. For the model is intended as a model. It should be used as an aid to understanding, not as a general theory of nature. In science, it is a good strategy to build up an understanding of nature's complexities by considering simple cases first and then building on them to understand the complex whole. Simple ideas rarely provide accurate, general theories; but they often provide powerful paradigms. The one-locus model is concrete and easy to understand and it is a good starting point for the science of population genetics. Indeed, population geneticists have constructed models of all the complications listed in the previous paragraph, and those models are all developments within the general method we have been studying.

5.11 A recurrent disadvantageous mutation will evolve to a calculable equilibrial frequency

The model of selection at one locus revealed how a favorable mutation will spread through a population. But what about unfavorable mutations? Natural selection will act to eliminate any allele that decreases the fitness of its bearers, and the allele's frequency will decrease at a rate specified by the equations of Section 5.6; but what about a recurrent disadvantageous mutation that keeps arising at a certain rate? Selection can never finally eliminate the gene, because it will keep on reappearing by mutation. In this case, we can work out the equilibrial frequency of the mutation: the equilibrium is between the mutant gene's creation, by recurrent mutation, and its elimination by natural selection.

A disadvantageous mutation may arise recurrently

To be specific, we can consider a single locus, at which there is initially one allele, a. The gene has a tendency to mutate to a dominant allele, A. We must specify the mutation rate and the selection coefficient (fitness) of the genotypes: define m as the mutation rate from a to A per generation. We will ignore back mutation (though actually this assumption does not matter). The frequency of a is q, and of A is p. Finally, we define the fitnesses as follows:

Genotype	aa	Aa	AA
Fitness	1	$1-s$	$1-s$

Evolution in this case will proceed to an equilibrial frequency of the gene A (we can write the stable equilibrium frequency as $p^\star$). If the frequency of A is higher than the equilibrium, natural selection removes more A genes than mutation creates and the frequency decreases; vice versa if the frequency is lower than the equilibrium. At the equilibrium, the rate of loss of A genes by selection equals their rate of gain by mutation.

We can use that statement to calculate the equilibrial gene frequency p^*. What is the rate per generation of creation of A genes by mutation? Each new A gene originates by mutation from an a gene and the chance that any one a gene mutates to an A gene is the mutation rate m. A proportion $(1 - p)$ of the genes in the population are a genes. Therefore:

We construct a model of the gene frequency of a recurrent disadvantageous mutation

Total rate of creation of A genes by mutation $= m(1 - p)$

And what is the rate at which A genes are eliminated? Each A gene has a $(1 - s)$ chance of surviving, or an s chance of dying. A proportion p of the genes in the population are A. Therefore:

Total rate of loss of A genes by selection $= ps$

At the equilibrium gene frequency (p^*):

$$\text{Rate of gain of } A \text{ gene} = \text{rate of loss of } A \text{ gene}$$
$$m(1 - p^*) = p^* s \tag{5.8}$$

Which can be multiplied out:

$$m - mp^* = p^* s$$
$$p^* = m/(s + m)$$

Of the two terms in the denominator, the mutation rate (maybe 10^{-6}, Section 2.6, p. 32) will usually be much less than the selection coefficient (perhaps 10^{-1} or 10^{-2}). With these values $s + m \approx s$ and the expression is therefore usually given in the approximate form:

$$p^* = m/s \tag{5.9}$$

The disadvantageous mutation has a low equilibrium frequency . . .

The simple result is that the equilibrium gene frequency of the mutation is equal to the ratio of its mutation rate to its selective disadvantage. The result is intuitive: the equilibrium is the balance between the rates of creation and elimination of the gene. To obtain the result, we used an argument about an equilibrium. We noticed that at the equilibrium the rate of loss of the gene equals the rate of gain and used that to work out the exact result. This is a powerful method for deriving equilibria, and we shall use an analogous argument in the next section.

The expression $p = m/s$ allows a rough estimate of the mutation rate of a harmful mutation just from a measurement of the mutant gene's frequency. If the mutation is rare, it will be present mainly in heterozygotes, which at birth will have frequency $2pq$. If p is small, $q \approx 1$ and $2pq \approx 2p$. N is defined as the frequency of mutant bearers, which equals the frequency of heterozygotes: i.e., $N = 2p$. As $p = m/s$, $m = sp$; if we substitute

$p = N/2$, $m = sN/2$. If the mutation is highly deleterious, $s \approx 1$ and $m = N/2$. The mutation rate can be estimated as half the birth rate of the mutant type. The estimate is clearly approximate, because it relies on a number of assumptions. In addition to the assumptions of high s and low p, mating is supposed to be random. We usually have no means of checking whether it is.

Chondrosdystrophic dwarfism is a dominant deleterious mutation in humans. In one study, 10 births out of 94,075 had the gene, a frequency of 10.6×10^{-5}. The estimate of the mutation rate by the above method is then $m = 5.3 \times 10^{-5}$. However, it is possible to estimate the selection coefficient, enabling a more accurate estimate of the mutation rate. In another study, 108 chondrodystrophic dwarves produced 27 children; their 457 normal siblings produced 582 children. The relative fitness of the dwarves was $(27/108)/(582/457) = 0.196$; the selection coefficient $s = 0.804$. Instead of assuming $s = 1$, we can use $s = 0.804$. Then the mutation rate is $sN/2 = 4.3 \times 10^{-5}$, a rather lower figure because with lower selection the same gene frequency can be maintained by a lower mutation rate.

For many genes, we do not know the dominance relations of the alleles at the locus. A similar calculation can be done for a recessive gene, but the formula is different, and it differs again if the mutation has intermediate dominance. We can only estimate the mutation rate from $p = m/s$ if we know the mutation is dominant. The method is therefore unreliable unless its assumptions have been independently verified. However, the general idea of this section — that a balance between selection and mutation can exist and explain genetic variation — will be used in later chapters.

5.12 Heterozygous advantage

5.12.1 Selection can maintain a polymorphism when the heterozygote is fitter than either homozygote

In some cases, heterozygotes have higher fitness than homozygotes

We come now to an influential theory. We are going to consider the case in which the heterozygote is fitter than both homozygotes. The fitnesses can be written:

Genotype	AA	Aa	aa
Fitness	$1 - s$	1	$1 - t$

t, like s, is a selection coefficient and has a value between 0 and 1. What happens here? There are three possible equilibria, but two of them are trivial. $p = 1$ and $p = 0$ are stable equilibria, but only because there is no mutation in the model. The third equilibrium is the interesting one; it has both genes present, and we can calculate the equilibrial gene frequencies by a similar argument to the one outlined in the previous section. The condition in which a population contains more than one gene is called *polymorphism*.

A genes and *a* genes are both removed by selection. The *A* genes are removed because they appear in the inferior *AA* homozygotes and the *a* genes because they appear in *aa* homozygotes. At the equilibrium, both genes must have the same chance of being removed by selection. If an *A* gene has a higher chance of being removed than an *a* gene, the frequency of *a* is increasing, and vice versa. Only when the chance is the same for both will the gene frequencies be stable.

What is the chance that an *A* gene will be carried by an individual who will die without reproducing? An *A* gene is either (with chance *q*) in a heterozygote and survives or (with chance *p*) in an *AA* homozygote and has a chance *s* of dying. Its total chance of dying is therefore *ps*. An *a* gene similarly is either (with chance *p*) in a heterozygote and survives or (with chance *q*) in an *aa* homozygote and has chance *t* of dying: its chance of death is *qt*. At the equilibrium,

Chance of death of an *A* gene = chance of death of an *a* gene

$$p^*s = q^*t \tag{5.10}$$

Substitute $\qquad p^*s = (1 - p^*)t$

and rearrange $\qquad p^* = t(s + t) \tag{5.11}$

Similarly if we substitute $q = (1 - p)$, $q^* = s/(s + t)$. Now we have derived the equilibrial gene frequencies when both homozygotes have lower fitness than the heterozygote. The equilibrium has all three genotypes present, even though the homozygotes are inferior and are selected against. They continue to exist because it is impossible to eliminate them. Matings among heterozygotes generate homozygotes. The exact gene frequency at equilibrium depends on the relative selection against the two homozygotes. If, for instance, *AA* and *aa* have equal fitness, then *s* = *t* and *p* = $\frac{1}{2}$ at equilibrium. If *AA* is relatively more unfit than *aa* then *s* > *t* and *p* < $\frac{1}{2}$; there are fewer of the more strongly selected against genotypes.

When heterozygotes are fitter than the homozygotes, therefore, natural selection will maintain a polymorphism. The result was first proved by Fisher in 1922 and independently by Haldane. We shall come later to consider in more detail why genetic variability exists in natural populations, and *heterozygous advantage* will be one of several controversial explanations to be tested.

5.12.2 *Sickle cell anemia is a polymorphism with heterozygous advantage*

Sickle cell anemia is the classic example of a polymorphism maintained by heterozygous advantage. It is a nearly lethal condition in humans, responsible for about 100,000 deaths a year. It is caused by a genetic variant of α-hemoglobin. If we symbolize the normal hemoglobin allele by *A* and the sickle cell hemoglobin by *S*, then people who suffer from sickle cell anemia are *SS*. Hemoglobin *S* causes the red blood cells to become curved and distorted (sickle shaped); they can then block capillaries and cause severe anemia if the blocked capillary is in the brain. About 80% of *SS* individuals die before reproducing. With such apparently strong selection against hemoglobin *S* it was a puzzle why it persisted at quite high frequencies (10% or even more) in some human populations.

If we compare a map of the incidence of malaria with a map of the gene frequency (Figure 5.9), we see that they are strikingly similar. Perhaps hemoglobin *S* provides

Figure 5.9
The global incidence of malaria coincides with that of the sickle cell form of hemoglobin. (a) A map of the frequency of the *S* allele of hemoglobin. (b) A map of malarial incidence. Redrawn, by permission of the publisher, from Bodmer & Cavalli-Sforza (1976).

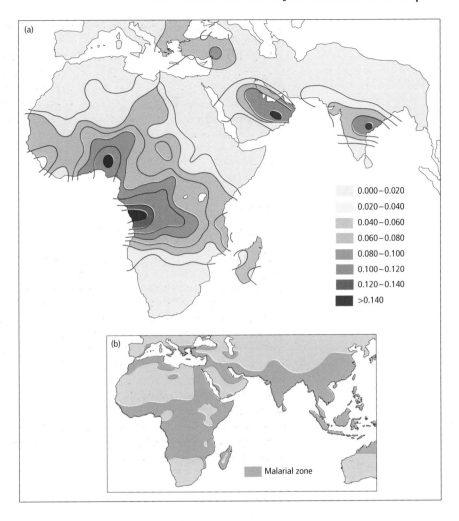

Sickle cell hemoglobin confers resistance to malaria

some advantage in malarial zones. Allison (1954) showed that, although *SS* is almost lethal, the heterozygote *AS* is more resistant to malaria than the homozygote *AA*. (Allison's was the first demonstration of natural selection at work in a human population.) The full reason was discovered later — *AS* red blood cells do not normally sickle, but they do if the oxygen concentration falls. When the malarial parasite *Plasmodium falciparum* enters a red blood cell it destroys (probably eats) the hemoglobin, which causes the oxygen concentration in the cell to go down. The cell sickles and is destroyed, along with the parasite. The human survives because most of the red blood cells are uninfected and carry oxygen normally. Therefore, where the malarial parasite is common, *AS* humans survive better than *AA*, who suffer from malaria.

Once the heterozygote had been shown physiologically to be at an advantage, the adult genotype frequencies can be used to estimate the relative fitnesses of the three genotypes. The fitnesses are:

Table 5.9
Estimates of selection coefficients for sickle cell anemia, using genotype frequencies in adults. The sickle cell hemoglobin allele is S, and the normal hemoglobin (which actually consists of more than one allele) is A. The genotype frequencies are for the Yorubas of Ibadan, Nigeria. One small detail is not explained in the text. The observed : expected ratio for the heterozygote may not be equal to 1. Here it turned out to be 1.12. All the observed : expected ratios are therefore divided by 1.12 to make them fit the standard fitness regime for heterozygote advantage. From Bodmer & Cavalli-Sforza (1976).

Genotype	Observed adult frequency (O)	Expected Hardy–Weinberg frequency (E)	Ratio O : E	Fitness
SS	29	187.4	0.155	$0.155/1.12 = 0.14 = 1 - t$
SA	2,993	2,672.4	1.12	$1.12/1.12 = 1.00$
AA	9,365	9,527.2	0.983	$0.983/1.12 = 0.88 = 1 - s$
Total	12,387	12,387		

Calculation of expected frequencies: gene frequency of S = frequency of SS + ½ (frequency of SA) = (29 + 2,993/2)/12,387 = 0.123. Therefore the frequency of A allele = $1 - 0.123 = 0.877$. From the Hardy–Weinberg theorem, the expected genotype frequencies are $(0.123)^2 \times 12,387$, $2(0.877)(0.123) \times 12,387$, and $(0.877)^2 \times 12,387$, for AA, AS, and SS, respectively.

Genotype	AA	AS	SS
Fitness	$1 - s$	1	$1 - t$

If the frequency of gene $A = p$ and of gene $S = q$, then the relative genotype frequencies among adults will be $p^2(1 - s) : 2pq : q^2(1 - t)$. If there were no selection ($s = t = 0$), the three genotypes would have Hardy–Weinberg frequencies of $p^2 : 2pq : q^2$.

Selection causes deviations from the Hardy–Weinberg frequencies. Take the genotype AA as an example. The ratio of the observed frequency in adults to that predicted from the Hardy–Weinberg ratio will be $(1 - s)/1$. The frequency expected from the Hardy–Weinberg principle is found by the usual method: the expected frequency is p^2, where p is the observed proportion of AA plus half the observed proportion of AS. Table 5.9 illustrates the method for a Nigerian population, where $s = 0.12$ ($1 - s = 0.88$) and $t = 0.86$ ($1 - t = 0.14$).

The method is only valid if the deviation from Hardy–Weinberg proportions is caused by heterozygous advantage and the genotypes differ only in their chance of survival (not their fertility). If heterozygotes are found to be in excess frequency in a natural population, it may indeed be because the heterozygote has a higher fitness. However, it could also be for other reasons. Disassortative mating, for instance, can produce the same result (in this case, disassortative mating would mean that aa individuals preferentially mate with AA individuals). But for sickle cell anemia, the physiological observations showed that the heterozygote is fitter and the procedure is well justified. Indeed, in this case, although it has not been checked whether mating is

We deduce selection coefficients of 0.12 and 0.14

random, the near lethality of *SS* means that disassortative mating will be unimportant; however, the assumption that the genotypes have equal fertility may well be false.

5.13 The fitness of a genotype may depend on its frequency

The next interesting complication is to consider selection when the fitness of a genotype depends on its frequency. In the models we have considered so far, the fitness of a genotype (1, 1 − s, or whatever) was constant, regardless of whether the genotype was rare or common. Now we consider the possibility that the fitness of a genotype goes up or down as the genotype frequency increases in the population (Figure 5.10). *Frequency-dependent selection* means that natural selection is acting and the fitnesses of the genotypes vary with the frequency of the genotypes. The two main kinds are *negative frequency dependence*, in which the fitness of a genotype goes down as its frequency goes up, and *positive frequency dependence*, in which the fitness of a genotype goes up as its frequency goes up.

In host–parasite relations, the fitness of a genotype may depend on frequency

Negative frequency dependence can arise in host–parasite interactions. For instance, two genotypes of a host may differ in their ability to keep out two genotypes of a parasite. This kind of set-up is like a lock and key. It is as if the two host genotypes are like two different locks, and the two parasite genotypes are like two different keys. One of the parasite keys fits one of the host locks and the other parasite key fits the other host lock. Then, if one of the host genotypes is in high frequency, natural selection will favor the parasite genotype that can penetrate that common kind of host. The result is that a high frequency automatically brings a disadvantage to a host genotype, because it creates an advantage for the kind of parasite than can exploit it. As the frequency of a host genotypes increases, its fitness soon decreases.

Snails and their parasite provide an example

Lively & Dybdahl (2000) recently described an example where the host is a snail, *Potamopyrgus antipodarum*, which (as its name hints at) lives in New Zealand, in freshwater habitats. The snail suffers from various parasites, of which a trematode called *Microphallus* is the most important (it is a parasitic castrator). The authors distinguished several strains (or clones) of the snail host and measured the frequency of each clone. They then measured, in an experiment, the ability of *Microphallus* to infect each clone. Figure 5.11 shows the infection rates achieved by parasites collected from two lakes, when experimentally exposed to snails taken from one of the two lakes. The local parasites infected the common clones better than the rare clones. It was the high

Figure 5.10
Frequency-dependent selection. (a) Negative frequency-dependent fitness means that the fitness of a genotype decreases as the frequency of the genotype increases. (b) Positive frequency-dependent fitness means that the fitness of a genotype goes up as its frequency increases. In general, frequency dependence refers to any case in which the graph is anything other than flat. A flat line, with fitness constant for all genotype frequencies, means that selection is not frequency dependent.

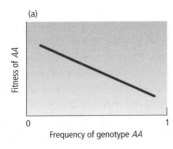

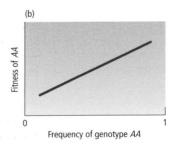

Figure 5.11

Parasites penetrate host genotypes more efficiently when they are locally abundant. Parasites from two lakes (Poerua and Ianthe) were experimentally put with snails of several genetic types (clones) from Lake Poerua. The four clones called 12, 19, 22, and 63 were common in the lake; several other clones were rare and they are all lumped together in the figure. The infection rates achieved by parasites taken from the two lakes were measured for each clone. (a) Infection rates achieved by parasites from Lake Poerua (sympatric parasites). (b) Infection rates achieved by parasites from Lake Ianthe (allopatric parasites). Note the higher infection rates achieved by the parasites on their local snails: the points are higher in (a) than in (b). But mainly note that the Poerua parasites in (a) infected the common snail clones more effectively than the rare clones; whereas the Ianthe parasites in (b) are no more effective with the common than the rare clones. From Lively & Dybdahl (2000). © 2000 Macmillan Magazines Ltd.

Frequency dependence can also arise in other circumstances

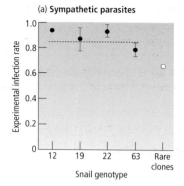

(a) Sympathetic parasites

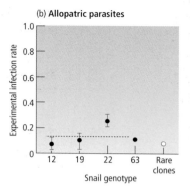

(b) Allopatric parasites

frequency of a clone that made it vulnerable to parasites. A clone that was common in one lake but rare in another was vulnerable to parasitism where it was common but not where it was rare.

Parasite–host relations are one important source of negative frequency-dependent selection (we return to this in Section 12.2.3, p. 323). Another important source is *multiple niche polymorphism*, a topic first discussed by Levene (1953). Suppose that a species contains several genotypes, and each genotype is adapted to a different set of environmental conditions. Genotypes *AA* and *Aa* might be adapted to the shade, and *aa* to sunny places (shady, and sunny, places then correspond to two "niches"). Then when the *A* gene is rare, *AA* and *Aa* experience less competition in their preferred areas, because there are fewer of them. As the frequency of *A* goes up, the shady areas become more crowded, competition increases, and fitness will tend to go down.

Frequency dependence is often generated by biological interactions. Competition and parasite–host relations are both biological interactions, and can generate negative frequency dependence. We shall meet some other examples, such as sex ratios (Section 12.5, p. 337) later in the book. Negative frequency-dependent fitnesses are important because they can produce stable polymorphisms within a species. As the frequency of each genotype goes up, its fitness goes down. Natural selection favors a gene when it is rare, but works against it when it is common. The result is that genotypes equilibrate at some intermediate frequency.

Positive frequency-dependent selection does not produce stable polymorphisms. Indeed it actively eliminates polymorphism, producing a genetically uniform population. For example, some species of insects have "warning coloration." They are brightly colored, and poisonous to eat. The bright coloration may reduce the chance of predation. When a bird eats the warningly colored insect, the bird is made sick and will remember not to eat an insect that looks like that again. However, the bird's lesson is not advantageous for the insect that made the bird sick; that insect is probably killed. When warningly colored insects are rare in a population mainly consisting of dull and cryptic individuals, the warningly colored genotypes are likely to have a low fitness. Few other insects exist to "educate" the local birds. This can create a problem in the evolution of warning coloration, because rare new mutants maybe selected against. The problem is not the point here, however. We are only considering it as an example of positive frequency dependence. The fitness of warningly colored genotypes will be higher at high frequencies, where the local birds are well educated about the dangers of eating the warningly colored forms.

The purpose of Sections 5.11–5.13 has been to illustrate the different mechanisms by which natural selection can maintain polymorphism. In Chapter 6 we look at another mechanism that can maintain polymorphism — genetic drift. Then, in Chapter 7, we tackle the question of how important the mechanisms are in nature.

5.14 Subdivided populations require special population genetic principles

5.14.1 *A subdivided set of populations have a higher proportion of homozygotes than an equivalent fused population: this is the Wahlund effect*

So far we have considered population genetics within a single, uniform population. In practice, a species may consist of a number of separate populations, each more or less isolated from the others. The members of a species might, for example, inhabit a number of islands, with each island population being separated by the sea from the others. Individuals might migrate between islands from time to time, but each island population would evolve to some extent independently. A species with a number of more or less independent subpopulations is said to have *population subdivision*.

Populations may be subdivided

Let us see first what effect population subdivision has on the Hardy–Weinberg principle. Consider a simple case in which there are two populations (we can call them population 1 and population 2), and we concentrate on one genetic locus with two alleles, A and a. Suppose allele A has frequency 0.3 in population 1 and 0.7 in population 2. If the genotypes have Hardy–Weinberg ratios they will have the frequencies, and average frequencies, in the two populations shown in Table 5.10. The average genotype frequencies are 0.29 for AA, 0.42 for Aa, and 0.29 for aa. Now suppose that the two

Table 5.10

The frequency of genotypes AA, Aa, and aa in two populations when A has frequency 0.7 in population 1 and 0.3 in population 2. The average genotypes are calculated assuming the two populations are of equal size.

	Genotype		
	AA	**Aa**	**aa**
Frequency	$(0.3)^2 = 0.09$	$2(0.3)(0.7) = 0.42$	$(0.7)^2 = 0.49$ population 1
	$(0.7)^2 = 0.49$	$2(0.7)(0.3) = 0.42$	$(0.3)^2 = 0.09$ population 2
Average	$0.58/2 = 0.29$	$0.84/2 = 0.42$	$0.58/2 = 0.29$

populations are fused together. The gene frequencies of A and a in the combined population are $(0.3 + 0.7)/2 = 0.5$, and the Hardy–Weinberg genotype frequencies are:

Genotype	AA	Aa	aa
Frequency	0.25	0.5	0.25

The Wahlund effect concerns the frequency of homozygotes in subdivided populations

In the large, fused population there are fewer homozygotes than in the average for the set of subdivided populations. This is a general, and mathematically automatic, result. The increased frequency of homozygotes in subdivided populations is called the *Wahlund effect*.

The Wahlund effect has a number of important consequences. One is that we have to know about the structure of a population when applying the Hardy–Weinberg principle to it. Suppose, for example, we had not known that populations 1 and 2 were independent. We might have sampled from both, pooled the samples indiscriminately, and then measured the genotype frequencies. We should find the frequency distribution for the average of the two populations (0.29, 0.42, 0.29); but the gene frequency would apparently be 0.5. There would seem to be more homozygotes than expected from the Hardy–Weinberg principle. We might suspect that selection, or some other factor, was favoring homozygotes. In fact both subpopulations are in perfectly good Hardy–Weinberg equilibrium and the deviation is due to the unwitting pooling of the separate populations. We need to look out for population subdivision when interpreting deviations from Hardy–Weinberg ratios.

Second, when a number of previously subdivided populations merge together, the frequency of homozygotes will decrease. In humans, this can lead to a decrease in the incidence of rare recessive genetic diseases when a previously isolated population comes into contact with a larger population. The recessive disease is only expressed in the homozygous condition, and when the two populations start to interbreed, the frequency of those homozygotes goes down.

5.14.2 *Migration acts to unify gene frequencies between populations*

When an individual migrates from one population to another, it carries genes that are representative of its own ancestral population into the recipient population. If it successfully establishes itself and breeds it will transmit those genes between the populations. The transfer of genes is called *gene flow*. If the two populations originally had different gene frequencies and if selection is not operating, migration (or, to be exact, gene flow) alone will rapidly cause the gene frequencies of the different populations to converge. We can see how rapidly in a simple model.

The spatial movement of genes is called gene flow

Consider again the case of two populations and one locus with two alleles (A and a). Suppose this time that one of the populations is much larger than the other, say population 2 is much larger than population 1 (2 might be a continent and 1 a small island off it); then practically all the migration is from population 2 to population 1. The frequency of allele a in population 1 in generation t is written $q_{1(t)}$; we can suppose that the frequency of a in the large population 2 is not changing between generations and

We construct a model of gene frequencies with migration

write it as q_m. (We are interested in the effect of migration on the gene frequency in population 1 and can ignore all other effects, such as selection.) Now, if we pick on any one allele in population 1 in generation $(t + 1)$, it will either be descended from a native of the population or from an immigrant. Define m as the chance that it is a migrant gene. (Earlier in the chapter, m was used for the mutation rate: now it is the *migration* rate.) If our gene is not a migrant (chance $(1 - m)$) it will be an a gene with chance $q_{1(t)}$, whereas if it is a migrant (chance m) it will be an a gene with chance q_m. The total frequency of a in population 1 in generation $(t + 1)$ is:

$$q_{1(t+1)} = (1 - m)q_{1(t)} + mq_m \qquad (5.12)$$

This can be rearranged to show the effect of t generations of migration on the gene frequency in population 1. If $q_{1(0)}$ is the frequency in the 0th generation, the frequency in generation t will be:

$$q_{1(t)} = q_m + (q_{1(0)} - q_m)(1 - m)^t \qquad (5.13)$$

(From $t = 1$ it is easy to confirm that this is indeed a rearrangement of the previous equation.) The equation says that the difference between the gene frequency in population 1 and population 2 decreases by a factor $(1 - m)$ per generation. At equilibrium, $q_1 = q_m$ and the small population will have the same gene frequency as the large population (Figure 5.12). In Figure 5.12, the gene frequencies converge in about 30 generations with a migration rate of 10%. Similar arguments apply if, instead of there being one source and one recipient population, the source is a set of many subpopulations, and p_m is their average gene frequency, or if there are two populations both sending migrants to, and receiving them from, another.

Gene flow binds biological species together

Migration will generally unify gene frequencies among populations rapidly in evolutionary time. In the absence of selection, migration is a strong force for equalizing the gene frequencies of populations within a species. Provided that the migration rate is greater than 0, gene frequencies will eventually equalize. Even if only one successful migrant moves into a population per generation, gene flow inevitably draws that population's gene frequency to the species' average. Gene flow acts, in a sense, to bind the species together.

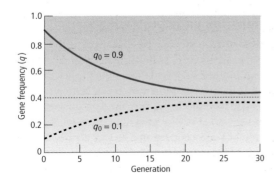

Figure 5.12
Migration causes the rapid convergence of gene frequencies in the populations exchanging migrants. Here a source population with gene frequency $q_m = 0.4$ sends migrants to two subpopulations, with initial gene frequencies of 0.9 and 0.1. They converge, with $m = 0.1$, onto the source population's gene frequency in about 30 generations.

5.14.3   *Convergence of gene frequencies by gene flow is illustrated by the human population of the USA*

The MN blood group is controlled by one locus with two alleles (Section 5.4). Frequencies of the *M* and *N* alleles have been measured, for example in European and African Americans in Claxton, Georgia, and among West Africans (whom we can assume to be representative of the ancestral gene frequency of the African American population of Claxton). The *M* allele frequency is 0.474 in West Africans, 0.484 in African Americans in Claxton, and 0.507 in the European Americans of Claxton. (The frequency of the *N* allele is equal to 1 minus the frequency of the *M* allele.) The gene frequency among African Americans is intermediate between the frequencies for European Americans and for the West African sample. Individuals of mixed parentage are usually categorized as African American and, if we ignore the possibility of selection favoring the *M* allele in the USA, we can treat the change in gene frequency in the African American population as due to "migration" of genes from the European American population. The measurements can then be used to estimate the rate of gene migration. In equation 5.13, q_m = gene frequency in the European American population (the source of the "migrant" genes), $q_0 = 0.474$ (the original frequency in the African American population), and $q_t = 0.484$. As an approximate figure, we can suppose that the black population has been in the USA for 200–300 years, or about 10 generations. Then:

The American population illustrates the model . . .

$$0.484 = 0.507 + (0.474 - 0.507)(1 - m)^{10}$$

. . . with a "migration" rate of about 3.5% per generation

This can be solved to find $m = 0.035$. That is, for every generation on average about 3.5% of the genes at the *MN* locus have migrated from the white population to the black population of Claxton. (Other estimates by the same method but using different gene loci suggest slightly different figures, more like 1%. The important point here is not the particular result; it is to illustrate how the population genetics of gene flow can be analyzed.) Notice again the rapid rate of genetic unification by migration: in only 10 generations, one-third of the gene frequency difference has been removed (after 10 generations the difference is $0.484 - 0.474$, against the original difference of $0.507 - 0.474$).

5.14.4   *A balance of selection and migration can maintain genetic differences between subpopulations*

If selection is working against an allele within one subpopulation, but the allele is continually being introduced by migration from other populations, it can be maintained by a balance of the two processes. We can analyze the balance between the two processes by much the same arguments as we used above for selection–mutation balance and heterozygous advantage. The simplest case is again for one locus with two alleles. Imagine selection in one subpopulation is working against a dominant *A* allele. The fitnesses of the genotypes are:

We construct a model of selection
and migration

	AA	Aa	aa
	$1-s$	$1-s$	1

The A allele has frequency p in the local population. Suppose that in other subpopulations, natural selection is more favorable to the gene A, and it has a higher frequency in them, p_m on average. p_m will then be the frequency of A among immigrants to our local population. In the local population, A genes are lost at a rate ps per generation. They are gained at a rate $(p_m - p)m$ per generation: m is the proportion of genes that are immigrants in a generation. Immigration increases the frequency in the local population by an amount $p_m - p$ because gene frequency is increased only in so far as the immigrating population has a higher frequency of A than the local population. If the immigrating gene frequency is the same as the local gene frequency, immigration has no effect.

There are three possible outcomes. If migration is powerful relative to selection, the rate of gain of A genes by immigration will exceed the rate of loss by selection. The local population will be swamped by immigrants. The frequency of the A gene will increase until it reaches p_m. If migration is weak relative to selection, the frequency of A will decrease until it is locally eliminated. The third possibility is an exact balance between migration and selection. There will be an equilibrium (with local frequency of $A = p^\star$) if:

Rate of gain of A by migration = rate of loss of A by selection

$$(p_m - p^\star)m = p^\star s \tag{5.14}$$

$$p^\star = p_m\left(\frac{m}{s + m}\right) \tag{5.15}$$

In the first case, migration unifies the gene frequencies in both populations, much in the same manner as Section 5.14.2: migration is so strong relative to selection that it is as if selection were not operating. In the second and third cases, migration is not strong enough to unify the gene frequencies and we should observe regional differences in the gene frequency; it would be higher in some places than in others. In the third case there is a polymorphism within the local population; A is maintained by migration even though it is locally disadvantageous.

Polymorphism or genetic unity can
result

This section has made two main points. First, a balance of migration and selection is another process to add to the list of processes that can maintain polymorphism. Second, we have seen how migration can be strong enough to unify gene frequencies between subpopulations, or if migration is weaker the gene frequencies of different subpopulations can diverge under selection. This theory is also relevant in the question of the relative importance of gene flow and selection in maintaining biological species (Section 13.7.2, p. 369).

Summary

1 In the absence of natural selection, and with random mating in a large population in which inheritance is Mendelian, the genotype frequencies at a locus move in one generation to the Hardy–Weinberg ratio; the genotype frequencies are then stable.

2 It is easy to observe whether the genotypes at a locus are in the Hardy–Weinberg ratio. In nature they will often not be, because the fitnesses of the genotypes are not equal, mating is non-random, or the population is small.

3 A theoretical equation for natural selection at a single locus can be written by expressing the frequency of a gene in one generation as a function of its frequency in the previous generation. The relation is determined by the fitnesses of the genotypes.

4 The fitnesses of the genotypes can be inferred from the rate of change of gene frequency in real cases of natural selection.

5 From the rate at which the melanic form of the peppered moth replaced the light-colored form, the melanic form must have had a selective advantage of about 50%.

6 The geographic pattern of melanic and light-colored forms of the peppered moth cannot be explained only by the selective advantage of the better camouflaged form. An inherent advantage to the melanic form, and migration, are also needed to explain the observations.

7 The evolution of resistance to pesticides in insects is in some cases due to rapid selection for a gene at a single locus. The fitness of the resistant types can be inferred, from the rate of evolution, to be as much as twice that of the non-resistant insects.

8 If a mutation is selected against but keeps on arising repeatedly, the mutation settles at a low frequency in the population. It is called selection–mutation balance.

9 Selection can maintain a polymorphism when the heterozygote is fitter than the homozygote and when fitnesses of genotypes are negatively frequency dependent.

10 Sickle cell anemia is an example of a polymorphism maintained by heterozygous advantage.

11 Subdivided populations have a higher proportion of homozygotes than an equivalent large, fused population.

12 Migration, in the absence of selection, rapidly unifies gene frequencies in different subpopulations; and it can maintain an allele that is selected against in a local subpopulation.

Further reading

There are a number of textbooks about population genetics. Crow (1986), Gillespie (1998), Hartl (2000), and Maynard Smith (1998) are relatively introductory. More comprehensive works include Hartl & Clark (1997) and Hedrick (2000). Crow & Kimura (1970) is a classic account of the mathematical theory. Dobzhansky (1970) is a standard study; Lewontin et al. (1981) contains Dobzhansky's most famous series of papers. Bell (1997a, 1997b) provides a comprehensive and a synoptic guide to selection.

For the peppered moth, Majerus (1998) is a modern, and Kettlewell (1973) a classic, account. Majerus (2002) is a more popular book, and contains a chapter on melanism. Grant (1999) is a review of Majerus (1998) and is also a good minireview of the topic in itself. Grant & Wiseman (2002) discuss the parallel rise and fall of the melanic form of the peppered moth in North America.

On pests and pesticides, see McKenzie (1996) and McKenzie & Batterham (1994). Lenormand et al. (1999) add further themes and molecular techniques, demonstrating seasonal cycles. The special issue of Science (4 October 2002, pp. 79–183) on the Anopheles genome has much background material on insecticide resistance and the various kinds of mosquito. See also Box 8.1 and Section 10.10, and their further reading lists.

See Endler (1986) on measuring fitness in general; Primack & Kang (1989) for plants; and Clutton-Brock (1988) for research on lifetime fitness.

The various selective means of maintaining polymorphisms are explained in the general texts. In addition, see Lederburg (1999) on the classic Haldane (1949a) paper and what it says about heterozygous advantage and sickle cell anemia. A recent possible example of heterozygote advantage in human HLA genes, providing resistance to HIV-1, is described by Carrington et al. (1999). Hori (1993) described a marvellous example of frequency dependence in the mouth-handedness of scale-eating cichlid fish. Another example is given by Gigord et al. (2001): the habits of naive bumblebees lead to a color polymorphism in an orchid.

Study and review questions

1 The following table gives genotype frequencies for five populations. Which are in Hardy–Weinberg equilibrium? For those that are not, suggest some hypotheses for why they are not.

	Genotype		
Population	AA	Aa	aa
1	25	50	25
2	10	80	10
3	40	20	40
4	0	150	100
5	2	16	32

2 For genotypes with the following fitnesses and frequencies at birth:

Genotype	AA	Aa	aa
Birth frequency	p^2	$2pq$	q^2
Fitness	1	1	$1 - s$

(a) What is the frequency of AA individuals in the adult population? (b) What is the frequency of the gene A in the adult population? (c) What is the mean fitness of the population?

3 What is the mean fitness of this population?

Genotype	AA	Aa	aa
Birth frequency	$1/3$	$1/3$	$1/3$
Fitness	1	$1 - s$	1

4 Consider a locus with two alleles, A and a. A is dominant and selection is working against the recessive homozygote. The frequency of A in two successive generations is 0.4875 and 0.5. What is the selection coefficient (s) against aa? (If you prefer to do it in your head rather than with a calculator, round the frequency of a in the first generation to 0.5 rather than 0.5125.)

5 What main assumption(s) is (or are) made in estimating fitnesses by the mark–recapture method?

6 Here are some adult genotype frequencies for a locus with two alleles. The polymorphism is known to be maintained by heterozygous advantage, and the fitnesses of the genotypes are known to differ only in survival (and not infertility). What are the fitnesses (or selection coefficients) of the two homozygotes, relative to a fitness of 1 for the heterozygote?

Genotype	AA	Aa	aa
Frequency among adults	$1/6$	$2/3$	$1/6$

7 There are two populations of a species, called population 1 and population 2. Migrants move from population 1 to 2, but not vice versa. For a locus with two alleles A and a, in generation n, the gene frequency of A is 0.5 in population 1 and 0.75 in population 2; in generation 2 it is 0.5 in population 1 and 0.625 in population 2. (a) What is the rate of migration, measured as the chance an individual in population 2 is a first-generation immigrant from population 1? (b) If the rate of migration is the same in the next generation, what will the frequency of A be in population 2 in generation 3? [Questions 8–10 are more in the nature of questions for further thought. They are not about things explicitly covered in the chapter, but are slight extensions.]

8 What is the general effect of assortative mating on genotype frequencies, relative to the Hardy–Weinberg equilibrium, for (a) a locus with two alleles, one dominant to the other; and (b) a locus with two alleles, and no dominance (the heterozygote is a distinct phenotype intermediate between the two homozygotes)? And (c) what is the effect on genotype frequencies of a mating preference, in which females preferentially mate with males of (i) the dominant, and (ii) the recessive phenotype?

9 Derive a recurrence relation, giving the frequency of the dominant gene A one generation on (p') in terms of the frequency in any generation (p) and of the selection coefficient (s) for selection against the dominant allele.

10 Derive the expression for the equilibrium gene frequency ($p*$) for the mutation–selection balance when the disadvantageous mutation is recessive.

6 Random Events in Population Genetics

The genotypes at a locus may all have the same fitness. Then the gene frequencies evolve by random genetic drift. This chapter starts by explaining why drift happens and what it means, and looks at examples of random sampling effects. We see how drift is more powerful in small than large popuations, and how in small populations it can counteract the effects of natural selection. We then see how drift can ultimately fix one allele. The Hardy–Weinberg ratios are not at an equilibrium once we allow for the effects of drift. We then add the effects of mutation, which introduces new variation: the variation observed in a population will be a balance between the drift to homozygosity and mutation that creates heterozygosity.

6.1 The frequency of alleles can change at random through time in a process called genetic drift

Imagine a population of 10 individuals, of which three have genotype *AA*, four have *Aa*, and three *aa*. There are 10 *A* genes in the population and 10 *a* genes; the gene frequencies of each gene are 0.5. We also imagine that natural selection is not operating: all genotypes have the same fitness. What will the gene frequencies be in the next generation? The most likely answer is 0.5 *A* and 0.5 *a*. However, this is only the most likely answer; it is not a certainty. The gene frequencies may by chance change a little from the previous generation. This can happen because the genes that form a new generation are a *random sample* from the parental generation. Box 6.1 looks at how genes are sampled from the parental gene pool, to produce the offspring generation's gene pool. In this chapter we look at the effect of random sampling on gene frequencies.

The easiest case in which to see the effect of random sampling is when natural selection is not acting. When the genotypes at a locus all produce the same number of offspring (they have identical fitness), the condition is called selective *neutrality*. We can write the fitnesses out in the same way as in Chapter 5, as follows:

Genotype	*AA*	*Aa*	*aa*
Fitness	1	1	1

Natural selection is not acting, and we might expect the gene frequencies to stay constant over time. Indeed, according to the Hardy–Weinberg theorem, the genotype frequencies should be constant at p^2, $2pq$, and q^2 (where p is the frequency of the gene *A* and q is the frequency of the gene *a*). But in fact random sampling can cause the gene frequencies to change. By chance, copies of the *A* gene may be luckier in reproduction, and the frequency of the *A* gene will increase. The increase is random, in the sense that the *A* gene is as likely by chance to decrease as to increase in frequency; but some gene frequency changes will occur. These random changes in gene frequencies between generations are called *genetic drift*, *random drift*, or (simply) *drift*. The word "drift" can be misleading if it is taken to imply an inbuilt bias in one direction or the other. Genetic drift is directionless drift.

Genetic drift is not confined to the case of selective neutrality. When selection is acting at a locus, random sampling also influences the change in gene frequencies between generations. The interaction between selection and drift is an important topic in evolutionary biology, as we shall see in Chapter 7. However, the theory of drift is easiest to understand when selection is not complicating the process and in this chapter we shall mainly look at the effect of drift by itself.

The rate of change of gene frequency by random drift depends on the size of the population. Random sampling effects are more important in smaller populations. For example (Figure 6.1), Dobzhansky and Pavlovsky (1957), working with the fruitfly *Drosophila pseudoobscura*, made 10 populations with 4,000 initial members (large populations) and 10 with 20 initial members (small populations), and followed the change in frequency of two chromosomal variants for 18 months. The average effect was the same in small and large populations, but the variability was significantly greater among the small populations. An analogous result could be obtained by flipping 10 sets of 20,

Genetic drift occurs because of random sampling

The power of drift depends on population size

Box 6.1
Random Sampling in Genetics

Random sampling starts at conception. In every species, each individual produces many more gametes than will ever fertilize, or be fertilized, to form new organisms. The successful gametes which do form offspring are a sample from the many gametes that the parents produce. If a parent is homozygous, the sampling makes no difference to what genes end up in the offspring; all of a homozygote's gametes contain the same gene. However, sampling does matter if the parent is a heterozygote, such as *Aa*. It will then produce a large number of gametes, of which approximately one-half will be *A* and the other half *a*. (The proportions may not be exactly one-half. Reproductive cells may die at any stage leading to gamete formation, or after they have become gametes; also, in the female, a randomly picked three-quarters of the products of meiosis are lost as polar bodies.) If that parent produces 10 offspring, it is most likely that five will inherit an *A* gene and five *a*. But because the gametes that formed the offspring were sampled from a much larger pool of gametes, it is possible that the proportions would be something else. Perhaps six inherited *A* and only four *a*, or three inherited *A* and seven *a*.

In what sense is the sampling of gametes random? We can see the exact meaning if we consider the first two offspring produced by an *Aa* parent. When it produces its first offspring, one gamete is sampled from its total gamete supply, and there is a 50% chance it will be an *A* and 50% that it will be an *a*. Suppose it happens to be an *A*. The sense in which sampling is random is that it is no more likely that the next gamete to be sampled will be an *a* gene just because the last one sampled was an *A*: the chance that the next successful gamete will be an *a* is still 50%. Coin flipping is random in the same way: if you first flip a head, the chance that the next flip will be a head is still one-half. The alternative would be some kind of "balancing" system in which, after an *A* gamete had been successful in reproduction, the next successful gamete would be an *a*. If reproduction was like that, the gene frequency contributed by a heterozygote to its offspring would always be exactly $\frac{1}{2}A$: $\frac{1}{2}a$. Random drift would then be unimportant in evolution. In fact reproduction is not like that. The successful gametes are a random sample from the gamete pool.

The sampling of gametes is only the first stage at which random sampling occurs. It continues at every stage as the adult population of a new generation grows up. Here is an imaginary example. Imagine a line of 100 pack horses are walking single file along a hazardous mountain path, but only 50 of them make it safely; the other 50 fall off the path and crash down the ravine. It could be that the 50 survivers were on average genetically surer of foot than the rest; the sampling of 50 survivers out of the original 100 would then be non-random. Natural selection would be determining which horses survived and which died. If we looked at the genotypic frequencies among the smashed horses at the bottom of the ravine they would differ from those among the survivers. Alternatively, death could be accidental: it could happen whenever a large rock bounced down the mountainside from above, and knocked one horse into the ravine. Suppose that the rocks come at unpredictable times and places and arrive so suddenly that defensive action is impossible; the horses do not vary genetically in their ability to avoid the falling rocks. The loss of genotypes would then be random in the sense defined above. If an *AA* horse had just fallen victim to a rock, that does not make it any more or less likely that the next victim will have the *AA* genotype. Now if we compared the genotype frequencies in the survivers and non-survivers, it is most likely that the two would not differ. The survivers would be a random genetic sample from the original population. They could, however, differ by chance. More *AA* horses might have been unlucky with falling rocks; more *aa* might have been lucky. Then there would be some increase in the frequency of the *a* gene in the population.

The sampling of pack horses is imaginary, but analogous sampling may happen at any time in a population, and at any life stage as juveniles develop into adults. Because there are many more eggs than adults, there is abundant opportunity for sampling as each new generation grows up. Random sampling occurs whenever a smaller number of successful individuals (or gametes) are sampled from a larger pool of potential survivers and the fitnesses of the genotypes are the same.

or 4,000, coins. On average, there would be 50% heads in both cases, but the chance of flipping 12 heads and 8 tails in the small population is higher than the chance of flipping 2,400 heads and 1,600 tails in the large.

If a population is small, it is more likely that a sample will be biased away from the average by any given percentage amount; genetic drift is therefore greater in smaller

Figure 6.1
Random sampling is more effective in small populations (a) than in large (b). Ten large (4,000 founders) and 10 small (20 founders) populations of the fruitfly *Drosophila pseudoobscura* were created in June 1955 with the same frequencies (50% each) of two chromosomal inversions, *AP* and *PP*. Eighteen months later the populations with small numbers of founders show a greater variety of genotype frequencies. Redrawn, by permission of the publisher, from Dobzhansky (1970).

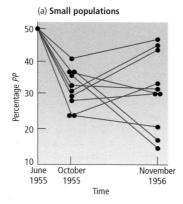
(a) Small populations

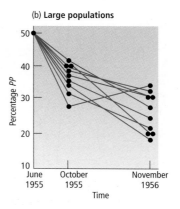
(b) Large populations

populations. The smaller the population, the more important are the effects of random sampling.

6.2 A small founder population may have a non-representative sample of the ancestral population's genes

A particular example of the influence of random sampling is given by what is called the *founder effect*. The founder effect was defined by Mayr (1963) as:

> the establishment of a new population by a few original founders (in an extreme case, by a single fertilized female) which carry only a small fraction of the total genetic variation of the parental population.

Population size may be reduced during founder events

We can divide the definition into two parts. The first part is the establishment of a new population by a small number of founders; we can call that a "founder event." The second part is that the founders have a limited sample of genetic variation. The full founder effect requires not only a founder event, but also that the founders are genetically unrepresentative of the original population.

Founder events undoubtedly happen. A population may be descended from a small number of ancestral individuals for either of two main reasons. A small number of individuals may colonize a place previously uninhabited by their species; the 250 or so individuals making up the modern human population on the island of Tristan da Cunha, for example, are all descended from about 20–25 immigrants in the early nineteenth century, and most are descended from the original settlers — one Scotchman and his family — who arrived in 1817. Alternatively, a population that is established in an area may fluctuate in size; the founder effect then occurs when the population passes through a "bottleneck" in which only a few individuals survive, and later expands again when more favorable times return.

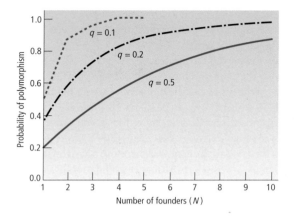

Figure 6.2
The chance that a founder population will be homozygous depends on the number of founders and the gene frequencies. If there is less variation and fewer founders, the chance of homozygosity is higher. Here the chance of homozygosity is shown for three different gene frequencies at a two-allele locus.

Founder events are unlikely to produce homozygosity

If a small sample of individuals is taken from a larger population, what is the chance that they will have reduced genetic variation? We can express the question exactly by asking what the chance is that an allele will be lost. In the special case of two alleles (*A* and *a* with proportions *p* and *q*), if one of them is not included in the founder population, the new population will be genetically monomorphic. The chance that an individual will be homozygous *AA* is simply p^2. The chance that two individuals drawn at random from the population will both be *AA* is $(p^2)^2$; in general, the chance of drawing *N* identical homozygotes is $(p^2)^N$. The founding population could be homozygous either because it is made up of *N AA* homozygotes or *N aa* homozygotes, and the total chance of homozygosity is therefore:

$$\text{Chance of homozygosity} = [(p^2)^N + (q^2)^N] \tag{6.1}$$

Figure 6.2 illustrates the relation between the number of individuals in the founder population and the chance that the founder population is genetically uniform. The interesting result is that founder events are not effective at producing a genetically monomorphic population. Even if the founder population is very small, with $N < 10$, it will usually possess both alleles. An analogous calculation could be done for a population with three alleles, in which we asked the chance that one of the three would be lost by the founder effect. The resulting population would not then be monomorphic, but would have two instead of three alleles. The general point is again the same: in general, founder events — whether by colonizations or population bottlenecks — are unlikely to reduce genetic variation unless the number of founders is tiny.

However, founder events can have other interesting consequences. Although the sample of individuals forming a founder population are likely to have nearly all the ancestral population's genes, the frequencies of the genes may differ from the parental population. Isolated populations often have exceptionally high frequencies of otherwise rare alleles, and the most likely explanation is that the founding population had a disproportionate number of those rare alleles. The clearest examples all come from humans.

Consider the Afrikaner population of South Africa, who are mainly descended from one shipload of immigrants who landed in 1652, though later arrivals have added to it.

Several human populations have otherwise rare genes in high frequency

The population has increased dramatically since then to its modern level of 2,500,000. The influence of the early colonists is shown by the fact that almost 1,000,000 living Afrikaners have the names of 20 of the original settlers.

The early colonists included individuals with a number of rare genes. The ship of 1652 contained a Dutch man carrying the gene for Huntington's disease, a lethal autosomal dominant disease. Most cases of the disease in the modern Afrikaner population can be traced back to that individual. A similar story can be told for the dominant autosomal gene causing porphyria variegata. Porphyria variegata is due to a defective form of the enzyme protoporphyrinogen oxidase. Carriers of the gene suffer a severe — even lethal — reaction to barbiturate anesthetics, and the gene was therefore not strongly disadvantageous before modern medicine. The modern Afrikaner population has about 30,000 carriers of the gene, a far higher frequency than in Holland. All the carriers are descended from one couple, Gerrit Jansz and Ariaantje Jacobs, who emigrated from Holland in 1685 and 1688, respectively. Every human population has its own "private" polymorphisms, which were probably often caused by the genetic peculiarities of founder individuals.

Both of the examples we have just considered are for medical conditions. The individual carriers of the genes will have lower fitness than average, and selection will therefore act to reduce the frequency of the gene to 0. For much of the time, the porphyria variegata gene may have had a similar fitness to other alleles at the same locus. It may have been a neutral polymorphism until its "environment" came to contain (in selected cases) barbiturates.

In contrast, the gene for Huntington's disease will have been consistently selected against. Thus its present high frequency suggests that the founder population had an even higher frequency, because it will have probably been decreased by selection since then. Any particular founder sample would not be expected to have a higher than average frequency of the Huntington's disease gene, but if enough colonizing groups set out, some of them are bound to have peculiar, or even very peculiar, gene frequencies. In the case of Huntington's disease, the Afrikaner population is not the only one descended from founders with more copies of the gene than average; 432 carriers of Huntington's disease in Australia are descended from the Miss Cundick who left England with her 13 children; and a French nobleman's grandson, Pierre Dagnet d'Assigne de Bourbon, has bequeathed all the known cases of Huntington's disease on the island of Mauritius.

6.3 One gene can be substituted for another by random drift

The frequency of a gene is as likely to decrease as to increase by random drift. On average the frequencies of neutral alleles remain unchanged from one generation to the next. In practice, their frequencies drift up and down, and it is therefore possible for a gene to enjoy a run of luck and be carried up to a much higher frequency — in the extreme case, its frequency could after many generations be carried up to 1 (become fixed) by random drift.

Evolution can occur by random drift

In every generation, the frequency of a neutral allele has a chance of increasing, a chance of decreasing, and a chance of staying constant. If it increases in one generation, it again has the same chances of increasing, decreasing, or staying constant in the next generation. A neutral allele thus has a small chance of increasing for two generations in a row (equal to the square of the chance of increasing in any one generation). It has a still smaller chance of increasing though three generations, and so on. For any one allele, fixation by random drift is very improbable. The probability is finite, however, and if enough neutral alleles, at enough loci, and over enough generations, are randomly drifting in frequency, one of them will eventually be fixed. The same process can occur whatever the initial frequency of the allele. A rare allele is less likely to be carried up to fixation by random drift than is a common allele, because it would take a longer run of "good" luck. However, fixation is still possible for a rare allele. Even a unique neutral mutation has some chance of eventual fixation. Any one mutation is most likely to be lost; but if enough mutations arise, one will be bound to be fixed eventually.

Random drift, therefore, can substitute one allele for another. What is the rate at which these substitutions occur? We might expect it would be faster in smaller populations, because most random effects are more powerful in smaller populations. However, it can be shown by an elegant argument that the neutral evolution rate exactly equals the neutral mutation rate, and is independent of population size. The argument is as follows. In a population of size N there are a total of $2N$ genes at each locus. On average, each gene contributes one copy of itself to the next generation; but because of random sampling, some genes will contribute more than one copy and others will contribute none. As we look two generations ahead, those genes that contributed no copies to the first generation cannot contribute copies to the second generation, or the third, or fourth . . . once a gene fails to be copied, it is lost forever. In the next generation some more genes will likewise "drop out," and be unable to contribute to future generations. Each generation, some of the $2N$ original genes are lost in this way (Figure 6.3).

If we look far enough forwards we eventually come to a time when all the $2N$ genes are descended from just one of the $2N$ genes now. This is because in every generation

For purely neutral drift, the rate of evolution is independent of population size

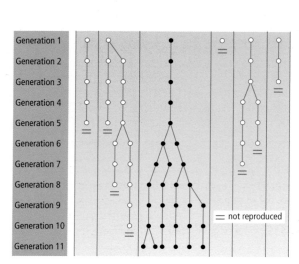

| Generation 1 |
| Generation 2 |
| Generation 3 |
| Generation 4 |
| Generation 5 |
| Generation 6 |
| Generation 7 |
| Generation 8 |
| Generation 9 |
| Generation 10 |
| Generation 11 |

= not reproduced

Figure 6.3

The drift to homozygosity. The figure traces the evolutionary fate of six genes; in a diploid species these would be combined each generation in three individuals. Every generation, some genes may by chance fail to reproduce and others by chance may leave more than one copy. Because once a gene has failed to reproduce its line is lost forever, over time the population must drift to become made up of descendants of only one gene in an ancestral population. In this example, the population after 11 generations is made up of descendants of gene number 3 (shaded circle) in generation 1.

Population size features in the workings . . .

some genes will fail to reproduce. We must eventually come to a time when all but one of the original genes have dropped out. That one gene will have hit a long enough run of lucky increases and will have spread through the whole population. It will have been fixed by genetic drift. Now, because the process is pure luck, each of the $2N$ genes in the original population has an equal chance of being the lucky one. Any one gene in the population, therefore, has a $1/(2N)$ chance of eventual fixation by random drift (and a $(2N-1)/(2N)$ chance of being lost by it).

Because the same argument applies to any gene in the population, it also applies to a new, unique, neutral mutation. When the new mutation arises, it will be one gene in a population of $2N$ genes at its locus (that is, its frequency will be $1/(2/N)$). The new mutation has the same $1/(2N)$ chance of eventual fixation as does every other gene in the population. The most likely fate of the new mutation is to be lost (probability of being lost $= (2N-1)/(2N) \approx 1$ if N is large); but it does have a small $(1/(2N))$ chance of success. That completes the first stage of the argument: the probability that a neutral mutation will eventually be fixed is $1/(2N)$.

The rate of evolution equals the probability that a mutation is fixed, multiplied by the rate at which mutations appear. We define the rate at which neutral mutations arise as u per gene per generation. (u is the rate at which new selectively neutral mutations arise, not the total mutation rate. The total mutation rate includes selectively favorable and unfavorable mutations as well as neutral mutations. We are here considering only the fraction of all mutations that are neutral.) At each locus, there are $2N$ genes in the population: the total number of neutral mutations arising in the population will be $2Nu$ per generation. The rate of neutral evolution is then $1/(2N) \times 2Nu = u$. The population size cancels out and the rate of neutral evolution is equal to the neutral mutation rate.

. . . and cancels out

Figure 6.3 also illustrates another important concept in the modern theory of genetic drift, the concept of coalescence (Box 6.2).

Box 6.2
The Coalescent

If we look forward far enough in time from any one generation, we must come to a time when all the genes at a locus are descended from one of the $2N$ copies of that gene in the current population (see Figure 6.3). The same argument works backwards. If we look far enough back from any one generation, we must come to a time when all the copies of the genes at one locus trace back to a single copy of that gene in the past. Thus, if we trace back from all the copies of a human gene, such as a globin gene, we must eventually come to a time in the past when only one gene gave rise to all the modern copies of the gene. (In Figure 6.3, look at generation 11 at the end. All copies of the gene trace back to a single gene in generation 5. Notice that the existence of a single ancestral gene for all the modern genes at a locus does not mean that only one gene existed at that time. Generation 5 has as many genes as every other generation.) The way all copies of a gene trace back to a single ancestral gene is called *coalescence*, and that single lucky ancestral gene is called *the coalescent*. Genetic coalescence is a consequence of the normal operation of genetic drift in natural populations. Every gene in the human species, and every gene in every species, traces back to a coalescent. The time when the coalescent existed for each gene probably differs between genes, but they all have a coalescent ancestor at some time. Population geneticists study how far back the coalescent exists for a gene, depending on population size, demography, and selection. A knowledge of the time back to the coalescent can be useful for dating events in the past using "gene trees," which we meet in Chapter 15.

Further reading: Fu & Li (1999), Kingman (2000).

6.4 Hardy–Weinberg "equilibrium" assumes the absence of genetic drift

Random drift has consequences for the Hardy–Weinberg theorem

Let us stay with the case of a single locus, with two selectively neutral alleles *A* and *a*. If genetic drift is not happening — if the population is large — the gene frequencies will stay constant from generation to generation and the genotype frequencies will also be constant, in Hardy–Weinberg proportions (Section 5.3, p. 98). But in a smaller population the gene frequencies can drift around. The average gene frequencies in one generation will be the same as in the previous generation, and it might be thought that the long-term average gene and genotype frequencies will simply be those of the Hardy–Weinberg equilibrium, but with a bit of "noise" around them. That is not so, however. The long-term result of genetic drift is that one of the alleles will be fixed. The polymorphic Hardy–Weinberg equilibrium is unstable once we allow for genetic drift.

Suppose that a population is made up of five individuals, containing five *A* alleles and five *a* alleles (that is obviously a tiny population, but the same point would apply if there were 500 copies of each allele). The genes are randomly sampled to produce the next generation. Maybe six *A* alleles are sampled and four *a* alleles. This is now the starting point to produce the next generation; the most likely ratio in the next generation is six *A* and four *a*: there is no "compensating" process to push it back toward five and five. Maybe in the next generation six *A* and four *a* are drawn again. The fourth generation might be seven *A* and three *a*, the fifth, six *A* and four *a*, the sixth, seven *A* and three *a*, then seven *A* and three *a*, eight *A* and two *a*, eight *A* and two *a*, nine *A* and one *a*, and then 10 *A*. The same process could have gone off in the other direction, or started by favoring A and then reversed to fix *a* — random drift is directionless. However, when one of the genes is fixed, the population is homozygous and will stay homozygous (Figures 6.3 and 6.4).

The Hardy–Weinberg equilibrium is a good approximation, and retains its importance in evolutionary biology. But it is also true that, once we allow for random drift, the Hardy–Weinberg ratios are not at an equilibrium. The Hardy–Weinberg ratios are for neutral alleles at a locus and the Hardy–Weinberg result suggests that the genotype (and gene) ratios are stable over time. However, random events cause gene frequencies to drift about, and one of the genes will eventually be fixed. Only then will the system be stable. The true equilibrium, incorporating genetic drift, is at homozygosity.

6.5 Neutral drift over time produces a march to homozygosity

Over the long term, pure random drift causes the population to "march" to homozygosity at a locus. The process by which this happens has already been considered (Section 6.4) and illustrated (Figure 6.3). All loci at which there are several selectively neutral alleles will tend to become fixed for only one gene. It is not difficult to derive an expression for the rate at which the population becomes homozygous. First we define the degree of homozygosity. Individuals in the population are either homozygotes or

Figure 6.4
Twenty repeat simulations of genetic drift for a two-allele locus with initial gene frequency 0.5 in: (a) a small population ($2N = 18$), and (b) a larger population ($2N = 100$). Eventually one of the alleles drifts to a frequency of 1. The other alleles are then lost. The drift to homozygosity is more rapid in a smaller population, but in any small population without mutation homozygosity is the final result.

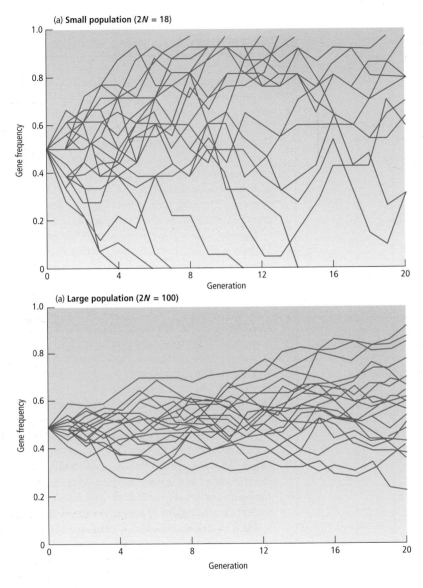

heterozygotes. Let f be the proportion of homozygotes, and $H = 1 - f$ is the proportion of heterozygotes (f comes from "fixation"). Homozygotes here includes all types of homozygote at a locus; if, for example, there are three alleles A_1, A_2, and A_3 then f is the number of A_1A_1, A_2A_2, and A_3A_3 individuals divided by the population size; H likewise is the sum of all heterozygote types. N will again stand for population size.

How will f change over time? We shall derive the result in terms of a special case: a species of hermaphrodite in which an individual can fertilize itself. Individuals in the population discharge their gametes into the water and each gamete has a chance of combining with any other gamete. New individuals are formed by sampling two

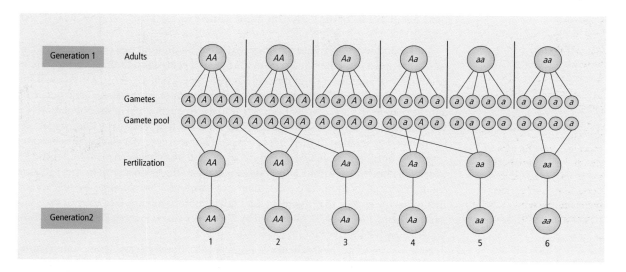

Figure 6.5

Inbreeding in a small population produces homozygosity. A homozygote can be produced either by combining copies of the same gene from different individuals, or by combining two copies of the same physical gene. Here we imagine that the population contains six adults, which are potentially self-fertilizing hermaphrodites, and each produces four gametes.

Homozygotes can then be produced by the kind of cross-mating assumed in the Hardy–Weinberg theorem (e.g., offspring number 2) or by self-fertilization (e.g., offspring number 1). Self-fertilization only necessarily produces a homozygote if its parent is homozygous (compare offspring 1 and 4).

gametes from the gamete pool. The gamete pool contains $2N$ gamete types, where "gamete types" should be understood as follows. There are $2N$ genes in a population made up of N diploid individuals. A gamete type consists of all the gametes containing a copy of any one of these genes. Thus, if an individual with two genes produces 200,000 gametes, there will be on average 100,000 copies of each gamete type in the gamete pool.

We construct a model of how homozygosity changes under drift

To calculate how f, the degree of homozygosity, changes through time, we derive an expression for the number of homozygotes in one generation in terms of the number of homozygotes in the generation before. We must first distinguish between a gene-bearing gametes in the gamete pool that are copies of the same parental a gene, and those that are derived from different parents. There are then two ways to produce a homozygote, when two a genes from the same gametic type meet or when two a genes from different gametic types meet (Figure 6.5); the frequency of homozygotes in the next generation will be the sum of these two.

The first way of making a homozygote is by "self-fertilization." There are $2N$ gamete types but, because each individual produces many more than two gametes, there is a chance $1/(2N)$ that a gamete will combine with another gamete of the same gamete type as itself: if it does, the offspring will be homozygous. (If, as above, each individual makes 200,000 gametes, there would be 200,000N gametes in the gamete pool. We first sample one gamete from it. Of the remaining gametes, practically 100,000 of them (99,999 in fact) are copies of the same gene. The proportion of gametes left in the pool that contain copies of the same gene as the gamete we sampled is 99,999/200,000N, or $1/(2N)$.)

The second way to produce a homozygote is by combining two identical genes that were not copied from the same gene in the parental generation. If the gamete does not combine with another copy of the same gamete type (chance $1 - (1/(2N))$) it will still form a homozygote if it combines with a copy made from the same gene but from another parent. For a gamete with an *a* gene, if the frequency of *a* in the population is *p*, the chance that two *a* genes meet is simply p^2. p^2 is the frequency of *aa* homozygotes in the parental generation. If there are two type of homozygote, *AA* and *aa*, the chance of forming a homozygote will be $p^2 + q^2 = f$. In general, the chance that two independent genes will combine to form a homozygote is equal to the frequency of homozygotes in the previous generation. The total chance of forming a homozygote by this second method is the chance that a gamete does not combine with another copy of the same parental gene, $1 - (1/(2N))$, multiplied by the chance that two independent genes combine to form a homozygote (f). That is, $f(1 - (1/(2N)))$.

Now we can write the frequency of homozygotes in the next generation in terms of the frequency of homozygotes in the parental generation. It is the sum of the two ways of forming a homozygote. Following the normal notation for f' and f (f' is the frequency of homozygotes one generation later),

$$f' = \frac{1}{2N} + \left(1 - \frac{1}{2N}\right)f \tag{6.2}$$

We can follow the same march to increasing homozygosity in terms of the decreasing heterozygosity in the population. A population's "heterozygosity" is a measure of its genetic variation. In formal terms, *heterozygosity* is defined as the chance that two genes at a locus, drawn at random from the population, are different. For example, a genetically uniform population (in which everyone is *AA*) has a heterozygosity of zero. The chance of drawing two different genes is zero. If half the individuals in the population are *AA* and half are *aa*, the chance of drawing two different genes is half, and heterozygosity equals one-half. Box 6.3 describes the calculation of heterozygosity. (Heterozygosity is symbolized by *H*.)

Heterozygosity can be shown, by rearrangement of equation 6.2, to decrease at the following rate (the rearrangement involves substituting $H = 1 - f$ in equation 6.2):

$$H' = \left(1 - \frac{1}{2N}\right)H \tag{6.3}$$

That is, heterozygosity decreases at a rate of $1/(2N)$ per generation until it is zero. The population size *N* is again important in governing the influence of genetic drift. If *N* is small, the march to homozygosity is rapid. At the other extreme, we re-encounter the Hardy–Weinberg result. If *N* is infinitely large, the degree of heterozygosity is stable: there is then no march to homozygosity.

Although it might seem that this derivation is for a particular, hermaphroditic breeding system, the result is in fact general (a small correction is needed for the case of two sexes). The march to homozygosity in a small population proceeds because two

Homozygosity can arise from crosses between different individuals

Heterozygosity is a measure of genetic variation

Box 6.3
Heterozygosity (*H*) and Nucleotide Diversity (*π*)

"Heterozygosity" is a general measure of the genetic variation per locus in a population. Imagine a locus at which two alleles (*A* and *a*) are present in the population. The frequency of *A* is *p*, the frequency of *a* is *q*. Heterozygosity is defined as the chance of drawing two different alleles if two random genes are sampled from the population (for one locus). The chance of drawing two copies of *A* is p^2, and the chance of drawing two copies of *a* is q^2. The total chance of drawing two identical genes is $p^2 + q^2$. The chance of drawing two different genes is 1 minus the chance of drawing two identical genes. Therefore, in this case $H = 1 - (p^2 + q^2)$.

In general, a population may contain any number of alleles at a locus. The different alleles can be distinguished by number subscripts. For instance, if a population has three alleles, their frequencies can be written p_1, p_2, and p_3. If a population has four alleles, their frequencies can be written p_1, p_2, p_3, and p_4, and so on for any number of alleles. We can symbolize the frequency of the *i*th allele by p_i (where *i* has as many values as there are alleles in the population). Now:

$$H = 1 - \sum p_i^2$$

The summation (symbolized by $\sum$) is over all values of *i*: that is, for all the alleles in the population at that locus. The term $\sum p_i^2$ equals the chance of picking two identical genes; $1 - t$ is the chance of picking two different genes.

If the population is in Hardy–Weinberg equilibrium, the heterozygosity equals the proportion of heterozygous individuals. But *H* is a more general definition of genetic diversity than the proportion of heterozygotes. The chance that two random genes differ measures genetic variation in all populations, whether or not they are in Hardy–Weinberg equilibrium. For example, *H* = 50% in a population consisting of half *AA* and half *aa* individuals (with no heterozygotes).

The term "heterozygosity" is meaningful for a diploid population.

However, the same measure of genetic diversity can be used for non-diploid genes, such as the genes in mitochondria and chloroplasts. It can also be used for bacterial populations. The word "heterozygosity" can sound rather odd for non-diploid gene loci, and population geneticists often call *H* "gene diversity."

The classic population genetic theory of diversity has been worked out in terms of heterozygosity at one locus. When talking about the theory, we usually refer to heterozygosity (*H*). However, most modern measurements of genetic diversity are at the DNA level. At this level, much the same index of diversity is referred to as "nucleotide diversity" and is symbolized by *π*.

Intuitively, the meaning of nucleotide diversity is as follows. Imagine picking out a stretch of DNA from two DNA molecules drawn at random from a population. Count the number of nucleotide differences between the two DNA stretches. Then divide by the length of the stretch. The result is *π*. *π* is the average number of nucleotide differences per site between a pair of DNA sequences drawn at random from a population. Here is a concrete example. Suppose a simple population has four DNA molecules. A comparable region of those four has the following set of sequences: (1) TTTTAGCC, (2) TTTTAACC, (3) TTTAAGC, and (4) TTTAGGC. We first count the number of differences between all possible pairs. Pair 1–2 has 1 difference, 1–3 has 2, 1–4 has 1, 2–3 has 1, 2–4 has 2, and 3–4 has 1. The average number of differences for all six pairwise comparisons is $(1 + 2 + 1 + 1 + 2 + 1)/6 = 1.33$. *π* is calculated per site, so we divide the average number of differences by the total sequence length (8). $\pi = 1.33/8 = 0.0166$. More formally,

$$\pi = \sum p_i p_j \pi_{ij}$$

where p_i and p_j are the frequencies of the *i*th and *j*th DNA sequence, and π_{ij} is the number of pairwise differences per site between sequences *i* and *j*. Some figures for *H* and *π* in real populations are given in Section 7.2 (p. 164).

The increase in homozygosity under drift is due to inbreeding

copies of the same gene may combine in a single individual. In the hermaphrodite, it happens obviously with self-fertilization. But if there are two sexes, a gene in the grandparental generation can appear as a homozygote, in two copies, in the grandchild generation. The process, by which a gene in a single copy in one individual combines in two copies in an offspring, is *inbreeding*. Inbreeding can happen in any breeding system with a small population, and becomes more likely the smaller the population. However, the general point in this section can be expressed without referring to inbreeding. With random sampling, two copies of the same gene may make it into an offspring in a future generation. Random sampling has then produced a homozygote. Genetic drift tends to increase homozygosity, and the rate of this increase can be exactly expressed by equations 6.2 and 6.3.

6.6 A calculable amount of polymorphism will exist in a population because of neutral mutation

Genetic variation for neutral genes is determined by a balance between drift and mutation

So far, it might appear that the theory of neutral drift predicts that populations should be completely homozygous. However, new variation will be contributed by mutation and the equilibrial level of polymorphism (or heterozygosity) will actually be a balance between its elimination by drift and its creation by mutation. We can now work out what that equilibrium is. The *neutral* mutation rate is equal to u per gene per generation. (u, as before, is the rate at which selectively neutral mutations arise, not the total mutation rate.) To find out the equilibrial heterozygosity under drift and mutation, we have to modify equation 6.2 to account for mutation. If an individual was born a homozygote, and if neither gene has mutated, it stays a homozygote and all its gametes will have the same gene. (We ignore the possibility that mutation produces a homozygote, for example by a heterozygote Aa mutating to a homozygous AA. We are assuming that mutations produce new genes.) In order for a homozygote to produce all its gametes with the same gene, neither of its genes must have mutated. If either of them has mutated, the frequency of homozygotes will decrease. The chance that a gene has not mutated equals $(1 - u)$ and the chance that neither of an individual's genes has mutated equals $(1 - u)^2$.

Now we can simply modify the recurrence relation derived above. The frequency of homozygotes will be as before, but multiplied by the probability that they have not mutated to heterozygotes:

$$f' = \left[\frac{1}{2N} + \left(1 - \frac{1}{2N} \right) f \right] (1 - u)^2 \tag{6.4}$$

Homozygosity (f) will now not increase to one. It will converge to an equilibrial value. The equilibrium is between the increase in homozygosity due to drift, and its decrease by mutation. We can find the equilibrium value of f from $f^* = f = f'$. f^* indicates a value of f that is stable in successive generations ($f' = f$). Substituting $f^* = f' = f$ in the equation gives (after a minor manipulation):

$$f^* = \frac{(1 - u)^2}{2N - (2N - 1)(1 - u)^2} \tag{6.5}$$

The equation simplifies if we ignore terms in u^2, which will be relatively unimportant because the neutral mutation rate is low. Then

$$f^* = \frac{1}{4Nu + 1} \tag{6.6}$$

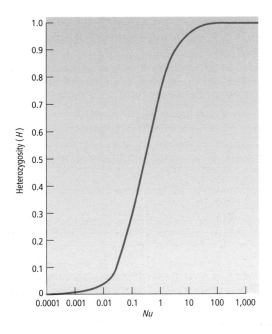

Figure 6.6
The theoretical relationship between the degree of heterozygosity and the parameter Nu (the product of the population size and neutral mutation rate).

The equilibrial heterozygosity ($H^\star = 1 - f^\star$) is:

$$H^\star = \frac{4Nu}{4Nu + 1} \qquad (6.7)$$

This is an important result. It gives the degree of heterozygosity that should exist for a balance between the drift to homozygosity and new neutral mutation. The expected heterozygosity depends on the neutral mutation rate and the population size (Figure 6.6). As the march to homozygosity is more rapid if the population size is smaller, it makes sense that the expected heterozygosity is lower if N is smaller. Heterozygosity is also lower if the mutation rate is lower, as we would expect. In sum, the population will be less genetically variable for neutral alleles when population sizes are smaller and the mutation rates lower.

6.7 Population size and effective population size

What is "population size"? We have seen that N determines the effect of genetic drift on gene frequencies. But what exactly is N? In an ecological sense, N can be measured by counting, such as the number of adults in a locality. However, for the theory of population genetics with small populations, the estimate obtained by ecological counting is only a crude approximation of the "population size," N, implied by the equations. What matters is the chance that two copies of a gene will be sampled as the next generation is produced, and this is affected by the breeding structure of the population. A population of size N contains $2N$ genes at a locus. The correct interpretation of N for the theoretical equations is that N has been correctly measured when the chance of drawing two copies of the same gene is $1/(2N)$.

Effective population size can differ from observed population size

If we draw two genes from a population at a locality, we may be more likely for various reasons to get two copies of the same gene than would be implied by the naive ecological measure of population size. Population geneticists therefore often write N_e (for "effective" population size) in the equations, rather than N. In practice, effective population sizes are usually lower than ecologically observed population sizes. The relation between N_e, the effective population size implied by the equations, and the observed population size N can be complex. A number of factors are known to influence effective population size.

1. *Sex ratio.* If one sex is rarer, the population size of the rarer sex will dominate the changes in gene frequencies. It is much more likely that identical genes will be drawn from the rarer sex, because fewer individuals are contributing genes to the next generation. Sewall Wright proved in 1932 that in this case:

$$N_e = \frac{4N_m \cdot N_f}{N_m + N_f} \tag{6.8}$$

Where N_m = number of males, and N_f = number of females in the population.

2. *Population fluctuations.* If population size fluctuates, homozygosity will increase more rapidly while the population goes through a "bottleneck" of small size. N_e is disproportionately influenced by N during the bottleneck, and a formula can be derived for N_e in terms of the harmonic mean of N.

3. *Small breeding groups.* If most breeding takes place within small groups, then the effective population size will differ from the total population size (made up of all the small breeding groups put together). N_e can be smaller or larger than N, depending on whether we look at the effective size of the local populations, or of all the local populations together. It also depends on the extinction rates of goups, and the migration rates between groups. Several models of population subdivision have been used to derive exact expressions for N_e.

4. *Variable fertility.* If the number of successful gametes varies between individuals (as it often does among males when sexual selection is operating, see Chapter 12), the more fertile individuals will accelerate the march to homozygosity. Again, the chance that copies of the same gene will combine in the same individual in the production of the next generation is increased and the effective population size is decreased relative to the total number of adults. Wright showed that if k is the average number of gametes produced by a member of the population and σ_k^2 is the variance of k (see Box 9.1, p. 233, for the definition of variance), then:

$$N_e = \frac{4N - 2}{\sigma_k^2 + 2} \tag{6.9}$$

For $N_e < N$, the variance of k has to be greater than random. If k varies randomly, as a Poisson process, $\sigma_k^2 = k = 2$ and $N_e \approx N$.

These are all quite technical points. The N_e in the equations for neutral evolution is an exactly defined quantity, but it is difficult to measure in practice. It is usually less than the observed number of adults, N. $N_e = N$ when the population mates randomly, is constant in size, has an equal sex ratio, and has approximately Poisson variance in fertility. Natural deviations from these conditions produce $N_e < N$. How much smaller N_e is than N is difficult to measure, though it is possible to make estimates by the formulae we have seen. Other things being equal, species with more subdivided and inbred population structures have a lower N_e than more panmictic species.

Summary

1 In a small population, random sampling of gametes to produce the next generation can change the gene frequency. These random changes are called genetic drift.

2 Genetic drift has a larger effect on gene frequencies if the population size is small than if it is large.

3 If a small population colonizes a new area, it is likely to carry all the ancestral population's genes; but the gene frequencies may be unrepresentative.

4 One gene can be substituted for another by random drift. The rate of neutral substitution is equal to the rate at which neutral mutations arise.

5 In a small population, in the absence of mutation, one allele will eventually be fixed at a locus. The population will eventually become homozygous. The Hardy–Weinberg equilibrium does not apply to small populations. The effect of drift is to reduce the amount of variability in the population.

6 The amount of neutral genetic variability in a population will be a balance between its loss by drift and its creation by new mutation.

7 The "effective" size of a population, which is the population size assumed in the theory of population genetics for small populations, should be distinguished from the size of a population that an ecologist might measure in nature. Effective population sizes are usually smaller than observed population sizes.

Further reading

Population genetics texts, such as those of Crow (1986), Hartl & Clark (1997), Gillespie (1998), or Hedrick (2000), and molecular evolution texts such as Page & Holmes (1998), Graur & Li (2000), and Li (1997), explain the theory of population genetics for small populations. Crow & Kimura (1970) is a classic account of the mathematical theory. Lewontin (1974) and Kimura (1983) also explain much of the material. Wright (1968) is more advanced. Beatty (1992) explains the history of ideas, including Wright's, about random drift. Kimura (1983) also contains a clear account of the parts of the theory most relevant to his neutral theory and discusses the meaning of effective population size. For the medical examples of founder events in humans, see Dean (1972) and Hayden (1981).

Study and review questions

1 A population of 100 individuals contains 100 *A* genes and 100 *a* genes. If there is no mutation and the three genotypes are selectively neutral, what would you expect the genotype and gene frequencies to be a long time, say 10,000 generations, in the future?

2 Review: (a) the meaning of "random" in random sampling, and the reason why drift is more powerful in smaller populations, and (b) the argument why all the genes at any locus (such as the insulin locus) in the human population are now descended from one gene in an ancestral population some time in the past.

3 What is the heterozygosity (*H*) of the following populations:

4 If the neutral mutation rate is 10^{-8} at a locus, what is the rate of neutral evolution at that locus if the population size is: (a) 100 individuals, or (b) 1,000 individuals?

5 What is the probability in a population of size *N* that a gene will combine (a) with another copy of itself to produce a new individual, and (b) with a copy of another gene?

6 Try to manipulate equation 6.2 into 6.3 and equation 6.6 into 6.7.

Population	Genotypes			H
	AA	Aa	aa	
1	25	50	25	
2	50	0	50	
3	0	50	50	
4	0	0	100	

7 Natural Selection and Random Drift in Molecular Evolution

This chapter discusses the relative importance of two processes in driving molecular evolution: random drift and natural selection. We begin by looking at what it means for drift to be a general explanation for molecular evolution. We then go on to some features of molecular evolution and, in particular, its relatively constant rate (the "molecular clock"). We see how certain details of molecular evolution have led to the development of the "nearly neutral" theory. We then look at the relation between functional constraint on molecules and their rate of evolution. Evolution in the non-coding parts of DNA, and for synonymous changes within genes, is probably mainly by drift. The relative contributions of selection and drift to non-synonymous (amino acid altering) changes are less clear. Natural selection can leave its signature in the statistical properties of DNA sequences, and the modern genomic era of biology has made it possible to study selection and drift in new ways. The chapter finishes by looking at four of these.

7.1   Random drift and natural selection can both hypothetically explain molecular evolution

Molecular evolution is studied in substitutions between species and polymorphisms within species

Evolution, at the molecular level, is observable as nucleotide (or base) changes in the DNA and amino acid changes in proteins. The word *substitution* is often used to refer to an evolutionary change. In particular, a gene (or a nucleotide) substitution means that one form of a gene (or a nucleotide) increases in frequency from being rare in the population to being common. Evolutionary substitutions are studied by comparing different species. If one species has nucleotide A at a certain site and another species has nucleotide G, then at least one substitution must have occurred in the evolutionary lineage connecting the two species. Molecular evolution is also studied by looking at polymorphisms within a species. A polymorphism exists if, for example, some individuals of a species have nucleotide A at a certain site while other individuals have G. A complete substitution has not occurred, because both A and G are in fairly high frequency, but some process must have driven up the frequency of one or both nucleotides in the past.

Polymorphism within a species, and evolutionary change between species, can be explained by two processes: natural selection and drift. This chapter will be looking at the contributions of drift and selection in molecular evolution. The subject hardly existed before the 1960s. Then gel electrophoresis (Section 4.5, p. 83) started to be used to study polymorphism, and the amino acid sequences of some proteins (such as cytochrome *c* and hemoglobin) became available for several species. The early evidence led Kimura (1968) and King & Jukes (1969) to suggest what Kimura called the *neutral theory of molecular evolution*. Motoo Kimura (who lived from 1924 to 1994) was a Japanese geneticist, and it was particularly him and his followers who promoted the neutral theory in the two decades after those original publications in 1968 and 1969.

Molecular evolution may be driven by selection or drift

The neutral theory does not suggest that random drift explains all evolutionary change. Natural selection is still needed to explain adaptation. It is, however, possible that the adaptations we observe in organisms required only a small proportion of all the evolutionary changes that have actually taken place in the DNA. The neutral theory states that evolution at the level of DNA and proteins, but not of adaptation, is dominated by random processes; most evolution at the molecular level would then be non-adaptive. We can contrast the neutral theory with its opposite: the idea that almost all molecular evolution has been driven by natural selection.

The difference between the two ideas can be understood in terms of the frequency distribution for the selection coefficients of mutations, or genetic variants. (It does not matter here whether we talk about new mutations or the set of genetic variants existing in a population at a genetic locus. "Genetic variant" could be substituted for "mutation" throughout this paragraph.) Given a mutation of a certain selection coefficient, the theory of random drift or selection (as described in Chapters 5 and 6) applies in a mathematically automatic way. If the selection coefficient is positive, the mutation increases in frequency; if it is negative, it is eliminated; if it is zero, the gene frequencies drift.[1]

---

[1]   This chapter uses a slightly different notation for selection coefficients from Chapter 5. In Chapter 5, the genotype with the highest fitness was given a fitness of 1 and the other genotypes were given fitnesses like $(1 - s)$. Here we shall be interested in whether one form of a molecule has a higher, lower, or equal, fitness with another form, and it will be more convenient to talk about selection coefficients that are +, 0, or −. A +ve selection coefficient means natural selection favors the variant; −ve means it is selected against; 0 means it is neutral.

What frequency of advantageous, disadvantageous, and neutral mutations do we expect there to be? Consider the nucleotide sequence of a gene in a living organism. The gene codes for a reasonably well adapted protein: the protein is unlikely to be a dud if the organism containing it is alive. Now consider all the mutations that can be made in the gene. You could work down the gene, altering one nucleotide at a time, and ask for each change whether the new version was better, worse, or equally as good as the original gene. In a population of organisms in nature, mutations will be occurring and causing these kinds of change, in certain frequencies.

Many mutational changes will be for the worse, and will have negative selection coefficients. Adaptation is an unlikely state of nature, and a random change in an adapted protein is likely to be for the worse. The disagreement has been about the relative frequencies of the other two classes of mutations: the neutral and the selectively advantageous. If natural selection has produced most evolutionary change at the molecular level, many advantageous mutations must have occurred, but few neutral mutations. If neutral drift has produced most evolutionary change at the molecular level, the relative frequencies are the other way round. Figure 7.1 illustrates two extreme views, in which most molecular evolution will be driven by selection (Figure 7.1a) or by drift (Figure 7.1b). The difference between the two is in the relative heights of the graph in the 0 and + regions. The high frequency of mutations in the – region is common to the two. Kimura's original neutral theory of molecular evolution implied something like Figure 7.1b.

At this point, it is worth pointing out two things that Kimura was not saying, and his modern followers are still not saying. The neutral theory says that the majority of molecular evolution is driven by neutral drift — but that does not mean the majority of *mutations* are neutral. Figure 7.1c illustrates what Kimura (1983) called "pan-neutralism," in contrast with his own ideas. Pan-neutralism mean that almost all mutations are neutral. Then, almost all evolution would be by neutral drift, just as in

Two extreme views — selectionist and neutralist — can be distinguished

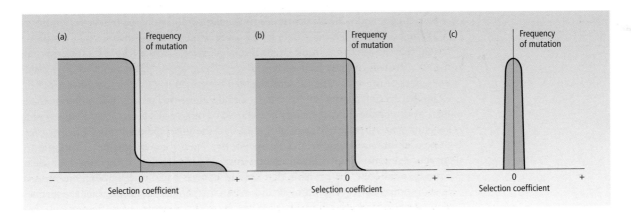

Figure 7.1
The neutral and selectionist theories postulate different frequency distributions for the rates of mutation with various selection coefficients. (a) According to the selectionists, exactly neutral mutations are rare and there are enough favorable mutations to account for all molecular evolution; whereas (b) neutralists believe there are more neutral, and hardly any selectively favored, mutations. (c) The theory of pan-neutralism, according to which all mutations are selectively neutral.

Neutralists do not claim that all mutations are neutral . . .

the neutral theory. But if most evolution is by neutral drift, that does not mean most mutations are neutral. Evolution is not the same as mutation. In Figure 7.1b, all the mutations that may end up contributing to evolutionary change are neutral, but the majority of mutations are disadvantageous and will be selected against. Disadvantageous mutations disappear from the population before they have any chance to show up as evolution. The neutral theory therefore does not rule out natural selection. It simply has a different use for it than has the selectionist theory of molecular evolution. The selectionist theory uses natural selection to explain both why mutations are lost (when they are disadvantageous) and are fixed (when they are advantageous). The neutral theory uses selection only to explain why disadvantageous mutation are lost; it uses drift to explain how new mutations are fixed.

Pan-neutralism is almost certainly false. We have strong evidence against it. For example, pan-neutralism has difficulty in explaining why different genes, and different parts of genes, evolve at different rates (Section 7.6 below). Nor is it theoretically plausible. It is absurd to suggests that hardly any mutations are disadvantageous. Organisms, including their molecules, are adapted to their environments; we only need reflect on the efficiency of digestive enzymes — or any other biological molecule — in supporting life to realize that. If molecules are adaptive, many (or most) changes in them will be for the worse.

. . . or that neutral drift explains adaptation

The other thing that the neutral theory of molecular evolution does not claim is that *all* molecular evolution is driven by neutral drift. It says that most molecular evolution is by neutral drift. An important fraction of molecular evolution is almost certainly driven by selection: the fraction of molecular evolution that occurs during the evolution of adaptations.

Biological molecules are well adapted for their functions. Hemoglobin carries oxygen; enzymes catalyze biochemical reactions. These adaptive functions did not evolve by accident. Random drift will not have contributed much, if at all, to adaptive evolution. The evolutionary events that gave rise to the adaptive functions of the modern molecules of life were almost all powered by selection.

Selectionist and neutral theories of molecular evolution agree that selection drives adaptive evolution. The disagreement is over what fraction of molecular evolution is adaptive. To see the point, imagine a gene of about 1,000 nucleotides (corresponding to a protein of about 300 amino acids). There are $4^{1,000}$ or about 10^{600} possible sequences of the gene. The protein encoded by the gene will have some function, for example carrying oxygen in the blood (actually done by hemoglobin, which is made up of four polypeptides of slightly less than 150 amino acids each). The neutral theory suggests that, of the 10^{600} possible molecules, the great majority would fail to carry oxygen at all, and many would do so poorly. Then there would be a minority, of maybe a few hundred different sequences, all very similar to one another, all of which would code for proteins that carried oxygen equally well. What we observe as evolution consists of shuffling round within this limited set of equivalent sequences. The selectionist alternative is that the few hundred variants are not equivalent, but that one works better in one environment, another in another environment, and so on. Evolution then consists of the substitution of one variant for another when the environment changes.

As the chapter unfolds, we shall see how the original neutral theory (illustrated in Figure 7.1b) has been modified in two ways. One is the development of the "nearly neutral" theory of molecular evolution. Kimura's original theory considered only purely neutral mutations, with a selection coefficient of zero. His modern followers also consider mutations with small positive or negative selection coefficients. Because drift is more powerful with small population sizes (Section 6.1, p. 138), these nearly neutral mutations are influenced more by drift in small populations and more by selection in large populations. The mutations become effectively neutral, or non-neutral, depending on population size.

The original neutral theory has been modified

Secondly, the original neutral theory made a global claim about all molecular evolution. The neutral theory suggested that almost all molecular evolution is driven by neutral drift. Now the theory has been refined. Some parts of the DNA appear to evolve by neutral drift, but the relative contributions of selection and drift in other parts of the DNA are less clear. The stark contrast between (a) and (b) in Figure 7.1 has been modified by 30 years of accumulated evidence.

The crucial difference between the selectionist and neutral theories of molecular evolution lies in the relative frequencies of neutral and selectively advantageous mutations. The direct way to test between them should simply be to measure the fitnesses of many genetic variants at a locus, and count the numbers with negative, neutral, or positive selection coefficients under certain environmental conditions. But the controversy has not been settled in this way. To measure the fitness of even one common genetic variant is a major research exercise, and to measure the fitnesses of many rare variants would be practically impossible.

In the first half of this chapter we shall look at three lines of less direct evidence that were originally used by Kimura, and King and Jukes, to argue for the importance of neutral drift in molecular evolution.

1. The absolute rate of molecular evolution and degree of polymorphism, both of which have been argued to be too high to be explained by natural selection.
2. The constancy of molecular evolution, which has been argued to be inconsistent with natural selection.
3. The observation that functionally less constrained parts of molecules evolve at a higher rate, which has been argued to be the opposite of what the theory of natural selection would predict.

Observation 1 is now of little influence. The molecular clock (observation 2) is not merely still influential, but has become the basis of a major research program in evolutionary biology. The relation between functional constraint and rate of evolution (observation 3) is also important. It has turned out that observation 3 can be studied more powerfully with DNA sequences, which have become increasingly available since the 1980s, than in protein sequences, which were used in the 1960s and 1970s.

In the second half of the chapter we shall look at some additional ways of testing between drift and selection that have become possible in the genomic era.

7.2 Rates of molecular evolution and amounts of genetic variation can be measured

Rates of evolution are estimated from the amino acid sequence of a protein, or nucleotide sequence of a region of DNA, in two or more species. For any two species,

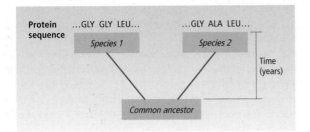

Rates of molecular evolution can be measured . . .

Figure 7.2

Imagine that some region of a protein has the illustrated sequences in two species. The evolutionary change has happened somewhere within the lineage connecting the two species via their common ancestor. The simplest interpretation is that either an alanine has been substituted for a glycine in the lineage leading to species 2, or a glycine for an alanine in the lineage to species 1. Either way, the amount of evolution is one change, and it has taken place in twice the time from the species back to their common ancestor; or, one change in $2t$ years. In practice, particularly with DNA data, the method of maximum likelihood is used to correct for multiple hits and the possibility that the ancestor had none of the states present in the modern species (Section 15.9.3, p. 442).

the approximate age of their common ancestor can be estimated from the fossil record. The rate of protein evolution can then be calculated as the number of amino acid differences between the protein of the two species divided by two times the time to their common ancestor (Figure 7.2). For example, if the species are humans and mice, their common ancestor probably lived about 80 million years ago. If we look at the sequence of a 100 amino acid protein in the two species and it differs at 16 sites, then the rate of evolution is estimated at $16/(100 \times 160 \times 10^6) \approx 1 \times 10^{-9}$ per amino acid site per year.

Much the same calculation can be made per nucleotide site for the rate of DNA evolution. But with DNA, a correction has to be made for "multiple hits." For instance, suppose that species 1 has nucleotide A at a certain site and species 2 has G at the equivalent site. Using the reasoning of Figure 7.2, we could deduce that one change has taken place in $2t$ years. However, more than one change may have occurred. The common ancestor might have had nucleotide A (the same reasoning applies if it had G). In the lineage leading to species 2, A changed to G. That requires at least one change, but there may have been more. Up that lineage, A may first have evolved to T and then T to G. In the lineage leading to species 1, A may have remained unchanged all the time. Alternatively, A may have evolved into C and then C evolved to A again. We see only one difference between the A and G in the modern species 1 and 2, but more than one change may underlie it.

The problem — that more than one substitution may underlie one observed difference between two species — is the problem of multiple hits. The problem is particularly acute for DNA, because DNA has only four states: the four nucleotides A, C, G, and T. Multiple evolutionary changes can easily end up leading to the same state in two species. For amino acids in proteins, there are 20 states (the 20 main amino acids) and multiple changes are less likely to result in the same state in two species. In Section 15.9.3 (p. 442) we look at how to correct for multiple hits in DNA data. Analogous corrections can be made for protein data. In this chapter, we simply assume that the necessary corrections have been made in estimates of evolutionary rates.

Table 7.1 gives some examples of evolutionary rate estimates, based on comparisons between humans and mice. As can be seen, different proteins evolve at different rates. Ribonuclease evolves slowly, albumin rapidly. Section 7.6 looks at why different

Table 7.1

Rates of evolution for amino acid changes in proteins, and for nucleotide changes in DNA. Rates are expressed as inferred number of changes per 10^9 years for an average amino acid site in the protein, or an average nucleotide site in the gene. Calculated using data in Li (1997).

Gene	Rate of amino acid evolution	Rate of nucleotide evolution
Albumin	0.92	6.08
α-globin	0.56	4.92
β-globin	0.78	3.36
Immunoglobin V_H	1.1	5.87
Parathyroid hormone	1.0	4.57
Relaxin	2.59	8.98
Ribosomal S14 protein	0.02	2.18
Average (45 proteins and genes)	0.74	4.25

proteins evolve at different rates. Here we are just looking at the approximate figures. An approximate, memorable figure suggested by Table 7.1 is that amino acids are substituted at a rate of a bit less than one per billion years at each amino acid site in a protein.

. . . and levels of polymorphism, . . .   Another important figure is for the amount of genetic variation within a species at a particular time. The amount of variation can be described by two main indexes. One is the chance that two randomly drawn alleles differ at an average locus, or heterozygosity (H, see Box 6.3, p. 149); we previously met H as a property of one locus. H can also be measured for a number of loci, and then expressed as an average for all of them. The other measure is the percentage of polymorphic loci. If, say, 20 loci are studied by gel electrophoresis, and 16 show no variation and four have more than one band on the gel, then the percent polymorphism would be $4/20 \times 100 = 20\%$. Gel electrophoretic evidence suggests that about 10–20% of loci are polymorphic in species in nature (Table 7.2).

Genetic variation has been measured at the DNA level in fewer species, because it requires sequencing a stretch of DNA in many individuals within one species. *. . . or nucleotide diversity*   DNA diversity within a species is expressed as the "nucleotide diversity" (π), which is mathematically equivalent to heterozygosity. In humans, is about 0.001. Thus, two randomly picked human DNA molecules (including two within any one human body) differ at about one in a 1,000 sites. Human DNA may be less diverse than that of many other species (Box 13.2, p. 365). *Drosophila* DNA has a nucleotide diversity almost 10 times higher than human DNA.

Kimura (1968, 1983) thought that the rate of molecular evolution, and the amount of molecular variation, was too high for a process driven by natural selection. His arguments are now mainly of historic importance and are outlined in Box 7.1.

Box 7.1
Genetic Loads and Kimura's Original Case for the Neutral Theory

Two of Kimura's (1968, 1983) three original arguments for the neutral theory made use of a general concept called "genetic load." Genetic load is a property of a population and is defined as follows. The population will contain a number of genotypes, and each genotype has a certain fitness. We identify the genotype, of those present in the population, that has the highest fitness, and assign that genotype a relative fitness of one. All the other genotypes will have fitnesses of less than one. We also measure the average fitness of the whole population; it is just the fitness of each genotype multiplied by its frequency and is called mean fitness (Section 5.6, p. 105). Mean fitness is conventionally symbolized by $\bar{w}$. The general formula for genetic load (L) is then

$$L = 1 - \bar{w}$$

If all the individuals in the population have the optimal genotype, $\bar{w} = 1$ and the load is zero. If all but one have a genotype of zero fitness, $\bar{w} = 0$ and $L = 1$. Genetic load is a number between 0 and 1 and it measures the extent to which the average individual in a population is inferior to the best possible kind of individual, given the range of genotypes in the population. To be exact, the genetic load equals the relative chance that an average individual will die before reproducing because of the disadvantageous genes that it possesses.

Genetic load can exist for several reasons. Kimura's original argument considered *substitutional load* and *segregational load*. Substitutional load arises when natural selection is substituting one (superior) allele for another (inferior) allele. While the inferior allele exists in the population, mean fitness is lower than if all individuals had the superior allele. The substitutional load is mathematically equivalent to another concept, defined by Haldane (1957), and called the "cost of natural selection."

Kimura, following Haldane, suggested that the rate of evolution has an upper limit. A favorable mutation might arise; initially it is a single copy in the population. At a very theoretical extreme, the favorable mutation could rise to a frequency of 100% in the population in three generations. In the first two generations, all individuals lacking a copy of the favorable mutation would have to die without breeding (except one in the first generation to provide a mate for the mutant). In the third generation, all individuals lacking two copies of the favorable gene would

have to die without breeding. The mutation would then have spread to a frequency of 100%.

Such rapid evolution is unlikely for various reasons, but the reason discussed by Haldane and Kimura was that the population would be driven down to such a low level that it would go extinct. A real population is unlikely to persist if it is cut down to one breeding pair. Moreover, a population certainly could not persist if two such mutations arose at separate loci, because then even the individuals who survived because they had one of the mutations would die for want of the other. Everyone would be dead. More realistic evolution will proceed at a lower rate, because the population must continue to exist in reasonable numbers while natural selection substitutes superior alleles. Haldane (1957) suggested an upper limit on the rate of evolution of about one gene substitution per 300 generations.

Molecular evolution proceeds at a far higher rate than this. When Kimura (1968) first estimated the total rate of molecular evolution in an average mammal species he derived a figure of one substitution every two generations. However, he only had evidence from amino acids. We now know that the rate of synonymous change is even higher. The full rate of DNA evolution is more like eight substitutions per year, or one substitution every 1.5 months (Hughes 1999, p. 41). The rate of molecular evolution is clearly far higher than Haldane's estimated upper limit. Kimura concluded that most molecular evolution could not be driven by natural selection. Molecular evolution must be driven instead by random drift. Random drift creates no genetic load, because all the genotypes concerned have equal fitness.

The argument with segregational load is similar. Segregational load arises when a polymorphism exists, maintained by heterozygous advantage (Section 5.12, p. 123). (Segregational load may or may not exist with polymorphisms maintained by frequency-dependent selection, but the original arguments considered heterozygous advantage.) With heterozygous advantage, the fitnesses of the genotypes are:

Genotype	AA	Aa	aa
Fitness	$1 - s$	1	$1 - t$

The population has a genetic load because the population cannot consist purely of heterozygotes. Even if a population did temporarily consist only of heterozygotes, they would produce homozygotes by normal Mendelian segregation in the next generation. For one locus, heterozygote advantage is plausible. A few individuals die because they are homozygotes, but the population continues to exist.

However, initial surveys suggested that about 3,000 loci might be polymorphic in fruitflies. Suppose all 3,000 were maintained by heterozygous advantage. The chance that an individual would be heterozygous at all 3,000 is essentially zero. All individuals will be homozygous at many hundreds of loci. If each such locus lowers fitness by a few percent, every individual will be dead several times over. (In terms of the example of sickle cell anemia, it is as if everyone has some such condition at hundreds of their loci. You might survive one of them, but not all of them.) Kimura concluded that it was impossible for natural selection to maintain all the genetic variation observed at the molecular level. The genetic variation must be maintained by random drift, which explains polymorphism by a balance of drift and mutation (Section 6.6, p. 150). Neutral variation does not create a genetic load.

Kimura's argument retains its interest, but is now generally thought to be inconclusive, for two main reasons. One is that the upper limits on the rate of evolution, and on the tolerable level of genetic variation, can be raised if we allow for *soft selection*. Haldane and Kimura's calculation assumed *hard selection*. Hard selection means that natural selection adds to the amount of mortality, decreasing the population size. We can distinguish between "background" mortality, due to normal ecological processes (Section 4.1, p. 72), and "selective" mortality, due to the action of natural selection. Organisms produce many more offspring than can survive, and many die without reproducing. If a cod produces 5,000,000 eggs, on average 4,999,998 die before reproducing, because of the operation of various ecological mortality factors. Natural selection is hard if it reduces the number of survivors below two. Natural selection is soft if converts some of the background ecological mortality into selective mortality. Population size is not reduced if selection is soft.

As a concrete example, imagine the population size is limited by the number of breeding territories. Only 100 territories exist in an area, and non-owners soon die of starvation. The 100 territory owners produce 10 eggs each, making 1,000 eggs in all. Half the eggs die before growing up into adults, such that 500 adults compete for the 100 territories each generation (400 will fail — though the numbers might need adjusting if gender introduces complexities). Consider first extreme soft selection. A new advantageous genotype arises, which increases juvenile survival, perhaps by 20%. Once the genotype is fixed, 600 juveniles will survive to become adults. However, the same 100 territories exist and the reproductive output of the population will not be altered.

Compare that with hard selection. A new disease arises that is only caught by territory holders. A new genotype arises, making the birds resistant to the disease; most of the birds initially have a disease-susceptible genotype. Until the disease-resistant genotype is being substituted by natural selection, the reproductive output of the birds will decrease. The mortality caused by the disease is additional. It comes on top of the ecological winnowing down, caused by the limited supply of territories.

Substitutional load ultimately limits the rate of evolution whether selection is hard or soft, but the limit is much lower with hard selection. Much selection in fact is probably soft, and does not reduce the reproductive output of a population. Evolution can then proceed at a higher rate than that calculated by Kimura and Haldane.

The second counterargument is that natural selection can act jointly on many loci. In the argument above about heterozgous advantage, we assumed that each homozygous locus in an individual reduces fitness by a few percent. Natural selection may not work like that. An individual may be able to survive equally well with one, two, three, or 100 homozygous loci, and only after the number of homozygous loci goes over some threshold, such as 500, will that individual's fitness seriously decrease. Then, many more heterozygous loci can be maintained in the population than if each locus contributes its own mortality. A similar argument can be made for the rate of evolution. A distinction is being made here between multiplicative fitnesses, in which each locus contributes its own independent effect on the organism's fitness, and epistatic fitnesses, in which the effects of different loci are not independent. Section 8.8 (p. 206) looks at the distinction more. It also features in the arguments about sex in Section 12.2.2 (p. 323).

A third counterargument is that genetic variation can be maintained by frequency-dependent selection without creating a genetic load. (The sex ratio, which maintains the X and Y chromosomes, is an example: see Section 12.5, p. 337.) Thus, even if Kimura's argument rules out heterozygous advantage as the explanation of much genetic variation, it does not rule out all forms of natural selection.

These counterarguments have not been shown to be correct in fact. They are hypothetical arguments, and reduce the theoretical force of Kimura's case. Neutral theory, for this reason, is now usually supported by arguments other than genetic load. However, the arguments are still worth knowing. They have been historically influential and also still constantly crop up, in one form or another, in many areas of evolutionary biology. Moreover, Williams (1992) suggested that the whole problem had been swept under the rug rather than solved, and that biologists should be paying more attention to the problem of loads.

Further reading: Lewontin (1974), Kimura (1983), Williams (1992), Gillespie (1998).

Table 7.2

Amounts of variation in natural populations. Variation can be measured as percentages of polymorphic loci (*P*) and average percent heterozygosity per individual (*H*). Also given is the number of loci used to estimate *P* and *H*. For meaning of *H*, see Box 6.3 (p. 149). Modified from Nevo (1988).

Species	Number of loci	P (%)	H (%)
Phlox cuspidata	16	11	1.2
Liatris cylindracea	27	56	5.7
Limulus polyphemus	25	25	5.7
Balanus eburneus	14	67	6.7
Homarus americanus	28–42	18	3.8
Gryllus bimaculatus	25	58	6.3
Drosophila robusta	40	39	11
Bombus americanorum	12	0	0
Salmo gairdneri	23	15	3.7
Bufo americanus	14	26	11.6
Passer domesticus	15	33	9.8
Homo sapiens	71	28	6.7

7.3 Rates of molecular evolution are arguably too constant for a process controlled by natural selection

The rate of molecular evolution can be measured for any pair of species by the method shown in Figure 7.2. Each pair of species needs a figure for the number of molecular differences and the time to their common ancestor. We can plot the point defined by these two numbers for many pairs of species; Figure 7.3 is an example for α-hemoglobin. The striking property of the graph is that the points for the different species pairs fall on a straight line. Molecular evolution appears to have an approximately constant rate per unit time; it is therefore said to show a *molecular clock*. Evolutionary change at the molecular level ticks over at a roughly constant rate, and the amount of molecular change between two species measures how long ago they shared a common ancestor. (Molecular differences between species can be used to infer the time of events in the evolutionary past, as we shall see in Parts 4 and 5 of this text.)

Molecular evolution seems to show a molecular clock

A graph such as Figure 7.3 requires a knowledge of the time to the common ancestor for each species pair. These times are estimated from the fossil record and are uncertain (Chapter 18); the results are therefore not universally trusted. However, we can also test the constancy of molecular evolution by another method, which does not require absolute dates, and this other test also suggests that molecular evolution is fairly clock-like (Box 7.2). There is empirical controversy as to how constant the molecular clock is, but the statistical details are involved and we shall not enter into them here. We can reasonably conclude at present that the rate of molecular evolution is constant enough to require explanation.

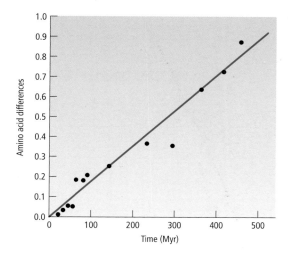

Figure 7.3

The rate of evolution of hemoglobin. Each point on the graph is for a pair of species, or groups of species, with the value for that pair being obtained by the method of Figure 7.2. Some of the points are for α-hemoglobin, others for β-hemoglobin. From Kimura (1983). Redrawn with permission of Cambridge University Press, © 1983.

Kimura argued that drift explains the molecular clock, whereas selection does not

What does a constant rate imply about whether molecular evolution is mainly driven by natural selection or neutral drift? Kimura reasoned that constant rates are more easily explained by neutral drift than selection. Neutral drift has the property of a random process and its rate will show the variability characteristic of a random process. Neutral mutations crop up at random intervals, but if they are observed over a sufficiently long time period the rate of change will appear to be approximately constant. Neutral drift will drive evolution at a fairly constant rate. Natural selection, Kimura argued, does not produce such constant change. Under selection, the rate of evolution is influenced by environmental change as well as the mutation rate; and it would require a surprisingly steady rate of environmental change, over hundreds of millions of years, in organisms as different as snails and mice and sharks and trees to produce the constant rate of change seen in Figure 7.3.

Moreover, if we look at characters, such as any adaptive morphological characters, that have undoubtedly evolved by natural selection, they do not seem to evolve at constant rates. Kimura (1983) discussed the evolution of the bird wing as an example. Before the wing evolved, there was a long period during which the vertebrate limb remained relatively constant (in the form of the tetrapod limb of amphibians and reptiles). Then came a shorter period when the wing originated and evolved. Finally, there was a long period of fine tuning a more or less finished wing form.

Morphological evolution is not clocklike

The wings of birds undoubtedly evolved by natural selection. The rate of change during wing evolution fluctuated between fast and slow. The rate of molecular evolution appears to be relatively constant, compared with morphological evolution. This observation is also Kimura's reason for confining the neutral theory to molecules, and not applying it to the gross phenotypes of organisms. Molecular evolution does appear to have a fairly constant rate, as would be expected for a random process. Morphological evolution has a different pattern, and is probably driven by the non-random process of selection.

Molecular evolution in "living fossils" provides a striking example both of the constant rate of molecular evolution and of the independence between molecular and

Box 7.2
The Relative Rate Test

The relative rate test is a method of testing whether a molecule (or, in principle, any other character) evolves at a constant rate in two independent lineages. It was first used by Sarich and Wilson in 1973. Suppose we know the sequence of a protein in three species, a, b, and c, and we also know the order of phylogenetic branching of the three species (Figure B7.1). We can now infer the amounts of change in the two lines from the common ancestor of a and b to the modern species (x and y in Figure B7.1). If the protein evolved at the same rate in the two lineages, the number of amino acid changes between the common ancestor and a (x) should equal the number of changes between the common ancestor and b (y); that is, $x = y$. x and y can be inferred by simple simultaneous equations. We know the differences between the protein sequences in a and b (k), b and c (l), and a and c (m). Thus

$$k = x + y$$
$$l = y + z$$
$$m = x + z$$

We have three equations with three unknowns and can solve for x, y, and z. We then test whether the rates were the same by seeing whether $x = y$. Notice that we do not need to know the absolute date (or the identity) of the common ancestors.

The relative rate test can only show that a molecule evolved at the same rate in the two lineages connecting the two modern species with their common ancestor. This does not prove that the molecule always has a constant rate; it does not, in other words, confirm the molecular clock. If identity of relative rate is shown for many pairs of species, with common ancestors of very different antiquities, that is suggestive of (and consistent with) a molecular clock, but it is not conclusive evidence. We can see why in a counterexample (Figure B7.2). Suppose that a molecule evolves at the same rate in all lineages at any one time, but that it has been gradually slowing down through evolutionary history. A pair of

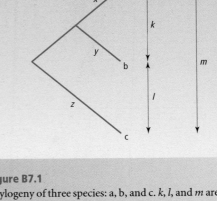

Figure B7.1
Phylogeny of three species: a, b, and c. k, l, and m are the observed number of amino acid differences between the three species. The amounts of evolution (x, y, z) in the three parts of the tree can be simply inferred, as the text explains.

species with a common ancestor 100 million years ago will then show rate constancy according to the relative rate test (because the molecule evolves at the same rate in all lineages at any one time); and any other species pair, for instance with a common ancestor 50 million years ago, will also show relative rate constancy. However, there is no molecular clock because the rate slows down through time. The relative rate test will not detect that the more recent species pair have a smaller absolute number of changes: absolute dates would be needed for that. The same point would apply if there were any trend in evolutionary rate with time, and it does not have to be directional. The molecule could speed up and slow down many times in evolution; but so long as the speeding up and slowing down apply to all lineages, the relative rate test will show equal rates of evolution in the two lineages. The relative rate test, therefore, cannot conclusively test the molecular clock hypothesis.

Figure B7.2
(a) The rate of evolution of a molecule has slowed down gradually through time; but the rate of evolution is always the same in all lineages at any one time. The molecule does not evolve like a clock (which would show up as a flat graph of rate against time). (b) Then, for any pair of species, with common ancestors at any time, the amount of change will be the same in both lineages.

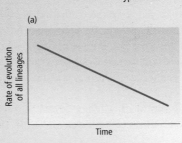

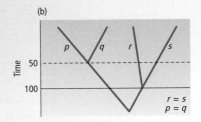

Table 7.3
Amino acid differences between the α- and β-hemoglobins for three species pairs. Adapted, by permission of the publisher, from Kimura (1983).

Species pairs	Number of amino acid differences
Human α vs human β	147
Carp α vs human β	149
Shark α vs shark β	150

Globin evolution in "living fossil" sharks illustrates the molecular clock

morphological evolution. The Port Jackson shark *Heterodontus portusjacksoni* is a living fossil — a species that closely resembles its fossil ancestors (some over 300 million years old). Its molecules have been evolving very differently from its morphology. Hemoglobin duplicated into α and β forms before the ancestor of mammals and sharks, at the beginning of the chordate radiation. We can count the amino acid differences between α- and β-hemoglobin as a measure of the rate of molecular evolution in the lineages leading to the modern species. Table 7.3 reveals that changes have accumulated in the Port Jackson shark lineage at the same rate as the human lineage. The rates of molecular evolution in the two lineages are roughly equal.

The constancy of molecular evolution in the shark and human lineages for the past 300 million years is in marked contrast with their rates of morphological evolution. The lineage leading to the modern Port Jackson shark has hardly had any change at all. But the lineage leading to humans has passed from an initial fish-like stage, through amphibian, reptilian, and several mammalian stages, before evolving into modern humans. Moreover, as Table 7.3 shows, human β-globin is as different from human α-globin as it is from carp α-globin. This is despite the fact that human α- and β-globin will have shared much more similar external selective pressures, as they have been locked in the same kind of organisms throughout evolution, than have human β-globin and carp α-globin.

The result suggests that the α- and β-globin molecules have been accumulating changes independently, at roughly constant rates, regardless of the external selective circumstances of the molecule. This in turn suggests that most of the evolutionary changes in the globin molecule have been neutral shifts among equivalent forms, of equal adaptive utility. While the rates of morphological change vary greatly among the various evolutionary lineages of vertebrates, the rates of molecular evolution all seem to have been more similar.

7.4 The molecular clock shows a generation time effect

The molecular clock seems to support the neutral theory of molecular evolution. However, when we examine the evidence in more detail, the support becomes less clear-cut. In particular, we should look at whether the clock runs relative to absolute time (in years) or generational time. Mice have shorter generations than elephants: but do molecules in mice show the same amount of evolutionary change per million years as equivalent molecules in elephants?

The prediction of the neutral theory depends on the mutational process. The rate of neutral evolution equals the neutral mutation rate (Section 6.3, p. 144). If species with short generation times have more mutations per year than species with long generation times, we expect species with short generations to evolve faster. We can distinguish three possibilities. One is that most mutations have external, environmental causes, such as UV-rays or chemical mutagens. Environmental mutagens probably hit organisms at an approximately constant rate through time. An organism that breeds after 1 year will have been hit by about 12 times as many mutagens as an organism that breeds after 1 month. The neutral theory then predicts the molecular clock will tick according to absolute time.

Secondly, at the opposite extreme, most mutations might occur during the disruptive events of meiosis. Meiosis happens only once per generation in all species, whether their generation times are long or short. The number of mutations per generation would then be similar in elephants and in shrews. The neutral theory predicts that the molecular clock should tick according to generational time.

Thirdly, mutations might mainly happen when DNA is replicated. The mutation rate would depend on the number of times DNA is replicated per generation, which equals the number of mitotic cell divisions in the cell lines that produce gametes. (The cell lines that produce the gametes are called the "germ line.") Species with long generation times do have more germ line cell divisions than species with short generation times, but the number is not proportional to generation time. For instance, a 30-year-old human female has 33 cell divisions behind each of her eggs, since the time when she was herself a zygote. A 30-year-old man has about 430 cell divisions behind each of his sperm. The average of the man and woman is about 230 cell divisions. A mature female rat has 29 cell divisions behind each egg, and a male rat about 58 cell divisions behind each sperm, giving an average of 43 cell divisions. The ratio of germ line cell divisions in a human to a rat is 230 : 43 or about five. The human generation length is about 30 years, the rat's about 1 year. The ratio of generation lengths in years is about 30, but humans have only about five times as many cell divisions in the germ line.

If mutations mainly happen at mitosis, the neutral theory predicts that the rate of evolution will be slower per year in species with longer generations than in species with shorter generation times, but not as slow as the ratio of their generation times (expressed in years) would predict.

For much of the twentieth century, mutations were thought mainly to have environmental causes. This belief followed from the discovery in the 1920s that X-rays and certain chemicals could cause mutations. But by the late twentieth century it had been established that most mutations are internal copying errors during DNA replication rather than externally caused. Thus, the third possibility is the most realistic. The neutral theory predicts there should be a generation time effect in the molecular clock.

Now let us turn to the evidence. What kind of time do real molecular clocks keep? For proteins, an important early paper by Wilson *et al.* (1977) strongly suggested that the clock runs relative to absolute time for protein evolution. Figure 7.4 shows their method. They picked a number of pairs of species. In each pair, one species had a short

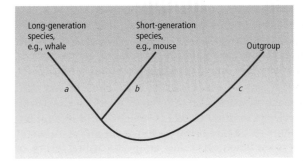

Figure 7.4

Wilson *et al.*'s (1977) method to test for a generation time effect on the rate of protein evolution. *a*, *b*, and *c* are the numbers of evolutionary changes in the three segments of the tree; they are estimated from the pairwise molecular differences between the species using the method of Box 7.2. The "outgroup" can be any species known to have a more distant common ancestor with the pair of species being compared. The evidence suggests that $a \approx b$ for many molecules and species pairs, whereas *a* would be less than *b* if generation time influenced evolutionary rate.

generation time and the other had a long generation time. Wilson *et al.* used a relative rate test (Box 7.2), and found that the amount of change was similar in the two lineages. The result now looked awkward for the neutral theory. At the time, a neutralist could easily argue that mutations occur at a probabilistically constant rate in absolute time, and the result was as expected.

When DNA evidence became available, it showed a different picture, at least for synonymous changes. (Synonymous changes are nucleotide changes that do not alter the amino acid. Nucleotide changes that do alter the amino acid are called non-synonymous. Synonymous changes are possible because of the redundancy in the genetic code — Section 2.5, p. 28.) Rodents, such as mice and rats, have shorter generation times than primates and artiodactyls (such as cows). For synonymous substitutions, evolution is faster in rodents than in artiodactyls, and faster in artiodactyls than in primates (Table 7.4). Synonymous substitutions occur faster in species with shorter generation times.

A generation time effect is seen in synonymous evolution . . .

Table 7.4

Rates of evolution in silent base sites are faster in groups with shorter generation times. There are estimates for various pairs of species, and each estimate is an average for a number of proteins; the number of sites is the total number of base sites (for all proteins) that have been used to estimate the rate. The divergence times, which are in millions of years, are uncertain; a range of estimates (in parentheses) have been made. Modified from Li *et al.* (1987).

Species pairs	Number of proteins	Number of sites	Divergence	Rate ($\times 10^{-9}$ years)	Generation time
Primates					
Man vs chimp	7	921	7 (5–10)	1.3 (0.9–1.9)	
Man vs orang-utan	4	616	12 (10–16)	2 (1.5–2.4)	} Long
Man vs OW monkey	8	998	25 (20–30)	2.2 (1.8–2.8)	
Artiodactyls					
Cow vs goat	3	297	17 (12–25)	4.2 (2.9–6)	} Medium
Cow/sheep vs goat	3	1,027	55 (45–65)	3.5 (3.0–4.3)	
Rodents					
Mouse vs rat	24	3,886	15 (10–30)	7.9 (3.9–11.8)	} Short

The DNA evidence for non-synonymous sites is more ambiguous. Some studies have borne out Wilson *et al.*'s finding, that the generation time effect is either absent or reduced in synonymous sites. Other studies have found that generation time influences the rate of evolution in non-synonymous sites much as in synonymous sites. Generation time may influence the rate of non-synonymous evolution in some genes, or some lineages, but not others.

The factual picture that has emerged is that DNA evolution is influenced by generation times for synonymous sites. For non-synonymous sites, where a substitution alters the amino acid, the generation time effect is less clear. Synonymous evolution fits the neutral theory. Non-synonymous evolution either does not fit the neutral theory, or does not fit it so well as synonymous evolution.

7.5 The nearly neutral theory

7.5.1 *The "purely" neutral theory faces several empirical problems*

The different effects of generation time in the molecular clocks for synonymous and non-synonymous evolution is one of several factual difficulties that had emerged in the neutral theory by the late 1980s. A related problem is that the molecular clock is not constant enough to fit the neutral theory. Molecular evolution does appear to be relatively constant. The exact degree of constancy in the rate of evolution is difficult to measure, for various statistical reasons, but by the time Gillespie (1991) wrote, many authors were claiming that the rate of molecular evolution is more erratic, or more episodic, than the neutral theory predicts. The molecular clock is not quite clock-like enough. One explanation might be the generation time effect that we have just looked at. If generation times fluctuate over evolutionary time, so will the rate of neutral evolution. Alternatively, some authors doubt whether generation times influence rates of evolution, and for them some other explanation is needed for inconstancies in the molecular clock.

A further problem emerged in the amounts of genetic variation. The neutral theory predicts a certain level of genetic variation, which can be expressed as heterozygosity. The heterozygosity is predicted to increase with population size (Figure 6.6, p. 151). Fruitflies, with large N, should have more genetic variation than horses, with small N. In fact it turned out that levels of heterozygosity are rather constant in all species, independent of N (Figure 7.5).

In all, the neutral theory put forward by Kimura (1968, 1983) seemed to have problems with several points:

1. The stronger influence of generation times on the rate of synonymous evolution than the rate of non-synonymous evolution.
2. The molecular clock, which is not constant enough.
3. Levels of heterozygosity, which are too constant between species and too low in species with large population sizes.
4. Observed levels of genetic variation and of evolutionary rates, which are not related in the predicted way.

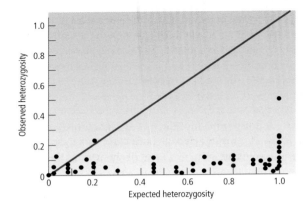

Figure 7.5
Observed levels of genetic variation (measured as heterozygosities) are too constant between different species, with different population sizes, than the neutral theory predicts. Each point gives the observed heterozygosity (y-axis) for a species (total 77 species), plotted against the "expected" heterozygosity from estimates of the population size and generation length of the species and assuming a neutral mutation rate of 10^{-7} per generation. Species with large population sizes appear to have too little genetic variation, relative to the neutral theory's prediction. Redrawn, by permission of the publisher, from Gillespie (1991).

Yet another problem for the neutral theory appeared in the "McDonald–Kreitman test," which we look at later in Section 7.8.3.

7.5.2 The nearly neutral theory of molecular evolution posits a class of nearly neutral mutations

In response to the factual difficulties we have just looked at, Ohta developed a modified version of the neutral theory. The modified version — the nearly neutral theory — grew in popularity until the 1990s. It is now a widely (though not universally) supported explanation for much of molecular evolution.

The nearly neutral theory invokes an effect for population size

Kimura's original, "purely" neutral theory explained molecular evolution by exactly neutral mutations. For exactly neutral mutations, we can ignore population size. For purely neutral mutations, the rate of evolution equals the neutral mutation rate. Population size cancels out of the equation (Section 6.3, p. 144). Population sizes are difficult to measure, and it is a great advantage if we can ignore it. However, the purely neutral theory appears not to fit all the facts. The nearly neutral theory can explain a greater range of facts, by bringing population size back into the theory.

Population size only cancels out for purely neutral mutations. For a nearly neutral mutation, the relative power of drift and selection depends on population size. Nearly neutral mutations behave as neutral mutations in small populations, and their fate is determined by random drift. They behave as non-neutral mutations in large populations, and their fate is determined by selection. To see why, consider a slightly disadvantageous mutation — one with a very small selective disadvantage. If it were purely neutral, its chance of eventually being fixed would be $1/2N$. If it is slightly disadvantageous, its chance of being fixed by random drift is slightly less than $1/2N$. In a small population, of 100 or so, the mutation has a fairly high chance (slightly less than one in 200) of ultimately being fixed by drift. But in a large population, of a million or so, the chance of being fixed by drift is negligible (slightly less than one in 1,000,000). This is just to restate the fact that drift is more powerful in small populations (Section 6.1, p. 138).

A slightly advantageous mutation, with a selective advantage of s relative to the other allele at the locus, has some chance of being lost by random accidents even though it is advantageous. The mutation might provide an advantage in the adult stage, but if the individual who contains the mutation has an accident while young the mutation will be lost. The chance that a slightly advantageous mutation is fixed by selection can be calculated and it is roughly $2s$. The mutation has a $1 - 2s$ chance of being lost by random factors. Thus, if a mutation increases the fitness of an organism by 1%, the chance that the mutation is lost by accident is 98%. (Graur & Li (2000, p. 54) give a simple derivation of this classic result.)

The 98% chance of being lost by accident is for any one copy of a mutation that has a selective advantage of 1%. An advantageous mutation is more likely to be present in one unique copy in a small, than a large, population. In a small population, an advantageous mutation may arise once but then be lost by chance. In a large population, the same mutation may occur several times and be present in multiple copies. (We assume the same mutation rate per gene in small and large populations.) Any one copy of the mutation may be lost by chance, but there are so many copies that one of them is likely to survive and be fixed by selection.

> Nearly neutral evolution is controlled by drift in small, and by selection in large, populations

Evolution, therefore, is arguably dominated by drift in small populations and by selection in large populations. We can be more exact. For mutations in populations where:

$$\frac{1}{2N} > 2s$$

random drift is more important than selection in deciding that mutation's evolutionary fate. Therefore mutations that satisfy the inequality:

$$s < \frac{1}{4N} \quad \text{or} \quad 4Ns < 1$$

behave as effectively neutral even if they have a non-zero selection coefficient.

The inequalities are often expressed in the approximate form:

$$s < \frac{1}{N} \quad \text{or} \quad Ns < 1$$

These are not strictly speaking accurate, but the four can often be dropped because the arguments in this area are often inexact.

> Nearly neutral mutations can be exactly defined

A mutation that satisfies the inequality $4Ns < 1$ (or $Ns < 1$) is a *nearly neutral mutation*. The class of nearly neutral mutations includes purely neutral mutations ($s = 0$), together with mutations that have small non-zero selection coefficients. The conceptual interest of nearly neutral mutations is that they evolve by random drift rather than natural selection.

The number of mutations that satisfy the inequality will depend on the population size. If N is large, only mutations with small s will satisfy the inequality and behave as neutral. As N decreases, more and more mutations, with higher and higher s, will satisfy the inequality and be dragged into the effectively neutral zone. The realized rate of neutral mutation therefore goes up as population size goes down. The number of mutations per gene is unchanged as population size decreases, but the fraction of them that behave as neutral will be higher if N is lower.

We can now distinguish two random drift theories of molecular evolution. According to Kimura's original neutral theory, most molecular evolution occurs as one purely neutral mutation ($s = 0$) is substituted for another. For the rest of this chapter I shall call this the *purely neutral theory* or *Kimura's neutral theory*. It should be distinguished from the *nearly neutral theory*, according to which most molecular evolution occurs as one nearly neutral mutation ($4Ns < 1$) is substituted for another.

7.5.3 · *The nearly neutral theory can explain the observed facts better than the purely neutral theory*

The nearly neutral theory can explain the observations on . . .

How can the nearly neutral theory explain the observations that did not fit the purely neutral theory? We can start with the observation that genetic variation is much the same within species with large population sizes as in species with small population sizes. For purely neutral mutations, species with larger population sizes should have more genetic variation; but they do not in fact. However, now suppose that many mutations are nearly, rather than exactly, neutral. Moreover, suppose that most of these nearly neutral mutations are slightly disadvantageous rather than slightly advantageous. (The assumption is probably correct, because random mutations in a well adapted molecule are more likely to make it worse than better.)

In a species with large populations, natural selection is more powerful than drift. The slightly disadvantageous mutations will be eliminated and not contribute to the observed genetic variation in that species. In species with small populations, natural selection is weak relative to random drift. Slightly disadvantageous mutations will behave as effectively neutral mutations. Some of them may drift up in frequency, contributing to the observed genetic variation. Genetic variation will be lower than the purely neutral theory predicts when population size is large. This is what is observed in reality (Figure 7.5).

. . . genetic variation . . .

Now we turn to the molecular clock. The easier problem to deal with is the relative inconstancy of the clock: the rate of molecular evolution is not as constant as the purely neutral theory predicts. However, if there is a large class of nearly neutral mutations, the rate of evolution will fluctuate over time when population sizes go up and down. As population size decreases, more slightly disadvantageous mutations will become effectively neutral. They may be fixed by drift, and the rate of evolution will increase. When population size increases, the slightly disadvantageous mutations will be eliminated by selection and the rate of evolution will slow down. The nearly neutral theory therefore predicts a more erratic rate of evolution than the purely neutral theory.

. . . unclocklike evolutionary rates . . .

The second problem we saw was that the molecular clock is more influenced by generation time for synonymous than for non-synonymous changes. Ohta's key argument here is the relation between population size and generation length. Species with long generation times tend to have smaller population sizes than species with short generation times (this relation was shown empirically by Chao & Carr (1993)). Whales, for example, live in smaller populations than fruitflies (even if we ignore the effects of humans on the two life forms).

. . . generation time effects . . .

Mutations at synonymous sites are probably mainly neutral. In Ohta's account, the rate of evolution at synonymous sites is influenced by generation length simply because the mutational process is influenced by generation length. DNA is copied fewer times per year in human gonads than in rat gonads. But why should there be less of a generation length influence (or even no influence) on the rate of evolution at non-synonymous sites? We begin by assuming that many amino acid-changing mutations are slightly disadvantageous. In a species with a long generation length, such as a whale, we now have two factors to consider: (i) DNA is copied slowly *per year*, which reduces the mutation rate per year; and (ii) population sizes are small, which makes drift more powerful than selection. Slightly disadvantageous mutations are less likely to be eliminated by selection, and are more likely to be fixed by drift. Factor (i) slows the rate of evolution; factor (ii) speeds it up.

Fruitflies, by contrast, have large population sizes but short generation times. They have a larger supply of mutations, because they copy their DNA more per year. But their population sizes are large, making fewer of the non-synonymous mutations effectively neutral. In all, generation length has two opposing influences on the rate of evolution for sites where many mutations are nearly neutral. Ohta suggests that the two effects could approximately cancel out, and the rate of evolution per year would be much the same whatever the generation length. That is her explanation for the possible absence of a generation time effect on the rate of amino acid substitutions. She may be right, but critics such as Gillespie argue that the two influences are unlikely to cancel each other out exactly. Then, some generation length effect would still be expected on the nearly neutral theory.

By this stage, we are at the frontiers of research, both for the facts and the theories. The nearly neutral theory can in principle account for what is known about molecular evolution, but that is not to say it has been shown to explain molecular evolution. The main conceptual difference between the nearly neutral theory and Kimura's original, purely neutral theory is in the use of population size. Population size does not affect the rate of evolution for purely neutral mutations. But it does affect the rate of evolution for nearly neutral mutations. This gives the nearly neutral theory great flexibility, because a wide variety of facts can be accounted for by assuming an appropriate history of population sizes. But the use of population sizes also make the theory difficult to test, because population sizes are difficult (and historic population sizes impossible) to measure. Kimura's original purely neutral theory, by contrast, was much more testable because its predictions did not require us to know anything about population sizes.

In summary, Ohta modified the purely neutral theory by positing a class of nearly neutral mutations. The relative power of selection and drift on these mutations depends on population sizes. The nearly neutral theory, by plausible arguments about population size, can account for several observations that present problems for Kimura's purely neutral theory.

7.5.4 *The nearly neutral theory is conceptually closely related to the original, purely neutral theory*

The nearly neutral theory makes use of natural selection. In some circumstances (large population size), the theory draws on natural selection; in other circumstances (small population sizes), it does not. Nearly neutral theory might be thought to blur the distinction between "selectionist" and "neutralist" explanations of molecular evolution.

However, a fundamental distinction remains. For any evolutionary change, in which one version of a gene is substituted for another, we can ask whether the force driving that change was natural selection or random drift. In the nearly neutral theory, just as in the original neutral theory, the force driving molecular evolution is neutral drift. Natural selection against disadvantageous mutations has a subtler, more flexible form in the nearly neutral theory than in the purely neutral theory. Drift and selection combine in different ways in the two theories to explain the observed facts of molecular evolution. But a crucial similarity remains: both theories explain evolution by drift. Natural selection has only a negative role, acting against disadvantageous mutations. This contrasts with all "selectionist" theories of molecular evolution, in which molecular evolutionary change occurs because natural selection favors advantageous mutations.

7.6 Evolutionary rate and functional constraint

7.6.1 *More functionally constrained parts of proteins evolve at slower rates*

Insulin illustrates the effect of functional importance

A protein contains functionally more important regions (such as the active site of an enzyme) and less important regions. The rate of evolution in the functionally more important parts of proteins is usually slower. For example, insulin is formed from a proinsulin molecule by excising a central region (Figure 7.6). The central region is discarded, and its sequence is probably less crucial than that of the outlying parts which form the final insulin protein. The central part evolves six times more rapidly than the outlying parts. The same result has been found by comparing evolutionary rates in the active sites and in other regions of enzymes; the surface of a hemoglobin, for example, may be functionally less important than the heme pocket, which contains the active site. The evolutionary rate is about 10 times faster in the surface region (Table 7.5).

A similar tendency may underlie differences in the rates of evolution of whole genes, or proteins. In Table 7.1 we saw that some proteins evolve faster than others. One

Figure 7.6
The insulin molecule is made by snipping the center out of a larger proinsulin molecule. The rate of evolution in the central part, which is discarded, is higher than that of the functional extremities. From Kimura (1983). Redrawn with permission of Cambridge University Press, © 1983.

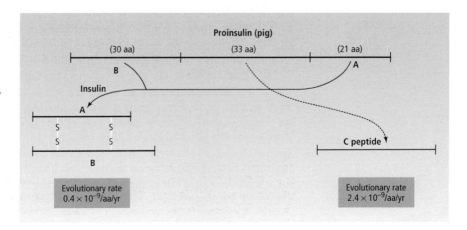

Table 7.5
Rates of evolution in the surface and heme pocket parts of the hemoglobin molecules. Rates are expressed as number of amino acid changes per 10^9 years. Reprinted, by permission of the publisher, from Kimura (1983). © 1983 Cambridge University Press.

Region	α-hemoglobin	β-hemoglobin
Surface	1.35	2.73
Heme pocket	0.165	0.236

Haemoglobin is another example . . .

generalization is that "house-keeping" genes, which control the basic metabolic processes of the cell, evolve slowly. The ribosomal protein, for instance, performs much the same function in the ribosome in almost all life forms. It evolves slowly. Other genes, such as the globins and immunoglobulins, have more specialized functions and only operate in specific cell types. They evolve more rapidly. The pattern is less clear-cut than the pattern we have just seen within a gene for insulin and for hemoglobin. However, the evidence does suggest that the degree of functional constraint is related to the rate of evolution for a large class of genes. A basic house-keeping gene may be more difficult to change during evolution than a gene with a more localized function.

The same relationships between functional constraint and evolutionary rate have been found for DNA as well as for proteins. Two properties of DNA sequences are particularly interesting: the relative rates of synonymous and non-synonymous changes in the DNA, and the evolutionary rate of pseudogenes.

. . . as is synonymous and non-synonymous evolution . . .

Synonymous base changes, which do not alter the amino acid, should be less constrained than non-synonymous changes. Kimura had predicted, before DNA sequences were available, that synonymous changes would occur at a higher rate. He was right: evolution in fact runs at about five times the rate in synonymous, as in non-synonymous, sites (Table 7.6).

. . . and pseudogenes

A pseudogene is a region of a DNA molecule that clearly resembles the sequence of a known gene, but differs from it in some crucial respect and probably has no function. Some pseudogenes, for example, cannot be transcribed, because they lack promotors and introns. (Promotors and introns are sequences that are needed for transcription, but are removed from the mRNA before it is translated into a protein. The pseudogene may have originated by reverse transcription of processed mRNA into the DNA.) Pseudogenes, once formed, are probably under little or no constraint and mutations will accumulate by neutral drift at the rate at which they arise. They will show pure neutral evolution in the "pan-neutral" (see Figure 7.1) sense that all mutations are neutral. The neutral theory predicts that pseudogenes should evolve rapidly. And they do — they evolve even more rapidly than synonymous sites in functional genes. The average rate of evolution at synonymous sites in Table 7.6 is 3.5 changes per 10^9 years. A comparable set of pseudogenes has evolved at about 3.9 changes per 10^9 years (Li 1997). A number of studies have shown the rate of pseudogene evolution to be about the same as, or somewhat higher than, the rate of synonymous evolution. (Box 7.3 describes how the rate of pseudogene evolution can be used to infer the total mutation rate in DNA.)

Table 7.6
Rates of evolution for synonymous and non-synonymous (that is, amino acid-changing) substitutions in various genes. Rates are expressed as inferred number of base changes per 10^9 years. These data were used to calculate the introductory figures in Table 7.1. Modified from Li (1997).

Gene	Non-synonymous rate	Synonymous rate
Albumin	0.92	5.16
α-globin	0.56	4.38
β-globin	0.78	2.58
Immunoglobin V_H	1.1	4.76
Parathyroid hormone	1.0	3.57
Relaxin	2.59	6.39
Ribosomal S14 protein	0.02	2.16
Average (45 genes)	0.74	3.51

Box 7.3
Using Pseudogenes to Infer the Total Mutation Rate

Nachman & Crowell (2000) estimated the rate of evolution in 18 pseudogenes that are present in both humans and chimpanzees. The average rate of evolution was about 2.5×10^{-8} per nucleotide site per generation. This can be multiplied by the diploid size of the human genome, about 6.6×10^9 nucleotides, to give the total number of mutations per human reproductive event. Nachman and Crosswell gave a range of estimates, and 175 mutations per generation is a representative figure.

The exact estimate for the total human mutation rate depends on what figure is used for the human generation length, the time since the common ancestor of humans and chimpanzees, and the ancestral population size. However, the results are usually somewhere in the 150–300 range for the human mutation rate. Numbers in this range were first deduced from pseudogene sequence data in the late 1980s (pseudogenes had not even been discovered before 1981), and the early estimates usually gave the figure as 200. When the number first appeared, it was much higher than had previously been supposed, but most human geneticists now accept that something like 200 mutations occur every time a new human being is reproduced. There may be some extra mutations, not estimated from the pseudogene sequence data — such as mutations in chromosome numbers or chromosome structure. But the 175–200 mutations estimated from pseudogenes probably includes most human mutations.

The inference assumes: (i) that the mutation rate in pseudogenes is representative of the genome as a whole; and (ii) all mutations in pseudogenes are neutral (that is, no selective constraints exist on them at all). The second assumption may not be valid (see Section 7.8.5 on codon bias).

7.6.2 *Both natural selection and neutral drift can explain the trend for proteins, but only drift is plausible for DNA*

The neutral explanation for the relation between evolutionary rate and functional constraint is as follows. In the active site of an enzyme, an amino acid change will probably change the enzyme's activity. Because the enzyme is relatively well adapted, the change is likely to be for the worse. It may well spoil the enzyme's function. In other parts of the molecule it may matter less what amino acid occupies a site, and a change is more likely to be neutral. The proportion of mutations that are neutral will be lower for the functionally constrained regions. Therefore, if the total mutation rate is similar throughout the enzyme, the number of neutral mutations will be lower in the active site. The evolutionary rate will then be lower too.

What is the selective explanation? The answer is usually expressed in terms of Fisher's (1930) model of adaptive evolution. We shall look at that model in Section 10.5.1 (p. 266). The model predicts that small, fine-tuning changes are more likely to improve the quality of adaptation than large changes. We could make an analogy with a radio. Biological molecules are fairly well adapted, but need to change from time to time to keep up with environmental change. This corresponds to a radio that is tuned to a station, but may wander out of tune from time to time as the signal changes. Most of the changes to the radio will be small, fine-tuning adjustments; a large jerk on the tuning knob would usually make things worse.

Mutations in a protein's active site will tend to have large effects; mutations in the outlying regions will have smaller effects. A change in amino acid in the active site is a virtual macromutation, which will almost always make things worse; natural selection will only rarely favor amino acid changes. But a similar change in the less functionally constrained parts may have more chance of being a small fine-tuning improvement which natural selection would favor. Selection will then more often favor changes in the less constrained regions of molecules, because there is more scope for fine tuning in those parts.

For amino acid changes in protein evolution, the neutralist and selectionist explanations are both possible. There was a controversy between the two in the 1970s and 1980s, and that controversy has never been settled. However, since the 1980s, interest has shifted more to DNA. For the DNA evidence — particularly the rapid evolution in synonymous sites and in pseudogenes — the selectionist explanation has few, or no, supporters. There is no evidence that the rapid evolution of these regions of DNA is due to exceptionally rapid, adaptive fine tuning. Pseudogenes are, after all, functionless and it is difficult to see what adaptation could be fine tuned within them. Some biologists favor the full neutralist view, according to which evolution at both synonymous and non-synonymous sites is mainly neutral. The slower evolution at non-synonymous sites is then because many amino acid changes are disadvantageous. Other biologists accept the neutralist view for synonymous sites and pseudogenes, but remain undecided whether amino acid changes are driven more by drift or positive selection.

Selectionists predict rapid fine-tuning adjustments in genes . . .

. . . but that theory is implausible for non-coding DNA

7.7 Conclusion and comment: the neutralist paradigm shift

Arguably, the view of evolutionary biologists about molecular evolution has shifted since the 1980s. When Kimura, King and Jukes first suggested the neutral theory in 1968 and 1969, they did so for protein evolution. The neutral theory was controversial in the 1970s. It was keenly discussed, but did not win widespread acceptance, nor did it inspire a big research program that assumed its validity. Indeed the neutral theory is controversial still for protein evolution. Natural selection may play a big part in the evolutionary change of, and genetic variation in, proteins — though this is far from confirmed.

Through the 1980s, DNA sequence data began to accumulate. The neutral theory was more, and the selectionist theory less, successful at predicting and explaining the patterns of evolution in DNA, particularly in synonymous and non-coding sites. Moreover, much of the DNA is non-coding. Perhaps 95% of human DNA does not code for genes. The nature of "non-coding" DNA has only slowly become clear — indeed biologists are still uncertain why non-coding DNA exists. During the 1970s, things were more uncertain than now and biologists could argue that the apparently excessive DNA could be informational in some way. Then selection would work on it. But much of the non-coding DNA is now generally accepted to have no function, though its nucleotide sequence may be partially constrained. It is difficult to see how selection could drive many changes in this sort of "junk" DNA. Most evolution in non-coding, non-genic DNA is thought to be neutral, though not pan-neutral. Therefore most of the substitutions that occur in the DNA as a whole are thought to be neutral too, because most of the DNA is non-coding. The conclusion is a little different from Kimura's original claim. He made it for proteins, that is for non-synonymous changes in the DNA. He "won" the argument, but not for the kind of evidence he originally discussed. It has turned out that most evolution is not in amino acid-changing parts of the DNA.

Biologists came to accept, through the 1980s, that most molecular evolution is by drift

The idea that most evolution in synonymous and non-coding DNA is neutral is now inspiring a huge research program: the reconstruction of the history of life using molecular evidence. Parts 4 and 5 of this book look at this kind of research. The research could have been built on the theory of natural selection, but it follows more easily from the neutral theory. Most of the biologists who are doing the work probably assume that the molecular changes they are studying occur by random drift.

As an interim conclusion, we can say that the neutral explanation for molecular evolution in synonymous sites within genes, and in non-coding parts of DNA, is widely accepted. This being so, the majority of molecular evolution proceeds by random drift rather than selection.

Natural selection is still evolutionarily important. It drives adaptive evolution, and we now turn to ways of looking for signs of adaptive evolution — or the signature of selection — in DNA sequence data.

7.8 Genomic sequences have led to new ways of studying molecular evolution

Genomic sequences have become available in large amounts recently, and they can be used to look for signs of selection and drift. We shall look at five examples of this current research trend, beginning with a classic result. They mainly make use of the distinction between synonymous and non-synonymous nucleotide changes.

7.8.1 *DNA sequences provide strong evidence for natural selection on protein structure*

The higher rate of synonymous than non-synonymous evolution . . .

When, in Chapter 4, we considered the evidence for biological variation, we noticed that many DNA sequence variants can be uncovered if proteins are sequenced at the DNA level (Section 4.5, p. 84). This observation has important implications for molecular evolution. At the alcohol dehydrogenase (*Adh*) locus in the fruitfly (*Drosophila melanogaster*), two alleles (fast and slow) are present. Kreitman (1983) sequenced the DNA of 11 different copies of the gene. He found that the proteins were uniform within each allelic class. He found only two amino acid sequences, corresponding to the two alleles. But he found a number of DNA sequences coding for each allele. Within an allelic class, he found synonymous, but not non-synonymous, variation. The combination of a fixed amino acid sequence and variable silent sites provides, as Lewontin (1986) emphasized, evidence that natural selection has been operating to maintain the enzyme structure.

There are two possible reasons why the enzyme sequence, at the amino acid level, should be fixed within each allelic class. One is "identity by descent": all the copies of each allele may be descended from an ancestral mutation, which had that sequence and has been passively passed from generation to generation. Eventually another amino acid-altering mutation may arise within one allelic class, and that allele will (at least temporarily) have become two alleles. The constant sequence within current fruitfly populations only means that not enough time has passed for such a mutation to occur. Alternatively, the gene copies that make up an allelic class may all have the same sequence because that sequence is maintained by natural selection; when a mutation arises, selection removes it.

. . . is evidence that natural selection acts against non-synonymous mutations

The observed variability distinguishes between these two hypotheses. The variability in the synonymous sites means that there has been time for mutations to arise in the molecule. If mutations have arisen in synonymous sites, they will surely have arisen in non-synonymous sites too. Therefore, we can reason that the identity in amino acid sequence is unlikely to be identity by descent. Mutations in non-synonymous sites have presumably not been retained because natural selection eliminated them.

If it had turned out that the *Adh-f* allele was fixed for one DNA sequence at all sites, synonymous and non-synonymous, we should not know whether the uniformity was due to selection or identity by descent. We should be in the same position as we were in before Kreitman's DNA-level study. The uniformity might mean only that no mutations had occurred. Kreitman's DNA sequencing thus provides evidence for selection that could not have been obtained with amino acid sequences alone.

The absence of amino acid sequence variation within the *Adh-f* (and *Adh-s*) allelic class is particularly striking because 30% of the enzyme is made up of isoleucine and valine, which are biochemically very similar (and indistinguishable by gel electrophoresis). A neutralist might have predicted that some of the valines could be changed to isoleucines, or vice versa. The only amino acid sequence variant is the one that causes the *Adh-f/Adh-s* polymorphism. That polymorphism is known to be maintained by natural selection. Therefore, none of the amino acids in the 255 amino acid alcohol dehydrogenase enzyme of the fruitfly can be changed neutrally. Interestingly, that means that we could almost construct Figure 7.1 for alcohol dehydrogenase at the amino acid level. The graph would be like Figure 7.1a for mutations that change an amino acid, but like

Figure 7.1b for synonymous mutations. Natural selection is powerfully maintaining the amino acid sequence, while synonymous changes evolve by drift.

7.8.2 *A high ratio of non-synonymous to synonymous changes provides evidence of selection*

When we compare the DNA sequence of a gene in two species, the usual result is for there to be more synonymous than non-synonymous nucleotide differences. Table 7.6 showed that synonymous evolution proceeds about five times as fast as non-synonymous evolution. The ratio of non-synonymous differences (dN) to synonymous differences (dS) will be about 1 : 5 or 0.2. As we have seen, synonymous evolution is faster because fewer synonymous mutations are disadvantageous, and more are neutral, than non-synonymous mutations. At least some amino acid changes are disadvantageous, and this slows down the rate of non-synonymous evolution.

Elevated dN/dS ratios are observed in some genes

However, some exceptional genes have been found in which the ratio of non-synonymous to synonymous evolution (the dN/dS ratio) is elevated. For example, Wyckoff *et al.* (2000) studied the protamine genes in the evolution of the great apes, including humans. The protamines function in the male reproductive system, and the genes evolve rapidly. Their evolution shows a high dN/dS ratio. The ratio for one protamine gene, *prm1*, for instance is 13.

What is the cause of elevated dN/dS ratios, such as we see in the protamine genes? One possibility is chance — the probability of a dN/dS ratio can be estimated statistically, and any one case may be a random blip in the data. What if we rule out chance? Two processes have been identified that increase the ratio of non-synonymous to synonymous evolutionary changes. One is positive selection in favor of a change in gene function. The other is relaxed selection.

They can be explained by natural selection . . .

The rate of amino acid-changing, non-synonymous evolution is usually low because change is disadvantageous. The protein that the gene codes for is probably well, or even perfectly, adapted and most or all non-synonymous change is for the worse. However, natural selection could favor a change in the protein. Then the rate of non-synonymous evolution will increase, while the rate of synonymous change will continue as normal, by random drift. Thus an elevated dN/dS ratio can result when natural selection has favored a change in the protein coded by a gene.

. . . or by relaxed selection

Alternatively, the dN/dS ratio can go up when natural selection is relaxed. Natural selection normally prevents amino acid changes. If natural selection is stopped from acting, the rate of amino acid evolution will increase. Changes that were disadvantageous become neutral in the absence of selection. Natural selection may be relaxed in humans, by medical care and other cultural practices that act against natural selection. More generally, a rapid increase in population size is a sign that selection has been relaxed. When a population colonizes some unexploited territory with abundant resources, there may be a phase of rapid population growth. Natural selection will probably be relaxed during this phase.

The two explanations for elevated dN/dS ratios are frustrating because they are conceptually almost opposite. The same data may mean either that positive selection, in favor of change, has been acting, or that negative selection, against change, has been relaxed. The rate of non-synonymous evolution could go up either way.

Wyckoff *et al.* thought of several ways round this dilemma. For instance, they looked for dN/dS ratios of more than one. Relaxed selection alone cannot take the ratio above one. When selection ceases to act on a DNA sequence, both non-synonymous and synonymous changes will be equally neutral and occur at the same rate. The dN/dS ratio will equal one. By contrast, positive selection in favor of change can take the dN/dS ratio much higher. If dN/dS ≪ 1, it is a strong sign that natural selection has been driving change.

The two explanations can be tested between

In summary, we have three zones of dN/dS ratio, and three associated evolutionary interpretations.

1. dN/dS low, perhaps 0.1–0.2 (though the actual value can vary down the DNA). Interpretation: synonymous change is neutral; there is no evidence that selection is driving the change in amino acids.
2. dN/dS between 0.2 and 1. Interpretation: either selection has been acting to change the amino acid sequence or selection has been relaxed; we do not know which.
3. dN/dS higher than 1. Interpretation: natural selection has been acting to change the amino acid sequence.

dN/dS ratios provide evidence of selection in some examples

Biologists have mainly been interested in using dN/dS ratios as evidence for positive selection. For them, relaxed selection is something to be ruled out. In the protamine gene, dN/dS > 1 and we have evidence of adaptive evolutionary change rather than relaxed selection. (Wyckoff *et al.* also presented other evidence for positive selection in protamine evolution, including evidence from the McDonald–Kreitman test that we discuss in the next section.)

High dN/dS ratios have been found in several genes. The genes concerned look like the sort of genes that may undergo rapid adaptive evolutionary change. The first genes to be found with high dN/dS were the HLA genes. HLA genes recognize parasite invaders in the body. They probably evolve fast to keep up with evolutionary changes in the parasites, which evolve to outsmart their host's immune systems. Other genes with high dN/dS are in signal–receptor systems and in the reproductive system.[2]

The relation between the two arguments in this section, and in the previous section, may be worth clarifying. It might seem that low dN/dS ratios were used as evidence of selection in the previous section and now high dN/dS ratios are being used as evidence of selection here. The answer is that the two sections are concerned with testing for different kinds of selection. Kreitman (1983) found synonymous, but no non-synonymous, variation between copies of one alcohol dehydrogenase allele in fruitflies. This shows that natural selection has been acting to prevent change. Wyckoff *et al.* (2000) found more non-synonymous than synonymous evolution in the protamine genes of apes. This shows, or at least suggests, that natural selection has driven adaptive evolutionary change. Kreitman's evidence by itself fits with all evolutionary change being by drift (there is evidence for selective changes in the *Adh* gene, but it comes from other research). Wyckoff *et al.*'s evidence challenges, and possibly refutes, random drift as the explanation of evolution in the protamine genes of humans and other apes.

Box 7.4 looks at a practical application of dN/dS ratios, in the genes coding for leptin.

---

[2] The possible rapid evolution of at least some reproductive systems' genes is a recurrent subtheme in this book. We return to it in Sections 12.4.7 (p. 336), 13.3.2 (p. 357), and 14.12 (p. 417). Swanson & Vacquier (2002) is a recent empirical review.

Box 7.4
Model Organisms for Biomedical Research

In late 1994, a hormone called leptin was discovered in mice. Leptin had the ability to make fat mice slim. Mice that are deficient in leptin are grossly obese; if those mice are fed leptin they become slim in a few weeks (and without noticeable side effects). A biotechnology company immediately snapped up the rights to leptin, for $20 million. Since then, there has been intense research on whether leptin influences human obesity, but little or no evidence of any effect has been found.

Leptin influences body weight in mice, but what about other species? Leptin genes have been found in several mammal species. We can measure the rates of non-synonymous to synonymous evolution (dN/dS) in various branches of the mammalian family tree (Figure B7.3). The ratio is generally low, at a typical level for genes in general; but it spurted up to more than two in the lineage between Old World monkeys and great apes.

The elevated ratio may be a chance blip, or artifact of the preliminary data, and meaningless. However, it may indicate a phase of adaptive evolution, when several new amino acids were established in the leptin molecule of apes. This could have caused a change in the function of leptin, such that leptin no longer regulates body weight in apes. Alternatively, leptin may have added or subtracted some functions, or changed its metabolic interactions. Many interpretations are compatible with the simple elevated dN/dS ratio. Further research would be needed to test between them. The point here is that the elevated dN/dS ratio alone is a clue that something happened to leptin in the evolution of the apes. If leptin simply evolved by random drift in all mammals, then leptin would probably have the same function in humans and mice. The elevated ratio is evidence of positive natural selection. In expensive and vital biomedical technologies, clues are valuable even when they are not decisive.

The result has several implications. One is that mice may not be a good model organism for research on human leptin. A second concerns the way related genes are identified by searching genomic databases. Leptin genes, related to the mouse leptin gene, were soon found in other mammalian genome libraries — but before jumping to conclusions about the function of the genes it is useful to know the dN/dS ratios in the phylogenetic branches connecting the species. Thirdly, the dN/dS ratios hint at an important change in leptin between mice and men. Such a change could explain why research has been slow to find an influence of leptin on human body weight. If the dN/dS ratios had been available in early 1995, that quick-draw biotechnology company might been slower with its wallet.

Further reading: Benner *et al.* (2002), newspiece in *Nature*, April 6, 2000, pp. 538–40.

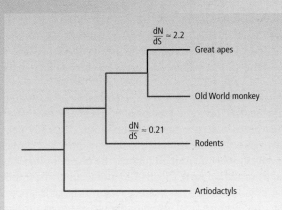

Figure B7.3
A spurt of meaningful evolution in the leptin gene during the origin of apes. The ratio of non-synonymous to synonymous evolution in the leptin gene has generally been low, such as the 0.2 figure for the rodent lineage. The ratio increased during the origin of apes, perhaps indicating a phase of adaptive modification. From Benner *et al.* (2002).

7.8.3 ### Selection can be detected by comparisons of the dN/dS ratio within and between species

A further test between drift and selection can be devised using the ratio of non-synonymous to synonymous evolution. The trick is to compare the ratio within one species and between two related species. Consider a gene like *Adh*, which we looked at in Section 7.8.1. Within the *Drosophila melanogaster* species, *Adh* is polymorphic — two alleles are present in most populations of the species. We can count the number of non-synonymous and synonymous differences between the two alleles and express the result as a dN/dS ratio *within* a species. We can also measure the number of differences between the *Adh* gene in *D. melanogaster* and in a related fruitfly species, to give the dN/dS ratio for evolutionary changes *between* the two species.

The McDonald–Kreitman test looks for selection by comparing dN/dS ratios within and between species

McDonald & Kreitman (1991) realized that, on Kimura's neutral theory, the dN/dS ratio should be the same both for polymorphism within a species and evolutionary divergence between species. In both cases, the dN/dS ratio equals the ratio of the non-synonymous neutral mutation rate to the synonymous neutral mutation rate.

The reason is as follows. The dN/dS ratio between species is the ratio of non-synonymous to synonymous evolutionary change. The rate of neutral evolution equals the neutral mutation rate (Section 6.3, p. 144). The ratio of non-synonymous to synonymous evolution should therefore, on Kimura's neutral theory, equal the ratio of the neutral mutation rates for non-synonymous and synonymous mutations. Within a species, the amount of neutral polymorphism is given by a more complex formula (Section 6.6, p. 151). But if we look at the ratio of polymorphism for non-synonymous to synonymous sites, everything in the formula cancels except the non-synonymous neutral mutation rate and the synonymous neutral mutation rate. The dN/dS ratio for polymorphism within a species is again the ratio of these two mutation rates.

If selection is at work, the dN/dS ratio is not expected to be the same within and between species. For instance, if natural selection favors a change in an amino acid in one species but not the other, the dN/dS ratio will be higher between than within a species. If natural selection favors a polymorphism, because of frequency-dependent selection or heterozygous advantage (Sections 5.12–5.13, pp. 123–8), the dN/dS ratio will be higher within a species than between. In summary, if the dN/dS ratio is similar for polymorphisms within a species and evolutionary change between species, that suggests random drift. If the ratio differs within and between species, that suggests natural selection.

Nearly neutral theory makes the test inconclusive for single genes

The McDonald–Kreitman test was initially used with individual genes such as *Adh*. The test seemed to rule out the neutral theory, at least in some cases. However, the test is not powerful for individual genes. The test can rule out Kimura's purely neutral theory; but it does not work against the nearly neutral theory. Once we allow for nearly neutral mutations as well as purely neutral mutations, the dN/dS ratios depend on population size as well as the mutation rate. The dN/dS ratio will only be the same within and between species if population size has been constant. In practice, population sizes fluctuate. Suppose, for instance, that the population size goes through a bottleneck while a new species originates. During that phase, more non-synonymous mutations may behave as nearly neutral mutations (for much the same reason as we met in Section 7.5.3). The ratio dN/dS will go up. The ratio for polymorphisms in the modern species will not be affected, because population sizes have been restored to normal. Only the dN/dS ratio for the comparison between species is affected. It is high because of the many substitutions that occurred during the population bottleneck.

For this reason, by the late 1990s, the McDonald–Kreitman test was thought to be interesting but not usually decisive. The test could be used against the purely neutral theory. However, the neutral theory had by then moved on to the nearly neutral theory, and the McDonald–Kreitman test did not work against that.

But the test has provided illuminating with genomic data, . . .

The McDonald–Kreitman test has enjoyed a revival as whole (or almost whole) genome sequences have become available. The dN/dS ratio could be calculated within and between species down the whole genome, if the whole genome had been sequenced for several individuals of two species. In practice, this kind of research has so far used parts of a genome, rather than whole genomes and has been confined to fruitflies (Fay *et al.* 2002; Smith & Eyre-Walker 2002). The dN/dS ratio is found to be larger between species rather than within. If that were true equally for all sites in the genome, the result could be explained either by positive selection for change or by the nearly neutral theory (with a population bottleneck during speciation). However, the excess non-synonymous substitutions are confined to only some sites in the genome. For many sites, the dN/dS ratio is equal within and between species. These sites have probably evolved by random drift. But at other sites, the amino acid has changed between related fruitfly species. It looks like selection has acted at those sites.

. . . and allows an estimate of how much non-synonymous evolution is driven by selection

More interestingly, the fraction of sites at which the dN/dS ratio is elevated between species can be used to estimate the fraction of evolutionary substitutions that have been driven by selection, as opposed to drift. In this way, Smith & Eyre-Walker (2002) estimated that 45% of non-synonymous substitutions between one pair of fruitfly species (*Dropsophila simulans* and *D. yakuba*) were fixed by positive selection.

The use of the McDonald–Kreitman test with genomic data avoids the problem of population sizes. A change in population size will influence the pattern of evolution across the whole genome. The new inferences use variation between sites within a genome. They focus on regions of the genome where the dN/dS ratio is abnormally high, between species. It cannot be argued that the sites with high dN/dS ratios have experienced one history of population sizes, and other sites (with lower dN/dS ratios) some other history of population sizes. All the sites in the genome must experience the same population size.

The results so far are preliminary. They are based on a limited genomic sample from one small group of species. However, the results have great interest. They suggest that natural selection may be a major force, at least for substitutions that change amino acids. They also show how genomic data may be used to estimate the relative importance of selection and drift in molecular evolution. In the future, the sequences of chimp and human genomes will become available. Evolutionary biologists can then scan down the sequences, to find sites where the dN/dS ratio is relatively high for comparisons between the species. Those sites may be the ones where selection has favored changes that have made us human.

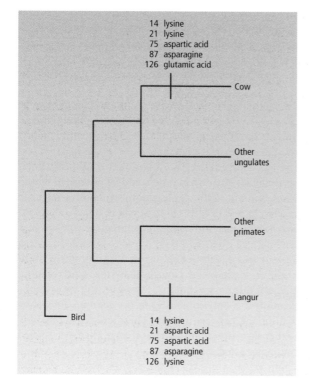

14 lysine
21 lysine
75 aspartic acid
87 asparagine
126 glutamic acid

Cow

Other
ungulates

Other
primates

Langur

Bird

14 lysine
21 aspartic acid
75 aspartic acid
87 asparagine
126 lysine

Figure 7.7
Convergent evolution of stomach lysozymes in langurs and ruminants. In the evolutionary lineages leading to langurs and cows, changes have occurred at the same five sites in the lysozyme protein, and the changes have been similar or identical. The numbers refer to amino acid sites in the protein.

7.8.4 *The gene for lysozyme has evolved convergently in cellulose-digesting mammals*

Lysozyme is a widespread enzyme, used in defense against bacteria. The enzyme breaks open the bacteria cell wall, causing lysis of the bacterial cell. Lysozyme is found in body fluids such as saliva, blood serum, tears, and milk. In two mammalian groups, ruminants (such as cows and sheep) and leaf-eating colobine monkeys (such as langurs), a new version of lysozyme has evolved in addition. Both taxa use lysozme to digest bacteria within their stomachs. The stomach bacteria themselves digest cellulose from plants, and the cow or langur obtains nutrients from cellulose by, in turn, digesting the bacteria.

Ruminants and colobine monkeys secrete lysozyme in their stomachs, which is a more acid environment than found in normal body fluids. When the sequences of stomach lysozymes in ruminants and colobines is compared with the sequence of standard lysozyme, we see that several identical amino acid changes have occurred independently in the two lineages (Figure 7.7). The amino acid changes allow the lysozyme to work better in acid environments, as well as providing other advantages.

The lysozymes of ruminants and colobine monkeys are a molecular example of convergent evolution (Section 15.3, p. 429). Convergence is usually due to adaptation

Convergence is evidence that selection has operated

to a common environment. In this case, convergence is good evidence that selection has been at work on the lysozyme gene. The case can be strengthened in two ways. One is that a third species, the South American bird called the hoatzin (*Opisthocomus hoazin*) has also independently evolved cellulose digestion. It also uses a lysozyme, secreted in its stomach, to digest cellulose-digesting bacteria. The hoatzin's lysozyme is a related but different gene from the one redeployed in ruminants and langurs, but it shows the same set of amino acid changes. Secondly, the evolution of lysozyme in ruminants and cows shows an elevated dN/dS ratio, which is suggestive of selection-powered adaptive evolution, as we saw in the previous section (Messier & Stewart 1997).

7.8.5   *Codon usages are biased*

The top part (green columns) of Figure 7.8 shows the relative frequency of the six leucine codons in two single-celled organisms, the bacterium *Escherichia coli* and the eukaryotic yeast *Saccharomyces cerevisiae*. The six codons are synonymous, and we

Evidence shows that synonymous codons are not used randomly

expect them to evolve by random drift. Notice two features of the figure: one is that the codon frequencies are unequal within a species. The other is that the species differ in which codons are abundant, and which rare. *E. coli* has more CUG; yeast has more UUG.

What is the explanation for codon biases? Two hypotheses have been suggested: selective constraint or mutation pressure. The mutation pressure hypothesis suggests that mutation is biased toward certain nucleotides (Section 4.8, p. 89). If A tended to mutate to G in *E. coli*, for instance, that might produce the excess of CUG and paucity of CUA codons.

Alternatively, some codon changes may be disadvantageous and selected against. Two possible reasons are the strength of DNA bonds and the relative abundance of transfer RNAs. The GC bond is stronger than the AT bond, because GC has three

Figure 7.8
Relative frequencies of codons match tRNA abundances. (a) The green columns (above) are the relative frequencies of six leucine codons in *Escherichia coli*; the gray columns (below) are the relative frequencies of the corresponding tRNA molecules in the cell. The two sets of codons joined by a + sign are recognized by a single tRNA molecule. (b) The same relation, but in *Saccharomyces cerevisiae*. Notice the different bias in codon usage in the two species, which reillustrates the point of Table 7.7. From Kimura (1983). Redrawn with permission of Cambridge University Press © 1983.

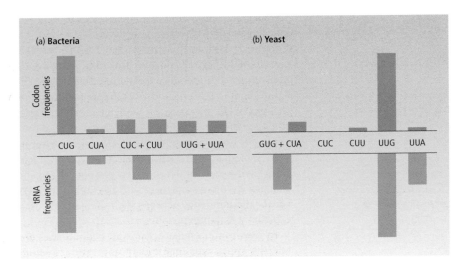

Table 7.7

Relative frequencies of six leucine codons in genes of *Escherichia coli* and yeast (*Saccharomyces cerevisiae*). The genes are divided into high usage genes and low usage genes: high usage genes are frequently transcribed, low usage genes are rarely transcribed. Note (i) codon biases are larger for highly used genes than low use genes, and (ii) the codon biases differ between the two species. The numbers are relative frequencies: they add up to six. A relative frequency of less than one means the codon is rarer than expected; of more than one means it is commoner than expected. Modified from Sharp *et al.* (1995).

Leucine codon	E. coli		S. cerevisiae	
	High	Low	High	Low
UUA	0.06	1.24	0.49	1.49
UUG	0.07	0.87	5.34	1.48
CUU	0.13	0.72	0.02	0.73
CUC	0.17	0.65	0.00	0.51
CUA	0.04	0.31	0.15	0.95
CUG	5.54	2.20	0.02	0.93

hydrogen bonds while AT has only two. Natural selection might work against GC to AT changes in regions of the DNA that need to be stably bonded. Secondly, different transfer RNAs are used by the different synonymous codons. (There are fewer kinds of tRNA than codons because of the phenomenon of "wobble." For some pairs of codons, one kind of tRNA can bind them both.) The different tRNAs have a certain frequency distribution in the cells: some tRNAs, among a synonymous set, are more frequent than others. Figure 7.8 shows tRNA abundances in the lower half (gray columns). A change in *E. coli* DNA from a CUG codon to a CUA codon might be selected against. The change might reduce the efficiency of protein synthesis, because the cell contains little leucine tRNA for the CUA codon.

In microbes, codon usage matches tRNA abundance, . . .

Figure 7.8 shows that the codon frequencies match the tRNA frequencies. The pattern makes sense if the two distributions evolve together, and changes from common to rare codons reduce translational efficiency. The argument can be strengthened. Some genes in bacteria and yeast are frequently translated. These can be called "high use" genes. Other genes are less often translated, and can be called "low use" genes. The efficiency of protein synthesis probably matters more for high use than low use genes. Table 7.7 shows that codon biases are much greater in high use than in low use genes. Thus, in high use genes natural selection works against codon changes. The cell benefits from having more of the codons corresponding to abundant tRNAs. In low use genes, changes are disadvantageous and the codon frequencies evolve by drift to be more equal. The difference between high use and low use genes in Table 7.7 is difficult to explain by mutation pressure.

. . . and the match is better for genes that are expressed more

At least in unicellular organisms, codon biases are thought to be caused more by selective constraints than mutation pressure. Evolution in synonymous sites still fits the neutral theory. Natural selection is a negative force, preventing certain changes. Evolutionary changes, when they do occur, are probably by neutral drift. However, the evidence for selective constraints means that evolution at synonymous sites is probably not "pan-neutral." Not all synonymous mutations are neutral. The rate of synonymous evolution will then be somewhat below the total mutation rate.

The argument we have looked at in this section is widely accepted for single-celled life forms. But the picture for multicellular life forms such as fruitflies and mammals may differ. The mutation-bias hypothesis may be more viable for mammals than for bacteria and yeast.

7.8.6 *Positive and negative selection leave their signatures in DNA sequences*

We have looked at five examples of the ways in which genomic sequences can be used to study natural selection. In the cases of the alcohol dehydrogenase gene and of codon bias, the effect of selection was negative: selection acted against disadvantageous mutations, preventing evolutionary change. Such evolutionary changes as do take place among synonymous codons are probably mainly driven by drift, but selection is acting to prevent some changes. The other three examples (elevated dN/dS ratios, different dN/dS ratios within and between species, and convergent evolution in lysozymes) illustrate positive selection: natural selecting actively favoring certain changes. The amino acid changes in the protamine and lysozyme genes have probably been driven by selection rather than drift.

The genomic era is allowing new tests of selection and drift, . . .

The examples illustrate two points. One is that the genomic era has opened up new ways to study selection. We saw earlier how natural selection can be studied ecologically, such as in the peppered moth or in insecticide resistance (Sections 5.7–5.8, pp. 108–18). The peppered moth has identifiable character states (light or dark coloration) and the fitnesses of these states can be measured in natural environments. This kind of ecological research is not the only way that selection has been studied, but it contrasts with research in the genomic era. When we look at dN/dS ratios, for instance, we are not looking at organismic character states, nor measuring fitnesses. We are counting large numbers of evolutionary changes, statistically, in a mass of sequence data. In Section 8.10 (p. 210) we shall see another statistical method for detecting selection in sequence data, in the phenomenon of selective sweeps.

. . . and identifying sites where selection appears to have operated

Secondly, the examples show that neutralism is not the whole story of molecular evolution. Random drift probably explains the majority of molecular evolution — provided we count "non-informational" changes. Evolution in non-coding regions of the DNA, and in synonymous sites within genes, looks neutral. But in the non-synonymous sites of genes, where DNA changes produce amino acid changes, selection is more important. Whole-genome analyses are being used to estimate the exact relative importance of selection and drift in amino acid substitutions. The lysozyme example shows how we can study the way selectin works in an identified gene. It makes sense that selection as well as drift should matter in molecular evolution. The molecules in living bodies are well adapted, and natural selection must work at least occasionally to keep those adaptations up to date.

7.9 Conclusion: 35 years of research on molecular evolution

In 1968, Kimura proposed the neutral theory of molecular evolution. His original argument was mainly based, in theory, on genetic loads and, in fact, on amino acid evolution. Neither his particular claim — that most molecular evolution proceeds by the random drift of neutral mutations — nor his argument using genetic loads, nor his evidence for proteins, has survived in its original form. However, he stimulated a huge area of research, which arguably led to a paradigm shift in our understanding of molecular evolution.

Kimura's neutral theory has developed into the nearly neutral theory. The nearly neutral theory shares with its predecessor the claim that most molecular evolution is by random drift — but the drift of nearly neutral ($4Ns < 1$ or $Ns < 1$) rather than exactly neutral ($s = 0$) mutations. Since Kimura first wrote, biologists have come to realize that DNA contains huge regions of non-coding sequences. If we use use the nearly neutral rather than the original, purely neutral theory, and confine it to substitutions in non-coding DNA and synonymous substitutions in coding DNA, then many (perhaps most) biologists accept a neutralist interpretation of molecular evolution. Most evolution at the DNA level is by random drift.

However, non-coding DNA is in some respects biologically less interesting than coding DNA. Non-synonymous substitutions, which alter amino acids, are biologically more important in the sense that they influence the form and functioning of the body. If we concentrate on non-synonymous evolution, in coding regions, then it is probably not true that most biologists are neutralists. We do not know the relative importance of drift and selection in driving amino acid change. Indeed, for most of the past 35 years we have lacked a decisive method to find out the relative importance of drift and selection. Molecular evolution is now entering the genomics era. Genomic data hold out the promise both of revealing the localities within the DNA where natural selection acts, and also of estimating the fractions of evolutionary substitutions that have been driven by natural selection and by random drift.

Summary

1 The neutral theory of molecular evolution suggests that molecular evolution is mainly due to neutral drift. On this view, the mutations that have been substituted in evolution were selectively neutral with respect to the genes they replaced. Alternatively, molecular evolution may be mainly driven by natural selection.

2 Three main observations were originally used to argue in favor of the neutral theory: molecular evolution has a rapid rate, its rate has a clock-like constancy, and it is more rapid in functionally less constrained parts of molecules.

3 One of the three arguments — from high rates of evolution (and high levels of polymorphism) — used the concept of genetic load and is no longer thought conclusive.

4 The constant rate of molecular evolution gives rise to a "molecular clock."

5 Neutral drift should drive evolution at a stochastically constant rate; Kimura pointed to the contrast between uneven rates of morphological evolution and the constant rate of molecular evolution and argued that natural selection would not drive molecular evolution at a constant rate.

6 For synonymous changes, evolution is faster in lineages with shorter generation times. For non-synonymous change, some evidence suggests that the rate of evolution is relatively constant independent of generation time, and other evidence suggests that the rate of evolution is faster in lineages with shorter generation times.

7 The original neutral theory had difficulty explaining certain observations, including: (i) the similar level of polymorphism in all species, independent of population size; (ii) the difference between synonymous and non-synonymous sites in whether the rate of evolution depends on generation times; and (iii) different ratios of non-synonymous to synonymous evolution for polymorphism within a species and divergence between species.

8 The nearly neutral theory of molecular evolution suggests that molecular evolution is driven by random drift, but includes the effect of drift on mutations with small disadvantageous (and advantageous) effects as well as on purely neutral mutations.

9 The nearly neutral theory can explain most observations about molecular evolution, including the observations that were problematic for the original purely neutral theory. Critics argue that the nearly neutral theory must invoke unrealistic assumptions about population size in order to explain all the observations.

10 The neutral theory explains the higher evolutionary rate of functionally less constrained regions of proteins by the greater chance that a mutation there will be neutral.

11 Pseudogenes and synonymous changes may be relatively functionally unconstrained. They have faster rates of evolution than non-synonymous changes, which alter the amino acid sequence of the protein. This high rate of evolution is probably due to enhanced neutral drift.

12 Genomic data can be used to study natural selection.

13 A high rate of non-synonymous evolution relative to synonymous evolution suggests that natural selection has been operating.

14 The McDonald–Kreitman test notes that the ratio of non-synonymous to synonymous evolution (dN/dS ratio) is equal between species and within a species when drift operates, but differs between and within species when selection operates.

15 The dN/dS ratio down the genome can be used to identify sites where selection has acted. It can also be used to estimate the fraction of sites where selection has acted. Preliminary work suggests that about half non-synonymous substitutions are by selection and half by drift.

16 Biases in codon usage can be caused by natural selection acting against certain codons in a synonymous set, and by biased mutation. For single-celled organisms, natural selection seems to explain the patterns of codon usage bias.

Further reading

The texts by Graur & Li (2000), Page & Holmes (1998), and Li (1997) introduce the topic. The classic works, in order of decreasing age, are: Lewontin (1974), Kimura (1983), and Gillespie (1991). Kimura (1991) up-dated his views.

The Festschrift for Lewontin (Singh & Krimbas 2000) contains a number of chapters on the topic. See Hardison (1999) on hemoglobin. Golding & Dean (1998) review studies of adaptation at the molecular level. Eanes (1999) reviews studies of enzyme polymorphism. The texts include material on levels of variation and rates of evolution. Przeworski *et al.* (2000) describe variation in human DNA. Mitton (1998) reviews classic, pregenomic studies of selection.

On molecular clocks, see Cutler (2000) on irregularity or overdispersion. The mutation process matters for the generation time effects in the molecular clock. For number of germ line divisions see Ridley (2001, p. 234). Mutations are mainly internal copying accidents: see Ridley (2001), and Sommer (1995), for example. However, Kumar & Subramanian (2002) provide evidence that some synonymous evolution rates, and therefore perhaps mutation rates, in mammals do not depend on generation time.

For nearly neutral theory, see the general texts. Ohta (1992) is a review, Ohta & Gillespie (1996) a historic perspective, and Ohta (2002) a recent update. The exchange between Ohta and Kreitman included in Ridley (1997) shows how various facts can be explained by the nearly neutral theory or selective evolution. Gillespie (2001) questions whether population size affects the rate of evolution, because the effect via hitch-hiking (Chapter 8) is the opposite of the effect on drift at one site.

On testing for selection in genomic data, Nielsen (2001) reviews the statistical tests. Brookfield (2001) introduces one case study. Hughes (1999) looks at ratios of non-synonymous to synonymous evolutionary rates.

Another test, similar to the test of McDonald and Kreitman (1991), was devised by Hudson, Kreitman, & Aguade (1987). The "HKA" test is also enjoying a revival with genomic data. It can be seen in action recently in Rand (2000) on mitochondrial genomics, and Wang *et al.* (1999) on genetic change during maize domestication. Bustamente *et al.* (2002) is another paper using the MK test with genomic data, like the two discussed in the text. They agree that fruitflies have substituted many advantageous non-synonymous changes, and add an inference that *Arabidopsis* has substituted more disadvantageous changes. The Ohta–Kreitman exchange cited above considers the MD and HKA tests further.

On codon biases, see Kreitman & Antezana (2000), Mooers & Holmes (2000), and Duret & Mouchiroud (1999).

Research on this topic can be followed in *Trends in Ecology and Evolution*, *Trends in Genetics*, *Bioessays*, and the December special issue each year of *Current Opinion in Genetics and Development*.

Study and review questions

1 Draw the frequency distribution for the selection coefficients of the genetic variants at a locus, according to the neutral and the selectionist theories of molecular evolution.

2 Why is the neutral theory confined to *molecular* evolution, rather than being applied to all evolution?

3 What facts about molecular evolution led to the proposal of the neutral theory of molecular evolution?

4 What facts about molecular evolution led to the proposal of the nearly neutral theory of molecular evolution?

5 Explain the relation between the degree of functional constraint on a molecule (or region of a molecule) and its rate of evolution by (a) the neutral theory, and (b) natural selection.

6 Do synonymous sites show pan-neutral evolution? What evidence is there bearing upon the answer?

7 Three genes have ratios of

$$\frac{\text{Rate of non-synonymous substitution}}{\text{Rate of synonymous substitution}}$$

of (a) 0.2, (b) 1, (c) 10. What inferences can we make about the evolution of these three genes?

8

Two-locus and Multilocus Population Genetics

We begin with an example of a character that is controlled by a multilocus genotype. The set of genes that an individual inherits from one of its parents form a "haplotype," and the theory of population genetics for multilocus systems traces haplotype frequencies through generations. The chapter has two main purposes. One is to introduce population genetic theory for multilocus systems, and the distinctive concepts that apply to it but not to one-locus models: we look at the multiple-locus concepts of linkage disequilibrium, together with its causes, of recombination, and of multiple-peaked fitness surfaces. The other purpose is to see the conditions under which single-locus models of evolution are inadequate, and multilocus models are necessary: the main condition is the existence of linkage disequilibrium. We also look at how multilocus genomic sequences can be used to test whether selection has recently been acting in a region of the DNA.

8.1 Mimicry in *Papilio* is controlled by more than one genetic locus

The characters we dealt with in earlier chapters have been characters controlled by single genetic loci. Enzymes, such as alcohol dehydrogenase, are encoded by a single gene, and it is not much of a simplification to treat the polymorphism in the peppered moth as a set of genotypes at one locus. We now move on to consider evolutionary changes at more than one locus.

The first example concerns a multilocus polymorphism. We can lead into it via a similar polymorphism that is controlled by a single locus. Both examples come from the same attractive group of butterflies called swallowtails. The swallowtails have a global distribution, and *Papilio* is the largest genus of them; their most striking characteristic is a "tail" on the hindwing. Swallowtail butterflies come in many colors — gorgeous greens, subtle shades of reds and orange, and marbled patterns in white and gray — but the commonest type has stripes of black and yellow. The North American tiger swallowtail *Papilio glaucus* is easy to recognize by its tiger stripes, as it flutters through woodland lanes or humid valleys. Or rather, *most* tiger swallowtails are easy to recognize in this way. In part of the species' range (roughly, to the southeast of a line from Massachusetts to south Minnesota and from east Colorado to the Gulf Coast) the standard form of the tiger swallowtail lives alongside another form of the same species. This second form is black, with red spots on its hindwings, and is called *nigra*; it is only found in females. The *nigra* form is not poisonous, but mimics another species, the pipevine swallowtail *Battus philenor*, which is poisonous. The *nigra* form's geographic distribution fits that of the pipevine swallowtail. The *nigra* form is well protected there from predatory birds that have learned by stomach-churning experience not to eat butterflies looking like pipevine swallowtails. The tiger swallowtail, therefore, has a *mimetic polymorphism*. It has both the standard non-mimetic tiger morph of yellow and black stripes, and a black mimetic morph.

The tiger swallowtail *P. glaucus* comes to look almost simple when compared with the amazing array of females in the species *P. memnon* (see Plate 3, between pp. 68 and 69). *P. memnon* lives in the Malay Archipelago and Indonesia; its male is again non-mimetic, though its color is deep blue rather than yellow and black stripes. However, instead of one mimetic female morph, *P. memnon* females come in almost numberless variety. Their forewings show different geometric patterns of black and white; their hindwings, as well as varying in shape, can be colored in yellow, orange, or blood red, and may or may not have a bright white spot; some have tails, others do not; the abdomen varies in color; and a spot at the butterfly's "shoulder" (i.e., at the base of the forewing near the head) called the epaulette, may be present in various shades of red.

Clarke & Sheppard (1969, also Clarke *et al.* 1968) suggested that each female form (or "morph") mimics a different model (Plate 3 shows six examples: notice that three of them have tails and three do not). Their evidence is not strong, as it comes only from the geographic ranges of mimic and model, and from superficial similarity of appearance (which is not exact in all cases). Good evidence for mimicry requires experimentally showing that birds that have learned to avoid the model will also then avoid the

The tiger swallowtail butterfly exists in two forms

Another swallowtail species exists in many forms

mimic; this has been done for the *nigra* morph of *P. glaucus*, but not for *P. memnon*. However, we can accept as a working hypothesis that the apparently mimetic morphs of *P. memnon* indeed are mimetic. (*P. memnon* has yet further, non-mimetic forms too, but they are not essential to the discussion here.)

Genetic crosses initially suggested one locus was at work . . .

Clarke and Sheppard were interested in the genetic control of this complex mimetic polymorphism. Crosses between the various morphs initially suggested that a single genetic locus, with many alleles, is at work. When two morphs are crossed, the offspring usually only contain individuals of one or other parental phenotype. This is the result expected if one locus is at work, with simple dominance relations among alleles. For instance, if one morph has genotype A_1A_1, another A_1A_2, and A_2 is dominant to A_1 then an $A_1A_1 \times A_1A_2$ cross produces the same two classes of offspring (A_1A_1 and A_1A_2) as were present in the parents.

But the genetic story soon grew more complicated. In addition to the mimetic and non-mimetic morphs of *P. memnon*, all of which exist in reasonable frequencies in nature, some much rarer morphs have been found. An example, in Java, is the rare morph called *anura* (Plate 3m). A specimen found in Borneo was sent to Clarke and Sheppard in Liverpool. When it was crossed with a known *P. memnon* morph, it behaved like another allelic form of the mimicry "locus"; but a closer look at *anura* suggests a different interpretation. *Anura's* morphology mixes patterns from two of the common morphs: it has the wing color pattern of the morph *achates* (Plate 3g–i), but it lacks *achates'* tail.

Clarke and Sheppard's interpretation is that *anura* is not an allelic variant, but a recombinant, and that the mimetic patterns of *P. memnon* are not controlled by one locus but by a whole set of loci. If *anura* is a recombinant, then there must be at least one locus (call it T) controlling the presence (allele T_+) or absence (T_-) of a tail and at least one other locus (C) controlling the color patterns (C_1 for *achates*, and C_2, C_3, etc., alleles for other color morphs). *Achates* would have a genotype made up of one or two sets (depending on whether the alleles are dominant) of the two-locus genotype T_+C_1, and *anura* would have T_-C_1, after recombination between a tailless morph and *achates*. The loci in question are so tightly linked that these recombinants practically never arise in the laboratory — which is why the different multilocus genotypes appear, when crossed, to segregate like single-locus genotypes. We can predict that if more than one locus really is involved, a sufficiently large number of crosses should be able to break one of the "alleles" (such as the *anura* "allele") into several real combinations of alleles at several loci.

. . . but it turned out that at least five loci were

From *anura* alone, at least two loci could be inferred to control the mimetic polymorphism of *P. memnon*; but other rare morphs have also been found. Some rare morphs, for example, combine the forewing color of one morph and the hindwing pattern of another, suggesting that separate loci control the color of the fore- and hindwings. When all the inferred recombinants are considered together, at least five loci seem to be at work: T, W, F, E, and B. They control, respectively, presence or absence of tail, hindwing pattern, forewing pattern, epaulette color, and body color. The *anura* morph is a recombinant between the T locus and the other four. The common morphs, which mimic natural models, should each consist of a particular set of alleles at the five loci. The morph mimicking model species no. 1, for example, might have genotype $T_+W_1F_1E_1B_1/T_+W_1F_1E_1B_1$, and another morph (mimicking a second model) might have

$T\_W_2F_2E_2B_2/T\_W_2F_2E_2B_2$ or $T\_W_2F_2E_2B_2/T_+W_1F_1E_1B_1$. The recombinant genotypes such as $T_+W_1F_1E_2B_2$ do not exist naturally, except as very rare forms like *anura*.

The point to remember is that each of the morphs of *P. memnon* is thought to be controlled by a multilocus genotype. The genetics of the mimetic polymorphism in swallowtails differs from the camouflage polymorphism in the peppered moth (Section 5.7, p. 108), in which the different morphs are controlled by genotypes at one locus. A whole set of one-locus genotypes is needed to produce each of the swallowtail butterfly morphs.

8.2 Genotypes at different loci in *Papilio memnon* are coadapted

Mimicry requires the correct combination of alleles at all the loci concerned

How will natural selection act on a rare recombinant morph of *Papilio memnon*, such as *anura* in Java? Successful mimicry requires as complete a resemblance as possible between a mimic and its model. A potential mimic that mixes the patterns needed to mimic two species will mimic neither as successfully as a mimic that resembles one model in all respects. It will probably be selected against. *Anura* has the color pattern of *achates*, but will not mimic the model species of *achates* because it lacks a tail on its hindwings. The models of tailless morphs, in turn, have different color patterns, and *anura* will not mimic them either.

In general, natural selection will act against any recombinants between the mimetic five-locus genotypes. A five-locus genotype that mimics one model species in all five respects will be favored. But a swallowtail collage, which mimics one model in three aspects and another model in two other aspects, will look like neither and will be selected against. The genes at the five loci in this situation are said to be *coadapted*, or to show *coadaptation*. Coadaptation means that a gene (or genotype), such as T_+ (or T_+/T_+), is favored by selection if it is in the same body as a particular gene (or genotype), such as W_1 (or W_1/W_1), at another locus, but is selected against when combined with other genes (or genotypes), such as W_2 (or W_2/W_2), at that other locus. For example, selection favors $T_+W_1F_1E_1B_1/T_+W_1F_1E_1B_1$ and $T\_W_2F_2E_2B_2/T\_W_2F_2E_2B_2$ individuals, but (if the alleles with the 2 subscript are dominant) works against $T_+W_2F_2E_2B_2/T_+W_1F_1E_1B_1$ individuals. Selection has not been empirically confirmed to work against the recombinant forms of *P. memnon*, but the argument is quite convincing.

8.3 Mimicry in *Heliconius* is controlled by more than one gene, but they are not tightly linked

The passion flower butterflies of the genus *Heliconius* make an interesting comparison with *Papilio memnon*. In South America, two species of *Heliconius*, *H. melpomene* and *H. erato*, have multiple mimetic forms (Figure 8.1). The color patterns are again controlled by many loci: 15 in *H. erato* and 12 in *H. melpomene*. However, in both species the loci are scattered at random among the chromosomes rather than being tightly

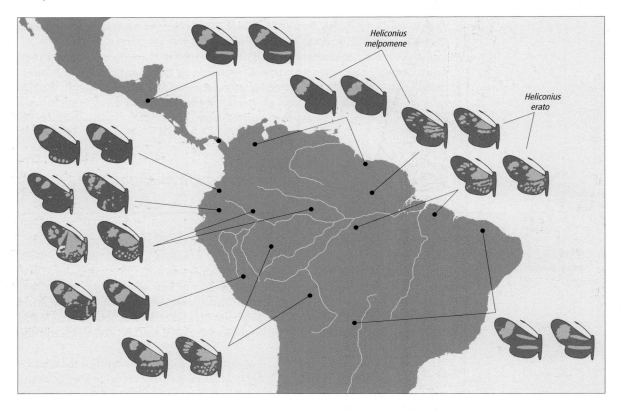

Figure 8.1

Two species of the butterfly *Heliconius* form parallel mimicry rings in South America. At each site indicated, both *H. erato* and *H. melpomene* are present and mimic each other; in different places the two species vary, in parallel, in appearance. Both species are poisonous. Redrawn, by permission of the publisher, from Turner (1976).

Heliconius also exists in multiple forms, . . .

linked. When two morphs of a *Heliconius* species are crossed, the offspring contain a kaleidoscopic variety of non-mimetic recombinant forms that resemble neither parent nor any known morph of the species.

Why do the genetics differ in *Heliconius* and *P. memnon*? The reason is probably geographic. At any one site in the range of *P. memnon*, several morphs are often living side by side. Crosses between them will happen with high frequency naturally. But in *Heliconius* only one morph is usually living at any one place. The different morphs are mainly geographically separated and will not interbreed in nature. Moreover, in *Heliconius*, the areas of overlap between neighboring morphs are probably due to recent range expansions: in the past the ranges were probably completely separate. The non-mimetic recombinant forms of *Heliconius* are usually generated only when morphs from different places are put together in the lab. In *Heliconius* it does not matter if the mimicry genes are scattered around the chromosomes, because the non-mimetic progeny are not usually produced. In *P. memnon* it does matter. If the mimicry genes were not linked, the recombinants would be produced — and be killed by predators.

. . . but its genetics differ from *Papilio memnon*

8.4 Two-locus genetics is concerned with haplotype frequencies

A haplotype is a haploid combination of genes at more than one locus

The theory of population genetics for a single locus is concerned with gene frequencies. The analogous variable in two-locus population genetics is *haplotype* frequency. (The term haplotype has two meanings. Here it refers to a combination of alleles at more than one locus. It is also used, in DNA sequencing, to refer to the base sequence of one of an individual's two sets of DNA.) For two loci with two alleles each (A_1 and A_2, B_1 and B_2) there are four haplotypes, A_1B_1, A_1B_2, A_2B_1, A_2B_2. A diploid individual's genotype will be something like A_1B_1/A_1B_2.[1] It has two haplotypes, one inherited from each parent, just as a one-locus genotype contains two genes from the two parents. If the A- and B-loci are on the same chromosome, each haplotype is a gene combination on a chromosome; but haplotypes can also be specified for loci on different chromosomes. The frequency of a haplotype in a population can be counted as the number of gametes bearing a particular combination of genes. A haplotype can be specified for any number of loci. We shall mainly discuss two-locus haplotypes, but the haplotypes in the *Papilio memnon* example had five loci, and the mimetic patterns of *Heliconius* are controlled by 12 or 15 gene loci. As this chapter will show, to understand the evolution of haplotype frequencies, we need some concepts that do not exist for gene frequencies. Two-locus population genetics is therefore not simply a doubled-up version of single-locus population genetics.

8.5 Frequencies of haplotypes may or may not be in linkage equilibrium

We can begin by asking a question like the one that led to the Hardy–Weinberg theorem for one locus. In the absence of selection, and in an infinite population with random mating, what will be the equilibrium frequencies of haplotypes? The question for multiple loci will lead us to another important concept, called *linkage equilibrium*.

The simplest case is for two loci with two alleles each. The crucial trick is to write the observed haplotype frequencies in terms of the gene frequencies at each locus, plus or minus a correction factor, called D. Let the gene frequency in the population of $A_1 = p_1$, $A_2 = p_2$, $B_1 = q_1$, and $B_2 = q_2$. Then:

---

[1] In this chapter, oblique strokes indicate diploid genotypes. Thus A_1/A_1 is a diploid genotype at one locus. The convention is to prevent confusion with haplotypes, which are written here without an oblique stroke, e.g., the A_1B_1 haplotype. A haplotype refers to the alleles at two (or more) loci that an individual received from one of its parents. A diploid individual has two haplotypes. Haplotypes have two different letters (for two loci), one-locus genotypes have only one letter. Diploid two-locus genotypes are also here written with an oblique stroke, e.g., A_1B_1/A_2B_2.

Haplotype	Frequency in population
A_1B_1	$a = p_1q_1 + D$
A_1B_2	$b = p_1q_2 - D$
A_2B_1	$c = p_2q_1 - D$
A_2B_2	$d = p_2q_2 + D$

Linkage disequilibrium is defined
as a deviation from a random
expectation

The total frequencies add up to one. That is, $a + b + c + d = 1$. (Also, $p_1q_1 + p_1q_2 + p_2q_1 + p_2q_2 = 1$, and the sum of the two $+D$ and two $-D$ factors is zero.) The important term to understand is D; it is a measure of "linkage disequilibrium." Linkage equilibrium is when $D = 0$ and means that the alleles at the two loci are combined independently. The two B alleles would then be found with any one A allele (such as A_1) in the same frequencies as they are found in the whole population. If we take all the A_1 genes, q_1 of them are with B_1 genes and q_2 with B_2 genes; likewise, q_1 of the A_1 genes are with B_1 genes and q_2 with B_2. At linkage equilibrium, the frequency of the A_1B_1 haplotype is p_1q_1. D measures the deviation from linkage equilibrium. If $D > 0$, A_1 is more often found with B_1 (and less often with B_2) than would be expected if alleles at the two loci were combined at random — the population contains an excess of $A_1 B_1$ (and of A_2B_2) haplotypes.

The butterflies provide an example

Papilio memnon is an example of high linkage disequilibrium. If Clarke and Sheppard are correct, the allele T_+ is almost always combined with the other alleles W_1, F_1, E_1, and B_1 rather than with W_2, W_3, or W_4 (and equivalent alleles at the other loci). There is a large excess of the haplotypes $T_+W_1F_1E_1B_1$, $T_-W_2F_2E_2B_2$, $T_-W_3F_3E_3B_3$, etc., while haplotypes such as $T_+W_2F_2E_2B_2$, $T_+W_1F_2E_2B_2$, or $T_+W_1F_1E_2B_2$ are almost absent. The linkage disequilibrium in *P. memnon*, as we have seen, is caused by selection. In this section, however, we are asking how a set of haplotype frequencies should change through time in the absence of selection.

We construct a model of haplotype
frequencies over time

Let us return again to the haplotype A_1B_1. It has frequency defined as a in one generation. What will its frequency be in the next generation? (We can use again the notation a' as the frequency of A_1B_1 one generation on.) In the absence of selection, the frequencies of each gene will be constant, but the frequencies of the haplotypes can be altered by recombination. The frequency of A_1B_1 cannot be altered by recombination in double or single homozygotes: the number of A_1B_1 haplotypes coming out of an A_1B_1/A_1B_1 individual, or of an A_1B_1/A_1B_2 individual, is the same as the number going in, whether or not there is recombination. The frequency can only be altered by recombination in the double heterozygotes A_1B_1/A_2B_2 and A_1B_2/A_2B_1. When recombination takes place in an A_1B_1/A_2B_2 individual, the number of A_1B_1 haplotypes is decreased. When it takes place in an A_1B_2/A_2B_1 individual, the number of A_1B_1 is increased. To be exact, half the genes of an A_1B_1/A_2B_2 double heterozygote are A_1B_1; when recombination hits between the loci the frequency of A_1B_1 decreases by an amount $-1/2$. Similarly, recombination in an A_1B_2/A_2B_1 individual increases the frequency of A_1B_1 by an amount $+1/2$.

The frequency of A_1B_1/A_2B_2 heterozygotes in the population is $2ad$ and of A_1B_2/A_2B_1 is $2bc$. The frequency at which the alleles at two loci are recombined per generation is defined as r. (r can theoretically have any value up to a maximum of 0.5, if the loci are on different chromosomes; r is between 0 and 0.5 for loci on the same chromosome depending on how tightly linked they are — see Section 2.8, p. 35.) So:

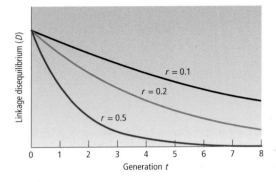

Figure 8.2
Non-random associations between genes at different loci are measured by the degree of linkage disequilibrium (D). Recombination between the loci breaks down the linkage disequilibrium, which decays at an exponential rate equal to the recombination rate between the loci.

$$a' = a - \frac{1}{2}r2ad + \frac{1}{2}r2bc$$

$$a' = a - r(ad - bc)$$

Now, the expression $(ad - bc)$ is simply equal to the linkage disequilibrium D. (This is easy to confirm by multiplying out $ad - bc$ from the definitions above of a, b, c, and d.) If $D = 0$, i.e., if the genes are randomly associated, the haplotype frequencies are constant: $a' = a$. But if there is an excess of A_1B_1 haplotypes, the excess decreases by an amount rD per generation. The same relation holds true for any successive pair of generations. We can see what is happening graphically if we substitute for a in the equation:

$$a' = p_1q_1 + D - rD$$

$$a' - p_1q_1 = (1 - r)D$$

The difference between a and p_1q_1 is the amount of "excess" of the A_1B_1 haplotype (i.e., the amount by which the frequency exceeds the random frequency). It is also equal to the linkage disequilibrium ($D = a - p_1q_1$). Therefore:

$$D' = (1 - r)D$$

Linkage disequilibrium tends to decay over time

In the absence of selection and in an infinite random mating population, the amount of linkage disequilibrium undergoes exponential decay at a rate equal to the recombination rate between the two loci (Figure 8.2). In other words, the difference between the actual frequency of a haplotype such as A_1B_1 (a) and the random proportion (p_1q_1) decreases each generation by a factor equal to the recombination rate between the loci.

Over time, any non-random genic associations will disappear; recombination will destroy the association. The higher the rate of recombination, the more rapid the destruction. The highest possible value of r is $1/2$, which is true when the two loci are on different chromosomes. Genic associations persist longer for tightly linked loci on the same chromosome, as we would intuitively expect.

The equilibrial haplotype proportions have $D = 0$. At equilibrium:

Haplotype	Equilibrial frequency
A_1B_1	$a = p_1q_1$
A_1B_2	$b = p_1q_2$
A_2B_1	$c = p_2q_1$
A_2B_2	$d = p_2q_2$

Recombination breaks down linkage disequilibrium

These are the haplotype frequencies we have met before and called linkage equilibrium. We can now see why it is called an "equilibrium." In the absence of selection, the action of recombination will drive the haplotypes to these frequencies and then keep them there.

Recombination randomizes genic associations over time. If an excess of one haplotype such as A_1B_1 exists, recombination will tend to break it down, and A_1 will end up with B_1 and B_2 in their population proportions (q_1 and q_2) and B_1 with A_1 and A_2 in their population proportions (p_1 and p_2). At linkage equilibrium, each of the two alleles at the A locus, A_1 and A_2, are then associated with B_1 in the same proportion.

Linkage equilibrium is, in a way, the analogy for a two-locus system of the Hardy–Weinberg equilibrium for the one-locus system. It describes the equilibrium that is reached in the absence of selection, and in an infinite, randomly mating population. Linkage equilibrium, however, is a property of haplotypes, not genotypes. A diploid individual has two haplotypes, and at equilibrium the genotypes at each locus will be in Hardy–Weinberg proportions while the haplotypes are at linkage equilibrium. Notice also that whereas the Hardy–Weinberg equilibrium for one locus is reached instantly in one generation (Section 5.3, p. 98), it takes several generations for linkage equilibrium to be reached.[2]

It is interesting to know if a population is in linkage equilibrium or disequilibrium

Linkage equilibrium has a three-fold interest. The Hardy–Weinberg theorem for one locus was the simplest model in single-locus population genetics and it illustrated how to construct a model with recurrence relations for gene frequencies. The model of linkage equilibrium is, likewise, the simplest model for two loci and shows us how to construct a recurrence relation for haplotype frequencies. Its second interest, also like the Hardy–Weinberg theorem, is that it provides a theoretical baseline telling us whether anything interesting is going on in a population. Deviations from Hardy–Weinberg proportions in a natural population suggest that selection, or non-random mating, or sampling effects may be operating. Likewise, if two loci are in linkage disequilibrium, we can also suspect that one or more of these variables are at work. If the first thing we had discovered about *Papilio memnon* had been its high linkage disequilibrium, we should have been led on to study how selection was operating on the loci, and perhaps

---

[2] The terms "linkage equilibrium" and "linkage disequilibrium" are not very satisfactory. They were first used by Lewontin and Kojima in 1960. "Linkage disequilibrium" can exist without linkage — among genes on different chromosomes — and it can also exist at equilibrium, as we shall see. It is, however, like the Hardy–Weinberg equilibrium, an equilibrium under certain specifiable conditions. The word linkage is avoided in certain other terms, such as "gametic phase equilibrium," which are also in use; but linkage disequilibrium is the commonest term. Also, there are other ways of measuring non-random associations between genes besides D, but all the points of principle can be made with D.

ended up discovering the mimetic polymorphism. In fact, the direction of research in *P. memnon* was the other way round; but the general point, that linkage disequilibrium indicates something interesting, holds true.

Linkage equilibrium can also tell us whether the more complex two-locus theory is needed in a real case. To a rough approximation, the theory of population genetics for a single locus is satisfactory for populations in linkage equilibrium. It is when genes become non-randomly associated that a two-locus model is needed. The case we have been discussing can show why. At linkage equilibrium, A_1 and A_2 are equally associated with B_1. To understand evolution at the A locus we can then ignore the relative fitnesses of B_1 and B_2, because if B_1 is fitter than B_2, the association with B_1 will benefit A_1 and A_2 equally (and B_2 will equally detract from them). But if A_1, for instance, is more associated with B_1 than is A_2 (that is, the population is in linkage disequilibrium), then any advantage of B_1 over B_2 will passively give rise to an advantage to A_1. To understand frequency changes of A_1 we then need to know the relative fitnesses of B_1 and B_2, and the degree of association A_1 has with them: we need a two-locus model.

The more complex multilocus models are only needed for populations in linkage disequilibrium

8.6 Human HLA genes are a multilocus gene system

The HLA system in humans is a set of linked genes on human chromosome 6; they control "histocompatibility" reactions. When an organ is transplanted from one individual to another, it is immunologically rejected by the recipient in a matter of days — skin grafts last about 2–15 days, for instance. The rejection implies that the immune system can distinguish between "self" and "foreign" cells, and the distinction is generally believed to be achieved by the products of the HLA genes. The HLA genes code for transmembrane proteins of immune system cells. Good evidence for their role comes from the time course of kidney transplant rejection among siblings that either have or have not been matched for their HLA genes. For kidney transplants between HLA-matched siblings, over 90% of transplants still survive after 48 months; but among HLA-unmatched siblings, 90% survive for 4 months and only about 40% for 48 months.

The HLA genes work in the immune response

The HLA system contains a number of genes (Figure 8.3). We shall concentrate on two of them, called HLA-*A* and HLA-*B*. Each HLA locus, in a human population, is highly polymorphic: at the *B* locus alone there will be maybe 16 alleles with frequencies of 1–10% and many more rare alleles; for example, a sample of 874 people in France contained 31 different alleles at the *B* locus and another 17 alleles at the *A* locus. These are exceptionally high degrees of variability. More typical loci (outside the HLA) might have one to five alleles, many less than the number found in the HLA. The reason for the high variability is uncertain: but it would allow the HLA genotype of an individual, even in a large population, to be unique, which is presumably important in the distinction of self from foreign cell types.

Some HLA loci are highly polymorphic

Particular HLA alleles are associated with particular diseases, and resistance to them. The strongest association found so far is between ankylosing spondylitis and the allele B27; 90% of people with the disease have the B27 allele, against only 7% in the population at large. On the other hand, allele B27 confers better than average resistance to

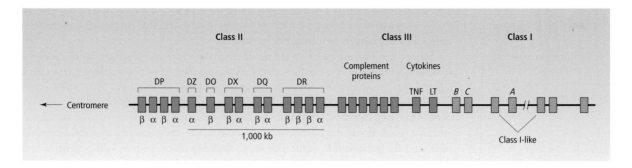

Figure 8.3
Genetic map of the human HLA loci on chromosome 6. The text concentrates on the *A* and *B* loci, but there are many more genes in the histocompatibility system (not drawn exactly to scale).

HIV. The full diversity of HLA types may reflect a history of coevolution between humans and disease agents. Disease agents may have tried to fool the immune system into treating the agent as part of the body, and the human population would then respond over evolutionary time by evolving new HLA alleles as new, reliable indicators of "self." This would provide a further advantage to variability in the HLA loci. A heterozygous individual with two HLA proteins can compare itself with a possible invader in two ways: the invader has to match a homozygote only in one respect, but a heterozygote has to be matched in two independent respects. The HLA loci therefore probably show heterozygous advantage (Section 5.12, p. 123), and the same process may have caused the exceptional pattern of evolution in silent and amino acid-changing bases within codon triplets (Section 7.8.2, p. 182).

Linkage disequilibrium exists between some HLA loci

The HLA system also provides examples of linkage disequilibrium. Particular combinations of genes are found in greater than random proportions. In North European populations, there is characteristically an excess of the A_1B_8 haplotype. Figure 8.4a is a more general picture. It shows the linkage disequilibrium values for all combinations of *B* alleles and the allele A_1. There could be an analogous graph for each *A* allele. In Figure 8.4a, $D = 0.07$ for A_1B_8. If A_1 and B_8 combined in their population proportions, A_1B_8 would have a frequency of about 0.023 (2.3%); but in fact it is found in about 9.3% of individuals ($0.093 - 0.023 = D = 0.07$). In all, the HLA-*A* and -*B* loci have about six clear cases of linkage disequilibrium; A_1B_8 and A_3B_7 are the most striking. The reason why these haplotypes are found in greater than random proportions is unknown, though it is generally believed to be due to selection in favor of the gene combinations. But selection is not the only possible reason for linkage disequilibrium, as the next section will reveal.

8.7 Linkage disequilibrium can exist for several reasons

Recombination breaks down non-random genic associations, and yet in some cases like *Papilio memnon* and the HLA genes, non-random associations exist. What is

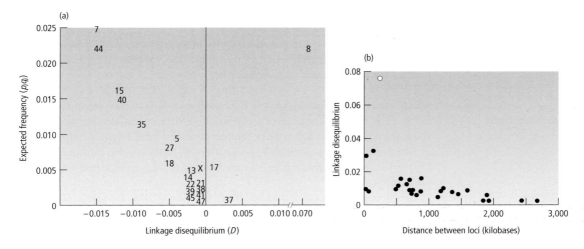

Figure 8.4

(a) One example of the pattern of degrees of linkage disequilibrium (*D*) in the HLA: the linkage disequilibrium between 21 *B* alleles and the A_1 allele. An analogous graph can be drawn for every *A* allele. The A_1B_8 haplotype occurs at a much higher than random frequency (note the gap in the *x*-axis). The *y*-axis is the expected frequency of the haplotype if the alleles were associated at random. Thus the observed frequency of a haplotype is the *y*-axis value plus (or minus) its *x*-axis value. (b) The linkage disequilibrium between eight HLA loci: more closely linked loci show higher linkage disequilibrium; the *y*-axis is a kind of average linkage disequilibrium for the multiple alleles (see Hedrick *et al.* (1991) for the exact measure). Figure 8.3 is a map of the loci, and compare with Figure 8.2 for the effect of recombination. Redrawn, by permission of the publisher, from Hedrick *et al.* (1991).

Linkage disequilibrium can be caused by selection . . .

causing the linkage disequilibrium? In *Papilio* and in at least some of the HLA associations, it is probably due to selection. If selection favors individuals with particular combinations of alleles, then it produces linkage disequilibrium. But selection is not the only possible cause for linkage disequilibrium, and a full study of a real case must examine three other factors.

The first factor is linkage. For linked loci, a number of generations are required for recombination to do its randomizing work (see Figure 8.2). Loosely linked loci will not show linkage disequilibrium for long. However, as the rate of recombination between two loci decreases the amount of time that alleles can be non-randomaly associated between them goes up. This may be one reason why, in the human HLA system, the average linkage disequilibrium is larger between more closely linked loci (Figure 8.4b). For tightly linked loci, some linkage disequilibrium can persist indefinitely.

. . . and by linkage, . . .

A second factor that can cause linkage disequilibrium is random drift. Random processes have the interesting property of being able to cause persistent, not just transitory, linkage disequilibrium. If random sampling produces by chance an excess of a haplotype in a generation, linkage disequilibrium will have arisen. This is true for all four haplotypes: random sampling that produces an excess of any of them will disturb the state of linkage equilibrium. Any haplotype could be "favored" by chance, so the disequilibrium is equally likely to have $D > 0$ or $D < 0$. As a population approaches linkage equilibrium, all random fluctuations in haplotype frequencies will tend to be away from the linkage equilibrium values. If a population is well away from the point of

. . . randon drift . . .

linkage equilibrium, random sampling is equally likely to move it towards, as away from, the equilibrium. Most natural populations are probably near linkage equilibrium (see below, Figure 8.5), and then the balance between the random creation of linkage disequilibrium and its destruction by recombination, in small enough populations, is such that linkage disequilibrium will persist.

The third factor is non-random mating. If individuals with gene A_1 tend to mate with B_1 types rather than B_2 types, A_1B_1 haplotypes will have excess frequency over that for random mating. (The exact effect depends on whether it is homozygous A_1/A_1 individuals that mate non-randomly, or the homozygotes and the A_1/A_2 heterozygotes, and on whether they mate preferentially only with B_1/B_1 homozygotes, or with B_1/B_2 heterozygotes too. But the general effect of non-random mating on linkage disequilibrium is not complicated.)

. . . and non-random drift

The three processes other than selection probably account for some cases of linkage disequilibrium in nature. The process that has most interested evolutionary biologists, however, is natural selection. Let us now consider how we can model the effect of selection on haplotype frequencies.

8.8 Two-locus models of natural selection can be built

The effect of natural selection on haplotype frequencies in two-locus models, like its effect on gene frequencies in single-locus models, depends on the fitnesses of the genotypes. We have to write down the fitness of each genotype, and there are many possible ways in which it can be done. In one of the simplest two-locus models, the fitness of a two-locus genotype is the product of the fitnesses of its two single-locus genotypes. The model is realistic if the fitness effect of one locus is independent of the genotype at the other. Suppose, for example, that the A locus influences survival from age 1 to 6 months, such that:

Natural selection may work on each locus independently . . .

Genotype	A_1/A_1	A_1/A_2	A_2/A_2
Chance of survival to age 6 months	w_{11}	w_{12}	w_{22}

and the other locus influences survival from age 6 to 12 months:

Genotype	B_1/B_1	B_1/B_2	B_2/B_2
Chance of survival from 6 to 12 months	x_{11}	x_{12}	x_{22}

The total chance of surviving from age 1 to 12 months would then be the product of the two genotypes that an individual possessed because selection at age 1–6 months is independent of selection at age 6–12 months:

. . . producing multiplicative fitnesses

	A_1/A_1	A_1/A_2	A_2/A_2
B_1/B_1	$w_{11}x_{11}$	$w_{12}x_{11}$	$w_{22}x_{11}$
B_1/B_2	$w_{11}x_{12}$	$w_{12}x_{12}$	$w_{22}x_{12}$
B_2/B_2	$w_{11}x_{22}$	$w_{12}x_{22}$	$w_{22}x_{22}$

These fitnesses are called *multiplicative*. An individual's fitness for its two genotypes is found by multiplying the fitnesses of each of its one-locus genotypes. The genotypes are independent, in the sense that the effect of one genotype on survival is independent of the other locus. An individual with the genotype A_1/A_1 has a chance of surviving from age 1 to 6 months of w_{11} whether its genotype at the other locus is B_1/B_1, B_1/B_2, or B_2/B_2.

A model of haplotype frequency changes can be built

The next step is to derive a recurrence relation between the frequency of a haplotype in one generation and in the next. However, we do not need to work through all the algebra here. In outline the procedure is the same as for the single-locus case, with the additional factor of recombination. The recurrence relation for haplotype frequency takes account of the frequency and fitness of all the genotypes that a haplotype is found in. It also has to add and subtract the number of copies gained and lost by recombination: we multiply by $(1 - r)$ the frequency of the double heterozygotes containing the haplotype and by r that of the double heterozygote that can generate it if recombination occurs. The Mendelian rules are then applied, and the frequency of the haplotype in the next generation results.

Which kinds of selection cause linkage disequilibrium? The question is important because, as we have seen, two-locus models are particularly needed when linkage disequilibrium exists. With multiplicative fitnesses, the haplotype frequencies almost always go to linkage equilibrium. (Linkage disequilibrium is only possible if both loci are polymorphic. If one gene is fixed at either locus, $D = 0$ trivially. The fitnesses, w_{11}, etc., as written above were frequency independent. A doubly heterozygous equilibrium then requires heterozygous advantage at both loci: $w_{11} < w_{12} > w_{22}$, $x_{11} < x_{12} > x_{22}$; see Section 5.12.1, p. 123.) If ever linkage disequilibrium exists between two loci that have multiplicative fitness relations, that disequilibrium will decay to zero as the generations pass.

Linkage disequilibrium arises with epistatic fitnesses, such as existed in the mimetic butterflies

The more interesting case is when the fitnesses of the two loci interact *epistatically* (the fitnesses are said to show *epistasis*). The selection in the mimetic polymorphism of *Papilio memnon* is epistatic. Epistatic interaction means that the fitness effects of a genotype depend on what genotype it is associated with at the other locus.

We can simplify the situation in *P. memnon* by imagining that one locus controls whether the butterfly has a tail on its hindwing and one other locus controls coloration. (In reality, at least four loci influence coloration.) Let T_+ (presence of tail) be dominant to T_- (absence). At the other locus, C_1 is dominant, and C_1/C_1 and C_1/C_2 individuals have a color pattern that mimics a model species with a tail, whereas C_2/C_2 individuals are colored like a model species that has no tail. The relative fitness of each genotype depends on what the genotype at the other locus is. For example, a T_+/T_+ genotype in the same butterfly as a C_2/C_2 will be less fit than a T_-/T_- with C_1/C_1. The fitnesses can be written as follows (the simplification relative to the earlier fitness matrixes arises because of dominance, and because there is one term for the fitness of both loci together rather than one term for each locus):

	T_+/T_+	T_+/T_-	T_-/T_-
C_1/C_1	w_{11}	w_{11}	w_{21}
C_1/C_2	w_{11}	w_{11}	w_{21}
C_2/C_2	w_{12}	w_{12}	w_{22}

In the case we discussed, w_{12} is the fitness of a butterfly with a tail and the color pattern of a tailless model, Therefore, $w_{12} < w_{11}$ and w_{22}. w_{21} is the fitness of a butterfly without a tail, but the color pattern of a tailed model. Therefore, $w_{21} < w_{11}$ and w_{22}. Selection now favors the $T_+/-$ genotypes when they are with $C_1/-$ but not when with C_2/C_2, and T_-/T_- when it is with C_2/C_2 but not when with $C_1/-$ (the dash implies it does not matter which gene is present, because of dominance). The fitness relations are epistatic. There can now be a doubly polymorphic equilibrium. All four alleles will be present, and the haplotypes T_+C_1 and T_-C_2 will have disproportionately high frequencies. T_+C_2 and T_-C_1 haplotypes are selected against, because they often find themselves in poorly mimetic butterflies. Linkage disequilibrium ($D > 0$ in this case) exists at the equilibrium.

In general, selection can only produce linkage disequilibrium at equilibrium when the fitnesses of the genotypes at different loci interact epistatically. Not all epistatic fitness interactions generate doubly polymorphic equilibria with linkage disequilibrium. But all (or nearly all) such equilibria do have epistatic fitnesses.

It is an empirical question, how often real fitness interactions are multiplicative or epistatic

We have been discussing the different sorts of fitness interactions — multiplicative or epistatic (and there are others too) — as properties of formal models. Real genes in real organisms will have fitness interactions too, and the more important question is what sort of interactions these are. There are cases like *Papilio* in which epistasis is present and powerful; but these may be isolated examples rather than representing a general condition. Evolutionary biologists are interested in whether fitness interactions between loci are generally epistatic and generate strong linkage disequilibrium, or whether they are generally independent and generate linkage equilibrium. These two extremes roughly correspond to a more "holistic" and a more "atomistic" (or "reductionist") school of thought, though that is not to say that they correspond to two clearly demarcated camps of biologists.

No general answer is yet available, but it is possible to make some observations. Different loci will tend to interact multiplicatively when they have independent effects on an individual's survival and reproduction. Some biologists suggest that loci which influence events at different times in an organism's life are more likely to show multiplicative fitness relations (though it is also possible for such events to interact). Epistatic interactions may be more likely for loci controlling closely interdependent parts of an organism. The extent to which we expect loci to interact epistatically or not then loosely depend on how atomistic or holistic a view we have of the organism (see also Section 8.12, below).

Fitness epistasis is not the same as genetic interaction

Notice that epistatic fitness interaction is not the same as mere physiological or embryological interaction. Fitness epistasis requires heterozygosity at two loci. Imagine a case in which the *A* locus controls, say, muscle strength and the *B* locus controls metabolic rate. Muscles and metabolism interact in a physiological sense: when muscles are put to work, the metabolic rate goes up. However, if the population is fixed for homozygotes at both loci (all individuals are A_1B_1/A_1B_1) then there cannot be any fitness epistasis. Epistatic fitness requires heterozygosity at both loci, and the kind of fitness relations we saw in the *Papilio memnon* example. This is a special condition. Though it is often called fitness "interaction," the term interaction is being used in a technical, not a colloquial, sense.

We can also test empirically how common epistatic fitness interactions are in nature. Linkage disequilibrium is produced by epistatic selection, and the degree of linkage

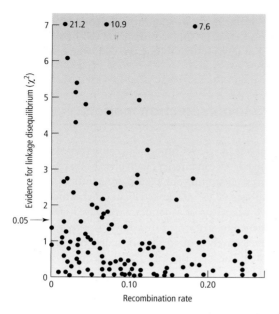

Figure 8.5

Linkage disequilibrium (on the y-axis) among gel electrophoretic samples of pairs of genes in *Drosophila*. It is plotted as a χ^2 value. The χ^2 value indicates how strong the evidence is for linkage disequilibrium: the higher the χ^2, the more linkage disequilibrium between that pair of genes. (The χ^2 value corresponding to a statistical significance of 0.05 is indicated.) Most of the gene pairs have insignificant or low linkage disequilibrium (i.e., low χ^2). The x-axis is the rate of recombination between the pairs of genes. Redrawn, by permission of the publisher, from Langley (1977).

disequilibrium in a population can be measured. If it is high, then epistatic selection may be common. The argument works in one direction but not the other: because there are several possible causes of linkage disequilibrium (Section 8.7), its existence does not demonstrate epistatic selection. However, if linkage disequilibrium is absent or low, we can infer that epistatic selection is unimportant in nature.

A few general surveys of the extent of linkage disequilibrium in natural populations have been made. One by Maynard Smith *et al.* (1993) for bacteria found high levels of linkage disequilibrium in some species, such as *Escherichia coli* (which lives in our, and other mammals', guts), but low levels in other species, such as *Neisseria gonorrhoeae*. The reason why many bacteria show linkage disequilibrium is that they reproduce asexually, and there is no recombination to break the linkage disequilibrium down. But some bacteria do sometimes exchange genes between individual cells, though not by the kind of sexual processes that eukaryotes use. *N. gonorrhoeae* presumably has enough genetic exchange between individuals to produce linkage equilibrium.

In eukaryotic organisms that are known to reproduce sexually, the evidence suggests that there is little deviation from linkage equilibrium in nature. The main evidence has historically come from surveys of protein polymorphisms, to see directly whether genes at different loci are associated. Figure 8.5 illustrates some comprehensive results for the fruitfly *Drosophila*. Some evidence of linkage disequilibrium is found, but the results suggest the level is low and most loci are in linkage equilibrium. DNA sequence evidence is now also becoming available and shows much the same pattern. Epistatic interactions are undoubtedly important in particular cases, like *Papilio*, but they may not be common for polymorphic loci in sexually reproducing species.

Not all biologists agree with this conclusion. They might be unconvinced by the evidence of Figure 8.5, perhaps calling it "limited," or "for a single species." The amount of

Linkage disequilibrium has been measured in microbes, . . .

. . . and in fruitflies

interaction between loci that must go on during the development of a complex, organic body is so high that they would expect epistatic fitness interactions to be common. Such is the assumption of the school of thought that follows Wright, whose ideas we shall discuss at the end of the chapter.

8.9 Hitch-hiking occurs in two-locus selection models

Natural selection at one locus can cause evolution at linked loci

When a gene is changing frequency at one locus over time, it can cause related changes at linked loci; conversely, events at linked loci can interfere with one another. Suppose, for instance, that directional selection is substituting one allele A' for another (A) at one locus, and there is a neutral polymorphism (B, B') at a linked locus. Then whichever of B and B' happened to be linked with A' when it arose as a mutant will have its frequency increased. If the new mutant A' happened to arise on a B-bearing chromosome, B will eventually be fixed together with the selectively favored allele A' unless recombination splits them before A has been eliminated. The increase in the B allele frequency is due to *hitch-hiking*.

Another possibility is for the polymorphism at the B locus to be a selectively "balanced" polymorphism, due to heterozygous advantage. Suppose again that a selectively favored mutation A' arises at a linked locus, and that it happens to arise on the same chromosome as a B allele. Now the polymorphism at the B locus will interfere with the progress of A'. As A' increases in frequency by directional selection it will increase the frequency of B with it. Because A' is linked to B, it will be more likely to be in a body with a B/B homozygote than will its allele A, and less likely to be in a B/b heterozygote. B/b has higher fitness than B/B and the selection against B/B individuals will also work against the A' gene. Depending on the selection coefficients at the two loci, and the rate of recombination between them, the heterozygous advantage at the B locus can slow the rate at which A' is fixed. The A' gene will then have to wait for recombination between the two loci before it can progress to fixation.

8.10 Selective sweeps can provide evidence of selection in DNA sequences

Natural selection at one locus tends to reduce diversity at linked loci

One consequence of hitch-hiking is that when natural selection fixes a new, favorable gene, the amount of genetic variation is reduced in the neighboring regions of the DNA. When a favorable mutation arises, it will initially be on a chromosome which has a particular sequence of nucleotides. As the mutation is fixed, it carries with it the nucleotides that are linked to it. Other nucleotide variants at neighboring sites in the DNA are eliminated, along with the inferior alleles at the locus where selection is acting. The result is reduced genetic diversity. (Genetic diversity can be measured by sequencing the DNA of many chromosomes from many fruitfly individuals, and counting the fraction of nucleotide sites that differ between two randomly picked chromosomes.)

The sweep reduces genetic diversity most at the locus where selection is acting. Nearby in the DNA, diversity will be reduced; further out, diversity will still be reduced

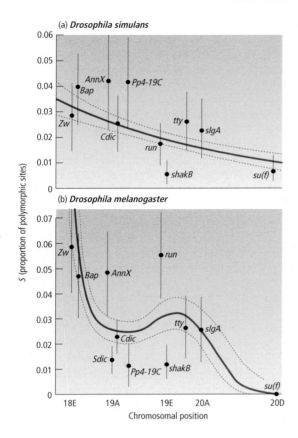

Figure 8.6

Selective sweep caused by recent substituion of the *Sdic* gene in *Drosophila melanogaster*. The *y*-axis gives the amount of genetic diversity. The *x*-axis is the position on the X chromosome. Diversity decreases toward the centromere (off to the right of the figure) where the recombination rate is decreased. The diversity near the *Sdic* gene in *D. melanogaster* is lower than we should expect from its position on the X chromosome. If there had not been a selective sweep of *Sdic*, the graph for *D. melanogaster* (b) would have looked much like the graph for *D. simulans* (a). The points are for various genetic loci, and the line around each point is the approximate 50% confidence interval. From Nurminsky *et al.* (2001)

but in decreasing amounts the further we go away from the selected locus. Recombination is more likely to have separated the favored mutation from its initially linked nucleotides at sites further away in the DNA. The homogenization (that is, reduction in diversity) of neighboring DNA when natural selection fixes a favorable new gene is called a *selective sweep*. As a mutation increases in frequency, it sweeps diversity out of the surrounding DNA.

Local reductions in genetic diversity
are a signature of selection

Local reductions in genetic diversity can be used as a "signature" of natural selection in DNA sequences. We can look down the DNA, and if we find a region of locally reduced diversity, one explanation is that natural selection has recently fixed a new gene somewhere in the region. Nurminsky *et al.*'s (2001) research on the gene called *Sdic* in *Drosophila melanogaster* is an example (Figure 8.6). The gene *Sdic* codes for a structure in the sperm. Figure 8.6b shows a trough in genetic diversity near *Sdic*, and this trough is part of Nurminsky *et al.*'s case that the version of *Sdic* in *D. melanogaster* has recently been fixed by natural selection.

A reduction in genetic diversity near a gene such as *Sdic* is not by itself strong evidence that selection has recently fixed a new version of the gene. Two alternative explanations need to be ruled out. One is that the mutation rate is locally depressed. This can be tested by McDonald & Kreitman's (1991) test (Section 7.8.3, p. 184). If the

Drosophila melanogaster has a local reduction of diversity near the gene Sdic . . .

mutation rate is low, we expect not only low diversity within a species, but also a low rate of evolutionary change. The rate of evolution can be found by comparing the gene in *D. melanogaster* and the closely related species *D. simulans*. In fact the rate of evolution is high, suggesting that the mutation rate has not been lowered near *Sdic*.

A second alternative is *background selection*. Deleterious mutations occur in the DNA (Section 12.2.2, p. 321, looks at the deleterious mutation rate). Natural selection acts against deleterious mutations, removing them from the population. As selection clears out deleterious mutations, it also reduces the local genetic diversity because any variants linked to a deleterious mutation will be removed along with it.

In some regions of the genome, the recombination rate is lower than in other regions. For example, recombination is less frequent near the centromere of a chromosome. Also, a whole chromosome may have a low recombination rate. Chromosome 4 in *Drosophila* is short and has a low recombination rate (Wang *et al.* 2002). In regions where the recombination rate is low, the diversity of DNA is known to be reduced: *D. melanogaster*'s fourth chromosome, and all chromosomes near their centromeres, show low genetic diversity. This reduction could be either because of selective sweeps or background selection. Both processes reduce genetic diversity, and both operate more powerfully where the recombination rate is low. Now, the *Sdic* gene is on the X chromosome and is near the centromere. The low local diversity could be due to background selection in a region of low recombination, rather than to a selective sweep.

. . . which is almost certainly due to a selective sweep

Figure 8.6 shows how Nurminsky *et al.* argue that the version of *Sdic* in *D. melanogaster* has caused a selective sweep. *D. simulans* (Figure 8.6a) shows a standard decrease in genetic diversity towards the centromere. The picture for *D. simulans* may well be due to background selection. If background selection caused the low diversity in *D. melanogaster* near the *Sdic* gene, we should expect much the same graph in both species. (There is no evidence that *Sdic* has undergone recent evolutionary change in *D. simulans*.) But Figure 8.6b shows that DNA diversity near *Sdic* in *D. melanogaster* is reduced relative to *D. simulans*. The reduced recombination rates near the centromere are not enough to explain the trough in diversity seen in *D. melanogaster*. The *Sdic* gene really does appear to have been fixed recently in *D. melanogaster*, and to have swept out the local diversity.

Selective sweeps, in which the local genetic diversity is reduced, can be added to the other signatures of selection that we looked at in Section 7.8 (p. 179 — signatures such as the relative rates of non-synonymous and synonymous evolution). The test has practical uses, and Box 8.1 describes how it can be used to detect which genes code for drug resistance in the malaria parasite. The test is most powerful if alternatives can be ruled out, and provides a further example of how DNA sequence data are allowing some novel tests of natural selection.

8.11 Linkage disequilibrium can be advantageous, neutral, or disadvantageous

Although linkage disequilibrium may be rare when we consider all the genes in a species, some examples still exist. We can distinguish between cases that are beneficial,

Box 8.1
A Genomic Hunt for Drug-resistance Genes

Chloroquine has been, since its introduction in 1946, one of the most effective and widely used drugs against malaria. Malaria is caused by the parasite *Plasmodium falciparum* (Section 5.12.2, p. 124), and *P. falciparum* that were resistant to chloroquine were first observed in 1957. Since then, chloroquine resistance has spread wherever the drug has been used. Medically, it is useful to know the mechanism of drug resistance. The genetic basis of drug resistance would classically have been identified by crosses; but now we can use genomic data and statistical tests for signs of selection. One recent study looked in the *P. falciparum* genome for regions of low genetic diversity and high local linkage disequilibrium — the signs of a selective sweep. The underlying idea is that any gene for drug resistance will have been selected for recently. Selection could have acted recently on other genes, but a sign of selection is at least a clue that could lead to the detection of a drug-resistance gene.

Wootton *et al.* (2002) looked at diversity at 342 marker sites in the *P.*

falciparum genome. They found locally reduced diversity in Asian, African, and South American *P. falciparum* in a region of chromosome 7 where a gene called *pfcrt* is located. The exact allele of *pfcrt* that has been selected for in each continent differs. Indeed, the sequences suggest that chloroquine resistance has originated independently four times — in Asia, Indonesia, South America, and Africa. However, the genomic site of lowered genetic diversity is the same in malaria parasites from all continents. At other regions of chromosome 7, diversity is low or high in inconsistent patterns between the continents. The *pfcrt* gene is one of a small number of sites where diversity is reduced in all populations.

The *pfcrt* gene is also a site of locally elevated linkage disequilibrium. A selective sweep produces linkage disequilibrium locally. As the frequency of the favored allele increases, the frequency of linked nucleotide variants will also be dragged up, producing linkage disequilibrium by hitch-hiking. The *pfcrt* locus is the only site on chromosome 7

at which there is both high local linkage disequilibrium and a local reduction in genetic diversity in all populations of *P. falciparum*. That locus shows a strong sign of recent selection. The genomic evidence alone would make us suspect that *pfcrt* influences drug resistance. As it happens, we have independent evidence that certain alleles of *pfcrt* do indeed code for drug resistance. But the genomic evidence shows how we could locate such genes even in the absence of independent evidence.

A local reduction in genetic diversity and a local high of linkage disequilibrium are both characteristics of a selective sweep. They can be used to find gene loci where selection has acted recently. Drug-resistance genes are medically important examples of gene loci where selection has recently acted. Any clue that enables us to find these genes is valuable. Selective sweeps can be used to hunt down, if not definitively identify, genes for drug resistance in disease organisms.

Further reading: *Science* October 4, 2002, pp. 79–183.

Advantageous, . . .

and cases that are not. The linkage disequilibrium in *Papilio memnon*'s mimetic polymorphism is advantageous. Natural selection favors individuals with genic associations like $T\_W_2F_2E_2B_2$, whereas it works against recombinants like $T_+W_2F_2E_2B_2$. An individual benefits from having the haplotypes that are in excess frequency in the population. Whole populations of *P. memnon* survive better than they would if the five loci were in linkage equilibrium.

. . . disadvantageous, . . .

In other cases, the opposite is true. We met an example in Section 8.9. It is where the spread of a favored allele interferes with a linked locus at which a heterozygote is advantageous. As the favored allele A' increases in frequency, the frequency of one of the alleles (such as B) at the linked polymorphic locus will also increased by hitch-hiking. Linkage disequilibrium builds up by selection on the A locus (creating an excess of the $A'B$ haplotype). This linkage disequilibrium is disadvantageous. The individuals on average have lower fitness than if linkage equilibrium existed between the A and B loci because the increase in the $A'B$ haplotype reduces the proportion of B/b heterozygotes. Natural selection will favor recombinant individuals that do not have the $A'B$ haplotype.

. . . and neutral linkage disequilibrium are all possible

A third possibility is for linkage disequilibrium to be selectively neutral. An example of this was provided by the hitch-hiking of an allele at a neutral polymorphic locus with a selectively advantageous mutant at a linked locus. While the mutant is being fixed, linkage disequilibrium temporarily builds up between it and the alleles it happens to be linked to at nearby loci. It disappears when the mutant reaches a frequency of one.

The distinction between advantageous and disadvantageous linkage disequilibrium is crucial to understanding one of the major problems of evolutionary biology: why recombination, and sexual reproduction, exists. We look at that problem in Sections 12.1–12.3 (pp. 314–27). We finish this chapter by looking at another influental multilocus population genetic concept — one that is so influential that it is part of the language of evolutionary biology.

8.12 Wright invented the influential concept of an adaptive topography

Wright's idea of an *adaptive topography* (or *adaptive landscape*) is particularly useful for thinking about complex genetic systems; but it is easier to begin with the simplest case. This is for a single genetic locus. The topography is a graph of mean population fitness ($\bar{w}$) against gene frequency (Figure 8.7). (Adaptive topographies can also be drawn for fitness in relation to genotype frequencies. They can even be drawn with phenotypic variables on the *x*-axis; see, for example, Raup's analysis of shell shape, in Figure 10.9 (p. 278). Figure 10.4 (p. 267), used in Fisher's theory of adaptation, is also similar.) We have repeatedly met the concept of mean fitness; it is equal to the sum of the fitnesses of each genotype in the population, each multiplied by its proportion in the population. In a case in which the genotypes containing one of the alleles have higher fitnesses than those of the alternative, the mean fitness of the population simply increases as the frequency of the superior allele increases and reaches a maximum when the gene is fixed (Figure 8.7a). That is fairly trivial. When there is heterozygous advantage, mean

Figure 8.7

A fitness surface, or adaptive topography, is a graph of the mean fitness of the population as a function of gene, or in some cases genotype, frequency. (a) If allele A (frequency p) has higher fitness than a (frequency $1 - p$), mean population fitness simply increases as the frequency of A increases. (b) With heterozygous advantage (fitnesses of genotypes $AA : Aa : aa$ are $1 - s : 1 : 1 - t$) mean population fitness increases to a peak at the intermediate frequency of A at which the proportion of heterozygotes is the maximum possible.

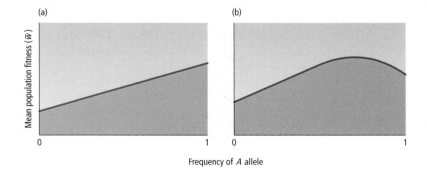

Adaptive topographies can be used to think about abstract evolutionary questions

fitness is highest at the equilibrium gene frequency given by the standard equation (Section 5.12.1, p. 123). Mean fitness declines either side of the equilibrium gene frequency, where more of the unfavorable homozygotes will be dying each generation than at the equilibrium (Figure 8.7b). The graph is also called a *fitness surface.*

In these two cases, natural selection carries the population to the gene frequency where mean fitness is at a maximum. With one favorable allele, the maximum mean fitness is where the allele is fixed — and natural selection will act to fix the allele. With heterozygous advantage, the maximum mean fitness is where the smallest number of homozygotes are dying each generation — and natural selection drives the population to an equilibrium where the amount of homozygote death is minimized.

A question of interest in theoretical population genetics is whether natural selection always drives the population to the state at which the mean fitness is the maximum possible. Frequency-dependent selection (Section 5.13, p. 127) is a case in which natural selection may not act to maximize mean fitness. When a polymorphism is maintained by frequency-dependent selection, the fitness of each genotype is highest when it is rare. But when a genotype is rare, natural selection acts to increase its frequency, making it less rare. The effect of selection can then be to reduce mean fitness.

If natural selection does not always maximize mean fitness, that opens up a further — and still unanswered — theoretical question of whether natural selection does act to maximize some other function, but we shall not pursue that question here. Whatever the answer to it, natural selection does still maximize simple mean fitness in many cases. For many purposes, we can safely think of natural selection as a hill-climbing process, by analogy with the hills in the adaptive topography (Figure 8.7).

Now consider a second locus. Selection can be going on here too, and the fitness surface for the two loci might look like Figure 8.8. Figure 8.8a shows a simple case in which one locus has heterozygous advantage and the other has a single favored allele. The idea of an adaptive topography can be extended to as many loci as interact to determine an organism's fitness, but further loci have to be imagined, rather than drawn, on two-dimensional paper.

Figure 8.8
Fitness surface for two loci.
(a) A combination of the patterns in Figures 8.7a and b: there is a heterozygous advantage at locus *A* and one allele has a higher fitness than the other at locus *B*.
(b) A two-locus fitness surface with two peaks.

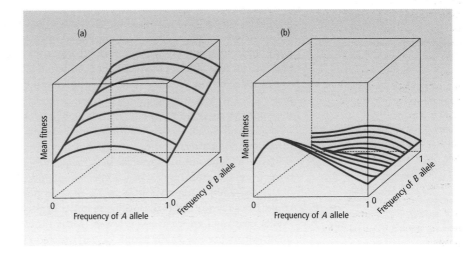

Wright believed that, because the genes at different loci interact, a real multidimensional fitness surface would often have multiple peaks, with valleys between them (Figure 8.8b). The kind of reasoning involved is abstract rather than concrete. We need to imagine a large number of loci, many with more than one allele, with the alleles at the different loci interacting epistatically in their effects on fitness. Epistatic interactions, we now imagine, are common because organisms are highly integrated entities compared with the atomistic chromosomal row of Mendelian genes from which organisms grow up: the genes will have to interact to produce an organism. As we saw above, developmental interactions among genes do not automatically generate epistatic fitness interactions among loci. The extent to which undoubted developmental interaction will produce a multiply peaked fitness surface is therefore open to question; but the possibility is plausible. (Wright called the genes that interact favorably to produce an adaptive peak an "interaction system.")

In the coadapted genes controlling mimicry in *Papilio memnon*, the mimetic genotypes occupy fitness peaks and the recombinants occupy various fitness valleys. The actual shape of the adaptive topography in nature is, however, a more advanced question than can be tackled here. The point of this section is to define what an adaptive topography is, and to point out that its visual simplicity can be useful in thinking about evolution when many gene loci are interacting.

8.13 The shifting balance theory of evolution

Wright used his idea of adaptive topographies in a general theory of evolution. He imagined that real topographies would have multiple peaks, separated by valleys, and that some peaks would be higher than others. When the environment changed, and competing species evolved new forms, the shape of the adaptive topography for a population would change too. The surface would also change shape when a new mutation arose. A new allele at a locus may interact with genes at other loci differently from the existing alleles, and the fitnesses of the genes at the other loci will then be altered; genetic changes will take place at other loci to adjust to the new mutant. All the time, natural selection will be a hill-climbing process, directing the population up toward the currently nearest peak. When the surface changes, the direction to the nearest peak may change, and selection will then send the population off in the new upward direction.

Natural selection, even in so far as it is a hill-climbing (i.e., mean fitness maximizing) process, is only a *local* hill-climbing process. In theory, the local fitness peak could be in the opposite direction from a higher, or global, peak (Figure 8.9). Natural selection, however, will direct the population to the local peak. Now suppose that the mean fitness of a population is a measure of the quality of its adaptations, such that a population with a higher mean fitness has better adaptations than a population with a lower mean fitness. Because natural selection seeks out only local peaks, natural selection may not always allow a population to evolve the best possible adaptations. A population could be stuck on a merely locally adaptive peak. Natural selection works against "valley crossing," where fitness is lower. (Mean fitness cannot always be equated with quality of adaptation. In the simple case in which one allele is superior to another (see

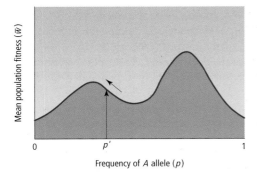

Figure 8.9

A two-peaked fitness surface with local and global maxima. Natural selection will take a population with gene frequency p' towards the local peak, away from the peak with highest average fitness.

Figure 8.7a), the organisms with the better genotype will also be better adapted. But when fitness is frequency dependent or when group and individual adaptations conflict (Chapter 11), the maximum of mean population fitness may not correspond to the best adaptation.)

Random drift may take a population away from a local peak

Wright was interested in how evolution could overcome the tendency of natural selection to become stuck at local fitness peaks. When fitness peaks correspond to optimal adaptations, the question is relevant to the evolution of adaptation; but when they do not, the question still has a technical interest in population genetics. Wright suggested that random drift could play a creative role. Drift will tend to make the population gene frequencies "explore" around their present position. The population could, by drift, move from a local peak to explore the valleys of the fitness surface. Once it had explored to the foot of another hill, natural selection could start it climbing uphill on the other side. If this process of drift and selection were repeated over and over again with different valleys and hills on the adaptive topography, a population would be more likely to reach the global peak than if it was under the exclusive control of the locally maximizing process of natural selection.

Wright's shifting balance theory is concerned with evolution on complex adaptive topographies

Wright's full shifting balance theory includes more than just selection and drift within a local population. He also suggested that populations would be subdivided into many small local populations, and drift and selection would go on in each. The large number of subpopulations would multiply the chance that one of them would find the global peak. If members of a subpopulation at the highest peak were better adapted, they could produce more offspring and more emigrants to the other subpopulations. Those other subpopulations would then be taken over by the superior immigrant genotypes. Thus the whole species would evolve to the higher peak. Wright's theory is thus an attempt at a comprehensive, realistic model of evolution. Everything is included: multiple loci, fitness interactions, selection within and between populations, drift, and migration. (The theory of adaptive peaks is also relevant to speciation: Section 14.4.4, p. 394.)

The question of how important the shifting balance process is in evolution is long standing, dating back to Wright's publications in the 1930s. Coyne *et al.* (1997) recently reopened the controversy, arguing that we have no good reason to think that the shifting balance process has contributed much to evolution. The full controversy has looked at many topics. Here are four of them.

1. What facts are better explained by the shifting balance process than by simple natural selection within one population? For instance, the passion flower butterflies (*Heliconius,* Section 8.2 above) have many morphs, each mimicking a different model. Each morph probably occupies an adaptive peak. An adaptive valley separates each peak, because intermediate forms would be poorly adapted to mimic any model and would be eaten. How can one morph evolve into another? On the shifting balance view, a morph can originate by drift within a local population, and then spread if it is advantageous. Alternatively, however, evolution in the mimetic form of the butterflies may be driven by changes in the abundance of local models. If a model with a certain coloration becomes locally common, perhaps because its resources are locally abundant, then the mimic species will evolve to match that coloration. Thus, although the species now shows a multipeaked fitness surface, the peaks may not have been separated by valleys in the past. The *Heliconius* example, as with all others that have been discussed in the controversy, is inconclusive.

2. Can genetic drift drive populations across real adaptive valleys? Genetic drift is powerful when it is not opposed by selection: that is, when drift is between different neutral forms. However, in Wright's theory, drift has to work in opposition to selection. This is a much more difficult process, and critics doubt whether it occurs. The selective disadvantages in the valleys between different morphs of *Heliconius,* for example, correspond to 50% fitness reductions. Random drift could not establish forms that have such large disadvantages.

3. Do populations have the structure proposed by Wright? Are populations subdivided into many small subpopulations? If populations are large, all the main possible genotypes will be present in it including the best genotype — the one corresponding to the highest adaptive peak. It can be fixed by normal natural selection within the population. The shifting balance process only helps if populations are so small that the best genotype happens never to have arisen in many local subpopulations. Supporters suggest that real populations are often as Wright suggested; critics doubt it.

4. Do real fitness surfaces have multiple peaks? Fisher, for instance, doubted whether natural selection would actually confine populations to local peaks. Fisher was preeminently a geometric thinker and he pointed out that, as the number of dimensions in an adaptive topography increases, local peaks in one dimension tend to become points on hills in other dimensions (Figure 8.10). In the extreme case, when there are an infinity of dimensions, it is certain that natural selection will be able to hill climb all the way to the global peak without any need for drift. Each one- (Figure 8.7) or two-dimensional peak (Figure 8.8) will be crossed at the peak by an infinity of other dimensions, and it is highly implausible that the fitness surface will turn downhill in all of them at that point. This is a highly interesting argument, though it is, of course, purely theoretical. It refutes Wright's theoretical claim that natural selection will get stuck at local peaks, but leaves open the empirical question of how important selection and drift have been in exploring the fitness surfaces of nature.

The importance of the shifting balance process remains undecided, but the controversy has a broader interest. Biologists distinguish between a "Fisher" and a "Wright" school of evolutionary thought. Fisher maintained that natural populations are generally too large for drift to be important, that epistatic fitness interactions do not interfere

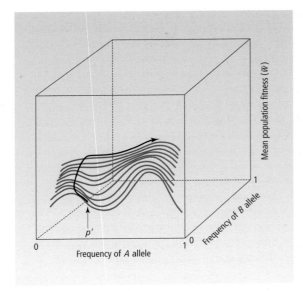

Figure 8.10
As extra gene loci are considered (extra axes in the adaptive topography) it becomes increasingly likely that what appeared to be a local maximum in fewer dimensions will turn out to be a hillside or saddle point in more dimensions. In this case, the fitness surface for the A locus at a gene frequency of zero for the B allele is the same as in Figure 8.9; but at other B gene frequencies, the local peak in the A locus fitness surface disappears. If the population started in the valley at B gene frequency = 0 and A gene frequency = p', natural selection would initially move gene frequencies to the local peak, but they would eventually reach the global peak by continuous hill climbing.

with the operation of selction, that adaptations evolve by selection within a population, and that adaptive evolution can proceed smoothly up to the highest fitness peak. Wright thought that populations are small, drift and epistatic fitnesses are important, and that adaptive evolution is liable to become stuck at a local optimum. Biologists today rarely count themselves simply as members of one school or the other, but the controversy between these two views has inspired, and continues to inspire, important evolutionary research.

Summary

1 Population genetics for two or more loci is concerned with changes in the frequencies of haplotypes, which are the multilocus equivalent of alleles.

2 Recombination tends, in the absence of other factors, to make the alleles of different loci appear in random (or independent) proportions in haplotypes. An allele A_1 at one locus will then be found with alleles B_1 and B_2 at another locus in the same proportions as B_1 and B_2 are found in the population as a whole. This condition is called linkage equilibrium.

3 A deviation from the random combinatorial proportions of haplotypes is called linkage disequilibrium.

4 The theory of population genetics for a single locus works well for populations in linkage equilibrium.

5 Linkage disequilibrium can arise because of low recombination, non-random mating, random sampling, and natural selection.

6 For selection to generate linkage disequilibrium, the fitness interactions must be epistatic: the effect on fitness of a genotype (such as A_1/A_2) must vary according to the genotype it is associated with at other loci.

7 Pairs of alleles at different loci that cooperate in their effects on fitness are called coadapted. Selection acts to reduce the amount of recombination between coadapted genotypes.

8 When selection works on one locus, it will influence gene frequencies at linked loci. The effect is called hitch-hiking.

9 The spread of a favorable mutation causes a reduction in diversity in neighboring regions of the DNA. The process is called a selective sweep. Local reductions in genetic diversity, in DNA sequence data, can provide evidence of selection.

10 The mean fitness of a population can be drawn graphically for two loci; the graph is called a fitness surface or an adaptive topography.

11 Wright suggested that real adaptive topographies will have many separate "hills," with "valleys" between them. Natural selection enables populations to climb

the hills in the adaptive topography, but not to cross valleys. A population could become trapped at a local optimum.

12 Random drift could supplement natural selection by enabling populations to explore the valley bottoms of adaptive topographies.

13 It is questionable whether real adaptive topographies do have multiple peaks and valleys. They might have a single peak, with a continuous hill leading to it. Natural selection could then take the population to the optimum without any random drift.

14 Two-locus population genetics uses a number of concepts not found in single-locus genetics. The most important are: haplotype frequency, recombination, linkage disequilibrium, epistatic fitness interaction, hitch-hiking, and multiple peaked fitness surfaces.

Further reading

Population genetics for two loci, as for one locus, is introduced in such standard textbooks as Hartl & Clark (1997) and Hedrick (2000).

The multilocus genetics of mimicry in *Papilio memnon* and *Heliconius* is explained by Turner (1976, 1984). Turner & Mallett (1996) discuss the puzzle of diversity in *Heliconius*, with the shifting balance process as one possible explanation. The HLA loci are introduced from a more evolutionary perspective by Hughes (1999), and a more molecular genetic perspective by Lewin (2000). Wolf *et al.* (2000) is a multiauthor book on epistasis and evolution. Wade *et al.* (2001) distinguish two meanings of epistasis, which differ between two-locus population and quantitative genetics.

Linkage disequilibrium in the human genome is described by Reich *et al.* (2001): for humans, linkage disequilibrium also matters for locating disease genes, and changes in population size need to be considered. Kohn *et al.* (2000) describe another example of a selctive sweep, like *Sdic* — the gene for warfarin resistance in rats. Gillespie (2001) looks at the effect of population size on hitch-hiking, and draws the subversive conclusion that population size may have little effect on evolution because its effect on hitch-hiking is the opposite of its effect at any one site.

On Wright's shifting balance theory, see Wright's four-volume treatise (1968–78), particularly volumes 3 and 4 (1977, 1978) and Wright (1986). Wright (1932) is a short and accessible paper from earlier on. See also Lewontin (1974, final chapter) and Provine (1986, chapter 9). For the modern controversy, see the exchange in *Evolution* (2000), vol. 54, pp. 306–27, including the references there.

Study and review questions

1 Here are the haplotype frequencies in four populations.

	Genotype								
Population	A_1B_1/A_1B_1	A_1B_1/A_1B_2	A_1B_2/A_1B_2	A_1B_1/A_2B_1	A_1B_1/A_2B_2	A_1B_2/A_2B_2	A_2B_1/A_2B_1	A_2B_1/A_2B_2	A_2B_2/A_2B_2
1	3/16	1/16	0	1/16	3/8	1/16	0	1/16	3/16
2	1/16	1/8	1/16	1/8	1/4	1/8	1/16	1/8	1/16
3	1/81	4/81	4/81	4/81	16/81	16/81	4/81	16/81	16/81
4	0	1/81	8/81	2/81	8/81	26/81	7/81	27/81	2/81

Calculate the linkage disequilibrium in each.

	Frequency of			
Population	A_1B_1	A_1	B_1	Value of D
1				
2				
3				
4				

2 What kinds of selection can be hypothesized to be operating in populations 1–4 of question 1?

3 In a large population, with random mating, and in the absence of selection, what haplotype frequencies would you expect in populations 1–4 of question 1 after a few hundred generations?

4 Distinguish three explanations for a local reduction in genetic diversity within the DNA of a species. How can you test between them?

5 Draw a fitness surface for a single locus with two alleles (*A* and *a*) and heterozygous advantage. Where is the equilibrium on the graph?

9

Quantitative Genetics

The Mendelian genetics of beak size is unknown, but the character shows evolutionary changes as the food supply changes through time. We start by looking at beak size in Darwin's finches as an example of the kind of character studied by quantitative genetics. We then move on to the theoretical apparatus used to analyze characters controlled by large numbers of unidentified genes. The influences on these characters are divided into environmental and genetic, and the genetic influences are divided into those that are inherited and influence the form of the offspring and those that are not. A number called "heritability" expresses the extent to which parental attributes are inherited by their offspring. With the theoretical apparatus in place, we can then apply it to a number of evolutionary questions: directional selection, in both artificial and natural examples, and stabilizing selection. We look at the effect of selection on heritability, and at mutation–selection balance. We end with some apparently puzzling observations, in which populations that are predicted to undergo evolutionary change in fact stay constant over time.

9.1 Climatic changes have driven the evolution of beak size in one of Darwin's finches

Fourteen species of Darwin's finches live in the Galápagos archipelago, and many of them differ most obviously in the sizes and shapes of their beaks. A finch's beak shape, in turn, influences how efficiently it can feed on different types of food. Peter and Rosemary Grant, together with a team of researchers, have been studying these finches since 1973, and they have evidence that beak size influences feeding efficiency. It comes from a comparison of two species (see Plate 4a, and b, between pp. 68 and 69), the large-beaked *Geospiza magnirostris* and the smaller *G. fortis*, feeding on the same kind of hard fruit.

Beak size influences feeding efficiency on different food types

The large-beaked *G. magnirostris* can crack the fruit (called the mericarp) of caltrop (*Tribulus cistoides*) transversely, taking on average only 2 seconds and exerting an average force of 26 kgf (255 N); it can then easily, in about 7 seconds, eat all the 4–6 seeds of the smashed fruit. The smaller *G. fortis* is not strong enough to crack *Tribulus* mericarps and instead twist open the lower surface, applying a force of only 6 kgf and taking 7 seconds on average to reach the seeds inside. Only one or two of the seeds can be obtained in this way and it takes an average of 15 seconds to extract them. *G. magnirostris* usually has an advantage with these large, hard types of food.

Smaller finches are probably more efficient with smaller types of food, but this is more difficult to show. Both large and small finches on the Galápagos do in fact eat small seeds, though there is an indirect reason (as we shall see) to believe that smaller finches do so more efficiently. From the evidence we have met so far, we can predict that natural selection would favor larger finches when large fruits and seeds are abundant. The prediction should apply both within and between species. A *G. magnirostris* finch looks like an enlarged *G. fortis*, and a larger individual *G. fortis* can probably deal with a large food item more efficiently than can a smaller conspecific, much as an average specimen of *G. magnirostris* is more efficient than an average *G. fortis*. When large seeds are common, we might expect the average beak size in a population of *G. fortis* to increase between generations, and to decrease when large seeds are rare — if beak size is inherited.

Beak size is inherited

If beak size is inherited . . . but is it? Beak size is inherited if parental finches with larger than average beaks produce offspring with larger than average beaks. The Grants measured the sizes of parental and offspring finches in several families and plotted the latter against the former (Grant 1986) (Figure 9.1). Large-beaked parental finches do indeed produce large-beaked offspring: beak size is inherited. It therefore makes sense to test the prediction that changes in beak size should follow changes in the size distribution of food items. The test was carried out on the species *G. fortis*, on one of the Galápagos islands, Daphne Major. Since the study began, this species has undergone two major, but contrasting, evolutionary events.

The weather in the Galápagos . . .

In the Galápagos, the normal pattern of seasons is for a hot, wet season from about January to May to be followed by a cooler, dryer season through the rest of the year. But in early 1977, for some reason, the rain did not fall. Instead of the normal progression, the dry season that began in mid 1976 continued until early 1978: one whole wet season did not happen. The finch population of Daphne Major collapsed from about 1,200 to

Figure 9.1
Parents with larger than average beaks produce offspring with larger than average beaks in *Geospiza fortis* on Daphne Major, showing that beak size is inherited. Results are shown here for 2 years in the 1970s. Grant & Grant (2000) show that the result persisted in future years. (0.4 in ≈ 10 mm.) Redrawn, by permission of the publisher, from Grant (1986).

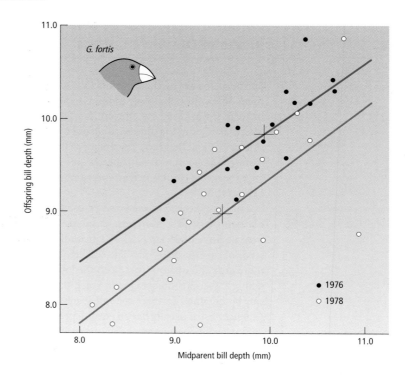

... influences the food supply ...

... leading to evolution in Darwin's finches

about 180 individuals, with females being particularly hard hit; the sex ratio at the end of 1977 was about five males per female. As the sex difference shows, not all finches suffered equally — smaller birds died at a higher rate. The reason, again, lies in the food supply. At the beginning of the drought, the various types of seeds were present in their normal proportions. *G fortis* of all sizes take small seeds, and as the drought persisted, these smaller seeds were relatively reduced in numbers. The average available seed size became larger with time (Figure 9.2). Now the larger finches were favored, because they eat the larger, harder seeds more efficiently. The average finch size increased as the smaller birds died off. (Females died at a higher rate than males because females are on average smaller.) Size, as we have seen, is inherited. The differential mortality in the drought therefore caused an increase in the average size of finches born in the next generation: *G. fortis* born in 1978 were about 4% larger on average than those born before the drought.

Four years later, in November 1982, the weather reversed. The rainfall of 1983 was exceptionally heavy and the dry volcanic landscape was covered with green in the periodic disturbance called El Niño (see Plate 4c and d, between pp. 68 and 69). Seed production was enormous. The theory developed for 1976–78 could now be tested. The conditions had reversed: the direction of evolution should go into reverse too. In the year after the 1983 El Niño event, there were more small seeds. If the smaller finches can in fact exploit small seeds more efficiently, the smaller finches should survive relatively better. The Grants again measured the sizes of *G. fortis* on Daphne

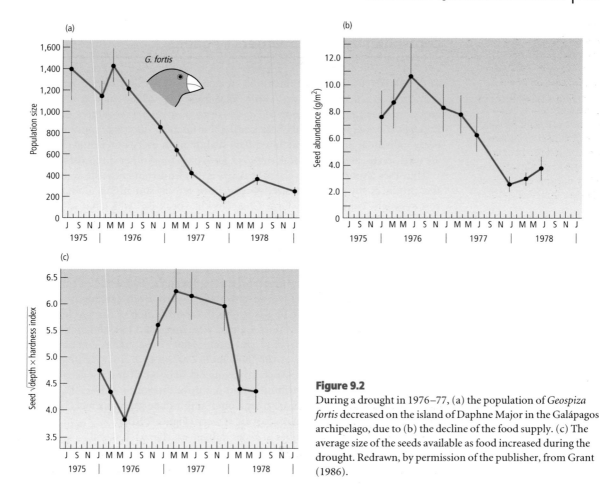

Figure 9.2

During a drought in 1976–77, (a) the population of *Geospiza fortis* decreased on the island of Daphne Major in the Galápagos archipelago, due to (b) the decline of the food supply. (c) The average size of the seeds available as food increased during the drought. Redrawn, by permission of the publisher, from Grant (1986).

Major in 1984–85 and found that the smaller birds were indeed favored. Finches born in 1985 had beaks about 2.5% smaller than those born before the El Niño downpours. The theory, that seed sizes control beak size in these finches, was confirmed. Further confirmation came at the next El Niño event in 1987. This time, the seed size distribution was hardly changed at all, leading to the prediction that the beak sizes of the finches would show no evolutionary change either; nor did they (Grant & Grant 1995).

Selection fluctuates over time

The fluctuations in the direction of selection on beak shape — with beaks evolving up in some years, down in other years, and staying constant in yet other years — probably results in a kind of "stabilizing" selection over a long period of time such that the average size of beak in the population is the size favored by long-term average weather. (Later in the chapter, we shall see how the degree of selection can be expressed more exactly; Figure 9.9 will show the results for 1976–77 and 1984–85.)

9.2 Quantitative genetics is concerned with characters controlled by large numbers of genes

Beak size is a continuous character

The beak size of Galápagos finches is an example which illustrates a large class of characters. It shows *continuous variation*. Simple Mendelian characters, like blood groups or the mimetic variation of *Papilio*, often have discrete variation; but many of the characters of species are like beak size in these finches — they vary continuously, and every individual in the population differs slightly from every other individual. There are no discrete categories of beak size in *G. fortis* or in most other species of birds.

The other important point about beak size is that we do not know the exact genotype that produces any given beak size. We can, however, say something about the general sort of genetic control it may have. Characters like beak size, which has an approximately normal frequency distribution (that is, a bell curve), are probably controlled by a large number of genes, each of small effect. The reason is as follows (Figure 9.3). Imagine first that beak size was controlled by a single pair of Mendelian alleles at one locus, with one dominant to the other, *AA* and *Aa* long and *aa* short. In this case, the population would contain two categories of individuals (Figure 9.3a). Imagine now that it was controlled by two loci with two alleles each. Beak size might now have a background value (say, 0.4 in or 1 cm) plus the contribution of the two loci, with an *A* or a *B* adding 0.04 in (0.1 cm). If *A* and *B* were dominant to *a* and *b*, then an *aabb* individual would have a 0.39 in (1 cm) beak; *AAbb*, *Aabb*, *aaBB*, and *aaBb* 0.43 in (1.1 cm); and *AABB*, *AaBB*, *AABb*, and *AaBb* 0.47 in (1.2 cm). Figure 9.3b is the frequency distribution if all alleles had a frequency of one-half and the two loci were in linkage equilibrium. The distribution now has three categories and has become more spread out. It becomes still more spread out if it is influenced by six loci (Figure 9.3c) and becomes normal when many loci are at work (Figure 9.3d).

When a large enough number of genes influence a character, it will have a continuous, normal frequency distribution. The normal distribution can result either if there are a large number of alleles at each of a small number of loci influencing the characters, or if there are fewer alleles at a larger number of loci. In this chapter, we shall mainly discuss the theory of quantitative genetics as if there were many loci, each with a small

Figure 9.3

(a) The phenotypic character, beak size for example, is controlled by one locus with two alleles (*A* and *a*); *A* is dominant to *a*. There are two discrete phenotypes in the population. (b) The character is controlled by two loci with two alleles each (*A* and *a*, *B* and *b*); *A* and *B* are dominant to *a* and *b*. There are three discrete phenotypes. (c) Control by six loci, with two alleles each. (d) Control by many loci with two alleles each. As the number of loci increase, the phenotypic frequency distribution becomes increasingly continuous.

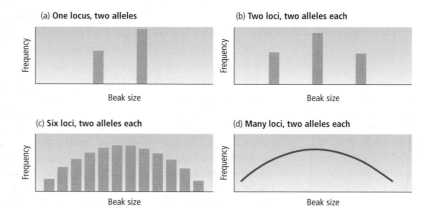

(a) **One locus, two alleles**

(b) **Two loci, two alleles each**

(c) **Six loci, two alleles each**

(d) **Many loci, two alleles each**

Continuous characters are influenced by many factors, each of small effect

number of alleles. This may well be the genetic system underlying many continuously varying characters. However, the theory applies equally well when there are a few (even one) loci and many alleles at each.

Mendel had noticed in his original paper in 1865 that multifactorial inheritance (i.e., the character is influenced by many genes) can generate a continuous frequency distribution; but it was not well confirmed until later work, particularly by East, Nilsson-Ehle, and others, in about 1910. Quantitative genetics is concerned with characters influenced by many genes, called *polygenic characters*. For a quantitative geneticist, 5–20 genes is a small number of genes to be influencing a character; many quantitative characters may be influenced by more than a hundred, or even several hundred, genes. For characters influenced by a large number of loci, it ceases to be useful to follow the transmission of individual genes or haplotypes (even if they have been identified) from one generation to the next. The pattern of inheritance, at the genetic level, is too complex.

There is an additional complication. So far we have only considered the effect of genes. The value of a character, like beak size, will usually also be influenced by the environment in which the individual grows up. Beak size is probably related to general body size and all characters to do with bodily stature will be influenced by the amount of food an organism happens to find during its life. If we take a set of organisms with identical genotypes and allow some to grow up with abundant food and others with limited food, the former will end up larger on average. In nature, each character will be influenced by many environmental variables, some tending to increase it, others to decrease it. Thus if we take a class of genotypes with the same value of a character before the influence of the environment and add the effect of the environment, some of the individuals of each genotype will be made larger and others smaller in various degrees. This produces a further "spreading out" of the frequency distribution. Any pattern of discrete variation in the genotype frequency distribution is likely to be obscured by environmental effects and the discrete categories converted into a smooth curve (Figure 9.4).

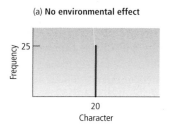

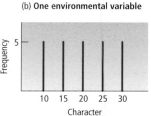

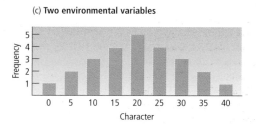

Figure 9.4

Environmental effects can produce continuous variation. (a) Twenty-five individuals in the absence of environmental variation all have the same phenotype, with a value for a character of 20. (b) Influence of one environmental variable. The variable has five states, and according to which state an organism grows up in its character becomes larger or smaller or is not changed. The five states change the character by +10, +5, 0, −5, and −10, and an organism has an equal chance of experiencing any one of them. (c) Influence of a second environmental variable. This variable also has five equiprobable states, and they change the character by +10, +5, 0, −5, and −10. Of the five individuals in (b) with character value 10, one will get another −10, giving a value of 0, a second will get −5, giving 5, etc. After the influence of both variables, the frequency distribution ranges from 0 to 40 and is beginning to look bell curved. With many environmental influences, each of small effect, a normal distribution will result.

Continuous characters are studied by quantitative genetics

The small effects of many genes and environmental variables are two separate influences that tend to convert the discrete phenotypic distribution of characters controlled by single genes into continuous distributions. If a character shows a continuous distribution, it in principle could be because of either process. Quantitative genetics is mainly concerned with characters influenced by both. Quantitative genetics employs higher level genetic concepts that are genetically less exact than those of one- or two-locus population genetics, but which are more useful for understanding evolution in polygenic characters. Instead of following changes in the frequency of genes or haplotypes, we now follow changes in the frequency distribution of a phenotypic character. Quantitative genetics is important because so many characters have continuous variation and multilocus control.

9.3 Variation is first divided into genetic and environmental effects

Quantitative genetics contains an unavoidable minimum of formal concepts that we need to understand before we can put it to use: those formalities are the topic of this section and the next. To understand how a quantitative character like beak size will evolve, we have to "dissect" its variation. We tease apart the different factors that cause some birds to have larger beaks than others. Suppose, for example, that all the variation in beak size was caused by environmental factors — that is, all birds have the same genotype and they differ in their beak sizes only because of the different environments in which they grew up. Beak size could not then change during evolution (except for non-genetic evolution due to environmental change). For the character to evolve, it has to be at least partly genetically controlled. We need to know how much beak size varies for environmental, and how much for genetic, reasons. However, even if different finches vary in their beak size for genetic reasons, that does not necessarily mean it can evolve by natural selection. As we shall see, we have to divide genetic influence into components that allow evolutionary change and those that do not.

The value of a character is expressed as a deviation from the mean

In quantitative genetics, the value of a character in an individual is always expressed as a deviation from the population mean. Beak size will have a certain mean value in a population, and we talk about environmental and genetic influences on an individual as deviations from that mean. The procedure is easy to understand if we think of the population mean as a "background" value, and then the influences leading to a particular individual phenotype are expressed as increases or decreases from that value. Let us see how it is done. Suppose there is one locus with two alleles influencing beak depth. AA and Aa individuals' beaks are 1 cm from top to bottom, and aa individuals' beaks 0.5 cm; the environment has no effect. If the population average was 0.875 cm (as it would be for a gene frequency of $a = \frac{1}{2}$), then we should write the beak phenotype of AA and Aa individuals as $+0.125$ cm and that of aa as -0.375 cm. In general, we symbolize the phenotype by P. In this case, $P = +0.125$ for AA and Aa individuals and $P = -0.375$ for aa.

Clearly, the value of P for a genotype depends on the gene frequencies. When the frequency of A is one-half, the genotypic effects are those just given. But when the frequency of A is one-quarter, the population mean would be 0.71875 cm. For aa, P now is -0.21875 and for AA and Aa, $P = +0.28125$. In this example the phenotype is controlled only by genotype. We can symbolize the effect of the genotype by G. G, like P, is expressed as a deviation from the population mean. In this case, for an individual with a particular genotype (because the environment has no effect):

Mean of population $+ P =$ mean of population $+ G$

Genetic and environmental influences on a character can be defined

The background population mean cancels from the equation, and can be ignored. We are then left (in this case in which the environment has no effect) with $P = G$.

The value of a real character will usually be influenced by the individual's environment as well as its genotype. If the character under study is something to do with size, for example, it will probably be influenced by how much food the individual found during its development, and how many diseases it has suffered. These environmental effects are measured in the same way as for the genotype, as a deviation from the population mean. If an individual grew up in an environment causing it to grow a bigger beak than average, its environmental effect will be positive; and vice versa if it grew up in an environment giving it a smaller than average beak. The phenotype can then be expressed as the sum of environmental (E) and genotypic influences:

$P = G + E$

This, simple as it is, is the fundamental model of quantitative genetics. For any phenotypic character, the individual's value for that character (expressed, remember, as a deviation from the population mean) is due to the effect of its genes and environment.

A simple Mendelian example illustrates the terms of quantitative genetics

We must look further into the genotypic effect. We need to consider both how to subdivide the genotypic effect, and why the subdivision is necessary. The main point can be seen in the one-locus example we have already used. The A gene is dominant, and both AA and Aa birds have 1 cm beaks ($P = +0.125$). (Because we are investigating the genotypic effect, it is simplest to ignore environmental effects, so $P = G$.) Suppose we take an AA individual and mate it to another bird drawn at random from the population. The gene frequency is $^1/_2$ and the random bird is AA with chance $^1/_4$, Aa with chance $^1/_2$, and aa with chance $^1/_4$; but whatever the mate's genotype all the offspring will have beak phenotype $P = +0.125$ because A is dominant. Now suppose we take an Aa individual and mate it to a random member of the population. The average phenotype P of their offspring is 0. (As can be confirmed by working out $(^1/_4 \times (+0.125)) + (^1/_2 \times 0) + (^1/_4 \times (-0.125))$ for the three offspring genotypes.) A P of 0 means that the average offspring beak size is the same as the population average. Thus, for two genotypes with the same beak size ($P = +0.125$), one produces offspring with beaks like their parent, the other produces offspring with beaks like the population average.

So some genotypic effects are inherited by the offspring and some are not. The next step is to divide the genotypic effect into a component that is passed on and a component that is not. The component that is passed on is called the *additive effect* (A) and the component that is not is called (in this case) the *dominance effect* (D). The full genotypic effect in an individual is the sum of the two:

$$G = A + D$$

The additive part of the genetic effect is most important . . .

The additive effect is the important one. The parent deviates from the population mean by a certain amount; its additive genotypic effect is the part of that deviation that can be passed on. However, when an individual reproduces, only half its genes are inherited by its offspring. The offspring inherit only half the additive effect of each parent. Thus the additive effect A for an individual is equal to twice the amount by which its offspring deviate from the population mean, if mating is random. For the AA parent, therefore, the additive effect is +0.25. (The full quantitative genetics of the AA individuals is $G = +0.125$, $A = +0.25$, and $D = -0.125$.) The offspring of Aa birds deviate by zero: their additive effect is twice zero, which is zero; the amount by which Aa heterozygotes deviate from the population mean is entirely due to dominance, and is not inherited by their offspring. (For Aa individuals, $G = +0.125$, $A = 0$, and $D = +0.125$.)

The division of the genotypic effect into additive and dominance components tells us what proportion of the parent's deviation from the mean is inherited, and reveals how the non-inheritance of the Aa individuals' genotypic effect is due to dominance. In practice quantitative geneticists do not know the genotypes underlying the characters they study; they only know the phenotype. They might, for instance, focus on the class of birds with 1 cm ($P = +0.125$) beaks. The additive component of their phenotypic value depends on the frequencies of the AA and Aa genotypes in this example: if all the birds with 1 cm beaks are Aa heterozygotes, then none of the offspring will inherit their parents deviation; if they are all AA, then half the offspring will.

Why is the additive effect of a phenotype so important? The answer is that once the additive effect for a character has been estimated, that estimate has much the same role in quantitative genetics as the exact knowledge of Mendelian genetics in a one- or two-locus case (Chapters 5 and 8). It is what we use to predict the frequency distribution of a character in the offspring, given a knowledge of the parents. In a one-locus genetic model, we know the genotypes corresponding to each phenotype, and can predict the phenotypes of offspring from the genotypes of their parents. In the case of selection, the gene frequency in the next generation is easy to predict if we know selection allows only AA individuals to breed. In two-locus genetics, the procedure is the same. If the next generation is formed from a certain mixture of Ab/AB and AB/AB individuals, we can calculate its haplotype frequencies if we know the exact mixture of parental genotypes.

. . . and is used to predict the character value in the next generation

In quantitative genetics, we do not know the genotypes. All we have are measurements of phenotypes, like beak size. But if we can estimate the additive genetic component of the phenotype, then we can predict the offspring in a manner analogous to the procedure when the real genetics are known. When we know the genetics, Mendel's laws of inheritance tell us how the parental genes are passed on to the offspring. When we do not know the genetics, the additive effect tells us what component of the parental phenotype is passed on. Estimating the additive effect is thus the key to understanding the evolution of quantitative characters. The estimates are practically made by breeding experiments. In the case of finches with 1 cm beaks in a population of average beak size 0.875 cm, the additive effect can be measured by mating 1 cm-beaked finches to random members of the population. The additive effect is then two times the offspring's deviation from the population mean.

The genetic partitioning that we have made so far is incomplete. It applies reasonably well for one locus: in that case, dominance is the main reason why the genotypic effect of a parent is not exactly inherited in its offspring. When many loci influence a charac-

Several non-additive genetic factors are known

ter, *epistatic interactions* between alleles at different loci can occur (Section 8.8, p. 207). Epistatic interactions are, like dominance effects, not passed on to offspring. They depend on particular combinations of genes and when the combinations are broken up (by genetic recombination) the effect disappears; nor are they are non-additive. An example of an epistatic interaction would be for individuals with the haplotype A_1B_2 to show a higher deviation from the population mean than the combined average deviations of A_1 and B_2; the extra deviation is epistatic. Other non-additive effects can arise because of gene–environment interaction (when the same gene produces different phenotypes in different environments) and gene–environment correlation (when particular genes are found more often than random in particular environments).

A full analysis can take all these effects into account. In a full analysis, just as in the simple one here, the aim is to isolate the additive effect of a phenotype. The additive effect is the part of the parental phenotype that is inherited by its offspring.

9.4 Variance of a character is divided into genetic and environmental effects

We return now to the frequency distribution of a character. We continue, as usual, to express effects as deviations from the population mean. If we consider an individual some distance from the mean, some of its deviation will be environmental, some genetic. Of the genetic component, some will be additive, some dominance, some epistatic. These terms have been defined so that they add up to give the exact deviation of the individual from the mean. Any individual has its particular phenotypic value (P) because of its particular combination of environmental experiences and the dominance, additive, and interaction effects in its genotype (Figure 9.5). The different combinations of E, D, and A in different individuals are the reason why the character shows a continuous frequency distribution in the population.

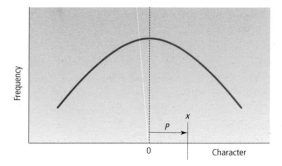

Figure 9.5

The x-axis of the continuous distribution for a character can be scaled to have a mean of zero. Consider the individuals (called x) with phenotype $+P$. Their phenotypic values are the sum of their individual combinations of the environmental, additive, and dominance effects ($E + A + D$). Individuals with character x can have any combination of E, A, and D such that $E + A + D = x$. Any individual's deviation from the mean is due to its individual combination of E, A, and D (as well as other effects, such as epistatic).

The variation seen for the character in any one population could exist because of variation in any one of, or any combination of, these effects. Thus the individual differences could be all due to different environmental effects, with every individual having the same value for G. Or it could be 25% due to the environment, 20% to additive effects, 30% to dominance effects, and 25% to interaction effects. The proportion of variation due to the different effects matters when we wish to understand how a population will respond to selection. If all the variation exists because different individuals have different values of E, there will be no response to selection; but if the variation is mainly additive genetic variation, the response will be large. The proportion of the variation that is due to different values of A in different individuals tells us whether the population can respond to selection.

The variability in the population due to any particular factor, such as the environment, is measured by the statistic called the variance. (Box 9.1 explains some statistical terms used in quantitative genetics.) Variance is the sum of squared deviations from the mean divided by the sample size minus one. So for all the values x of a character like beak size in a population, the variance of x is (see Box 9.1 for the notation):

$$V_X = \frac{1}{n-1} \sum (x_i - \bar{x})^2$$

We have seen how the total phenotype, genetic effect, environmental effect, and so on, can be measured for an individual; the measurements (P for phenotypic effect, etc.) are expressed as deviations from the mean. We can therefore easily calculate, for a population, what their variances are:

Phenotypic variance $\quad = V_P = \dfrac{1}{n-1} \sum P^2$

Environmental variance $= V_E = \dfrac{1}{n-1} \sum E^2$

Genetic variance $\quad\quad = V_G = \dfrac{1}{n-1} \sum G^2$

Dominance variance $\quad = V_D = \dfrac{1}{n-1} \sum D^2$

Additive variance $\quad\quad = V_A = \dfrac{1}{n-1} \sum A^2$

The phenotypic variance for a population, for example, expresses how spread out the frequency distribution for the character is. If the frequency distribution is wide, with different individuals having very different values of the character, the phenotypic variance will be high. If it is a narrow spike, with most individuals having a similar value for the character, phenotypic variance will be low.

Box 9.1
Some Statistical Terms Used in Quantitative Genetics

The text mentions three main statistical terms. This box explains variance, but serves mainly as a reminder of the exact definitions of covariance and regression; reference should be made to a statistical text for fuller explanation.

Variance

The variability of a set of numbers can be expressed as a variance. Take a set of numbers, such as 4,3,7,2,9. Here is how to calculate their variance.

1. Calculate the mean:

$$\text{Mean} = \frac{4 + 3 + 7 + 2 + 9}{5} = 5$$

2. For each number, calculate the square of its deviation from the mean. For the first number, 4, it is $(5 - 4)^2 = 1$. We do likewise for all five numbers.
3. Add up the sum of the squared deviations from the mean. For the five numbers, it is $1 + 4 + 4 + 9 + 16 = 34$.
4. Divide the sum by $n - 1$; $n = 5$ in this case.

$$\text{Variance} = 34/4 = 8.5$$

The general formula for the variance of a character X is:

$$V_X = \frac{1}{n - 1} \sum (x_i - \bar{x})^2$$

where $\bar{x}$ is the mean and x_i is a standard notation for a set of numbers. Here we have five numbers. In terms of the notation, that means that i can have any value from 1 to 5 and is the value of the character for each i. Thus $x_1 = 4$, $x_2 = 3$, and $x_5 = 9$. The summation in the general formula is for all values of i: here it is for all five numbers. If there had been 50 numbers, i would have varied from 1 to 50 and we should proceed as in the example for all 50 numbers. The variance describes how variable the set of numbers is: the higher the variance, the greater the differences among the numbers. If all the numbers were the same (all $x_i = \bar{x}$) then their variance is zero.

Standard deviation

This is the square root of the variance.

Covariance

Now imagine the individuals have been measured for two characters, X and Y. The covariance between the two is defined as:

$$\text{cov}_{XY} = \frac{1}{n - 1} \sum (x_i - \bar{x})(y_i - \bar{y})$$

Covariance measures whether, if an individual has a large value of X, it also has a large value for Y. If the x_i and y_i of an individual are both large, the product $x_i y_i$ will also be large, but if y_i is not large when x_i is, then the product will be smaller. Generally, if X and Y covary, the product (and so the covariance) is large, and if they do not, the sum of the products will come to zero.

Regression

The regression, symbolized by b_{XY}, between characters X and Y is their covariance divided by the variance of X:

$$b_{XY} = \frac{\text{cov}_{XY}}{V_X}$$

Regressions are used to describe the slopes of graphs and are therefore useful in describing the resemblance between classes of relatives. If X and Y are unrelated, $\text{cov}_{XY} = 0$ and $b_{XY} = 0$; if they are related, the covariance and regression can be positive or negative (Figure B9.1).

Figure B9.1
The relation between two variables (x and y): (a) negative regression coefficient ($b < 0$); (b) no relation ($b = 0$); and (c) positive regression coefficient ($b > 0$).

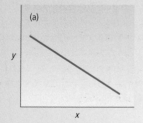
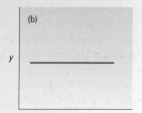
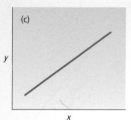

9.5 Relatives have similar genotypes, producing the correlation between relatives

Figure 9.1 above was a graph of beak depth in many parent–offspring pairs in *Geospiza fortis*. Each point is for one offspring and the average of its parents. Beak depth in *G. fortis*, like many characters in most species, shows a correlation between parent and offspring. This is an example of the correlation between relatives, for just as parents and offspring are similar to each other, so are siblings and more distant relatives to some extent.

Similarity among relatives may be either environmental or genetic, and the two effects may have different relative importance in different kinds of species. In humans, where parents and offspring live together in social groups, much of the similarity will be due to the family's common environment (i.e., there is gene–environment correlation). At the other extreme, in a species like a bivalve mollusk in which eggs are released into the sea at an early stage, relatives will not necessarily grow up in similar environments. The environment may not have such a strong influence on similarity among relatives. Darwin's finches are probably somewhere between these two extremes. It is easy to understand the similarity among relatives that is caused by similar environments: in so far as relatives grow up in correlated environments, and there is environmental variation in a character, relatives will be more similar than non-relatives.

The similarity between relatives due to their shared genes is evolutionarily more important. It is possible to deduce the correlation due to shared genes among any two classes of relatives from the variance terms we have already defined. We shall consider only one case, the correlation between parents and offspring, to see how it is done. We can keep things simple by assuming that the environments of parent and offspring are uncorrelated so environmental effects can be ignored (because any environmental effect in the parent will not show up in the offspring: if the parent is larger than average because it chanced on a good food supply, that does not mean its offspring will too). Any correlation between parent and offspring will then be due to their genetic effects.

The genetic value of the character in the parent is, we have seen, made up of several components of which only the additive component is inherited by the offspring. When mating is random, half that additive component of the individual parent is diluted. At a locus, a parent has an additive deviation from the population mean in both its genes. When an offspring is formed, one of the parental genes goes into the offspring together with another gene drawn at random from the population (because we are assuming random mating). The average value of the character in the offspring is, as we saw above, half the additive value of the parent ($\frac{1}{2}A$); the average genetic value in the parent is $A + D$. The correlation between parent and offspring is the covariance between the two (see Box 9.1):

$$\text{cov}_{\text{OP}} = \sum \frac{1}{2} A(A + D)$$

where the sum is over all offspring–parent pairs. The covariance can be re-expressed as:

Relatives, such as siblings or parents and offspring, are similar because of shared environments . . .

. . . and shared genes

We construct a quantitative genetic model to predict parent–offspring similarity

Table 9.1
The covariances between several different classes of relatives.

Relatives	Covariance
One parent and offspring	$\frac{1}{2}V_A$
Mid-parent and offspring	$\frac{1}{2}V_A$
Half sibs	$\frac{1}{4}V_A$
Full sibs	$\frac{1}{2}V_A + \frac{1}{4}V_D$

$$\text{cov}_{OP} = \frac{1}{2}\sum A^2 + \frac{1}{2}\sum AD$$

$^{1}/_{2}\sum AD = 0$ because A and D have been so defined as to be uncorrelated. If an individual has a big value of A, we know nothing about whether its value of D will be big or small. That leaves:

$$\text{cov}_{OP} = \frac{1}{2}\sum A^2 = \frac{1}{2}V_A$$

Parent–offspring similarity depends on the additive genetic variance

In words, the covariance of an offspring and one of its parents is equal to half the additive genetic variance of the character in the population.

The expression for the covariance between one parent and its offspring is true for each parent. It is a small step (though we shall not go into it here) to show that the same expression also gives the covariance between offspring and the midparental value: it is also $^{1}/_{2}V_A$. Other expressions can be deduced, by similar arguments, for the covariance between other classes of relatives (Table 9.1). The formulae are useful for estimating the additive variances of real characters. However, the estimates become most interesting, for the evolutionary biologist, when expressed in terms of the statistic called heritability.

9.6 Heritability is the proportion of phenotypic variance that is additive

The similarity between relatives in general, and between parents and offspring in particular, is governed by the additive genetic variance of the character. If a character has no additive genetic variance in a population, it will not be inherited from parent to

Heritability is a measure of the genetic influence on a character

offspring. For instance, many of the properties of an individual phenotype are accidentally acquired characters, such as cuts, scrapes, and wounds; if we measure these in parent and offspring they will show no correlation: $V_A = 0$. Moreover, some characters, such as the number of legs per individual in a natural population of, say, zebra, show practically no variation of any sort and for them V_A trivially is zero. Additive variance is therefore often discussed as a fraction of total phenotypic variance, and it is this fraction that is called the *heritability* (h^2) of a character:

$$h^2 = \frac{V_A}{V_P}$$

Heritability is a number between zero and one. If heritability is one, all the variance of the character is genetic and additive. Given that $V_P = V_E + V_A + V_D$, all the terms on the right other than V_A must then be zero. In so far as the factors other than additive variance account for the variance of a character, heritability is less than one.

Heritability has an easy intuitive meaning. Consider two parents that differ from the population by a certain amount. If their offspring also deviate by the same amount, heritability is one; if the offspring have the same mean as the population, heritability is zero; if the offspring deviate from the mean in the same direction as their parents but to a lesser extent, heritability is between zero and one. Heritability, therefore, is the quantitative extent to which offspring resemble their parents, relative to the population mean.

Heritability can be measured by several methods

How can we estimate the heritability of a real character? One method is to cross two pure lines. This is mainly of interest in applied genetics, where the problem might be to breed a new variety of crops; it has little interest in evolutionary biology. The two other main methods are to measure the correlation between relatives and the response to artificial selection. Figure 9.1 is an example which uses the correlation between relatives. The slope of the graph, which shows the beak size in offspring finches in relation to the average beak size of the two parents, is equal to the heritability of beak size in that population. The reason is as follows. The slope of the line is the regression of offspring beak size on mid-parental beak size. The regression of any variable y on another variable x equals $\text{cov}_{xy}/\text{var}_x$ (Box 9.1). The covariance of offspring and mid-parental value equals $^1/_2 V_A$ (Table 9.1) and the variance of the mid-parental beak size is equal to $^1/_2 V_P$. (It is half the total population variance because two parents have been drawn from the population and their values averaged: if Figure 9.1 had the value for one parent on the x-axis, its variance would be V_P.) The regression slope simply equals V_A/V_P, which is the character's heritability. For beak depth in *Geospiza fortis* on Daphne Major, the regression and therefore heritability is 0.79.

9.7 A character's heritability determines its response to artificial selection

How can quantitative genetics be applied to understand evolution? There are many ways, and we shall consider two of them here: directional selection and stabilizing selection. As we have seen (Section 4.4, p. 76), three main kinds of selection are usually

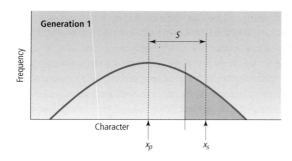

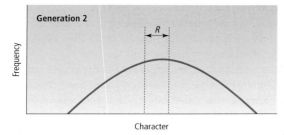

Figure 9.6

Truncation selection: the next generation is bred from those individuals (shaded area) with a character value exceeding a threshold value. The selection differential (S) is the difference between the whole population mean (x_p) and the selected subpopulation's mean (x_s); $S = x_s - x_p$. Because R is about 0.4 of S we could deduce in this case that heritability, h^2, ≈ 0.4.

distinguished; disruptive selection is the third category, which we shall not discuss further here. This section will be concerned with *directional selection*, which has particularly been studied through artificial selection experiments. Artificial selection is important in applied genetics, as it provides the means of improving agricultural stock and crops.

If we wish to increase the value of a character by artificial selection, we can use any of a variety of selection regimes. One simple form is truncation selection: the selector picks out all individuals whose value of the character under selection is greater than a threshold value, and uses them to breed the next generation (Figure 9.6). What will be the value of the character in the offspring generation? First, we can define S as the mean deviation of the selected parents from the mean for the parental population; S is also called the *selection differential*. The response to selection (R) is the difference between the offspring population mean and the parental population mean. In this case, calculating the response to selection is found by regressing the character value in the offspring on that in the parents, where the parents are the individuals that were selected to breed: we plot the offspring's against the parental deviation from the population average to produce a graph like Figure 9.1. The slope of the graph for parents and offspring is symbolized by b_{OP} and we saw in the previous section that for any character $b_{OP} = h^2$; the parent–offspring regressional slope equals the heritability. Therefore:

$$R = b_{OP}S \quad \text{or}$$

$$R = h^2 S$$

This is an important result. The response to selection is equal to the amount by which the parents of the offspring generation deviate from the mean for their population

We construct a quantitative genetic model of directional selection

multiplied by the character's heritability. (The response to selection or the parent–offspring regression can be used to estimate the heritability of a character; for a selected population, they are two ways of looking at the same set of measurements.)

A real example of directional selection may not have the form of truncation selection. In truncation selection, all individuals above a certain value for the character breed and all individuals below do not breed. All the selected individuals contribute equally to the next generation. It could be instead that there is no sharp cut off, but that individuals with higher values of the character contribute increasing numbers of offspring to the next generation. However, the same formula for evolutionary response works for all forms of directional selection. The difference between the mean character value in the whole population and in those individuals that actually contribute to the next generation (if necessary, weighted by the number of offspring they contribute) is the "selection differential" and can be plugged into the formula to find the expected value of the character in the next generation.

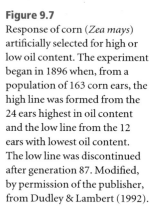

Sustained directional selection reduces heritability

A population can only respond to artificial selection for as long as the genetic variation lasts. Consider, for example, the longest running controlled artificial selection experiment. Since 1896, corn has been selected, at the State Agricultural Laboratory in Illinois, for (among other things) either high or low oil content. As Figure 9.7 shows, even after 90 generations the response to selection for high oil content has not been exhausted. However, the oil content finally became negligibly low in the line selected for low oil content. The seeds had become difficult to maintain and the "low oil" experiment was discontinued after 87 generation.

The "high oil" experiment continues, but it too will eventually come to a stop. As the corn is selected for increased oil content, the genotypes encoding for high oil content will increase in frequency and be substituted for genotypes for lower oil content. This process can only proceed so far. Eventually all the individuals in the population will come to have the same genotype for oil content. At the loci controlling oil content, no

Figure 9.7

Response of corn (*Zea mays*) artificially selected for high or low oil content. The experiment began in 1896 when, from a population of 163 corn ears, the high line was formed from the 24 ears highest in oil content and the low line from the 12 ears with lowest oil content. The low line was discontinued after generation 87. Modified, by permission of the publisher, from Dudley & Lambert (1992).

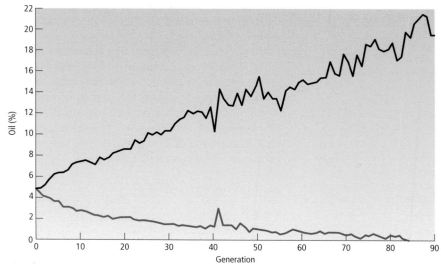

Table 9.2

The heritability of oil content in corn populations after different numbers of generations of artificial selection for high or low oil content, in the Illinois corn experiment (see Figure 9.7). The heritability declines as selection is applied. From Dudley & Lambert (1992).

Generation	Heritability of oil content	
	High line	Low line
1–9	0.32	0.5
10–25	0.34	0.23
26–58	0.11	0.1
59–90	0.12	0.14

New variation can be introduced by recombination or mutation

additive genetic variance will then be left; heritability will have been reduced to zero and the response to artificial selection will come to a stop. In the Illinois corn experiment, the process has not yet run its full course. The heritability of oil content in both the high and low selected lines decreased early in the experiment (Table 9.2), but since then it has been constant at about 10–15% for about 65 generations. The population continues to respond to selection because heritability continues to be above zero.

In other artificial selection experiments, the full process has been recorded. Figure 9.8 shows the response of a population of fruitflies to consistent directional selection for increased numbers of scutellar chaetae (i.e., bristles on a dorsal region of the thorax). Initially the population responded; then, as the additive genetic variation was used up (or, as its heritability declined), the rate of change slowed down to a stop in generations 4–14. It also appeared that, if selection was still continued after the response stopped,

Figure 9.8

Artificial selection for increased numbers of scutellar chaetae in *Drosophila melanogaster*. The response took place in two rapid steps that coincided with observable changes in the form of chromosomes 2, and 2 and 3, respectively; the changes are thought to have been recombinational events. Redrawn, by permission of the publisher, from Mather (1943).

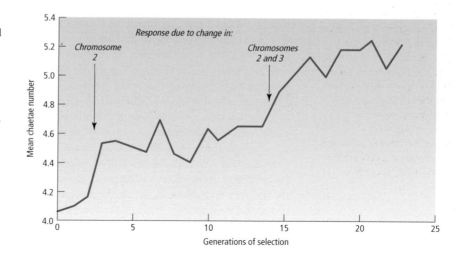

the population suddenly started to respond again after an interval (in generations 14–17). The renewed bout of change is attributed to a rare recombinant or mutation that reinjected new genetic variation into the population.

The quantitative genetic model has broad application

The relation between response to selection (R), heritability, and selection differential (S) enables us to calculate any one of the three variables if the other two have been measured. For example, we saw in Chapter 4 that fishing has selected for small size in salmon, because larger fish are selectively taken in the nets. The selection differential S can be estimated from three measurements: the average size of salmon caught in the nets, the average size of salmon in the population at the mouth of the river (before they are fished), and the proportion of the population that is taken by fishing. All three have been measured and lead to the estimates that the salmon who survive to spawn are about 0.4 lb (0.18 kg) smaller than the population average. Figure 4.3 shows that response (R) — the average size of the salmon — decreased by about 0.1 lb between each 2-year generation. We can therefore estimate the heritability, $h^2 = 0.1/0.4 = 0.25$.

9.8 Strength of selection has been estimated in many studies of natural populations

A character such as beak size may be experiencing directional selection in a bird population. We can estimate the response to selection (R) by measuring the average size over a number of years. Standard quantitative genetic techniques can be used to estimate heritability. We can then use the two numbers to estimate the selection differential. The selection differential expresses how strongly selection is acting (in the case of directional selection, but not stabilizing selection). If the successful individuals are very different from the average individuals in the population, selection is strong, and the selection differential (S) will be large. If selection is weak, the successful individuals will be more like a random sample from the population as a whole and S will be small.

The strength of selection has been measured in Darwin's finches . . .

In Darwin's finches, Gibbs & Grant (1987) measured the response to selection (R), and heritability, for several characters related to body size, and used these to estimate selection differentials. We saw that in *Geospiza fortis* heritability of beak size is about 80%; and after the bout of selection for large size in 1976–77, the finches were about 4% larger. We can estimate the selection differential as $S = 0.04/0.8 = +5\%$. The results for several characters in three periods is shown in Figure 9.9. As the direction of selection reversed from favoring larger beaks between 1976 and 1978 and smaller beaks between 1983 and 1985, the selection differentials reversed from positive values in 1976–77 to negative values in 1984–85. The next El Niño event came along after Gibbs and Grant's paper. This time the changed weather had little effect on the seed size distribution, and the selection differential was round about zero (Grant & Grant 1995).

In Darwin's finches the measured relations between the selection differential, heritability, and response to selection all fit with the predictions of quantitative genetic theory. Any two of the three can be measured, and the third accurately predicted (Grant & Grant 1995). However, Section 9.12 below will look at some more puzzling cases — in which a character is subject to directional selection (the value of S is non-zero), and has been shown to be genetically heritable, but shows no evolutionary

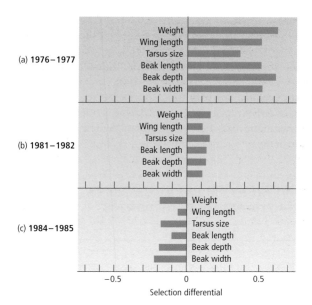

Figure 9.9

(a) In the drought of 1976–77, *Geospiza fortis* individuals with larger beaks survived better and the average size of the finch population increased. (b) In the normal years of 1981–82 there was a slight advantage to having a larger beak, but this was much smaller than during the drought. (c) After the 1983 El Niño, in 1984–85, finches with smaller beaks survived better and the average size of the finch population decreased. The x-axis expresses the selection differential (S) in standardized form: the mean for the survivors after selection minus the population mean before selection all divided by the standard deviation of the character. (The standard deviation is the square root of the variance; see Box 9.1 for the meaning of variance.) A value of S of about 5% for beak depth in the text corresponds here to a standardized S of about 0.6. Redrawn, by permission, from Gibbs & Grant (1987). © 1987 Macmillan Magazines Ltd.

response. When we come to those more puzzling results, it is worth keeping in mind the "successful" results for Darwin's finches.

... and in other studies

Kingsolver *et al.* (2001) compiled the results of 63 studies of directional selection, on 62 species, performed by many different biologists and published between 1984 and 1997. Figure 9.10 shows the distribution of selection differentials found in the studies. For a survey of many characters, the selection differentials need to be "standardized." The selection differentials we have looked at so far have been "non-standardized." They were absolute measurements (0.4 lb or 0.18 kg) in salmon and percentages (5%) in the finches. The equation $R = h^2S$ works in any one study with absolute numbers or percentages. A standardized selection differential expresses the deviation from the mean of the successful individuals as a fraction of the phenotypic standard deviation in the population. (Box 9.1 formally explains standard deviation, but intuitively it is a

Figure 9.10
Frequency distribution of selection differentials (S) found in 63 studies of directional selection in 62 species. Drawn from the database of Kingsolver *et al.* (2001).

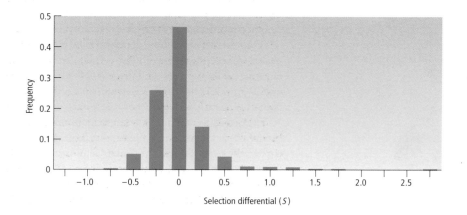

measure of the amount of variation that is independent of the units used to do the measuring. The equation $R = h^2S$ works with standardized selection differentials too, and gives the response as a fraction of the standard deviation of the population.)

Figure 9.10 shows that standardized selection differentials are mainly in the range −0.25 to +0.25. Kingsolver *et al.* (2001) and Hoekstra *et al.* (2001) use the results of their survey to make some tentative further deductions, but for our purposes their survey shows that we have a large amount of evidence, in which quantitative genetic techniques have been used to study directional selection in nature. It also shows the general range of results of those studies.

9.9 Relations between genotype and phenotype may be non-linear, producing remarkable responses to selection

Figure 9.11 illustrates a remarkable artificial selection experiment. Scharloo selected a population of fruitflies for increased relative length of the fourth wing vein (Figure 9.11a). The figure shows the frequency distribution of vein lengths in the population for 10 generations. A length of 60–80 has been reached by about generation 5. At this stage the frequency distribution (amid the scatter that is often seen in real experiments) starts to show a consistent bimodality: it is clearest in generations 5–7, with only the high peak being maintained in generations 8–10. The experiment suggests that more complicated things can occur in artificial selection experiments than we have seen so far. What is going on?

The key to understanding the shape of the response is the relation between genotype and phenotype. A simple response, such as that for oil content in Figure 9.7 or bristle number in Figure 9.8, results when there is an approximately linear relation between genotype and phenotype (Figure 9.12a). Genotype here is expressed as a metrical variable. The easiest way to think of this is to imagine that the character is controlled by many loci; at each, some alleles (+) cause the phenotypic character to increase, and others (−) to decrease. The more positive genes an individual has, the higher its genotypic value (Figure 9.12a). Then when we select for an increase in the character, we pick the individuals with more positive genes, and the value of the character will increase smoothly between generations in the manner of the Illinois corn oil experiment.

The approximately linear form of Figure 9.12a is not the only possible relation between genotype and phenotype (cf. Figure 9.12b and c). The bimodal response in Figure 9.11 is thought to result from a threshold relation between genotype and phenotype (Figures 9.12c and 9.13). In Figure 9.13, the graph has been rotated through 180° relative to the form in Figure 9.12; the *x*-axis (genotype) is drawn down the page on the left. The genotype is thought to control the amount of some vein-inducing substance. Vein length is shown across the top of the graph. The relation between substance and vein length is hypothesized to contain a jump at vein length 60–80 (where the artificial selection response goes bimodal).

Imagine the course of selection for longer wing veins with this threshold relation between genotype and phenotype. The population starts at the top left of the graph with

Look at the bimodal response in the figure opposite!

Relations between genotype and phenotype can be linear or non-linear

Figure 9.11

(a) The main veins in the wing of *Drosophila*. The relative length of the fourth vein was measured by the ratio $L_5 : L_3$. (b) Artificial selection for relative length of the fourth vein. The series on the left is for nine generations in females, with 10 generations of males on the right. Each graph has, as a green line, a frequency distribution for the selected population. In both males and females, a bimodal distribution appears at vein lengths of about 60–80. The black lines are controls. Redrawn, by permission of the publisher, from Scharloo (1987). © 1987 Springer-Verlag.

(a)

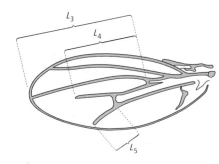

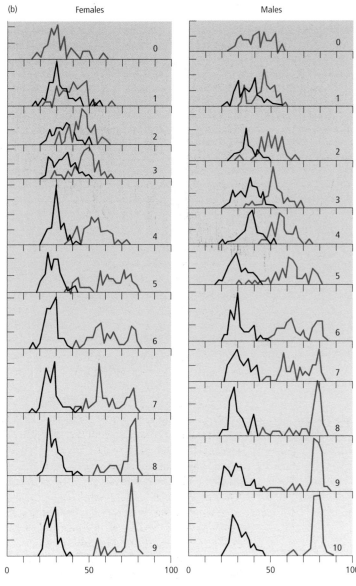

(a) **Linear relation**

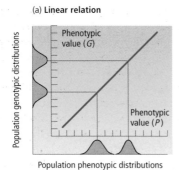

(b) **"Canalized" relation**

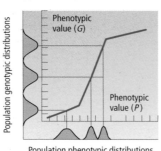

(c) **Threshold relation**

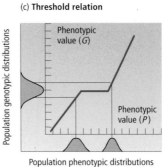

Figure 9.12

The relation between genotype and phenotype. The genotypic value is for a polygenic character, in which higher values might be produced by more positive alleles. The shaded frequency distributions for the phenotypes shown below the *x*-axes are what would be observed, given the shape of the graphs, if the genotypic frequency distributions were as illustrated. (a) Linear relation. (b) "Canalized" relation. (c) Threshold relation. In the canalized relation, the phenotypic variance of a population is reduced relative to the variance in the causal genotypic factor, and vice versa in the threshold relation.

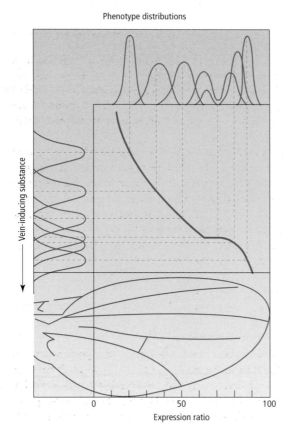

Figure 9.13

A model of the relationship between genotype and phenotype for the *Drosophila* fourth wing vein. The genotype produces an amount of vein-inducing substance. The phenotype (vein length) is drawn along the top; six different population frequency distributions are shown. The graph shows the relation between genotype and phenotype, and between the genotypic and phenotypic frequency distributions of a population. The phenotype in the expression ratio 60–80 range matches a threshold jump in the genotype–phenotype relation; a bimodal phenotypic frequency distribution arises there. Redrawn, by permission of the publisher, from Scharloo (1987). © 1987 Springer-Verlag.

Certain non-linear relations can produce the bimodal response

relatively short veins. Initially, there is some variation in the population for genotype, and an associated normal distribution of vein lengths. Artificial selection produces flies with a higher concentration of the vein-inducing substance and with correspondingly longer wing veins. This process can continue until the population reaches a vein length of about 60–80. At that point, the normal (unimodal) variation for the amount of vein-inducing substance generates a bimodal distribution of vein lengths. The bimodal distribution will later disappear, as the population passes beyond the jump in the genotype–phenotype mapping function. Hence the observed response to selection. The relation of genotype and phenotype for vein length in Figure 9.13 is a hypothesis only; but it does show how in theory a bimodal response to selection could arise.

The main points are that when the genotype–phenotype relation has the linear form of Figure 9.12a, there is a simple response to artificial selection. The population changes until the genetic variation is used up. However, we have no reason to think that this is the typical genotype–phenotype relation. When the relation is more complex, the response to artificial selection can be interestingly different, as the bimodal response to selection on wing vein length in fruitflies illustrates.

9.10 Stabilizing selection reduces the genetic variability of a character

We saw earlier that directional selection reduces the amount of genetic variation for a character, and this can be measured as a decrease in heritability (see Table 9.2). But what about stabilizing selection? In nature, many (perhaps most) characters are subject to stabilizing selection, in which the extremes on either side of some optimum are selected against. (See Section 4.4, p. 78, where Figure 4.4 illustrates how birth weight in humans is an example of stabilizing selection.)

A polygenic, polymorphic character, subject to stabilizing selection . . .

Stabilizing selection will also tend to reduce heritable variation. Consider a character that is influenced by a large number of genes. Some of the genes increase the value of the character and some of them decrease it. Suppose that the character is influenced by 10 loci, and at each two alleles are present. One of the two alleles increases (+) the value of the character, the other decreases (−) it. An individual's haplotypes will then each be a series of alleles, and might for example be symbolized by −++−+−−++. Natural selection favors individuals with an intermediate phenotype, produced by any genotype made up of half + genes and half − genes. Here are three examples:

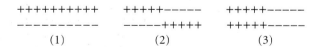

Four different haplotypes are found in these three individuals. In a population that contains these four haplotypes, the three genotypes (1), (2), and (3) will all have the same fitness. We might expect the population to retain considerable genetic variation as these genotypes interbreed and produce a variety of offspring types. However genotypes like (3) that breed true have a small advantage. All the offspring of genotype (3)

have the optimal phenotype, whereas some of the offspring of genotypes (1) and (2) do not. In a population made up of these three genotypes, selection slightly favors genotype (3). If the environment were constant for a long time, always favoring the same phenotype, selection should eventually produce a uniform population with a genotype like (3).

. . . will tend to become uniform, with homozygotes at all loci, . . .

We can take the argument a stage further. Genotype (3) is not the only true-breeding homozygote that can produce the intermediate optimal form. All the following do too:

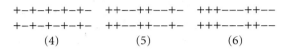

Suppose there was a population made up of genotypes (3) to (6), and selection still favors the intermediate phenotype. What will happen now? Evolution will again tend toward a population with only one genotype, and that genotype should be a multiple homozygote.

The reason is that any one of the homozygotes that happens to have a slightly higher frequency than the others has an advantage. Suppose, for example, that genotype (4) had a higher frequency than (3), (5), and (6). All the genotypes will now be most likely to mate with genotype (4). When genotype (4) mates with genotype (4), all their offspring have the favored phenotype, identical to their parents. But when genotype (3), (5), or (6) mates with genotype (4), the offspring contain potentially disadvantageous genotypes. The offspring of a mating between genotype (3) and genotype (4) will be +−+−+−+−+/++−−++−−+− and has the favored intermediate phenotype. However, *its* offspring will contain disadvantageous recombinants. The end result is for selection to produce a uniform population, in which minority genotypes are selected against because they do not fit in with the majority form. Selection eventually reduces the genetic variability to zero, even with stabilizing selection.

In conclusion, whether a character is subject to directional or stabilizing selection, the effect of selection is to reduce the amount of genetic variation, and the heritability. If selection were the only factor at work, and it worked steadily for a period of time, heritability would be reduced to zero.

9.11 Characters in natural populations subject to stabilizing selection show genetic variation

. . . and yet real characters show much genetic variation

The conclusion of the previous section is contradicted by observable facts. Heritabilities can be measured for real characters, and many show significant genetic variation. Figure 9.14 summarizes some measurements for *Drosophila*. It suggests that typical values for heritability are in the range 20–50%. Heritabilities have been measured in other species too, such as the Galápagos finch, and the results fit the same pattern. Real characters have heritabilities of more than zero.

If selection, whether directional or stabilizing, eliminates genetic variation, why does all this genetic variation exist? Until now, in this chapter, we have been on fairly solid

Figure 9.14

Heritabilities of quantitative characters in *Drosophila*. Hoffmann (2000) compiled field and laboratory estimates of heritabilities for many characters in fruitflies. He divided the characters into four categories. Each datapoint in the figure is an average for a number of estimates for one character. For instance, heritabilities have been estimated for four life history characters in fruitflies. Any one of those characters may have been studied more than once, and the results of the different studies produce a range of estimates, but only the average is shown here. The 29 morphological characters are things like "wing length" and "tibia width." The heritabilities of morphological and physiological characters tend to be higher than those for behavioral and life history characters, but it is questionable whether the difference is biologically meaningful. Drawn from data in Hoffmann (2000).

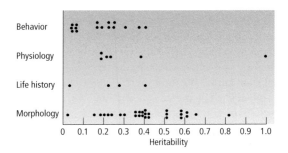

ground. We have been looking at widely accepted theories and results. The results in Figure 9.14 are also widely accepted, as is the reasoning in Section 9.10. But when we move on to think about what maintains the genetic variation we are moving on to a frontier research problem. The question does not yet have a generally accepted answer.

9.12 Levels of genetic variation in natural populations are imperfectly understood

For characters subject to stabilizing selection, two processes can explain the existence of heritable genetic variation. One is mutation–selection balance. The character may have some optimum value, and natural selection eliminates genes that cause deviations from that optimum. But mutations will continually arise, causing no deviations from the optimum. The result is an equilibrium, at which some genetic variation exists because selection cannot clear out mutations instantly with 100% efficiency.

For any one locus, the amount of variation maintained by this selection–mutation balance is low because mutation rates are low (Section 5.11, p. 122); but for a polygenic character, mutation rates should be approximately multiplied by the number of loci influencing the character. A character controlled by 500 loci will have 500 times the mutation rate of a one-locus character. The amount of variability that can be maintained is proportionally increased.

Genetic variation has been studied by two theoretical approaches

How much genetic variation will exist? The question has been thought about in two main ways. One, revived and developed by Lande (1976), considers stabilizing selection on a continuous character (such as body size) controlled by many loci. Mutations at any of the loci can influence the character; because a genotype may be above or below the optimal value for the trait, a small random mutation has a 50% chance of being an improvement. The other, revived and developed by Kondrashov & Turelli (1992), does not consider stabilizing selection on a phenotypic character, but supposes mutations are occurring at many loci and the great majority (many more than 50%) are deleterious. The result is a balance between selection and deleterious mutation at many loci.

We do not have space to go far into either theoretical system, or their relative merits. However, the research so far suggests that selection–mutation balance can explain some heritable genetic variation, but not the high levels typically seen in nature (Figure 9.14). Something else is probably at work too.

Natural selection can favor the maintenance of genetic variation. We look at the evidence in more detail in a later chapter about biological species (Sections 13.6–13.7, pp. 366–73). A simple version of the argument is as follows. Suppose, for example, that the members of a finch population have a range of beak sizes. The beaks may be adapted to a range of seed sizes: finches with larger beaks are better at eating larger seeds, and finches with smaller beaks are better at eating smaller seeds. (In terms of Chapter 5, this corresponds to multiple niche polymorphism: Section 5.13, p. 128.) If the seeds in the local environment are all the same size, then natural selection will produce a bird population with beaks of one size. If the seeds are a range of sizes, natural selection will favor a range of beak sizes in the birds. The actual seed size distribution available to the bird population will depend on whether any competitors are present, as well as what seeds are produced by the local plants. However, although this "ecological" kind of selection can theoretically maintain genetic variation, we do not know whether it is in fact causing the genetic variation observed in the heritability measurements of Figure 9.14.

Recent research has uncovered another puzzle in the observations we have on heritabilities. A number of species have been the subjects of long-term studies in nature. Each generation, the action of natural selection is measured, as a selection differential. For example, in European collared flycatchers, the tarsus lengths of reproductively successful birds and of average birds have been measured from 1980 to 2000 (Figure 9.15a). (The tarsus is part of the bird's leg.) In most years the selection differential is positive: the average number is about 0.2. (This is a "standardized selection differential," and means that the successful birds have tarsus lengths 0.2 of a standard deviation longer than average birds.) The character is also heritable, with a heritability of about 0.35. Natural selection favors longer tarsuses in collared flycatchers, and collared flycatchers have genetic variation for tarsal size.

Some force of natural selection is needed to explain observed levels of genetic variation

Figure 9.15
(a) Natural selection favors collared flycatchers (*Ficedula albicollis*) with longer legs. The results are expressed as a selection differential. The average selection differential over this 20-year period was just below +0.2 (indicated by the dotted line). (b) Observed changes in tarsal length. No net change is statistically detectable over the 20-year period. Year 0 is 1980, year 20 is 2000. From Kruuk *et al.* (2001).

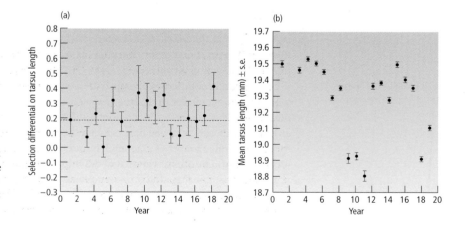

Some heritable characters do not respond to selection . . .

In the same way as we did with salmon and finches (Sections 9.7 and 9.8), we can use the standard equation $R = h^2 S$ to predict how the flycatchers will evolve. The evolutionary response should be a bit less than 0.35×0.2 per generation. This is a "standardized" response, expressed as a fraction of 1 standard deviation. In absolute terms, it corresponds to a predicted increase of about 0.04 mm per generation (+0.022 mm per year). Kruuk *et al.* (2001) tested the prediction by measuring tarsus length over the 20-year period in a large sample of birds. The result is shown in Figure 9.15b. Tarsus length has been net constant. (In fact it shows a fractional decrease, though the decrease is statistically and probably biologically insignificant.)

. . . and the reason is the subject of unfinished research

The collared flycatcher is not the only species in which a character is apparently subject to directional selection, and apparently shows heritability, but is not showing an evolutionary response. Biologists are puzzled by these results and have a number of hypotheses about them. The measurements of heritability may be erroneous or inappropriate. Migration, or hybridization with other species, may be confusing the picture.

A biologically more interesting possibility is that the effect of selection, to increase (for example) the tarsus length each generation, is balanced by some other force that decreases tarsus length by approximately the same amount. For instance, the population might be in selection–mutation equilibrium (Section 5.11, p. 122, but allowing for multiple loci). If a large tarsus is advantageous, the effect of disadvantageous mutations each generation will be to reduce average tarsus length. Then, at equilibrium, the increase in tarsus length each generation by directional selection could be balanced by a decrease due to mutation. The character would be both constant over time and heritable. Mutation is not the only factor that could balance directional selection in this way; intraspecific competition could too. However, these ideas are all hypotheses to be tested. The lack of evolutionary response in heritable characters apparently subject to directional selection is not understood. Quantitative genetics is one of the oldest topics in evolutionary biology, and contains many solid findings. But it has its share of unsolved problems for the future too. Non-evolution, in species that are predicted to show evolutionary change, is one of them.

9.13 Conclusion

One- and two-locus population genetics are used for characters controlled by one or two loci and whose genetics are known. Quantitative genetics provides the techniques to understand evolution in characters that are influenced by a large number of genes, and for which the exact genotype (or genotypes) producing any given phenotype are unknown. It is possible that the majority of characters have this kind of genetics, in which case quantitative genetics would be appropriate for understanding the majority of evolution; at any rate, it is a highly important set of techniques. We have seen in this chapter how quantitative genetics divides up the variation in a character, to recognize the component — the additive genetic effect — that controls how offspring resemble their parents. The additive genetic effect plays the same role in quantitative genetics as a knowledge of Mendelian genetics in one- and two-locus population genetics. The response to selection can be analyzed by means of the heritability of a character, which is the fraction of its variation due to additive genetic effects. However, even with simple directional selection, the exact response depends on the underlying genetic control. For example, the possible threshold relation between the genotype and phenotype for the wing veins of the fruitfly generates an interesting bimodal response to selection. Here, the heritability of the character would show strange changes as the character evolved. Directional selection unambiguously should continue to alter a character until its heritability is reduced to zero. With stabilizing selection, it might be thought that many genotypes could be maintained if they all produce the same intermediate phenotype. However, even here it can be argued that all but one of the genotypes should eventually be eliminated by selection. The argument appears to be contradicted by the facts, and biologists do not yet fully understand the observed values of heritabilities in natural populations.

Summary

1 Quantitative genetics, which is concerned with characters controlled by many genes, considers the changes in phenotypic and genotypic frequency distributions between generations, rather than following the fate of individual genes.

2 The phenotypic variance of a character in a population can be divided into components due to genetic and environmental differences between individuals.

3 Some of the genetic effects on an individual's phenotype are inherited by its offspring; others are not. The former are called additive genetic effects and the latter are due to such factors as dominance and epistatic interaction between genes.

4 The heritability of a character is the proportion of its total phenotypic variance in a population that is additive.

5 The heritability of a character determines its evolutionary response to selection.

6 The additive genetic variance can be measured by the correlation between relatives, or by artificial selection experiments.

7 The strength of selection can be estimated as a selection differential, either directly by measuring reproductively successful individuals and average individuals in a population, or indirectly, using measurements of heritability and of observed evolutionary change in a population. Selection differentials often have values in the range -0.2 to $+0.2$ in populations that are experiencing directional selection.

8 The response of a population to artificial selection depends on the amount of additive genetic variability and on the relation between genotype and phenotype. If the relation is non-linear, strange bimodal responses can arise.

9 Stabilizing selection acts to reduce the amount of genetic variability in a population. However, polygenic characters show non-trivial values for heritability.

10 Biologists do not completely understand the observed levels of heritability for characters in natural populations. Genetic variation is maintained by some mix of mutation and selection. Some characters do evolve by the amount predicted from their heritability and the strength of selection; others, however, seem not to.

Further reading

Falconer & Mackay (1996) is the standard introduction. Lewontin's chapters on the subject in Griffiths *et al.* (2000) are also introductory. Roff (1997) and Lynch & Walsh (1998) are recent comprehensive texts. Volume 2 of Wright (1969) is classic and advanced.

On Darwin's finches, see Grant (1986, 1991), Grant & Grant (1995, 2000, 2002), and the popular book by Weiner (1994).

I have described quantitative genetics as being concerned with characters for which the genes are unknown. One current research topic is to identify the genes contributing to quantitative characters. See the textbooks and Beldade *et al.* (2002a) for a recent example that connects with Chapter 20 in this text about the gene *Distal-less*, which contributes to variation in butterfly eyespots.

The various debates about heritability, genetic variation, and response to selection can be traced through the general texts I refer to above. The review of rates of evolution by Hendry & Kinnison (1999) mainly connects with Chapter 21 in this book, but the material in it is similar to that in the reviews by Kingsolver *et al.* (2001) and Hoekstra *et al.* (2001): this allows a line of synthesis between this chapter and Chapter 21. Scharloo (1987, 1991) reviews non-linear selection responses and canalization. Gibson & Wagner (2000) also discuss canalization. See Chapter 10 for more on canalization and then Chapter 20 for references on the breakdown of canalization and on "evolvability."

Study and review questions

1 Review the reasons why characters that are influenced by a large number of genetic and/or environmental effects will show a normal distribution.

2 Suppose the average spine length in a population of porcupines is 10 in and we take a number of individuals with 12 in spines and mate each with a random member of the population. The offspring grow up with spines of average length 10.5 in. What is the additive effect of those 12 in porcupines?

3 Here are measurements, or estimates, of the phenotypic values and their additive genetic components in a population of nine porcupines. (a) Calculate V_P, V_A, and h^2. (b) What would you predict the average spine length would be in the next generation if porcupines eight and nine were used to produce it?

4 Imagine we are selecting porcupines to make them pricklier. The heritability of prickliness is 0.75. Average prickliness before selection is 100 (and the standard deviation is also 100). The average prickliness of the porcupines who produce the next generation is 108. What will the prickliness of the next generation be?

5 If a character is subject to stabilizing selection, how might the genotype–phenotype relationship for it evolve?

6 Suppose there is stabilizing selection on a character controlled by eight gene loci, and the optimum individual should have six positive genes and 10 negative genes. What will happen through time to a population that initially contains individuals of two sorts: those that are +++−−−−−/+++−−−− and those that are ++++++−−/−−−−−−−−?

Individual	1	2	3	4	5	6	7	8	9	V_P	V_A	h^2
P	−10	−5	−5	0	0	0	+5	+5	+10			
A	−4	−2	−2	0	0	0	+2	+2	+4			

Part three

Adaptation and Natural Selection

The three chapters of this part are about adaptation — the fit of organisms to life in their environments. We begin with a conceptual chapter. Adaptation was known about long before Darwin lived, and pre-Darwinian thinkers had tried to explain the existence of adaptation. Chapter 10, however, argues that only natural selection can explain adaptation. Some characters, particularly molecular characters, have evolved by processes other than natural selection, but they are not adaptations. Not all evolution proceeds by natural selection, but all adaptive evolution does. The chapter also looks at "gradualism" in Darwin's theory — it looks at the way new adaptations evolve by modifications of previously existing parts, and at the size of the genetic changes that occur during adaptive evolution. The chapter discusses how perfect the adaptations of living species are, and what constraints there are on adaptive perfection. The chapter ends by considering various definitions of adaptation.

In Chapter 11, we move on to ask what the entity is that adaptations evolve for the benefit of. Evolution by natural selection happens because adaptations benefit something, but what is it exactly — genes, whole genomes, individual organisms, groups of organisms, species, or what? This is the question of "What is the unit of selection?" Adaptations, the chapter suggests, usually benefit organisms, but there is a deeper criterion that can be used to understand the exceptions as well as the rule: more fundamentally, adaptations evolve for the benefit of genes. Only genes last long enough for natural selection to be able to adjust their frequencies over evolutionary time. Organismal adaptations usually result because gene reproduction is more closely tied to the reproduction of organisms than any other entity, and gene reproduction is maximized if adaptations are at the organismal level.

Chapter 12 uses three examples from sexual reproduction to illustrate the practical study of adaptation. The deepest problem — of why sex exists — is still mainly at the stage of developing a hypothesis. The theory of sexual selection is well developed, and the crucial empirical work is now beginning. In the theory of sex ratios, there is a good match of theoretical prediction and empirical tests. The existence of sex itself is one of the biggest unsolved problems in biology. The problem of sex ratios is not only solved, but provides a powerful "model system" for the analysis of adaptation.

10 Adaptive Explanation

This chapter considers a series of points about adaptation. Many of them are matters of controversy in the professional literature and this chapter aims to explain the main positions in those controversies, and to provide the background to understanding them. We look first at the argument to show that natural selection is the only known explanation of adaptation; and second at whether natural selection can explain all adaptations, including such complex organs as the eye. Then we move on to the genetics of adaptation, particularly Fisher's model, which suggests that adaptations evolve in many small genetic steps. We briefly look at the main methods of studying adaptation: the relation between the predicted and observed forms of a character, experiments, and interspecies comparisons. Most of the rest of the chapter is about various reasons why adaptations may not be perfect. Adaptations may be out of date or may be constrained by genetics, developmental mechanisms, historic origins, or trade-offs between multiple functions. We finish by looking at the definitions of adaptation.

10.1 Natural selection is the only known explanation for adaptation

Before Darwin, adaptation was explained theologically, . . .

The fact that living things are adapted for life on Earth is sufficiently obvious that philosophers did not have to wait for Darwin to point it out. In Section 1.2 (p. 6) we looked at a classic example of adaptation, the woodpecker's beak. In later chapters we have met many more examples, such as camouflage in moths, mimicry in butterflies, and drug resistance in HIV. Living creatres are, in many ways, well adjusted for living in their natural environments. Adaptation was a crucial concept in *natural theology* — a school of thought that was highly influential from the eighteenth century until Darwin's time. Natural theologians explained the properties of nature, including adaptation, theologically (that is, by the direct action of God). John Ray and William Paley were two important thinkers of this type. In our time, the ideas of natural theology are still used by certain kinds of modern creationist.

Darwin himself was much influenced by the examples of adaptation, such as the vertebrate eye, discussed by Paley. Paley explained adaptation in nature by the creative action of God: when God miraculously created the world and its living creatures, he or she miraculously created their adaptations too. Natural theology was influential as a way of understanding adaptations in nature, but its main influence — beyond biology — was as an argument to prove that God exists, called the "argument from design." This is one of several classic philosophical arguments for the existence of God. Part of the reason why Darwin's theory was so controversial was that it wrecked one of the most popular (at that time) arguments for the existence of God. The key difference between natural theology and Darwinism is that the former explains adaptation by supernatural action, and the latter by natural selection.

. . . or by directed variation

Natural theology and natural selection are not the only explanations that have been put forward for adaptation. The inheritance of acquired characters ("Lamarckism") suggests that the hereditary process produces adaptations automatically. Other theories suggest that the hereditary mechanism itself produces designed, or directed, mutations and adaptation results as the consequence. These theories differ from Darwinism. In Darwinism, variation is not directed toward improved adaptation. Instead, mutation is undirected and selection provides the adaptive direction in evolution (Section 4.8, p. 88).

It is one of the most fundamental claims in the Darwinian theory of evolution that natural selection is the only explanation for adaptation. The Darwinian, therefore, has to show that the alternatives to natural selection either do not work or are scientifically These alternatives can be ruled out philosophically, . . . unacceptable. Let us consider the natural theologians' supernatural explanation first. We can accept that an omnipotent, supernatural agent could create well adapted living things: in that sense the explanation works. However, it has two defects. One is that supernatural explanations for natural phenomena are not used in science (Section 3.13, p. 67). The second is that the supernatural Creator is not explanatory. The problem is to explain the existence of adaptation in the world, but the supernatural Creator already possesses this property. Omnipotent beings are themselves well designed, adaptively complex, entities. The thing we want to explain has been built into the explanation. Positing a God begs the question of how such a highly adaptive and well designed thing could in its turn have come into existence. Natural theology is therefore arguably non-explanatory, and its use of supernatural causes is unscientific.

The "Lamarckian" theory — the inheritance of acquired characters — is not unscientific.[1] It posits a hereditary mechanism that can be tested for, and that could give rise to adaptations. Biologists generally reject Lamarckism for two reasons. One is factual. Since Weismann, in the late nineteenth century, it has generally been accepted that acquired characters are, as a matter of fact, not inherited. More than a century of genetics since Weismann has supported this view. (A few minor exceptions are known, but they do not challenge the general principle.)

The second objection is theoretical. Lamarckism by itself arguably cannot account for the evolution of adaptation. Consider the adaptations of zebras to escape from lions. Ancestral zebras would have run as fast as possible to escape from lions. In doing so, they would have exercised and strengthened the muscles used in running. Stronger legs are adaptive as well as being an individually acquired character: if the acquired character was inherited, the adaptation would be perpetuated. Superficially, this looks like an explanation, whose only defect is that acquired characters happen not to be inherited.

Now let us imagine (for the sake of argument) that acquired characters are inherited, and look more closely at the explanation. The adaptation arises because zebras, within their lifetimes, become stronger runners. However, muscles do not by some automatic physical process become stronger when they are exercised. The muscles might just as well become weaker, because they are used up. Muscle strengthening in an individual zebra requires explanation and cannot be taken for granted. Muscles, when exercised, grow stronger because of a pre-existing mechanism which is adaptive for the organism. But where did that adaptive mechanism come from? The theoretical defect in Lamarckism is that it has no good answer to this question. To provide a complete explanation for adaptation, it would have to fall back on another theory, such as God or natural selection. In the former case it would run into the difficulties we discussed above. In the latter case it is natural selection, not Lamarckism, that is providing the fundamental explanation of adaptation. Lamarckism could work only as a subsidiary mechanism; it could only bring adaptations into existence in so far as natural selection had already programed the organism with a set of adaptive responses. Pure Lamarckism does not by itself explain adaptation.

All theories of directed or designed mutation have the same problem. A theory of directed mutation, if it is to be a true alternative to natural selection, must offer a mechanism for adaptive change that does not fundamentally rely on natural selection to provide the adaptive information. Most alternatives to natural selection do not explain adaptation at all. For example, in the early twentieth century, some paleontologists, such as Osborn, were impressed by long-term evolutionary trends in the fossil record. The titanotheres are a classic example. Titanotheres are an extinct group of Eocene and Oligocene perissodactyls (the mammalian order that includes horses). In a number of

---

[1] I put "Lamarckian" in quotes because, as we saw in Chapter 1, the inheritance of acquired characters was not especially important in Lamarck's own theory; nor did he invent the idea. However, the inheritance of acquired characters has generally come to be called Lamarckism and we can conveniently follow normal usage, outside purely historic discussion.

Margin notes:

. . . factually, . . .

. . . or theoretically

Directed variation assumes, rather than explains, adaptation

Figure 10.1
Two lineages of titanotheres showing parallel body size increase and the evolution of horns. Only two of many lineages are illustrated. Reprinted, by permission of the publisher, from Simpson (1949).

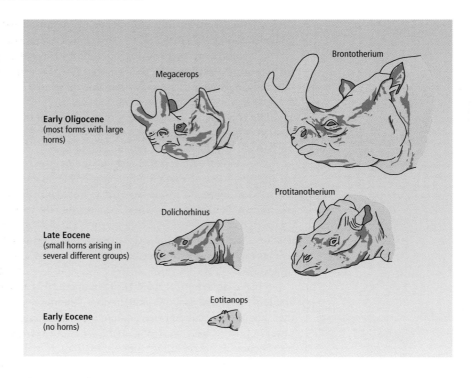

The fossil record shows some apparently directed trends . . .

. . . but they are unlikely to have been driven by directed variation

lineages, the earlier forms lacked horns whereas later ones had evolved them (Figure 10.1). Osborn, and others, believed that the trend was *orthogenetic*: that it arose not because of natural selection among random mutations but because titanotheres were mutating in the direction of the trend.

Directed mutation could explain a simple, adaptively indifferent trend. If a titanothere was equally well adapted no matter what size its horns were, then a trend toward larger horns might be generated by directed mutation. In fact, the horns are thought to be adaptive, and that makes directed mutation an implausible explanation. Mutation is random with respect to adaptation (Section 4.8, p. 88). If mutation is directed, it is in a non-adaptive way. Thus, if someone explains a trend by orthogenesis (or directed mutation) we can ask how the "orthogenetic" mutations could keep on occurring in the direction of adaptive improvement. If the reply is that variation just happens to be that way, then adaptation is being explained by chance — and chance alone cannot explain adaptation, almost by definition.

This objection is not all that strong for titanothere horns, because their adaptive function is little understood. The trends might have been possible by simple increases in size. However, for other known trends in the fossil record, such as the evolution of mammals from mammal-like reptiles (Section 18.6.2, p. 542), the objection is much more powerful. Mammals evolved over about 100 million years, during which time changes occurred in the teeth, jaws, locomotion, and physiology. Almost every feature of the animals was altered in an integrated way. Directed mutation alone would be highly unlikely to drive a complex, multicharacter, adaptive trend of this kind. A random process alone will not explain adaptation. For this reason, directed mutation on its own, like Lamarckism, is ruled out as an explanation for adaptation.

In conclusion a strong argument can be made that natural selection is the only currently available theory of adaptation. The alternatives variously rely on chance, on unscientific causes, on processes that do not operate in fact, or are non-explanatory.

10.2 Pluralism is appropriate in the study of evolution, not of adaptation

Not all evolution is adaptive

So natural selection is our only explanation for adaptation. This statement, however, applies only to adaptation and not to evolution as a whole. Biologists, such as Gould & Lewontin (1979), have pointed out that Darwin did not himself rely exclusively on natural selection, but admitted other processes too; and they urge that we should accept a "pluralism" of evolutionary processes, rather than relying exclusively on natural selection. For evolution as a whole, this is a sensible idea. In Chapter 7, for instance, we saw that many evolutionary changes in molecules may take place by random drift. The molecular sequences among which drift takes place are not different adaptations. They are different variants of one adaptation, and natural selection does not explain why one organism has one sequence variant, and another organism has another. We need drift as well as selection in a full theory of evolution.

The fact that processes beside natural selection can cause evolutionary change does not alter the argument of Section 10.1. It just goes to show that not all evolution need be adaptive. This being so, we should be pluralists about evolution; but when we are studying adaptation, it is sensible to concentrate on natural selection.

10.3 Natural selection can in principle explain all known adaptations

The argument so far has been negative: we have ruled out the alternatives to natural selection, but we have not made the positive case for natural selection itself. We have seen before (Chapter 4) that natural selection can explain adaptation, but we can also ask a stronger question: can it explain *all* known adaptations?

The question is important historically, and it still often rises in popular discussions of evolution. The case against selection would run something like this. There is no doubt that natural selection explains some adaptations, such as camouflage. However, the adaptation in this case, as well as in other famous examples of natural selection, are all simple. In the peppered moth it is just a matter of adjusting external color to the background. The problem arises in complex characters that are adapted to the environment in many interdependent respects. Darwin's explanation for complex adaptations is that they evolved in many small steps, each analogous to the simple evolution in the peppered moth; that is what Darwin meant when he called evolution gradual. Evolution has to be gradual because it would take a miracle for a complex organ, requiring mutations in many parts, to evolve in one sudden step. If each mutation arose separately, in different organisms at different times, the whole process becomes more probable (we look at this further in Section 10.5).

In Darwin's theory, complex adaptations evolve in many small steps

Darwin's "gradualist" requirement is a fundamental property of evolutionary theory. The Darwinian should be able to show for any organ that it could, at least in principle, have evolved in many small steps, with each step being advantageous. If there are exceptions, the theory is in trouble. In Darwin's (1859) words, "if it could be demonstrated that any complex organ existed which could not possibly have been formed by numerous successive slight modifications, my theory would absolutely break down."

Darwin argued that all known organs could have evolved in small steps. He took examples of complex adaptations and showed for them how they could have evolved through intermediate stages. In some cases, such as the eye (Figure 10.2), these intermediates can be illustrated by analogies with living species, in other cases they can only be imagined. Darwin only had to show that the intermediates could possibly have existed. His critics had the more difficult task of showing that the intermediates could not have existed. It is very difficult to prove negative statements. Nevertheless, many critics suggested, for various adaptations, that natural selection cannot account for them. These types of adaptation can be considered under two headings.

Coadaptations

Critics suggest that complex organs cannot evolve by natural selection

Coadaptation here refers to complex adaptations, the evolution of which would have required mutually adjusted changes in more than one of their parts. (Coadaptation is a popular word: it has already been used in a different sense in Chapter 8, and will be used in a third sense in Chapter 20!) In a historic dispute in the 1890s, Herbert Spencer and August Weismann discussed the giraffe's neck as an example. Spencer supposed that the nerves, veins, bones, and muscles in the neck were each under separate genetic control. Any change in neck length would then require independent, simultaneous changes of the correct magnitude in all the parts. A change in the length of the neck-bones would malfunction without an equal change in vein length, and evolution by natural selection on one part at a time would be impossible. The example is unconvincing now because of the obvious retort that the lengths of all the parts could be under common genetic control.

The other standard example of a complex coadaptation is the eye. When one eye part, such as the distance from the retina to the cornea, changes during evolution, changes in other parts, such as lens shape, would (it is said) be needed at the same time. Because of the improbability of simultaneous correct mutations in both parts at the same time, a complex, finely adjusted engineering device like the eye could not therefore have evolved by natural selection. The Darwinian reply (illustrated in Figure 10.2) is that the different parts could evolve independently in small steps: it is not necessary for all the parts of an eye to change at the same time in evolution.

A computer model study by Nilsson & Pelger (1994) illustrates the power of Darwin's argument. Although the eye of a vertebrate or an octopus looks so complex that it can be difficult to believe it could have evolved by natural selection, in fact light-sensitive organs (not all of them complex) have evolved 40–60 times in various invertebrate groups — which suggests either that the Darwinian explanation faces a 40- to

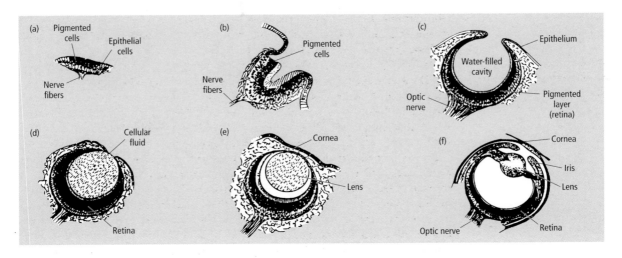

Figure 10.2
Stages in the evolution of the eye, illustrated by species of mollusks. (a) A simple spot of pigmented cells. (b) A folded region of pigmented cells, which increases the number of sensitive cells per unit area. (c) A pin-hole camera eye, as is found in *Nautilus*. (d) An eye cavity filled with cellular fluid rather than water. (e) An eye is protected by adding a transparent cover of skin and part of the cellular fluid has differentiated into a lens. (f) A full, complex eye, as found in the octopus and squid. Redrawn, by permission of the publisher, from Strickberger (1990).

60-fold more difficult problem than the vertebrate eye alone presents, or that it may not be so difficult for the things to evolve after all.

Nilsson and Pelger simulated a model of the eye to find out how difficult its evolution really is. Their simulation began with a crude light-sensitive organ consisting of a layer of light-sensitive cells sandwiched between a darkened layer of cells below and a transparent protective layer above (Figure 10.3). The simulation, therefore, does not cover the complete evolution of an eye. To begin with it takes light-sensitive cells as given (which is an important but not absurd assumption, because many pigments are influenced by light) and at the other end it ignores the evolution of advanced perceptual skills (which are more a problem in the evolution of the brain than the eye). It concentrates on the evolution of eye shape and the lens; this is the problem that Darwin's critics have often pointed to, because they think it requires the simultaneous adjustment of many intricately related parts.

From the initial simple stage, Nilsson and Pelger allowed the shape of the model eye to change at random, in steps of no more than 1% change at a time. One percent is a small change, and fits in with the idea that adaptive evolution proceeds in small, gradual stages. The model eye then evolved in the computer, with each new generation formed from the optically superior eyes in the previous generation; changes that made the optics worse were rejected, just as selection would reject them in nature.

The particular optical criterion used was visual acuity or the ability to resolve objects in space. The visual acuity of each eye in the simulation was calculated by methods of optical physics. The eye is particularly well suited to this kind of study because optical

A simulation suggests that the eye could easily evolve in gradual advantageous stages . . .

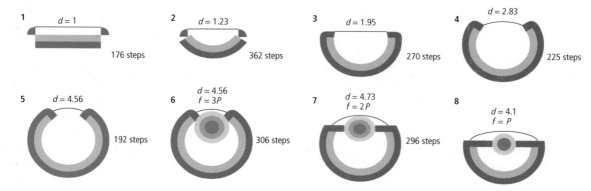

Figure 10.3

Eight stages in the evolution of the eye in a computer model. The initial stage has a transparent cell layer, a light-sensitive cell layer, and a dark pigmented bottom cell layer. It first improves its optical properies by buckling in (up to stages 4–5); by stage 5 it approximately corresponds to the pin-hole camera eye (see Figure 10.2c). It then improves by the evolution of a lens (stage 6). The lens shape then changes, and the iris flattens, to improve the focusing properties. f is the focal length of the lens; it has the best optical properties when f equals the distance from the lens to the retina (P): this feature gradually improves in the final three phases (stages 6–8). d indicates the change in shape and is the normalized diameter of the eye. Redrawn, by permission of the publisher, from Nilsson & Pelger (1994).

qualities can readily be quantified: it is possible to show objectively that one model eye would have better acuity than another. (It is not so easy to imagine how to measure the quality of some other organs, such as a liver or a backbone.) The simulated eye duly improved over time, and Figure 10.3 shows some of the phases along the way. After 1,000 or so steps the eye had evolved to be rather like a pin-hole camera eye (Figure 10.2c shows a real example). Then, the lens started to evolve by a local increase in the refractive index of the layer that had started out simply as transparent protection. The lens to begin with had poor optical qualities but its focal length improved until it equaled the diameter of the eye, at which point it could form a sharply focused image.

. . . in less than half a million generations

How long did it take? The complete evolution of an eye like that of a vertebrate or octopus took about 2,000 steps. What had looked like an impossibility actually turns out to be possible in a short interval of time. Nilsson and Pelger (1994) used estimates of heritability and strength of selection (Section 9.7, p. 236) to calculate how long the change might take; their answer was about 400,000 generations. With a generation time of 1 year, the evolution of an eye from a rudimentary beginning would take less than 0.5 million years. Far from being difficult to evolve, the model shows that it is rather easy.

The work also illustrates the value of building models to test our intuitions. Darwin himself referred to the evolution of complex organs by natural selection as presenting a problem for the imagination, not the reason. Nilsson and Pelger's computer study supports his remark.

Functionless, or disadvantageous, rudimentary stages

An organ has to be advantageous to its bearer at all stages in its evolution if it is to be produced by natural selection. Some adaptations, it is said, although undoubtedly

Some critics suggest that the initial stages of a character would be disadvantageous

advantageous in their final form, could not have been when in a rudimentary form: "What is the use of half a wing?" is a familiar example. The anatomist St George Jackson Mivart particularly stressed this argument in his *The Genesis of Species* (1871). The Darwinian reply has been to suggest ways in which the character could have been advantageous in its rudimentary form. In the case of the wing, partial wings might have broken the force of a fall from a tree, or protowinged birds might have glided from cliff tops or between trees — as many animals, such as flying foxes, do now. These early stages would not have required all the muscular back up of a full, final wing. The concept of preadaptation (see below) provides another solution to the problem.

Evolutionary biologists are sometimes challenged with arguments about functionless rudimentary stages or the impossibility of complex adaptive evolution. It is impossible to imagine, someone will insist, how such-and-such a character could have evolved in small, advantageous steps. In reply, the evolutionary biologist may offer a possible series of stages by which the character might have evolved. We need to keep in mind the status of the evolutionary biologist's argument here. The series of stages may in some cases not be particularly plausible, or well supported by evidence, but the argument is put forward solely to refute the suggestion that we cannot imagine how the character could have evolved.

A second order of critics may latch on to the argument at this point and accuse evolutionists of making speculative, even fanciful, suggestions about the stages through which individual adaptations could have evolved. But the critics overlook the original point of the discussion. The speculations are not the prize specimens of evolutionary analysis. It is not being claimed that the series is particularly profound or realistic, or even very probable. The long evolutionary history that precedes any complex modern adaptation will appear, with hindsight, to be an improbable series of accidents: the same point is as true for human history as evolution. Given the state of our knowledge at any one time, for some characters we can reconstruct their evolutionary stages with some rigor (Chapter 15), but for others we cannot — and for these it is only possible to make guesses to illustrate possibilities, not conduct a careful scientific investigation.

It is fair to conclude that there are no known adaptations that definitely could not have evolved by natural selection. Or (if the double negative is confusing!), we can conclude that all known adaptations are in principle explicable by natural selection.

10.4 New adaptations evolve in continuous stages from pre-existing adaptations, but the continuity takes various forms

10.4.1 *In Darwin's theory, no special process produces evolutionary novelties*

We saw in the previous section that Darwin's theory of adaptation is "gradualist." New adaptations evolve in small stages from pre-existing organs, behavior patterns, cells, or molecules. Another way of saying the same thing is to say that there is continuity between all the forms of adaptation that we see in the world today. This view of continuity contrasts with, for example, a creationist view of life in which the adaptations of different species originate separately and there is no continuity between them.

Gradual change produces novelties

The continuity of adaptive evolution can challenge our understanding of novelty. During evolution, organs do arise that can be described as evolutionary novelties. The vertebrate eye, for example, exists in vertebrates including ourselves, but is not found in all life. It was in some sense newly evolved during vertebrate ancestry. It is eventually recognizable as a new structure that did not exist before. However, as we saw in the previous section, the eye evolved in continuous small stages ultimately from ancestral photoreceptive cells on the body surface. There is no distinct stage at which the "eye" sudenly and distinctly came into existence. The vertebrate eye evolutionarily blurs out through multiple ancestral stages. Thus something that we recognize as a novelty can arise even though it evolved through the modification of previously existing structures.

In Darwin's theory, no special evolutionary process operates to create new structures. The same evolutionary process of adaptation to the local environment is at work throughout. The cumulative effect of many small modifications can be such that something "new" has arisen. (This view of evolutionary novelty is not universally agreed by biologists. Some biologists do argue that evolutionary novelty is a special process: however, they would probably agree that theirs is a minority view.)

10.4.2 *The function of an adaptation may change with little change in its form*

During the evolution of the eye, the function of the organ was relatively constant throughout. From simple photoreceptive cells to full eyes, the organ was a sense organ — sensitive to light. Probably, many organs evolve in this way, by gradual transformation of a structure that has a constant function. In other cases, organs can change their function with relatively little change in structure. Feathers are an example, suggested by dramatic, recently excavated, evidence from fossils in China. Feathers are found in modern birds and mainly function in flight. Birds likely evolved from a group of dinosaurs, and dinosaur fossils typically lack feathers. We might therefore infer that feathers evolved along with flight during the origin of birds.

Feathers preceded flight in the evolution of birds

However, in the past 5 years or so a series of fossils have been described from China (Prum & Brush 2002). The fossils are described as non-avian dinosaurs, but they have feathers or rudimentary feathers. Feathers probably originally evolved for some function other than flight — thermoregulation, perhaps, or display. Later on, flight evolved and feathers turned out to be useful aerodynamically. Feathers then took on their modern function. (Feathers are still used in display and thermoregulation so it might be more accurate to say a function was added, rather than changed. Alternatively, we could say that a function of flight plus display is a change from a function of display alone.)

The classic Darwinian term for a case such as the feathers in non-avian dinosaurs is *preadaptation*. A preadaptation is a structure that happens to be able to evolve some new function with little change in structure. A second example is the tetrapod leg. Fish lack legs, which evolved during the evolution of amphibians and are now used for walking on land. Fossil evidence, such as from *Acanthostega*, suggests that legs originally evolved for underwater swimming. The bone structure of swimming paddles in one

Box 10.1
Molecular Cooption

The term *cooption* is often used to describe the evolutionary process in which a molecule takes on a new function, but with little change in structure. Cooption is conceptually much the same as preadaptation. One term (cooption) happens to be used more about molecules, and the other (preadaptation) about morphology.

The crystallins are a remarkable example. These are the molecules that make up the lens in the eye. Many unrelated molecules appear to be able to function as lens proteins, and the exact molecule that is found in the lens can change during evolution. The lens of human eyes, like those of many vertebrates, contains α-crystallin, which is very similar to a heat-shock protein and probably evolved by gene duplication from a gene coding for a heat-shock protein. The eye lenses of some other vertebrate taxa contain other crystallins that are unrelated to the heat-shock protein. A few birds, and crocodiles, use ε-crystallin, which has much the same sequence as (and indeed is) lactate dehydrogenase. The usual crystallin of birds and reptiles is δ-crystallin, which is arginosuccinate lyase. Other odd crystallins are found in individual taxa such as elephant shrews. Apparently all that is needed for a molecule to serve as a crystallin is that it forms a certain globular shape. Many enzymes meet this requirement, and during evolution the molecules that have been used as lens proteins have chopped and changed while the lens itself has remained much the same. (The crystallins make an interesting case study in homology. The lens of human eyes is homologous with the lens of a crocodile eye. The molecules that make up the lenses are not. On homology, see Section 15.3, p. 427.)

The emerging subject of "evo-devo" (Chapter 20) is documenting many examples of molecular cooption. In embryonic development, certain regulatory genes code for subroutines that can be useful in many circumstances. A gene that regulates how far a limb grows before the feet start to develop may also prove useful in regulating the size of an "eyespot" pattern on a butterfly wing. In both cases, some embryonic process has to operate for a certain time, or across a certain space, and then come to a stop. Much the same genetic instructions may be able to control the development of both limbs and eyespots.

Further reading: Raff (1996), Carroll *et al*. (2001), Gould (2002b).

group of creatures turned out to be appropriate for a leg that could walk on land. (Section 18.6.1, p. 540, describes the fish–amphibian transition.)

Many further examples of preadaptation are being discovered in molecular evolution and Box 10.1 gives an example.

10.4.3   *A new adaptation may evolve by combining unrelated parts*

One enzyme used in milk synthesis evolved from two unrelated enzymes

So far we have seen how new adaptations may evolve by changes in structures that have a constant function, or by changes in the function of a structure. A third possibility is that a novelty may result when two pre-existing parts are combined. For example, the use of milk to feed the young is a unique feature in mammals. Mammals evolved from reptiles, who did not produce milk. The full story of the evolution of lactation has many components. One of them is the evolution of the enzymatic machinery to synthesize milk. Milk contains large amounts of a sugar, lactose, and mammals have evolved a new enzyme — lactose synthetase — to manufacture it. Lactose synthetase catalyzes the conversion of glucose into lactose and is made up of modified versions of two pre-existing enzymes, galactosyl transferase and α-lactalbumin. Galactosyl transferase functions in the golgi apparatus of all eukaryotic cells and α-lactalbumin is related to the enzyme lysozyme that all vertebrates use in their antibacterial defenses. In this example, an evolutionary novelty resulted from the combination of two pre-existing parts with unrelated functions. A lactose-manufacturing enzyme evolved by combining a golgi enzyme and an antibacterial enzyme.

The evolution of milk digestion is a molecular example in which a new enzyme evolved by combining two pre-existing enzymes. A related process operates at a higher level when two whole species merge by symbiosis and evolve into a new species with a new combined physiology. For example, the mitochondria and chloroplasts in eukaryotic cells each originated when one bacterial cell engulfed another bacterial cell. In the case of mitochondria, the combined cell was capable (or soon evolved to be capable) of burning carbohydrates in oxygen — a process that releases more energy than anerobic respiration. The new cell had a more complex metabolism than either ancestral cell by itself.

Evolution can proceed by symbiosis

Evolution by symbiosis, or combining several genes into new composite genes, can violate the letter, but not the spirit, of Darwinian gradualism. According to the gradualist requirement, new adaptations evolved in many small, continuous stages. When two cells merge, there may be a relatively sudden transition to a new adaptation in one big step. However, no deep principle in Darwinism has been violated because the adaptive information within each ancestral cell was built up in gradual stages.

10.5 Genetics of adaptation

10.5.1 *Fisher proposed a model, and microscope analogy, to explain why the genetic changes in adaptive evolution will be small*

Adaptations have been suggested to evolve in few, large genetic steps, or many, small ones

Evolutionary biologists distinguish between a "Fisherian" and a "Goldschmidtian" view of the genetic steps by which adaptations evolve. Goldschmidt (1940) argued that new adaptations, and new species, evolve by macromutations (or "hopeful masters"). A macromutation is a mutation with a large phenotypic effect, such that the individual carrying the mutation is outside the normal range of variation for its population (Figure 1.7, p. 14). Fisher doubted whether macromutations contribute much to evolution, and argued that adaptive evolution mainly proceeds in many small steps. The mutations that contribute to adaptive evolution have small phenotypic effects.

Fisher's argument begins by noting that living things are fairly well adapted to their environments. They must be at least reasonably well adjusted, or they would be dead. Next, Fisher assumes that most characters are in an optimally adapted state. If the character is larger or small than the optimum, the organism's fitness declines (Figure 10.4a). Because living organisms are at least fairly well adapted, they are somewhere near the peak in Figure 10.4a. We now assume that the direction of mutations is random with a mutation having a 50% chance of increasing the character state, and a 50% chance of decreasing it. A small mutation therefore has a 50% chance of improving the adaptation. But a large mutation would make things worse either way. It either is directed away from the optimum, or shoots past the optimum down the slope on the other side (Figure 10.4a). Fisher calculated, on the assumption that the organism is near the adaptive peak, that an indefinitely small mutation has a half chance of improving the adaptation, and the

(a)

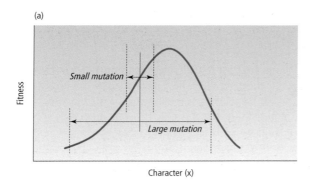

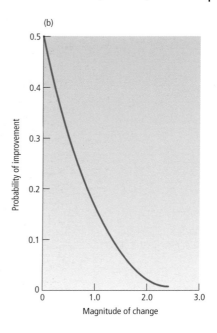

(b)

Figure 10.4

(a) A general model of adaptation. For a trait (*x*), the fitness of an individual has an optimum at a certain value of *x*, and declines away from that point. There is then a hill of fitness values. A mutation which changes the value of *x* also changes its bearer's fitness. A mutation of small effect is more likely to improve its bearer's fitness if the bearer is somewhere near the adaptive peak. (b) Fisher's calculations concerning the chance that a mutation improves fitness, depending on the magnitude of the mutation's phenotypic effect. The *y*-axis units refer to a particular model, but the general shape of the graph will be the same in any model like (a).

Fisher's microscope analogy

Wright's shifting balance theory is another alternative

chance of improvement decreases as the phenotypic effect of the mutation increases (Figure 10.4b). A macromutation has zero chance of being advantageous.

Fisher's argument can be explained less formally. Any well adjusted machine is more likely to be improved by fine tuning than by gross insult. If your radio wanders out of tune, you can recover the station by a small adjustment to the tuning knob. Taking a hammer to the machine is much less likely to help. Fisher used the analogy of focusing a microscope. If the microscope is already fairly well focused on its object, most focusing movements will be small, fine adjustments, taking the lens up or down a tiny amount. A big jerk on a randomly picked part of the microscope is unlikely to improve the focus.

The main assumption of Fisher's argument is that the organism is near the adaptive optimum. We should also notice a second, related assumption, which is that the adaptation has a single peak. If Figure 10.4a had multiple peaks, separated by valleys, a macromutation might have some chance of improving things by taking the organism to another peak. In Section 8.13 (p. 216), we saw that fitness surfaces with multiple peaks were the basis of Wright's shifting balance theory of evolution. Wright's theory is, with Goldschmidt's, a second alternative to Fisher's. Wright did not invoke macromutations. He argued that adaptive evolution was facilitated by random drift in small, subdivided populations. As we saw, Fisher doubted whether real adaptive surfaces have multiple peaks and judged Wright's shifting balance process unnecessary. But if real adaptive surfaces are sometimes multipeaked, it could complicate the details of Fisher's calculations. Fisher's basic point, however, that large changes in well adjusted systems are usually for the worse, still stands. Adaptive evolution is usually by piecemeal reform, not revolution.

10.5.2 *An expanded theory is needed when an organism is not near an adaptive peak*

For populations not close to their adaptive optima . . .

Fisher's assumption that living things are close to their adaptive optima may not always be met. Then, mutations of larger phenotypic effects have a better chance of improving adaptation. Small mutations are still more likely to be improvements than are large mutations, but Kimura (1983) pointed out that this may be overridden by a second factor. A large mutation, when it is advantageous, may have a larger selective advantage than a small mutation, because it moves higher up towards the peak (Figure 10.4a). The full contribution of any class of mutations (e.g., the class of mutations with small phenotypic effect or the class of mutations with large phenotypic effect) to evolution depends on three factors:

. . . three factors influence the magnitude of their genetic steps

Chance that the mutations are substituted	=	chance that the mutations arise	×	chance that the mutations are advantageous	×	selective advantage of the mutations

(The equation is not mathematically correct. For instance, we saw in Section 7.5.2 (p. 172) that the chance that a selectively advantageous mutation is fixed is about $2s$, where s is the selective advantage. The third term in the equation here should be $2s$, not simply s. However, the equation identifies the three factors at work, and the three are approximately multiplicative.)

To find the relative contribution of small and large mutations to adaptive evolution, we need to know the relative size of all three factors for the two classes of mutations. Fisher's argument only looks at the second factor.

No rigorous results are available for the first factor, but theory and evidence both indicate that mutations of small effect are more frequent than mutations of large effect. Orr (1998) looked mathematically at the second and third factors together, combining Fisher's argument and Kimura's conjecture. In Orr's model, the mutations that were substituted showed a certain frequency distribution of phenotypic effects (negative exponential, to be exact). A small number of mutations of large effect were substituted initially, followed by an increasing number of mutations with smaller effects. The reason is that, in a population away from a peak, somewhat large mutations are initially fixed. Then there is a phase of fine tuning in which many small mutations are fixed. Therefore, it is theoretically possible that some adaptive evolution is by large mutations, particularly in poorly adapted populations.

10.5.3 *The genetics of adaptation is being studied experimentally*

Genetic evidence bearing on Fisher's theory . . .

So far we have been looking at theory. What do the facts tell us about the genetics of adaptation? Two kinds of evidence are available. One comes from crosses between different forms within a species, or between closely related species. Orr & Coyne (1992) reviewed eight such crosses for cases in which the two crossed forms differed in an unambiguously adaptive character. In five or six of these, the difference was controlled by a single gene with major effect. Orr and Coyne concluded that Fisher's theory is poorly supported.

Figure 10.5

A cross between the two modern species would show that the large beak size difference is controlled by a single gene of large effect (A vs A_4). However, the modern difference has evolved by a number of small stages in the past. The Fisherian evolution is invisible in a modern cross. (0.5 in ≈ 1.25 cm.)

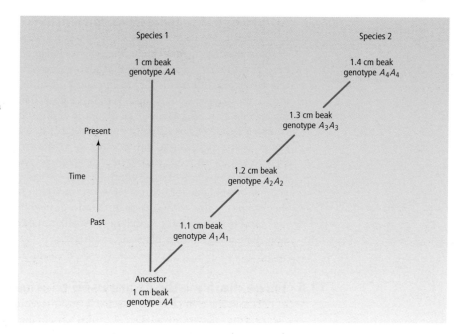

There is an inherent difficulty in testing theories about the evolutionary past using genetic crosses between modern forms. The genes in modern species will only reflect the way evolution proceeds if no genetic change has occurred since the adaptations originally evolved. Figure 10.5 illustrates the problem. Two species have diverged in a series of small steps, but the modern species differ by one gene with large effect. The problem does not invalidate Orr and Coyne's conclusion that the evidence for Fisher's theory is poor. But it would be a mistake to turn their evidence round and count it against Fisher's theory. In practice, more extensive genetic crosses are needed. In the African swallowtail *Papilio dardanus*, for instance, an initial cross suggests mimetic polymorphism is due to a single gene of large effect. But further crosses between apparently similar morphs from different regions of Africa show that several genes are at work (Turner 1977, p. 184).

. . . is hard to interpret

Burch & Chao (1999) pioneered a second kind of experiment. They knocked a bacteriophage away from its adaptive peak by allowing deleterious mutations to accumulate. They then measured the mutational steps by which the phage population evolved back to its former level of adaptation. The result depended on population size. In small populations, the phage evolved back to its peak in many small mutational steps. In large populations, they evolved back in some large, and some small, mutational steps. Their explanation is that large advantageous mutations are rarer than small advantageous mutations. In a small population, no large advantageous mutations may arise and adaptive evolution proceeds using mutations of small effect. In large populations, a few large advantageous mutations may be present, and they contribute to adaptive evolution. Burch and Chao's results fit with the basic theory of Sections 10.5.1 and 10.5.2 here, and show that population size also matters. Their work also shows how microbial systems can be used to test themes about the genetics of adaptation.

Some experimental work supports and expands it

10.5.4   *Conclusion: the genetics of adaptation*

We have met four theories about the genetic changes that occur during adaptive evolution. The "Goldschmidt" theory, that adaptations evolve by macromutations, has been rejected because of its theoretical implausibility. Macromutations will almost always reduce the quality of adaptation. Wright's theory, that adaptations evolve by the shifting balance process, has not been the topic of this section; but should be included for completeness. We saw in Section 8.13 (p. 216) that the shifting balance theory continues to inspire research, but no one has yet shown it to be important in evolution. Fisher's original theory suggested that adaptive evolution proceeds only by many mutational steps each of small effect. This theory has never been ruled out (or ruled in), and has been highly influential. However, modern research is looking at an expanded theory, that builds additional factors onto Fisher's basic model. Experimental work may be able to test what mix of large or small mutations contribute to adaptive evolution, depending on the ecological conditions.

10.6   Three main methods are used to study adaptation

"Whether" and "how" a character is adapted are different questions

We should distinguish two questions about any character of an organism. One is *whether* it is adaptive. The other is (if the character is an adaptation) *how* it is an adaptation. The first question is complicated, because the answer will depend on what definition of adaptation is used. Several definitions exist, and the methods of recognizing adaptations vary from definition to definition. We shall return to the question in Section 10.8 below. Here we can look at the methods used to study adaptations, to work out how the attribute in question is adaptive.

The study of adaptation proceeds in three conceptual stages. The first is to identify, or postulate, what kinds of of genetic variant the character could have. Sometimes, as in peppered moths (Section 5.7, p. 108) for instance, this is done empirically. Other characters do not vary genetically and for them it is necessary to postulate appropriate theoretical mutant forms. For example, when we come in Chapter 12 to look at why sex exists, we shall postulate mutant forms that reproduce clonally, or asexually.

Adaptation can be studied . . .

The second stage is to develop a hypothesis, or a model, of the organ or character's function. The original hypothesis for peppered moths was that coloration functioned as camouflage. Hypotheses are of varying quality, but they can be improved on as work proceeds. As we saw in Section 5.7 (p. 108), melanic coloration in peppered moths seems to have some other advantage in addition to camouflage in polluted areas. Another example comes from beak shape in birds. In this book, we shall often consider beak shape as an adaptation to the food supply. Larger beaks are better at eating larger and tougher food items, as we saw from the Grant's research on Darwin's finches (Section 9.1, p. 223). However, beaks have other functions too, including lice preening, and beak shape matters for those other functions (Clayton & Walther 2001).

. . . by engineering models, . . .

A good hypothesis is one that predicts the features of an organ exactly, and makes testable predictions. In morphology, these predictions are often derived from an engineering model. For example, hydrodynamics is used to understand fish shape, while

Table 10.1
The wing stripe of some butterflies was painted over, and controls were painted with transparent paint that did not affect their appearance. The number with intact wings at different times after the treatment was measured. The frequency distributions are not significantly different. From Silberglied *et al.* (1980).

Age at capture (week)	Painted butterflies		Controls	
	n	%	*n*	%
0	81	83.5	88	90.7
1	14	14.4	6	6.2
2	2	2.1	2	2.1
3	0	0	1	1.0
4	0	0	0	0
5	0	0	0	0

construction engineering is used for shell thickness in a mollusk: the costs of building a thicker shell have to be weighed against the benefits of reduced breakage, by wave action or predators. This sort of research can be carried out at all levels, from the simple and qualitative through to sophisticated algebraic modeling.

Stage three is to test the hypothesis's predictions. Three main methods are available. One is simply to see whether the actual form of an organ (or whatever character is under investigation) matches the hypothetical prediction; if it does not, the hypothesis is wrong somehow.

. . . by experiment, . . .

A second method is to do experiments. It is only useful if the organ, or behavior pattern, can be altered experimentally. Almost any hypothesis about adaptation will predict that some specified form of an organ will enable its bearer to survive better than some other forms, but the alternatives are not always feasible. We cannot, for example, make an experimental pig with wings to see whether flight would be advantageous. When they are possible, experiments are a powerful means of testing ideas about adaptation. Animal coloration, for instance, has been studied in this way. Color patterns in some butterfly species are believed to act as camouflage by "breaking up" the butterfly's outline. Silberglied *et al.* (1980), working at the Smithsonian Tropical Research Institution at Panama, experimentally painted out the wing stripes of the butterfly *Anartia fatima*. The butterflies with their wing stripes painted out showed similar levels of wing damage (which is produced by unsuccessful bird attacks) and survived equally as well as control butterflies (Table 10.1); the wing stripes, therefore, are not in fact adaptations to increase survival. They may have some other signaling or reproductive function, though that would need to be tested by further experiments.

. . . and by the comparative method

The *comparative method* is the third method of studying adaptation. It can be used if the hypothesis predicts that some kinds of species should have different forms of an adaptation from other kinds of species. Darwin's classic study of the relation between sexual dimorphism and mating system is an example we shall discuss below. Some hypotheses predict that different kinds of species will have different adaptations, others do not. Darwin's theory of sexual dimorphism does; but, for example, an optical engineer's model of how the eye should be designed might specify just a single best design, with the implication that all animals with eyes should have that design. The comparative method would in that case be inapplicable.

In summary, the three main methods of studying adaptation are to compare the predicted form of an organ with what is observed in nature (and perhaps also to measure the fitness of different forms of organism), to alter the organ experimentally, and to compare the form of an organ in different kinds of species.

10.7 Adaptations in nature are not perfect

Natural selection has brought into existence creatures that are in many respects marvellously well designed. The designs, however, are generally imperfect, and for a number of reasons. We shall look at several reasons in this chapter. In Chapter 11 we shall see another reason: that it may not be possible for an adaptation to be simultaneously perfect at all levels of organization. For example, birth control may be good for the population but not the individual. Most of the familiar examples of adaptation benefit the organism. They will therefore be (at best) imperfect at other levels, such as the genic, cellular, and group levels. However, we can still ask whether organismal adaptations are perfect even for the organism.

The quality of adaptation will progressively improve for as long as there is genetic variation to work on. If some genetic variants in the population produce a better adaptation than others, natural selection will increase their frequency. Although this process must always operate in the direction of improvement, it has never reached the final state of perfection. As Maynard Smith (1978) remarked, "if there were no constraints on what is possible, the best phenotype would live for ever, would be impregnable to predators, would lay eggs at an infinite rate, and so on." What are the constraints that prevent this kind of perfection from evolving?

10.7.1 *Adaptations may be imperfect because of time lags*

Fruits coevolve with animals

Many flowering plants produce fruits, in order to induce animals to act as dispersal agents. The fruits of different species are adapted in various ways to the particular animals they make use of. They must be attractive to the relevant animal, but also protect the seed from destruction by the animal's digestive system; they also must remain in the animal's gut for about the right amount of time to be dispersed an appropriate distance from the parent and then be properly deposited, which can be achieved by laxatives in the fruit. Many details are known about the ways in which individual fruits are adapted to the habits and physiology of the animal species that disperse them. Over evolutionary time, plants presumably have adapted the form of their fruits to whatever animals are around, and when the fauna changes, the plants will evolve (or rather coevolve — see Chapter 22), in time, to produce a new set of adapted fruits.

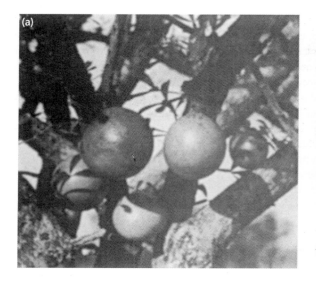

Figure 10.6
The fruits of (a) *Crescentia alata* (Bignoniaceae) and (b) *Annona purpurea* (Annonaceae) are two examples of fruits that were probably eaten by large herbivores that recently went extinct. The larger fruits in (a) are about 8 in (20 cm) long; the fruit in (b) is nearer 12 in (30 cm) long. Both trees were photographed in Santa Rosa National Park, Costa Rica. (Photos courtesy of Dan Janzen.)

But some fruits appear adaptively out of date . . .

Natural selection, however, takes time, and there will be a period after a major change in the fauna during which the adaptations of fruits will be out of date, and adapted to an earlier form of dispersal agent. Janzen & Martin (1982) have argued that the fruits of many trees in the tropical forests of Central America are "neotropical anachronisms" (see Figure 17.2, p. 495, for the geographic term neotropical). The fruits are anachronistically adapted to an extinct fauna of large herbivores (Figure 10.6).

Until about 10,000 years ago North and Central America had a fauna of large herbivores comparable in scale to that of Africa in recent times. Just as Africa has elephants, giraffes, and hippopotamuses, in Central America there were giant ground sloths, a giant extinct bear, a large extinct species of horse, mammoths, and a group of large relatives of mastodons called gomphotheres. These mammals now are all gone, but the species of trees that they used to walk beneath still remain. In the tropical forests of Costa Rica, some trees still drop large and hard fruits in great quantities. It accumulates, and much of it rots, at the base of the trees, and those that are moved by small mammals such as agoutis are not moved far. Here is how Janzen & Martin (1982) describe the fruiting of the large forest palm *Scheelea rostrata*: "in a month as many as 5000 fruits accumulate below each fruit-bearing *Scheelea*-palm. The first fruits to fall are picked up by agoutis, peccaries, and other animals that are soon satiated. . . . The bulk of the seeds perish directly below the parent." The fruits seem overprotected with their hard external coverings, they are produced in excessive quantities, and they are not adapted for dispersal by small animals like agoutis. It looks like a case of maladaptation: "a poor adjustment of seed crop size to dispersal guild." However, the fruits make sense if they are anachronistic adaptations to the large herbivores that have so recently gone extinct. The large size would have been appropriate for a gomphothere, and the hard external cover would have protected the seeds from the gomphothere's powerful crushing teeth. Ten thousand years has not been long enough for the trees to evolve fruits appropriate to the more modestly sized mammals that now dwell among them.

The principle illustrated by the fruits of these Central American plants is a general one. Adaptations will often be imperfect because evolution takes time. The environments of all species change more or less continually because of the evolutionary fortunes of the species they compete, and cooperate, with. Each species has to evolve to keep up with these events, but at any one time they will lag some distance behind the optimal adaptation to their environment. Adaptation will be imperfect when natural selection cannot operate as fast as the environment of a species changes. (Box 22.1, p. 624, contains further discussion of fruit coevolution.)

10.7.2 *Genetic constraints may cause imperfect adaptation*

When the heterozygote at a locus has a higher fitness than either homozygote, the population evolves to an equilibrium at which all three genotypes are present (Section 5.12, p. 123). A proportion of the individuals in the population must therefore have the deleterious homozygous genotypes. This is an example of a *genetic constraint*. It arises because the heterozygotes cannot, under Mendelian inheritance, produce purely heterozygous offspring: they cannot "breed true." In so far as heterozygous advantage exists, some members of natural populations will be imperfectly adapted. The importance of heterozygous advantage is controversial, but there are undoubted examples such as sickle cell hemoglobin, which is indeed a practical manifestation of imperfect adaptation due to genetic constraint.

50% of European crested newt
offspring die, because of a genetic
peculiarity

The balanced lethal system of the European crested newt *Triturus cristatus* is a more dramatic example. Members of the species have 12 pairs of chromosomes, numbered from 1 to 12, 1 being the longest and 12 the shortest. Macgregor & Horner (1980) found that all individual crested newts of both sexes are "heteromorphic" for chromosome 1: an individual's two copies of chromosome 1 are visibly different under the microscope. They named the two types of chromosome 1, 1A, and 1B (the same two types are found in every individual). Meiosis, they found, is normal so that an individual produces equal numbers of gametes with 1A chromosomes as with 1B chromosomes. There is also little, if any, recombination between the two chromosomes.

The puzzle is why there are no chromosomally homomorphic newts, with either two 1A or two 1B chromosomes. Macgregor and Horner carried out breeding experiments, in which they crossed two normal individuals, and counted the proportion of eggs that survived. In every case, approximately half the offspring died during development. It is almost certainly the homomorphic individuals that die off, leaving only the heteromorphs.

The reason why half the offspring die is as follows. The adult population has two types of chromosome, each with a frequency of one-half. If we write the frequency of the 1A chromosome as p and of the 1B chromosome as q, $p = q = \frac{1}{2}$. By normal Mendelian segregation, and the Hardy–Weinberg principle, the proportion of homozygotes (or homomorphs) is $p^2 + q^2 = \frac{1}{2}$. In each generation, therefore, the heterozygous newts mate together and produce half homozygous offspring and half heterozygous offspring, and then all the homozygotes die. The system looks incredibly inefficient, because half the reproductive effort of the newts each generation is wasted; but the same sort of inefficiency exists, to some extent, at any genetic locus with heterozygous advantage. If in humans a new hemoglobin arose that was resistant to malaria and viable in double dose, or in the crested newt a new chromosome 1 arose that was viable as a homomorph, it should spread through the population. Presumably the inefficiency remains only because no such mutations have arisen.

Could a system with heterozygous advantage easily evolve into a pure breeding genotype with the same phenotypic effect? It probably could by gene duplication (Section 2.5, p. 30). Imagine that the relevant hemoglobin gene duplicated in an Hb^+/Hb^s individual, to become Hb^+Hb^+/Hb^sHb^s. Genetic recombination could then produce a Hb^+Hb^s chromosome, and that chromosome should be able to achieve anything that an Hb^+/Hb^s heterozygote can. The chromosome would also breed true, once it had been fixed. We might expect therefore that the existing Hb^+/Hb^s system would evolve to a pure Hb^+Hb^+/Hb^sHb^s system. Some "dosage compensation" might be needed after the gene had duplicated, but that should be no difficulty because regulatory devices are common in the genome. The apparent ease of this evolutionary escape from heterozygous advantage and segregational load is one possible explanation for the (apparent) rarity of heterozygous advantage. However that may be, the existence of some cases of heterozygous advantage suggests that natural populations can be imperfectly adapted because a superior mutation has not arisen.

Heterozygous advantage may lead to gene duplication

10.7.3 *Developmental constraints may cause adaptive imperfection*

Developmental systems influence the course of evolution

A nine-penned discussion (Maynard Smith *et al.* 1985) of *developmental constraints* gave the following definition: "a developmental constraint is a bias on the production of variant phenotypes or a limitation on phenotypic variability caused by the structure, character, composition, or dynamics of the developmental system." The idea is that different groups of living things that evolved distinct developmental mechanisms, and that the way an organism develops will influence the kinds of mutation it is likely to generate. A plant, for example, may be likely to mutate to a new form with more branches than would a vertebrate, because it is easier to produce that kind of change in the development of a plant (indeed it is not even clear what a new "branch" would mean in the vertebrate — perhaps it might be extra legs, or having two heads). The rates of different kinds of mutation — or of "production of variant phenotypes" in the quoted definition — therefore differs between plants and vertebrates.

Developmental constraints can arise for a number of reasons. *Pleiotropy* is an example. A gene may influence the phenotype of more than one part of the body. A trivial instance would be that genes influencing the length of the left leg probably also influence the length of the right leg. The growth of legs probably takes place through a growth mechanism controlling both legs. This mechanism does not have to be inevitable for a constraint to exist. Perhaps some rare mutants do affect the length only of the right leg. A developmental constraint exists whenever there is a tendency for mutants (in this example) to affect both legs, and the tendency is due to the action of some developmental mechanism.

Pleiotropy exists because there is not a one-to-one relation between the parts of an organism that a gene influences and the parts of an organism that we recognize as characters. The genes divide up the body in a different way from the human observer. Genes influence developmental processes, and a change in development will often change more than one part of the phenotype. Much the same reasoning lies behind a second sort of developmental constraint. New mutations often disrupt the development of the organism. A new mutant, with an advantageous effect, may also disrupt other parts of the phenotype and these disruptions will probably be disadvantageous; but if the mutant has a net positive effect on fitness, natural selection will favor it. In some cases, the disruption can be measured by the degree of asymmetry in the form of the organism. In a species with bilateral symmetry, any deviation from that symmetry in an individual is a measure of how well regulated its development was. Mutations can therefore cause *developmental asymmetry*.

New mutations may disrupt development . . .

The Australian sheep blowfly *Lucilia cuprina* provides an example. It is a pest, and farmers spray it with insecticides. The flies, as we would expect (Section 5.8, p. 115), soon respond by evolving resistance. This evolutionary pattern has been repeated with a series of insecticides and resistance genotypes in the flies, and McKenzie has studied a number of cases. When the resistance mutation first appears, it produces developmental asymmetry as a by-product. Presumably, the disruption of development is deleterious, though not so deleterious that the mutation is selected against. The advantage in insecticide resistance more than makes up for a little developmental disruption. The mutation therefore increases in frequency. Selection will then start to act at other loci, to favor genes there that reduce the new mutation's deleterious side effects while maintaining its advantageous main effect. That is, selection will make the new mutation fit in with the blowfly's developmental mechanism. The genes at the other loci that restore symmetric development, while preserving the insecticide resistance, are called *modifier genes*, and the type of selection is called *canalizing selection*. Over time, in the sheep blowfly, the resistance mutation was modified such that it no longer disrupted development (Figure 10.7).

. . . but selection over time reduces the disruptive effect

McKenzie was able to show that the modification was caused by genes at loci other than the mutation-carrying locus. (This is important because, just as there is selection at other loci to reduce the deleterious side effects of the mutation, so selection at that locus will favor other mutations that can produce insecticide resistance without harmful side effects.) It is probably common, given the extent of genetic interaction in development, for new mutations to disrupt the existing developmental pattern. Canalizing selection, to restore developmental regulation with the new mutation, is therefore likely to be an important evolutionary process.

Another sort of developmental constraint can be seen in the "quantum" growth mechanism of arthropods. Arthropods grow by molting their exoskeleton and then growing a new, larger one. They do not grow while the exoskeleton is hard. The arthropod growth curve shows a series of jumps, often with a fairly constant size ratio of 1.2–1.3 before and after the molt. Now, there are various models of how body size can be adaptive: body size, for example, influences thermoregulation, competitive power, and what food can be taken. But none of these factors can plausibly explain the jumps in the arthropod growth curve. If, for example, the body size of an arthropod was adapted to the size of food items it fed on, it would hardly be likely that the distribution of sizes of food items in its environment set up a selection pressure for quantum

Figure 10.7

Developmental asymmetry in genotypes of the Australian sheep blowfly (*Lucilia cuprina*) that are, or are not, resistant to the insecticide malathion. (a) Developmental asymmetry in genotypes when the resistant gene *RMal* first appeared, soon after malathion was first used. + is the original, non-resistant genotype. *RMal* disrupts development, producing greater average asymmetry; and is selectively disadvantageous in the absence of malathion. (b) Developmental asymmetry of *RMal* flies after modifiers (*M*) have evolved to reduce the developmental disruption; it is now reduced near to the level of the original +/+ flies, and in the absence of malathion *RMal* has little selective disadvantage or is neutral relative to +. The sample size is 50 flies for each genotype. Redrawn, by permission of the publisher, from McKenzie & O'Farrell (1993).

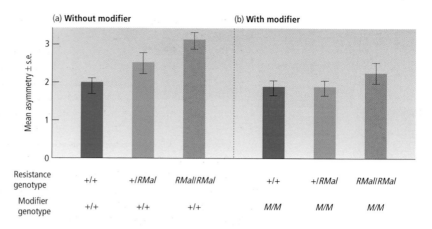

A morphospace for shells shows all the shell forms that could possibly exist

growth. The explanation for the quantum jumps is a developmental constraint: growth, by molting, is dangerous and to grow with a smooth curve would require frequent risky molts. It is better to molt more rarely and grow in jumps.

Developmental constraints have been suggested as an alternative explanation to natural selection for two main natural phenomena. One is the persistence of fossil species for long periods of time without showing any change in form (Section 21.5, p. 606). The other is the variety of forms to be found in the world. We can imagine plotting a *morphospace* for a particular set of phenotypes and then filling in the areas that are and are not represented in nature.

Raup's analysis of shell shapes is an elegant example. Raup found that shell shapes could be described in terms of three main variables: translation rate, expansion rate, and distance of generating curve from the coiling axis (Figure 10.8). Any shell can be represented as a point in a three-dimensional space, and Raup plotted the regions in this space that are occupied by living shells (Figure 10.9).

Large parts of the shell morphospace in Figure 10.9 are not occupied. There are two general hypotheses to explain why these forms do not exist: natural selection and constraint. If natural selection is responsible, the empty parts of the morphospace are regions of maladaptation. When these shell types arise as mutations, they are selected

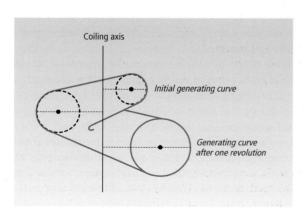

Figure 10.8

The shape of a shell can be described by three numbers. The translation rate (T) describes the rate at which the coil moves down the coiling axis: $T = 0$ for a flat planispiral shell, and is an increasingly positive number for increasingly elongated shells. The expansion rate (W) describes the rate at which the shell size increases; it can be measured by the ratio of the diameter of the shell at equivalent points in successive revolutions; $W = 2$ in the figure. The distance from the coiling axis (D) describes the tightness of the coil; it is the distance between the shell and the coiling axis, and in the figure it is half the diameter of the shell. See Figure 10.9 for many theoretically possible shell shapes with different values of T, W, and D. Redrawn, by permission of the publisher, from Raup (1966).

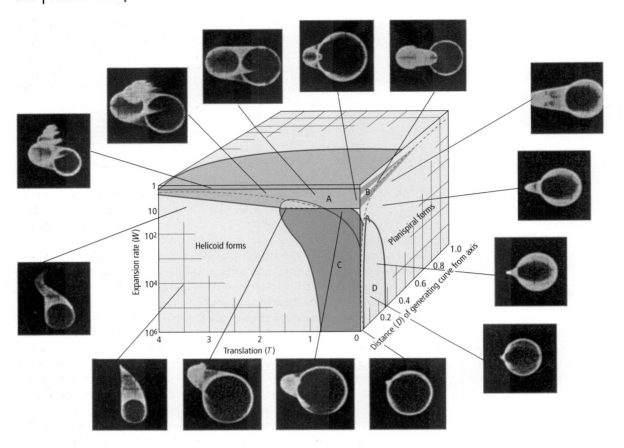

Figure 10.9

The three-dimensional cube describes a set of possible shell shapes. Around the outside of the figure, 14 possible shell shapes are illustrated as drawn by a computer. Only four regions in the cube are actually occupied by natural species: A, B, C, and D. All other regions in the cube represent theoretically possible but naturally unrealized shell shapes. The space is called a morphospace. reprinted, by permission of the publisher, from Raup (1966).

against and eliminated. Alternatively, the empty parts could be regions of constraint: the mutations to produce these shells have never occurred. If the constraint was developmental, it would mean that for some reason it is developmentally impossible (or at least unlikely) for these kinds of shells to grow. The non-existent shells would be embryological analogies for animals that disobey the law of gravity — they are shells that break the (unknown) laws of embryology. The absence of these shells would then be no more due to natural selection than is the absence of animals that break the law of gravity.

Just as natural selection and constraint are hypotheses to explain the absence of any form from nature, so they can both hypothetically explain the forms that are present. Faced with any form of organism, we can ask whether it exists because it is the only form that organism possibly could have (constraint), or whether selection has operated

Constraint and selection can be alternatives . . .

in the past among many genetic variants and the form we now observe was the one that was favored. If the form of an organism is the only one possible, an analysis that treated it as an adaptation would be misdirected. In some cases we can be more certain that variation is strongly constrained than in others. If the constraint is the law of gravity, adaptation is a fanciful hypothesis; but if the constraint is a conjectural piece of embryology, adaptation is much more worth investigating.

How can we test between selection and constraint? Maynard Smith and his eight coauthors listed four general possibilities: adaptive prediction, direct measures of selection, heritability of characters, and cross-species evidence.

The first test is the use of adaptive prediction. If a theory of shell adaptation predicted accurately and successfully the relation between shell form and environment — which forms should be present, and which absent, in various conditions — then, in the absence of an equally exact embryological theory, that would count in favor of adaptation and against developmental constraint. Conversely, a successful, exact embryological theory would be preferred to an empty adaptive theory.

The second test is a direct measure of selection. In the case of the shell morphospace, this would mean somehow making the naturally non-existent shells experimentally, and testing how selection then worked on them (Section 10.6). We then find out by observation whether there is negative selection against these forms.

Thirdly, we can measure the character's heritability. If a constraint is preventing mutation in a character, it should not be genetically variable. Genetic variability can be measured, and the constraint hypothesis will be refuted for any character that shows significant heritability. As it happens, this kind of evidence suggests that the gaps in the shell morphospace are not caused by developmental constraint. The heritability of a number of shell properties has been measured, and significant genetic variation found. Shell shape, therefore, is unconstrained to some extent.

Finally, cross-species evidence may be useful. It has particularly been used for pleiotropic developmental constraints. When more than one character is measured, and the values for the two characters in different organisms are plotted against each other, a relation is nearly always found. (This is true whether the different organisms are all in the same species, or from different species.) The graphs have been plotted most often for body size together with another character, and the relations are then called *allometric* (Darwin referred to it as the "correlation of growth"). Allometric relations are found almost whenever two aspects of size are plotted against each other graphically. A graph of brain size against body size for various species of vertebrates, for example, shows a positive relation. Graphs like these are two-dimensioned morphospaces, and are analogous to Raup's more sophisticated analysis for shells.

The observed distribution of points might, once again, be due either to adaptation or to constraint. It might be adaptive for an animal with a large body to have a large brain. Or it might make no difference what size an animal's brain is, and changes in brain size would simply be the correlated consequences of changes in body size (or vice versa). Mutations altering one of the characters would in that case be constrained also to alter the other. Huxley was an influential early student of allometry, and he liked to explain allometric relations by the hypothesis of constraint: "whenever we find [allometric relationships], we are justified in concluding that the *relative size* of the horn, mandible, or other organ is automatically determined as a secondary result of a single common

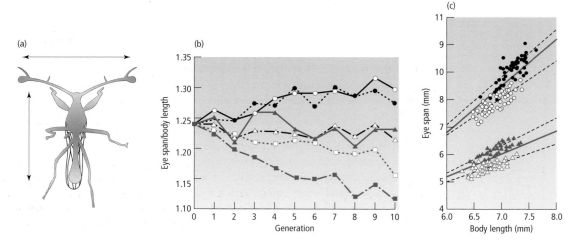

Figure 10.10

Artificial selection to alter the allometric shape of the stalk-eyed Malaysian fly *Cyrtodiopsis dalmanni*. (a) A silhouette of a fly, with arrows to indicate how eye span and body length were measured. (b) Results of one set of experiments on males. Circles are experimental lines in which males with high ratios of eye span to body length were selected to breed; squares are experimental lines in which males with low ratios of eye span to body length were selected to breed; and triangles are unselected control lines. Two replicates were done in each condition and they are distinguished by whether or not the symbol is filled in. (c) Another illustration of the allometric change; there are four sets of points. The top two (circles) are for males; the bottom two (triangles) for females. The filled in symbols are individuals of the high line after 10 generations of selection for increased relative eye span; the open symbols are individuals of the low line after 10 generations of selection for decreased relative eye span; and the dashed lines indicate the allometry in the unselected control lines. The male points correspond to replicate 1 (open circle) in (b). Note the response to selection, showing allometric relations are changeable, with the important change being in the slope of the lines in (c), which is more easily visible as a change in the ratio in (b). (0.25 in ≈ 6 mm.) Redrawn, by permission of the publisher, from Wilkinson (1993).

growth-mechanism, and *therefore is not of adaptive significance*. This provides us with a large new list of non-adaptive specific and generic characters" (Huxley 1932).

Some kinds of evidence are more persuasive than others. Allometric relations, in particular, are not strong evidence of developmental constraint. We can use the third kind of evidence (genetic variability) to see whether allometric relations are embryologically inevitable, or whether they can be altered by selection. Whenever anyone has looked, allometric relations have been found to be as malleable as any other character.

Figure 10.10 illustrates an artificial selection experiment by Wilkinson (1993) on the weird Malaysian fly *Cyrtodiopsis dalmanni*. These flies have their eyes at the ends of long eye stalks (Figure 10.10a, and Plate 5, between pp. 68 and 69). The eye stalks are particularly elongated in males and the character probably evolved by sexual selection. The important point here is that body and eye-stalk lengths are found to be correlated when they are measured in a number of individuals (Figure 10.10c). The ratio of eye span to body length in the natural population was 1.24 (yes, that is not a misprint: the eye stalks really are longer than the entire length of the body!). Wilkinson selected for increases or decreases in eye span relative to body length in two experimental lines and was able to alter the allometric relation in both directions (Figure 10.10). The

Allometric relations have been treated as non-adaptive

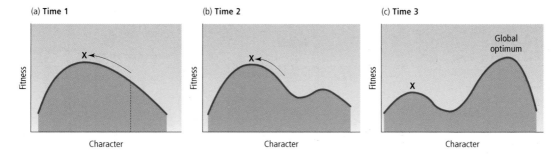

Figure 10.11

A historic change in adaptive topography has left a species stranded on a local peak. (a) Initially, there is a single optimum state for a character, and the population (X) evolves to that peak. (b) As the environment changes through time, the adaptive topography changes. The species has now reached the optimum. (c) The topography has changed, and a new global peak has arisen. The species is stuck at the local peak, because evolution to the global peak would traverse a valley: natural selection does not favor evolution toward the global peak.

allometric relation, therefore, is not a fixed law of embryonic development. Results like Wilkinson's suggest that allometric relations will have been tuned by natural selection in the past, to establish a favorable shape in each species.

In conclusion, not much is known about how embryology constrains mutation, but the general idea is plausible. The way an organism develops will influence the mutations that can arise in some of its characters. The interesting problems begin when we try to move from this general claim to an exact demonstration in a real case. The attempts so far, as in the example of allometry, have not been finally convincing. In particular cases, we can test between the alternatives of selection and constraint.

10.7.4 *Historic constraints may cause adaptive imperfection*

A population may be stuck at a local optimum

Evolution by natural selection proceeds in small, local steps and each change has to be advantageous in the short term. Unlike a human designer, natural selection cannot favor disadvantageous changes now in the knowledge that they will ultimately work out for the best. As Wright emphasized in his shifting balance model (Section 8.13, p. 216), natural selection may climb to a local optimum, where the population may be trapped because no small change is advantageous, though a large change could be. As we saw, selection itself (when considered in a fully multidimensioned context), or neutral drift, may lead the population away from local peaks; but it also may not. Some natural populations now may be imperfectly adapted because the accidents of history pointed their ancestors in what would later become the wrong direction (Figure 10.11).

The recurrent laryngeal nerve provides an amazing example. The laryngeal nerve is, anatomically, the fourth vagus nerve, one of the cranial nerves. These nerves first evolved in fish-like ancestors. As Figure 10.12a shows, successive branches of the vagus nerve pass, in fish, behind the successive arterial arches that run through the gills. Each nerve takes a direct route from the brain to the gills. During evolution, the gill arches have been transformed; the sixth gill arch has evolved in mammals into the ductus arteriosus, which is anatomically near to the heart. The recurrent laryngeal nerve still

Figure 10.12

Evolution of the recurrent laryngeal nerve. (a) In fish, the vagus nerve sends direct branches between successive gill arches. (b) In mammals, the gill arches have evolved into a very different circulatory system. The descendant nerve of the fish's fourth vagus now passes from the brain, down to the heart (in the thorax) and back up to the larynx. Redrawn, by permission of the publisher, from Strickberger (1990), modified from de Beer (1971).

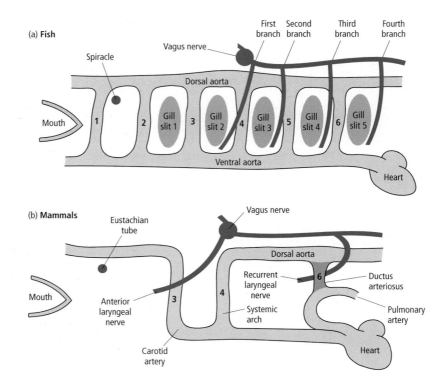

The recurrent laryngeal nerve is probably maladapted in giraffes

follows the route behind the (now highly modified) "gill arch": in a modern mammal, therefore, the nerve passes from the brain, down the neck, round the dorsal aorta, and back up to the larynx (Figure 10.12b).

In humans, the detour looks absurd, but is only a distance of a foot or two. In modern giraffes, the nerve makes the same detour, but it passes all the way down and up the full length of the giraffe's neck. The detour is almost certainly unnecessary and probably imposes a cost on the giraffe (because it has to grow more nerve than necessary and signals sent down the nerve will take more time and energy). Ancestrally, the direct route for the nerve was to pass posterior to the aorta; but as the neck lengthened in the giraffe's evolutionary lineage the nerve was led on a detour of increasing absurdity. If a mutant arose in which the nerve went directly from brain to larynx, it would probably be favored (though the mutation would be unlikely if it required a major embryologic reorganization); the imperfection persists because such a mutation has not arisen (or it arose and was lost by chance). The fault arose because natural selection operates in the short term, with each step taking place as a modification of what is already present. This process can easily lead to imperfections due to historic constraint — though most will not be as dramatic as the giraffe's recurrent laryngeal nerve.

A similar historic contingency may produce not actual imperfection, but differences between populations or species that are not adaptively significant. In an adaptive topography with several adaptive peaks, there may be more than one of similar height. The giraffe's laryngeal nerve looks like a case in which a local peak is clearly lower than

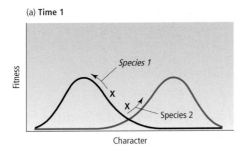

(a) Time 1

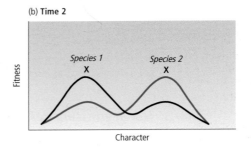

(b) Time 2

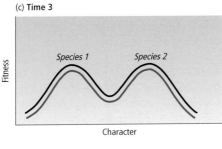

(c) Time 3

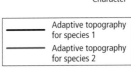

Figure 10.13
Different starting conditions lead to two species occupying different, but equivalent, adaptive peaks. (a) The adaptive topographies for two species differ, and each evolves to its own peak. (b) The adaptive topographies now change, until (c) they become identical for the two species; but each species remains on its own peak. At stage (b) the species difference was adaptive, thus it was better for species 1 to be on its peak, and species 2 on its. By (c) the species difference is non-adaptive as either species would be equally well adapted on either peak.

the global peak, and it is therefore recognizably an imperfect adaptation. If there were several peaks of similar height, one would not be recognizably inferior to the others. Imagine now that the ancestors of a number of different populations started out near different future peaks. If they then experienced the same external force of selection, each one would still evolve to its nearest peak. The different populations would then evolve different adaptations. But they have evolved different adaptations because of their different starting conditions, not because they have adapted to different environments (Figure 10.13).

Species could have non-adaptive differences

Kangaroos and placental herbivores such as gazelles are possible examples. The two forms are ecologically analogous, but have different methods of locomotion. Kangaroo hopping is no better or worse for moving than running on four legs. The lineage leading to kangaroos improved one method of moving, while that leading to gazelles concentrated on another. The difference is probably mainly a historic accident. If the argument is right, the distant ancestors of kangaroos faced different selective conditions from those of gazelles. The adaptations fixed in those ancestors then influenced subsequent evolution such that now, even though kangaroos and gazelles occupy similar ecological niches, the mutations influencing locomotion that are favored in the two groups are completely different.

Kangaroos and gazelles may be an example

The example illustrates a different idea from the giraffe's recurrent laryngeal nerve. Neither kangaroo nor gazelle is claimed to be imperfectly adapted; it is only the difference between the two that may be a historic accident. In the giraffe lineage, a similar kind of historic accident has generated an actual imperfection in its laryngeal nerve.

Whether historic accident leads to imperfection, or a neutral difference between lineages, depends on whether a global peak stays during evolution as a global peak or evolves into a local peak. In either case, past evolutionary events can lead to the establishment of forms that cannot be explained by a naive application of the theory of natural selection. Adaptation has to be understood historically.

10.7.5 *An organism's design may be a trade-off between different adaptive needs*

Many organs are adapted to perform more than one function and their adaptations for each are a compromise. If an organ is studied in isolation, as if it were an adaptation for only one of its functions, it may appear poorly designed.

Mouth design is a trade-off between feeding and eating

Consider how the mouth is used for feeding and breathing in different groups of tetrapods (amphibians, reptiles, birds, mammals). In mammals, the nose and mouth are separated by a secondary palate, and the animal can chew and breathe at the same time. The earliest tetrapods, some modern reptiles, and all modern amphibians, lack a secondary palate and have only a limited ability to eat and breathe simultaneously. A boa constrictor, for example, has to stop breathing while it goes through the complex motions of swallowing its prey — a process that can take hours. The mouth of any species that cannot breathe while it is feeding may, if it is judged only as an adaptation for feeding, appear inefficient compared with the mammalian system; the snake's mouth is a compromised adaptation for feeding. Of the reptilian groups, only crocodiles have a full secondary palate like mammals (it is presumably useful in crocodiles as it enables them to breathe air through the nose while the mouth is under water), and reptilian feeding systems can be understood as compromised in varying degrees by the need to breathe.

Trade-offs do not only exist in organ systems. In behavior, an animal has to allocate its time between different activities, and the time allocated to foraging (for example) might be compromised by the need to spend time on other demands. Trade-offs exist over the whole lifetime too: an individual's life history of survival and reproduction from birth to death is a trade-off between reproduction early in life and reproduction later on. At any one time, an animal may appear to be producing less offspring than it could, but that does not mean it is poorly adapted as it may be conserving its energies for extra reproduction later.

In summary, the adaptations of organisms are a set of trade-offs between multiple functions, multiple activities, and the possibilities of the present and future. If a character is viewed in isolation it will often seem poorly adapted; but the correct standard for assessing an adaptation is its contribution to the organism's fitness in all the functions it is employed in, throughout the whole of the organism's life.

10.7.6 *Conclusion: constraints on adaptation*

Evolutionary biologists are concerned to understand both why different species have different adaptations and how adaptations function within each species. They use different methods to analyze adaptive differences between species and adaptations within a species. As we have looked at the sources of adaptive imperfection, we have seen some that produce adaptively insignificant difference between species and others that produce imperfect adaptation within one species. Let us finish by summarizing how the various kinds of imperfection could upset the methods of analyzing adaptation (Section 10.6).

The comparative method could be misled by cases of adaptively insignificant differences between species. If the different forms of the adaptation are selectively neutral, or are equivalent locally adaptive peaks that different species evolved by historic accident, then attempts to correlate the differences with ecological circumstances should be unsuccessful. However, the fact that an adaptation can have several equivalently good forms does not disturb the study of the character by itself. The possibility of multiple adaptive forms should emerge from the analysis. If an enzyme has an optimal form, then it is no less an optimal form if 100 different amino acid sequences can realize it in practice. The problem for studies of particular adaptations within a species comes from the other source of imperfection. If the perfect form of the character has not arisen for reasons of history, embryology, or the genetic system, or because the environment has changed recently, then the character itself will be imperfectly adapted. If we try to predict the form of the character by an analysis of optimal adaptation, the prediction will be wrong.

What should the investigator do when a prediction turns out wrong? An analysis purely in terms of adaptation may produce spurious results. Any particular character could have evolved as an adaptation for any of a large number of reasons. Body size, for example, may be adaptive for thermoregulation, storing food, subduing prey, fighting other members of the same species, or other factors. If we assume that body size is an adaptation, we begin research by picking on one factor, such as thermoregulation, build a model relating thermoregulation to body size, and see whether the model predicted body size correctly. If the model fails, we could move on to another factor, such as diet. We build a model relating diet to body size, and see if that predicts body size any better. If this model fails, we could move on to a third factor . . . and so on. This method, however, if carried far enough, will almost inevitably find a factor that "predicts" body size correctly. Eventually, by chance, a relation will be found if enough other factors are studied, even if body size is a neutral character.

The solution to the problem can be stated in a conceptually valid, but not always practically useful, form. The methods of studying adaptation work well *if we are studying an adaptation*. If the character under study is an adaptation then it must exist because of natural selection. We are right to persist in looking for the particular reason why natural selection favors it. If body size is an adaptation, there will be an adaptive model for it that is correct. However, if the character (or different forms of it) is not favored by natural selection, the method breaks down. Methods of studying adaptation should therefore be confined to characters that are adaptive, which in practice they mainly are. Adaptation can be a self-evident property of nature, and it would be absurd to claim that no properties of living things are adaptive. While research concentrates on obvious adaptations, it should be philosophically non-controversial.

However, that leaves plenty of room for controversy. Biologists do not all agree on how widespread, and how perfect, adaptations are in nature. Some biologists believe

Imperfect adaptations may or may not cause problems for the methods of studying adaptation

The methods are foolproof if the character under study is adaptive that natural selection has fine tuned the details, and established the main forms, of organic diversity. Others think that the main forms may be historic accidents and the fine details due to random drift. Not surprisingly, the evolutionary biologists who study adaptation tend to be among the former and those who criticize it among the latter. But this difference of opinion is not about the fundamental coherence of the methods; it is about the range of their application. The controversy is unlikely to disappear in the absence of an objective, universally applicable criterion by which we can recognize which characters are adaptations. That brings us back to the problem of defining adaptation.

10.8 How can we recognize adaptations?

10.8.1 *The function of an organ should be distinguished from the effects it may have*

Obeying the law of gravity is not an adaptation

A character of an organism can have beneficial effects that are not strictly speaking adaptive. Some consequences follow from the laws of physics and chemistry without any need for shaping by natural selection. Here is an example discussed by Williams (1966).

> Consider a flying fish that has just left the water to undertake an aerial flight. It is clear that there is a physiological necessity for it to return to the water very soon; it cannot long survive in the air. It is, moreover, a matter of common observation that an aerial glide normally terminates with a return to the sea. Is this the result of a mechanism for getting the fish back into water? Certainly not; we need not invoke the principle of adaptation here. The purely physical principle of gravitation adequately explains why the fish, having gone up, eventually comes down.

The flying fish is not adapted to obey the law of gravity. When evolutionary biologists seek to understand how a character is adaptive, they consider the likely reproductive success of mutant, altered forms of the character. We can imagine many changes in the shape of the flying fish, but none of them will prevent it from returning to the sea. Even though returning to the sea is a "biological necessity," natural selection in the past has not acted between some types of fish that did return to the sea and some types that did not, with the former surviving and reproducing better.

A thought experiment about alternative forms of a character is only sensible if the alternatives are plausible. Fish that disobey gravity are not. Imagining alternative forms of a character is not absurd, but it can be taken to absurd extremes. In real cases, the alternatives are usually plausible and may even be known to exist. For example, postulating a melanic form of the peppered moth is not absurd, because it can be seen in nature.

In addition, not all the beneficial consequences of a character are properly called adaptations. A character is an adaptation in so far as natural selection is maintaining its form in modern populations. Beneficial consequences that are independent of natural selection are not adaptations. The point is obvious in practice, but must be borne in mind in conceptual discussion.

10.8.2 *Adaptations can be defined by engineering design or reproductive fitness*

The "design" of an eye for seeing is evidence the eye is an adaptation

We can distinguish between concepts of adaptation that define it in terms of the inherent design of a character and those that look at its reproductive consequences. The vertebrate eye is a good example to explain the "design" concept. Almost everyone would accept that the eye is an adaptation. We could recognize that it is adapted by describing its inherent design. From the principles of optical physics, we can tell that the eye is correctly shaped to form optical images. Likewise, the heart is designed to pump blood and the skeleton to support muscles. On the "design" concept, we recognize adaptations as characters that are, on some appropriate engineering principle, fitted for life in the environment of the species.

Adaptation has also been defined in terms of fitness measurements

Alternatively, we could define adaptations using measurements of reproductive success. If a character is an adaptation, then natural selection will work against genetic alternatives. Natural selection will act against mutant forms of the eye that produce inferior images. Reeve & Sherman (1993) define an adaptation as that form of a character, among a set of variants, that has the highest fitness.

The two concepts — the "design" and "fitness" concepts of adaptation — are closely related. A well designed form of an organ such as the eye will also have high fitness. Both concepts are concerned with much the same underlying facts. However, they have different strengths and weaknesses. One strength of defining adaptation by measurements of fitness is that it is objective and unambiguous. A mutant version of a character either will or will not spread.

One weakness of the "fitness" concept is that it cannot always be used. Even in a character that does exist in many variant forms, it takes a lot of work to measure reproductive success in all the variants. Moreover, some characters do not vary in an easily measurable way. The vertebrate eye is undoubtedly an adaptation, but nobody has ever correlated variation in its optical properties with survival and reproductive success. A third problem is that a character could still be adaptive even if its relation with reproductive success was statistically undetectable. Natural selection can theoretically work on a character over millions of years and produce major changes through selection coefficients of 0.001 or less. It would be practically impossible to detect this amount of selection in a modern population with the normal resources of an evolutionary biologist. Forces that are important in evolution can in some cases be impossible to study directly because they are so small. A direct measurement of reproductive success is most likely to demonstrate that a character is adaptive if the selection coefficient is large; but these will tend to be the "obvious" characters in any case. The method will be less useful for characters whose adaptive status is controversial.

The two concepts have strengths and weaknesses

The strength of the design concept is that it is widely applicable. We can study any character to see whether it is designed for some purpose. The weakness of the concept is that it can be ambiguous. For example, the brain is surely an adaptation. However, brain size might be 15 in^3 (250 cm^3) in one species and 18 in^3 (300 cm^3) in another species. Is the difference between the two species adaptive? The design criterion alone may not tell us the answer.

We might make an analogy with the uncertainty in the definition of "design" in human fabrications. If we were to travel round the world and guess which objects were

The recognition of adaptations can be uncertain in some cases

brought about by human design, we would see many obvious cases, such as architecture and engineered objects, and many non-obvious cases, such as heaps of earth. However, earth could have been heaped up for a special purpose, such as for a burial mound, or it could have just accumulated there by natural accident. We cannot always tell which cause operated just by looking at the result. The two causes are objectively distinct, but the distinction is historic: either the heaps of earth were constructed by human agency or they were not. However, the history is unobservable, and when we have to make the distinction purely using modern observable evidence, there will be difficult border-line cases. We should not, therefore, expect the distinction between designed and non-designed entities to always be clear in either the case of natural adaptation or of human fabrications.

Likewise, body coloration may be a simple adaptation, brought about by natural selection, or it may be non-adaptive and brought about by chance, as may be the case for the red color of the sediment-dwelling worm *Tubifex* (visual factors are not important in the sediment at the bottom of the water column). Again, either natural selection is favoring the body coloration or it is not; but if we try to decide whether it is just from looking at the character, the answer may not be clear. We have a clear theoretical concept of what an adaptation is, but that concept implies that adaptation cannot have a universal, foolproof, practical definition.

Summary

1 Three theories have been put forward to explain the existence of adaptation: supernatural creation, Lamarckism, and natural selection. Only natural selection works as a scientific theory.

2 Natural selection is not the only process that causes evolution, but is the only process causing adaptation.

3 Natural selection, at least in principle, can explain all known adaptations. Examples of coadaptation and useless incipient stages have been suggested but they can be reconciled with the theory of natural selection. The vertebrate eye could have evolved rapidly by small advantageous steps.

4 Some new organs (and new genes) evolve by continuous modification of a previously existing organ (or gene), while the function is constant. Others evolve by continuous modification, but with a change in function. Yet others evolve when previously existing but separate parts are combined.

5 Fisher proposed a model in which adaptation evolves in many small genetic steps. His model contrasts with Goldschmidt's, in which adaptations evolve by sudden macromutations. Fisher's model is being modified theoretically, and tested experimentally.

6 Adaptations may be imperfect because of time lags: a species may be adapted to a past environment because it takes time for natural selection to operate.

7 Adaptations are imperfect because the mutations that would enable perfect adaptation have not arisen. The imperfections of living things are due to genetic, developmental, and historic constraints, and to trade-offs between competing demands.

8 For particular characters, adaptation and constraint can be alternative explanations. Likewise, differences in the form of a character between species may be due to adaptation to different conditions or to constraint. Forms that are not found in nature may be absent because they are selected against or because a constraint renders them impossible.

9 Adaptation and constraint can be tested between by several methods: by the use of predictions from a hypothesis of adaptation or constraint, by direct measures of selection, by seeing whether the character is variable and whether the variation is heritable and can be altered by artificial selection, and by examining comparative trends.

10 The methods of analyzing adaptation are valid when applied to adaptive characters and interspecific trends; they might be misleading for non-adaptive characters and trends.

11 Not all the effects of an organ will have evolved as adaptations by natural selection. Some will be inevitable consequences of the laws of physics.

12 Biologists disagree about how exact, and how widespread, adaptation is in nature.

13 There are criteria to distinguish adaptive from non-adaptive characters. Measurement of selection provides an objective criterion, but is not always practical. The inherent engineering design of a character is not always an objective crtierion, but is widely applicable. The two criteria are closely related.

Further reading

Williams (1966) is a classic work on adaptation. Gould & Lewontin (1979) is an influential paper that criticizes the way adaptation has often been studied; Cain (1964) argues the opposite. Pigliucci & Kaplan (2000) look at 20 years of discussion about Gould & Lewontin (1979). Lewontin (2000) and Gould (2002b) variously update their viewpoints. Reeve & Sherman (1993) is a stimulating paper about adaptation. Dawkins (1982, 1986, 1996) argues that only natural selection can explain adaptation; the 1986 and 1996 books are written for a wide audience. Dennett (1995) is also written for a broad audience and discusses several of the topics covered in this chapter.

Allen *et al.* (1998) have compiled an anthology of classic papers about adaptation. My evolution anthology contains a section of extracts about adaptation (Ridley 1997) and Rose & Lauder (1996) have edited a multiauthor volume on the topic.

The natural theologian's argument from design was philosophically undermined by Hume in his *Dialogues Concerning Natural Religion*, which are in print in various paperback editions and (unlike some of Hume's other philosophical writings) readily intelligible. I include the passage in Ridley (1997). However, Hume's abstract argument did not convince people and it was Darwin's mechanistic theory of natural selection that historically toppled that long tradition of thought. See Simpson (1944, 1953) on orthogenesis.

Dawkins (1996) includes a popular account of Nilsson & Pelger's (1994) paper about eye evolution. Land & Nilsson (2002) is a book about animal eyes. Nitecki (1990) is a multiauthor book about evolutionary innovations. On feathers, see Prum & Brush (2002) and their references. On preadaptation in general, see also the popular essay by Gould (1977b, chapter 12). Gerhart & Kirschner (1997) discuss the lactose example.

On the genetics of adaptation, Leigh (1987) includes an account of Fisher's argument. Travisano (2001) discusses the emerging research program with microbial experimental systems.

The methods of studying adaptation are discussed (in addition to the multiauthor volumes referred to above) by Orzack & Sober (1994), Harvey & Pagel (1991), Parker & Maynard Smith (1990), Maynard Smith (1978), and Rudwick (1964). For the experimental method, see the special issue of *American Naturalist*, a supplement to vol. 154 (July 1999).

On constraints, Antonovics & van Tienderen (1991) look at terminology. Barton & Partridge (2000) look at the topic in general. On "ghost" adaptations like the neotropical fruit, see the popular book by Barlow (2000). Byers (1997) is an example discussing the social behavior of the American pronghorn and Macgregor (1991) reviews the remarkable genetic constraint in the crested newt and refers to earlier work.

On developmental constraint, Maynard Smith *et al.* (1985) and Gould (2002b) are major reviews. McKenzie & Batterham (1994) and McKenzie (1996) discuss the insecticide resistance example (see also the further reading in Chapter 5, p. 135). Antibiotic resistance in microbes is a related topic. Levin *et al.* (2000) discuss how compensatory mutations that reduce the harmful side effects of the initial resistance mutations may influence the persistence of antibiotic resistance. The arguments are related to those in Box 5.2 (p. 119). On developmental stability in general, see Lens *et al.* (2002). Harvey & Pagel (1991) contains an account of, and references to, recent work on allometry. Chapter 9 has further references for canalizing selection. Chapter 20 looks at evolutionary development, which probably provides the concepts for future studies of developmental constraint. Galis *et al.* (2001) discuss the special case of constraints on digit numbers.

Certain human genes confer resistance to disease, but are otherwise disadvantageous. These genes probably illustrate constraints due to history (they evolved recently) and to trade-offs (disease resistance is so important that other adaptations are compromised). Schliekman *et al.* (2001) give some calculations for three such genes: $CCR5^-$ (resistance to HIV), hemoglobin S, and $\Delta 32$ (resistance to bubonic plague).

On definition, see the references already given to Williams (1966) and Reeve & Sherman (1993). I have extracted them, along with another good discussion by Grafen, in Ridley (1997). A further distinction is between historic and non-historic definitions. Gould has argued that only characters that retain a constant function should be called adaptations. See Gould (2002b) for a thorough recent statement of his view, and Reeve & Sherman (1993) for problems with it.

Study and review questions

1 What difficulties do theories of Lamarckian inheritance and directed mutation encounter when they are used as general theories of evolution, independent of (or in the absence of) natural selection?

2 Outline the main stages by which the vertebrate or octopus eye might have evolved, with successive stages showing improvements in the optical properties of the eye.

3 Some new adaptations evolve by symbiosis. For instance, plant cells acquired photosynthesis when cyanobacteria with photosynthetic abilities were incorporated into a larger cell. Do events of this kind falsify Fisher's model of the genetic evolution of adaptation, and Darwin's principle of gradualism?

4 What magnitude of genetic steps is expected in the adaptive evolution of a species that is: (a) near an adaptive peak and subject to slow environmental change; and (b) distant from an adaptive peak, or subject to rapid environmental change?

5 Feathers seem to have originally evolved for some function other than flight — perhaps display or thermoregulation. Feathers are called a preadaptation for flight. Does this imply that evolution has some anticipatory, futuristic ability, in which characters evolve because they will be useful for some function in the future?

6 Consider a morphospace, such as the one for shell morphologies, or a brain–body size allometric graph. (a) There are regions in the space in which there are no natural representatives. What are the two main theories to explain the absence of these forms? (b) How can we test whether a particular interspecies pattern in morphospace is caused by one theory or the other?

11 The Units of Selection

*A*daptations clearly benefit something in the living world, but to understand exactly why adaptations evolve we need to know in theory what it really is that adaptations evolve for the benefit of. This chapter begins by explaining the problem. The first main section of the chapter considers a series of adaptations which benefit increasingly higher levels of organization of life: we start with adaptations that benefit only a small clusters of genes, and move through the cell, organism, and family levels, up to possible adaptations to benefit whole groups. The examples illustrate the conditions for adaptations to evolve to benefit the different hierarchical levels, and reveal why adaptations at most levels other than the organism (and family group) are rare, though not non-existent. We finish this section with a general criterion that an entity must satisfy in order to evolve adaptations: it must show heritability. The second main section of the chapter asks the more fundamental question of what entity natural selection operates on, and describes an argument to suggest that the entity is the gene, though defined in a special sense.

11.1 What entities benefit from the adaptations produced by selection?

An adaptation may benefit one level of organization but not another

It is a familiar idea that life can be divided into a series of levels of organization, from nucleotide to gene, through cell, organ, and organism, to social group, species, and higher levels. Which, if any, of these levels does natural selection act on and produce adaptations for the benefit of? In a fairly superficial analysis, the answer does not matter. If an adaptation benefits an individual organism, it will often also benefit its species at a higher level and, at a lower level, all the parts that make up the individual. But there can be conflicts between these levels. In some cases, what benefits an organism may not also benefit its species, and in these cases the evolutionary biologist needs to know which level natural selection most directly benefits. The question therefore matters when we are studying particular adaptations. If we are seeking to understand why an adaptation evolved, we need to know what entities adaptations in general evolve for the benefit of. The question also has a more general, almost philosophical, interest: the theory of evolution should include a precise, and accurate, account of why adaptations evolve.

The issue can be made clearer in an example. Let us consider the adaptations that can be seen when lions go on a hunt. Lions often hunt alone, but they can improve their chance of success by hunting in a group. Here is part of a description of a group of hunting lions by Bertram (1978).

> When prey have been detected, a wildebeest herd perhaps, the lions start to stalk towards them. As they get close, they take different routes, some going on straight ahead, and some to the sides, so the prey herd is approached by lions stalking them from different directions. . . . Eventually one lion gets close enough to make a rush at a wildebeest, or else a lion is detected by the prey.

Lion hunts benefit lions but not mammals as a whole

Then the trap is sprung. Panicked wildebeest run in all directions, some of them into the reach of other lions. The cooperative behavior of the hunting party is here an adaptation for catching food, but it is not the lion's only feeding adaptation. The lions' muscular jaws and limbs, their teeth and five senses, all contribute to the success of the hunt. Lions are well adapted for feeding: although some hunts are unsuccessful, and individual lions may starve to death, the lions in the Serengeti Plains of Kenya spend about 20 hours a day in rest or sleep, and only 1 hour a day on average in hunting. Visitors tend to think lions are lazy.

When a lion hunt is successful, there are benefits for all but the highest biological levels of organization. The individual lions obviously benefit, as does the pride. Each time a hunt is successful, there will be a small incremental increase in the species' chance of survival, or avoiding extinction. The survival probability will also be increased, if by a smaller amount, for the genus *Felis* and the cat family Felidae. The hunt's effect at higher levels will depend on exactly what prey the lions caught. Almost all the Serengeti lions' food is made up of other mammals, so when we reach the class Mammalia, the effect of the hunt has probably become neutral. The lion's gain is the wildebeest or the zebra's loss and the chance of survival of the class Mammalia is more or less unaffected. The beneficial effect of the hunt spreads downwards as well as up from the individual lion. As the survival of the lion is increased, so is the survival of its constituents: the organs, cells, proteins, and genes. (Though if we trace the effect down through the nucleotides and their constituent atoms, it again disappears. A lion's atoms survive just as well whether it is alive and well fed or dead due to starvation.)

The levels of organization, from gene through individual lion to Felidae, are to a large extent bound together in their evolutionary fate, and what benefits one level will usually also benefit the others. However, this is not always so. Male lions can only join a pride by forcibly evicting the incumbent males. In the fight, lions may get killed or wounded, and in any case lions have a low rate of survival after they have been evicted from a pride. These fights have losers as well as winners: here the benefit of winning is confined to the individual (or male coalition) level and below. The lion species does not benefit. The survival of the species may be little affected by the death of male lions, because the mating system is polygynous and has plenty of males to spare; but the effect is clearly not positive.

Different adaptations, therefore, have different consequences for different units in nature. At one extreme, there may be adaptations — an improved DNA replication mechanism perhaps — that could benefit all life, but most adaptations will benefit only a smaller subsample of living things. Because the levels of living organization are bound together, if natural selection produces an adaptation to benefit one level, many other levels will benefit as a consequence. The question in this chapter is whether natural selection really acts to produce adaptations to benefit one level, with benefits at other levels being incidental consequences, or whether it acts to benefit all levels. And if it benefits one level, which is it? In evolutionary biology, this question is expressed as "What is the *unit of selection*?"

We define the question of "what is the unit of selection?"

We shall seek to answer it by looking at a series of adaptations that appear to benefit different "levels" in the hierarchy of biological organization. Some adaptations seem to be in the interests of individual genes, at the organism's expense, others benefit organisms at the group's expense, others may benefit higher levels. When we have seen the example, we can discuss generally which of the types of adaptation we should expect to see most often in nature.

11.2 Natural selection has produced adaptations that benefit various levels of organization

11.2.1 *Segregation distortion benefits one gene at the expense of its allele*

With normal Mendelian segregation at a genetic locus, on average half of an organism's offspring inherit one of the alleles and the other half the other allele. Mendelian segregation is so to speak "fair" in its treatment of genes: genes emerge from Mendelian segregation in the same proportions as they went in. There are, however, some curious cases in which Mendel's laws are broken in which one of the alleles, instead of being inherited by 50% of a heterozygote's offspring, is consistently overrepresented. The *segregation distorter* gene of *Drosophila melanogaster* is an example of this phenomenon, which is also called *meiotic drive*.

The segregation distorter gene breaks Mendel's laws

The segregation distorter gene was first found in *Drosophila* stocks from Wisconsin and Baja California. The gene is symbolized by *sd*, and we can call the other, more normal, alleles at the locus "+". A heterozygote for the segregation distorter is then *sd*/+. The majority (90% or more) of offspring from male heterozygotes have the *sd* gene because the sperm containing the + gene fail to develop. Female heterozygotes have normal Mendelian segregation. (The segregation distorter gene in fruitflies is really a pair of closely linked genes. However, we can discuss it as if it were a single locus: the points of principle will be clearer that way and the facts not badly misrepresented.)

The *sd* gene gains an advantage for itself, at a cost to the rest of the body it is in

A segregation distorter gene can have a great selective advantage. The allele that gets into more than half of a heterozygote's offspring will automatically increase in frequency and should spread through the population. Once the allele is fixed, the effect would disappear, provided that segregation is normal in homozygotes. In the case of the segregation distorter gene in *Drosophila*, however, other things are not equal. The abnormal sperm are infertile, and the total fertility of male heterozygotes is accordingly lowered. The fertility of an *sd*/+ male is about half that of a normal male. The effect of the lowered fertility on selection at the *sd* locus is complex, and depends on whether the reduction in fertility is more or less than 50% and what effect the reduced fertility has on the number of offspring produced. There are, however, at least some segregation distorter alleles that produce enough copies of themselves in heterozygotes to have an automatic selective advantage. They produce more copies of themselves than would a normal Mendelian heterozygote; their increase in frequency up to fixation is then inevitable.

Segregation distortion sets up an interesting selection pressure in the rest of the genome. On average, all other genes at other loci suffer a disadvantage because of segregation distortion. In any case, one gene has only a 50% chance of getting into a particular gamete because gametes are haploid whereas the individual is diploid. But segregation distortion produces a further reduction of about 50% in the chance that genes at other loci are passed on. In an *sd*/+ heterozygous fruitfly, a gene on another chromosome has a 50% chance of entering an *sd* sperm, which will be fertile, and a 50% chance of entering a + sperm, which will be infertile. Genes on other chromosomes from the *sd* locus are all net losers. If they are in the same sperm as the favored *sd* allele, things are normal; if they are in the shrivelled disfavored sperm, they die. Selection at other loci will favor genes that suppress the distorters and restore the *status quo*.

The *sd* gene illustrates intragenomic conflict

When selection acts in conflicting ways on different genes in the same individual body, it is called *intragenomic conflict*. The *sd*/+ fruitfly has intragenomic conflict, because selection on the *sd* gene favors segregation distortion and selection on other genes favors restoring normal segregation. Which genes win out can depend on many factors, but the point of the example here is to show what it means for natural selection to favor an adaptation that is the interest of a single gene (such as *sd*) within a body.

11.2.2 *Selection may sometimes favor some cell lines relative to other cell lines in the same body*

In organisms like ourselves, a new individual develops from an initial single-cell stage and that single cell derives from a special cell line, the germ line, in its parents. This kind of life cycle is called *Weismannist*, after the German biologist August Weismann, who first expounded the distinction between germ and somatic cell lines. In a "Weismannist" organism, most cell lines (the soma) inevitably die when the organism dies; reproduction is concentrated in a separate germ line of cells.

The separation of the germ line limits the possibilities for selection at the suborganismic level, between cell lines. One cell may mutate and become able to out-reproduce other cell lines and (like a cancer) proliferate through the body. But this "adaptation" will not be passed on to the next generation unless it has arisen in the germ line. Any somatic cell line comes to an end with the organism's death. For this reason, cell selection is not important in species like ourselves.

Many multicellular life forms do not have Weismannish development

However, Buss (1987) pointed out that Weismannist development is relatively exceptional among multicellular organisms (Table 11.1). We tend to think of it as usual because vertebrates, as well as the more familiar invertebrates like arthropods, develop in a Weismannist manner. However, more than half the taxa listed in Table 11.1 have the capacity for somatic embryogenesis — a new generation may be formed from cells other than those in specialized reproductive organs. The most striking examples are from plants. Steward, for instance, in a famous experiment in the 1950s, grew new carrots from single phloem cells taken from the root of an adult plant.

In a species in which new offspring can develop from more than one cell lineage, selection between cell lines becomes possible. When the organism is conceived it will be a single cell, and for the first few rounds of cell division the organism will probably remain genetically uniform. No selection can take place between cell lines if they are all genetically identical. Eventually a mutation may arise in one of the cells. If the mutation increases the cell's rate of reproduction, the cell line will cancerously proliferate at the expense of other cell lines in the organism. In a Weismannist species, that cell line will die when the organism dies and any selection between cell lines will be unimportant.

One cell line may proliferate within the body

However, if any cell line in the body has some chance of giving rise to the next organismal generation, the mutant cell line would increase its chance of being in an offspring and be favored by selection. Explained in this way, selection between cell lines within the body is detrimental to the organism. However, the process could also be advantageous for the organism. Whitham & Slobodchikoff (1981) argued that in plants selection between cell lines enables the individual to adapt to local conditions more rapidly than would be possible with strictly Weismannist inheritance.

The process is at present more of a theoretical possibility than a confirmed empirical fact, but it may well be important in non-Weismannist species. It may also have been important in the non-Weismannist ancestors of such modern Weismannist forms as arthropods and humans. Buss has developed the idea that cell selection can explain certain features of embryology in Weismannist species.

11.2.3 *Natural selection has produced many adaptations to benefit organisms*

We do not need to consider an example of organismal adaptation here: most of the adaptations described elsewhere in the book, from the beaks of the woodpecker

Table 11.1

The modes of development in different groups of living things. In the cellular differentiation column, + means that it is present in all the species that have been studied in the group and +/− means it is present in some species and absent in others. In the developmental mode column, s means new organisms can develop from the "somatic" cells of their parent, e means epigenetic development, p means preformationistic development, and u means unknown. From Buss (1987).

Taxon	Cellular differentiation	Developmental mode	Taxon	Cellular differentiation	Developmental mode
Protoctista			Animalia (continued)		
Phaeophyta	+/−	s	Mesozoa	+	p
Rhodophyta	+/−	s	Platyhelminthes	+	s, e, p
Chlorophyta	+/−	p	Nemertina	+	e
Ciliophora	+/−	s	Gnathostomulida	+	u
Labyrinthulamycota	+/−	s	Gastrotricha	+	p
Acrasiomycota	+/−	s	Rotifera	+	p
Myxomycota	+/−	s	Kinorhyncha	+	u
Oomycota	+	s	Acanthocephala	+	p
Fungi			Entoprocta	+	s
Zygomycota	+	s	Nematoda	+	p
Ascomycota	+	s	Nematomorpha	+	u
Basidiomycota	+	s	Bryozoa	+	s
Deuteromycota	+	s	Phoronida	+	s
Plantae			Brachiopoda	+	u
Bryophyta	+	s	Mollusca	+	e, p
Lycopodophyta	+	s	Priapulida	+	u
Sphenophyta	+	s	Sipuncula	+	u
Pteridophyta	+	s	Echiura	+	u
Cycadophyta	+	s	Annelida	+	s, e, p
Coniferophyta	+	s	Tardigrada	+	p
Angiospermophyta	+	s	Onychophora	+	p
Animalia			Arthropoda	+	e, p
Placozoa	+	s	Pogonophora	+	u
Porifera	+	s	Echinodermata	+	e
Cnidaria	+	s	Chaetognatha	+	p
Ctenophora	+	p	Hemichordata	+	s, e
			Chordata	+	e, p

(Section 1.2, p. 6) and the Galápagos finches (Section 9.1, p. 223), to the color patterns of the peppered moth *Biston betularia* (Section 5.7, p. 108) and mimetic butterflies (Section 8.1, p. 195), are all adaptations that benefit the individual organism. It can hardly be doubted, therefore, that organismal adaptations exist, and natural selection can favor them.

11.2.4 *Natural selection working on groups of close genetic relatives is called kin selection*

In species in which individuals sometimes meet one another, such as in social groups, individuals may be able to influence each other's reproduction. Biologists call a behavior pattern *altruistic* if it increases the number of offspring produced by the recipient and decreases that of the altruist. (Notice that the term in biology, unlike in human action, implies nothing about the altruist's intentions: it is a motive-free account of reproductive consequences.) Can natural selection ever favor altruistic actions that decrease the reproduction of the actor? If we take a strictly organismic view of natural selection, it would seem to be impossible. And yet, in a growing list of natural observations, animals behave in an apparently altruistic manner. The altruism of the sterile "workers," in such insects as ants and bees, is one undoubted example; here the altruism is extreme, as the workers do not reproduce at all in some species.

Altruistic behavior often takes place between genetic relatives, and when it does the most likely explanation is the theory of *kin selection*. Let us suppose for simplicity that there are two types of organism, altruistic and selfish. A hypothetical example might be that, when someone is drowning, an altruist would jump in and try and save him or her whereas the selfish individual would not. The altruistic act decreases the altruist's chance of survival by some amount, which we can call c (for cost), as the altruist runs some risk of drowning too. It increases the chance of survival of the recipient by an amount b (for benefit). If the altruists dispensed their aid indiscriminately to other individuals it would be received by other altruists and selfish individuals in the same proportion as they exist in the population. Natural selection would then favor the selfish types, because they receive the benefits but do not pay the costs.

For altruism to evolve, it has to be directed preferentially to other altruists. Suppose, to begin with, that acts of altruism were only ever given to other altruists; what would be the condition for natural selection to favor altruism? The answer is that the altruism must take place only in circumstances in which the benefit to the recipient exceeds the cost to the altruist. This will be true if the altruist is a better swimmer than the recipient, but it does not logically have to be true (if, for instance, the "altruist" were a poor swimmer and the "recipients" were capable of looking after themselves, the net result of the altruist's heroic plunge into the water might merely be that the altruist would drown). If the recipient's benefit does exceed the altruist's cost then the average fitness of the altruistic types as a whole will increase. This condition is only of theoretical interest. In practice altruism usually (maybe always) cannot be directed only to other altruists, because they cannot be recognized with certainty. However, altruism can be directed only at a class of individuals that contains a disproportionate number of altruists relative to their frequency in the population. This is true when altruism is directed toward genetic relatives: if a gene for altruism is in an individual, it is also likely to be in its relatives. Define r (for relatedness) as the probability that a new rare gene that is in one individual is also in another individual. The probability is between zero and one, depending on the other individual concerned. The appropriate r can be deduced from Mendel's rules. If the new mutation is in a parent, there is a half chance it will be in its offspring; and there is likewise a half chance that a gene in an individual is also in its brother or sister.

Under what condition will natural selection favor altruism? The altruist still pays a cost of c for performing the act, and the recipient receives a benefit b. However, the chance that the altruistic gene is in the recipient is r. When rb exceeds c there will be a net increase in the average fitness of the altruists. The number of copies of the gene for altruism will increase because the loss of copies from the excess death of the individuals who actually perform acts of altruism is more than made up for by the excess survival of the individuals who receive it (and contain the gene for altruism). The condition for natural selection to favor altruism among relatives is that it should be performed if:

We define Hamilton's rule

$$rb > c$$

This is the theory of kin selection. It states that an individual is selected to behave altruistically provided that $rb > c$. The condition itself is called Hamilton's rule, after W.D. Hamilton, who mainly invented the theory of kin selection.

Hamilton's rule is testable. For an act of altruism we can measure the benefit and the cost, and r can be deduced if the pedigree relationship between the altruists and recipients is known. The details of how b and c are estimated depends on the example. Here we shall look at one example: "helpers at the nest" in the Florida scrub jay (*Aphelocoma coerulescens*, see Plate 6, between pp. 68 and 69).

The scrub jay is distributed widely across the western USA, and also has an isolated population that breeds in the shrinking areas of oak scrub in central Florida. It has been continuously studied there since 1969 by Woolfenden and his colleagues. A breeding pair of Florida scrub jays may be helped by up to six other birds. Woolfenden knows the pedigrees of the birds and therefore has been able to show that the majority of these helpers are either full or half sibs of the young they are helping. r is therefore known, but how can we estimate b and c?

Hamilton's rule has been tested in the Florida scrub jay

"Benefit" and "cost" properly refer to the change in lifetime reproductive successes of the altruist and recipient, relative to if the act had not been performed. The true values of b and c are therefore unmeasureable, because they refer to a situation that does not exist (namely, if the act had not been performed). They can, however, be estimated. Mumme (1992) experimentally removed the helpers from 14 nests in 1987 and 1988, and measured the reproductive success both in these experimental nests and in 21 untreated control nests in the same area. There were an average of almost 1.8 helpers at the experimental nest, which we can round up to two helpers per nest in some approximate calculations below; there were a similar number of helpers at the control nests.

The removal of the helpers significantly reduced the survival of the offspring (Table 11.2). Mumme found that the contribution of helpers is mainly in defending the nest against nest predators such as snakes and other birds. A nest with a helper is more likely to have a sentinel bird present at the nest at any time than is a nest without helpers, and nests with helpers can "mob" predators more effectively. ("Mobbing" is a kind of group defense of birds against predators, in which the birds dive at and harrass the predator. It is most commonly seen in birds such as crows mobbing domestic cats.) The young in nests with helpers are also fed more and (probably in consequence) survive better after fledging.

Mumme's result enables us to estimate the benefit (b) of helping as the difference between the survival rate of young in nests with and without helpers. In Table 11.2, the

Table 11.2

The survival of young Florida scrub jays in nests where the helpers had either been experimentally removed or left undisturbed. The offspring in the experimental groups had lower survival during and immediately after the period of parental care, but not at the egg stage. These results are for 1987; the post-fledgling difference was similar in 1988, but the pre-fledging difference was not. Modified from Mumme (1992).

	Experimental groups (helpers removed)	Control groups (helpers present)
Initial sample size	45	63
% survival from egg to hatching	67	68
% survival from hatching to fledging	30	63
% survival from fledging to day 60	33	81
% survival from egg to day 60	7	35

survival rate was increased somewhere between two- and fivefold if helpers are present. Let us use the total figure, for survival to day 60, to calculate whether natural selection favors helping. The survival of an average young scrub jay to day 60 is increased from 7% to 35% if helpers are present: the difference is $35 - 7 = 28\%$. We divide this by two to find the benefit of helping per helper: $28/2 = 14\% = b$.

The cost of helping is more difficult to estimate. The cost is equal to the reproductive success that a helper would have had if it had not helped. We can make an upper and lower bound estimate. The lower bound estimate is zero if the helper had been unable to breed independently. This may be close to the true value of c in the saturated habitat of Florida scrub jays. It is thought that one of the main advantages of staying at the parental nest is the chance either of inheriting the territory or budding off another territory at the edge of it: most new territories are formed in this way. A young jay may have to stay at home in order ever to be able to breed independently. An alternative, higher bound estimate of the cost is the reproductive success of pairs without helpers. The justification of this estimate is that if the helper had bred by itself it would lack helpers (which are mainly derived from earlier clutches) and thus achieve the success of an unhelped pair. The cost of helping is then 7%, the chance of survival to day 60 of an egg in a nest without helpers.

To apply Hamilton's rule in this case we have to notice that the helper's choice is between producing sibs and producing its own offspring. It should help its sibs if:

$$r_{sib}b > r_{off}c$$

where r_{sib} is the relatedness to a sib and r_{off} is the relatedness to its own offspring, both of which are $1/2$ if the sibs are full sibs. (The small difference from the $rb > c$ version given above arises because there we imagined changes to the survival of altruists and recipients. The relatedness of an altruist to itself is necessarily 1 and the cost is implicly multiplied by this. Here the altruism affects the numbers of two kinds of offspring — its own

Costs are hard to measure empirically

and its parents' — and we have to weigh each kind by the chance they share a gene with the helper.)

For the two methods of estimating cost, the inequalities are approximately:

Lower bound estimate: $1/2 \times 14 > 1/2 \times 0$
Higher bound estimate: $1/2 \times 14 > 1/2 \times 7$

Kin selection is at work in this example

Either way, natural selection favors helping behavior in young Florida scrub jays. The estimates of both b and c are fallible, however, and the test is uncertain. Despite these uncertainties, the test does illustrate how we can attempt a quantitative test of the theory of kin selection.

11.2.5 *Whether group selection ever produces adaptations for the benefit of groups has been controversial, though most biologists now think it is only a weak force in evolution*

Some adaptations may benefit the group, but not the individual . . .

A group adaptation is a property of a group of organisms that benefits the survival and reproduction of the group as a whole. Adaptations produced by kin selection — such as helping in family groups of birds — will satisfy that definition, but we are concerned here with group adaptations that did not evolve by kin selection. If any exist, they will have come into existence by selection between groups: groups possessing the group adaptation would have gone extinct at a lower rate, and sent out more emigrants, than groups lacking it. The group adaptation would have been favored by the differential reproduction of whole groups.

Many characters are beneficial at the group level, but also benefit all the individuals in the group. This would trivially be true, for example, of an improvement in the hunting skill of a lion: after the improvement has spread by individual selection, all the individuals in the group, and the group as a whole, will be better adapted. That is just to restate the earlier point that adaptations that have evolved for the benefit of one level of organization can incidentally benefit higher levels. The controversial group adaptations are those that benefit the group but not the individual. A hypothetical example is Wynne-Edwards theory, put forward in *Animal Dispersion in Relation to Social Behaviour* (1962), that animals restrain their reproduction in order not to overeat the local food supply. If all the individuals in a group reproduce at the maximum rate, their offspring might overeat the food supply, and the group would then go extinct. They could avoid this fate by collectively restraining their reproduction, to maintain the balance of nature. Natural selection on individuals does not favor reproductive restraint. An individual that increases its reproduction will automatically be favored relative to individuals that produce fewer offspring. Within a group, if some individuals produce more offspring than others, the former will proliferate. But can individual selection within the group be overcome by selection between groups?

. . . such as reproductive restraint

The question is highly important, both conceptually and historically. It is important historically because vague group selectionist thinking — particularly in the form of statements like "adaptation X exists for the good of the species" — was once common. It is now more usual (though by no means universal) for biologists to believe that group

The evidence for group selection has been criticised . . .

selection is a weak and unimportant process. There are both theoretical and empirical reasons. Empirically, there are no definite examples of adaptations that need to be interpreted in terms of group advantage: Williams (1966) argued that the characters that Wynne-Edwards had suggested evolved to regulate population size can all be explained as adaptations that benefit individuals. Individuals generally reproduce at the maximum rate they can. The only obvious exceptions concern genetically related individuals, and can be explained by kin selection. Moreover, living things have characteristics that contradict the theory of group selection. The 50 : 50 sex ratio, which we discuss in Section 12.5 (p. 337), is a case in point. In polygynous species, it is inefficient for the population to produce 50% males, most of whom are not needed. The widespread existence of the 50 : 50 sex ratio suggests that group selection has been ineffective on this trait.

Group selection is also implausible in theory. Consider a population containing two genotypes. One codes for an altruistic, or group adaptive, trait like reproductive restraint. The other codes for a selfish trait, like reproducing as fast as possible. The population is made up of a number of groups, which can contain any proportion of altrusitic and selfish individuals; groups with mainly altruistic members we call altruistic groups and those with mainly selfish members, selfish groups. The altruistic groups will go extinct at a lower rate as group selection favors altruism. Individual selection favors the selfish individuals within all groups so within each group the selfish individuals increase in frequency. An altruistic group may temporarily contain no selfish members, but as soon as it is "infected" with one the selfish trait will proliferate and become fixed in the group. What result should we expect to find? It depends on the balance between the two processes. In theory, we can imagine a rate of group extinction so high that altruists will predominate: just imagine, for sake of argument, what would happen if all groups with more than 10% selfish types instantly went extinct. All the groups we should see would clearly have at least 90% altruists. But that is only a thought experiment. The interesting question is what we should expect to happen naturally.

. . . as has the theory

The reason most biologists suppose that group selection is a weak force in opposition to individual selection stems from the slow life cycles of groups as compared to individuals. Individuals die and reproduce at the rate of once per generation, and many individuals can move between groups within a generation. Groups go extinct at a much slower rate. The amount of time it will take for selfish individuals to infect and proliferate in a group is a small part of the group's lifespan; at any one time, therefore, individual adaptations will predominate.

Migration between groups undermines group selection

Many models of group and individual selection exist, but they can mainly be reduced to a common form (Figure 11.1). The groups are supposed to occupy "patches" in nature. As before, some patches are occupied by altruistic and others by selfish groups. There are also empty patches. A selfish group in the model drives itself extinct by overeating its patch's resources. The result of the model depends on whether a selfish group can infect an empty or altristic patch before going extinct. Maynard Smith (1976) defines the number m as the number of successful migrants produced by one selfish group on average between its origin and extinction. (Successful means that the migrant establishes itself in another group and breeds.) If $m = 1$ the system will be stable; if $m < 1$ the selfish groups decrease in number, and if $m > 1$ they increase. In other words, a selfish group only needs to produce more than one successful emigrant during

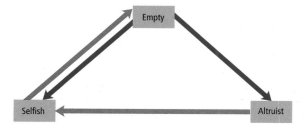

Figure 11.1
Maynard Smith's formulation of group selection models. A patch may change state in the direction of the arrows. Redrawn, by permission of the publisher, from Maynard Smith (1976). © 1976 University of Chicago Press.

its existence for the selfish trait to take over. This is a small number. So small that we can expect selfish individual adaptations to prevail in nature. Group selection, we conclude, is a weak force. It only works if migration rates are implausibly low and group extinction rates implausibly high. It is also not needed to explain the facts.

The case against group selection is presented here in stark terms, but only to make the arguments clear. The matter has not been settled finally, and group selection probably operates sometimes. Moreover, group selection can have evolutionary consequences even if it never overrides individual selection. In Section 23.6 (p. 658) we look at a process called "species selection." Species selection operates when different species (or even higher taxa) possess different individual-level adaptations, and their different adaptations have different consequences for the rate of extinction or speciation. Taxa with lower extinction, or higher speciation, rates tend to proliferate. Much the same could be true of groups within a species.

The theories of group selection and of species selection are distinct

In species selection, there is no conflict between selection at lower (individual) and higher (species, or even group) levels. In all the species (or groups), individuals act in their own selfish interest. Species selection is theoretically uncontroversial, though its empirical importance is open to doubt. The controversy about group selection that we looked at above was theoretical as well as empirical. Critics of group selection doubt whether group selection could be strong enough to cause individuals to sacrifice their own reproductive interests to those of its group.

In nature, group selection is rarely likely to override individual selection, and to establish individually disadvantageous behavior. In the laboratory, however, conditions can be made extreme enough for it to do so. Let us finish by looking at such a case: Wade's experiment on flour beetles, *Tribolium castaneum*. The life cycle of the flour beetle takes place, from egg to adult, in stored flour. They are pests, but they have also become a population biologist's standard experimental animal, particularly in Chicago. Wade set up an experiment to illustrate Wynne-Edwards' hypothesis of reproductive restraint.

Group selection can be experimentally simulated

Figure 11.2a illustrates the experimental design. In each of three experimental treatments, there were 48 different colonies of *Tribolium*. Each colony was allowed to breed for 37 days. Then 48 new colonies, of 16 beetles each, were set up from the progeny of the old ones. Wade (1976) artificially selected for groups that had showed a low (or high) fecundity (the third treatment was a control). He selected for low fecundity by forming a new generation of colonies from *Tribolium* colonies that had a low population density at the end of the 37 days; and for high fecundity by forming each new generation from colonies that had high population densities. He repeated a number of rounds of the process.

(a)

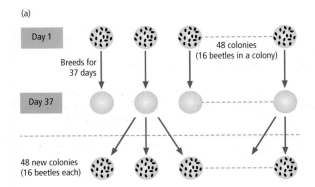

(b)

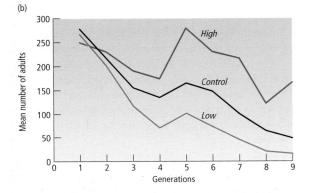

Figure 11.2
Wade's experiment with *Tribolium* beetles. (a) Experimental design: 48 colonies were bred for 37 days, and then a new round of colonies was formed from the colonies that had grown to a low (or high) population density. (b) Results showing population densities in lines selected for high or low population densities and in unselected controls. The results are the means for 48 colonies. Redrawn, by permission of the publisher, from Wade (1976).

Group-selected beetles evolve lower fecundity

Not surprisingly, the population density in the "low" lines decreased relative to the "high" lines (Figure 11.2b). The decrease in the low lines is due to group selection. Presumably, within the 37 days of any one cycle, the beetle types with high fecundity were increasing within each colony relative to the less fecund beetles. However, between cycles, Wade's group selection for low fecundity more than out-weighed the individual selection and average fecundity declined. The group selection was strong enough to work.

In a way, the group selective structure of the experiment is superfluous. We could simply breed from beetles with lower fecundity. Artificial selection of this kind would reduce beetle fecundity without their being kept in groups for 37 days. But Wade's purpose was to illustrate group, not artificial, selection and his experiment does so. It has alternating rounds of individual and group selection and the experimental group selection is strong enough to produce the effect that Wynne-Edwards thought to be common in nature. Box 11.1 shows how the group selection design of Wade's experiment has had a practical application.

The fact that group selection can be implemented in an experiment does not mean that group selection is important in nature. Biologists doubt group selection for theoretical reasons, and because of the kinds of adaptations seen in nature. The experiments are instructive, however. They show what group selection means, and how individual selection can decrease the efficiency of a group. Muir's experiment (Box 11.1) also has a commercial interest.

Box 11.1
Group Selection for Egg Laying in Hens

Muir (1995) found a practical application of the basic experimental set up used by Wade to illustrate group selection. Muir repeated the experiment on egg-laying chickens. Farmers have selected chickens to lay the maximum possible number of eggs, but in farms chickens are often kept in groups. The individual chicken that lays the most eggs may then be an "antisocial" individual who lays extra eggs by reducing the productivity of other chickens in the group. The bird that lays the most eggs may, for instance, take a disproportionate amount of resources, or be aggressive to other birds. Muir kept chickens in groups of nine birds, and selected the most productive group. The consequence was that the chickens evolved to be less selfish, and mortality declined. Productivity, in terms of numbers of eggs laid, went up dramatically from less than 100 to over 200 a year per

hen, in five generations (Figure B11.1). The experiments show how individual and group selection can conflict. A selective regime that prevents individual selection can improve the average output of the whole group. The improved

egg laying, combined with the improved quality of social life for the chickens, has considerable agricultural interest.

Further reading: Sober & Wilson (1998, pp. 121–3).

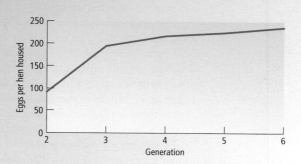

Figure B11.1
Group selection for increased egg laying. Hens were kept in nine-hen groups, and the groups that laid more eggs contributed more to the next generation, in an experimental design similar to Figure 11.2. From Muir (1995).

11.2.6 *Which level in the hierarchy of organization levels will evolve adaptations is controlled by which level shows heritability*

Adaptations can exist for the benefit of genes, cells, organisms, kin, or groups of unrelated individuals. Genic adaptations like segregation distortion are rare; cell line adaptations are very rare, at most, in Weismannist species, but may be found in non-Weismannist species such as plants; organismic adaptations are common; the number of examples of kin-selected adaptations is increasing; and group adaptations are probably rare.

Most adaptations appear to benefit the individual organism

Why do adaptations mainly appear at the organismic level, with a few additional cases for groups of kin? Earlier in this chapter, we have already discussed the answer to this question, but a more general answer can be given. The units in nature that show adaptations are the units that show heritability (Section 9.6, p. 235). Mutations that influence the phenotype of a unit (whether a cell, organism, or group) must be passed on to the offspring of that unit in the next generation; if this happens natural selection can act to increase the mutation's frequency.

Organisms show heritability in this sense. A finch with an improved beak shape, caused by a genetic change, will on average produce offspring with the improved beak shape. Natural selection can work on individual finches.

But groups do not show this sort of heritability when group and individual advantage are in conflict.[1] A genetic variant that increases a group's chance of success tends not to be inherited by future groups. Immigration contaminates the group's genetic composition, such that heritability from generation to generation is low. Thus altruistic groups do not exclusively generate descendant altruistic groups, and selfish groups generate selfish groups. Migration from selfish groups causes altruistic groups to become selfish. Thus a group in one generation will only be genetically correlated with the group of its offspring in the next generation when there is practically no migration; then group selection works.

The same point can be made about kin selection and selection among cell lines. Kin selection operates because an "offspring" kin group genetically resembles the "parental" kin group. Cell selection, in Weismannist species, tends not to operate because somatic cells, although they are inherited down a cell line during one organism's brief life, are not passed on from an organism to its offspring; but when they are (in non-Weismannist species), cell selection and the evolution of cell line adaptations becomes theoretically more plausible. For genic adaptations such as segregation distortion, the same basic argument applies.

In summary, we should expect to find adaptations existing for the benefit of those units in nature that show heritability. Adaptations will, therefore, usually benefit organisms. The cases of adaptations that benefit higher or lower levels of organization can be understood in the same general terms, because they only evolve in circumstances when groups, or parts, of organisms show heritability from one generation to the next. We can now give our first answer to the question of what is the unit of selection. The general answer is "that entity that shows heritability"; more specifically, it is usually the organism, with some interesting exceptions. This first answer specifies the units in nature that should possess adaptations.

Heritability is the key to which levels of organization show adaptations

11.3 Another sense of "unit of selection" is the entity whose frequency is adjusted directly by natural selection

Natural selection over the generations adjusts the frequencies of entities at all levels. We have implicitly seen this in the example of the lion hunt. If the lions of one pride become more efficient at hunting, perhaps because of some new behavioral trick, natural selection will favor them. If the trick is inherited, that type of lion will increase in frequency relative to other types of lion. All things associated with the trick will increase in frequency too. The type of lion, its type of neurons and proteins, and their encoding genes would all increase in frequency relative to their alternatives. When the hunting

---

[1] Species-level heritability does exist with species selection, because individual and species advantage are not in conflict (Section 23.6.3, p. 665). That is why species selection is more widely accepted, in theory, than group selection.

success of the lions as a whole increases, the frequency of lions in the ecosystem will increase too, and lions might, over geological time, come to replace other competing predators on the plains. The question in this section is whether natural selection directly adjusts the frequency of any of these units — nucleotides, genes, neurons, individual lions, lion prides, lion species?

The unit of selection, in a second sense, is the gene . . .

The answer was most clearly given by Williams in *Adaptation and Natural Selection* (1966) and Dawkins in *The Selfish Gene* (1989a). It is at least implicit in all theoretical population genetics and, indeed, in the previous section of this chapter. For natural selection to adjust the frequency of something over the generations, the entity must have a sufficient degree of permanence. You cannot adjust the frequency of an entity between times t_1 and t_2 if between the two times it has ceased to exist. A character that is to increase in frequency under natural selection therefore has to be inherited.

We can work through the argument in terms of the example of an improvement in lion hunting skill. (We shall express it in terms of selection on a mutation: the same arguments apply when gene frequencies are being adjusted at a polymorphic locus.) When the improvement first appeared, it was a single genetic mutation. At a physiological level, the mutation would produce its effect by making some minor change in the lion's developmental program. After the mutation has appeared, there is a "pool" of two types — the new mutation, and all the rest (i.e., all alleles of the mutation, and the behavior patterns they produce). There will, of course, be genetic variation at loci other than the one where the mutation arose, but that variation can be ignored because it will be randomly distributed among the mutant and non-mutant types. The lions with the mutation will survive better and produce more offspring. Natural selection is starting to work. Now we can ask what natural selection is adjusting the frequency of. Is it lions? Lion genomes? Or the mutation?

. . . because only genes last long enough for natural selection to adjust their frequencies

Williams and Dawkins' answer is the gene — the particular mutation that produces improved hunting. Natural selection cannot work on whole lions because lions die: they are not permanent. Nor can it work on the genome. The mutant lion's offspring inherit only genetic fragments, not a copy of a whole genome, from their parents. Meiotic recombination breaks the genome. In Williams' expression, "meiosis and recombination destroy genotypes [i.e. genomes] as surely as death." What matters, in the process of natural selection, is that some of the lion's offspring inherit the mutation. These offspring in turn produce more offspring, and the gene increases in frequency. The gene can increase in frequency because it is not (like the genome) fragmented by meiosis or (like the phenotype) returned to dust by death. The gene, in the form of copies of itself, is potentially immortal, and is at least permanent enough for it to be possible to alter its frequency in successive generations.

It may be objected that recombination breaks genes as well as genomes. Recombination strikes at almost random intervals in the DNA and therefore could strike within the mutation we are concerned with. A little reflection, however, shows that is irrelevant. The information of the gene, not its physical continuity, is what matters. Consider the length of chromosome containing the gene and its mutant form; there will usually be a number of polymorphic loci around the mutant locus (Figure 11.3a). Now consider what happens when recombination strikes either in a neighboring gene or in the gene itself. Nearby recombination breaks the information in the chromosome — which is just to repeat the point already made, that recombination destroys the genome

Figure 11.3
(a) Three genes along a chromosome. Loci *A* and *B* are heterozygous. The * indicates where the nucleotide differs from the other allele. Now consider the effect of recombination on the structure of the gene at the *B* locus. (b) Intragenic recombination does not affect the structure of the *A* genes; (c) nor does recombination in the neighboring *B/b* locus. The same pair of *A* and *a* gene sequences come out of recombination as were present before.

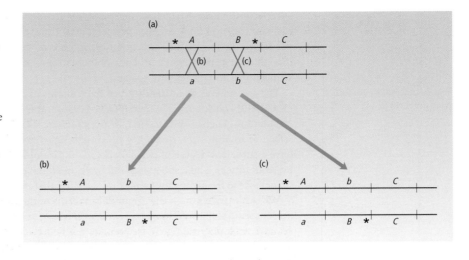

The longevity of a genetic unit matters relative to the time evolutionary change takes

(Figure 11.3c). Recombination within the gene does not usually alter the outcome (Figure 11.3b). If the locus was homozygous before the mutation, all the gene except for the mutant base pair will be identical in the original and mutant forms. Intragenic recombination therefore produces exactly the same result within the gene as no recombination; it only alters the combinations of genes.

Intragenic recombination can destroy the heritable information in a gene in one special circumstance. If the locus was heterozygous before the mutation and recombination occurs between the mutant site and the other site that differs between the two strands, the products of recombination differ from the initial strands (Figure 11.4). Clearly, this could happen. When it does, the length of DNA whose information is inherited is shorter than a gene. For this reason, if we take a long enough view, the only finally permanent units in the genome are nucleotide bases; recombination does not alter them. However, this long view is of little interest. We are concerned with the timescale of natural selection. It takes a few thousand generations for a mutation's frequency to be significantly altered (Section 5.6, p. 107) and, over this time, genes, but not genomes or phenotypes, will be practically unaltered. Genes will then act as units of selection and will be permanent enough to have their frequency altered by natural selection.

Williams defined the gene to make it almost true by definition that the gene is the unit of selection. He defined the gene as "that which segregates and recombines with appreciable frequency." The gene in this definition need not be the same as a cistron (i.e., the length of DNA encoding one protein, or polypeptide). It is instead the length of chromosome that has sufficient permanence for natural selection to adjust its frequency: longer lengths are broken by recombination and shorter lengths have no more permanence that the gene (for the reason shown in Figure 11.3). The gene in Williams' definition is what Dawkins calls the *replicator*. In practice, the replicator (or Williams' gene) does not consistently correspond to any particular length of DNA.

When selection is taking place at one locus, a cistron at a neighboring locus will to some extent (depending on the amount of recombination) have its frequency adjusted as a consequence. In a population genetic sense, this is hitch-hiking (Section 8.9,

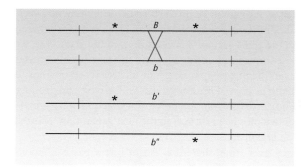

Figure 11.4
When intragenic recombination happens between two alleles with different heterozygous nucleotides, it breaks the gene structure. The * indicates where the nucleotide sequence differs from the allele. The gene sequences coming out of recombination differ from the initial sequences.

p. 210) and builds up linkage disequilibrium between genes. The same will be true of loci further down the DNA from the selected locus. The hitch-hiking effect is gradually reduced with distance by recombination, but there is no clear cut-off. This poses no problem for Williams' definition of the gene. The neighboring allele that is hitch-hiking with the selected mutation is, in Williams' definition, part of the gene that is having its frequency altered.

The word "gene" is being used in a technical sense

Williams' "gene" has a statistical reality, because shorter lengths of DNA are more permanent and longer lengths are less permanent. The random hits of recombination will generate a frequency distribution of genome lengths lasting for different periods of evolutionary time. The average length that survives long enough for natural selection to work on has been defined by Williams and Dawkins as the gene. Population geneticists have scolded them from time to time for assuming a one-locus, zero linkage disequilibrium view of evolution, but the dispute is a matter of definition, not substance. The critics are identifying "gene" with "cistron." It would be interesting to know whether the gene in Williams' sense is also a physical cistron; but it is a secondary question and has nothing to do with the fundamental logic of Williams and Dawkins' argument.

We must discuss one further matter before considering the significance of the genic unit of selection. Critics, such as Gould (2002b), have objected that gene frequencies change between generations only in a passive, "book-keeping" sense. The frequency changes provide a record of evolution, but are not its fundamental cause. True natural selection, the critics would say, happens at the level of organismic survival and reproduction. The actual selection in the lion example happens when a lion catches, or fails to catch, its prey. The differential hunting success drives the gene frequency changes, and it is a mistake to identify the gene frequency changes as causal. Williams and Dawkins, however, do not deny that whatever ecological processes are causing differential organismic survival produce gene frequency changes within a generation. What they deny is that this ecological interaction of organisms means that natural selection directly adjusts the frequencies of organisms over the evolutionary timescale of many generations.

Criticisms have been made

There is an easy philosophical method of deciding whether natural selection works on genes or larger phenotypic units. We can consider a phenotypic change such as a new hunting skill, and ask whether natural selection can work on it if it is produced genically and if it is produced non-genically. The case we discussed above was genic: the advantageous new hunting behavior was caused by a genetic mutation. Now suppose that the same advantageous phenotypic change was caused by a non-heritable phenotypic change instead, such as individual learning or some developmental accident in the lion's nervous system. The thought experiment provides a test case between the organismic, phenotypic and the genic accounts of evolution. In the genic case, we know, natural selection favors the improved hunting type and the gene for it increases in frequency. But what happens in the phenotypic case? The answer is too obvious to labor over. The individual lion with improved hunting ability will survive and produce more offspring than an average lion, but no evolution or natural selection, in any interesting sense, will occur. The trait will not be passed on to the next generation. Natural selection cannot directly work on organisms.

Genes are actively, not passively, part of natural selection

The change in gene frequency over time, therefore, is not just a passive "bookkeeping" record of evolution. Genes are crucial if natural selection is to take place. The need for inheritance, and the fact that acquired characters are not inherited, gives the gene a priority over the organism as a unit of selection. Whenever a gene is being selected, it produces a phenotypic change and the frequency of different organismal types will change along with the gene frequency. But the change in organism frequency is a consequence of the change in gene frequency: it is the gene frequency that natural selection is actually working on and this is why Williams and Dawkins maintain that the gene is the unit of selection.

Why does the argument matter? Its importance is to tell us what entities adaptations exist for the good of. Evolutionary biologists work on particular characters (like banding patterns in snails and sex), trying to work out why the characters exist. The ultimate, abstract answer is that any adaptation exists because it increases the reproduction of the genes encoding it, relative to that of the alleles for alternative characters. The genes that exist in nature are the genes that in the past have out-reproduced alternative alleles. Natural selection will always favor a character that increases the replication of the genes encoding it.

The issues are scientifically important

It is important to know what the ultimate beneficiaries of adaptations are. When we are trying to explain the existence of particular characters, we need to know whether a proposed explanation is correct. The argument that genes are the units of selection provides the fundamental logic that is used to find out. We imagine different genetic forms of the character, and the correct explanation must specify how the genes for the observed form of the character will out-reproduce other genetic types. In practice, there may be several possible hypotheses, and they can be tested between using the methods of Chapter 10, but before those methods are applied we have to insure that the hypotheses make theoretical sense. We can rule a hypothesis about adaptation out before the practical testing stage if it contradicts the theory of gene selection.

11.4 The two senses of "unit of selection" are compatible: one specifies the entity that generally shows phenotypic adaptations, the other the entity whose frequency is generally adjusted by natural selection

We have now specified what the unit of selection is in two different senses. They have sometimes been confused, but many evolutionary biologists now appreciate the distinction. The two have been given names; Hull (1988), for instance, distinguishes between interactors and replicators, and Dawkins (1982) between vehicles and replicators. It is most important, however, to realize that there are two distinct issues and to understand the arguments used in the two cases.

Adaptations evolve because the genes encoding them out-reproduce the alternative genes. In this sense, adaptations can only evolve if they benefit replicators. Genes do not, however, exist nakedly in the world, and the kinds of adaptations that evolutionary biologists seek to understand, such as social behavior, beak shape, or flower coloration, are not simple properties of genes. They are phenotypic properties of higher level entities (whole organisms, or societies). We therefore also have to ask which higher level entities should benefit from the natural selection of replicating genes. The answer is usually organisms, but in some cases it is a family of genetically related organisms.

Summary

1 Adaptations evolve by means of natural selection. When natural selection acts, it alters the frequencies of entities at many levels in the hierarchy of biological levels of organization. It also produces adaptations that benefit entities at many levels.

2 The discussion of units of selection aims to find out which level natural selection directly acts on, and which ones it affects only incidentally.

3 Evolutionary biologists are interested in what the unit of selection is both in order to understand why adaptations evolve and also in order that, when they study adaptations, they can concentrate on theoretically sensible hypotheses.

4 We can find out which level of organization shows adaptations by considering a series of adaptations at genic, cellular, organismic, and group levels and asking which evolves most often.

5 Segregation distortion is an adaptation of a gene against its allelic alternatives. Examples of this kind are rare.

6 In Weismannist organisms, with separate germ and somatic cell lines, selection between cell lines is a weak force. But many species do not have separate germ lines and in these we expect cell lines to evolve adaptations enabling them to proliferate at the expense of other cell lines. No clear examples are known, but Buss has suggested that the embryology of modern Weismannist species can be explained by a history of cell selection.

7 Adaptations are common at the level of organisms. When genetic relatives interact, adaptations may evolve for the benefit of kin groups (kin selection).

8 Group selection, in which selection produces adaptations for the benefit of groups of unrelated individuals, is thought to be a weak force.

9 Adaptations are possessed by the levels in the hierarchy of life that show heritability, in the sense that genetic changes are inherited by the progeny at that level. Group selection is weak because of the low genetic correlation (heritability) between succeeding generations of groups.

10 Natural selection only adjusts the frequencies of entities that are sufficiently permanent over evolutionary time. It therefore fundamentally adjusts the frequency of small genetic units. This small genetic unit is called the replicator. The gene can be so defined to be the unit of selection; but it is then not necessarily always a cistron in length.

11 Adaptations evolve because they increase the replication of genes. The replication of genes, in the real world, is enhanced by adaptations that benefit entities that show heritability.

12 The question of whether natural selection adjusts the frequencies of genes or of organisms is distinct from the question of the relative power of individual, kin, and group selection.

Further reading

The multiauthor book edited by Keller (1999) contains chapters by expert authors on most of the themes in this chapter.

On segregation distorters, I have written a popular book (Ridley 2001) that includes an account of them and the reasons that Haig and others have suggested to explain why they are rare. The book has references to the original literature. A further example has been found in the eye-stalk flies, illustrated in Plate 5 (between pp. 68 and 69), where a driving sex chromosome shrinks the eye stalks (Wilkinson *et al.* 1998). Various subcellular entities provide a level between the "gene" and "cell" levels in this chapter. Wolbachias are an example, and they are the subject of a newspiece in *Nature* July 5, 2001, pp. 12–14. Mitochondria are another example, and they enjoy an amazing system of multilevel selection, discussed by Rand (2001). (I also discuss selection in mitochondria in Ridley (2001).) On kin selection, the fundamental works are included in volume 1 of Hamilton's (1996) collected papers; Dawkins (1989a) is more introductory; Clutton-Brock (2002) is a review; and Woolfenden & Fitzpatrick (1990) is about the Florida scrub jay. Sober & Wilson (1998) is about group selection.

For replicator selection, and the relation between the two senses of selection unit, see Dawkins (1982, 1989a), Gould (2002b), Maynard Smith (1987), and Williams (1966, 1992), who also refer to the prior literature.

Study and review questions

1 Give examples of adaptations that benefit: (a) both the individual organism and the species that the organism belongs to; (b) the individual organism, but at a cost to its species; (c) a local group of organisms, at a cost to its individual members; and (d) a small genetic system, at a cost to the organism containing it.

2 What is (are) the main theoretical factor(s) in models of group versus individual selection that determine whether individual or group adaptations tend to evolve?

3 In the measurements of Woolfenden and Fitzpatrick, the average number of young birds produced by a nest of scrub jays with helpers is 2.2 and the average number by a nest without helpers is 1.24. The average number of helpers present, for the nests with helpers, is 1.7. What values for b and c can be estimated from this data? If the helpers are brothers or sisters of the individuals they are helping, does kin selection favor helping?

4 In both kin selection and pure group selection, adaptations often evolve that benefit the local group. What key difference is there between the kinds of groups, and the plausibility of the two processes?

5 Does the fact that individual selection is normally more powerful than group selection, benefit the *average* individual in a group?

6 What is the unit of selection, in the sense of a replicator, in: (a) a species that reproduces asexually; and (b) a species in which there is no recombinational crossing-over at meiosis?

12 Adaptations in Sexual Reproduction

This chapter mainly concentrates on three related research questions: the questions of how sex, sex differences, and sex ratios are adaptive. Biologists do not understand why sexual, as opposed to asexual, reproduction exists and we look at four hypotheses — genetic constraint, group selection, deleterious mutation, and parasite–host coevolution. Once sex had evolved, natural selection favored different sets of adaptations in males and females. The theory of sexual selection aims to explain male–female differences. We look at female choice for bizarre male traits, such as the peacock's tail, and at evolutionary conflicts between the sexes. Thirdly, we turn to the sex ratio (the ratio of males to females). The sex ratio is one of the best understood adaptations, and we look at a recent study of a successfully predicted deviation from a 50 : 50 sex ratio in a bird species with helpers at the nest.

12.1 The existence of sex is an outstanding, unsolved problem in evolutionary biology

12.1.1 Sex has a 50% cost

Sex poses a problem . . .

In asexual (or clonal) reproduction, a parent produces an offspring that is a genetic copy of the parent. In sexual reproduction, a parent combines half its DNA with half the DNA of another individual and the offspring is only a half genetic copy of each parent. Sexual reproduction poses an evolutionary problem because it seems to be half as efficient a method of reproducing as its alternative, asexual reproduction.

Figure 12.1 imagines a simple population with one asexual individual, one sexual female, and one sexual male. (If the numbers seem unrealistically low, they can be multiplied up by any amount. Each individual in Figure 12.1 could stand for 1,000 individuals, for instance.) We assume that the members of the two groups are identical in all other respects: sexual and asexual individuals are equally good at finding food, avoiding enemies, and staying alive; they produce the same number of offspring, and those offspring have an equal chance of survival. We are considering only whether natural selection favors sexual or asexual reproduction.

. . . because it seems to be less efficient than cloning

Suppose, for simplicity, that each female produces two offspring. After one generation, the asexual group will have grown to two individuals. The sexual female will also produce two offspring, but only one of these will be a daughter. Now we have four individuals in all, and the proportion of asexual females has increased from one-third to one-half. After another generation, there will be four asexual females, one sexual female and one male; the proportion of asexual females has grown to two-thirds. Asexual reproduction will soon take over completely from sexual reproduction. The clone of offspring from an asexual female multiplies at twice the rate of the progeny descended from a sexual female, and a sexual female has only 50% of the fitness of an asexual female.

In some species, the cost of sex may be less than 50%. For instance, in some single-celled organisms sex is not associated with reproduction. In *Paramecium*, two cells may

Figure 12.1

The 50% cost of sex. A population initially contains equal numbers of asexual and sexual females. The females have identical survival and fecundity (two offspring per parent). Asexual reproduction rapidly takes over, because it doubles the rate of reproduction.

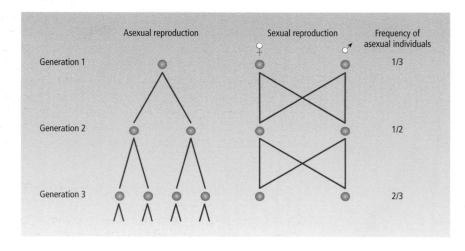

Figure 12.2

Non-reproductive sex in *Paramecium*. *Paramecium* normally contains one micronucleus and one macronucleus. When it prepares for sex, the macronucleus dissolves and the micronucleus is duplicated. Two such cells can then conjugate, swapping one of their micronuclei. Meiosis then occurs within each cell. The sex act is non-reproductive. *Paramecium* cells reproduce by binary fission. Thus, sex and reproduction are not associated in *Paramecium*. The same is true in many single-celled life forms.

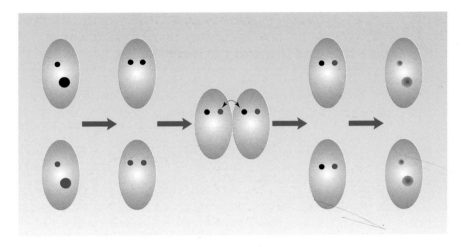

Some kinds of sex are cost-free

conjugate (Figure 12.2). The two cells swap copies of their DNA and then separate. Meiosis then occurs within each cell. Sex is non-reproductive: there were two cells before conjugation and there are two after it. Sex has no cost as in Figure 12.1. Sex acquired its cost as sex became associated with reproduction, perhaps around the time of the evolution of multicellular life. Sex probably originated in single-celled life, and had little cost at that time. The origin of sex therefore poses no deep evolutionary problem. But in many life forms today, sex does have a 50% cost and its existence is a problem.

Fifty percent is a large cost. The problem of explaining sex is to find a compensating advantage of sexual reproduction that is large enough to make up for its cost. We are on the look out for an extraordinarily large selective advantage. Typical evolutionary events are thought to involve selective advantages of a few percent at most, and more often 1% or less. Consider this: a female who has survived to adulthood and is about to reproduce must be fairly well adapted to her environment. If she were to reproduce asexually, she would just make a copy of herself and produce a daughter as well adapted to the conditions of the next generation as she would be herself. If she reproduces sexually instead, she discards half her genes and produces an offspring by mixing the remaining half with other genes drawn from a stranger. If sex is to outweigh its twofold cost, the sexual female must by this procedure expect to produce a daughter who will be twice as fit as a simple copy of herself. The problem, therefore, is not trivial. Indeed, G. C. Williams has described it as "the outstanding puzzle in evolutionary biology." The puzzle is still a puzzle, but we can look at some possible solutions to it. Box 12.1 discusses how the puzzle has taken on a practical importance, with the rise of cloning technologies.

But a big unsolved problem remains

12.1.2   *Sex is unlikely to be explained by genetic constraint*

One possibility is that life uses sexual reproduction because it is "stuck with it." That is, the mutations to produce asexual reproduction have not occurred. (In terms of

Box 12.1
The Ethics of Human Cloning

The question of why sex exists has until recently been scientifically rather than practically or ethically important. That has been changed by developments in cloning technology. Cloning still has to overcome some technical problems, and the theory of evolution is of little relevance there. But the evolution of sex is highly relevant for the ethics of reproductive cloning. Sexual reproduction probably only exists because it is advantageous. Given the 50% cost of sex, it probably at least doubles the fitness of an average sexual offspring relative to a cloned offspring. As we shall see, the two most plausible theories at present suggest that sex helps organisms limit the effects of genetic or infectious disease. If so, then doing away with sex would increase the chance that offspring would succumb to disease. A decision to produce cloned offspring might be ethically analogous to producing sexual offspring and taking them to a plague-ridden city where the chance of dying of infectious disease was twice the normal rate — or to damaging enough of their genes to double the chance of dying from genetic disease. This argument could be wrong. The parasitic and mutational theories of sex may both be incorrect. But then sex probably has some other advantage, which could be lost by cloning. Cloning would be unproblematic if sex exists because we are stuck with it, or because its evolutionary advantage no longer matters in modern human society. However, we need research results before we could draw any such conclusion. The general point — that we need to understand the evolution and function of a character before medically altering it — is the principle of *Darwinian medicine* (Nesse & Williams 1995). The argument here particularly applies to reproductive cloning. However, cloning may also be used to produce new cells for a single individual, to replace faulty cells or organs, and in this case theories about sex may or may not be relevant.

Section 10.7.2, p. 274, this explains sex by genetic constraint.) This hypothesis is unlikely, for two reasons.

Mutations to remove sex can probably occur easily

A mutation to produce asexual reproduction in a sexual form is not a biologically difficult mutation. All that the mutation has to do is eliminate the meiotic cell division at the end of the cell line that produces the gametes. The reproductive cells would then be produced by mitosis rather than meiosis. This is a "loss" mutation, in which a piece of biological information (that is, all the cellular processes of meiosis) is lost. Nothing new is being created. The reproductive cell division will become mitotic; but mitosis already exists — all the other cell divisions in the body are by mitosis.

Secondly, asexual reproduction exists in many forms of life. Asexual reproduction has evolved many times within sexual branches of the tree of life, showing that the necessary mutations can occur. Mutations to produce asexual reproduction are therefore plausible in theory and occur in fact. Their absence is probably not the reason for persistence of sex.

12.1.3   *Sex can accelerate the rate of evolution*

A population of sexually reproducing organisms can, under some conditions, evolve faster than a similar number of asexual organisms. Sexual reproduction can greatly increase the rate at which beneficial mutations, at separate loci, can be combined in a

Figure 12.3

Evolution in (a) asexual and (b) sexual populations. The mutations *A*, *B*, and *C* are all advantageous. In the asexual population, an *AB* individual can only arise if the *B* mutation arises in an individual that already has an *A* mutation (or vice versa). In the sexual population, an *AB* individual can be formed by the breeding of a *B* mutation-bearing individual with an *A* mutation-bearing individual; the second mutation of *B* is not needed. (c) If favorable mutations are rare, each will have been fixed before the next arises, and sexual populations do not evolve faster. The relative rates of evolution in asexual and sexual populations depends on the rate at which favorable mutations arise.

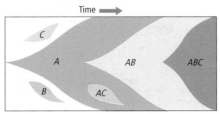

(a) **Asexual: high rate of favorable mutation**

Time ➡

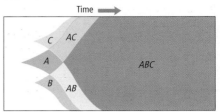

(b) **Sexual: high rate of favorable mutation**

Time ➡

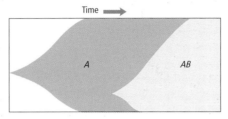

(c) **Sexual or asexual: low rate of favorable mutation**

Time ➡

Sexual populations can evolve faster than asexual populations, . . .

. . . if the rate of favorable mutation is high

single individual (Figure 12.3). Suppose, for example, that a sexual and an asexual population are both fixed for genes *A′* and *B′* at two loci. In the environment where the two populations are living, mutations *A* and *B* are advantageous. *A* and *B* mutations would be likely to arise initially in different individuals. The asexual population will then come to consist of *A′B* and *AB′* individuals, because the *A* mutant cannot spread into the *A′B* clone or vice versa. *AB* individuals cannot appear until an *A′* gene mutates to *A* within the *A′B* clone (or *B′* to *B* in the *AB′* clone).

In the sexual population, evolution proceeds much faster. After *A* and *B* have arisen in different individuals, they can soon combine in a single individual by sex without waiting for the mutations to occur twice. Natural selection can therefore take the population from the state *A′B′* to *AB* faster than under asexual reproduction. This argument was first put forward by Fisher and by Muller in the 1930s. They concluded that sexual populations have a more rapid rate of evolution than would an otherwise equivalent group of asexual organisms.

However, subsequent research has shown that the rate of evolution in sexual populations is not necessarily faster than in equivalent asexual populations. For instance, the result depends on the rate of mutation. If favorable mutations are rare, each one will have been fixed in the population before the next one arises (Figure 12.3c). Sexual and asexual populations then evolve at the same rate. New favorable mutations will always arise in individuals that already carry the previous favorable mutation: they must, because the previous favorable mutation is already in every member of the population. In terms of the example, the *B* mutation will arise in an *AB′* individual in both sexual and asexual populations. However, if favorable mutations arise more frequently, Fisher and Muller's argument works: the sexual population evolves faster. Each new favorable mutation will usually arise in an individual that does not already possess other

Figure 12.4
The taxonomic distribution of asexual reproduction is spindly and is found in odd isolated taxa.

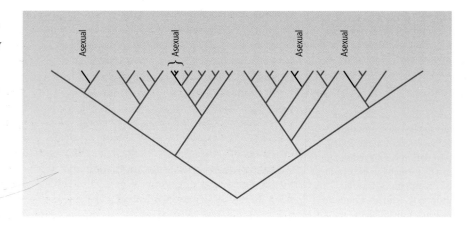

favorable mutations; the greater speed with which the different favorable mutations combine together causes the sexual population to evolve faster. The higher the rate at which favorable mutations are arising, the greater the evolutionary rate of a sexual relative to an asexual population.

Other factors can also influence the relative rate of evolution in sexual and asexual populations. However, the basic Fisher–Muller result remains valid in many, if not all, circumstances. Rice & Chippindale (2001) experimentally demonstrated that the Fisher–Muller theory can be realistic. They found that the rate of evolution was faster in the presence, than in the absence, of sexual recombination.

12.1.4   *Is sex maintained by group selection?*

Sex may exist despite an individual disadvantage

Perhaps the commonest answer to the question of why sex exists is that it speeds up the rate of evolution. This is the "group selection" theory of sex. It accepts that sex is disadvantageous for the individual, because of its 50% cost, but claims that the cost is more than made up for by the reduced extinction rate of populations, or groups, of sexually reproducing organisms. Sexual populations, or groups, can accumulate superior adaptations more rapidly than asexual populations, or groups. The asexual population will then be out-competed and go extinct faster. Each sexual female is in a sense sacrificing herself (she could produce more offspring by reproducing asexually) in order to save the group from extinction.

The main argument for group selection, as an explanation for sex, comes from the taxonomic distribution of asexual reproduction. In multicellular life, exclusively asexual reproduction is mainly confined to small twigs of the phylogenetic tree (Figure 12.4).[1] A few exceptions have been suggested, but it is difficult to be sure that an

---

[1]   The spindly taxonomic distribution of asexuality only applies to multicellular life. In single-celled and viral life, there probably are large chunks of the phylogenetic tree in which asexual reproduction prevails. Many (and arguably all) bacteria, for instance, may do without sex, though this is far from confirmed. But we are not concerned with those single-celled forms here.

apparently asexual form does not use sex in some rare or cryptic circumstances. The bdelloid rotifers are the best documented exception. The Bdelloidea is an entire sub-order of rotifers, containing 300 or so species. Mark Welch & Meselsohn (2000) used a new method, in which they reconstructed gene trees (Section 11.5, p. 457) to show that bdelloid rotifers indeed are exclusively asexual.

Asexual reproduction has a spindly phylogenetic distribution

Despite an exception or two, asexual reproduction does mainly have a spindly taxonomic distribution in multicellular life. The spindly taxonomic distribution of asexual reproduction suggests that asexual lineages have a higher extinction rate than sexual lineages — that asexual lineages usually do not last long enough to diversify into a genus or larger taxonomic group. The higher extinction rate could be because asexual populations do not evolve fast enough to keep up with environmental change, as discussed in the previous section. Alternatively, it could be because asexual forms accumulate more deleterious mutations than sexual populations, as we shall discuss in Section 12.2 below. Either way, according to the group selection theory, sexual reproduction prevails despite its cost for the individual because sexually reproducing groups have a lower extinction rate.

The argument for group selection is not watertight, . . .

The argument is not completely convincing. To say that sexual populations have a lower extinction rate than asexual populations is to say one thing: to say that sex exists because of its lower extinction rate is to say something much stronger. Sex might exist in sexual species because sex is advantageous to the individuals of those species, and asexual reproduction exists in asexual species because it is advantageous to the individuals in those species. The different extinction rates would then be species-level consequences of different individual adaptations in the two types of species. By analogy, carnivores could have higher extinction rates than herbivores, but that would not mean that herbivory was disadvantageous to individual herbivores and only maintained because of its advantage to the group. The taxonomic distribution of asexuality, therefore, although it is consistent with the group selectionist theory of sex, does not confirm it. The same pattern could have arisen if sex had an individual advantage.

. . . and there are general . . .

There are also arguments against group selection. As we saw in Section 11.2.5 (p. 301), biologists are generally suspicious of group selectionist theories. When individual and group advantages conflict, individual selection is usually more powerful. Adaptations that are disadvantageous for the individual are not expected to evolve even if they do benefit the group. Although sexual populations last longer than asexual ones, sexual *individuals* reproduce more slowly than asexual individuals. Asexuality, once it has arisen, will tend to take over sexual groups. Asexuality can arise in a group by either mutation or immigration, and neither of these processes is likely to be, on an evolutionary timescale, all that rare. Asexual reproduction probably arises at a fairly high rate. The reason to be suspicious of group selection is that it requires the rate at which asexual females arise in sexual groups to be very low.

. . . and specific arguments against group selection

Williams (1975) also put forward a specific objection against group selection in the case of sex. His objection has come to be called the *balance argument*. Some species, such as many plants, aphids, sponges, rotifers, and water fleas (Cladocera), can reproduce both sexually or asexually according to the conditions. These species are called heterogonic. Many heterogonic species time their sexual reproduction for periods of environmental uncertainty, and reproduce asexually when conditions are more stable; but that is not the important point here. What matters is that an individual can reproduce in either way. Therefore, when an aphid reproduces sexually, it must be advantageous to the individual, because if it was not the aphid could have reproduced asexually. Both sexual and asexual reproduction must have "balanced" advantages to maintain them in the species' life cycle, otherwise the inferior one would be lost.

The group selectionist proposes that sex is disadvantageous to the individual, and only advantageous to the group. But in aphids and other heterogonic species in which individuals have a "choice," sex almost has to have an individual advantage. The argument can be extended. If sex is advantageous in aphids, it is probably also advantageous to the individual in non-heterogonic species too. We have no good reason to think that sex is exceptional in aphids, or that special factors favor sex in heterogonic species. If we must find an individual advantage for sex in aphids, that same advantage will probably also exist in other species. If group selection can be ruled out for aphids, it can probably also be ruled out for other species.

The balance argument is not decisive

Williams' argument is powerful, but not decisive. In most heterogonic species, the asexual and sexual propagules differ in other respects besides being asexual and sexual. For example, the cladoceran sexual offspring form special winter eggs that are adapted for winter survival. Any cladoceran that gave up sex would also lose its overwintering stage: in practice, the loss of sex while retaining the winter egg would need two mutations, one for the loss of sex and the other for transferring the winter egg phenotype to asexual eggs. So the balance argument is not perfectly clear-cut.

Group selection may or may not explain sex

In summary, group selection will tend to favor sexual over asexual reproduction because sexual populations will have a lower rate of extinction. The taxonomic distribution of asexuality suggests that asexual populations tend to go extinct relatively quickly in evolution. However, biologists doubt whether group selection is the reason why sex exists, for two main reasons. One is a general disbelief in group selection; the other is Williams' balance argument. Neither of these objections is completely convincing, and group selection cannot finally be ruled out. However, the objections are strong enough to have inspired biologists to look for a short-term, individual advantage to sex.

12.2 There are two main theories in which sex may have a short-term advantage

12.2.1 *Sexual reproduction can enable females to reduce the number of deleterious mutations in their offspring*

Sex may help clear deleterious mutations, . . .

A certain number of deleterious mutations arise every generation, and every individual contains some defective genes. Selection acts to remove these deleterious mutations. Here we shall consider how effectively selection removes them, depending on whether reproduction is sexual or asexual. The theory that sex exists because it enhances the power of selection against deleterious mutation, was proposed by Kondrashov (1988). It is sometimes called the *mutational theory* of sex. Maynard Smith has given an analogy to explain Kondrashov's theory. Imagine you have two cars, with two different defects. One is broken down because it has faulty brakes, and the other because it has a faulty ignition. What should you do? One thing you could do is swap the components

. . . as illustrated by a motor car analogy

between the cars, creating one car with two good components, at the expense of the second car with two (rather than one) bad components. This is an improvement. You have created a car that goes out of two that did not. If a car is a wreck, it does not much matter whether it contains one, two, or 20 defects. So you can load a second bad component into an already broken-down car without making things worse.

In genetic terms, imagine a simple haploid model with two loci and two alleles. At each locus, there is a good version of the gene, symbolized by 1, and a bad (deleterious) version, symbolized by 0. Four haplotypes are possible: 11, 01, 10, 00. (Remember these are combinations of alleles at two loci, not the more familiar diploid genotypes at one locus. See Section 8.4, p. 199, on haplotypes.) Sex, as in the car analogy, helps when two individuals with single complementary defects interbreed: that is, an 01 × 10 mating. That will produce some 11 offspring (or grandchildren) at the expense of some 00 offspring. The advantage of sex is that it increases the number of deleterious mutations removed in one death. If an 01 individual clones itself, one death among its offspring removes one bad gene. If it reproduces sexually, one death of an 00 offspring removes two bad genes. The average quality of the surviving offspring can be increased.

Kondrashov's theory requires two conditions in order for natural selection to favor sex despite the 50% cost. We can look at them in turn.

12.2.2 *The mutational theory predicts* U > 1

The first condition is that the deleterious mutation rate is high enough. If deleterious mutations are rare, any advantage of sex will be minor. If they are common, sex may be more advantageous. The deleterious mutation rate is expressed as a genomic figure, that is the average number of new deleterious mutations that occur in each offspring. It is the sum of the deleterious mutations carried into that offspring by the sperm plus the number carried by the egg. The genomic deleterious mutation rate is symbolized by U.

A controversial prediction about the genomic deleterious mutation rate

Sex becomes advantageous relative to cloning if U is more than about one. This is the most controversial prediction of Kondrashov's theory, because deleterious mutation rates have historically been thought to be much lower. Measurements of deleterious mutation rates are being attempted by two methods at present, though neither has yet yielded a conclusive result.

One method is the mutation–accumulation experiment, pioneered by the Japanese geneticist Terumi Mukai. The experimenter attempts to create conditions in which selection does not act against mutation. Mutations will then accumulate over time at the same rate as they occur. From time to time, the fitness of individuals in the experimental population are measured relative to control individuals. Any decline in fitness of the experimental line can be used to estimate the deleterious mutation rate. Mukai's original experiment produced a dramatic decline (Figure 12.5), suggesting that the deleterious mutation rate (U) could be one, or even more than one, in fruitflies.

Mutation–accumulation experiments are so far ambiguous

Since then, these experiments have been created several times in several species. Some experiments produce high U, like Mukai's; others produce negligibly low U. Mutation–accumulation experiments are an active area of research, but currently ambiguous.

The second method uses rates of DNA sequence evolution. We begin with a region of DNA, such as a pseudogene, that evolves in a completely neutral manner. This DNA

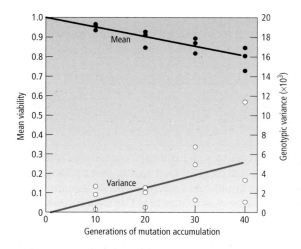

Figure 12.5

The mutational meltdown in fruitflies protected from selection, showing a decrease in viability. Viability is measured in flies that are homozygous for a chromosome that has experimentally accumulated mutations relative to flies that are heterozygous for the same chromosome. The decline is due to the accumulated deleterious mutations. There were 104 lines and the variance in viability among lines increased through time (see Box 9.1, p. 233, for the definition of variance). Redrawn, by permission of the publisher, from Mukai *et al.* (1972).

Table 12.1

Mutation rates in various life forms. More complex life forms have more cell cycles per generation and more DNA. The mutation rate per nucleotide is probably approxiamtely constant in all eukaryotes. The total number of errors, including harmful errors, increases from bacteria to human beings. Two methods of estimating the fraction of all mutations that are deleterious give different results, hence the columns labelled (1) and (2) for *U*. All numbers are approximate. After various sources; see Ridley (2001).

Creature	Mutation rate per nucleotide	DNA length	Cell cycles per generation	Total number of mutations	Number of deleterious mutations 1	2
Bacteria	10^{-9}–10^{-10}	10^6	1	≪1	≪1	≪1
Fruitfly	10^{-9}–10^{-10}	3.6×10^8	20	4	<1	~1
Human being	10^{-9}–10^{-10}	6.6×10^9	200	200	~2	~20

Sequence analysis gives estimates of U . . .

will evolve at a rate equal to the total mutation rate. We can extrapolate this figure to estimate the total mutation rate for the whole genome. Box 7.3 (p. 177) showed that the resulting figure is 200 or so in humans. This (high) number is uncontroversial but it is not the number we need. It is the total mutation rate, whereas we need the deleterious mutation rate. Most of the 200 mutations are probably neutral so we need to know what fraction are deleterious. Two estimates are available, and they are frustratingly inconsistent. Eyre-Walker & Keightley (1999) estimate that about 1% of mutations are deleterious whereas Shabalina *et al.* (2001) estimate more like 15% are.

The total mutation rate increases from bacteria to fruitflies, and from fruitflies to humans (Table 12.1). The mutation rate per nucleotide per copying event is approximately constant in all eukaryotes, but the genome size and total number of DNA replications per generation goes up. On both the high and low estimates of

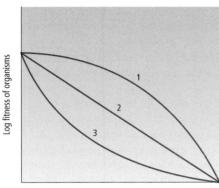

Number of mutations in organism

Figure 12.6
Three different relations between the fitness of an organism and the number of deleterious mutations it carries. The *y*-axis is logarithmic. 1, Synergistic epistasis: multiple mutations have an increasingly damaging effect on the organism. 2, Independent, or multiplicative, fitness effects. 3, Multiple mutations have a decreasingly damaging effect. Deleterious mutations are purged more (slope 1), equally (slope 2), and less (slope 3), efficiently with sexual, than asexual, reproduction. The relative positions of lines 1, 2, and 3 up the *y*-axis is irrelevant; only the slope matters. If slope 1 was drawn below 2 it would still be synergistic epistasis.

. . . that may or may not confirm the mutational theory of sex

deleterious mutation rates, U is less than one in bacteria and more than one in great apes, including humans. Kondrashov's theory correctly predicts the absence of sex in bacteria. The problematic area is around flies and worms. They reproduce sexually and Kondrashov's theory predicts $U > 1$. On the high estimate, the prediction is upheld; on the low estimate, it is falsified. Thus, further research is needed on the fraction of mutations that are deleterious. However, it is worth noting that U is > 1 in humans whether the correct figure is 2 or 30 deleterious mutations per generation. If Kondrashov's theory turns out to be wrong, and sex does not help the selective purge of deleterious mutations, we shall be left with a paradox — how can humans exist, given their high deleterious mutation rate?

Work on a second prediction is also inconclusive

The second prediction of Kondrashov's theory concerns the relation between the fitness of an organism and the number of deleterious mutations it contains. Three sorts of relation are theoretically possible (Figure 12.6). Kondrashov's theory only works if the graph slopes down — a condition called *synergistic epistasis*. Experimenters are also trying to test this prediction, but no conclusive results are yet available.

In conclusion, the mutational theory suggests that sex exists to help life cope with its load of deleterious mutations. The theory has been worked out, and is internally consistent. It makes two predictions about real sexual creatures: they should have deleterious mutation rates of one or more, and their fitness relations should show synergistic epitasis. These predictions have inspired a major research programme — one of the most active and important in modern evolutionary biology — but it is currently inconclusive. The next few years of work should tell us whether U is > 1 in fruitflies, but we do not yet know.

12.2.3 Coevolution of parasites and hosts may produce rapid environmental change

The second theory we shall look at ignores the effect of deleterious mutations and concentrates on external environmental change. Sex is more likely to be advantageous if environments change rapidly: the problem is to work out *how* environments could

Sex may be advantageous in
changeable environments . . .

possibly be changing rapidly enough. It is not difficult to believe that environments might change fast enough to make sex advantageous every few hundred years, but how could they be changing fast enough to make it advantageous every generation? Remember, the environment would have to be changing so rapidly that an average sexual female's daughters must be twice as fit as those of an average asexual female. We cannot take it for granted that ordinary environmental change will be enough. If we are to explain the existence of sex by environmental change, we have some work to do.

One promising suggestion is that the coevolution between parasites and hosts may generate fast enough environmental change to make sex advantageous in the short term. The "environment" here, for the parasite, is the host's resistance mechanism and, for the host, the parasite's method of penetrating its defenses. Several authors have sug-

. . . such as in parasite–host
coevolution

gested that *parasite–host coevolution* may be important in the maintenance of sex, and Hamilton is the best known of them.

The theory can be made more exact by a simple model. Some parasite–host relationships have gene-for-gene matching systems such that one host genotype is adapted for resisting one parasite genotype, another host genotype for another parasite genotype, and so on. The best understood example is from wheat and parasitic rusts, and similar selection may operate in the human HLA system (Section 8.6, p. 203).

The simplest genetic model for host–parasite coevolution is haploid, with two alleles in each of the host and parasite species. One parasite allele is adapted to penetrate hosts with one of the host alleles, the other parasite allele penetrates the other (Table 12.2).

Table 12.2
A simple model of gene-for-gene matching in a pair of host and parasite species. The numbers in the table are the fitnesses of the genotypes.

(a) Fitness of parasite genotype in two types of host.

	Host genotype	
	H_1	H_2
Parasite genotype		
P_1	0.9	1
P_2	1	0.9

(b) Fitness of host genotype against two types of parasite.

	Parasite genotype	
	P_1	P_2
Host genotype		
H_1	1	0.9
H_2	0.9	1

Figure 12.7

Frequency changes of host and parasite genotypes. (a) As H_2 becomes commoner, there is selection to increase the frequency of P_2, which in turn selects against H_2, and H_1 increases in frequency. (b) Plotted against time, the frequency of each genotype oscillates cyclically.

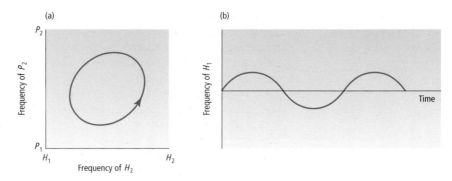

The theory predicts cyclical changes in gene associations

Selection of this sort generates cyclic changes in gene frequency (Figure 12.7). As a genotype increases in frequency, its fitness (after a time lag) decreases. If parasite genotype P_1 is commoner, host genotype H_1 will be favored and will increase in frequency; the fitness of P_1 then goes down as more hosts are resistant to it. Then, as P_2 becomes commoner, the fitness of H_1 decreases. When H_1 becomes rarer, the frequency of P_1 will in turn increase again. Cycles of gene frequency are driven by corresponding cycles of gene fitness.

We need a more complex, and probably more realistic, model to produce an advantage for sex. Imagine now that resistance and counterresistance are controlled by two loci. Again, a haploid model is simplest. With two loci and two alleles at each, there are four haplotypes, AB, Ab, aB, and ab. There will be complimentary sets in the host and parasite; if $A_H B_H$, $A_H b_H$, $a_H B_H$, and $a_H b_H$ are the host genotypes, then we could write the parasite genotypes as $A_P B_P$, $A_P b_P$, $a_P B_P$, and $a_P b_P$. $A_H B_H$ and $A_P B_P$ are analogous to H_1 and P_1 in the previous model. As a concrete example, A_H and B_H might control two cell surface molecules used by the parasite to penetrate the host. Hosts with allele A_H are efficiently penetrated by parasites with a_P, but not A_P; B_H hosts are penetrated by b_P parasites but not B_P. b_P parasites are therefore favored if the hosts are mainly B_H; likewise, b_H is a gene for resistance to b_P parasites. Haplotype frequencies at both loci will oscillate for the same reasons as did H_1 and P_1 in the simpler model. In the sexual parasites and the sexual hosts, alleles at the two loci can recombine, whereas in asexual parasites and hosts, they cannot. A third locus determines whether reproduction is sexual or asexual.

How can sex be advantageous? As the frequencies of the four haplotypes oscillate through time, there will be some chance that any one of them will be lost at the trough of its frequency cycle. Suppose, for example, that the frequency of $A_H b_H$ is driven so low in one cycle that it is lost from both the asexual and sexual populations. In the asexual population it has been lost forever, whereas in the sexual population it will be recreated by recombination between the other three genotypes. As the frequency of the parasites that specialize in attacking $A_H b_H$ hosts increase again, sexual reproduction will be an advantage as it is more often associated with the resistant genotype. Thus sex has an advantage because it maintains in reserve an ability to recreate multilocus genotypes that have been disadvantageous, but may be needed again. The cycles of host–parasite coevolution are exactly the kind of circumstances in which this ability is favored. If the

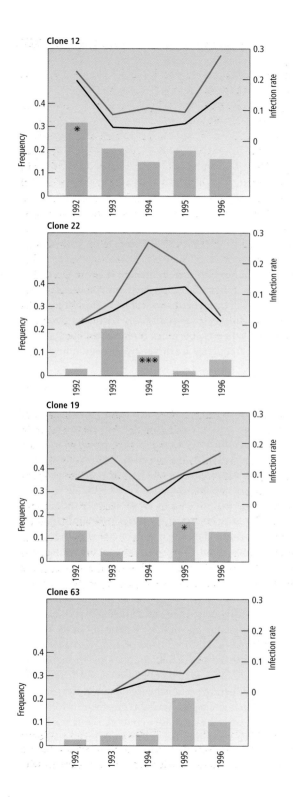

Figure 12.8

Genetic cycles in snails, possibly driven by parasite–host coevolution. Data are shown for the four commonest clones of the snail *Potamopyrgus antipodarum* in a lake in New Zealand. For each, the histogram at the bottom of each part shows the frequency of the clones over time. Note that all four undergo frequency cycles, but not synchronously: clone 12, for instance, peaks in 1992 and clone 63 in 1995. The line graphs at the top of each part shows the infection rate of each clone: that is, the fraction of snails of that clone that are parasitized. The total infection rate, by all parasites, is the green line. The infection rate due to *Microphallus* is the black line. Note that the infection rate tends to peak 1 year after the clone's frequency peak. Clone 22, for example, peaked in 1993 and its peak of infection was in 1994. Clones 19 and 63 show the same pattern, but clone 12 is exceptional. Asterisks indicate that the clone is significantly more infected by *Microphallus* than would be expected by chance (*, $P < 0.05$; ***, $P < 0.0001$). From Dybdahl & Lively (1998).

environmental change is more erratic, or open ended, such that once a genotype has been eliminated it is unlikely to be useful again, sex is not advantageous.

The parasitic theory has not inspired such a large research program as the mutational theory. The parasitic theory is plauible, not least because parasite–host coevolution is widespread in life. But we do not know whether the specific prediction of the theory — cycles in the associations between resistance genes in hosts — is correct. The most direct test so far is by Dybdahl & Lively (1998), which is part of a long-term study of sex in New Zealand snails by Curt Lively, where several genetic clones of the aquatic snail *Potamopyrgus antipodarum* have been distinguished. The snail exists in both sexual and asexual forms. The main parasite of the snails in a trematode (*Microphallus*) which, as its name suggests, is a parasitic castrator.

Genetic cycles in snails are consistent with the theory

The clones underwent frequency cycles in the 1990s (Figure 12.8). Moreover, Dybdahl and Lively showed experimentally that the parasites were best able to infect the snail clone that had the highest frequency in the previous year. This suggests that the parasites are adapting to penetrate the commonest host genotypes. (In Section 5.13, p. 127, we saw evidence for frequency-dependent selection in this system.) The results are all consistent with the parasite theory of sex. However, further work is needed to show that the genetic cycles are of the right type to explain the existence of sex.

12.3 Conclusion: it is uncertain how sex is adaptive

Both deleterious mutation and parasite–host coevolution are reasonable theories of sex, but it has not been conclusively shown for either of them that they really maintain sex in nature. They are afloat in evolutionary biology today as stimulating hypotheses that are inspiring much research. They are not mutually exclusive ideas, and both factors could turn out to be contributing to the selective advantage of sex. Other hypotheses exist too, and some of them are highly ingenious.

Today, the question of why sex exists remains an "outstanding puzzle." Evolutionary biologists are not confident the question has been satisfactorily answered. Maybe we need some radically new idea that has not yet been put forward or lies unappreciated. Alternatively, the gist of the answer may lie in the theories we have discussed and the problem is more one of showing how they apply in nature. Whatever the answer turns out to be, it is likely to tell us something about the safety, or otherwise, of cloning technology (Box 12.1).

12.4 The theory of sexual selection explains many differences between males and females

12.4.1 *Sexual characters are often apparently deleterious*

For the most part, the characters of organisms are adaptive: they increase the organisms' chances of surviving to reproduce. However, there are some characters that do the opposite, and (as Darwin was well aware) natural selection does not explain why these characters exist. If a population contains some types with higher survival than other types, natural selection will fix the former and eliminate the latter.

Characters that reduce survival can be called "deleterious" or "costly." One large class of apparently costly characters are those found usually only in males and which Darwin called secondary sexual characters. The primary sexual characters are things like genitalia that are needed for breeding. The secondary sexual characters are not actually needed for breeding, but they function during reproduction. The peacock's "tail" (or, more exactly, train) is an example. In many other bird species too, the males have tails or other extravagantly developed and brightly colored structures. A peacock could inseminate a female just as well without his remarkable tail, and in that sense it is a secondary, not a primary, sexual organ. The peacock's tail almost certainly reduces the male's survival (though this disadvantage has never actually been demonstrated) as the tail reduces maneuverability, powers of flight, and makes the bird more conspicuous; its growth must also impose an energetic cost. Why are these costly characters not eliminated by selection?

The peacock's tail is an example of a costly sexual character

12.4.2 *Sexual selection acts by male competition and female choice*

Darwin's solution was his theory of sexual selection. He defined the process by saying that it "depends on the advantage which certain individuals have over other individuals of the same sex and species, in exclusive relation to reproduction." A structure produced by sexual selection in males exists not because of the struggle for existence, but because it gives the males that possess it an advantage over other males in the competition for mates. Darwin's idea is that the reduced survival of peacocks with long, colorful tails is more than compensated by their increased "advantage in reproduction."

Sexual selection works mainly by . . .

Darwin discussed two kinds of sexual selection. One is for males to compete among each other for access to females. *Male competition* can take the form of direct fighting, or it can be more subtle. Some male insects, for instance, can remove sperm from females they are copulating with — sperm that was stored from matings with previous males. However, we shall not discuss adaptations of sperm competition here, or other adaptations of male competition, because they do not pose deep theoretical questions. The situation is different for Darwin's other mechanism: *female choice*.

. . . male competition . . .

A structure like the peacock's tail cannot plausibly be explained by male competition. It would be no use in fighting — indeed it would reduce the male's fighting power — and no one has ever thought up a more subtle competitive function for the tail. Darwin suggested that the tail exists instead because females preferentially mate with males that have longer, brighter, or more beautiful tails. If they do, the mating advantage of males with longer tails will compensate a corresponding amount of reduced male survival.

. . . and female choice

Darwin's main argument for the importance of sexual selection was comparative. Sexual selection should operate more powerfully in polygamous than in monogamous species. In a polygynous species, in which several females mate with one male (and other males do not breed at all), a single male can potentially breed with more females than under monogamy; selection in favor of adaptations that enable males to gain access to females (whether by male competition or female choice) is proportionally stronger. Darwin therefore reasoned that secondary sexual characters would be more developed in polygynous, than monogamous, species. Polygynous species should have stronger *sexual dimorphism*.

Darwin's book *The Descent of Man, and Selection in Relation to Sex* (1871) contains a long review of sexual dimorphism in the animal kingdom. It is still the best (and classic) demonstration that sexual dimorphism is indeed mainly found in polygynous species. In polyandrous birds, such as phalaropes, sexual selection is reversed: females compete for males, and it is the females that are the larger and more brightly colored sex. There are exceptions, such as monogamous ducks that are sexually dimorphic; Darwin had an additional theory for them. However, the main point is that Darwin's principal evidence for sexual selection came from a comparison of large numbers of species that showed that species with brightly colored, large, or dangerously armed males are more often polygynous and species in which males and females are more similar are more often monogamous.

Darwin provided comparative evidence for sexual selection

12.4.3 *Females may choose to pair with particular males*

For Darwin, female choice among males was an assumption; he was mainly concerned to show that, if it exists, it can explain extraordinary phenomena like the peacock's tail. He did not have much to say about the prior question of why the female preference should ever evolve to begin with. Selection can work on a female preference just like on any other character. If females with one type of preference produce more offspring than females with another, selection will favor the more productive preference. The difficult case is in an extreme case like the peacock, in which the form of female choice appears to be disadvantageous to the female. Females are picking males that possess a costly character that will be passed on to their sons; the female preference therefore seems to be causing the females to produce inferior sons.

The advantage to a female of mating with a male that carries a costly character is not obvious

We can spell the problem out more fully in terms of selection on a mutant, non-choosy female. Suppose that peahens do prefer peacocks with dazzling tails, and a mutant female, who does not prefer these males, arises; she might mate at random, or prefer some other sort of male. What does selection do to this mutation? The mutant female will produce sons that do not possess the costly character, or at least in a less extreme form. Her sons will therefore survive better than average. So the mutant should be favored, the female preference should be lost, and the extreme male forms should disappear.

Or should it? The mutant female will indeed produce sons that survive better than the population's average. But that, as Fisher (1930) first realized, is not enough to guarantee that the mutation will spread. When the mutant female's sons grow up, with their inferior tails, they will be rejected as mates. The mutant female is a rare mutant, in a population where the majority of females prefer males with long tails, and this majority preference will work against the mutant's sons. Despite their superior survival, they will be condemned to celibacy. The randomly mating mutation, therefore, may not spread.

But Fisher suggested an answer

Fisher also discussed how the preference for a costly character could evolve to begin with. After the long male tail has evolved it is costly, but at an earlier evolutionary stage,

Figure 12.9
(a) The early stage in the evolution of a bizarre character such as the peacock's tail. Before females preferred to mate with long-tailed males, there might have been a positive correlation between tail length (then much shorter than in their descendants) and male fitness. (b) Full relation between degree of exaggeration of character (tail length) and survivorship. There is an intermediate optimum. Modern species like the peacock occur toward the right of the graph.

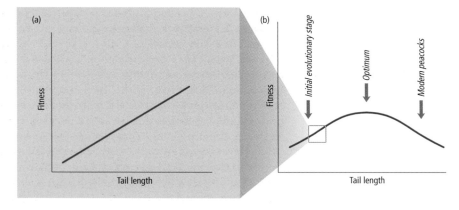

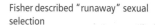

Fisher described "runaway" sexual selection

before the female preference arose, things might have been different. Male tails would have been shorter then. Suppose that, before some mutant female arose who picked long-tailed males, most females picked their mates at random; suppose also that there was at that time a positive correlation between male tail length and survival (Figure 12.9a). Selection would then favor a mutant female with a preference for males with longer tails as she would produce sons with longer than average tails, with an associated higher survival. Then, as the mutation spread, the males with longer tails would start to acquire a second advantage. There are increasing numbers of females in the population who prefer to mate with longer tailed males, and the males so endowed will not only survive better but also enjoy an advantage in mating. The evolution of longer tails in males, and a mating preference for them in females, thus come to reinforce each other, in what Fisher called a *runaway* process.

Technically, they reinforce each other because the genes encoding them are in linkage disequilibrium (Section 8.5, p. 199). The offspring of a female who mates preferentially with longer tailed males will possess both their mother's genes for choice and their father's genes for long tails. These two kinds of genes thus become non-randomly associated, and the genes for female choice increase in frequency by hitch-hiking with the advantageous genes for long tails in males.

If we consider a sufficiently wide range of tail lengths, the full relation between it and male survival presumably shows some increase and decrease on either side of an optimum (Figure 12.9b). Eventually, powered by female choice, the average tail length in the population will reach the optimum; but evolution does not stop there. As the population evolves towards the optimal tail length, the longer tailed males are still preferred. By now the female preference will have spread through the population and the majority of females will prefer longer tailed mates. Now the mating preference alone drives the evolution of longer tails. The preference may have become strong enough to compensate lower male survival, and evolution will proceed into the interesting zone in which the male character, in a complete reversal of the original selective forces, evolves to become increasingly costly to its bearers.

The evolution of the long tail therefore proceeds through three stages. Long tails initially have only a survival advantage. Then the survival advantage is supplemented by a mating advantage. As female choice grows commoner and the tail length grows past the optimum for survival, the relative importance of the two advantages shifts over, until we reach a third stage at which further elongation is driven purely by female choice.

As the population evolves past the point of optimum tail length, the selective forces at work have become almost absurd. The runaway process will only come to a stop when the death rate of males, due to their feathery excess, is so high that their success in mating no longer makes up for it. The tail length will then reach an equilibrium. That equilibrium, according to Fisher, is what we are now observing in birds like peacocks and birds of paradise.

The original problem was to explain the evolution of a set of apparently deleterious characters. Darwin's solution was that they could be maintained by female choice. He did not, however, explain why females should come to choose males with deleterious characters, nor why the choice would not be lost by natural selection. In Fisher's theory, when the choice first evolved, the male character was much smaller and choice then favored males with higher survival. Genes for choice could thus increase in frequency from being rare mutants to being the majority form in the population. Once nearly all the females in a population choose mates in a certain way, mutant females that pick some other sort of male are selected against (because of the effect on the kind of sons they produce). The cost of the male character at the final equilibrium serves no function for the female. The male character is maintained by female choice but is the useless end product of an initially useful process.

12.4.4 *Females may prefer to pair with handicapped males, because the male's survival indicates his high quality*

The "handicap" theory is a second evolutionary theory of female choice

We now turn to a second theory, in which the costliness of the male character is positively useful to the female, called the *handicap theory* (Zahavi 1975). ("Handicap" is Zahavi's term for what we have been calling a costly or deleterious character that reduces survival.) The argument runs like this. Suppose that the males in the population vary in their quality. We shall concentrate on species, like peacocks, in which males contribute only sperm; in them, quality must mean genetic quality, because nothing else is transferred. We are thus assuming that some males have genes that confer higher fitness ("good genes") than do other males (who have "bad genes"). In practice there could be all degrees between good and bad, but the point can be explained more easily in the simple dichotomous case.

If a female mates at random, her mates will have good and bad genes in the same proportions as the good and bad genes have in the whole population; if half the males in the population have good genes and half have bad, then 50% of her mates will have good genes and 50% bad. Now suppose that some of the males in the population possess a handicap or character that reduces their survival. If only males with good genes can survive possessing this handicap, then a female who mates preferentially with such handicapped males will only mate with males with good genes (Table 12.3). The choice will be favored by selection if the advantage through the superior genes outweighs the cost of the handicap: then the net quality of the choosy female's offspring will be higher than those of the randomly mating female.

Table 12.3

The handicap principle. If only males with good genes can survive the possession of a handicap, females who mate with handicapped males will mate only with males who possess good genes. A female who mates with males lacking a handicap will mate with males possessing good and bad genes in their population proportions.

	Males with bad genes	Males with good genes
No handicap	Alive	Alive ⎤ in population
Handicap	Dead	Alive ⎦ proportions

The costliness of a signal makes it reliable

The handicap acts as an indicator of genetic quality. But why does the indicator have to be costly? The reason is that the cost guarantees that the indicator will be reliable. A male's genetic quality does not come written on him: it has to be inferred, and if females inferred it from an an inexpensive signal, there would be selection on males to cheat. If females preferentially mated with males who merely said "I have good genes" (or rather, in a non-human species, something analogous to saying this) and rejected those that said "I have poor genes," mutant males who said the former independently of their true genetic quality would be favored. Words (and their analogs) are cheap. But if the criterion favored by females is costly, as growing a long and ostentatious tail is, then selection will less automatically favor cheats. In particular, if the cost of growing a handicap is less for a truly high quality male than for a low quality male, handicaps will be grown only by high quality males and will be reliable signals for females to use. (This condition was met in the simple example in Table 12.3: the cost of the handicap for the males with bad genes was far higher than for males with good genes.)

So the reason for the costliness of the male character is completely different in Fisher and Zahavi's theories. In Fisher's theory, the cost arose as the end product of a runaway process. To begin with, long tails were not costly, but as an open-ended female preference for males with longer tails was selected into the population, the tails evolved past their optimum and ended up reducing the survival of their bearers. In Zahavi's theory, the male character had to be costly from the start, and to remain costly as the female preference spreads. The function of the chosen male character is to indicate genetic quality at other loci, and it has to be costly in order to be reliable.

12.4.5 *Female choice in most models of Fisher's and Zahavi's theories is open ended, and this condition can be tested*

There are two crucial means by which Fisher and Zahavi's ideas can be tested. The first concerns the exact kind of female preference that they require. The preference is open-ended. We can distinguish between absolute preferences, which are of the form "mate preferentially with males whose tails are 12 in long," and open-ended preferences, such as "mate preferentially with the male who has the longest tail you can find."

In Fisher's theory, at the initial stage when the male character was positively corre- lated with survival, either an absolute or an open-ended preference could be favored. The average tail length might then have been 2 in (5 cm), and the longest tails in the population might have been 12 in (30 cm). If the mutant that was selected happened to be one encoding an absolute preference for males with 12 in tails, then evolution would proceed until the average tail length was 12 in and then come to a stop. Only if the preference is open-ended can there be an equilibrium with a costly male character. At equilibrium, the lower survival of males with longer than average tails has to be com- pensated by a higher frequency of mating. It is not enough for the females to prefer average males, or to mate at random. The females must actively prefer males with longer than average tails.

Likewise, in the handicap theory, females must prefer males with the most costly handicaps. If the greater cost paid by the higher quality males is not compensated by higher mating success, a less costly handicap will evolve. Therefore, in both theories, the female choice must be open-ended in a species with a costly male ornament. If we find evidence for such preferences, it suggests the male character is indeed maintained by female choice. However, it does not tell us whether Fisher's runaway theory or Zahavi's handicap theory are at work.

The prediction has been tested in more than one species. One example to illustrate the procedure is Møller's (1994) study of barn swallows (*Hirundo rustica*). The two sexes are similar in the swallow, except for the outermost tail feather which is about 16% longer in the male than in the female. Møller tested whether females open-eddedly prefer males with longer tails by experimentally shortening the tails of some males, by cutting them off with a pair of scissors, and elongating the tails of others, by sticking those severed tail feathers on to other intact males with superglue that hardened in less than 1 second. He then measured how long it took the different males to find a mate. Males with elongated tails mated faster (Figure 12.10a), resulting in higher repro- ductive success (Figure 12.10b).

Møller also confirmed that the male character is costly. Swallows molt in the fall and grow a new tail for the following breeding season. A male's new tail is on average about 0.2 in (5 mm) longer than in the previous year, but the males whose tails were elongated grew a tail in the following year that was shorter than before the experimental treat- ment (Figure 12.10c). (Møller did not tamper with their tails in the year after the experiment.) Those males had enjoyed a good year during the experiment, but the extra effort of flying with an elongated tail exacted a physiological cost. Next year the cost was paid: the males took longer to find a mate and their reproductive success decreased. In summary, Møller has shown that the sexually dimorphic tail feathers of swallows are maintained by female choice, that the choice is open-ended, and that the character chosen is costly.

12.4.6 *Fisher's theory requires heritable variation in the male character, and Zahavi's theory requires heritable variation in fitness*

In Fisher's runaway theory, the reason why females choose males with long tails at the final equilibrium point is that a mutant female who mated at random would have lower

Figure 12.10

Barn swallows with longer tails are preferred by females, but the character is costly. Møller experimentally shortened some males' tails and elongated others; as one control he cut the males' tails off and then immediately stuck them back on again (control 1) and as another he left the males untreated (control 2). (a) Males with elongated tails obtain mates more quickly, (b) have higher reproductive success, but (c) next year grow a shorter tail while the other males grow a longer tail. (Møller also measured both the mating advantage and the cost of longer tails by other criteria too, and those results support the results illustrated here.) (0.25 in ≈ 6 mm.) Redrawn, by permission, from Møller (1994). © 1994 Macmillan Magazines Ltd.

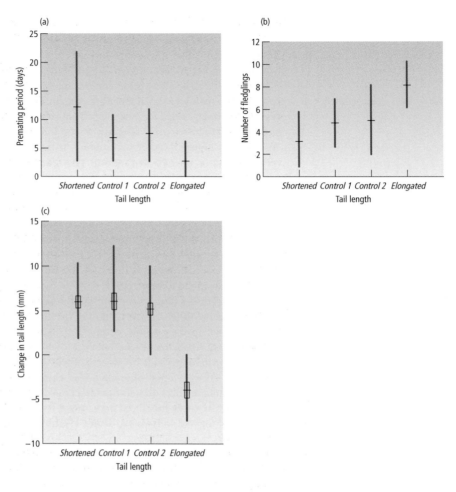

Theory predicts the male character is heritable

fitness because her sons would have shorter tails and be rejected as mates. This is only true if male tail length is heritable. If all the variation in tail length were environmental, and its heritability were zero (Section 9.6, p. 235), the tail length of the mutant female's sons would be no shorter on average than those of the choosy females. If mate choice imposed any cost on a female at all, the randomly mating mutant would spread. Tail length, therefore, must be heritable or selection will favor the female who mates at random. This condition is testable, but has never been tested in a species with a costly and extravagant character like the peacock's tail.

In Zahavi's theory, the advantage of female choice does not depend on the inheritance of the male character. Choice could be maintained even if all members of the population had the same genes for tail length. But the theory has an analogous condition. In species in which males transfer only sperm, female choice is for male genetic quality. There must be variation in male genetic quality: some males must have good genes, others bad. This is a condition called *heritability of fitness*. Heritability of fitness means that individuals of higher than average fitness (that is, they produce more offspring than average) produce offspring who also have higher than average fitness. If high quality males do not produce high quality offspring, there is no point in picking them as mates.

But selection may have removed heritable variation

The conditions in the two theories face a common difficulty. While selection operates on any character, it reduces its heritability (see, for example, the Illinois maize experiment, Table 9.2, p. 239, and Figure 9.7, p. 238). In a population in which some individuals possess good genes and others bad genes, selection acts to fix the good genes — and once it has done so there will be no variation in genetic quality left. Zahavi's theory then would not work.

In fact, the amount of variation in genetic quality for species in nature is unknown, and no firm conclusion can be drawn. However, there are three arguments to be aware of. One is the possibility just noted, that Zahavi's theory may not work in species in which males transfer only sperm because there is not enough variation in genetic quality.

Other processes may recreate variation

Alternatively, enough genetic variation may exist for the handicap process to operate. The variation in genetic quality is likely to exist because of one or other of the two factors we looked at earlier in the chapter: deleterious mutation and host–parasite coevolution. At any one locus, mutation contributes little genetic variation. But if the highest estimates of the genomic deleterious mutation rate (see Table 12.1) turn out to be correct, then a substantial amount of variation in genetic quality will exist. Females, by picking males with long tails, may be picking mates with relatively few bad genes. Likewise, if the cyclic gene association proposed in the parasitic theory are correct (see Figure 12.7), a population will contain variation in genetic quality for this reason. Females, by picking males with long tails, may be picking mates with good parasite-resistance genes; may be only healthy males are able to grow long tails.

Some evidence exists that female choice is influenced by genetic quality. For instance, Welch *et al.* (1998) did a particularly clear experiment with gray tree frogs (*Hyla versicolor*) in Missouri. Welch *et al.* fertilized half a female's eggs with sperm of preferred males, and half with sperm of unpreferred males. The offspring of the preferred males had higher fitness. Research at present aims to test how widespread these female preferences for good genes are, and what the explanation is for the differences in genetic quality between males.

12.4.7 Natural selection may work in conflicting ways on males and females

The evolution of sex opened up the possibility of future evolutionary conflicts between males and females. Consider an animal such as a fruitfly, in which females usually mate with a number of different males over a period of time. The female stores sperm after each mating. She produces eggs steadily, at a certain rate, over time, and draws on her store of sperm to fertilize the eggs as she lays them. One male, therefore, fertilizes most of the eggs that a female lays between when he mates with her and when the next male mates with her.

Several selective forces will be at work in male and female fruitflies. For instance, selection favors a male who can accelerate a female's egg production immediately after

A hypothesis about conflicting selection pressures on males and females . . .

mating with him, because that male will then fertilize more eggs. Male fruitflies seem to transfer chemicals with their sperm that act as hormones in the female and accelerate egg production. The accelerated egg production may not be in the interest of the female. Her optimal rate of egg production will be some trade-off between her survival and reproduction. If she produces extra eggs now, it will be by allocating less energy to maintaining her body. Survival will decrease, and her total lifetime output of eggs will also decrease. The male gains extra eggs in the short term, at the cost to the female of reduced lifetime fitness. The cost is not paid by the male, because the later eggs that a female loses because she dies younger would have been fertilized by another male. Natural selection on females favor resistance to the male techniques of accelerating egg production. Females may evolve counterhormones or other methods of restoring the optimal egg production rate.

The selective forces differ if there is lifetime monogamy. Now the "interests" of the male and female are identical. If a male causes a female to accelerate her short-term egg production, but reduces her lifetime fitness, his own fitness will go down by the same amount.

Holland & Rice (1999) tested this reasoning experimentally with fruitflies. Fruitflies usually mate with several members of the opposite gender. Holland and Rice allowed some fruitflies to mate normally, as controls. In their experimental lines, they imposed monogamy by selecting at random one individual to be the only mate of another individual (of the opposite gender). They bred lines of these experimentally monogamous fruitflies for 47 generations.

. . . is supported by ingenious experimental results

As predicted, the male fruitflies in the monogamous lines evolved to be less harmful to their mates and females evolved to be less resistant to males. Figure 12.11 shows some of Holland and Rice's results. After 47 generations of monogamy, males had evolved lower rates of courting (Figure 12.11a). Also, the total reproduction output per female increased under monogamy (Figure 12.11b). This result suggests that the conflict between males and females in normal fruitflies is reducing the fitness of the average individual by 20% or so. The main interest of the results is to illustrate the evolutioanry theory of intersexual conflict, but the results have other interests too. Intersexual conflict is a further factor that can be added to the list of causes of adaptive imperfection in Chapter 10. Sexual selection, including intersexual conflict, may also underlie the relatively rapid evolution of genes that are expressed in the reproductive system — a phenomenon noted in Sections 7.8.2 (p. 182) and 14.12 (p. 417).

12.4.8 *Conclusion: the theory of sex differences is well worked out but incompletely tested*

The theory of sexual selection is at a more advanced stage than the theory of why sex exists. The models, such as those of Fisher and Zahavi, may be correct, and some work has been done to test them. The tests, however, are at an early stage. There are several pieces of evidence for open-ended female choice in species with extravagant, costly male characters. This suggests Darwin was right to explain those characters by female choice. But less work has been done on the other crucial theoretical variable: the inheritance of genetic quality.

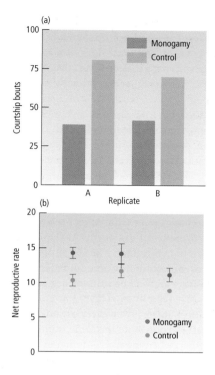

Figure 12.11

Experimentally imposed monogamy causes the evolution of reduced reproductive conflict in fruitflies. (a) Individual males were put with individual females and the amount of courtship behavior was measured. Courtship rates were reduced in the monogamous fruitflies. Results are shown for two replicate lines (A and B) of males sampled from the experimental line after 45 generations of monogamy and from a control line. (b) The total reproductive output per female increased (this was measured as the number of mature progeny per female). It is shown here for the experimental and control lines for the final three generations of the 47-generation experiment. Redrawn, by permission of the publisher, from Holland & Rice (1999).

Many other consequences of sexual selection are also being investigated. One hot topic is the experimental study of intersexual conflict. The evolutionary forces of intersexual conflict depend on the mating system. By experimentally altering the mating system from polygamy to monogamy, for example, it is possible to produce a predictable reduction in male–female conflict over evolutionary time.

12.5 The sex ratio is a well understood adaptation

12.5.1 *Natural selection usually favors a 50 : 50 sex ratio*

The sex ratio is one of the most successfully understood adaptations. The main idea is again due to Fisher. In most species, the sex ratio at the zygote stage is about 50 : 50. Fisher explained the 50 : 50 sex ratio as an equilibrium point: if a population ever comes to deviate from it, natural selection will drive it back.

Group selection favors a female-biased sex ratio

At first glance, the 50 : 50 sex ratio might seem inefficient. Most species do not have parental care and are not monogamous, meaning one male can fertilize several females. It would be more efficient for the species to produce more females than males. The extra males are not needed to fertilize the females of the species and do not increase its reproductive rate. (This is another "group selection" argument, see Section 11.2.5, p. 301.) However, imagine what would happen to a population with a persistently female-biased sex ratio — one with four females for every one male, for instance. Each male in the population will fertilize on average four females. This condition could not be stable for long in evolution, because an average male is producing four times as many off-spring as an average female. There is an advantage to being a male, and an advantage to a female who produces extra sons — because sons have a higher reproductive success than daughters.

Individual selection favors a 50 : 50 sex ratio . . .

If a mutant female arose who produced only sons, the total reproductive success of her offspring would be 20/8 times that of an average female (the mutant produces five males, each with a relative reproductive success of 4, for every one male and four females produced by the average female). The mutant would spread. As it spread, the population sex ratio would become less and less female biased. The same argument works in reverse for a population with a male-biased sex ratio. The reproductive suc-cess of the average female is then higher than that of a male, and natural selection will favor mutant females that produce more daughters than sons. Only when the sex ratio is equal are the relative reproductive successes of the two sexes equal. At that point there is no advantage in producing more of one sex than the other. The 50 : 50 sex ratio is the equilibrium that the population moves to, over evolutionary time, and then stays at. Any population that deviates from the 50 : 50 sex ratio will be shifted back to it by natural selection.

The fundamental reason why the 50 : 50 sex ratio is stable is that every organism has one father and one mother. All the females together contribute the same number of genes to the next generation as all the males together; when members of one sex are in short supply, their average success must increase.

To be exact, Fisher's theory predicts a 50 : 50 ratio of investment by the parents in male and female offspring. This usually translates into a 50 : 50 sex ratio in the zygotes. The sex ratio among adults may be biased away from 50 : 50 if males have higher or lower mortality than females. This does not mean that selection favors any compensat-ing bias at the early stages to produce more of the high-mortality gender.

. . . unaffected by sex differences in adult mortality

Suppose, for instance, that males have higher mortality than females on average. The adult sex ratio will be female biased. Among the individuals that do survive to repro-duce, the average male will have a higher reproductive success than the average female. But as far as a mother is concerned, the extra reproductive success of her surviving sons exactly balances the zero reproductive success of those that die before reproducing. When she produces a son, she cannot "know" in advance whether he will be a survivor who will have higher than average success, or die and not reproduce at all. At birth, a male can only be expected to have the success of an average male, and the average male has the same reproductive success as the average female. The offspring of any one par-ent will suffer any sex differences in mortality in the same proportion as the population as a whole and there is nothing to be gained from producing more of the sex that will (in the adult stage) be in the minority. Any parent who did so would simply increase the average mortality rate among her progeny.

However, Fisher's argument makes various assumptions. When the assumptions are altered, the predicted sex ratio is altered too. In the past 30 years or so, biologists have used Fisher's basic theory to test for biased sex ratios in many peculiar circumstances. In some cases, the predictions are quantitative and the tests experimental. The sex ratio has proved to be a remarkably fertile testing ground for theories of adaptation, and the sex ratio is among the best understood adaptations in life. Here is one example to illustrate this area of research.

12.5.2 Sex ratios may be biased when either sons or daughters disproportionately act as "helpers at the nest"

We looked in Section 11.2.4 (p. 298) at "helpers at the nest." In some bird species, some offspring remain at their parents' nest after fledging. These offspring do not themselves breed, but help their parents to rear the next brood of offspring. In some cases, mainly male offspring act as helpers; in others, mainly female offspring do. Natural selection can then favour a sex ratio other than 50 : 50, depending on the exact circumstances.

The Seychelles warbler (*Acrocephalus sechellensis*) is a bird that lives in the Seychelles Islands, an archipelago 60 miles (100 km) or so north of Madagascar. Helpers at the nest are seen in this species, and the helpers are mainly daughters of the reproducing pair (88% of helpers are female). Sons mainly disperse to other territories after fledging. Komdeur (1996) found that helpers had opposite effects on the reproductive success of the nest, depending on the quality of the territory. (Komdeur measured territory quality by counting samples of insects of the sort eaten by the warblers.)

On territories of high quality, the presence of 1–2 helpers increases the reproductive success of the nest. But on territories of low quality, the presence of any number of helpers decreases the reproductive success of the nest. The reason is likely that food is in short supply and the food consumption of the helper herself reduces the food available for the breeding pair and the nestlings. Even on a high quality territory, too many helpers reduces nest success — nests with three or more helpers had a lower success than if the helpers were absent. Again, the reason is probably competition for food. In technical language, the two factors are called local resource competition and local resource enhancement: the former refers to the case in which one gender of offspring decreases parental reproductive success, and the latter to any case in which one gender of offspring improves the local resources, for instance by bringing food to the young, and increases parental reproductive success.

When helpers are net beneficial to the parents, natural selection favors parents who produce more of the helping sex (daughters in the Seychelles warbler). When helpers are net disadvantageous to the parents, natural selection favors the production of more of the non-helping sex (sons in the Seychelles warbler). Komdeur (1996) found that these predictions were found in reality. More sons were produced on low quality territory and more daughters on high quality territory (Table 12.4a).

Komdeur also tested the theory by a translocation experiment. Certain pairs on either low or high quality territories were moved to other islands, to territories of high quality. The control pairs (moved from one high quality territory to another) continued to produce extra daughters. But the experimental pairs moved from high to low quality territories dramatically shifted from daughter production to son production (Table 12.4b).

Komdeur did further experiments, and they provide further support for the theory. However, the results of Table 12.4 are enough to illustrate the kind of evidence available, although they do raise further questions. For instance, what is the mechanism by

Theory predicts deviations from 50 : 50 sex ratios in certain special cases

Seychelles warblers produce more sons in some conditions, and more daughters in others

Table 12.4

The remarkable adjustable sex ratio of the Seychelles warbler. (a) The sex ratio depends on territory quality (based on 118 nests in 3 years). (b) Parents adjust their offspring sex ratio after experimental translocation from low to high quality territories. Control pairs did not adjust their sex ratio after translocation between similar quality territories. The sex ratios in (a) and (b) were measured both molecularly, in eggs, and in nestlings. Territory quality was measured by insect sampling. Simplified from data in Komdeur (1996).

(a) The sex ratio and territory quality.

Territory quality	Number of sons	Number of daughters	Percent sons
Low	44	13	77
Medium	14	13	55
High	4	32	12.5

(b) Translocation experiment.

Before translocation			After translocation			
	Sex ratio			Sex ratio		
Territory quality	Female	Male	Territory quality	Female	Male	Number of pairs
Low	2	18	High	29	5	4
High	15	4	High	16	4	3

which parents adjust the sex ratio of their offspring? Molecular evidence suggests that the sex ratio biases are already present when the eggs are laid. What is going on at earlier stages, when the sex ratio bias is established, is unknown.

In summary, when one gender of offspring enhance parental reproduction, natural selection favors parents who produce more offspring of that gender. When one offspring gender reduces parental reproduction, natural selection favors parents who produce less of that gender. Both these predictions have been successfully tested in the Seychelles warbler.

The sex ratio is a successfully understood adaptation

Local resource competition, and local resource enhancement, are two examples in which deviations from a 50 : 50 sex ratio have been successfully predicted. Some other examples are even more detailed. For instance, quantitative differences in the sex ratio produced by different ants nests can be predicted from the genetic relatedness within each nest. However, the example of the Seychelles warbler is enough to illustrate how the theory of adaptive sex ratios can be remarkably successful in explaining both the normal 50 : 50 sex ratio and deviations from it. The basic theory has inspired various different kinds of test. The theory is quantitative in its predictions; and the key variable

— sex ratio — is easy to measure. The sex ratio will therefore likely stay in the vanguard of evolutionary research for some time to come.

12.6 Different adaptations are understood in different levels of detail

We have looked at the function of sex, sexual selection, and the sex ratio, as three related examples of research on adaptation. In each case, the research has advanced to a different stage.

The problem of sex is still unsolved. Until recently, the main work has been theoretical, aiming to build a model in which some hypothesized advantage to sex is large enough to outweigh the 50% cost. We now have two reasonably well worked out theories (the mutational and parasitic theories) and research is moving on to an empirical phase.

In the case of sexual selection, the main theories of female choice have been around for some time. They provide a satisfactory abstract explanation for organs such as the peacock's tail. The full repertoire of techniques — model building, experiment, comparative methods — are being used. At a detailed natural history level, many questions remain unanswered. We do not know whether the abstract ideas correctly explain the full natural variety of sexual behavior and dimorphism, and the ideas are not easy to test.

The theory of sex ratio is still further advanced. The relation between facts and theories is good. We have not only a general abstract theory as for sexual selection, but the theory also makes quantitative predictions and suggests a number of types of test in special cases. Several of the tests have been followed up, and the fit of results to predictions suggests that the theory stands a good chance of being correct.

Summary

1 For many characters, it is not obvious how (or whether) they are adaptive.

2 Adaptation can be studied by comparing the observed form of an organ with a theoretical prediction, by experimentally altering the organ, and by comparing the form of the organ in many species.

3 Sex has a 50% fitness disadvantage relative to asexual reproduction.

4 Sexually reproducing populations will evolve faster than a set of asexual clones, provided that the rate of favorable mutation is high enough.

5 The taxonomic distribution of asexual reproduction suggests that asexual forms have a higher extinction rate than sexual forms. However, it is generally doubted that sex is maintained by group selection.

6 Two modern theories of why sex exists propose that it is favored by: (i) the large numbers of deleterious mutations, which are more efficiently removed by sexual than asexual reproduction; and (ii) the coevolutionary arms race of parasites and hosts. The problem of why sex exists has not been finally solved.

7 Males in many species have bizarre and deleterious secondary sexual characters; the peacock's train is an example.

8 Darwin explained the evolution of strange secondary sex characters by sexual selection: the characters reduce their bearers' survival, but increase their success in reproduction; sexual selection in most species works by male competition and by female choice.

9 The greater sexual dimorphism of polygynous species than monogamous species suggests the importance of sexual selection.

10 The preference of females for males with deleterious characters is theoretically puzzling. It may be explained by Fisher's theory, in which deleterious characters were formerly advantageous and are maintained by majority preference, or by a Zahavi's handicap theory, in which the costly character indicates superior genetic quality.

11 There can be conflicting forces of selection on males and on females. The conflict depends on the mating system, and can be studied by experimentally altering the mating system and allowing the population to evolve to a new adaptive state.

12 The sex ratio is usually 50 : 50 because the reproductive success of all the males in a population must equal the reproductive success of all the females. If the population sex ratio deviates from 50 : 50, natural selection favors individuals that produce more offspring of the rarer sex.

13 The theory of sex ratio has correctly predicted when the ratio should differ from 50 : 50. It has been tested experimentally in the case of "helpers at the nest" in the Seychelles warbler.

14 The functions of sex, sexual selection, and sex ratio are three of the most important areas of research on adaptation. They have reached different stages of theoretical advance.

Further reading

On sex, my popular book (Ridley 2001) explains the basic problem and Kondrashov's theory. The Fisher–Muller theory can be traced through Barton & Charlesworth (1998), Burt (2000), and Otto & Lenormand (2002). Butlin (2002) reviews ancient asexuals, and the Meselsohn test. Rice (2002) reviews experimental work on the evolution of sex.

For the mutational theory, two thorough reviews of the research on U are by Keightley & Eyre-Walker (1999) and Lynch *et al.* (1999). (Note that Lynch uses U for the genomic rate, Keightley for the gametic rate. The latter is half the former.) These can now be updated by the exchange between Kondrashov and Eyre-Walker & Keightley in *Trends in Genetics* (2001), vol. 17, pp. 75–8, and by Shabalina *et al.* (2001).

One further related theme is the possibility of destroying HIV by enhancing its mutation rate: Holmes (2000a) is a popular piece on it.

The other main theory is parasitic. Hamilton (2001) is volume 2 of his collected papers and contains all his key papers on sex, together with introductions that update the literature reviews. Otto & Lenormand (2002) discuss the theory, as do Barton & Charlesworth (1998). Lively (1996) introduces his research. Chapter 22 contains further references on parasite–host coevolution generally; and see Simmons (1996) on the genetics of parasite–host relations in plants.

The evolution of sex was the topic of special issues of three journals recently: *Science* September 25, 1998 (vol. 281, pp. 1979–2008), *Journal of Evolutionary Biology* (vol. 12, no. 6, 1999), and *Nature Reviews Genetics* (vol. 3, no. 4, 2002). The *Journal of Evolutionary Biology* (sometimes informally referred to as the journal of evil biology) special issue contains a "target" article by West *et al.*, together with commentaries by many expert authors. West *et al.* argue that the mutational and parasitic theories may act jointly to maintain sex, rather than being alternatives.

For both sexual selection and the sex ratio, Dawkins (1989a) is a good introduction. For sexual selection, Andersson (1994) is a comprehensive review; Cronin (1991) is another clear introduction and is good on history and the broader context. Møller (1994) describes his work on swallows. On female choice for good genes, see also Wilkinson *et al.* (1998)'s work on stalk-eyed flies. On antagonistic coevolution, further results in the same vein as Holland & Rice (1999) are reported by Hosken *et al.* (2001) and by Civetta & Clark (2000).

On sex ratios, Fisher (1930) is the classic source. West *et al.* (2000) is a short review and Hardy (2002) is an edited volume on modern research. Hewison & Gaillard (1999) review another deviation — the Trivers & Willard (1973) effect — in ungulates.

Study and review questions

1 What is the cost of sex in a species in which the sex ratio at birth is: (a) 1 male : 2 females, and (b) 2 males : 1 female?

2 (a) What condition is required of the rate of deleterious mutation for natural selection to favor sexual over asexual reproduction? Does reason and evidence suggest that the condition is met naturally? (b) Draw the relation between the number of deleterious mutations in an organism and its fitness that is required for sex and recombination to be advantageous (include a specification of the *y*-axis). What form is arguably true in reality?

3 What would you investigate in order to determine whether sex is favored by host–parasitic coevolution?

4 Why does a male character have to be costly (a handicap) in order to signal genetic quality?

5 In Fisher's runaway theory, what maintains the female preference for extreme males — why does the preference not evolutionarily disappear?

6 If one male can fertilize several females in a species, why do parents not produce a sex ratio of many daughters per son?

7 In terms of the modes of selection (or fitness regimes) discussed in Chapter 5, what kind of selection operates in Fisher's model of: (a) female choice, and (b) the sex ratio?

8 [This question draws on material in all three chapters in Part 3.] If organisms in a polygynous species produce a 50: 50 sex ratio in their offspring, is it a perfect adaptation from the viewpoint of: (a) the individual organism, and (b) the group of organisms? What general moral about the perfection of adaptation does the example illustrate?

Part four

Evolution and Diversity

Darwin closed *The Origin of Species* with the following words:

> There is grandeur in this view of life, with its several powers, having been originally breathed into a few forms or into one; and that, whilst this planet has gone cycling on according to the fixed law of gravity, from so simple a beginning endless forms most beautiful and most wonderful have been, and are being, evolved.

Part 4 of this book is about how the theory of evolution can be used to understand the diversity of life or, in Darwin's words, the "endless forms most beautiful." The units in which biologists measure the endless forms are species. We begin this set of chapters by looking at what biological species are, and also at diversity within a species. In evolutionary biology, species can be understood as gene pools — sets of interbreeding organisms — and these are important units because, in the theory of population genetics, natural selection adjusts the frequency of genes in gene pools.

The millions of species now inhabiting this planet have, as Darwin said, evolved from a common ancestor, and the multiplication in the number of species has been generated as single species have split into two. Speciation (Chapter 14) has probably often occurred when two populations have evolved independently, and accumulated incompatible genetic differences. Much is understood about this process, but we also look at some other, less well understood, ways in which new species may arise.

Chapter 15 describes how the phylogenetic relations of species, and higher taxonomic groups, can be reconstructed. The history of species cannot be simply observed, and phylogenetic relations have to be reconstructed from clues in the molecules, chromosomes, and morphology of modern species (and in the morphology alone of fossils). Phylogenetic reconstruction is a crucial part of modern taxonomy, which we look at in Chapter 16. Arguably, phylogeny provides a better principle for biological classification than any alternatives. In order to classify species, therefore, we need to know their phylogenetic relations and Chapter 16 logically follows Chapter 15.

Finally, the theory of speciation, as well as classification and phylogenetic reconstruction are all needed in evolutionary biogeography (Chapter 17) — the use of evolutionary theory to understand the geographic distribution of species.

13 Species Concepts and Intraspecific Variation

Evolutionary theorists have suggested a number of reasons why biological species exist, and there has been a controversy about which of the reasons is most important. This chapter is about species concepts, and the controversy among them. We begin by seeing how species are recognized in practice, and then move on to the theoretical ideas. We take, in order, phenetic, reproductive (biological and recognition), and ecological concepts, which all aim to define species at a point in time. We concentrate on two properties of each species concept: (i) whether it theoretically identifies natural units; and (ii) whether it explains the existence of the discrete phenetic clusters we recognize as species. While looking at the biological species concept, which defines species by interbreeding, we also consider the topic of isolating mechanisms that prevent interbreeding between species. We examine some test cases from asexual organisms and from genetic and phenetic patterns in space. We then turn to cladistic and evolutionary species concepts that can supplement the non-temporal concepts and define species through time. We finish by considering the philosophical question of whether species are real categories in nature, or nominal ones.

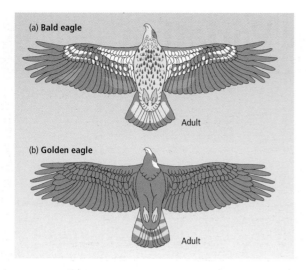

Figure 13.1
(a) Adult bald eagle (*Haliaeetus leucocephalus*) and (b) adult golden eagle (*Aquila chrysaetos*), seen from underneath. The species can be distinguished by their pattern of white coloration.

13.1 In practice species are recognized and defined by phenetic characters

Species are formally defined and practically recognized by phenetic characters

Biologists almost universally agree that the species is a fundamental natural unit. When biologists report their research, they identify their subject matter at the species level and communicate it by a Linnaean binomial such as *Haliaeetus leucocephalus* (bald eagle) or *Drosophila melanogaster* (fruitfly). However, biologists have not been able to agree on exactly how species should be defined in the abstract. The controversy is theoretical, not practical. No one doubts how particular species are defined in practice. Taxonomists practically define species by means of morphological or phenetic characters.[1] If one group of organisms consistently differs from other organisms, it will be defined as a separate species. The formal definition of the species will be in terms of characters that can be used to recognize members of that species. The taxonomist who describes the species will have examined specimens of it and of related species, looking for characters that are present in specimens of the species to be described, and absent from other closely related species. These are the characters used to define the species.

Almost any phenetic character may end up being useful in the practical recognition of species. Figure 13.1 for example shows the adults of the bald eagle (*Haliaeetus leucocephalus*) and the golden eagle (*Aquila chrysaetos*), seen from below. A bird guide will

[1] Phenetic characters are all the observable, or measurable, characters of an organism, including microscopic and physiological characters that may be hard work in practice to observe or measure. Morphological characters are characters of the shape or observable form of the whole organism or a large part of it. Behavioral and physiological characters are part of the phenetic description of an organism, but not part of its morphology. However, taxonomic descriptions are usually made from dead specimens in a museum, and the phenetic characters that are specified in taxonomic descriptions are usually morphological characters. The words "phenetic" and "morphological" are therefore practically almost interchangeable here. Also, the word "phenotypic" could be used instead of "phenetic."

Box 13.1
Description and Diagnosis in Formal Taxonomy

The point of the example of the two eagle species is intended merely to demonstrate that species are defined in practice by observable phenetic characters. We should also notice a terminological formality, distinguishing between a formal *description* of a species and a *diagnosis*. The formal definition is the description of the species — in terms of phenetic characters — that a taxonomist originally supplied when naming the species. Certain rules exist about the naming of new species, and the characters specified in the formal definition are the "defining" characters of the species in a strictly formal sense.

The formally defining characters of a species may be difficult to observe in practice. They might, for instance, be some fine details of the creature's genitalia, which can be recognized only by an expert using a microscope. Taxonomists do not on purpose pick obscure characters to put in their definitions, but if the only distinct characters that the species' first taxonomist noticed were obscure ones then they will provide its formal definition. If the formally defining characters are inconvenient to observe, subsequent taxonomists will try to find other characters that are more easily

observable. These useful characters, if they are not in the formal description, provide what is called a "diagnosis." A diagnosis does not have the legalistic power of a description to determine which names are attached to which specimens, but it is more useful in the day to day practical taxonomic task of recognizing which species specimens belong to. As research progresses, better characters (i.e., more characteristic of the species and more easily recognized) may be found than those in the first formal description. The formal definition then loses its practical interest, and the characters given in a work like Peterson's *Birds* are more likely to be diagnostic than formally defining.

When an evolutionary biologist discusses the definition of species, the formal distinction between description and diagnosis is beside the point. All that matters is that phenetic characters are used to recognize species, as in the eagles. The distinction is worth knowing about, however: both in order to avoid unnecessary muddles, and for other reasons — taxonomic formalities are important in the politics of conservation, for instance.

give a number of characters by which the two species can be told apart. In the adult, the bald eagle has a distinctive white head and tail, and a massive yellow bill. In North America, a bald eagle can therefore be recognized by the color of its feathers and bill. (Strictly speaking, the characters used to recognize species are often "diagnostic" rather than "defining" characters. Box 13.1 explains the distinction.)

In practice, the characters that define a species will not be present in all members of that species and absent from all members of other species. Nature is too variable. A perfectly defining character cannot usually be found, because the individuals of a species do not all look the same. One bald eagle will differ in color from another bald eagle. Real species form a "phenetic cluster": the individuals in the species show a range of appearances, but they tend to be more similar to one another than to members of other species. Bald eagles tend to have one color pattern, golden eagles another. The defining characters are not perfectly discriminatory, but they do indicate how most members of the species differ from most members of other, related species.

In the most difficult cases, two species may blur into each other (Figure 13.2). Two species that only recently evolved from a common ancestor, or two populations that have not yet separated into two full species will be particularly likely to blur into each other. Ring species are an example (Section 3.5, p. 50, and Plate 1, opposite p. 68). In a ring species, two species appear to be present at one place, but those two "species" are connected by a series of forms that are geographically arranged in a ring. No phenetic character could be used, except arbitrarily, to divide the ring into two species. Such a

Interindividual variation causes problems

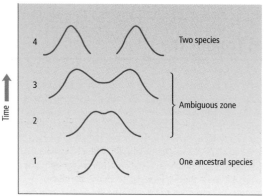

Time

4   Two species

3

    Ambiguous zone

2

1   One ancestral species

Form, or character state (e.g., beak size)

Figure 13.2
Difficulties in species recognition are expected in the theory of evolution, because variation exists within each species and new species evolve by the splitting of ancestral species. During the evolution of new species, the distinction between the species will be ambiguous during times 2 and 3. At stage 3, for instance, no phenotypic character can unambiguously distinguish between two species; indeed two species do not yet exist.

division of the ring would also be theoretically meaningless: there really is a continuum, not a number of clear-cut, separate species. Problems of this kind are exactly what we should expect given that species originated by an evolutionary process. We should not expect clear-cut defining characters to exist for all species; that is not the way nature is.

The evolutionary controversies about species are not mainly concerned with practical or formal issues

Species are in practice mainly recognized by phenetic characters, more or less successfully. However, when evolutionary biologists discuss species concepts, they are not usually discussing how species are recognized in practice. They are discussing deeper, theoretical concepts of species, concepts that may lie beneath the practical procedures that are used to recognize particular species. Is the bald eagle just the set of eagles that have white heads and tails? Imagine that a parental pair of bald eagles with good white heads and tails produced a nest of eagles of some different color pattern. Would they have given birth to a new species? If the color of the head and tail was all there was to being a member of *Haliaeetus leucocephalus*, then the answer would clearly be yes. However, if the species have a more fundamental definition, and the coloration was picked only as a practically useful marker, then the answer would be no. Indeed, the new eagles without the white coloration would render that taxonomic character out of date, and it would be time to start looking for some other characters to recognize the species. Most of the discussion of species concepts that follows assumes that species definition has some deeper meaning than the phenetic characters used to recognize the species in practice. When biologists argue about species concepts they are not arguing about how species are defined in practice.

13.2 Several closely related species concepts exist

A first distinction among species concepts is between horizontal and vertical concepts (Figure 13.3). A horizontal concept aims to define which individuals belong to which species at any instant in time. A vertical concept aims to define which individuals belong to which species at all times. Vertical concepts are mentioned here mainly for completeness; most of the interest in species concepts is in horizontal concepts.

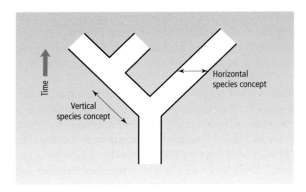

Figure 13.3
Horizontal and vertical species concepts. A horizontal concept aims to define species at a time instant and specifies which individuals belong to which species at one time. A vertical concept aims to define species through time and specifies which individuals belong to which species through all time.

Biologists are mainly concerned with defining species in the present, and this requires a horizontal concept. We need to know which eagles are *Haliaeetus leucocephalus* now, and are less interested in eagles a million years in the past or the future. This chapter concentrates on horizontal concepts.

13.2.1 *The biological species concept*

Species may be defined by interbreeding

The *biological species concept* defines species in terms of interbreeding. Mayr (1963), for instance defined a species as follows: "species are groups of interbreeding natural populations that are reproductively isolated from other such groups." The expression "reproductively isolated" means that members of the species do not interbreed with members of other species, because they have some attributes that prevent interbreeding. The species concept that is now called the biological species concept actually predates Darwin — it was the species concept used by John Ray in the seventeenth century, for instance — but it was strongly advocated by several influential founders of the modern synthesis, such as Dobzhansky, Mayr, and Huxley, and it is the most widely accepted species concept today, at least among zoologists.

The biological species concept is important because it places the taxonomy of natural species within the conceptual scheme of population genetics. A community of interbreeding organisms make up, in population genetic terms, a gene pool. In theory, the gene pool is the unit within which gene frequencies can change. In the biological species concept, gene pools become more or less identifiable as species. The identity is imperfect, because species and populations are often subdivided, but that is a detail. The species, in this concept, is the unit of evolution. Organisms do not evolve but species do, and higher taxonomic groups such as phyla only evolve in so far as their constituent species are evolving.

The biological species concept explains why the members of a species resemble one another, and differ from other species. When two organisms breed within a species, their genes pass into their combined offspring; as the same process is repeated every generation, the genes of different organisms are constantly shuffled around the species gene pool. Different family lineages (of parent, offspring, grandchildren, and so on) soon become blurred by the transfer of genes between them. The shared gene pool gives the species its identity. By contrast, genes are not (by definition) transferred to other species, and different species therefore evolve a different appearance. The movement of genes through a species by migration and interbreeding is called *gene flow*. According to the biological species concept, gene flow explains why each species forms a phenetic cluster.

Moreover, the constant shuffling around of genes sets up a selection pressure favoring genes that interact well with genes at other loci to produce an adapted organism; a gene that does not fit in with the workings of other genes will be selected against. When we look at organisms today, we are looking at the effects of selection in the past. We should expect to see genes that interact well together within a species. The same is not true of genes in two separate species. These genes have not been tried out together and sifted by selection, and we have no reason to expect them to interact well. When combined in a single body, they may produce a genetic snarl-up. (Section 14.4, p. 389, further develops the theory of gene interactions within, and between, species.) Sexual interbreeding within a species produces what Mayr (1963) calls "cohesion" (and others call "cohesiveness") in the species' gene pool.

And how, in this concept, should the taxonomist's method of defining species be interpreted? Taxonomists actually identify species by morphology, not interbreeding. On the biological species concept, the taxonomist's aim should be, as far as possible, to define species as interbreeding units. The justification for defining species morphologically is that the morphological characters shared between individuals are indicators of interbreeding. When taxonomists can study interbreeding in nature they should do so and define the arrays of interbreeding forms as species. With dead specimens in museums, taxonomists should use the interbreeding criterion to guide their analysis of morphological criteria. Taxonomists should seek morphological criteria which define a species as a set of forms that appears to have the kind of variation that an interbreeding community would have. The morphological characters of species are then indicators of interbreeding, as estimated by the taxonomist. Eagles with white heads and tails are one interbreeding unit; eagles with the color pattern of the golden eagle are another.

A closely related species concept is the *recognition species concept* of Paterson (1993). Paterson defines a species as a set of organisms with a shared specific mate recognition system (SMRS). The specific mate recognition system is the sensory method by which organisms recognize potential mates. For example, as many as 30 or 40 different species of crickets may be breeding within a single habitat in the USA. The male crickets broadcast their songs and are approached by females. Interbreeding is confined within a species because each species has its own distinctive song and females only approach males that are singing their species song. The system of a male song and a female acoustic system that leads females to approach some songs and not others is an example of what is meant by an SMRS. The set of organisms that are defined as a species by the biological and recognition species concepts will be very similar, because organisms that interbreed will usually also have a shared SMRS.

Another closely related concept has been developed to make use of the increasing quantities of data from molecular genetic markers, which can be used to recognize which sets of organisms belong to the same evolutionary lineage (Howard & Berlocher 1998). In all, several species concepts exist that are inspired by the underlying idea that species exist because of interbreeding among the individual organisms within each species. The biological species concept is the most influential of these reproductive species concepts.

13.2.2   *The ecological species concept*

The forms and behavior of organisms are, at least to some extent (Chapter 10), adapted to the resources they exploit and the habitats they occupy. According to the *ecological species concept*, populations form the discrete phenetic clusters that we recognize as species because the ecological and evolutionary processes controlling how resources are divided up tend to produce those clusters. About half a century of ecological research, particularly with closely related species living in the same area, has abundantly demonstrated that the differences between species in form and behavior are often related to differences in the ecological resources the species exploit. The set of resources and habitats exploited by the members of a species form that species' ecological niche and the ecological species concept defines a species as the set of organisms exploiting a single niche. (In some cases, a full definition would have to be more long-worded. If, for instance, the juvenile stage of an organism lives in plankton while the adult stage is attached to rocks, then the different life stages exploit different ecological niches. However, the definition could be expanded to define a species as a set of organisms who exploit a certain set of niches, where the set includes the niches exploited by different life stages, genders, or other forms within the species.)

Why should ecological processes produce discrete species? Parasite–host relations provide a clear example. Imagine the parasites exploiting two host species. The host species will differ in certain respects, perhaps in where they live, or the times of day they are active, or their morphology. The parasites will evolve appropriate adaptations to live in one or the other host. The parasites then tend to become two discrete species, because their environmental resources (hosts in this case) come in two discrete kinds.

Two host species can clearly provide two discrete sets of ecological resources. But in other cases, the ecological resources may not come in such discrete units. Consider, for example, the five species of warblers in Maine that were the subject of a classic study by MacArthur (1958). MacArthur showed that each species mainly exploited a particular subregion of the trees they all lived in. Some species foraged higher, some lower; some foraged near the ends of branches, others nearer the center of the tree. These variables, like height in tree, are continuous. The warblers form five distinct, discrete species, but they divide up resource variables that are continuous. In this case, an ecological explanation for the existence of discrete species mainly comes from the principle of *competitive exclusion*. Only species that are sufficiently different can coexist. The result is that even with a continuous resource distribution, species may evolve into a series of discrete forms along the continuum. If the species blurred into one another, superior competitors could drive inferior competitors extinct, and gaps between species would appear. (The theory of speciation (Chapter 14) suggests some further reasons why discrete species evolve on continuous resources. Also, Section 13.7.2 discusses further evidence that ecological factors influence the array of phenetic forms in a species.)

The ecological and biological species concepts are closely related. Life, according to the ecological species concept, comes in the form of discrete species because of

Species may be defined ecologically, by a shared ecological niche

The ecological force of competitive exclusion maintains species differences

adaptation to exploit the resources in nature. Interbreeding is shaped by the same process. Natural selection will favour organisms who interbreed with other organisms that have a similar set of ecological adaptations. For instance, the ecological adaptation might be the size of the beak, if the beak is adapted to eat seeds found locally. Natural selection favors individual birds that interbreed with other birds that have similar beaks. Then they will on average produce offspring that are well adapted to eat the local seeds. Natural selection works against birds that interbreed with mates that have very different beaks as their offspring will tend to have maladapted beaks. The patterns of interbreeding and the ecological adaptations in a population are therefore shaped by common evolutionary forces. Notwithstanding the close relations between the concepts, some controversy still exists between them (Section 13.7 below).

13.2.3 The phenetic species concept

Species may be defined by shared phenetic attributes

The phenetic species concept can be understood as an extension of the way taxonomists define species (Section 13.1). Taxonomists define each species by a particular defining character, or characters, that is shared by its members. In general we could define a species as a set of organisms that are phenetically similar, and distinct from other sets of organisms. This would be a "phenetic" species concept: it defines species in general by shared phenetic attributes. One noteworthy feature of the phenetic concept is that it is not based on a theory of why life is organized into discrete species. The biological and ecological concepts are both theoretical, or explanatory, concepts. They define species in terms of processes that are thought to explain the existence of species: interbreeding or ecological adaptation. The phenetic species concept is non-theoretical, or descriptive. The concept simply notes that species do in fact exist, in the form of phenetic clusters. Why species exist in this form is a separate question.

The classic version was the typological, . . .

The classic version of the phenetic species concept is the "typological species concept" (the term "morphological species concept" has also been used to refer to much the same concept). The word "typological" comes from the word "type," which is used in formal taxonomy. When a new species is named, its description is based on a specimen called the type specimen, which has to be deposited in a public collection. According to the typological species concept, a species consists of all individuals that look sufficiently similar to the type specimen of the species. We shall look further at "typological thinking" in Section 13.5, where we shall see why typology is thought to be invalid in modern evolutionary theory.

. . . a later version was the numerical, . . .

A later version of the phenetic concept was developed by the school of numerical taxonomy in the 1960s. (On numerical taxonomy, see Section 16.5, p. 476.) Numerical taxonomists developed statistical techniques for describing the phenetic similarity of organisms. Those techniques could be applied to recognize species. A species could then be defined as a set of organisms of sufficient phenetic distinctness (where the word "sufficient" could be made precise by the statistical methods used to describe phenetic similarity). The numerical taxonomists' phenetic species concept has nothing to do with the typological concept, but belongs to the same family of concepts.

Some versions of a more recently proposed *phylogenetic species concept* also define species by a kind of phenetic similarity. For instance, Nixon & Wheeler (1990) define a species as "the smallest aggregation of populations (sexual) or lineages (asexual) diagnosable by a unique combination of character states in comparable individuals."

The various phenetic species concepts are closely related to the biological and ecological concepts. All these concepts will recognize much the same species in nature. A set of organisms that are adapted to a similar niche are likely to be phenetically similar, because they share a set of phenetic characters that are used to exploit the ecological resources. A set of organisms that interbreed are also likely to be phenetically similar. The ancestors of the modern members of the species have interbred, resulting in genetic (and therefore phenetic) similarity among the members of the species now. In Section 13.7 we look at controversies among species concepts. It will be worth keeping in mind that all the concepts agree most of the time both about what species exist in nature and about what the biological forces are that explain those species.

13.3 Isolating barriers

13.3.1 *Isolating barriers prevent interbreeding between species*

Isolating barriers evolve between species

Why is it that closely related species, living in the same area, do not breed together? The answer is that this is prevented by *isolating barriers*. An isolating barrier is any evolved character of the two species that stops them from interbreeding.[2] The definition specifies "evolved characters" to exclude non-interbreeding due to simple geographic separation. Interbreeding between two geographically separate populations of a species is impossible, but the geographic separation is not an isolating barrier in the strict sense. Geographic separation alone does not have to be an evolved character, and is unlikely to be an evolved character when it is between two populations of a species. One subpopulation can colonize a new area without any genetic change, or the populations may have been separated by a geographic accident, such as the formation of a new river. Courtship, however, is an example of an isolating barrier. If two species do not interbreed because their courtship differs, then the courtship behavior of at least one of those species must have undergone evolutionary change.

There are several kinds of barrier

Several kinds of isolating barrier are distinguished; Table 13.1, based on Dobzhansky (1970), gives one classification. The most important distinction is between prezygotic and postzygotic isolation. Prezygotic isolation means that zygotes are never formed, for instance because members of the two species are adapted to different habitats and never meet, or have different courtships and do not recognize each other as potential

---

[2] What is here called an "isolating barrier" has until recently (following Dobzhansky (1970)) usually been called an "isolating mechanism." Some biologists have criticized the word "mechanism" because it might imply that the character that causes isolation evolved in order to prevent interbreeding — that the isolating mechanism is an adaptation to prevent interbreeding. As we shall see in Chapter 14, the characters that cause reproductive isolation certainly sometimes, and perhaps almost always, evolve for other reasons and prevent interbreeding only as an evolutionary by-product. The use of the term "isolating barrier" is becoming common now, and I follow this usage. However, the older expression could be defended. In biology, a mechanism of X is not always something that evolved to cause X. Compare, for instance, "population regulation mechanism," "mechanism of mutation," "mechanism of speciation," and "mechanism of extinction." Isolating mechanism could mean only a mechanism that isolates, not a mechanism that evolved in order to isolate.

> **Table 13.1**
> Dobzhansky's classification of reproductive isolation barriers. From Dobzhansky (1970).
>
> 1. *Premating* or *prezygotic* mechanisms prevent the formation of hybrid zygotes
>   (a) *Ecological* or *habitat isolation*. The populations concerned occur in different habitats in the same general region
>   (b) *Seasonal* or *temporal isolation*. Mating or flowering times occur at different seasons
>   (c) *Sexual* or *ethological isolation*. Mutual attraction between the sexes of different species is weak or absent
>   (d) *Mechanical isolation*. Physical non-correspondence of the genitalia or the flower parts prevents copulation or the transfer of pollen
>   (e) *Isolation by different pollinators*. In flowering plants, related species may be specialized to attract different insects as pollinators
>   (f) *Gametic isolation*. In organisms with external fertilization, female and male gametes may not be attracted to each other. In organisms with internal fertilization, the gametes or gametophytes of one species may be inviable in the sexual ducts or in the styles of other species
>
> 2. *Postmating* or *postzygotic* isolating mechanisms reduce the viability or fertility of hybrid zygotes
>   (g) *Hybrid inviability*. Hybrid zygotes have reduced viability or are inviable
>   (h) *Hybrid sterility*. The F_1 hybrids of one sex or of both sexes fail to produce functional gametes
>   (i) *Hybrid breakdown*. The F_2 or backcross hybrids have reduced viability or fertility

mates. Alternatively, the members of two species may meet, mate, and form zygotes, but if the hybrid offspring are inviable or sterile then the two species have postzygotic isolation.

13.3.2 Sperm or pollen competition can produce subtle prezygotic isolation

Over evolutionary time, differences accumulate between species and the result is that they become fully isolated by both prezygotic and postzygotic isolating barriers. They will evolve different appearances, different courtships, different ecological adaptations, and different and incompatible genetic systems. However, closely related and recently evolved species may be only partly isolated, and then research can reveal which isolating barriers are at work.

Gametic isolation is a kind of isolating barrier . . .

One factor that has been investigated recently in several species is "gametic isolation" (Table 13.1). The simplest kind of gametic isolation occurs when the sperm and eggs of two species do not fertilize each other. But a process called "sperm competition" can cause a subtler kind of gametic isolation. Two species may not interbreed because the sperm, or pollen, of species 1 outcompetes that of species 2 when fertilizing the eggs of species 1, but the sperm, or pollen, of species 2 outcompetes that of species 1 when fertilizing the eggs of species 2. Wade *et al.* (1993), for instance, studied reproductive isolation between two beetles, *Tribolium castaneum* and *T. freemani*. *T. castaneum* is a worldwide pest of stored flour called the flour beetle, and *T. freemani* is a closely related species that lives in Kashmir. The two species are not isolated at the premating stage: males of both species copulate indiscriminately with females of both species. Wade *et al.* quote a remark about the mating propensities of male flour beetles, who "will attempt copulation with other males, dead beetles of both sexes, or with any object, such as a lump of flour or frass, which looks like a beetle."

. . . that has been shown by recent experiments

Wade *et al.* (1993) did an experiment in which they put female *T. freemani* in one of three situations: (i) with two successive males of *T. freemani*; (ii) with two successive males of *T. castaneum*; or (iii) with one male *T. freemani* and then one male *T. castaneum*. The female beetles laid a similar number of eggs in all three cases, and a similar percentage of the eggs hatched and grew up (though the interspecies hybrid offspring are sterile). This shows that the sperm of male *T. castaneum* are capable of fertilizing *T. freemani* eggs. When female *T. freemani* were put with males of both species (condition (iii)), less than 3% of the offspring were hybrids — over 97% of the eggs had been fertilized by the *T. freemani* male's sperm. The reason is that when two males inseminate the same female, their sperm compete inside the female to fertilize her eggs. In this case, when no *T. freemani* sperm are present, *T. castaneum* sperm can fertilize the eggs, but when *T. freemani* sperm are present they outcompete the *T. castaneum* sperm. Sperm competition is causing reproductive isolation. (Sperm competition is a form of sexual selection, discussed in Section 12.4, p. 327. It is a form of male competition, and its outcome may well be influenced by female choice. In this case, the "choice" would be effected by the female's internal reproductive physiology. Section 14.11, p. 413, discusses how sexual selection may contribute to speciation, and provides further contexts for these observations.)

The experiment matters not only for revealing the nature of reproductive isolation in this pair of beetles, but also shows what needs to be done in research on prezygotic isolation. An experiment in which males of one species are simply crossed with females of another species is inadequate to measure prezygotic isolation. When male *T. castaneum* are put with female *T. freemani* they produce hybrid offspring. We might falsely conclude that these two species are not prezygotically isolated. But if the females are put with male *T. freemani* and *T. castaneum*, hardly any hybrid offspring are produced and the prezygotic isolation is revealed. Isolation by sperm, or pollen, competition has recently been found in many species (Howard 1999).

13.3.3 *Closely related African cichlid fish species are prezygotically isolated by their color patterns, but are not postzygotically isolated*

Cichlid fish are found globally in warm freshwater environments, but they are famous for the huge numbers of species that have evolved in the East African lakes. They are also famous as a conservation disaster, as a large but unknown number of species have been lost following the introduction of a predatory fish, the Nile perch, into the lakes, together with increasing lake eutrophication. Here we concentrate on the reproductive isolation between two cichlid species that live in Lake Victoria.

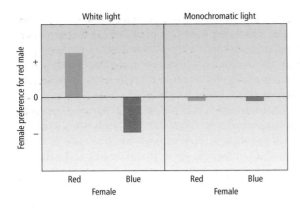

Figure 13.4

Mating preferences (a form of prezygotic isolation) in two cichlid species from Lake Victoria, Africa. The two species are referred to as the "red" and "blue" species: see text for details and Plate 7 (between pp. 68 and 69) for illustration. Individual females of each species were given a choice of two males, one from each species. A preference for males of the red species was arbitrarily defined as a positive preference; a negative preference indicates a preference for males of the blue species. Females preferred conspecific males in normal white light, but the preference disappeared in monochromatic light, where the two species were visually indistinguishable. From Seehausen & van Alphen (1998).

Experiments show that cichlid fish are isolated by color pattern . . .

Cichlids often have beautiful color patterns, and *Pundamilia nyererei* and *P. pundamilia* are related species that differ in color (see Plate 7, between pp. 68 and 69). For simplicity, we can refer to *P. nyererei* as red and to *P. pundamilia* as blue, but the color illustrations show that the words red and blue hardly describe the gorgeous colors of the two species. Seehausen & van Alphen (1998) performed a laboratory experiment on the mating preferences of the two species. They first tested the preferences of females of both species for males of one species or the other, in normal light. The result was that the females of both species preferred conspecific males (Figure 13.4). The two species show prezygotic isolation by mating behavior. Seehausen and van Alphen then repeated the experiment, but in monochromatic light, in which the color difference between the two species was invisible (Plate 6). Now the females of both species show no preference between red and blue males. The experiment shows that the prezygotic isolation is due to the color patterns of the two fish species.

. . . and not postzygotically, . . .

Seehausen's lab has also measured postzygotic isolation (Seehausen *et al.* 1997). The two species will interbreed in the lab and produce hybrids. The hybrids are fertile, and by 2001 five generations of hybrids had been successfully bred: the two species are not postzygotically isolated. In conclusion, *P. nyererei* and *P. pundamilia* are isolated prezygotically by color pattern but not postzygotically.

. . . which has consequences for the effect of pollution

The main point of Seehausen's experiment here is to show how isolating barriers can be investigated, but the results have two other interests. One is in relation to conservation. The color differences between the two species become less visible in cloudy, eutrophic waters. Pollution in Lake Victoria is making it more likely that the two species hybridize. Pollution is leading to a loss of biodiversity, not by the normal mechanism of extinction but by removing the isolating barrier between closely related species. The other interest is in relation to speciation, and illustrates a similar point to the study of flour beetles. Mate preference, like sperm competition, is a form of sexual selection. Sexual selection is thought to drive speciation, particularly sympatric speciation (Section 14.11, p. 414). The African lake cichlids provide some of the strongest evidence for sympatric speciation (Section 14.10.3, p. 413). Seehausen's experiments, which show that mating preferences are the first kind of isolation to evolve in these fish, fits in with the broad idea that sexual selection has contributed to the spectacular radiation of cichlids in East Africa.

In conclusion, over evolutionary time the amount of isolation between two species will increase and the species will eventually be isolated by most of the barriers listed in Table 13.1. (Think of how humans are isolated from a distant species, such as a baboon — we are probably isolated from them by everything in the list except habitat and breeding season.) Experiments can be done to reveal what the particular isolating barriers are between closely related species. These experiments can reveal what isolating barriers are at work in the early stages of speciation. We return to this topic in Chapter 14.

13.4 Geographic variation within a species can be understood in terms of population genetic and ecological processes

Intraspecific variation exists both at any one place and between different places. If we sample a number of individuals belonging to one species at any one locality, those individuals may differ — variation within a population — often showing a normal distribution (Section 9.2, p. 226). Also, if we sample individuals belonging to one species, from different places, they may differ — variation between populations, or geographic variation.

We need to examine intraspecific variation both in order to understand the nature of species and also to understand how new species evolve. As Chapter 14 will discuss, the evolution of new species consists of the conversion of variation within one species into differences between species. Chapters 5–9 looked at the factors that control variation within a population: variation may be maintained by natural selection, or a balance of selection and mutation, or a balance of drift and mutation. Here we shall look at variation between populations (geographic variation), and its relation with variation within each population. (The theory in Section 5.14, p. 129 is related to the topic here.)

13.4.1 *Geographic variation exists in all species and can be caused by adaptation to local conditions*

Johnston & Selander (1971)measured 15 morphological variables in 1,752 house sparrows (*Passer domesticus*) sampled from 33 sites in North America. The 15 characters can be reduced to a single abstract character of "body size" (to be statistically exact, this character was the first principal component). In Figure 13.5 the average body size of house sparrows is plotted on a map and two things are immediately important.

We have good evidence of geographic variation

First, and more important for our purposes, is simply that the characters vary in space: house sparrows from one part of the continent differ from those in other parts. Almost every species that has been studied in different places has been found to vary in some respect. Not all characters vary (for instance, humans have two eyes everywhere), but populations always differ in some characters. Different populations have been found to differ in morphology, in the amino acid sequences of their proteins, and the base sequence of their DNA. Geographic variation is ubiquitous. Mayr, most

Figure 13.5
Size of male house sparrows in North America. Size is measured as a "principal component" score, derived from 15 skeletal measurements. The score of 8 is for the largest birds, the score of 1 is for the smallest. The study described in Section 3.2 (pp. 46–7) is a precursor of this research. Redrawn, by permission, from Gould & Johnston (1972), corrected from Johnston & Selander (1971). © 1972 Annual Reviews Inc.

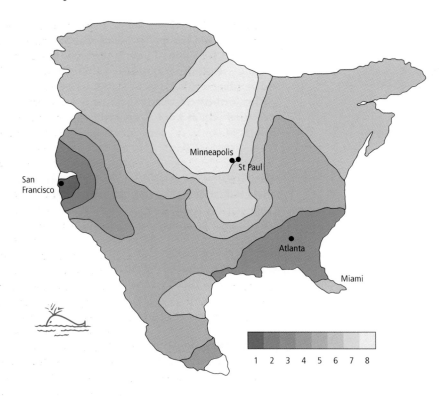

powerfully in his book *Animal Species and Evolution* (1963, chapter 11), has collected more evidence about geographic variation than anybody else and he concludes that "every population of a species differs from all others," and "the degree of difference between different populations of a species ranges from almost complete identity to distinctness almost of species level."

The second point to notice in Figure 13.5 is that the form of the geographic variation is explicable. House sparrows are generally larger in the north, in Canada, than in the center of America. The generalization is imperfect (compare, for instance, the sparrows of San Francisco and Miami); but in so far as it applies, it illustrates *Bergman's rule*. Animals tend to be larger in colder regions, presumably for reasons of thermoregulation. Geographic variation in these two species is therefore adaptive: the form of the sparrows differs between regions because natural selection favors slightly differing shapes in different regions.

Sparrows illustrate Bergman's rule

13.4.2 *Geographic variation may also be caused by genetic drift*

House mice (*Mus musculus*) have a standard diploid chromosomal set of 40 chromosomes ($2N = 40$). The centromeres of all 20 chromosomes are near the chromosomes ends and, perhaps for this reason, chromosomal fusions often take place in this species. In a chromosomal fusion, two chromosomes join together at their terminal centromeres. They form a new, longer chromosome with its centromere nearer the center. A fused chromosome often becomes established in a local population of house mice. The result is that the local population has less than 40 chromosomes per mouse.

Geographic variation in mice chromosomes . . .

Britton-Davidian *et al.* (2000) described a remarkable example recently, in the mice on the island of Madeira (see Plate 8, between pp. 68 and 69). They found that different chromosomal fusions were fixed in local mouse populations only 5–10 km apart. One local population might have 28–30 chromosomes per mouse, because five or six chromosomal fusions had occurred. In another population, three further fusions reduced the numbers to 2N = 22. According to Britton-Davidian *et al.* "house mice are thought to have been introduced onto Madeira following the first Portuguese settlement during the fifteenth century." If this is correct, the geographic variation illustrated in Plate 8 has all evolved in under 500 years. Mice, and rodents generally, show rapid chromosomal evolution. By way of contrast, all human populations have the same set of chromosomes, except for rare mutants.

. . . looks like an example of genetic drift . . .

What is the cause of this chromosomal evolution? The answer is uncertain, but it is thought to be random drift. A mouse containing a fused chromosome contains the same genes as a mouse with the two separate chromosomes. The mouse may grow up identically either way. However, a chromosomal mutation will initially exist in heterozygous form, and such heterozygotes tend inherently to be disadvantageous. A fusion between chromosomes 1 and 2 can be represented as 1+2. The heterozygote can be written 1,2/1+2. The heterozygote is disadvantageous during cell division, particular meiosis. For instance, the fused chromosome 1+2 may pair with chromosome 1, leaving chromosome 2 unpaired. The unpaired chromosome 2 may then segregate with chromosome 1, producing viable offspring. Or it may be segregated with chromosome 1+2, producing offspring with too many, or too few, chromosomes.

When a new chromosomal fusion mutation arises, it will be selected against because of its disadvantage in heterozygous form. But if it drifts up to a locally high frequency, as may easily happen in a local, small, and perhaps inbreeding, mouse population, natural selection will favor it. Natural selection favors whichever chromosomal form is locally common (this is an example of positive frequency-dependent selection, Section 5.13, p. 127). Natural selection alone cannot explain the geographic variation observed by Britton-Davidian *et al.* Natural selection alone would cause all the mice to have the same chromosome numbers. The variation is more likely to be explained by drift, with different individual chromosomal fusions drifting up in frequency in different localities. Natural selection may also be at work, depending on the frequency of the chromosomes. But whatever the cause of the pattern in Plate 8, it is a further example of geographic variation.

. . . though other factors may contribute

Geographic variation is probably rarely caused only by drift or only by selection. Also, more than one selective factor is likely to operate. In the case of the mouse chromosomes, natural selection probably interacts with drift, depending on the chromsomes' frequency. But other kinds of natural selection can act, such as meiotic drive (Section 11.2.1, p. 294) and a full account of mouse chromosomal evolution is complex (Nachman & Searle 1995). Moreover, very thorough research is needed to test between selection and drift.

Linanthus parryae is a small desert flower, living on the edge of the Mojave Desert in California. Local populations vary according to the frequency of the white and blue flowers. Wright (1978) considered it to be the best example of how drift causes differences between local populations (the first stage in Wright's shifting balance theory of evolution — see Section 8.13, p. 216). However, a long-term study by Schemske & Bierzychudek (2001) measured the fitness of blue and white flowers and found that selection is at work in a complex way that differs from year to year. A small study, over 1 or 2 years, might have supported Wright's interpretation but Schemske and Bierzychudek counted more than 710,000 seeds from more than 42,000 flowers over an 11-year period and they have effectively refuted drift as the explanation of variation in this particular species.

Hard work is needed to measure the contributions of drift and selection in particular species. But in general, patterns of geographic variation can be explained by some mix of selection, as seems to explain body size variation in sparrows, and of drift, as seems to explain chromosomal variation in house mice.

13.4.3 *Geographic variation may take the form of a cline*

If we drew a line on Figure 13.5 from Atlanta to Minneapolis and St Paul, or from the twin cities to San Francisco, and looked at the size of sparrows along it, we should have an example of a cline. A *cline* is a gradient of continuous variation, in a phenotypic or genetic character, within a species. Clines can arise for a number of reasons. In the house sparrows, the reason is likely that natural selection favors a slightly different body size along the gradient; sparrows are continuously adapted to an environment that changes continuously in space (Figure 13.6). For instance, body size may be adapted to environmental temperature. Temperature gradually decreases to the north, and body size in the sparrows increases as we go north. Alternatively, the environment may change discontinuously in space and different genes may be adapted to the two regions (Figure 13.6b). A cline can then arise because of gene flow: the movements of individuals, or their pollen in the case of plants.

> A cline is a continuous gradient of variation, within a species

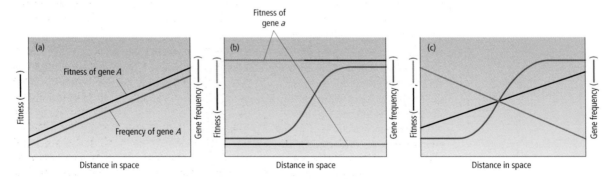

Figure 13.6

A cline can arise in various forms. (a) It can occur in a continuous environmental gradient. The house sparrow example (see Figure 13.5) probably has polygenic inheritance; the *y*-axis would more appropriately express the proportion of genes for larger body size than the average for the USA. (b) A cline can also arise when natural selection favors different genotypes in different discrete environments and there is gene flow (migration) between them. (c) A situation like (b) except that the environment changes gradually rather than suddenly.

Clines may be smooth or "stepped" (Figure 13.6c), depending on how suddenly gene frequencies change in space. If the environment varies smoothly, the cline will also be smooth. If the environment changes more suddenly, the cline may be more stepped. The shape of the step depends on the fitness difference between the genotypes in the two regions, the fitness of any intermediate genotypes (such as heterozygotes or recombinants), and the amount of gene flow. A sudden change in the environment is called an "ecotone" (Section 13.7.2 below contains an example from the grass *Agrostis*). However, ecotones are not the only explanation for stepped clines. Stepped clines may also result when the ranges of two formerly separate populations expand and the two populations meet up (Section 17.4, p. 500). Or they may result from genetic drift. When biologists see a stepped cline, they are interested to know whether it corresponds with an ecotone or has some other explanation. The main point here, however, is that geographic variation often takes the form of a cline. Clinal variation contrasts with a case such as that of the mice of Madeira, where the local populations do not show a gradient of variation.

13.5 "Population thinking" and "typological thinking" are two ways of thinking about biological diversity

Species show variation, both between individuals at any one place (often this has the form of a "bell curve" or normal variation) and geographic variation between individuals from different places. This variation has been thought about in two main ways: "population thinking" and "typological thinking" (Mayr 1976). We have already met the typological species concept (Section 13.2.3). A "type" specimen has to exist in order for a species to be defined. However, variation will exist in the species with some individuals more like the type specimen, and others less like it. By typological thinking, Mayr meant the idea that the type individual, and other individuals like it, are in some sense "better" examples of their species — they are more real, or more representative, members of their species. We can see what this means if we think about the classification of many non-biological entities.

Suppose we are classifying objects as chairs or non-chairs. Some objects will be better specimens of chairs than others. If an object has four equal length legs and a horizontal surface to sit on, it is a "good" chair. By calling something a good chair, or a better specimen of a chair, we mean that it is easily recognized as a chair, not that it is morally superior to other objects that are less easily recognized as chairs. Some other object may look rather like a chair, but have two legs missing and a third broken, making it is less representative of the category of chairs. Other objects may be so smashed up that we might hesitate to call them chairs at all. The variation between objects consists of some objects that are good chairs, and others that are less good chairs. The "less good" chairs mainly exist because of some kind of accident or environmental error, such as an accident in which a leg is broken off. We think to some extent typologically about chairs: some entities are typical chairs, others are less typical because there is something wrong with them.

Creationism could give an account of biological species that is rather like the typological account of chair classification. Each species might have a "best" form, perhaps corresponding to the optimal adaptation for the local environment. Individuals that deviated from that optimum might be less clearly recognized as members of their species; and they are also adaptively inferior. They might deviate from the optimum because of mutational error, or environmental accidents, that carried the phenotype away from the optimum.

Typological thinking is inappropriate in biological classification

Typological thinking means dividing variation into good, type specimens that are more real members of their category, and accidental deviants that are less good members of the category. The example we have just looked at, in terms of optimal adaptation and mutational and environmental error, is only one version of typological thinking. Historically, typological thinking has been based on ideas that no longer look scientific. For instance (in an extreme case) the nineteenth century taxonomist Louis Agassiz said that species are thoughts in the mind of God. The "good" specimens of the species, near the type specimen, would then exactly correspond to God's thoughts and the other specimens, away from the center of of the bell curve, would be inferior approximations. *Any* theory in which some versions of a species are better representatives of that species than are other versions is likely to be a case of typological thinking.

In modern evolutionary thinking, however, variation is non-typological. All the individuals of a species are equally good specimens of that species and they are equally representative of it. The species does not have some individuals that are more typical of it than others. We can see the case for population thinking in the evidence for geographic variation. Sparrows vary in size through North America: the variation is partly due to temperature differences between places, with sparrows evolving larger sizes where it is colder. It is not true that one size of sparrow is better, or more real, or more representative of sparrowness, than any other size within the species' range. All the sparrows are equally good sparrows.

Both selection . . .

The same can be said about the chromosomal forms of mice, if they are indeed caused by genetic drift. One chromosomal form is as good as another. The variation is neutral, and no one form of mouse can be recognized as a truer type of mouse than the others. Even if the variation within a species is partly due to mutation–selection balance (and some individuals are better adapted than others), the environment could change and the currently less fortunate individuals would improve in fitness. That is how evolutionary change occurs. Variation is essential for the evolutionary process. It is true that one individual of the species is used to define each species, and that individual is called the type specimen; but the use of type specimens is now just a legalistic naming procedure. It does not imply that individuals with the exact set of characters used to define the species are in any way better or more representative members of the species than are other individuals who happen to have variant forms of the defining characters.

. . . and drift . . .

. . . cause biological populations to show variation

Mayr (1976) argued that the replacement of typological by population thinking was one of the key features of the Darwinian revolution. And the main point here has been that population thinking makes more sense than typological thinking given what we understand about evolution. However, the distinction has some wider implications. Typological thinking can easily complement racist or other illiberal ideologies in which some humans, or kinds of humans, are regarded as superior, or fuller, specimens of humanity than are others. Box 13.2 looks at human variation, and at evidence that humans have an exceptionally low amount of interracial difference relative to other species.

Box 13.2
Human Variation and Human Races

Imagine a species made up by a number of geographic populations. How can we describe the amount of genetic divergence between the local populations? A number of statistics exist, of which G_{ST} is one of the clearest.

$$G_{ST} = \frac{H_T - H_S}{H_T}$$

H stands for heterozygosity (Box 6.3, p. 149); the subscript "S" implies subpopulation and "T" implies total population. We can see how G_{ST} behaves by looking at two extreme cases. Imagine first a case of maximal geographic divergence. Imagine there are two local populations, equal in size, and allele A is fixed in one and allele a is fixed in the other. We first compute the heterozygosity of the total population (H_T). Because the two subpopulations are equal in size, the gene frequency of A is 0.5, and of a is 0.5: $H_T = 0.5$. We now compute the heterozygosity in each subpopulation (H_S). Only one allele is present in each case and $H_S = 0$. $G_{ST} = (0.5 - 0)/0.5 = 1$.

Now imagine that the same two alleles are present, but the two local subpopulations are identical. The frequency of A is 0.5 and of a is 0.5 in both populations. H_T again is 0.5 because the gene frequencies are 0.5 in

the combined population as a whole. H_S also is 0.5 within each subpopulation. $G_{ST} = (0.5 - 0.5)/0.5 = 0$. With no genetic divergence between local populations, G_{ST} is 0; with complete divergence, G_{ST} is 1; with intermediate levels of divergence, G_{ST} has a value between 0 and 1.

What values does G_{ST} have for real species? Table B13.1 lists some figures. We can notice two features. One is that different species show a range of degrees of divergence between local populations. The other is that the figure for humans is low, relative to the majority of other species; the genetic difference between the major human races is lower than for the geographic races of most other species. The figure of $G_{ST} = 0.07$ means that 93% of human genetic variation is present within each racial group. Only 7% of human genetic variation is due to genetic differences between races. The figures in Table B13.1 are based on protein data, but much the same results have been obtained with human DNA (Barbujani et al. 1997). DNA data, however, are not available for enough species to enable an interspecies comparison possible.

Why is racial divergence relatively low in humans as compared with other species? The answer is unknown, but one reason may be that the human

Table B13.1

The fraction of genetic variation within, and between, races of a species, as expressed in the statistic G_{ST}. From data in Crow (1986).

Species	G_{ST}
Horseshoe crab	0.07
Humans	0.07
Drosophila equinox	0.11
Mouse	0.12
Club moss	0.28
Kangaroo rat	0.67

species has evolved only recently. All modern humans may share a common ancestor who lived in Africa as recently as 100,000 years ago (and at any rate less than 500,000 years ago). The genetic differences between human races have accumulated since then. Maybe human races are too recent for much genetic difference to have evolved. In other species, races may have been longer established and G_{ST} has built up to a larger number. Whatever the interpretation, G_{ST} and the other statistics like it provide useful ways of describing geographic variation within a species.

Further reading: Cavalli-Sforza (2000).

In summary, we have seen two concepts of intraspecific variation. One is typological, and supposes that some individuals within a range of variation are better representatives of a species than are other individuals. The other concept is population thinking and treats variation as real and important: no one individual within the range of variation is privileged in any way and all specimens are equally good members of a species.

13.6 Ecological influences on the form of a species are shown by the phenomenon of character displacement

Ecological competition can influence the form of a species (as we mentioned, theoretically, in Section 13.2.3 above). The range of a morphological character, such as beak size, within a species may be limited because the extreme forms suffer competition from neighboring species. In this section we shall look at some evidence for the influence of ecological competition on species. The clearest evidence is provided by *character displacement*.

Character displacement can arise in the following conditions. Two closely related species exist — species that may be ecological competitors. The two species must have a special kind of geographic distribution: it must be the case that both species are present in some places, but only one of the species is present at other places. That is, the two species must have partly overlapping ranges. Character displacement means that individuals of the two species differ more if they are sampled from a place where both species are present (*sympatry*, same place) than do individuals sampled from places where only one of the species is present (*allopatry*, other place). In these terms, character displacement means that sympatric populations of two species differ more than do allopatric populations of the same two species.

Character displacement is difficult to detect because it requires two competing species to have partly overlapping ranges. Many pairs of species either have completely separate ranges, or ranges that are very similar; in either case, it is impossible to study character displacement.

An example of character displacement comes from two species of salamander, *Plethodon cinereus* and *P. hoffmani*. *P. cinereus* lives throughout much of northeastern USA, except for parts of Pennsylvania and Virginia, whereas *P. hoffmani* lives in parts of Pennsylvania where *P. cinereus* is absent. The two species also live together, sympatrically, in a small region of overlap in Pennsylvania. The two species differ in the shape of their heads and jaws: *P. hoffmani* has a jaw that is relatively weak but can be closed fast and *P. cinereus* has a stronger jaw but is slow to snap it shut. *P. hoffmani* is better adapted to eat large prey items, which are caught by immediately closing the mouth on them, whereas *P. cinereus* is better adapted to eat smaller prey, which are eaten by pressing them between the tongue and teeth.

Figure 13.7 shows that the two species differ more in locations where both species are present, that is they show character displacement. The standard interpretation of character displacement is that, where only one species is present, it is released from competition with the other species and it evolves to exploit resources that would be taken by its competitor if it were present. All the allopatric populations evolve to have a similar array of forms. Where both species are present (in sympatry), each species evolves to exploit the resources that it is better adapted to. Competition forces each species to become more specialized. Character displacement shows how ecological competition results in a discrete array of forms within each species.

However, it takes rigorous research to show conclusively that a result such as Figure 13.7 is really caused by ecological competition. Taper & Case (1992), Losos (2000), and Schluter (2000) discuss six criteria that a full study would need to satisfy. For instance,

Figure 13.7
Character displacement in North American salamanders. (a) Character displacement can only be studied in two species with partly overlapping ranges, such that in some places both species are present (sympatry) and in other places only one species is present (allopatry). (b) Where only one of the species of *Plethodon cinereus* or *P. hoffmani* is found (allopatric populations) the form of the species is similar. Where both species are found together (sympatric populations) they differ more. (c) Measurements were made of the skull form, which is related to diet. Redrawn, by permission of the publisher, from Adams & Rohlf (2000).

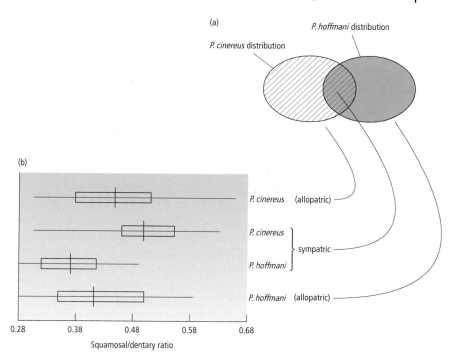

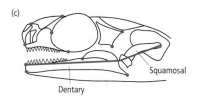

the pattern could be caused by differences in resources, for example if the insect prey differed between sites, or it could be caused by chance. Adams & Rohlf (2000) came close to ruling out all the alternatives to competition: they met five of the six criteria in their study of the salamanders *P. hoffmani* and *P. cinereus*. These salamanders are about the best example we have of character displacement, and its explanation by ecological competition.

13.7 Some controversial issues exist between the phenetic, biological, and ecological species concepts

We saw in Section 13.2 that the phenetic, ecological and biological species concepts are closely related. Most species probably exist in a phenetic, ecological, and biological (that is, interbreeding) sense. However, the three factors do not exactly coincide in nature. The cases in which they do not can be used as test cases, to test whether one species concept is superior to another. The controversies have mainly been between the phenetic and the biological species concepts, or between the ecological and the biological species concepts.

13.7.1 *The phenetic species concept suffers from serious theoretical defects*

The phenetic species concept defines a species as a certain set, or cluster, of phenotypic forms. But why should one set of phenotypic forms rather than another be recognized as a species? The classic version of the phenetic species concept was the typological species concept. It defined a species by reference to the "type" of the species. The trouble with this idea is that, as we saw in Section 13.5, types do not exist in Darwinian theory. Typological theories of species are mainly rejected. A more modern version of the phenetic species concept was developed by the numerical taxonomists. They tried to define species simply as phenetic clusters. The trouble with this (as discussed in Section 16.5, p. 476) is that several statistical methods exist for recognizing phenetic clusters, and those methods can disagree about what the clusters are. The definition of species then requires an arbitrary choice between the different statistical procedures. The underlying problem is that distinct phenetic species do not simply exist "out there" in nature. Some species form obvious phenetic units, but others do not and then we need some other criterion to fall back on.

But the criteria that the phenetic concept might fall back on are unable to save the phenetic species concept as a general stand-alone species concept. The phenetic concept might, for example, fall back on the biological species concept, which defines a species as a set of interbreeding organisms. A set of interbreeding organisms often forms a phenetic cluster, but it does not always. If a set of interbreeding organisms always evolved to differ by x phenetic units from the next such set of interbreeding organisms, we could recognize phenetic species as differing x units from the nearest species. But in fact the two biological species may differ by almost any phenetic amount. *Sibling species* are one case in which phenetic and reproductive units do not coincide. Sibling species are pairs of species that differ reproductively but not morphologically. The classic example is the species pair *Drosophila persimilis* and *D. pseudoobscura*. The two species are separate interbreeding units: if flies from a *persimilis* line are put with flies from a *pseudoobscura* line, they do not interbreed. But they are phenetically almost indistinguishable. Sibling species are an extreme example, to illustrate the general point that phenetic and interbreeding units are not the same in nature. Far from saving the phenetic species concept by providing a measure of phenetic distinctness, the biological species concept shows that the phenetic species concept is trying to do something impossible. Phenetic clusters alone do not satisfactorily divide all of life up into species.

The same point can be illustrated by examples at the opposite extreme: a single species (in the biological sense) that contains a huge array of distinct phenetic forms. Some highly "polytypic" species contain many forms, each of which would be distinct enough to count as a separate species on the classic typological definition of a species.

The phenetic species concept is ambiguous in theory

Sibling species are phenetically almost identical

Polytypic species have diverse phenetic forms within one species

Some butterfly species, such as *Heliconius erato* (Section 8.3, p. 197), contain a number of forms that differ more than do most butterfly species. But the forms can interbreed and are all included in the same species. Species like *H. erato* are called "polytypic": they cannot be defined by reference to one type specimen because they have many distinct forms. Taxonomic practice in sibling species and highly polytypic species follows the biological species concept where sibling species are split into pairs of formally named species, and the many forms of a species such as *H. erato* are all formally named as one species. Many, perhaps most, species form phenetic clusters. But not all do and the phenetic procedures for defining species can only be justified by falling back on the biological species concept. That ultimate reliance on the biological species concept is made clear in difficult test cases such as sibling and highly polytypic species.

13.7.2 *Ecological adaptation and gene flow can provide complementary, or in some cases competing, theories of the integrity of species*

The reproductive and ecological aspects of species are probably usually correlated in nature. Interbreeding among the members of a species results in a set of organisms with shared adaptations to an ecological niche, as we saw in Section 13.2.2. The ecological and biological species concepts are therefore usually not in conflict. However, there are some test cases in which the two concepts make different predictions. For instance, gene flow (migration) can rapidly unify the gene frequencies of separate populations if selection is weak (Section 5.14.4, p. 132). On the other hand, a strong selection force can in theory keep two populations distinct despite gene flow. The relative importance of adaptation to the local ecological conditions and gene flow is an empirical question in cases where the two forces conflict.

Selection can produce divergence despite gene flow

Bradshaw (1971) carried out a major ecological genetic study of plants, particularly the grass *Agrostis tenuis*, on and around spoil-tips in the UK. Spoil-tips are deposited from metal mining and contain high concentrations of such poisonous heavy metals as copper, zinc, or lead. Only a few plants have been able to colonize them, and of these the grass *A. tenuis* has been studied most closely. It has colonized these areas by means of genetic variants that are able to grow where the concentration of heavy metals is high; around a spoil-tip, therefore, there is one class of genotypes growing on the tip itself, and another class in the surrounding area. Natural selection works strongly against the seeds of the surrounding forms when they land on the spoil-tip: the seeds are poisoned. Selection also acts against the metal-tolerant forms off the spoil-tips. The reason is less clear, but the detoxification mechanism may cost something to possess. Where the mechanism is not needed the grass is better off without it.

Metal-tolerant grasses show spatial divergence . . .

Populations of *A. tenuis* show divergence, in that there are markedly different frequencies of genes for metal tolerance on and off the spoil-tips. The pattern is clearly favored by natural selection — but what about gene flow? The biological species concept predicts that gene flow will be low, otherwise the divergence could not have taken

. . . despite gene flow

place. In fact, gene flow is large. Pollen blows in clouds over the edges of the spoil-tips and interbreeding between the genotypes is extensive. In this case, selection has been strong enough to overcome gene flow.

The situation in *A. tenuis* fits better with the ecological species concept than the biological species concept. Ecological adaptation, not reduced gene flow, explains the divergence between the grass on and off the spoil-tips. However, the conditions on the spoil-tips are exceptional and recently established. The selective conditions may soon be removed, for instance if the spoil-tips are cleaned up. If the selective conditions do persist, the conflict between gene flow and ecological adaptation may disappear over time. The grass might evolve a flexible genotype that could switch a metal-tolerating mechanism on or off, depending on where the grass grew up. Or a cost-free detoxification mechanism might evolve (in much the same way as pesticide resistance has evolved in insect pests, see Section 10.7.3, p. 276). Alternatively, gene flow may be reduced. The flowering times of the tolerant and normal types already differ in *A. tenuis*, and that will reduce the gene flow between them. In the future the two forms could evolve into two separate species. One way or another, the conflict between gene flow and selection will be short lived. Either the gene flow pattern, or the selection regime, will change. *A. tenuis* is a partial exception to the rule that biological and ecological species concepts usually agree, but the exception is likely to be minor and short lived relative to evolutionary time.

Selection can produce uniformity in the absence of gene flow

In other cases, different populations of a species have similar gene frequencies even though no gene flow seems to occur between the populations. For instance, Ochman *et al.* (1983) studied the snail *Cepaea nemoralis* in the Spanish Pyrenees. The snail rarely lives above 4,600 feet (1,400 m) in the mountains, and never above 6,500 feet (2,000 m) because of the cold. In the Pyrenees, it lives in neighboring river valleys separated by mountains: where those mountains are higher than 4,600 feet (1,400 m), gene flow between valleys will be absent — and there is probably little gene flow even between the valleys in lower mountains. If gene flow is required to maintain the integrity of the species (that is, the similarity of gene frequencies), populations in different valleys should have diverged.

Snails show genetic uniformity in the absence of gene flow

Ochman *et al.* (1983) measured several characters, including the frequencies of four alleles of the gene coding for an enzyme, indophenol oxidase (*Ipo-1*), in 197 populations (shown as dots in Figure 13.8a). As Figure 13.8c shows, the *Ipo-1* alleles

Figure 13.8 (*opposite*)
(a) Map of the Pyrenees showing sites where the snail *Cepaea nemoralis* was sampled and the river valleys. The rivers are separated by high ground and mountains, and the shaded gray area running from left to right indicates regions where the altitude exceeds 4,900 feet (1,500 m). The stippled green area in the middle indicates the area around which gene frequencies are differentiated: see (c) below. (b) Shell morphology (in this case, background color) shows little geographic variation.

(c) Protein polymorphism, however, falls into three main areas. The map is for the four alleles of one enzyme, indophenol oxidase (*Ipo-1*). Three or so regions can be seen from left to right, with characteristic gene frequencies: to the left, allele 130 is more frequent, in the center, allele 100 is more frequent, and to the right allele 80 is more frequent. These regions transcend the high grounds shown in (a). Similarity within an area is unlikely to be maintained by gene flow. Redrawn, by permission of the publisher, from Ochman *et al.* (1983).

(a)

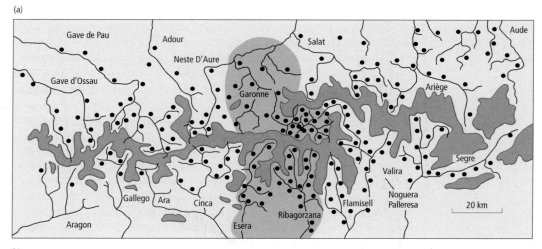

(b)

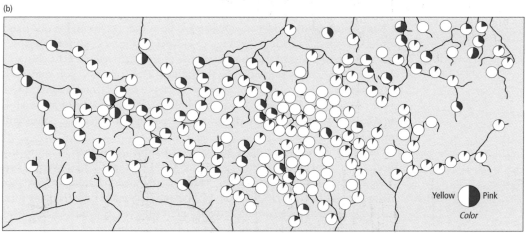

(c)

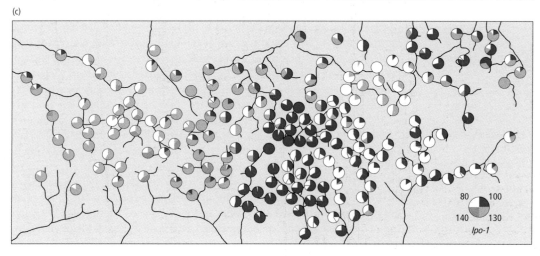

divide the snails into three main regions. From the left, the first main region has a relatively high frequency of allele 130; the second has a high frequency of allele 100; and the right-hand region has a higher frequency of allele 80. These regions transcend the mountian barrier to gene flow, which is shown as a gray region from left to right in Figure 13.8a. The similarity of the populations within each of the three regions, on both sides of a barrier to gene flow, is difficult to explain by the biological species concept. The ecological species concept might be able to explain the pattern, but we need further research on how the different alleles are adapted to different regions across the map.

Asexual forms . . .

A further test case comes from asexual species. The ecological species concept predicts equally clear-cut species in both sexual and asexual forms. There is no reason why only sexual, and not asexual, forms should inhabit niches, and selection should therefore maintain asexual species in integrated clusters much like sexual species. But the biological species concept predicts a difference. Asexual forms do not interbreed and no gene flow occurs. If gene flow holds species together, then asexual forms should have blurred edges; nothing will stop asexual species from blurring into a continuum. Sexual forms should be more clearly discrete than their asexual relatives.

. . . may . . .

Unfortunately, the evidence published so far is indecisive. On the one hand, many authors — especially critics of the biological species concept, Simpson (1961b) being an example — have asserted that asexual species form integrated phenetic clusters just like sexual species. This is supported by the study of Holman (1987) on rotifers. Bdelloid rotifers are a large asexual taxon (Section 12.1.4, p. 319). The monogont rotifers are the sister taxon of the bdelloids, but monogonts are at least sometimes sexual. Holman showed that species in bdelloids have been recognized at least as consistently as in monogonts.

. . . or may not . . .

On the other hand, examples also exist of asexual forms that do not form distinct species. Maynard Smith (1986) has pointed to the example of hawkweed (*Hieracium*). It reproduces asexually and is highly variable, such that taxonomists have recognized many hundreds of "species," and no two taxonomists agree on how many forms there are. In all, asexual species are a potentially interesting test case, but the evidence that has so far been assembled does not point to a definite conclusion.

. . . exist as distinct species

Bacteria, and other microbes, illustrate much the same point. Bacteria mainly reproduce asexually, and yet distinct species of bacteria are named just like in multicellular, sexual life forms. This could mean that the biological species concept is inadequate because it is unable to account for bacterial species. However, genetic exchange does take place between bacterial cells. The units recognized as species in bacteria may then be maintained by gene flow. Alternatively, bacteria may not really form distinct species, and the habit of naming bacterial "species" may be misleading. The evidence about genetic variation in bacteria is too limited to allow a broad conclusion about bacteria

Microbes show clearer species in some cases than in others

as a whole. Much is known about the population genetics of a few bacteria, such as *Escherichia coli*, but the population genetics of most microbes remains obscure. One popular interim conclusion is that some bacteria have extensive genetic exchange between cells and form good species, but other bacteria have little genetic exchange and the application of species concepts in them may be problematic. Cohen (2001) and Lan & Reeves (2001) discuss microbial species. Maynard Smith *et al.* (1993) look at the kind of data that are needed. Meanwhile bacteria, like asexual forms generally, pose a problem for the future, rather than contributing decisive evidence in the present, for the controversies about biological species.

13.7.3 *Both selection and genetic incompatibility provide explanations of reduced hybrid fitness*

Ecological factors can influence the fitness of hybrids

When closely related species can produce hybrids, the hybrid offspring often have low fitness. The hybrids may be sterile (for example, mules) or have reduced viability. The reduced fitness of the hybrids is an example of postzygotic isolation (see Table 13.1), and may be explained by either or both of two processes. One is that the hybrids may have a form that is intermediate between the two parental species and be maladapted because few resourses exist for an intermediate form. In an area where seeds are large or small, one species may have large breaks and another species have small beaks. Hybrids between the two species may have low fitness because few medium-sized seeds are available. This is an ecological theory of low hybrid fitness. It can be illustrated by a study of Darwin's finches by Grant & Grant (2002).

The medium ground finch *Geospiza fortis* lives on the island of Daphne Major, in the Galápagos, and it eats relatively large, hard seeds. The small ground finch *G. fuliginosa* is an occasional immigrant. It eats smaller seeds, and has a lower survival rate than *G. fortis* in normal conditions, when the supply of small seeds is low. The immigrant *G. fuliginosa* hybridizes with the resident *G. fortis*, producing hybrids with intermediate-sized beaks. The hybrids also mainly eat small seeds, and have relatively low survival in normal conditions (Table 13.2). But following the El Niño event, the supply of small seeds increased massively (see Section 9.1, p. 223, and Plate 4, between pp. 68 and 69). The fitness of the hybrids now increased, to at least as high a level as *G. fortis*. The degree of postzygotic isolation between *G. fortis* and *G. fuliginosa* depends on the food supply.

Table 13.2
Hybrid fitness (and therefore postzygotic isolation) between two species of Darwin's finches depends on the food supply. In normal years, small seeds are rare and pure *Geospiza fortis* individuals have higher fitness; following El Niño, the supply of small seeds increases and hybrid fitness improves. Fitness is here measured by survival from egg to first year. (Other measures of fitness showed the same trend.) From Schluter (2000), from data of Grant & Grant.

	Survival to first year
Normal years	
fortis × *fuliginosa* hybrids	0.16
fortis × *fortis*	0.32
El Niño year	
fortis × *fuliginosa* hybrids	0.84
fortis × *fortis*	0.82

Most of the time, small seeds are rare and the hybrids have low fitness. The Grants' measurements show that the reason for the low fitness is that they are poorly adapted ecologically.

Low hybrid fitness may also be due to genetic incompatibilities

Alternatively, hybrids may have low fitness because the two parental species contain genes that do not work well when put together in a hybrid offspring. Section 14.4 (p. 389) will look at this theory further. Suppose that one member of species 1 contains genes *A* and *B* at two loci, and members of species 2 contain genes *a* and *b*. *A* and *B* work well together and produce a good, functioning body, as do genes *a* and *b*. But a hybrid may contain genes *A* and *b*. These two genes may be incompatible. (A crude example would be for *A* and *B* to code for a long left and right leg, and *a* and *b* for a short left and right leg. The unfortunate hybrids would then have one long and one short leg.) This is a genetic explanation for low hybrid fitness. The mule (a hybrid between a male ass *Equus africanus* and a female horse *E. caballus*) is probably explained by some incompatiblity between the genes of asses and horses.

Ecological maladaptation and genetic mismatches can be competing hypotheses to explain any one case of low hybrid fitness. They can be tested between, but the conflict should not be exaggerated. Both factors probably operate in nature, and both can be incorporated into our understanding of species. For instance, the ecological explanation of low hybrid fitness may apply more to closely related species living in the same area. They may hybridize sufficiently often for their genes to remain compatible. Such may be the case in Darwin's finches. The genetic explanation may become more important over time, as two species diverge and their genes become increasingly different.

In summary, nature has supplied us with certain test cases to examine the processes invoked by the biological and ecological species concepts. The processes (ecological adaptation and gene flow) probably usually act together to produce the same result. In some cases, the two processes appear to be in conflict. The test cases may be short lived (as in *Agrostis tenuis*) and of little evolutionary importance; or the results may be ambiguous (as in asexual species); or the test cases may suggest that both processes should be incorporated in the two concepts (as in the theories of low hybrid fitness).

Ecological and reproductive factors are likely both at work

The evidence seems to suggest that both ecological adaptation and interbreeding are needed to explain the sets of forms that we recognize as species. Some biologists, therefore, have suggested that we need a more general species concept. Templeton (1998), for instance, favours a "cohesion species concept," in which all species show "cohesion" (that is, species exist as discrete phenetic clusters) but the reason may differ from one species to another. Some species may exist because of ecological adaptation, others because of gene flow, others because of a mix of the two.

13.8 Taxonomic concepts may be nominalist or realist

13.8.1 *The species category*

When we classify the natural world into units such as species, genera, and families, are we imposing categories of our own devising on a seamless natural continuum, or are the categories real divisions in nature? The problem is an old one. It applies to all taxonomic categories, but has particularly been discussed in the case of the species category. The idea that species are artificial divisions of a natural continuum is called *nominalism*; the alternative, that nature is itself divided into discrete species, is called *realism*.

Biological species are real, not nominal units

On the biological species concept, species are real rather than nominal units in nature. If we take the set of all organisms currently classified as human beings and as chimpanzees, then these organisms do divide into two discrete reproductive units. A human being can interbreed with any other human (subject to provisos, such as that the two humans are of opposite sex and of reproductive age), but with no chimpanzee. Interbreeding between species does not blur out. Here is a thought experiment to illustrate what "blurring out" would mean. Take the set of all human plus all chimpanzee individuals. Then pick an individual at random. Now experimentally place that individual with a range of potential mates from across the entire set of other individuals. If reproductive output varied continuously from 100% to 0% across the full set of mates, then interbreeding could be said to blur out. In fact reproductive output would jump between 100% (or a high figure) and 0% with nothing in between. Human and chimpanzee interbreeding does not blur out. In a way, the strangeness of imagining what blurring out would mean illustrates how humans and chimpanzees form real, not nominal, reproductive units. In any case, humans in fact form a real reproductive unit. So too do most species.

Species are likely to form phenetic units in consequence. Because interbreeding is confined to a certain set of individuals, an advantageous new mutation will spread through that set of individuals, but not into other such sets (that is, other species). If chimpanzees gain a favorable mutation, it will not spread to us even if we would benefit from it. For this reason, biological species often form real, rather than nominal, phenetic clusters. The most striking evidence that species exist as phenetic clusters comes

Folk taxonomy often matches formal taxonomy

from "folk taxonomy." People working independently of Western taxonomists usually have names for the species living in their area, and we can look at whether they have hit on the same division of nature into species as have Western taxonomists working with the same raw material. Some people, it seems, do use much the same classification of species. The Kalám of New Guinea, for instance, recognize 174 vertebrate species, all but four of which correspond to species recognized by Western taxonomists.

As we saw (Section 13.7.1), phenetic and reproductive units do not always coincide. In polytypic butterfly species, there are many discrete phenetic forms and "folk taxonomies" of these butterflies tend to recognize many forms rather than the single biological species. Likewise, folk taxonomies would probably not distinguish sibling species, though most sibling species are too obscure for this question even to have been asked. In summary, species in nature are real rather than nominal interbreeding units in most cases, but not in all.

13.8.2 *Categories below the species level*

Species in many cases form discrete phenetic units. This contrasts with subspecific units such as "subspecies" and "races." (I put the words in quotes because, although the categories are sometimes used, biologists are skeptical about their utility for the reason we are about to look at.) Subspecies and races — the two terms are almost

Figure 13.9
Different species form relatively discrete genetic (and usually phenetic) units; different subspecific units such as races do not. (a) Evolution in two species. Successive genes spread within each species. Species 1 forms a cluster with genes *A* and *B*; species 2 is a distinct cluster with genes *C* and *D*. No individuals have a discordant gene combination such as *AbcD*. (b) Evolution within one species. Advantageous and neutral genes spread locally. Different genes may spread in different places, partly depending on the local conditions. Discordant gene combinations can easily arise, and in area 1 some individuals have the gene (*C*) found in area 2 and other individuals do not. To produce discordant gene combinations between species, a gene (such as *C*) would need to spread not only through species 2 but also through part of species 1. This is usually impossible because of isolating barriers between species. The argument here applies to phenetic characters as well as genes, in so far as the genes code for distinct phenetic characters.

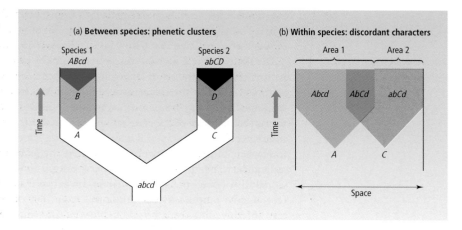

interchangeable — are defined as geographic populations within a species that have a distinct phenetic appearance. The trouble is that variation within a species does not form discrete genetic phenetic clusters in the way that differences between species often do. Sparrows in North America, for example, form a cline in body size from south to north (see Figure 13.5). Northern sparrows are bigger, because of adaptation to temperature. But if we looked at a second character, such as their vocalizations (song) or the frequency of a gene, there is no reason to expect it to form a cline along the same gradient. It might form a complicated gradient related to the rainfall or another factor. Different characters form different spatial patterns, related to different adaptive factors or random drift.

Thus the distributions of different characters within a species are "discordant." Nothing forces the sparrows in one area to form a discrete genetic or phenetic cluster. Interbreeding and non-interbreeding cause different species to form phenetic clusters. The distributions of different characters tend to be concordant, as one mutation after another is fixed within a species (Figure 13.9). Within a species, any character distribution is possible. This is part of the reason why discrete races cannot be recognized in the human species. (The problem is compounded by the low genetic variation within our species; see Box 13.2.) When different people have tried to classify human races, they have found as few as six or as many as 60 races. An objective classification is impossible because different characters vary independently within a species. Skin color, eye shape, and blood groups form independent and discordant clines. This does not show that race is a meaningless concept for human beings. It has cultural and political meanings, but race has a nominal rather than a real meaning in evolutionary biology.

13.8.3 Categories above the species level

The reality of taxonomic categories above the species level in part depends on how those categories are defined, and that is a later topic (Chapter 16). However, one point

Biologists disagree about how real the higher taxonomic units are

can be made here. Evolutionary biologists who support the biological species concept characteristically differ from those who support the ecological species concept in their attitude to the reality of taxa above the species level. The biological species concept can apply to only one taxonomic level. If species are defined by interbreeding, then genera, families, and orders must exist for some other reason. Mayr has been a strong supporter of the biological species concept and (in 1942, for example) duly reasoned that species are real, but that higher levels are defined more phenetically and have less reality; that is higher levels are relatively nominalistic. Dobzhansky and Huxley held a similar position.

Simpson, however, favored a more ecological theory of species. The ecological concept can apply in much the same way at all taxonomic levels. If the lion occupies an adaptive zone corresponding to a single ecological niche, then the genus *Felis* may occupy a broader adaptive zone, and the class Mammalia an even broader adaptive zone. Adaptive zones could have a hierarchical pattern corresponding to (and causing) the taxonomic hierarchy. All taxonomic levels could then be real in the same way. The relative reality of the species, and of higher taxonomic levels, is therefore part of the larger controversy between the ecological and reproductive species concepts.

13.9 Conclusion

In evolutionary biology, the interesting questions about species are theoretical. The practical question of which actual individuals should be classified into which species can on occasion be awkward, but biologists do not tie themselves in knots about it. The majority — perhaps over 99.9% of specimens can be fitted into conventionally recognized species and do not raise even practical problems. Other specimens can be identified after a bit of work — or even left on one side until more is learned about them.

The more interesting question is why variation comes in nature arranged in the clusters we recognize as species. There are several possible answers, as we have seen. Different species concepts follow from different ideas about the importance of interbreeding (or gene flow) and natural selection. It is sometimes possible to test between them, but the results so far have not been enough to confirm any one concept (or any plurality of concepts) decisively. However, there is general agreement that phenetic distinction alone is not an adequate concept, and that the key explanatory processes are interbreeding and the pattern of ecological resources.

Summary

1 In practice, species are defined by easily recognizable phenetic characters that reliably indicate what species an individual belongs to.

2 The biological species concept defines a species as a set of interbreeding forms. Interbreeding between species is prevented by isolating mechanisms.

3 The ecological species concept defines a species as a set of organisms adapted to a particular ecological niche.

4 The phenetic species concept defines a species as a set of organisms that are sufficiently phenetically similar to one another.

5 The biological, ecological, and phenetic (and several other) species concepts are all closely related, and are concerned to explain or describe much the same fact: that life seems to come in the form of distinct species.

6 Individuals mainly interbreed with other members of their own species because of isolating barriers that prevent interbreeding with other species. Isolating barriers can be prezygotic or postzygotic.

7 Geographic variation can be adaptive or neutral. The amount of genetic variation among geographic races of a species can be described quantitatively and is low in human beings relative to other species.

8 The theory of evolution justifies population thinking rather than typological thinking about intraspecific variation: all individuals in a population are equally good members of a species, rather than some being better specimens than others.

9 Character displacement occurs when two species have partly overlapping geographic ranges and the two species differ more in sympatry than in allopatry. Character displacement maybe caused by ecological competition.

10 The biological species concept explains the integrity of species by interbreeding (which produces gene flow), the ecological concept by selection. The two processes are usually correlated, but it is possible to test between them in special cases. Selection can be strong enough to overcome gene flow, and selection can maintain a species' integrity in the absence of gene flow.

11 Taxonomic entities such as biological species may be real or nominal. According to the biological species concept, species can be real, but lower and higher taxonomic levels are nominal. According to the ecological species concept, all taxonomic levels can have a similar degree of realism.

Further reading

Mayr (1963) is the classic account of the species in evolutionary biology; see also Mayr (1976, 2001) and Mayr & Ashlock (1991). Coyne (1994) discusses species concepts, particularly in relation to Mayr's ideas. Dobzhansky (1970), Huxley (1942), Cain (1954), and Simpson (1961b) also contain classic material. Ereshefsky's (1992) anthology contains many of the important papers on species concepts.

More recent books include the volume edited by Howard & Berlocher (1998), which has good chapters on species concepts by Harrison, Templeton, Shaw, and de Queiroz that discuss the use of molecular markers and coalescence. See Levin (2000) on plants. Two other recent books are by Hey (2001) and by Ereshefsky (2001), both of whom question whether species, as recognized in conventional Linnaean clasification, correspond to species as fundamental evolutionary units. The practical problems of species definition are dealt with in most general books about classification, which I list in the further reading for Chapter 16.

On the biological species concept, see almost every source in the previous paragraphs, particularly those by Mayr. On the ecological concept, see Van Valen (1976). On the phenetic species concept, see Sneath & Sokal (1973) and many of the references in Chapter 16. Paterson (1993) is the main source for the recognition species concept, as well as the authors in Lambert & Spencer (1994). For criticism, see Coyne *et al.* (1989). Ritchie & Philips (1998) provide evidence of intraspecific variation in SMRS, in contrast with the theory that stabilizing selection acts on SMRS. See also the material on antagonistic sexual selection in Section 12.4.7 of this text.

For isolating mechanisms see the books by Mayr and Dobzhansky above. On plants, see Grant (1981) and Levin (2000). For background on the African cichlids, see Stiassny & Meyer (1999). See also Fryer (2001).

For geographic variation the classic source is again Mayr (1963), and the topic is covered in population genetic texts such as those listed in Chapter 5 in this book. For more on the *Linanthus* example, see Wright (1978) for background and Turelli *et al.* (2001b) for the cutting edge of modern research. Huey *et al.* (2000) is a nice example of recently evolved geographic variation. On population versus typological thinking, see Ghiselin (1997) and Hull (1988) in addition to Mayr (1976 — of which I extracted one classic essay in Ridley (1997)). Pre-Darwinian taxonomists have, since Mayr, often been criticized as typologists. However, the distinction between population and typological thinking is better used conceptually than historically — Winsor (2003) argues that essentially no pre-Darwinian taxonomists were typologists, though they did not appreciate variation in the way we now do.

Character displacement is well reviewed by Schluter (2000), most recently, and by Taper & Case (1992). Brown & Wilson (1958) is the original source. Schluter (2000, pp. 166–8) has a table of other examples such as the salamanders, together with information on how well they have been studied. Another classic example comes from Darwin's finches, and chapter 10 of Weiner (1994) is a popular account, while Grant (1986) contains a more authoritative discussion of it.

The difficulties in the phenetic species concepts are a special case of the difficulties in all phenetic classification: see the references in Chapter 16 later in this book. On heavy metal tolerance in plants, see Bradshaw (1971) and Ford (1975), and Palumbi (2001b) on human-driven evolution in general.

European oaks are a further good case study in ecological versus biological (gene flow) species concepts: see Van Valen (1976) again, and Muir *et al.* (2000). Other recent studies of selection and gene flow include Blondel *et al.* (1999) on blue tits in Corsica, and Smith *et al.* (1997) on rainforest biodiversity. The ecological and genetic explanations of hybrid fitness are discussed in Schluter (2000) and many of the papers about reinforcement, hybrid speciation in plants, and the Dobzhansky–Muller theory that are referred to in Chapter 14.

Berlin (1992) is a book about folk taxonomy, and Gould (1980) contains a popular essay on the subject.

Study and review questions

1 Review the main arguments for and against the phenetic, biological, and ecological species concepts.

2 In a pair of "sibling species," how many species are there in the: (i) phenetic, (ii) biological, and (iii) ecological species concepts?

3 Review the kinds of prezygotic and postzygotic isolating barriers that exist.

4 Calculate the statistic G_{ST}, which describes the amount of geographic differentiation within a species, for species 1–3 below.

Species	H_T	H_S	G_{ST}
1	0.5	0.5	
2	0.5	0.25	
3	0	0	

What biological factors might cause G_{ST} to be lower in some species than in others?

5 Do asexual organisms form species like sexual organisms, and what consequences does the answer have for our concept of species?

6 Is population or typological thinking more appropriate in classifying the following entities? (The answer is not certain in all cases, and they are as much topics to think about and discuss, as to provide final answers about.) (i) Chemical elements (such as atoms of carbon, hydrogen, gold, etc.); (ii) human cultures; (iii) biological species; (iv) human emotions (such as fear, anger, etc.); (v) mechanisms of transport (such as cars, walking, airplanes, and so on); and (vi) scientific theories (such as evolution, gravity, quantum theory, etc.).

7 How can we test between the ecological and genetic theories of postzygotic isolation?

8 In the (i) phenetic, (ii) biological, and (iii) ecological species concepts, are (a) species, (b) subspecies/races, and (c) higher taxonomic categories, real or nominal entities in nature?

14 Speciation

Speciation means the evolution of reproductive isolation between two populations. Two main processes have been suggested by which reproductive isolation can evolve. Reproductive isolation may evolve as a by-product of evolutionary divergence between two populations. Or it may be directly favored, in a process called reinforcement. This chapter begins by showing that we have extensive evidence for, and a good theoretical understanding of, the "by-product" theory of speciation. The evidence comes from laboratory experiments and biogeographic observation. The chapter then goes on to look at the theory of reinforcement. This theory is controversial: the evidence is inconclusive and we cannot show either that it is important or that it is trivial. The chapter also looks at the special case of hybrid speciation in plants, at the possibility of speciation between populations that are not geographically separated, and at two current research trends — the influence of sexual selection on speciation, and the use of modern genomic techniques to identify genes that cause reproductive isolation.

14.1 How can one species split into two reproductively isolated groups of organisms?

Reproductive isolation is the main topic in research on speciation

The crucial event for the origin of a new species is reproductive isolation. As we saw in Chapter 13, the members of a species usually differ genetically, ecologically, and in their behavior and morphology (that is, phenetically) from other species, as well as in who they will interbreed with. Some biologists prefer to define species not by reproductive isolation but by other properties, such as genetic or ecological differences. Probably no single property can provide a universal species definition, applicable to all animals, plants, and microorganisms. However, many species do differ by being reproductively isolated, and even if the evolution of reproductive isolation is not always the crucial event in speciation, it is certainly the key event in research on speciation. The topic of this chapter is the evolution of reproductive isolation. The aim is to understand how a barrier to interbreeding can evolve between two populations, such that one species evolves into two.

Reproductive isolation can be caused by many features of organisms (see Table 13.1, p. 356). However, for most of the research in this chapter, we only need a distinction between prezygotic and postzygotic isolation. Prezygotic isolation exists when, for instance, two species have different courtship or mate choices, or different breeding seasons. Postzygotic isolation exists when two species do interbreed, but their hybrid offspring have low viability or fertility. Some of the theories of speciation apply only to prezygotic isolation, some only to postzygotic isolation, and some to both.

14.2 A newly evolving species could theoretically have an allopatric, parapatric, or sympatric geographic relation with its ancestor

Speciating populations can have various kinds of geographic relations

We can start with a distinction between different geographic conditions in the speciating populations. If a new species evolves in geographic isolation from its ancestor, the process is called *allopatric speciation*. If the new species evolves in a geographically contiguous population, it is called *parapatric speciation*. If the new species evolves within the geographic range of its ancestor, it is called *sympatric speciation* (Figure 14.1). The distinctions between these three kinds of speciation can blur, but we shall begin the chapter with the most important of the three processes: allopatric speciation. Almost all biologists accept that allopatric speciation occurs. The importance of parapatric and sympatric speciation are more in doubt, and we shall come on to them later.

In allopatric speciation, new species evolve when one (or more) population of a species becomes separated from the other populations of the species, in the manner of Figure 14.1a. This kind of event often happens in nature. For example, a species could split into two separate populations if a physical barrier divided its geographic range. The barrier could be something like a new mountain range, or river, cutting through the formerly continuous population. Or the intermediate populations of a species may be driven extinct, perhaps by a local disease outbreak, leaving the geographically

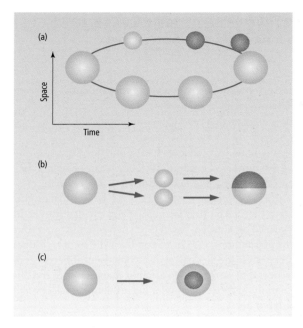

Figure 14.1
Three main theoretical types of speciation can be distinguished according to the geographic relations of the ancestral species and the newly evolving species. (a) In allopatric speciation the new species forms geographically apart from its ancestor; (b) in parapatric speciation the new species forms in a contiguous population; and (c) in sympatric speciation a new species emerges from within the geographic range of its ancestor.

extreme populations cut off from each other. Or a subpopulation may migrate (actively or passively) to a new place, outside the range of the ancestral species, such as when a few individuals colonize an island away from the mainland. Such a population, at the edge of the main range of a species, is called a "peripheral isolate."

Geographic separation alone is not reproductive isolation

One way or another, a species can become geographically subdivided, consisting of a number of populations between which gene flow has been cut off. This is not, in itself, an isolating barrier in the sense of Table 13.1 (p. 356). An isolating barrier is an evolved property of a species that prevents interbreeding. When two populations are geographically cut off, gene flow ceases but only because members of the population do not meet. The two populations have not yet evolved a genetic difference. The evolution of an isolating barrier requires some new character, such as a new courtship song, to evolve in at least one of the populations — a new character that has the effect of preventing gene flow. In the theory of allopatric speciation, the cessation of gene flow between allopatric populations leads, over time, to the evolution of intrinsic isolating barriers between the populations. Let us see what happens to the reproductive isolation between these populations over evolutionary time.

14.3 Reproductive isolation can evolve as a by-product of divergence in allopatric populations

We have two main kinds of evidence that reproductive isolation evolves when geographically separate populations are evolving apart. One comes from laboratory experiments and the other comes from biogeographic observations.

14.3.1 *Laboratory experiments illustrate how separately evolving populations of a species tend incidentally to evolve reproductive isolation*

When two geographically separate populations are evolving independently, different genes will be fixed in each, whether by drift or adaptation to different environments. The theory of allopatric speciation suggests that two such populations will also, at least sometimes, evolve some degree of reproductive isolation in consequence.

Populations can be kept on different resources . . .

The idea has been tested experimentally. We can keep two populations apart, allowing them to evolve indendently for a number of generations. Then we test whether they have evolved any degree of reproductive isolation. Dodd (1989), for example, performed the experiment with fruitflies (*Drosophila pseudoobscura*). The flies had originally been caught in Utah and were then taken to a lab at Yale and divided into eight populations: four of them were placed on a starch-based food medium; the other four on a maltose-based medium. The populations were reared on these different resources for a number of generations. After a while the flies had evolved detectable differences in their digestive enzymes — differences that were almost certainly adaptations to the different resources. Thus, the populations had diverged under the influence of selection to live on different resources in the laboratory.

. . . and prezygotic isolation evolves between them

Dodd exploited these populations to see whether any reproductive isolation had evolved as an incidental consequence of the divergence. She placed recently emerged males and females from the starch and maltose populations in a cage, after marking all the individuals of one of the populations. She then measured who mated with whom, and found that the "starch" flies preferred a "starch" mate, and the "maltose" flies a "maltose" mate (Figure 14.2). Some reproductive isolation had evolved — in this case it is prezygotic isolation. It presumably evolved because the changes that had occurred in the population influenced reproductive behavior in some way.

The result is general, and remarkable

Dodd's is only one experiment among many. Rice & Hostert (1993) listed 14 experiments that measured whether prezygotic isolation emerged between populations that had been experimentally isolated, and found that in 11 of them it did; in the other three there was no significant change. The evolution of reproductive isolation is a general result in experiments in which two populations are evolving separately, in different environmental conditions. The result is quite striking because it might not have been predicted. Would you expect that if you kept one population of humans on a starch diet and another on a sugar diet for a number of generations, that at the end of the period they would have evolved a mating preference for their own dietic type? The evolution of adaptations to the food supply (starch in one population, sugar in the other) is predictable. The evolution of reproductive isolation might not be, but it does seem to happen. The result is also interesting because reproductive isolation is not, at least directly, being selected for. The adaptations to the environment (such as diet) though are being selected for, and the flies in Dodd's experiment duly evolved appropriate digestive enzymes. However, the reproductive isolation just "drops out," as an incidental consequence of the experimental procedure. The experiment did not selectively breed from individuals that showed some mating preference, such that mating preferences evolved upwards over the generations. The mating preference somehow evolves as a correlated response when selection favors new adaptations to the environment.

Figure 14.2

Prezygotic isolation emerges between populations that have adapted to different conditions. (a) The experimental design: four populations of fruitflies were kept on starch medium and four others on maltose medium. After a number of generations the tendency of the flies to mate with others like themselves was measured. In the experimental series, 12 females from a maltose population and 12 more from a starch population were put in a cage with 12 males from a starch and 12 males from a maltose population; the numbers of the four kinds of mating couple that formed were counted. One such experiment was done for each of the four starch populations with each of the four maltose populations, making 16 experiments in all. (b) An example of the results. (c) The average isolation for all 16 experiments. In the control series, 12 females and 12 males from one of the four starch populations were put with another 12 males and 12 females from another of the four starch populations; the same controls were done with the maltose populations too. Again, one example is given in (b): a pair may be formed between a male and a female from the same starch population, or a male from one of the four starch populations may pair with a female from a different starch population. Notice the higher value of the isolation index for the experimental crosses than the control ones in (c). Prezygotic isolation had evolved between the populations that had experienced different media, but not among populations that were isolated but on the same media. Drawn from data in Dodd (1989).

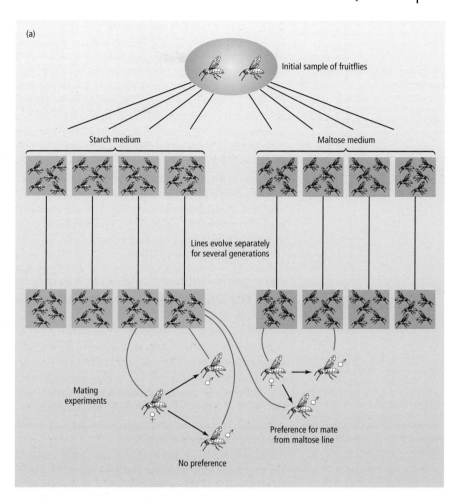

(a)

Initial sample of fruitflies

Starch medium

Maltose medium

Lines evolve separately for several generations

Mating experiments

Preference for mate from maltose line

No preference

(b) **Examples of results**

Experimental cage

Male	Female	
	Starch	Maltose
Starch	22	9
Maltose	8	20

Isolation index $= \left(\dfrac{42 - 17}{59}\right) = 0.42$

Control cage

Male	Female	
	Same	Different
Same	18	15
Different	12	15

Isolation index $= \left(\dfrac{33 - 27}{60}\right) = 0.1$

(c) **Average isolation indexes for all 16 crosses**

	Average isolation index
Starch × maltose population crosses	0.33
Control crosses	0.014

The isolation index is calculated as (number of matings to same type − number of matings to different type)/total number of matings. It varies from 1 (for complete reproductive isolation) to 0 (for random mating, or zero isolation) through to the theoretical extreme of −1 if matings were exclusively between opposite typed flies.

Divergence by drift alone may not
be enough

Two other results of the experiments are worth noticing. One is that they suggest, though they do not prove, that speciation normally requires natural selection; genetic drift alone is not enough. Look at the controls in Dodd's results, for instance (Figure 14.2). No reproductive isolation evolved between populations that were evolving separately but in the same environment. These populations would have evolved apart by drift, but not by selection. Reproductive isolation only evolved between lines kept on different foods, and selection would have been acting differently between them. Templeton (1996), however, has argued that this experimental design is inappropriate for testing the influence of drift in speciation. Secondly, experiments have usually measured the evolution of prezygotic, not postzygotic, isolation. This likely only reflects what the experimenters happened to do. Postzygotic isolation would probably evolve by the same process in experimental populations, but this has not been properly shown. In conclusion for the experiments on allopatric speciation, we have strong evidence that prezygotic isolation tends to evolve in populations that are kept separately, in different conditions, for many generations.

14.3.2 Prezygotic isolation evolves because it is genetically correlated with the characters undergoing divergence

In an experiment such as Dodd's (Figure 14.2), the experimenter is not directly selecting for reproductive isolation. The experimenter selects for an ecological adaptation: some populations of flies are selected to live on one food type, other populations to live on another food type. The prezygotic isolation between the populations evolves somehow as a by-product. Here we shall look at the genetic reason. It is probably that the characters influencing the ecological adaptation are genetically correlated with characters influencing prezygotic isolation. The correlation could exist for two reasons: pleiotropy and hitch-hiking.

Two genetic factors may contribute to the evolution of prezygotic isolation, pleiotropy . . .

Pleiotropy means that one gene influences more than one phenotypic character of the organism. Consider, for example, a gene that influences the shape of a bird's beak. Beak size is related to the food that the bird eats: smaller beaks are adapted to eat smaller seeds, larger beaks to eat larger seeds (Section 9.1, p. 223). If two populations occupied two islands with different-sized seeds, the populations would evolve apart as the birds adapted to the local food supply. Beak shape, in this sense, is an ecological adaptation.

Beak shape can also influence reproductive behavior. Some birds may choose their mates by direct physical inspection of their beaks, but the influence may often be less direct. Figure 14.3 shows an example, from the research of Podos (2001), in Darwin's finches. Beak shape is associated with the kind of song the bird sings. Species with large beaks, for instance, do not produce rapid trills, whereas species with small beaks do. This may be a direct physical consequence of the beak size as it may be physically harder for a bird with a large beak to sing a rapid trill than for a bird with a small beak. Then, when two populations adapt to different food supplies, their songs will change too. Darwin's finches partly choose their mates according to the songs they sing. Thus a change in diet can incidentally cause a change in reproductive isolation. The genetic mechanism is pleiotropy: a gene that is favored because it improves ecological

Figure 14.3
Song form may be pleiotropically correlated with beak shape in Darwin's finches. The beaks and sound spectrograms are shown for eight species from one of the Galápagos Islands. The spectrogram has time on the *x*-axis (the bar represents 0.5 seconds) and frequency on the *y*-axis. Notice that the species with larger beaks produce slower trills (less of the recognizable units per unit time) and with a lower range of frequencies. Statistical analysis shows that the effect is not due to phylogenetic relationship. Redrawn, by permission of the publisher, from Podos (2001).

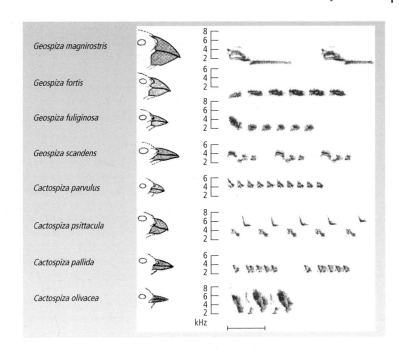

adaptation will also cause some reproductive isolation. The pleiotropy arises because the same morphological character (the beak) influences both feeding and mating.

... and hitch-hiking

Hitch-hiking means that when natural selection favors a gene at one locus, genes at linked loci may also increase in frequency (Section 8.9, p. 210). In Dodd's experiment, natural selection increased the frequency of genes coding for appropriate digestive enzymes. Maybe a closely linked locus influences the fruitfly's courtship dance. Then, when a gene for an ecological adaptation (the digestive enzyme) increases in frequency, it may drag along with it a linked gene for a new courtship dance step. Prezygotic isolation could again evolve as a by-product, but the genetic mechanism is hitch-hiking rather than pleiotropy.

14.3.3 *Reproductive isolation is often observed when members of geographically distant populations are crossed*

The populations of a species in different geographic areas tend to evolve genetic differences (Section 13.4, p. 359). We usually do not know whether the members of different populations can interbreed because they live in separate places and the opportunity for interbreeding does not arise. But in some cases, a biologist has brought samples from distant populations into the lab, and measured the amount of reproductive isolation. Figure 14.4 shows an example from the Californian flower *Streptanthus glandulosus*. The flower lives on serpentine soils, which are found in discrete, local areas. The flower therefore has a discontinuous distribution, with many small local populations. The

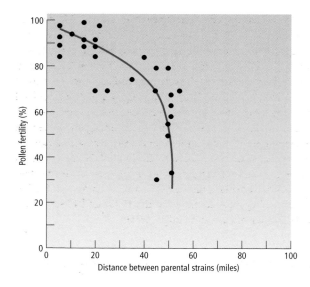

Figure 14.4

Postzygotic isolation between populations of the Californian flower *Streptanthus glandulosus*. Flowers taken from different populations were crossed, and produced hybrids. The amount of good pollen produced by the hybrid offspring was measured and expressed as a percentage of the pollen produced by the parents. Hybrids from more distant crosses tended to be less fertile. (50 miles ≈ 80 km.) Redrawn, by permission of the publisher, from Kruckeberg (1957).

results show that crosses between members of nearby populations are usually fertile, but fertility decreases for crosses between more distant populations.

Geographically distant populations have often evolved isolating barriers

Research of this kind has been mainly done with plants, and has measured postzygotic isolation. The usual result is that some crosses show postzygotic isolation while others do not, and that more distant populations are more strongly isolated. Some research of this kind has also been done on animals, and on prezygotic isolation. Korol *et al.* (2000), for example, showed that fruitflies from different slopes of Mount Carmel in Israel preferentially mated with flies from their own locality. In all, we have extensive evidence that isolating barriers tend to evolve between geographically distant populations of a species in nature.

The members of distant populations normally have to be brought together in the lab by an experimenter in order to measure the amount of reproductive isolation between them. They do not usually naturally meet up. However, there is one exceptional circumstance in which members of "distant" populations come together in nature: this is the phenomenon of *ring species* (Section 3.5, p. 50). In California, the salamander *Ensatina eschscholtzii* spread down from the north, and sent separate colonizing populations down the coast and inland to the east of the central valley (Jackman & Wake 1994). These two lines of populations meet up in the south, in San Diego County, where they are reproductively isolated and effectively two species (Wake *et al.* 1986). The two reproductively isolated populations, or species, in the south are connected by a continuously interbreeding series of populations looping round north California and Oregon.

Ring species are striking examples

Ring species provide dramatic evidence that normal genetic divergence within one species can build up to a sufficient level to generate two species. We do not see it, in most cases, because the genetic extremes within a species are living far apart. But in a ring species the extremes exist side by side and the resulting reproductive isolation is directly observable in nature. Examples are rare because the populations of a species are rarely arranged in a ring. But several examples do exist (Irwin *et al.* 2001b). The herring gull (*Larus argentatus*) and lesser black-backed gull (*L. fuscus*) of northern Europe are a ring species, connected by a ring of populations around the North Pole. In central Asia, the warbler *Phylloscopus trochiloides* is distributed in a ring around the treeless Tibetan Plateau. The species appears to have originated in the south, in India. It then colonized northwards, either side of the Tibetan Plateau. The colonizing populations evolved slight genetic differences, and the songs diverged between the eastern and western populations. Eventually, these two lines of populations meet up again in central Siberia, but here the songs are so different that the members of the east population do not interbreed with members of the western population (Irwin *et al.* 2001a). And on the tiny Pacific island of Moorea, certain species of snail in the genus *Partula* had evolved ring species around individual small mountains (Murray & Clarke 1980). Alas, the *Partula* of Moorea are extinct. They all succumbed in the 1980s and 1990s, in yet another unnecessary manmade ecological disaster (Tudge 1992).

14.3.4 *Speciation as a by-product of divergence is well documented*

In summary, we have abundant evidence that reproductive isolation evolves as a by-product of divergence between geographically separate populations. If populations are experimentally kept in different conditions in the lab, those populations evolve adaptations to their conditions — and reproductive isolation also evolves as a consequence. If separate populations adapt to different local conditions in nature, they also evolve reproductive isolation as a consequence. For prezygotic isolation, we saw that the genetic basis is probably pleiotropy or hitch-hiking. However, that explanation is mainly hypothetical at present, because little genetic research has been done on prezygotic isolation. We can turn now to the genetics of postzygotic isolation. It has been the subject of much genetic research, and is well understood both empirically and theoretically.

14.4 The Dobzhansky–Muller theory of postzygotic isolation

14.4.1 *The Dobzhansky–Muller theory is a genetic theory of postzygotic isolation, explaining it by interactions among many gene loci*

Could postzygotic isolation evolve by changes at one locus?

What kind of genetic changes give rise to postzygotic isolation? Postzygotic isolation means that hybrid offspring are produced but they either die before breeding or live and are sterile. In this section, we shall see that postzygotic isolation is likely to be caused, genetically, by an interaction between the genotypes at multiple loci, rather than by genotypes at a single locus. The simplest hypothetical genetic control would have one genetic locus, and the fitnesses of the genotypes would be as follows:

	Species 1	Hybrid	Species 2
One-locus genotype	*AA*	*Aa*	*aa*
Fitness	High	Zero	High

Here we have supposed that each species is fixed for a different allele at the locus. The hybrids are heterozygotes, and if there is postzygotic isolation then those heterozygotes must have low fitness. However, there is a theoretical argument to suggest that the genetics underlying postzygotic isolation is unlikely to have this form. The problem is: how could it have evolved in the past? Species 1 and 2 exist now, with genotypes *AA* and *aa*. In the past, an ancestral species split into two to give rise to modern species 1 and 2. What genotype did that ancestral species have for this locus? We do not know, but two simple possibilities are that it had either *AA* or *aa*. Suppose, for instance, that the ancestor had *AA*. The genotype has been retained in species 1. In the evolution of species 2, *AA* has evolved into *aa*. The allele *a* was, we might reason, advantageous in species 2. The allele *a* appeared as a new mutation in the *AA* ancestor — creating an *Aa* heterozygote. But we know that *Aa* heterozygotes are lethal or sterile (this is shown in the postzygotic isolation between modern species 1 and 2). The one-locus model of postzygotic isolation contains a paradox. The modern set up (with *AA* in one species and *aa* in the other) must have evolved somehow. But evolution has to pass through a disadvantageous, or even deadly, stage — which is improbable, if not impossible. The same problem arises if the ancestor was *aa*. It also arises, in a more convoluted form, if the ancestor had some third allele, such as *A**. The paradox is unavoidable if postzygotic isolation is controlled by one-genetic locus.

The solution to the paradox was suggested by Dobzhansky and by Muller in the 1930s, and is often called the *Dobzhansky–Muller theory*. They realized that postzygotic isolation could evolve without difficulty if it was controlled by interaction among more than one genetic locus. The simplest case has two loci (Figure 14.5): the ancestor has a two-locus genotype such as *AABB*. It splits into two allopatric populations. In the environmental conditions of population 1, the allele *a* is advantageous. Two copies of *a* are better than one, and the population will evolve from *AABB* to *AaBB* to *aaBB*; natural selection fixes the *a* allele. This is simple evolution by natural selection. In the environmental condition of population 2, a change at the other locus is advantageous. Natural selection drives the population from *AABB* to *AABb* to *AAbb*, and fixes the *b* allele.

Now suppose we cross members of the two populations. One is *AAbb*, the other *aaBB*; the hybrid offspring will be double heterozygotes *AaBb*. These hybrids may have low fitness, without creating the paradox we met in the one-locus model. The double heterozygote has never existed before. The two new alleles *a* and *b* have never been together in the same body. The *a* gene may be advantageous in a *BB* body but not in a

Probably not

How about more than one locus?

Figure 14.5

The Dobzhansky–Muller theory for the evolution of postzygotic isolation. An ancestral species splits into more than one population, between which gene flow is absent. Each population adapts to its local conditions by genetic change. The genetic changes are likely to be at different gene loci in the different populations. If the two populations later meet up, the genetic changes in each will probably be incompatible and the hybrids sterile or inviable. The genotypes shown are for two loci. *A* and *a* are alleles at one locus; *B* and *b* are alleles at a second locus.

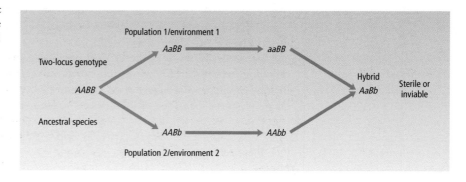

b-containing body. The *a* and *b* genes may be incompatible. In formal terms, the fitness interaction between the two loci are epistatic (Section 8.8, p. 207). The fitness of a genotype, such as *Aa*, depends on the genotype at the *B/b* locus. In informal terms, the *a* and *b* genes cause some kind of genetic snarl-up when they are combined in the same body. Therefore, postzygotic isolation can evolve by interactions between genes without the paradox that we met with only one locus. The evolution of postzygotic isolation is theoretically more likely to be caused by multilocus genetic interactions than by single-locus ones.

14.4.2 *The Dobzhansky–Muller theory is supported by extensive genetic evidence*

The main prediction of the Dobzhansky–Muller theory is that postzygotic isolation is caused by multilocus interaction, not by single loci. The prediction is in principle easy to test. Two closely related species can be forced to interbreed in the laboratory, and classic genetic methods used to estimate the number of gene loci that contribute to sterility or inviability of the hybrid offspring. Many such experiments have been performed, particularly with fruitflies. Coyne & Orr (1998) reviewed evidence from 38 experiments on 26 pairs of species (or near species). Only in two species pairs was low hybrid fitness due to a genotype at one locus. In the other 24 species the problems in the hybrid were due to epistatic interactions at multiple loci. It is a well supported generalization about speciation, that postzygotic isolation is due to multilocus gene interactions.

Hybrid fitness is influenced by many loci

Coyne & Orr (1998) also note that the Dobzhansky–Muller theory makes two subtler, more specific predictions. One is that the amount of postzygotic isolation should "snowball" as the number of loci differing between two population goes up. If new alleles have evolved at only two loci, they may be incompatible and cause postzygotic isolation, as shown in Figure 14.5. But, alternatively, *a* may not be incompatible with *b* and the hybrids then do not have reduced fitness. Now suppose a third locus also undergoes evolutionary change. Hybrids now contain three new genes, *a*, *b*, and *c*. The new *a* gene was compatible with *b*, but the hybrid may now suffer because *a* is incompatible with *c*, or *b* with *c*. The increase from two to three loci has increased the number of gene interactions from one to three. The number of possible gene interactions (any one of which may be incompatible) goes up faster than the number of gene loci that differ between the species. In the Dobzhansky–Muller theory, postzygotic isolation is caused by gene interactions. Hence the prediction that postzygotic isolation will "snowball" as two populations diverge genetically.

Two subtler predictions can be made

A second subtle prediction is that there should be asymmetry in the gene interactions in the hybrid. In Figure 14.5, low hybrid fitness is caused by the snarl-up between the two new alleles *a* and *b*. If this is true, we can see that the other pair of alleles (*A* and *B*) ought not to cause a problem. The reason is that the *A* and *B* alleles are the ancestral combination, and the *AABB* ancestral organisms were good, functioning creatures. Thus if we can identify which gene combinations are causing problems in the hybrid, we can predict that the complementary sets of genes at those loci will not cause problems.

These two subtler predictions of the Dobzhansky–Muller theory have not been as extensively tested as the basic prediction that postzygotic isolation is caused by multi-locus interactions. However, the Dobzhansky–Muller theory has been rather neglected until recently. Biologists are beginning to explore its rich implications for the genetic changes that cause speciation, and these two subtle predictions are examples of the kinds of hypotheses being tested now.

14.4.3 The Dobzhansky–Muller theory has broad biological plausibility

The account of the Dobzhansky–Muller theory that we looked at in Figure 14.5 was abstract. We considered two loci with two alleles each (A/a and B/b), but said nothing about what the genes code for. We simply deduced that *if* the new alleles (a and b) were incompatible, then there is postzygotic isolation. In reality, whether or not the alleles are incompatible will depend on the biological details of what they code for. How general the Dobzhansky–Muller process is will depend on how common it is for newly evolved alleles in different populations to be incompatible.

A first concrete example concerns genes that interact in a metabolic pathway. Imagine two genes (G_1 and G_2) coding for two enzymes (E_1 and E_2) that successively process a substrate ($S_1 \rightarrow S_2 \rightarrow S_3$):

Metabolic pathways . . .

$$
\begin{array}{cc}
G_1 & G_2 \\
\downarrow & \downarrow \\
\end{array}
$$
$$
S_1 \xrightarrow{\;E_1\;} S_2 \xrightarrow{\;E_2\;} S_3
$$

In the environment of one population, the food resources may differ from those in the environment of the other population. The two populations will evolve different enzymes to digest their differing local food supplies. We can symbolize the enzymes by E_1 and E_2 for population 1 and $E_1{}^*$ and $E_2{}^*$ for population 2. The enzyme pairs work within each population, as E_1 processes the substrate into a form that can be tackled by E_2. But in a hybrid there will be E_1 from one population and $E_2{}^*$ from the other. E_1 may process the substrate into a form that $E_2{}^*$ does not bind to, resulting in a metabolic inefficiency. Some biologists doubt whether multiple genes in metabolic pathways in fact underlie postzygotic isolation (Orr & Presgraves 2000), but the example at least illustrates how the Dobzhansky–Muller process could operate.

. . . signal–receptor systems . . .

As a second example, consider genes coding for an egg receptor protein and a sperm lysin in abalone (Swanson & Vacquier 1998). The sperm lysin breaks a hole in the egg by binding to a particular receptor molecule on the egg membrane. The sperm lysin gene and egg receptor gene evolve in concert within a species; the sperm can recognize eggs of the right species. But different forms of the genes evolve in different species. Now imagine that the ancestral abalone had an allele L_1 of the sperm lysin gene and R_1 of the egg receptor gene. Its genotype was $L_1L_1R_1R_1$. It split into two population, each of which evolved different genotypes: $L_2L_2R_2R_2$ in one and $L_3L_3R_3R_3$ in the other. If we cross members of the two populations the hybrid offspring are L_2R_2/L_3R_3. They may have normal viability, but if they are bred their fertility may be approximately halved.

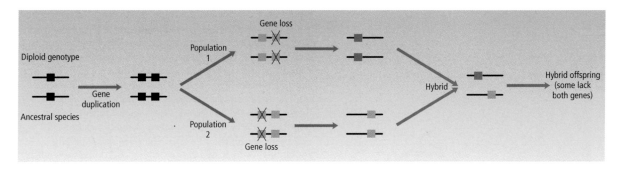

Figure 14.6

Gene duplications can provide an example of the Dobzhansky–Muller process (which was illustrated in Figure 14.5). A gene duplicates in an ancestral species. The species then splits into two populations, and different copies of the duplicated gene are lost in each population. If members of the two populations later interbreed, the hybrids will have reduced fertility because some of their offspring (one-sixteenth of them in a simple case) lack both copies of the gene.

Half the sperm–egg encounters among hybrids will be incompatible and something may go wrong in them.

. . . parasite–host relations . . .

Thirdly, consider parasite–host coevolution. In Section 12.2.3 (p. 323) we looked at the way in which hosts may evolve specific resistance mechanisms against locally abundant parasites. Population 1 may evolve a set of resistance genes (R_1R_2) that work against parasites in its environment; population 2 may evolve another set of resistance genes (R_3R_4) that work against its parasites. But hybrids may contain combinations of resistance genes (R_1R_3 or R_2R_4) that do not work against any parasites.

. . . and gene duplication and loss . . .

Fourthly, consider gene duplication and loss (Figure 14.6). (Sections 2.5, p. 30, 10.7.2, p. 275, and 19.3, p. 559, provide background on gene duplications.) A gene may be duplicated, or a species may split into two and the two new populations may lose different copies of the gene. If members of the two populations then meet, hybrids may initially be viable because they contain copies of the gene from both parental populations. However, recombination within hybrids can produce offspring with no copies of the gene, as Figure 14.6 illustrates. This is the phenomenon of "hybrid breakdown," which is a form of postzygotic isolation (Table 13.1, p. 356). The hybrids themselves are healthy but subsequent generations have reduced fitness.

. . . are four (of many) examples that illustrate the Dobzhansky–Muller set-up

These are only four of the many ways in which simple evolutionary change in separate populations may fit the Dobzhansky–Muller process. Almost any genetic changes in the DNA have the potential to prove incompatible with genetic changes elsewhere in the DNA. There are extensive interactions between gene loci within a body, and these interactions will not proceed smoothly by chance. The genes in a human body interact well because natural selection has been acting, over millions of years, to favor versions of our genes that interact well. No such force constrains our genes to interact well with the genes of other species, such as chimpanzees. Human–chimpanzee hybrids would be likely to show a great genetic snarl-up, and be inviable. The Dobzhansky–Muller theory has good biological plausibility as well as being theoretically coherent and empirically supported.

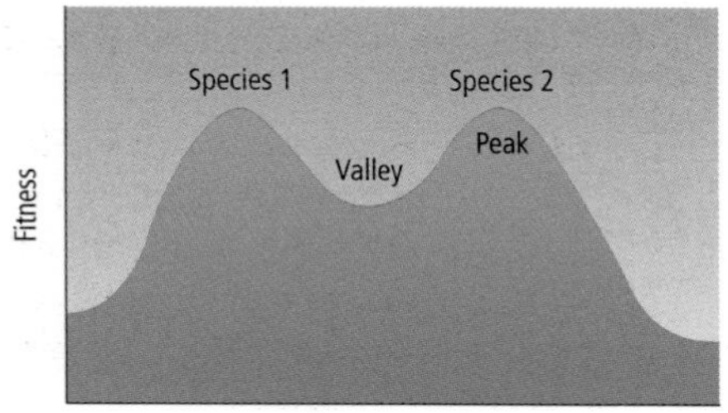

Figure 14.7

Valley crossing during speciation. The figure shows an adaptive landscape (see Figure 8.7, p. 214): quality of adaptation, or fitness, is on the *y*-axis; character state, or genotype, is on the *x*-axis. Related species are adapted to somewhat different environments, and each is well adapted to its own environment, Intermediate forms are less well adapted with a fitness valley lying between the two species. Natural selection acts against valley crossing. If the landscape has the illustrated shape then speciation is an evolutionarily difficult process, perhaps requiring special conditions in which the action of natural selection is suspended. Some theories of speciation, such as the Dobzhansky–Muller theory, do not require valley crossing.

14.4.4 *The Dobzhansky–Muller theory solves a general problem of "valley crossing" during speciation*

We began this section on the Dobzhansky–Muller theory by looking at a hypothetical one-locus genetic model of postzygotic isolation. We can now look at a more general version of that argument, and use it to explore a general question about speciation. Is speciation an "easy" evolutionary process that follows almost automatically from normal evolutionary change, or is it an evolutionarily "difficult" process that requires extraordinary mechanisms?

Speciation might be thought to be a difficult evolutionary process . . .

The members of a species are usually fairly well adapted to their environments, and the genes at different loci work well together — they interact well enough to produce viable, fertile bodies. Species probably lie near, if not on, the peaks of an adaptive landscape, and different species occupy different peaks (Figure 14.7). The problem in speciation is that it seems to require "valley crossing." For species 1 to evolve into species 2, or vice versa, the population has to pass through a disadvantageous phase. The one-locus model illustrates the difficulty — the fitness valley in the one-locus model corresponds to the heterozygous hybrid (Section 14.4.1). However, many other genetic models could also have a valley between two adaptive peaks.

. . . requiring drift . . .

It is hard, if not impossible, for a population to cross an adaptive valley. Natural selection and random drift are the two main forces of evolution. Natural selection almost always acts to drive species toward a peak on an adaptive landscape. Natural selection opposes valley crossing as it requires genotypes of lowered fitness to somehow spread through the species. Random drift is only a powerful force when the alternative genotypes are selectively neutral. For drift to drive a population across a valley, it has to work contrary to selection, and that is unlikely. Therefore, if speciation requires valley crossing, speciation is a difficult evolutionary process and will not normally happen; it will require some special conditions.

. . . or a genetic revolution . . .

For instance, evolutionists have argued that speciation happens in small stressed populations where a "genetic revolution" occurs (Mayr 1963, 1976). Or that it happens by a special process of "peak shifts." Or that it happens when the action of natural selection is temporarily suspended, perhaps when a colonizing population exploits abundant resources in the absence of competitors (the "founder flush" model: see

Templeton 1996). Without entering into the details of these models, we can note that they all invoke peculiar evolutionary mechanisms. Speciation requires the normal action of selection and drift to be suspended. The inspiration of these ideas is that speciation is a difficult process, because of the need for valley crossing. This is one view of speciation.

. . . in exceptional conditions

The Dobzhansky–Muller model offers a different view of speciation. It has no valley crossing. The fitness valley is generated as a consequence of the separate evolution of the two species. In the Dobzhansky–Muller view, speciation happens as an almost automatic consequence of ordinary selection and drift within a population, as each population evolves in its own environmental conditions. Speciation does not require special conditions, in which normal evolutionary processes are suspended.

In the Dobzhansky–Muller theory, speciation is evolutionarily easy

The Dobzhansky–Muller theory applies only to postzygotic isolation, but a similar argument can be made for prezygotic isolation. We saw earlier that prezygotic isolation evolves as a by-product of normal evolutionary change, by genetic processes such as pleiotropy (Section 14.3.2 above).

We have theories for both prezygotic and postzygotic isolation that are well validated in fact and in theory, and in both cases the evolution of reproductive isolation does not require valley crossing. Speciation instead is an almost automatic consequence of evolutionary change. The special mechanisms proposed in the alternative, valley crossing, view are little supported or unsupported by facts and are at best questionable in theory (Turelli *et al.* 2001a). That could change in the future, but many evolutionists currently prefer the view that speciation is an evolutionarily "easy" process, requiring no more than the most commonplace of evolutionary mechanisms.

14.4.5 *Postzygotic isolation may have ecological as well as genetic causes*

We distinguish two theories of hybrid fitness

The Dobzhansky–Muller theory is not the only theory of postzygotic isolation. In Section 13.7.3 (p. 373), we looked at an ecological theory of postzygotic isolation, closely related to the ecological species concept. We looked at an example from Darwin's finches (Table 13.2, p. 373): the fitness of hybrids between *Geospiza fortis* and *G. fuliginosa* on Daphne Island in the Galápagos archipelago depends on the food supply. When, in the 1982–83 El Niño event, the supply of small seeds increased, hybrid fitness increased too. When small seeds were rare, hybrid fitness was lower.

The ecological theory of postzygotic isolation differs from the Dobzhansky–Muller theory. In the Dobzhansky–Muller theory, hybrids are inferior because of incompatibilities between their genes. The internal working of the hybrid's body will be defective. In the ecological theory, the internal workings of the hybrid's body are just as good as in any member of a pure species. Any hybrid inferiority is due to external conditions. The ecological and Dobzhansky–Muller theories are potentially competing theories in any one case; but they could also combine to explain the full amount of postzygotic isolation. The ecological theory may work better for very closely related species that repeatedly hybridize; their gene pools will contain few incompatible genes. The Dobzhansky–Muller process may become more powerful as two species evolve separately for a longer time. But it will take further research to establish the influence of each theory. Meanwhile, the Dobzhansky–Muller theory has been extensively tested and supported and has almost undoubtedly contributed to speciation, but only limited work has been done on the ecological theory so its contribution is more uncertain.

14.4.6   *Postzygotic isolation usually follows Haldane's rule*

In 1922, J.B.S. Haldane identified the following pattern in postzygotic isolation:

Hybrids of the heterogametic sex have lower fitness than hybrids of the homogametic sex

> When in the F_1 offspring of the two different animal races one sex is absent, rare, or sterile, that sex is the heterozygous one.

We should now say "heterogametic" instead of "heterozygous." In mammals and in fruitflies males are heterogametic (XY — whereas females are XX). In birds and in butterflies, it is the other way round and the females are heterogametic. Haldane found that in crosses in which one gender of hybrid offspring has lower fitness than the other gender, the gender with lower fitness is male in mammals and fruitflies (and female in birds and butterflies). Eighty years on, the facts continue to support Haldane remarkably well (Table 14.1). His generalization has come to be called *Haldane's rule*.

The rule has also gained an extra interest. As stated, Haldane's rule only says that *when* fitness differs between the sexes of hybrid offspring, the heterogametic sex has lower fitness. But it could be that in most cases the sexes do not differ; the rule would then only be something of a curiosity. However, we now know that in fact most speciation events do go through a "Haldane rule" phase. Coyne & Orr (1989) quantified this fact as follows. We can define the amount of postzygotic isolation (I) as the average fractional reduction in fitness of the hybrid offspring in a cross between the two species, or near species. I equals one minus hybrid fitness. Thus if we cross two members of a species their offspring will have high fitness, and $I = 0$. If we cross individuals from two different species, usually the hybrid fitness is zero, and $I = 1$. The isolation (I) increases from 0 to 1 during speciation. For Haldane's rule we are interested in pairs of "species"

Table 14.1
Support for Haldane's rule. "Asymmetry" in the column "Hybridizations with asymmetry" means that one sex is affected more than the other with respect to the trait such as fertility. Many species of butterflies, moths, and mosquitos are also known to follow the same rule. From Coyne & Orr (1989).

Group	Trait	Hybridizations with asymmetry	Number obeying Haldane's rule
Mammals	Fertility	20	19
Birds	Fertility	43	40
	Viability	18	18
Drosophila	Fertility and viability	145	141

(a) **It could have been …**

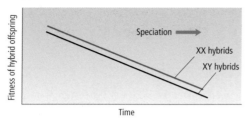

(b) **… but in fact it normally is**

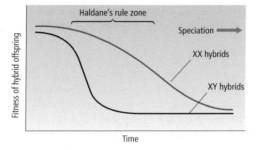

Figure 14.8

Haldane's rule. (a) We might naively expect that the fitness of male and female hybrid offspring would on average decrease in much the same way as speciation proceeds. (b) In fact sterility or inviability evolve first in the heterogametic hybrid offspring. This gives rise to an intermediate stage in speciation, during which XX hybrids have higher fitness than XY hybrids; this stage is where Haldane's rule operates. In fruitflies and mammals, males are XY and females XX. In birds and butterflies males are XX and females XY. (In (a) and part of (b) the fact that the XY line has been drawn just below the XX line is insignificant: they are meant to be one on top of the other. Also, the fact that the line is drawn straight in (a) is insignificant. What matters is that it goes down the same for male and for female hybrid offspring; whether the line goes down smoothly or suddenly or in a curve is unimportant here.)

(or near species), that are in the gray area on the way to full speciation. They have I between 0 and 1. In the simplest case, all the male hybrid offspring (in a mammal, for instance) would be dead or sterile and all the female offspring would be perfectly alright. Then $I = 0.5$, averaging over all offspring.

Now it might be thought that during speciation, the degrees of isolation would increase in some fashion from 0 to 1 (Figure 14.8a). But Coyne & Orr (1989) found that of 43 *Drosophila* species pairs with intermediate degrees of isolation ($0 < I < 1$), 37 showed a sex difference and fitted Haldane's rule. It is a normal fact about speciation, at least in fruitflies, that low male hybrid fitness evolves earlier than low female fitness. The true course of speciation looks something like Figure 14.8b. Haldane's rule is a general property of speciation, not a curiosity.

Modern genetic techniques have been used to test Haldane's rule in new ways. For instance, snazzy genetic tricks can be used to introduce various numbers of genes from one fruitfly species into another fruitfly species. True *et al.* (1996) introduced genes from *Drosophila mauritiana* into *D. simulans*. They found that if they introduced the same number of random genes into males and females, the males were six times as likely to be sterilized. The result can be dramatized in an anthropomorphic analogy. It is as if we introduced a certain number of chimpanzees genes into human males and females, randomly scattering the chimp genes in the human DNA. True *et al.*'s result would then imply that the men were more likely to be sterilized by the experience than the women. The evolutionary interpretation is that male hybrid sterility evolves faster, at a lower level of genetic divergence between species, than female hybrid sterility. That is another way of expressing Haldane's rule.

Haldane's rule is a big generalization about speciation. Whatever the explanation is for the rule, we can conclude that speciation often proceeds in the manner of

Haldane's rule describes a normal stage in speciation, not a curiosity

Box 14.1
Haldane's Rule is Probably (at Least in Part) Explained by the Dobzhansky–Muller Theory

Postzygotic isolation in the Dobzhansky–Muller theory is caused by interactions between genes at many loci. Let us see what happens if one of the loci is on the X chromosome and one on an autosome (Figure B14.1). The two parental species have internally satisfactory combinations of genes, but the hybrid offspring contain gene

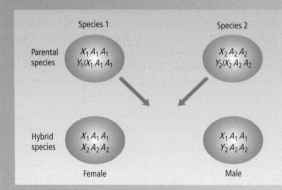

Figure B14.1

The "dominance theory" of Haldane's rule. Some gene combinations in hybrids will be new and incompatible: for example, X_1/A_2 combinations (where X_1/A_2 refers to a combination of one gene on the X chromosome from species 1 and another gene on an autosome of species 2). If the incompatible gene on the X chromosome is recessive then the gene combination will not be expressed in female hybrids (the X_2 gene is dominant) and female hybrids are alright. But in male hybrids the X_1/A_2 combination is expressed because the male lacks an X_2 chromosome. Males have reduced fitness. The theory is illustrated as if for a fruitfly, in which males are heterogametic. In a species in which females are heterogametic, "males" and "females" would need to be reversed.

combinations that are incompatible. A gene on the X chromosome of species 1 (X_1) may be incompatible with a gene on an autosome of species 2 (A_2). What matters is whether the defect caused by the incompatibility between X_1 and A_2 is dominant or recessive. If it is dominant, it will damage both male and female hybrid offspring; if it is recessive it will damage male but not female offspring. Thus the standard Dobzhansky–Muller theory can explain Haldane's rule with the added assumption that some of the genes at work on the X chromosome are recessive. In all, we explain Haldane's rule by two genetic properties: (i) postzygotic isolation is due to interactions between many gene loci (that is, epistasis), and some of these gene loci will be on the X chromosome; and (ii) some of these X-linked genes are recessive.

This is called the "dominance" theory of Haldane's rule. It works well for genes causing inviability in fruitflies. However, it is probably not a complete explanation. The genetic explanation for cases in which the heterogametic sex of hybrid offspring are inviable probably differs from cases in which they are sterile. Genes that influence viability usually affect males and females equally, because they are genes that influence the well-being of the whole body. Male bodies do differ in some respects from female bodies, and some genes do influence viability in only one gender; but these genes are exceptional. The genes that influence fertility, by contrast, mainly differ in male and female bodies. Genes influencing female fertility are expressed in the ovaries and influence oogenesis; these genes are switched off in male bodies. Some explanation other than the simple dominance theory given here may be needed for cases of sterility that fit Haldane's rule. Also, the theory as given here works for the kind of X chromosome dosage compensation that is found in fruitflies, but the theory needs to be extended to explain mammals.

Therefore, the Dobzhansky–Muller theory can be used to explain, at least in part, the longstanding and well documented generalization known as Haldane's rule. Haldane's rule has been a topic of active research recently, particularly since a classic experiment by Coyne (1985). Haldane's rule has proved to be an excellent route to understanding the genetic changes that cause speciation, or at least that cause postzygotic isolation.

Figure 14.8b, not Figure 14.8a. Postzygotic isolation evolves faster in the heterogametic gender. But what is the explanation for Haldane's rule? The question has been the topic of active research recently, and many detailed genetic hypotheses have been tested. Box 14.1 describes how the basic Dobzhansky–Muller theory can explain Haldane's rule in at least some cases. The Dobzhansky–Muller theory may not be a universal, complete theory of Haldane's rule, but it is an important influence. Thus, the Dobzhansky–Muller theory not only explains the multilocus, epistatic genetic control of postzygotic isolation, it also helps to explain a general sex difference in the time course of speciation.

14.5 An interim conclusion: two solid generalizations about speciation

Speciation often occurs as a by-product of evolutionary divergence

We have abundant evidence, from experiments and biogeographic observations, that speciation evolves as a by-product when two geographically separate populations evolve apart. For prezygotic isolation, we have some hypotheses, but few research results, on the genetic changes that underlie it. For postzygotic isolation, we have extensive theory and evidence about the genetic changes that underlie it. We can conclude the chapter so far by saying that there are two solid results in the study of speciation: reproductive isolation evolves as a by-product of allopatric divergence, and postzygotic isolation is caused by epistatic interactions among multiple genetic loci.

These generalizations are worth keeping in mind as we move on. We are now going to turn to some less solid, more controversial areas of research on speciation. If we concentrated on these controversial areas alone, it might appear that little is known about speciation and that it is a permanently confused area of evolutionary biology. John Herschel, a senior scientific figure when Charles Darwin was starting in scientific research, described the question of how a new life form could appear on Earth as "the mystery of mysteries." It is a haunting phrase, and Darwin remembered it. The theory we have looked at so far does not provide a complete account of how all new species evolve. But it does help to demystify the origin of species, and provides a clear scientific searchlight for future research.

14.6 Reinforcement

14.6.1 Reproductive isolation may be reinforced by natural selection

So far we have been looking at one of two main theories about how reproductive isolation evolves: that it evolves as a by-product when natural selection favors different genetic changes in separately evolving populations. The second theory suggests that natural selection can act directly to increase the amount of isolation between two populations. The process is called *reinforcement* and the general precondition for reinforcement to operate is as follows. We assume there are two genetic types within a population, and hybrids between them have lower fitness than the offspring of matings within each type. The genetic difference between the two types could be in a multilocus genotype, or in alleles at one locus, or in chromosomal form. The symbols A and A' are general, and stand for any of these kinds of genetic differences. The condition for reinforcement is then:

Reinforcement means . . .

	Type 1	Hybrid	Type 2
Genetic type	AA	AA'	$A'A'$
Fitness	High	Low	High

The two types are partially, but not completely, postzygotically isolated from each other. More formally, we could say the fitnesses of the two pure forms AA and $A'A'$ are 1, and the fitness of the hybrids is more than 0 but less than 1.

One way in which this set up could arise would be if two populations initially occupied separate (allopatric) ranges and diverged, but their ranges then changed and the two populations met up again. One population has one genetic type (A), the other A'. These two populations might have evolved some postzygotic isolation by the Dobzhansky–Muller process, but the isolation might not be complete. (We shall meet some other ways in which the same basic set up can arise, in Sections 14.9 and 14.10, when we look at parapatric and sympatric speciation.) What will be the next evolutionary step?

. . . natural selection to increase reproductive isolation

Reinforcement is one possibility. Natural selection may increase the amount of prezygotic isolation. If an AA individual mates with another AA individual, they produce offspring with high fitness. If an AA individual mates with an $A'A'$ individual they produce hybrid AA' offspring who have low fitness. Natural selection favours individuals who mate with others who are genetically like themselves — that is, assortative mating.[1] The theory of reinforcement assumes that some postzygotic isolation exists, and argues that prezygotic isolation will increase. Natural selection cannot, except in strange circumstances, favor increases in postzygotic isolation. Natural selection favors increased prezygotic isolation, because the individuals save themselves from producing inferior hybrid offspring. But an increase in postzygotic isolation means that the fitness of hybrids goes down. The hybrids become more likely to die. Natural selection cannot favor genes that make their bearers more likely to die (except in special conditions described in the theory of kin selection, Section 11.2.4, p. 298). Indeed the main effect of natural selection on postzygotic isolation will be to decrease it, by favoring fitter hybrids. Thus reinforcement is really only a theory of prezygotic isolation, not postzygotic isolation.

How important is reinforcement in speciation? The initial condition for it looks simple, and probably arises quite often. All we need is the evolution of two genetic forms between which crosses are disadvantageous. The argument, that natural selection then favors prezygotic isolation (or assortative mating), looks simple and inevitable. We might therefore expect reinforcement to occur quite often during speciation, as a supplement to the "by-product" theory we have looked at. Many evolutionary theorists, from Wallace to Dobzhansky, have supported the theory of reinforcement. However, when we look in more detail at the theory, and the evidence that has been put forward for it, we find that the case is unconvincing. The arguments are well worth knowing,

The theory of reinforcement has had supporters

---

[1] Assortative mating means like mates with like. It can be contrasted with disassortative mating, in which individuals preferentially mate with the other type from themselves, and with random mating. The theory of reinforcement is usually concerned with evolution from random mating toward stronger and stronger assortative mating. When assortative mating is absolute — an individual will never mate with someone of the other genetic type — prezygotic isolation is complete and speciation has occurred.

however, because the theory of reinforcement has not been definitively falsified. Reinforcement remains a topic of active research, and biologists hold a range of views on how important it is in evolution.

14.6.2 *Preconditions for reinforcement may be short lived*

The theory of reinforcement has problems with . . .

The precondition for reinforcement is that two genetic types exist, and hybrids produced by crosses between those types are disadvantageous. Natural selection favors assortative mating. However, other evolutionary forces will also be acting, and may remove the preconditions before reinforcement has increased reproductive isolation to the point of full speciation.

. . . the loss of rare forms . . .

1. *Natural selection may eliminate the rarer genotype.* The precondition for reinforcement is inherently unstable. Imagine that 90% of the population are *AA* and 10% are *A'A'*. Initially the two types mate randomly. For simplicity, we can asume that *AA* and *A'A'* individuals have equal chances of survival, and *AA'* hybrids have a much lower chance of survival. With random mating, an *AA* individual has a 90% chance of mating with another *AA* individual and producing *AA* offspring. *AA* individuals mate with *A'A'* individuals only 10% of the time, producing inferior hybrids. *A'A'* individuals mate among themselves, producing high quality *A'A'* offspring only 10% of the time; they mate with *AA* individuals and produce low quality hybrid offspring 90% of the time. The rarer genotype has an automatic disadvantage, and natural selection acts to eliminate it. It may be driven extinct before full assortative mating has evolved. (The precondition for reinforcement is an instance of positive frequency-dependent selection: Section 5.13, p. 127.)

. . . gene flow . . .

2. *Gene flow merges the two genetic types.* Imagine that the two genetic types in the population are multilocus sets of genes. *A* might stand for one set of genes at five loci (*BCDEF*) and *A'* stand for another set (*bcdef*). Hybrids (*BCDEF/bcdef*) have low fitness, but some recombinants will be formed (*Bcdef*, *bcdEF*, and so on), provided hybrid fitness is more than zero. Over time the two distinct types will blur into a continuous population. The rate of blurring will depend on the fitnesses of the different gene combinations. Again, the precondition for reinforcement may disappear before speciation takes place.

. . . and recombination

3. *Recombination between gene loci may disrupt reinforcement.* A model of reinforcement usually has three gene loci (or three sets of gene loci). One controls adaptation: that is, the *A/A'* locus (or loci) in the case that we have been looking at. This locus might control a digestive enzyme, with *A* types able to eat one kind of food, *A'* another kind of food, and *AA'* hybrids not able to eat either kind of food. A second locus controls the degree of assortative mating. A third locus controls a character that is used in mating decisions. Mating may be decided by coloration, and a color locus may have two types, blue and green. Reinforcement can work if *AA* individuals are usually blue and *A'A'* individuals are usually green. Natural selection favors assortative mating based on color. The problem is that recombination may generate green *AA* individuals (and blue *A'A'* individuals). A blue *AA* who mates assortatively may now have an *A'A'* partner, and produce inferior hybrid offspring. The process of reinforcement will then collapse. Reinforcement requires tight linkage

between the character used in mating decisions and the character influencing hybrid fitness; but that requirement often may not be met.

These three objections considerably weaken the theory of reinforcement. But they do not show that it is impossible, and counterarguments can be made. For instance, the preconditions can be stabilized if the two genetic types are a polymorphism that is actively maintained by natural selection (by any of the standard mechanisms of Sections 5.11–5.14, pp. 121–33). If the rarer genotype can eat a food type that the common type cannot, its advantage in feeding may balance its disadvantage in more often producing hybrid offspring. The same process can prevent the two genetic types from being blurred away by gene flow. Reinforcement then has more time to act. There are also ways of configuring the various loci such that recombination does not disrupt reinforcement (Schluter 2000, p. 192). The theory about reinforcement is therefore inconclusive. We can identify weaknesses in the theory, but they are not enough to show that it is impossible. We need to turn to the facts to find out how important reinforcement has been in nature.

14.6.3 *Empirical tests of reinforcement are inconclusive or fail to support the theory*

Evidence for reinforcement from artifical selection is . . .

Two kinds of evidence have been used to test for reinforcement, one experimental and the other biogeographic. The experimental evidence consists of artificial selection experiments, in which the experimenter creates the preconditions for reinforcement. For instance, Kessler (1966) put two closely related fruitfly species together in the laboratory, in conditions in which they interbred. Over a number of generations, any hybrids were prevented from breeding. The experimenter gave the hybrids low (indeed zero) fitness. The tendency of the fruitflies to mate assortatively was measured, and it increased over time (Figure 14.9). Natural selection favored assortative mating, which duly increased. Many other experiments have obtained similar results. The problem with these experiments, for our purposes here, is that arguably they do not test the theory of reinforcement. Reinforcement is a process that drives speciation. But the experimenter made hybrid fitness zero, meaning that speciation was effectively complete. Gene flow between the lines was experimentally prevented. Rice & Hostert (1993) called experiments of this kind "destroy the hybrids" experiments.

. . . inadequate . . .

However, the experiments do have value. They show, for instance, how natural selection can increase prezygotic isolation once postzygotic isolation is complete. But they do not provide much of a test of reinforcement. A good test would make the hybrid fitness low, but not zero, with some gene flow continuing during the experiment. Hostert (1997) did this experiment and found no increase in assortative mating when hybrid fitness was allowed to be anything above zero. But one experiment is not enough to prove that reinforcement never works. Another species, in some other conditions, might show a different result. However, at present the evidence from artificial selection either fails to test, or fails to support, the theory of reinforcement.

. . . or negative

The second main kind of evidence comes from biogeography. We require a special biogeographic set up, in which two closely related species have partly overlapping ranges. (This is the same set up we met in Section 13.6, p. 366, when looking at ecological

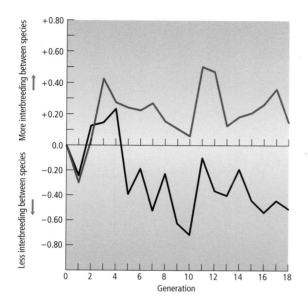

Figure 14.9

Artificial selection in female *Drosophila pseudoobscura* for increased (low isolation, green line) and decreased (high isolation, black line) tendencies to mate with male *D. persimilis*. The *y*-axis is an index of frequency of mating with a member of the other species (heterospecific mating). When the index is positive, females are more likely to mate with heterospecific males than are control females; when it is negative they are less likely to. Redrawn, by permission of the publisher, from Kessler (1966).

Evidence from biogeography . . .

character displacement.) For example, *Drosophila mojavensis* and *D. arizonae* are two closely related species of cactus-eating fruitfly that coexist in Sonora, Mexico. But elsewhere in the southwest, each species can be found living without the other: *D. mojavensis*, for example, lives in Baja California where *D. arizonae* is absent, and *D. arizonae* lives in other regions of Mexico without *D. mojavensis* (Figure 14.10).

The key result concerns the amount of prezygotic isolation between the species in sympatric and allopatric populations. When a male of one species is put with a female of the other, they are less likely to mate than are a pair from the same species. Wasserman & Koepfer (1977) measured the degree of mating discrimination in populations taken both from where the two species co-occur and from where only one of the species lives. They found that discrimination against potential mates from the other species was stronger in the flies from regions where both species are found (Figure 14.10c).

The result is an example of character displacement. Character displacement occurs when two species differ more in sympatry than in allopatry. The term can refer to any character, and *D. mojavensis* and *D. arizonae* show reproductive character displacement, or to be exact character displacement for prezygotic reproductive isolation. These two species are one example among several in which this result has been found.

. . . is consistent with reinforcement . . .

One interpretation of reproductive character displacement is that prezygotic isolation has been reinforced in sympatry. When the two species do not encounter each other (that is, allopatrically), natural selection will not have favored discrimination against mates from the other species. In sympatry, where interbreeding may produce hybrids of reduced fitness, selection will have favored mechanisms to prevent crossbreeding. Reinforcement has acted to increase prezygotic isolation only where the two species coexist.

(a) Simplified system to test for character displacement

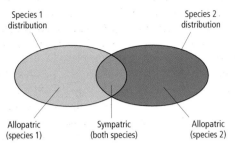

Species 1 distribution

Species 2 distribution

Allopatric (species 1)

Sympatric (both species)

Allopatric (species 2)

(b) **Distribution of** *D. arizonae* **and** *D. mojavensis* **in the southwest**

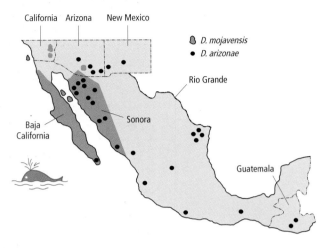

California   Arizona   New Mexico

🐚 *D. mojavensis*
● *D. arizonae*

Rio Grande

Baja California

Sonora

Guatemala

(c) **Isolation between** *D. arizonae* **and** *D. mojavensis*

D. mojavensis female with *D. arizonae* male		*D. arizonae* female with *D. mojavensis* male	
D. mojavensis female		*D. arizonae* female	
Allopatry	$I = 0.3$	Allopatry	$I = 0.9$
Sympatry	$I = 0.94$	Sympatry	$I = 0.8$
D. arizonae male		*D. mojavensis* male	
Allopatry	$I = 0.54$	Allopatry	$I = 0.78$
Sympatry	$I = 0.6$	Sympatry	$I = 0.92$

Figure 14.10

(a) A study of character displacement requires two species with partly overlapping ranges. (b) Distributions of *Drosophila mojavensis* and *D. arizonae* in southwest America. They coexist in part of Sonora, Mexico, and are found alone in other areas, including large regions of Mexico for *D. arizonae* and Baja California for *D. mojavensis*. The dots on the map for *D. arizonae* are the collecting sites for the experiment in (c), within a fairly continuous distribution. (c) Experimental demonstration that reproductive isolation is higher between the two species in sympatry than in allopatry. The experiments give (i) a female of one species a choice of mating with males of either of the species, and (ii) a male of one species a choice of mating with females of either species. In the experiment, the number of matings with members of the same species (H_s) and with the other species (H_o), and total number of matings (N) was measured, and the isolation index was calculated, $I = (H_s - H_o)/N$, as explained in Figure 14.2. In (c) the top left number means that *mojavensis* females, taken from a place where *arizonae* does not live (Baja California) — allopatric *D. mojavensis* — when put with males of both species, show an isolation index of only 0.3. The same applies for the other seven conditions. Redrawn, by permission of the publisher: (b) from Koepfer (1987) and (c) from Wasserman & Koepfer (1977).

The problem is that reinforcement is not the only explanation for the observations. The same observations could also arise without reinforcement. The reason is that sympatric species pairs with low levels of isolation may be lost, by fusion or extinction. To see the problem, imagine that a number of populations, all descended from one ancestral species, are evolving allopatrically. They will evolve various degrees of isolation, depending on how the Dobzhansky–Muller process happens to influence postzygotic isolation, and how pleiotropy and hitch-hiking happen to influence prezygotic isolation. Some pairs of populations will evolve high isolation, other pairs will evolve low isolation. We could measure the average amount of isolation between allopatric populations by bringing individuals sampled from them into the lab. The figure would be some average for the range of values for particular pairs of populations.

. . . but also open to alternative interpretations

Now imagine also that the geographic distributions of some of the populations change, and some of the formerly allopatric populations become sympatric. If two populations that become sympatric had already evolved a high amount of reproductive isolation, then the two will propably continue to coexist. But if they happen to have evolved a low level of isolation, the processes that we looked at in Section 14.6.2 will start to operate. Either the rarer of the two populations will be lost, or gene flow between the two populations will cause them to fuse into one. Either way, the population pair is lost from the dataset. The only population (or species) pairs we are left with in sympatry are the pairs with high isolation. Then the amount of isolation between pairs of species living in sympatry will be high because the pairs with low isolation have been lost, not because reinforcement has increased isolation.

The argument does not show that reinforcement has not operated in cases such as Figure 14.10, it only shows that the evidence is inconclusive. Coyne & Orr (1989) ingeniously subdivided the evidence in various ways, making a stronger case for reinforcement; but their evidence is still explicable without it (Gavrilets & Boake 1998). Therefore, the biogeographic evidence, like the evidence from artificial selection, is currently inconclusive. Evolutionary biologists remain undecided about reinforcement. Few would say that it never operates, but the theoretical and empirical case for its importance is unconvincing. Of the two processes that can drive the evolution of reproductive isolation — (i) divergence with isolation as a by-product, and (ii) reinforcement — the first is well documented and is almost certainly important in speciation, but the second is not well documented and its influence in speciation is indeterminate.

14.7 Some plant species have originated by hybridization

We encountered the origin of plant species by hybridization in Section 3.6 (p. 53), where we saw how a new species of primrose, and a natural species of *Galeopsis*, were artificially produced by hybridization. Interspecies hybrids are largely sterile, usually because the chromosome pairs, which consist of one chromosome from one species and another chromosome from the second species, do not segregate regularly at meiosis. In order for a new species to evolve, this sterility has to be overcome. One famous mechanism is polyploidy. If the chromosome numbers are doubled, each chromosome pair at meiosis contains two chromosomes from one species, and regular segregation is restored. Polyploids arise naturally, by mutation, and may lead to the evolution of a new species. The polyploid hybrids are interfertile among themselves, but reproductively isolated (by the mismatch in chromosome numbers) from the parental species; they are therefore well defined new species.

Hybrids, after polyploidy, may form a new species

The simplest cases to identify are those, like *Primula kewensis*, in which the new species is a simple 50 : 50 hybrid, produced from two parental species, with 50% of its genes coming from one parental species and 50% from the other. Within the past century, the natural evolution of four new species of this sort has been recorded, two in

Britain and two in North America. The latter two examples belong to the genus *Tragopogon*. *Tragopogon* is an Old World genus, but three species have been introduced to North America: *T. dubius*, *T. pratensis*, and *T. porrifolius* (whose common name is salsify, and whose roots can be eaten as a vegetable). All three species are found together in regions of east Washington and Idaho, and they all first became established there in the first two or three decades of the twentieth century. By 1950, Ownbey discovered that two new species had appeared in this region, *T. mirus* and *T. miscellus*. Both of them continue to thrive, and samples taken 40 years later by Novak *et al.* (1991) showed that *T. miscellus* had become a common weed of roadsides and vacant lots in and around Spokane, Washington, and to the east.

Tragopogon has evolved new hybrid species in the past century

Ownbey showed that *T. mirus* and *T. miscellus* (each with 12 pairs of chromosomes) are tetraploid hybrids of pairs of the three introduced species (which are diploid and have six pairs of chromosomes). The forms of the chromosomes in the species, as well as other characters such as flower color, revealed that *T. mirus* is derived from *T. dubius* and *T. porrifolius*, and *T. miscellus* from *T. dubius* and *T. pratensis*.

Ownbey found many interspecies hybrids in nature, but they were all diploid and sterile. Presumably tetraploid mutants occur in the hybrids from time to time in nature, and have given rise to the new species. The tetraploid hybrids are fertile, and reproductively isolated from the parental species. Subsequent work has used more discriminating genetic markers, and has shown that the new *Tragopogon* species have originated more than once. The parental species have hybridized (and the hybrids then tetraplodized) independently in different areas. The hybrids from the different origination events are interfertile, and all belong to the same species. Soltis & Soltis (1999) remarked that *T. miscellus* may have "originated" as many as 20 times, and *T. mirus* 12 times, in eastern Washington in the past 60–70 years.

The diploid hybrids of *Tragopogon* are sterile, and the origin of a new species could not occur until a polyploid mutant arose. In other pairs of species, the initial hybrids are partly interfertile with one or both parental species. The hybrids then backcross to the parents; this gene flow from parental species into hybrid population is called *introgression*. Many outcomes are possible from introgression, depending on the degree of interfertility with the parents. Often, the hybrids and parental species interbreed to some extent for a number of generations, and the hybrid population builds up a complex mixture of the genes of the two parental species. At some point, the hybrid population becomes reproductively isolated from the parental species. It has then evolved into a new species (Figure 14.11).

Hybrid speciation may proceed via introgression

Many cases of hybrid speciation in plants probably involve a number of generations of introgression, rather than an instantaneous speciation event. The difference between introgression and simple hybridization is that in introgression the new species will have a complex mix of parental genes, according to the history of backcrosses between hybrid and parental species during the origin of the new species, whereas in a simple hybrid it has 50% of its genes from one parental species and 50% from the other. Rieseberg & Wendel (1993) reviewed introgressive speciation in plants: they listed 155 cases in which it had been suggested, and they judged that the evidence for introgression was good in 65 of them.

One of the best examples comes from the wetlands of south Louisiana (see Plate 9, between pp. 68 and 69). There, a number of species of attractive irises grow in the

Figure 14.11
Hybrid speciation by introgression. The initial hybrid individuals are interfertile with one or both parental species and backcross with them, producing a hybrid population with various mixtures of genes from the two parental species. At some stage, the hybrid population may evolve far enough to be reproductively isolated fom the parental species; it is then a new species. From Rieseberg & Wendel (1993).

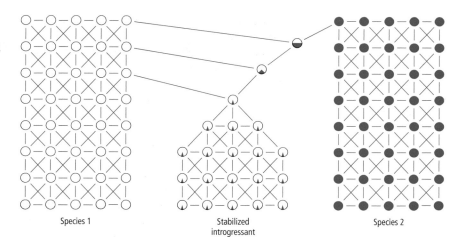

Species 1   Stabilized introgressant   Species 2

Iris has also evolved new hybrid species recently, but without polyploidy

swamps and rivers, and Arnold and his colleagues have been using genetic markers to reconstruct their origin. Plate 9a illustrates the three parental species in this example. Two of them — *Iris fulva*, with tawny colored flowers, and *I. hexagona*, whose flowers are colored violet with yellow crests — are widespread in streams in the southeast, and in southern Louisiana they live in the water channels called bayous that are derived from the Mississippi River. The third parental species is *I. brevicaulis*, which is colored like *I. hexagona* but has a different growth habit. *I. hexagona* grow up to 4 feet (1.2 m), whereas I. *brevicaulis* tend to lie flatter on the ground and curve upwards. *I. brevicaulis* lives in drier habitats such as hardwood forest. Where a bayou happens to flow into an *I. brevicaulis* habitat, the three species come into near contact and hybridize (Figure 14.12). Natural hybrids are not uncommon, though the rate at which they are formed is low. Hybrids may, to some extent, cross back to the parental species, producing a complicated mix of genotypes in the populations where the species meet.

In the 1960s, a new species of iris was detected in the region where hybrids are found, and was named *I. nelsonii* (Plate 9b). It has a morphology, including flower color,

Figure 14.12
Where bayous flow into swamps in southern Louisiana, the bayou-dwelling *Iris fulva* and *I. hexagona* come into contact with the swamp-dwelling *I. brevicaulis*. In the intermediate regions, the hybrid species *I. nelsonii* has evolved by introgression. Redrawn, by permission, from Arnold & Bennett (1993), after Viosca. © 1993 Oxford University Press Inc.

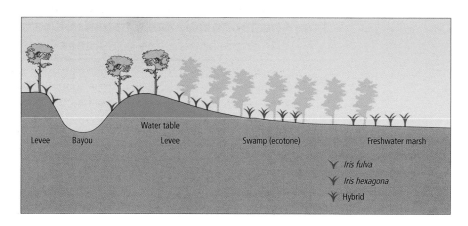

Levee   Bayou   Water table   Levee   Swamp (ecotone)   Freshwater marsh

Iris fulva
Iris hexagona
Hybrid

and chromosomal complement that shows similarities to *I. fulva* and *I. hexagona*, but *I. brevicaulis* also contributed genes to its origin. Genetic markers suggest that *I. nelsonii* mainly resulted through repeated backcrossing into *I. fulva*, rather than from one simple hybridization event in the manner of the Kew primrose (*Primula kewensis*). *I. nelsonii* is not polyploid.

We have concentrated on the problem of how a new reproductively isolated hybrid genotype can evolve. But some further problems are likely to arise in the evolutionary transition from a rare new hybrid genotype to a full hybrid species. One is finding a mate. When a fertile polyploid hybrid first arises, it is one hybrid (or perhaps one of a small number) within two large populations of the parental species. It may simply be infertile with both parental species because of the chromosomal difference; or the situation may be worse if the parental species' pollen fertilizes the hybrid's eggs and they then fail to develop or reproduce. The hybrid's interfertility with other hybrids like itself can only be expressed if other hybrids exist. Natural selection on the hybrid therefore has a kind of positive frequency dependence (Section 5.13, p. 127): when it is rare its fitness is lower because of the difficulty of finding a mate. It may have to reach some threshold of abundance before natural selection favors it. (Strictly speaking, this is number, rather than frequency, dependence; but there is frequency-dependent selection in at least an informal sense.)

This problem is probably the reason why hybrid speciation has been much commoner in some groups of plants than others. A new hybrid can more easily cross the difficult transition stage, in which it is rare, if it has alternative reproductive options besides sexual cross-fertilization. Stebbins (1950) has shown that hybrid speciation is commoner in groups in which asexual reproduction or self-fertilization are possible. *Iris nelsonii*, for example, can reproduce asexually by rhizome runners, in addition to sexual cross-fertilization via pollen that is carried by bumblebees.

Hybrid speciation is a distinctive contribution to evolutionary biology that has come from the study of plants. Hybrid speciation is probably commoner in plants than in animals (though animal examples do exist, as Arnold's (1997) book shows). It is certainly much better understood in plants than in animals, and practically all our understanding of the process has come from plants.

A new hybrid species must overcome reproductive problems

14.8 Speciation may occur in non-allopatric populations, either parapatrically or sympatrically

The Dobzhansky–Muller process works in allopatry

In the theory we have looked at so far, reproductive isolation can evolve either as an incidental consequence (or by-product) of divergence between two populations or by reinforcement. What is the relation between these theories and the allopatric, parapatric, and sympatric theories of speciation (see Figure 14.1)? Both prezygotic and postzygotic isolation can evolve as by-products of divergence. Postzygotic isolation evolves according to the Dobzhansky–Muller theory, and that theory is closely tied to the allopatric theory of speciation. The Dobzhansky–Muller theory requires that separately advantageous, but jointly disadvantageous, genes be fixed in two populations. This is only likely to happen in separately evolving (and therefore allopatric) populations. Within one population, natural selection will not favor a genetic change that is incompatible with genes at other loci.

Prezygotic isolation, however, does not require incompatible genetic change at several loci. Prezygotic isolation can evolve as a by-product of divergence if the characters that have diverged between populations are genetically correlated with characters causing prezygotic isolation. This theory is less strongly tied to the theory of allopatric speciation. The process can indeed occur between populations that are separately evolving in different places. But adaptive divergence can also occur within one population, as we shall see, and that at least raises the possibility that speciation could occur non-allopatrically.

The other theory was reinforcement. Reinforcement only occurs in sympatry. Natural selection only favors discrimination among potential mates for the range of mates that are present in a particular place. The theory of reinforcement is only weakly tied to the theory of allopatric speciation. Indeed, it is hardly an allopatric theory of speciation at all. Reinforcement was only used in the allopatric theory to "finish off" speciation that was incomplete in allopatry.

Thus, in the theories we have met so far, speciation in non-allopatric populations is relatively unlikely. One well supported theory, the Dobzhansky–Muller theory, is allopatric. Reinforcement is a sympatric process, but (as we saw) little supported by evidence and problematic in theory. However, non-allopatric speciation has not been ruled out, and in the next two sections we shall look some more at whether speciation could occur parapatrically or sympatrically.

14.9 Parapatric speciation

14.9.1 *Parapatric speciation begins with the evolution of a stepped cline*

In parapatric speciation, the new species evolve from contiguous populations, rather than completely separate ones, as in allopatric speciation (see Figure 14.1). The full process could occur as follows. Initially, one species is distributed in space. The species evolves a "stepped cline" pattern of geographic variation (Section 13.4.3, p. 363). The stepped cline could exist because of an abrupt environmental change: one form of the species would be adapted to the conditions on one side of the boundary, the other form to the conditions on the other side of the boundary.

A *hybrid zone* is a stepped cline in which the forms on either side of the boundary are sufficiently different that they can easily be recognized. The two forms may have been given different taxonomic names, as subspecies or races, or they may be different enough to have been classified as separate species.

European crows provide an example of a hybrid zone

The carrion crow (*Corvus corone*) and hooded crow (*C. cornix*) in Europe are a classic example of species round a hybrid zone (Figure 14.13). The hooded crow is distributed more to the east, the carrion crow to the west, with the two species meeting along a line in central Europe. At that line — the hybrid zone — they interbreed and produce hybrids. The hybrid zone for the crows was first recognized phenotypically,

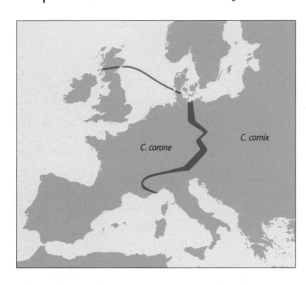

Figure 14.13
Hybrid zone between the carrion crow (*Cornix corone*) and
hooded crow (*C. cornix*) in Europe. Here the two crows are
shown as two separate species, but some taxonomists classify
them as subspecies. Redrawn, by permission of the publisher,
from Mayr (1963). © 1963 President and Fellows of Harvard
College.

because the hooded crow is gray with a black head and tail, whereas the carrion crow is
black all over. The two species (or near species) are now known to differ in many other
respects too. The fact that the crows interbreed in the hybrid zone means that speciation between them is incomplete. We shall meet some more examples of hybrid zones
in Section 17.4 (p. 497).

The conditions in a hybrid zone (or a stepped cline) are particularly ripe for
speciation if it is a *tension zone*. A tension zone exists when the hybrids between the
forms on either side of the boundary are selectively disadvantageous. (A hybrid zone
is not a tension zone if the hybrids have intermediate, or superior, fitness to the
pure forms.) For instance, if one homozygote (*AA*) is adapted to one environment,
and another homozygote (*aa*) to another environment, heterozygotes (*Aa*) will be
produced where the two environments meet up. If the heterozygotes are disadvantageous, the meeting place is an example of a tension zone. Most known hybrid zones
are in fact tension zones (see, for example, Barton & Hewitt's (1985) review of
170 hybrid zones).

In a tension zone, the conditions are exactly the preconditions for reinforcement
(Section 14.6.1). Matings within a type are advantageous, and matings between types
produce disadvantageous hybrids. Natural selection favors assortative mating. We can
therefore imagine a sequence where a stepped cline initially evolves, and then becomes
distinct enough to count as a hybrid zone. We are near the border of the origin of a new
species. Reinforcement could then finish speciation off, eliminating hybridization
from the hybrid zone. That sequence of events constitutes parapatric speciation.

The strong point of the theory of parapatric speciation is that the environment
"stabilizes" the preconditions for reinforcement. We saw that these conditions are
liable to autodestruct, as the two forms interbreed, or as one eliminates the other. But
if the environment varies in space, the clinal variation will be maintained. Parapatric
speciation could work, in theory.

Many hybrid zones are tension
zones . . .

. . . in which reinforcement may
operate

14.9.2 Evidence for the theory of parapatric speciation is relatively weak

Most hybrid zones are due to secondary contact

The theory of parapatric speciation has two main weak points in the evidence. One is the evolutionary history of hybrid zones. Hybrid zones can be "primary" or "secondary." A hybrid zone is primary if it evolved while the species had approximately their current geographic distribution. It is secondary if in the past the species was subdivided into separate populations, where the differences between the forms evolved, and the populations later expanded and met up at what is now the hybrid zone. Real hybrid zones only illustrate a stage in parapatric speciation if they are primary. The abundance of hybrid zones in nature would only be evidence that parapatric speciation is a plausible process if those hybrid zones are mainly primary. If most hybrid zones are secondary, the difference between the forms evolved allopatrically not parapatrically. In fact the evidence suggests that most hybrid zones are secondary. Hooded and carrion crows, for instance, have met up after their ranges expanded following the most recent ice age. Indeed, range expansion following the ice age is a common explanation of hybrid zones (Section 17.4, p. 497). Hybrid zones provide little support for the theory of parapatric speciation.

Secondly, if reinforcement operates in hybrid zones, we predict that prezygotic isolation will be stronger in the hybrid zone than between the two forms away from the hybrid zone. The prediction is a special case of the general biogeographic test of reinforcement (Section 14.6.3). The evidence does not support the prediction: we have little good evidence that prezygotic isolation is reinforced in hybrid zones.

Thus, the process of parapatric speciation is possible in theory. The theory solves one key problem in reinforcement. Most (but not all) stages of parapatric speciation can be illustrated by evidence. But parapatric speciation lacks the solid weight of supporting evidence and the theoretical near inevitability of allopatric speciation. Parapatric speciation cannot be ruled out, and probably operates in some cases. But the case that it is important has still to be made.

14.10 Sympatric speciation

14.10.1 Sympatric speciation is theoretically possible

In sympatric speciation, a species splits into two without any separation of the ancestral species' geographic range (see Figure 14.1). Sympatric speciation has been a source of recurrent controversy for a century or so. Mayr (1942, 1963) particularly cast doubt on it, and in doing so has stimulated others to look for evidence and to work out the theoretical conditions under which it may be possible.

In the theory of parapatric speciation, the initial stage in speciation is a spatial polymorphism (or stepped cline). In sympatric speciation, the initial stage is a polymorphism that does not depend on space within a population. For instance, two forms of a species may be adapted to eat different foods. If matings between the two are disadvantageous, because hybrids have low fitness, reinforcement will operate between them. Most models of sympatric speciation suppose that natural selection initially establishes a polymorphism, and then selection favors prezygotic isolation between the polymorphic forms. "Host shifts" in a fly called *Rhagoletis pomonella* provide a case study that may illustrate part of the process.

14.10.2 *Phytophagous insects may split sympatrically by host shifts*

The apple maggot fly has only recently moved on to apples

Rhagoletis pomonella is a tephritid fly and a pest of apples. It lays its eggs in apples and the maggot then ruins the fruit, but this was not always so. In North America, *R. pomonella*'s native larval resource is the hawthorn. Only in 1864 were these species first found on apples. Since then it has expanded through the orchards of North America, and has also started to exploit cherries, pears, and roses. These moves to new food plants are called *host shifts*. In the host shift of *R. pomonella*, speciation may be happening before our eyes.

The *R. pomonella* on the different hosts are currently different genetic races. Females prefer to lay their eggs in the kind of fruit they grew up in: females isolated as they emerge from apples will later choose to lay eggs in apples, given a choice in the laboratory. Likewise, adult males tend to wait on the host species that they grew up in, and mating takes place on the fruit before the females oviposit. Thus there is assortative mating: male flies from apples mate with females from apples, males from hawthorn with females from hawthorn.

The races on apples show some isolation from the ancestral races on hawthorn

The races are presumably about 140 generations old (given that they first moved on to apples nearly one and a half centuries ago). Is this long enough for genetic differences between the races to have built up? Gel electrophoresis shows that the two races have evolved extensive differences in their enzymes. They also differ genetically in their development time: maggots in apples develop in about 40 days, whereas hawthorn maggots develop in 55–60 days. This difference also acts to increase the reproductive isolation between the races, because the adults of the two races are not active at the same time.

Apples and hawthorns differ and selection will therefore probably favor different characters in each race; this may be the reason for their divergence. If it is, selection may also favor prezygotic isolation and speciation. If flies from the different races are put together in the lab, however, they mate together indiscriminately. Either reinforcement has not operated when it might have been expected, or, alternatively, the differences in behavior and development time in the field may be enough to reduce interbreeding to the level natural selection favors. Selection would then not be acting to reinforce the degree of prezygotic isolation. We do not know which interpretation is correct; we need to know more about the forces maintaining the genetic differences between the races. Once again, the evidence for reinforcement is the weak point in a theory of speciation.

But the example is incomplete

In the case of host shifts, we can be practically certain that the initial host shift, and formation of a new race, has happened in sympatry. The shift took place in historic time. However, it is not a full example of sympatric speciation because the races have not fully speciated. Indeed, we do not know whether they will, or whether the current situation, with incomplete speciation, is stable.

How general a process is sympatric speciation by host shifts? A definite answer cannot be given as it has not even been confirmed that sympatric speciation ever does take place by host shifts. But there are interesting hints that the process might be important (Section 22.3.3, p. 620). Several phytophagous insect taxa have undergone extensive phylogenetic radiations on plant host taxa. There are, for example, about 750 species of fig wasps, and each breeds on its own species of fig; in Britain alone there are 300 species of leaf miners in the dipteran family Agromyzidae, and 70% of them each feed on only one plant species. It is easy to imagine how these groups could have radiated from a single common ancestor, as successive new species arose by host shifts like the one taking place in the apple maggot fly in the USA. If phytophagous insect species consisted of an occasional odd species scattered through the phylogeny of insects, and feeding on unrelated kinds of food plants, the process would probably have not been operating; but the existence of whole large taxa of host plant-specific phytophages does suggest that speciation by host shifts could have contributed to their diversification.

14.10.3 *Phylogenies can be used to test whether speciation has been sympatric or allopatric*

Direct attempts to test the theory of sympatric speciation, such as in *Rhagoletis pomonella*, are only one way to test whether sympatric speciation occurs. Recently, a new kind of evidence has been put forward for sympatric speciation. The evidence suggests that sympatric speciation occurs, but tells us nothing about how it occurs. The evidence comes from the shape of phylogenetic trees, and was first obtained for cichlid fish in African lakes (Schliewen *et al.* 1994). As we saw (Section 13.3.3, p. 357), many species of cichlid fish have evolved in the East African lakes. Did they originate by sympatric, or allopatric, speciation?

Figure 14.14 shows the argument. If a new species arises by allopatric speciation, its nearest relative will usually live in a different geographic area, such as in a nearby lake or river. If the species evolved sympatrically, the nearest related species will usually live in the same lake. In the case of a number of fish species, including the African cichlids, the phylogenetic evidence supports sympatric speciation. Similar studies for other taxa usually suggest allopatric speciation (Barraclough & Vogler 2001).

In conclusion, few biologists would rule out non-allopatric mechanisms of speciation. Speciation probably occurs non-allopatrically, though it may only be rare. Sympatric and parapatric speciation are more controversial theories than allopatric speciation, except for special cases such as hybrid speciation in plants, because they are not supported by such an impressive range of evidence.

Phylogenetic tests for sympatric speciation have recently been devised

14.11 The influence of sexual selection in speciation is one current trend in research

We can finish this chapter by looking briefly at two big themes in current, and possibly future, research on speciation. One is the possibility that sexual selection is important.

Figure 14.14
Phylogenetic test between sympatric and allopatric speciation. The test has been used most for lake-dwelling fish species. (a) With allopatric speciation, new species evolve in separate lakes and for any one species, its closest relative is predicted to live in a different lake. (b) With sympatric speciation, new species evolve alongside their ancestors. Related species are predicted to live in the same lake. Evidence for some fish species, including African lake cichlids, shows the phylogenetic pattern expected with sympatric speciation.

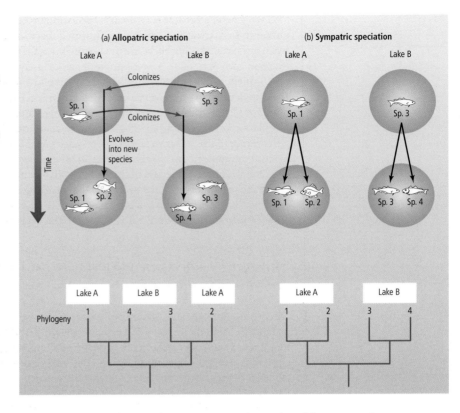

Sexual selection is discussed in Section 12.4 (p. 327) and has two main components: male competition and female choice. The mechanisms that females use to the choose mates may influence speciation because they can contribute to, or even wholly determine, prezygotic isolation.

Sexual selection may contribute to the evolution of prezygotic isolation

The way natural selection acts on mate choice may help explain the evolution of prezygotic isolation in both allopatric and sympatric populations. Consider again those experiments in which some populations of a species are allowed to evolve in two environmental conditions, such as a diet of maltose or of starch (see Figure 14.2). We saw that prezygotic isolation evolves as a by-product and its genetic basis may be pleiotropy or hitch-hiking. Now let us think some more about how natural selection will work in each experimental population. On a starch medium, selection favors individuals who can eat, digest, and thrive on starch. But it also favors female flies who choose as mates those males that are better than average at living on a starch diet. Over time, females may evolve a preference for males with adaptations for life on starch. In the maltose population, females evolve preferences for males who are adapted for life on maltose. Then, at the end of the experiment we give females a choice of males and find the females from the starch lines prefer males from starch lines.

One way of seeing this argument is as providing an explanation for pleiotropy. Prezygotic isolation evolves as a by-product when the character concerned with ecological adaptation happens, perhaps coincidentally, also to be concerned with mate choice

(Section 14.3.2). But it may be less of a coincidence. Females may evolutionarily seize on those male characters that contribute to superior adaptation. Natural selection works on mate choice mechanisms as well as ecological adaptation, and the two may become associated.

A similar association arises in some recent models of sympatric speciation (Dieckmann & Doebeli 1999; Higashi *et al.* 1999; Kondrashov & Kondrashov 1999). One theoretical problem in reinforcement is that recombination tends to break down any association between genes for assortative mating and genes for ecological adaptation (Section 14.6.2). But sexual selection can help to strengthen the association, making sympatric speciation more plausible.

These two arguments are only two of several ways in which sexual selection has recently been suggested to drive speciation. (Schluter (2000, p. 195) gives a table with six or so additional ideas. For instance, evolutionary conflict between males and females (Section 12.4.7, p. 336) may contribute to speciation.) Most of the arguments are hypothetical. Sexual selection has not yet been shown to drive the evolution of prezygotic isolation in any case of speciation, though good suggestive evidence exists. We do not know that sexual selection is a general force of speciation. But much research on this topic is being done.

This is a "hot topic"

14.12 Identification of genes that cause reproductive isolation is another current trend in research

We have discussed the genetics of both prezygotic and postzygotic isolation, the latter extensively in the Dobzhansky–Muller theory. Prezygotic isolation may be due to pleiotropy and hitch-hiking; postzygotic isolation to epistatic interactions among multiple gene loci. The discussion, however, has been abstract. Genetic crosses do provide evidence for the Dobzhansky–Muller theory, and it is possible to explain why biological systems fit the theory (Section 14.4.3). But in none of the work did we look at particular examples of genes. Relatively little research has been done yet to identify genes that contribute to pre- or postzygotic isolation. But research of this sort may provide one way forward in studying speciation. If we can identify genes that cause prezygotic isolation, we can see what (if anything) their pleiotropic, and hitch-hiked, effects are. If we can identify genes causing postzygotic isolation, we can investigate what their epistatic interactions are with other genes and why those interactions arise. Our understanding of speciation should improve as we move from abstract theory to concrete examples. Moreover, modern genetics has powerful techniques for identifying genes — techniques that were not available before the "genomics" era.

The *Odysseus* gene controls reproductive isolation in two fruitfly species

As an example, consider the work of Ting *et al.* (1998) on a gene called *Odysseus*. *Drosophila simulans* and *D. mauritiana* are two closely related fruitfly species and an interspecific cross between them conforms to Haldane's rule — that the male hybrids are sterile. Ting *et al.* used genetic techniques to insert bits of *D. mauritiana* DNA into *D. simulans*. They were able to show that male hybrid sterility is caused by a gene on the X chromosome. If they inserted only this gene (*Odysseus*) into *D. simulans* males, those males were sterilized as in an interspecies cross (Figure 14.15).

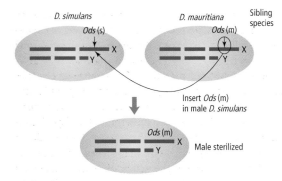

D. simulans   D. mauritiana   Sibling species

Ods (s)   Ods (m)

Insert Ods (m) in male D. simulans

Ods (m)   Male sterilized

Figure 14.15

A gene that has been identified and that causes reproductive isolation. Ting *et al.* (1998) experimentally inserted the *Drosophila mauritiana* version of the gene *Odysseus* into a *D. simulans* genetic background. The result was that the males were sterilized, just as in normal hybrid crosses between these two species. *Odysseus* probably causes reproductive isolation between these two species. *Odysseus* is a gene on the X chromosome in fruitflies. It is a homeobox gene, likely expressed in spermatogenesis. It has also evolved exceptionally fast between these two (sibling) species: it is more different between the two than it is between rodents and nematode worms.

They looked into *Odysseus* some more. *Odysseus* contains a "homeobox," a sequence found in genes that regulate development (Section 20.6, p. 582). It is expressed in the development of the male reproductive system. The sterility of *D. mauritiana* × *simulans* hybrids may be caused by an incompatibility between the *mauritiana* form of the *Odysseus* gene and a *simulans* gene that is also expressed in the male reproductive system.

Odysseus has evolved fast in these species

One striking feature of *Odysseus* is its rate of evolution. *Odysseus*, like most homeobox genes, normally evolves slowly. But it has undergone a sudden spurt of evolution in these fruitflies. Indeed *Odysseus* is more different between *D. mauritiana* and *D. simulans*, which share a common ancestor about half a million years ago, than it is between a worm and a mouse, which share a common ancestor at least 700 million years ago. The rate of evolution of this gene has zoomed up over 1,000-fold in these fruitflies. And associated with that, it causes postzygotic isolation.

We can fit these observations in with a general idea about speciation: the idea of "speciation genes." Speciation genes can be defined as genes that differ between a pair of species, and cause reproductive isolation between them. (A more demanding definition would be that speciation genes are genes that differ between a pair of species and drove speciation between them. However, empirical research can usually only show that a gene causes reproductive isolation — and we remain uncertain whether the gene also drove speciation.)

Biologists discuss various hypotheses about speciation genes. We can distinguish a strong and weak claim. The strong claim would be that some genes in the genome may be particularly likely to drive speciation. That is, we can look at the genome in advance of speciation and say "if gene X changes, speciation will follow." For instance, changes in the genes concerned with courtship or mate choice might be more likely to drive speciation than changes in other genes. If true, genes that influence courtship and mate choice would be "speciation genes." Other possible examples include genes on the X chromosome, or genes such as the segregation distorter genes (Section 11.2, p. 294), or chromosomal mutation. But none of these kinds of genes have been shown to drive speciation in general, and the strong claim about speciation genes may well be false.

Alternatively, changes in almost any gene might be able to drive speciation. Then we can talk about speciation genes in a weaker sense — simply to refer to the genes that happen to cause reproductive isolation in a particular pair of species. In the

Dobzhansky–Muller theory, any gene can cause isolation, provided it can have an epistatic interaction with other genes in the genome. However, the genes that drive speciation will be the genes that have changed in evolution. An unchanging, conserved gene cannot cause isolation between two species. The genes driving speciation will be the first genes to change — that is, the genes that evolve fastest. Maybe they will be genes like *Odysseus*, which does not normally evolve fast but happened to in one population. One gene may have an evolutionary spurt in one lineage, and cause speciation there. Another gene may spurt in another lineage, and cause speciation there. The "speciation genes" will be those that happened to evolve fast in a particular lineage. Or it could be that some genes in the genome evolve faster than average in all life forms. Then, these fast evolving genes may be the speciation genes. One suggestion of this sort is that genes expressed in the reproductive system may evolve faster than other genes (see Swanson & Vacquier (2002) for the facts). Then speciation will more often be caused by evolution in the genes of the reproductive system than in genes of (for example) the nervous or digestive system.

These ideas about speciation genes are currently conjectural. However, they are an example of the kind of general idea about speciation that we should be able to investigate as modern genetic techniques are used to identify the genes that are causing reproductive isolation in particular species.

Biologists are interested in theories of "speciation genes"

14.13 Conclusion

At the beginning of the chapter we saw that there are two theories of how reproductive isolation evolves: the "by-product" theory and reinforcement. When Darwin discussed the topic in *On the Origin of Species* (1859) he favored what is here called the by-product theory, saying that the sterility of interspecies hybrids is "incidental on other acquired differences." He devoted a chapter to arguing the point. He was less interested in the geographic circumstances of speciation, but argued for something like what we would now call sympatric speciation rather than allopatric speciation. Competition between forms within an area would force them to diverge, he reasoned.

Our modern understanding of speciation shows several advances since Darwin

The impressive evidence that we now have from artificial selection experiments (Section 14.3.1) plugs one hole in Darwin's case. Darwin had no evidence that reproductive isolation evolved between domestic varieties that had been selected apart. "The perfect fertility [he wrote] of so many domestic varieties, differing widely from each other in appearance, for instance of the pigeon or the cabbage, is a remarkable fact." It is not so remarkable now, because more careful, and better controlled, measurements have shown that reproductive isolation often evolves between artificially selected varieties.

After Darwin, the evolutionary biologists of the "modern synthesis" added four or five main claims, in the period from about 1930 to 1950. One claim, argued for by Mayr (1942), was that new species arise allopatrically rather than sympatrically. Associated with this was a second claim, that speciating populations tend to be small and that genetic drift is particularly important in speciation. Thirdly, Dobzhansky and others argued that reinforcement also contributes to speciation. (Dobzhansky (1970) gives a later update of his views.) Fourthly, new species often arise by hybridization, particularly in plants. Fifthly, Darwin's idea that isolation evolves as a by-product was expanded and explained genetically in the Dobzhansky–Muller theory.

Now, 50 or more years later, the allopatric theory of speciation still stands up. Many biologists would allow some contribution from sympatric speciation, but most accept that allopatric speciation is the main process. In this respect, biologists now agree with the modern synthesis rather than Darwin. The second claim, that speciation is often powered by genetic drift now has few supporters. It is the least important of the five claims listed above, and may not have been strongly believed in even during the period from the 1930s to the 1950s. In the 1920s, biologists often suggested that the characters that differ between species are non-adaptive. This partly inspired "non-adaptive" theories of speciation, but few biologists now argue that species differences are non-adaptive. The experimental evidence and theory of speciation suggest the genetic drift is not all that important in speciation. Speciation is probably more often a by-product of normal adaptive divergence between populations.

The theory of reinforcement has had its ups and downs. Reinforcement continues to tantalize biologists, but a compelling case for its importance has yet to be made. The theory of hybrid speciation in plants, by contrast, has held up well. New genetic techniques have enabled biologists to trace the ancestry of modern species, providing a detailed description of hybrid speciation.

Finally, the genetics of postzygotic isolation has become a major field of research. Darwin seems to have been right that postzygotic isolation evolves as an incidental by-product of divergence. The Dobzhansky–Muller theory improved our understanding of the genetic events by which postzygotic isolation can incidentally drop out of normal evolutionary change. Evidence for the theory has accumulated, not least as the evidence for Haldane's rule has been incorporated (if partly) in the general theoretical scheme. The Dobzhansky–Muller theory looks as if it may continue to inspire research as the techniques of modern genomics are imported into the study of speciation.

Summary

1 The evolution of a new species happens when one population of interbreeding organisms splits into two separately breeding populations.

2 Two theories of how reproductive isolation evolves have been suggested: it evolves as a by-product of divergence between two populations, or it evolves by reinforcement.

3 The "by-product" theory is well supported by experimental and biogeographic evidence.

4 Experiments have demonstrated that reproductive isolation tends to arise incidentally between two populations that are kept separate from each other and allowed to evolve in different environments for a number of generations.

5 Members of a species can be sampled from different parts of the species' biogeographic range, brought into the laboratory, and crossed. Reproductive isolation is often found between the individuals from distant parts of the species' range.

6 Prozygotic isolation evolves when it is genetically correlated, by pleiotropy or hitch-hiking, with the characters undergoing divergent evolution between populations.

7 The Dobzhansky–Muller theory explains the genetics of postzygotic isolation. When two populations diverge, they may evolve new genes that are incompatible when put together. Postzygotic isolation between two species is usually caused by epistatic interactions among multiple genes, and not by a single genetic locus.

8 Haldane's rule is a generalization about postzygotic isolation. It states: "When in the F_1 offspring of the two different animal races one sex is absent, rare, or sterile, that sex is the heterozygous one." Postzygotic isolation evolves first in the heterogametic gender of the hybrid offspring. The Dobzhansky–Muller theory can partly explain Haldane's rule.

9 Reinforcement is the enhancement of reproductive isolation by natural selection: forms are selected to mate with their own, and not with the other, type.

10 The theory and evidence for reinforcement are both problematic. Reinforcement may contribute to the evolution of reproductive isolation, but a compelling case for it has not yet been made.

11 Many new plant species have originated following hybridization of two existing species.

12 Speciation may occur in parapatric (that is, geographically contiguous) populations. Parapatric speciation begins with a stepped cline, and prezygotic isolation then evolves between the forms on either side of the step.

13 Speciation may occur in sympatry. The process can begin with the establishment of a polymorphism, and reproductive isolation then evolves between the different forms. The shape of phylogenies, for instance in lake-dwelling fish, provides evidence that sympatric speciation has occurred.

14 Two current trends in research are: (i) to look at the influence of sexual selection in speciation; and (ii) to identify particular genes that cause reproductive isolation between species.

Further reading

The July 2001 issue of *Trends in Ecology and Evolution* (vol. 16, pp. 325–413) is a special issue on speciation, and introduces most of the modern research trends. The opening paper, by Turelli *et al.* (2001a), is an overview of the whole subject. Coyne & Orr (2003) authoritatively review speciation at book length. Howard & Berlocher (1998) is a multi-author research-level book about speciation. Schiltuizen (2001) is a single-author, more popular, book about speciation. The recent research monographs by Arnold (1997), Levin (2000), and Schluter (2000) contain much material about speciation, as does the conference proceedings edited by Magurran & May (1999). The special issue of *Genetica* (2001), vol. 112/113 contains several papers on speciation; it was also issued as a separate book (Hendry & Kinnison 2001). Also see the supplement (edited by Via) to vol. 159 of *American Naturalist* (2002); it is a special issue on the ecological genetics of speciation.

Rice & Hostert (1993) review the experimental research on speciation, including the evolution of reproductive isolation as a by-product of divergence. Meffert (1999) relates experimental work of this kind to conservation. For the biogeographic evidence we have no similar review, but there are many further studies like Kruckeberg (1957). Levin (2000) lists several. Vickery (1978) is a particularly thorough study of North American monkey flowers. Ring species illustrate the same point: see Chapter 3 in this book and Irwin *et al.* (2001b). Nosil *et al.* (2002) take this line of research further. They not only show that more distant populations of a walking-stick insect have higher prezygotic isolation, but also that isolation is influenced by ecological similarity — specifically, host plant similarity.

Other recent examples, like Podos (2001), in which reproductive isolation is an almost automatic consequence of change in some character or other, include Keller & Gerhardt's (2001) study of polyploidy and call structure in frogs. An excellent related example is provided by flower morphology and pollinator specialization. See Schemske & Bradshaw's (1999) work on monkey flowers, and Waser (1998) generally; Section 22.3 will pick up this theme and has further references.

The genetics of postzygotic isolation has been well reviewed recently. See Orr (2001) and Turelli *et al.* (2001a) in the special issue of *Trends in Ecology and Evolution*. See also Orr & Presgraves (2000) and Coyne & Orr (1998). See Johnson (2002) for a historic perspective. Other biological examples that fit the basic Dobzhansky–Muller scheme include segregation distorters (see Section 11.2, p. 294, and the paper by Tao *et al.* (2001)). On parasites, see Hamilton (2001). Fishman & Willis (2001) show that Dobzhansky–Muller incompatibilities are at work in monkey flowers. Wolbachias are another special case: see Breeuwer & Werren (1990) for an example, and *Nature* (2001), vol. 409, p. 675 for a picture of how it fits the Dobzhansky–Muller scheme. (Wolbachias are worth looking into in their own right for dramatic experiments, such as Breeuwer and Werren's, in which antibiotic treatment "cures" speciation. Werren (1997) is a review.)

Two other excellent case studies in the genetics of speciation are the work of Schemske & Bradshaw (1999) on monkey flowers, in which genes influence flower coloration, which influences pollinators, and of Rieseberg on sunflowers (see the hybrid speciation references given below). Rieseberg also has a piece in the special issue of *Trends in Ecology and Evolution* (2001) that introduces the role of chromosomal change in speciation — which is a further big historic theme in the speciation literature. Noor *et al.* (2001) is a recent study of a pair of *Drosophila* species in which a chromosomal inversion influences reproductive isolation.

For Haldane's rule, see Turelli *et al.* (2001a) and Orr (2001) in the special issue of *Trends in Ecology and Evolution*, and Orr & Presgraves (2000). The recent literature is huge, and they introduce it.

On reinforcement generally, Noor (1999) is a recent review and also look at Howard (1993). Servedio (2001) expands the topic, looking at other ways that natural selection can act on prezygotic isolation. More specific studies include Saetre *et al.* (1997) on Corsican birds, and Higgie *et al.* (2000) on Australian fruitflies. Coyne & Orr's (1989) study (updated in 1997) is also important.

On hybrid speciation, see Arnold (1997), Rieseberg (1997, 2001), and Rieseberg & Wendel (1993). Soltis & Soltis (1999), Ramsey & Schemske (1998), and Leitch & Bennett (1997) discuss polyploidy in plants, a closely related topic. Grant (1981) is a classic and covers plant speciation in general. See also general books on plant evolution, such as Niklas (1997). Arnold (1997) and Dowling & Secor (1997) discuss evidence for animals too. A further case study that I did not cover in the text is the sunflower *Helianthus* in southwest USA; Rieseberg & Wendel (1993) and Arnold (1997) both discuss it, and see Rieseberg *et al.* (1996) and Ungerer *et al.* (1998) for marvellous results on the genetics. Hybrid fitness is a further topic. The classic theory is the Dobzhansky–Muller theory, suggesting that hybrid fitness will be low. But Veen *et al.* (2001) have interesting results on hybrid fitness, showing how hybrids are not as unfit as might naively be thought. See Arnold (1997) on this generally, as well as Grant & Grant's (2002) work on Darwin's finches (discussed in Chapter 13).

On parapatric speciation see Endler (1977), which includes an important discussion of the biogeographic evidence for hybrid zones. Harrison (1993) is a multiauthor book about hybrid zones. See also Chapter 17 of this text, and the Hewitt references in it. On the European crows, see Cook (1975).

On sympatric speciation, Mayr (1942, 1963) is the classic critic, though see Mayr (2001) for his current view. Guy Bush has inspired much work, and the book edited by Howard & Berlocher (1998) was a *Festschrift* for Bush: it includes several papers on host shifts and *Rhagoletis*, as well as on other topics in sympatric speciation. The issues of *Nature* (1996), vol. 382, p. 298, and of *Science* for September 13, 1996 have news features on a conference again mainly about Bush's work. Via (2001) reviews sympatric speciation, and Barraclough & Nee (2001) discuss the use of phylogenetic evidence, in the special issue of *Trends in Ecology and Evolution*. For the cichlids see Stiassny & Meyer (1999) and Fryer (2001).

Panhuis *et al.* (2001) have a piece about sexual selection in the special issue of *Trends in Ecology and Evolution*. See also Turelli *et al.* (2001) therein, and Schluter (2000), as well as the references I give in the text. On the second modern trend, identifying individual genes, *Period* is also worth looking into besides *Odysseus* as it may influence prezygotic isolation in fruitflies — see Ritchie & Phillips (1998).

The classic treatises on speciation by Mayr (1942, 1963) and Dobzhansky (1970) remain good, if dated, introductions. See Mayr (2001, and Mayr & Ashlock 1991) for his more recent ideas; Coyne (1994) discusses speciation, particularly in relation to Mayr's ideas. The multiauthor book edited by Otte & Endler (1989) is becoming dated, but introduces many themes in speciation.

Study and review questions

1 When two populations are kept experimentally in different conditions for a number of generations, reproductive isolation is found to have evolved between them. What is the genetic reason for the evolution of reproductive isolation? Give an example.

2 How can the amount of (a) postzygotic, and (b) prezygotic isolation be expressed quantitatively? For (b) imagine a mating experiment, where an individual female of species 1 is given a choice between a male of species 1 and a male of species 2. The experiment is repeated with 100 females. The numbers of females who mated with each kind of male are given below. Calculate an index of prezygotic reproductive isolation (I) for the three cases.

	Species 1 male	Species 2 male	I
(i)	100	0	
(ii)	75	25	
(iii)	50	50	

3 How does the phylogeny of the greenish warblers around the Tibetan Plateau help us to understand the evolution of ring species?

4 Why is postzygotic isolation theoretically unlikely to be due genetically to a single locus?

5 (a) What is Haldane's rule? (b) If humans split in the future into two species, for instance following colonization into the galaxy, do you expect the sons or daughters of the hybrids bertween the two emerging species to evolve sterility first? (c) How can the Dobzhansky–Muller theory explain Haldane's rule?

6 What is meant by "valley crossing" in the origin of species? Is there valley crossing when: (a) prezygotic isolation evolves by pleiotropy; (b) postzygotic isolation evolves by the Dobzhansky–Muller process; and (c) postzygotic isolation evolves by changes at a single genetic locus?

7 Two species have partly overlapping ranges. Females of species 1, taken from an area where only species 1 lives and given an experimental choice between males of the two species, mate indiscriminately. Females of species 1 taken from the area where both species live and given the same choice mate preferentially with males of species 1. What is the phenomenon called? And what are the two main evolutionary explanations for it?

8 What reasons suggest that reinforcement may be a weak evolutionary force in nature?

9 Explain why hybrid zones exist according to the theory of (a) allopatric and (b) parapatric speciation.

10 How can we test between the allopatric and sympatric theories of speciation, using phylogenetic trees?

15

The Reconstruction of Phylogeny

A knowledge of the phylogenetic relations among species is essential for many other inferences in biology, and a proportionally large effort has been put into reconstructing the tree of life. Classically, phylogenies were inferred using morphological evidence from living and fossil species. Phylogenies are now increasingly inferred from molecular sequence evidence. The principles of phylogenetic inference from morphological and molecular evidence are fundamentally the same, but the techniques used differ in many ways and the chapter looks at the two separately. We begin with "cladistic" techniques, which are used with morphological evidence. We then move on to molecular evidence, looking at three classes of statistical procedure. We also look at when these statistical procedures lead to the right, and the wrong, inference. We finish with a classic case study from human evolution in which different kinds of evidence came into conflict.

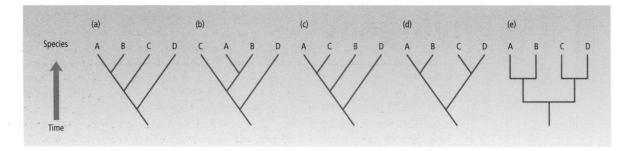

Figure 15.1

A phylogeny shows, for a group of species, the order in which they share common ancestors with one another. (a) Species A and B share a more recent common ancestor with each other than either does with C; the group of species A, B, and C share a more recent common ancestor with one another than any of them do with species D. (b) In a phylogeny, any of the nodes can be rotated without altering the relation shown: (a) and (b) are identical, but (c) and (d) differ from (a) and (b) because the order or the pattern of branching is altered. (e) Phylogenies may be drawn with either right-angled or diagonal lines; the information is identical: (d) and (e) are the same phylogeny.

The only information in the phylogeny is the order of branching: the x-axis does not necessarily represent phenetic similarity. In particular, (d) does not imply that species B and C show convergent evolution. Sometimes a phylogenetic diagram does also display phenetic similarity (e.g., Figure 15.6), but then it is explicitly drawn on. The vertical axis expresses the direction of time, which goes up the page. However, the axis is not usually exactly proportional to time: (a) does not imply that the time between the successive branching was constant. Some phylogenetic diagrams do display absolute time, and then it is again made explicit (e.g., Figure 15.12).

15.1 Phylogenies express the ancestral relations between species

What a phylogeny is

A *phylogenetic tree*, or *phylogeny*, or *tree*, for a group of species is a branching diagram that shows the relationships between species, according to the recency of their common ancestors. For each species, or group of species, a phylogeny shows which other species (or group of species) it shares its most recent common ancestor with. A phylogeny implicitly has a time axis, and time usually goes up the page. In Figure 15.1a, for instance, species A and B share a more recent common ancestor with each other than either shares with any other species (or group of species). There are many possible phylogenies for the four species, A, B, C, and D, in Figure 15.1. Maybe A shares its most recent common ancestor with B, as shown in Figure 15.1a. Or maybe A shares its most recent common ancestor with C, as shown in Figure 15.1c. Figure 15.1d is another possibility. In all, any set of four species have 15 possible phylogenies.[1] The problem of phylogenetic inference is to work out which of those 15 is correct, or most likely to be correct. The answer has to be found by inference, rather than by direct observation or experiment. The splitting events occured, and the common ancestors lived, in the past. They cannot be directly observed.

---

[1] We are assuming that the phylogenies only contain two-way splits, or bifurcations, and no higher order splits, such as three-way splits, or trifurcations. The assumption is likely to be valid, given the number of species that exist and the amount of time they evolved in.

Shared characters are used to infer phylogenies

Phylogenies are inferred using characters that are shared between species. The characters may be at the level of gross morphology. For instance, humans and chimpanzees share such vertebrate characters as brains and backbones, mammalian characters such as lactation, and great ape characters such as their distinctive molar teeth and absence of a tail. Or the characters used may be at the level of chromosomes, such as the number or structure of the chromosomes in the species under study. Much phylogenetic inference in biology today uses molecular sequences, particularly the nucleotide sequence of DNA in different species. In this chapter we shall look first at how phylogenies are inferred with morphological evidence, and then move on to molecular evidence. The methods used for morphology and molecules both rely, at an abstract level, on the same logic. However, the detailed implementation of that logic differs so much between morphological and molecular evidence that it is convenient to look at them separately. We can also look at some examples where the two kinds of evidence have come into conflict.

15.2 Phylogenies are inferred from morphological characters using cladistic techniques

Phylogenetic inference using morphological characters proceeds in the same way with both living and fossil species. For fossil species, we usually have evidence only from hard parts, such as bones in vertebrates or shells in mollusks. For living species, we have further evidence from the soft parts. We also have evidence from characters that are not morphological in a narrow sense, but can be included with morphological characters in phylogenetic research. For instance, mammals are viviparous (produce live young) and lactate, whereas birds are oviparous (lay eggs). Reproductive and physiological characters of this kind are all good evidence for phylogenetic inference. In this chapter, "morphological" evidence refers to all observable characters in the whole organism, as distinct from molecular characters.

Cladism uses morphological characters . . .

The techniques used with morphological characters are called *cladistic* techniques. (The word cladistics comes from the Greek word for a branch.) The techniques were mainly formalized in a book, *Phylogenetic Systematics*, by the German entomologist Willi Hennig (1966). The book is not an easy read, but has been highly influential, with good reason. Hennig had thought through the problem of phylogenetic inference more thoroughly than most of his predecessors. Subsequent work (with morphological characters) mainly follows on Hennig's lead.

. . . divided into discrete states

For cladistic analysis, the evidence consists of a number of characters, each with a number of discrete character states. For example, one character might be "mode of reproduction" and it might have the states "viviparity" or "oviparity." Another character might be "structure of forelimb," and its states might be "wing" and "arm." The particular characters and character states will depend on the species that are being studied. They may also be revised during research: the character state "wing" might have to be replaced by "bird wing" and "bat wing" if both birds and bats were included in the study. The division of an organism's morphology into characters, and the division of characters into discrete states, can itself be problematic. However, in this chapter we shall take the characters and character states as the starting point. They are usually represented by symbols, such as *a* and *a'* (where *a* might stand for oviparity and *a'* for viviparity); *a* and *a'* are two states of one character. The states of a second character might be symbolized by *b* and *b'*.

Different characters conflict with each other

Phylogenetic inference is not simple, mainly because not all the characters for which we have evidence will point to the same phylogeny. In an easy case, all the characters will agree. For example, suppose we want to know the phylogeny of three species — humans, chimpanzees, and a species of worm. Some character states are shared between humans and chimpanzees; many character states are shared between all three species; practically no character states are shared between worms and either chimpanzees (but not humans) on the one hand, or between worms and humans (but not chimpanzees) on the other. Humans and chimpanzees, we conclude, share a more recent common ancestor than either does with the worm. If all cases were this easy, we could simply read phylogenetic relations from the character states. Cladistics would hardly need to have been invented.

But suppose now that we are studying the phylogeny of humans, a bat, and a bird. Some character states are similar in birds and bats: both have wings and other skeletal adaptations for flight. Other character states are similar in humans and bats: both are viviparous and lactate. Which evidence should we rely on? Figure 15.2 shows another famously problematic example, from the relations of birds and reptiles. Suppose we are studying the phylogeny of a crocodile, a bird, and a lizard. The crocodile and lizard share many similarities: they have scales and walk on four legs, whereas birds have feathers and walk with two of their appendages and fly with the other two. But a detailed study of the skull shows that birds and crocociles have important similarities there, whereas lizards have a different skull anatomy. Which evidence should we rely on? These two examples illustrate a general problem. In most phylogenetic research, different characters point to different phylogenies. (I should stress the word *research* in the previous sentence. Easy cases — such as humans, chimpanzees, and worms, or humans, gorillas, and oak trees — have all been solved. We know their phylogeny. The cases left for research are the ones that are not easy. They are not easy either because we have practically no knowledge of the character states in the species, and phylogenetic research has yet to begin, or because of conflict among characters.)

Cladism aims to distinguish reliable from unreliable characters

When different characters point to conflicting phylogenies, we can be sure that at least some of the characters are misleading. A set of species has only one phylogeny: the phylogeny that represents the ancestral relations that those species possess. A set of species can no more have multiple phylogenetic relations than a human family can have more than one family tree. If a human family has two conflicting family trees in its possession, at least one of them must be wrong. Likewise, if two characters suggest incompatible phylogenies, something is wrong with at least one of them.

The techniques of cladistics work by distinguishing between reliable and unreliable characters. The unreliable characters, once identified, can be discarded. The amount of character conflict in the shortened list of reliable characters should be reduced — and in a good case the conflict will be reduced to zero, and all the reliable characters will agree on the same phylogeny. The analysis of characters, to distinguish reliable from unreliable characters, proceeds in two stages: we first distinguish homologies from homoplasies, and then distinguish derived homologies from ancestral homologies.

Figure 15.2
Character conflict in the phylogeny of birds and reptiles. The gait and the anatomy of the skull link crocodiles and birds; leg number, physiology, and external surface group link the reptilian groups. The anapsid skull has no openings, apart from the eye socket; the key feature of the diapsid skull is a single upper temporal opening, though most diapsids have an additional lower opening too. Archosaurs and lepidosaurs differ in their skulls (to be exact, lepidosaurs lack a lower temporal arch).

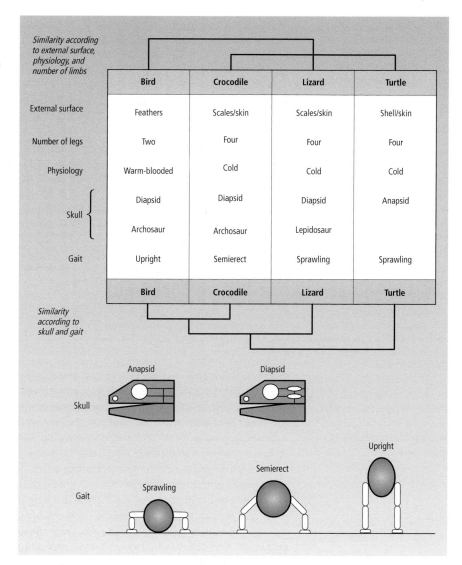

	Bird	Crocodile	Lizard	Turtle
External surface	Feathers	Scales/skin	Scales/skin	Shell/skin
Number of legs	Two	Four	Four	Four
Physiology	Warm-blooded	Cold	Cold	Cold
Skull	Diapsid	Diapsid	Diapsid	Anapsid
	Archosaur	Archosaur	Lepidosaur	
Gait	Upright	Semierect	Sprawling	Sprawling
	Bird	Crocodile	Lizard	Turtle

Similarity according to external surface, physiology, and number of limbs

Similarity according to skull and gait

Skull — Anapsid / Diapsid

Gait — Sprawling / Semierect / Upright

15.3 Homologies provide reliable evidence for phylogenetic inference, and homoplasies provide unreliable evidence

The first stage of cladistic analysis is to distinguish homologies from homoplasies. The difference between *homologies* (or a homologous character) and *homoplasies* (or a homoplasious character) is as follows. A homology is a character shared between two or more species that was present in their common ancestor. A homoplasy is a character

Figure 15.3

(a) A homology is a character state shared between two species that was present in their common ancestor.
(b) A homoplasy is a character state shared between two species that was not present in their common ancestor. *A* and *A′* are two character states.
(c) The wings of birds and bats are an example of a homoplasy. They are structurally different as the bird wing is supported by digit number 2, and the bat wing by digits 2–5. The bird wing is also covered with feathers, the bat's with skin.

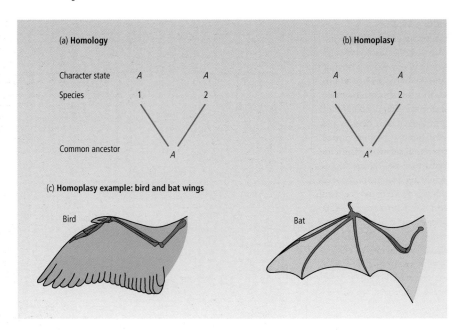

(a) **Homology**

Character state	*A*	*A*
Species	1	2

Common ancestor — *A*

(b) **Homoplasy**

A   *A*
1   2

A′

(c) **Homoplasy example: bird and bat wings**

Bird

Bat

We distinguish homologies from homoplasies

shared between two or more species that was not present in their common ancestor (Figure 15.3). Thus we start with a character that is similar in two species. We then trace all the way back to their most recent common ancestor. If the common ancestor had that same character, then the character in the two descendant species is similar by common evolutionary descent and is a homology. If the common ancestor had some different character state, then the character in the two descendant species evolved independently and is a homoplasy. The distinction matters because homologies may reveal phylogenetic relationships, whereas homoplasies do not.

A homologous character such as the heart, or lungs, of a human and a chimpanzee is easily recognized as the same character, presumably shared from a common ancestor that also possessed that character. In other cases, the similarity is less obvious. The five-digit limb of tetrapods is homologous even though its form varies (Figure 3.6, p. 58), and in extreme cases homologies are so subtle that it takes clever detective work to reveal them. The ear bones of mammals, for example, do not superficially resemble the skull and jaw bones of reptiles. But a classic piece of comparative anatomic research in the nineteenth century traced a series of intermediates that can be found between three skull and jaw bones of reptiles and three ear bones of mammals. The bones also have a common embryonic origin. A homologous character does not have to be the same in all the species possessing it — there only has to be some shared morphological information among them.

Homoplasies can arise for a number of reasons. In DNA evidence, as we discuss later, homoplasy can easily arise by chance. In morphological evidence, chance is unlikely to

Figure 15.4

Convergence in marsupial and placental carnivores. (a) The reconstructed bodies and skulls of *Thylacosmilus*, a saber-toothed marsupial carnivore that lived in South America in the Pliocene and of *Smilodon*, a saber-toothed placental carnivore from the Pleistocene in North America. (b) *Prothylacynus patagonicus*, a borhyaenid marsupial from the early Miocene in Argentina; *Thylacinus cynocephalus*, the extinct marsupial Tasmanian wolf; and *Canis lupus*, the modern placental wolf. From Strickberger (1990). © 1990 Jones & Bartlett Publishers.

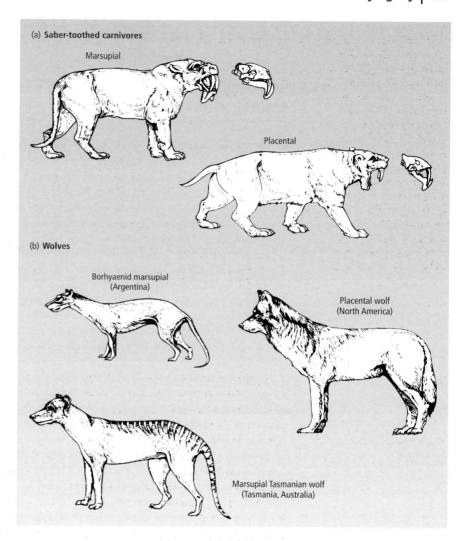

(a) Saber-toothed carnivores

Marsupial

Placental

(b) Wolves

Borhyaenid marsupial
(Argentina)

Placental wolf
(North America)

Marsupial Tasmanian wolf
(Tasmania, Australia)

Convergent evolution produces homoplasies

cause homoplasy; the most important cause is convergent evolution, when the same selection pressure has operated in two lineages. A classic example of convergence is seen in the two major groups of mammals, the marsupials and placentals (Figure 15.4). The marsupial and placental saber-toothed carnivores both evolved long, gashing canine teeth and there are also striking similarities in skull shape and body form in the marsupial and placental wolves. If we inferred the phylogeny of the marsupial wolf, placental wolf, and kangaroo from the pattern of phenetic similarity, we would obtain the wrong answer. The two wolves are phenetically more similar, even though the marsupial wolf is phylogenetically closer to the kangaroo than to the placental wolf.

The phenetic similarity between the wolves is homoplasious and is not due to a close phylogenetic relationship.

15.4 Homologies can be distinguished from homoplasies by several criteria

Homoplasies do not indicate phylogenetic relationships, and the first task is to recognize homoplasies, as opposed to homologies. How can we recognize them? The crude answer is that homologies are identifiably the same character in two species, but homoplasies differ in some way that suggests the character has evolved independently in the species that possess it. Research therefore starts with a character that shows some similarity in two (or more) species, and then examines the character in detail to find out whether it really is the same in all the species.

First, if a character is homologous it is likely to have the same fundamental structure. The wings of birds and bats, for example, are superficially similar; but they are constructed from different materials and supported by different limb digits (Figure 15.3c). The differences suggest that the bird and bat wings are homoplasies, and evolved independently from a common ancestor that lacked wings.

Second, homologies usually have the same relations to surrounding characters. Homologous bones, for example, are usually connected in a similar way with their surrounding bones.

Third, the character is likely to have the same embryonic development in different groups. A character that looks similar in the adult forms, but develops by a different series of stages, is unlikely to be homologous. One example, which we meet again in Chapter 16, is the relation between a barnacle, a mollusk such as a limpet, and a crab (Figure 16.1, p. 474). At least superficially, the adult form of a barnacle is more like a limpet than a crab. The relations of barnacles had been uncertain for centuries until John Vaughan Thompson discovered their larvae in 1830. The barnacle larva is very like the larva of several groups of Crustacea, and unlike those of mollusks. Barnacles therefore share a more recent common ancestor with crabs than with limpets. The similarities between the adult barnacle and limpet, such as their hard external armour, attachment to rocks, and feeding through a hole in the shell, are all homoplasious.

Finally, some other criteria can sometimes be useful. Convergence is caused by natural selection, when organisms in different evolutionary lineages face similar functional requirements (such as flying in birds and bats). We have grounds for suspecting that a shared morphological structure may be homoplasious when the species that share it clearly need it for their way of life.

The criteria in this section are not the only ones that can be used to distinguish homologies from homoplasies. However, the criteria discussed here do illustrate that we have techniques to analyze characters shared between species to distinguish homologies from homoplasies. A homology can be recognized as a character that has fundamentally the same structure, relations with surrounding parts, and development, in a set of species. Once the homologies are (often tentatively) identified, they can be retained in the list of evidence used to infer the phylogeny. The homoplasies are discarded.

Homologies are recognized by . . .

. . . structural similarity . . .

. . . relations between parts . . .

. . . embryonic development . . .

. . . and other criteria

15.5 Derived homologies are more reliable indicators of phylogenetic relations than are ancestral homologies

The next stage is to divide the homologies into ancestral and derived homologies. Consider the number of digits on the feet of a frog, a dog, and a horse. The frog and dog have standard tetrapod feet, with five digits. This is the ancestral state for all tetrapods. (Tetrapods are the group of amphibians, reptiles, birds, and mammals.) Horses have reduced the number of their digits, and have only one of the five digits left. The similarity of the dog and frog is not evidence that they share a more recent common ancestor with each other than either does with a horse. Indeed, both the dog and the horse are mammals, and share a more recent common ancestor than either does with a frog. In the group of the frog, dog, and horse, the state of having five digits per foot is a homology in the dog and frog, but not evidence of a phylogenetic relationship.

We distinguish ancestral from derived homology

We therefore need to distinguish ancestral from derived homologies (Figure 15.5). To see the distinction, first take the set of species under analysis. A homology that is present in the common ancestor of that group is an ancestral homology and is useless for determining phylogenetic relations within the group. The character state A' in Figure 15.5 is like the five-digit tetrapod foot in the group of the frog, dog, and horse. However, if we are studying instead the relations of a frog, a dog, and a fish, the five-digit foot is no longer the ancestral state. It was not present in the common ancestor of the three species. For these three species, the five-digit foot is a derived homology. It evolved within the group of species that we are studying and tells us something about the phylogeny. It tells us that the frog and dog share a more recent common ancestor with each other than either does with the fish.

Ancestral homologies are characters that were present in the common ancestor of the group of species under study. Derived homologies are homologies that evolved after the common ancestor, within the group of species under study. The distinction between ancestral and derived homologies is meaningless if we are only talking about two species by themselves: any homology for the two species is simply a homology. The

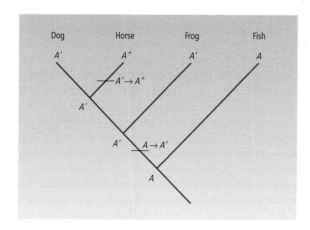

Figure 15.5

Ancestral and derived homologies. A' is an ancestral homology if we are studying the phylogeny of the dog, horse, and frog. A' is a derived homology if we are studying the phylogeny of the dog, horse, and fish. An ancestral homology was present in the common ancestor of the group of species under study; a derived homology evolved more recently that the common ancestor of the group of species being studied. The distinction between derived and ancestral homologies is relative to the group of species. Derived homologies reliably indicate phylogenetic relations; ancestral homologies do not.

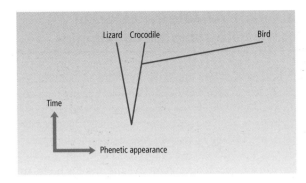

Figure 15.6
The evolution of birds, crocodiles, and lizards illustrates how if one lineage undergoes rapid evolution, members of the other two lineages are left looking relatively similar even though they are phylogenetically distant. A crocodile looks more like a lizard than a bird; but a crocodile has a more recent common ancestor with a bird than with a lizard.

distinction implies that we are comparing the two species with at least one other species. Then whether the homology is ancestral or derived depends on what that third species is.

Ancestral homologies are most dangerous for phylogenetic inferences in cases such as a bird, a crocodile, and a lizard (Figure 15.6). Here one lineage within a group of species has undergone rapid evolution. Birds have evolved wings and other skeletal and physiological adaptations for flight. The lineages to the crocodiles and lizards have evolved slowly in comparison. They have both retained ancestral reptilian characters such as scales and walking on four legs. The crocodile and lizard have been left looking relatively similar compared to birds because of the evolutionary spurt of the latter. But the similarity between crocodiles and lizards is ancestral similarity. It is for characters that were present in the common ancestor of all three groups. The similarity is not evidence that crocociles and lizards share a more recent common ancestor with each other than either does with birds.

The complete analysis of a character has two stages, first distinguishing homoplasies from homologies and then distinguishing ancestral from derived homologies. In all, a character can belong to any one of three types (Figure 15.7). The distinction matters because, of the three kinds of shared character, only derived homologies are evidence that the two species share a more recent common ancestor with each other than with any other species under investigation.

Phylogenies, therefore, should not be inferred from simple phenetic similarity. Phenetic similarity mixes reliable similarity (in derived homologies) with unreliable similarity (in homoplasies and ancestral homologies). That phenetic similarity is misleading in the case of convergence is widely appreciated; but ancestral homologies cause the same problem, and more insidiously. We have seen it in the reptiles (though they are not the only example): a crocodile looks more like a lizard than a bird, but is phylogenetically closer to a bird than to a lizard. The point of these two examples is not that phenetic similarity never indicates phylogenetic relationships, but that it is unreliable. If we sift the evidence, and concentrate on derived homologous similarity, we should make fewer mistakes in phylogenetic inference.

In Section 15.2 we saw that phylogenetic inference faces the problem of character conflict, that different characters suggest different phylogenies. The conflict is caused by homoplasies and ancestral homologies, both of which can fall into incompatible sets

Ancestral homology can be misleading

Derived homologies are reliable evidence

Figure 15.7
Shared characters divide into homoplasies, ancestral homologies, and derived homologies. (a) *a'* is a homoplasy: it is not in the common ancestor of the species that share it. (b) *a* is an ancestral homology: it is in the common ancestor of the species that share it, but has been lost in some descendants of that common ancestor. (c) *a'* is a derived homology: it is in the common ancestor of the species that share it, and in all its descendants. Notice that only derived homologies always indicate phylogenetic relationships. Figure 16.4 (p. 480) and Table 16.1 (p. 475) show the kinds of taxa that are defined by these three kinds of character.

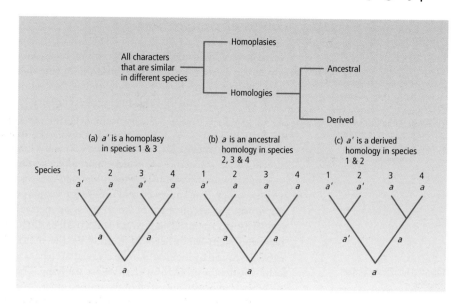

of species. If we successfully identify the homoplasies and ancestral homologies, and discard them, the problem of character conflict should be removed. Correctly identifed derived homologies must all agree on the same phylogeny. They all evolved in the same phylogenetic tree, and should all fall into the same pattern of groups ("horizontal" transfer of characters between lineages is an exception). If, in a number of characters, the homologies and character polarities have been correctly identified, it is impossible for different derived homologies to suggest incompatible phylogenies. The same is not true for homoplasies and ancestral homologies. Ten different homoplasies, or 10 different ancestral homologies, can fall into up to 10 different, and conflicting, groupings of species.

In summary, we have divided characters into three kinds and considered theoretically how each relates to groups of species in a phylogeny. Only shared derived homologies consistently reveal phylogenetic groups. But how can we distinguish in practice between ancestral and derived states of characters?

15.6 The polarity of character states can be inferred by several techniques

Cladists infer character polarities

The question of how to distinguish derived from ancestral homologies has the following general form. A character has two states, which we can call *a* and *a'*: we need to know whether *a* evolved from *a'*, or the other way round. In this section, we discuss two of the methods. The distinction between ancestral and derived character states is sometimes referred to as *character polarity*. Analyzing a character to work out which of its states are ancestral and which derived is also working out the "polarities" of the character states.

15.6.1 *Outgroup comparison*

Amniotes are the group made up of reptiles, birds, and mammals; all these animals possess an egg membrane, called the amnion, during their development. It is known that amniotes are a monophyletic group, that is they all share a unique common ancestor. Here, we will assume that the amniotes are indeed known to be a good phylogenetic group but that we do not know the relations among the different amniotes. For instance, in a set of six amniotic species (such as a mouse, a kangaroo, a bird of paradise, a robin, a crocodile, and a tortoise), does the kangaroo share a more recent common ancestor with a mouse, a bird of paradise, or what?

Suppose we have established homologies in various characters, including reproductive physiology. The kangaroo and mouse are viviparous, and the other four species are oviparous. Did the ancestor of the group of six species breed viviparously, in which case viviparity was ancestral and oviparity derived, or did it breed oviparously, in which case evolution went the other way round? By the method of *outgroup comparison*, the answer is found by looking at a closely related species which is known to be phylogenetically outside the group of species we are studying. The character state in that outgroup is likely to have been ancestral in the group under consideration.

In this case, we might look at a salamander, a frog, or even a fish. They are all near relatives of the amniotes, but are not amniotes themselves. These "outgroup" species almost all breed oviparously. The inference by outgroup comparison, therefore, is that oviparity is ancestral in the amniotes. Viviparity, in the kangaroo and mouse, would then be a shared derived character and oviparity in the other four a shared ancestral character.

In the abstract, there could be two species, species 1 and 3, sharing homology *a* and two others, species 2 and 4, with homology *a'* (Figure 15.8). We wish to know whether

Outgroup comparison infers character polarity from the states in related species

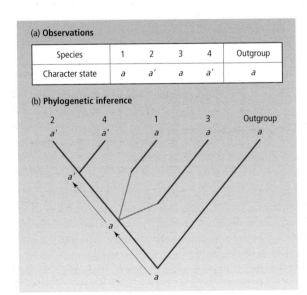

Figure 15.8
(a) Species 1–4 have the character states as given. We wish to know whether *a* or *a'* was the state in their common ancestor. (b) We look at a closely related species, the outgroup. It has state *a*, and we infer that was the state in the ancestor of species 1–4. The gray lines for species 1 and 3 indicate their branching relations remain uncertain.

character *a* evolved into *a'*, or *a'* into *a*. We look at a closely related species and infer that the state there is ancestral in the group of four. If the outgroup had *a* we should infer that species 2 and 4 share a more recent common ancestor with each other than with any of the other species; the relations of 1 and 3 remain uncertain (as is spelled out further below).

The underlying assumption of outgroup comparison is that evolution proceeded via the least possible steps. This is the assumption of "parsimony," which we look at in more detail later (Section 15.9.4). In Figure 15.8, if the character in the outgroup (*a*) is ancestral in the group of species 1–4, there must have been at least one evolutionary event in the phylogeny: a transition from *a* to *a'* before the ancestor of species 2 and 4. If, having observed *a* in the outgroup, we had reasoned that *a'* was the ancestral state of species 1–4, we should need at least two events: a change from *a'* to *a* somewhere between the outgroup and species 1–4, and then a change from *a'* back to *a* in species 1 and 3. If the character state in the outgroup is ancestral, the fewest evolutionary events are required.

Outgroup comparison, like all techniques of phylogenetic inference, is fallible. Sometimes, one possible outgroup will suggest that one character state is ancestral, but another outgroup will suggest that a different character state is ancestral. The result will then depend on which outgroup we rely on. The method is most reliable when the closely related species that could be used as outgroups all suggest the same inference, but it is possible to be led astray by the method in particular cases. The inference should be treated with caution, and if possible tested against other evidence.

Before we can use outgroup comparison, we need to know something about the phylogeny. We needed to know that fish and amphibians were outside the Amniota in order to use them as "outgroups." In practice this is not a major problem. Outgroup comparison cannot be used when we are absolutely ignorant, but if we know something about the phylogeny of a group (for example that amphibians are not amniotes, but are closely related to them) we can build on that knowledge to find out more (in this case, more about the phylogeny within the amniotes).

Parsimony underlies outgroup comparison

15.6.2 *The fossil record*

In the evolution of mammals from mammal-like reptiles, many characters changed (Section 18.6.2, p. 542). Posture evolved from a "sprawling" to an "upright" gait, and jaw articulation and circulatory physiology also changed. Some, though not all, of these characters leave a fossil record, and we can infer which character states were ancestral and which derived by seeing which is found in the earlier fossils.

The reasoning could hardly be easier. The ancestral state of a character must have preceded the derived states in fact, and therefore the earlier state in the fossil record is likely to be ancestral. In the case of the mammal-like reptiles, the criterion is reliable, because the fossil record is relatively complete. If the record is less complete, a derived character could be preserved earlier than its ancestral state (Figure 15.9), and the paleontological inference will be the opposite of the truth.

For a whole fossil series like the mammal-like reptiles, we can be reasonably sure which states are ancestral. At the other extreme, where there are few fossils and a highly

Character polarities can be inferred from the fossil record

Figure 15.9
(a) The ancestral state of a character (*a*) must have evolved before its derived state (*a'*). (b) If the fossil record is relatively complete, the ancestral state will be preserved in earlier fossils; (c) but if it is incomplete, the derived state may (ii) or may not (i, iii) be preserved earlier than the ancestral state.

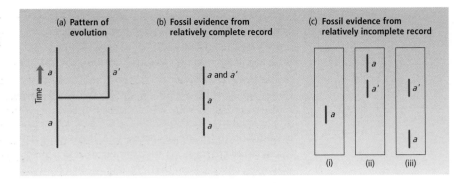

imperfect record, the evidence may be practically worthless. Most real cases somewhere between, and an intermediate level of confidence is appropriate.

15.6.3 Other methods

Outgroup comparison and the fossil record are not the only ways to determine character polarity. A third classic technique uses embryonic development, and we shall encounter a fourth (and recently invented) technique in Section 15.13 when we look at paralog rooting.

15.7 Some character conflict may remain after cladistic character analysis is complete

Cladistic analysis reduces character conflict

The cladistic techniques are intended to infer the phylogenetic relations of a group of species from conflicting evidence. The conflict in the raw evidence arises because some of the characters are homoplasies, some ancestral homologies, and some derived homologies. The cladistic analysis boils down the initial evidence to a list of derived homologies and should reduce the conflict relative to an unanalyzed list of characters, for the theoretical reasons given in Section 15.5. In an ideal case, the conflict should be reduced to zero, because real derived homologies cannot conflict. However, the actual level of conflict is likely to be reduced to something more than zero because the techniques can all make mistakes. Convergence can be deceptively exact, and homoplasies can be mistaken for homologies. The criteria for determining character polarities may be inapplicable (if the character lacks a fossil record, or its phylogenetic relations with nearby outgroups are obscure), and even when they can be used they are still fallible. Moreover, the existence of more than one criterion may increase the uncertainty. If a character can be studied by more than one of the criteria, they can be played against each other: if the criteria agree, that increases the plausibility of the conclusion, but if they disagree, we have another problem of deciding which evidence to trust.

Suppose, for instance, we began with a list of 100 characters, of which 30 pointed to one phylogeny (which we can call *a*), 30 to a second phylogeny *b*, 20 to a third phylogeny *c*, and 20 to other idiosyncratic arrangements. We then study the characters, identify the homoplasies and ancestral homologies, and discard them. Maybe 30 of the 100 initial characters remain. If all 30 point to the same phylogeny, our job is done. But in practice, 20 of the derived homologies may support phylogeny *a*, six support *b*, and four support *c*. The reason is probably that some of the characters we think are derived homologies are in fact homoplasies or ancestral homologies.

We have four options when faced with conflict in the evidence: we can scrutinize, and rescrutinize, the contradictory results to test their reliability; we can suspend judgment; we can collect more evidence; or we can infer that the phylogeny supported by the most evidence is the correct one. If 20 of the 30 characters support phylogeny *a*, then we could infer *a* is the correct answer. The viability of the four options will vary from problem to problem.

15.8 Molecular sequences are becoming increasingly important in phylogenetic inference, and they have distinct properties

The sequences of proteins and DNA are both used in phylogenetic inference. Proteins blazed the trail. The first protein to have its amino acid sequence worked out was insulin, which was sequenced by Sanger in 1954. Protein sequencing became an automated process through the 1960s, and the sequences of some proteins, such as cytochrome *c* and hemoglobin, became available in enough species for large-scale phylogenies to be inferred. DNA sequences followed on, about 20 years later. It was Sanger again who sequenced the first decent-sized sequence of DNA, in this case the whole genome (containing 5,375 bases) of the bacteriophage ϕX164, in 1977. DNA sequencing since then has expanded, almost explosively, and most current molecular phylogenetic work is concerned with DNA sequences. Many of the methods and concepts of molecular phylogenetics were established for proteins, however, and here we consider the two kinds of molecules together.

The deep logic of phylogenetic inference is identical for molecular and morphological characters, but the two have distinct properties and the methods and concepts used for each can appear very different. The homology/homoplasy distinction, in particular, differs for the two. When confronted by apparently conflicting homologies for morphological characters (like the wings of birds and bats), the first thing to do is to re-examine the organs, and their embryology, in detail to see whether their similarity really is fundamental, or superficial and homoplasious. Homology is a powerful concept for morphological organs such as wings. Wings are complex in structure and can take on an almost infinite variety of shapes; they have an embryonic development and morphological relations with the rest of the body. If the information in the structure and development of a wing in two species is the same, those wings are highly likely to evolved from a common ancestor who had similar wings.

The homology/homoplasy distinction is much less powerful for molecular evidence. Suppose a nucleotide is identical in two species. Evolutionary changes take place among a very limited set of alternatives (the four bases A, C, G, and T) and it is fairly probable that the same informational state could independently evolve in the two

Protein and DNA sequences have been transcribed in recent decades

Molecules are used in phylogenetic inference . . .

species. The argument is relaxed for proteins, because there are 20 amino acid states, but it still applies because 20 fixed states is still a small number compared with the variety of morphological forms. Thus for molecules it is not so unlikely that similarity in the states of two species could have evolved independently. Moreover, the morphologist's methods are absurd for molecules. The amino acid at site 12 of cytochrome c is methionine in humans, chimps, and rattlesnakes, but glutamine in all other species — including many mammals and birds — that have been studied. We cannot dissect the rattlesnake's methionine, or trace its embryonic development, to see whether it is only "superficially" methionine and "more fundamentally" glutamine. It is a methionine molecule, and that is that.

Nor can we usually assess the reliability of different pieces of molecular evidence by thinking about how natural selection could have acted on them. When morphologists examine a similarity between the organs of two species, they keep a look out for functional convergences — such as the evolution of wings in species that fly. This kind of analysis is impossible if we do not understand the relation between the structure (the wing) and its function (flight). For molecules, we usually lack this understanding. If we knew, for example, that a change from glutamine to methionine at site 12 of cytochrome c made functional sense in certain kinds of animals, then the same kind of arguments as appear in morphology could be used for the protein. Otherwise, we have to treat molecules in the way a morphologist would treat an organ of unknown function.

Molecular sequences have other distinctive properties. The amount of evidence they provide is large; cytochrome c alone, for example, has 104 amino acids, which can be treated as 104 pieces of phylogenetic evidence. A typical morphological study might be based on perhaps 20 or so characters, and it is exceptional for many more than about 50 characters to be used.

In addition, the recognition of independent units of evidence appears to be straightforward. With morphological evidence, two apparently separate organs may really be a single evolutionary unit. At one extreme, non-independence is obvious; no one would think of treating the right leg and the left leg as two pieces of evidence. But less obvious correlations can also arise as a consequence of developmental processes, which makes the recognition of independence tricky. For nucleotides, the mutations down the DNA molecule are effectively independent as each site can evolve independently of each other site.[2]

Evolution at different amino acid and nucleotide sites is easily comparable: one change at one site is equivalent to one change at another. This is a huge advantage when we are weighing up conflicting evidence. Suppose the nucleotides at 10 sites support one phylogeny for a group of species, and the nucleotides at five other sites support a different phylogeny. Each of the 10 sites in one set is approximately equivalent to each of the five sites in the conflicting set. We can assume that the phylogeny supported by the 10 nucleotides is the better estimate of the true phylogeny. However, if one

---

[2] However, not all sites may in fact evolve independently. For instance, a change at one site may set up selection for a compensatory change at another site. How much of a problem, if any, this creates for phylogenetic inference is unsettled. Genomic analyses are starting to reveal the amount of non-independent change at different sites. Averof *et al.* (2000) found non-independence in one sequence comparison; Silva & Kondrashov (2002) did not in another. As genomic analyses proliferate, understanding should deepen.

phylogeny is supported by 10 morphological characters and another by five other morphological characters, the comparison is less straightforward. It is not easy to say what amount of evolution in a knee bone is equivalent to any given change in a skull bone. Although the phylogeny supported by five characters has less characters in support, the evolution in those five characters may be somehow weightier, or more reliable. In most phylogenetic inference, we have to weigh one set of characters against another — because different characters sets will often support different phylogenies. However, no general method exists for comparing evolution between different morphological characters. Molecular characters are readily comparable and therefore easier to use.

These four properties of DNA and protein sequence data — the impossibility of any deeper analysis of the character, the large amounts of evidence, the recognizability of independent characters, and the comparability of evidence — have encouraged the development of statistical techniques to infer phylogenies. The same techniques are in principle just as applicable to morphological evidence, though here it is always tempting to try to pre-empt statistical analysis and resolve the apparent conflicts by ever-deepening character analysis. Morphological data are also less readily divisible into neat character states for statistical analysis.

Molecular phylogenetics uses statistical techniques

15.9 Several statistical techniques exist to infer phylogenies from molecular sequences

A full review of the statistical techniques that can be used to infer phylogenies from molecular evidence would have to cover dozens of techniques. Instead we shall concentrate on the basic principles of the three main classes of techniques that are currently in use. But before we come to these three, we need to know about "unrooted" as opposed to "rooted" trees.

15.9.1 *An unrooted tree is a phylogeny in which the common ancestor is unspecified*

The phylogenies that we have been concerned with so far (such as Figure 15.1) are all rooted trees. If you look at the phylogeny of species A–D in Figure 15.1 you can see the common ancestor (or "root") at the bottom of the tree. A rooted tree has a time axis on it, and successively more distant ancestors are successively lower on the page. A rooted tree is the goal of phylogenetic research. It is the way biologists think about the evolutionary relationships between species.

We establish the relation between rooted and unrooted trees

However, most molecular phylogenetic techniques first work out what is called an *unrooted tree* (Figure 15.10). An unrooted tree is like a rooted tree but with the time axis taken off; it shows the branching relationships between a set of species, but not the location of their common ancestor. Figure 15.10 illustrates the relation between a rooted and unrooted tree for four species. An unrooted tree is a less informative statement about phylogenetic relations. For four species, one unrooted tree is compatible with five rooted trees. We need extra information (the location of the root) to say which

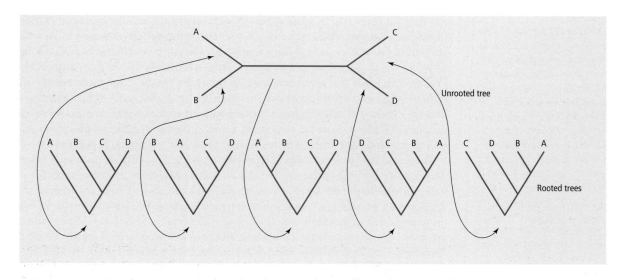

Figure 15.10
Unrooted and rooted trees. One unrooted tree for four species is compatible with five rooted trees. An unrooted tree is a timeless picture of branching relations and does not specify where the ancestor (or root) of the tree is. The root could be anywhere in it, and there are five topological possibilities, as drawn below. In general, any one unrooted tree of s species has $2s - 3$ internal branches and therefore $2s - 3$ possible rooted trees. (Here, as elsewhere in the chapter, we confine ourselves to strictly bifurcating trees.)

rooted tree is correct. In general, the root could be in any one of the internal branches. A four-species unrooted tree has five internal branches. A five-species unrooted tree has seven internal branches and is compatible with seven rooted trees.

Unrooted trees can be rooted cladistically

Unrooted trees can be thought of as part of the internal workings of molecular phylogenetic techniques. The unrooted tree links the species according to the evidence that is used to infer the phylogeny, but it does not show ancestral relations. Once a technique has found the unrooted tree for a set of species, some further evidence is used to find the root, and thus the ancestral relationships between the species. This further evidence often consists of one of the cladistic techniques for determining character polarity (Section 15.6 above). For instance, in Figure 15.10 we could look at the molecular sequence in some closely related species (or "outgroup"). If it was most similar to species A, that would suggest the root of the tree lies in the branch leading to A. The rooted tree would be the one at the left in Figure 15.10. In some cases however, the location of the root cannot be found, or an analysis can proceed with an unrooted tree alone. Then the unrooted tree is the final product of the molecular phylogenetic study.

15.9.2 **One class of molecular phylogenetic techniques uses molecular distances**

Imagine that we know the sequences of a particular 100-nucleotide stretch of DNA in four species, A, B, C, and D. For any pair of species, such as A and B, the nucleotides will

Figure 15.11

Distance methods. (a) The data consist of a matrix of distances between species. Here we have four species (A, B, C, and D) and the matrix shows pairwise distances between all the species. If distance is measured as percent difference between the DNA of two species, for example, then the DNA of species A and B would differ by 4%. The shaded region of the matrix is either meaningless or redundant. (b) Each species is grouped with the other species that it has the shortest distance to. The numbers on the branches are the implied amounts of evolutionary change, and add up to the total distances in (a).

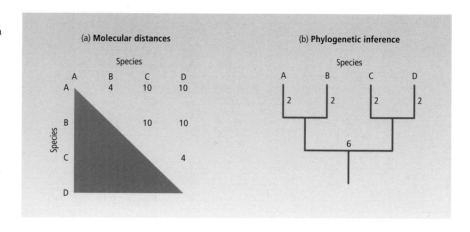

Molecular distances between species can be measured

be the same at some sites and different at others. Maybe it is the same at 96 sites and different at four. The two sequences are then 4% different. This figure is a simple example of a *molecular distance*. The simplest kind of molecular phylogenetic inference uses the matrix of molecular distances between species to infer the phylogeny. The species with shorter distances between them are inferred to be more closely related (Figure 15.11). This is a quick and dirty method of phylogenetic inference. The method assumes a "molecular clock" (Section 7.3, p. 164).[3] If the molecular distances between species increases constantly with time, the species pairs with shorter distances will indeed share more recent common ancestors.

Some classic molecular phylogenetic inferences have been made by what are essentially "distance" methods. For example, the molecular distance between two whole DNA molecules, from two species, can be measured by DNA hybridization. This method begins with DNA from a number of species. The DNA of any pair of species is "denatured": the double-stranded molecule is made into two single strands, usually by heating the molecule up. The single strands of DNA from the two species are allowed to join up and form double-stranded hybrid DNA. This hybrid molecule is then in turn denatured by heating it up. The crucial measurement is how hot you have to make the hybrid DNA before it will separate into its two single strands. The more similar the DNA of the two species is, the stronger the bond between them, and the higher the temperature required to separate them. The same procedure is followed for all pairs of species, producing a matrix of distances for all the species. The matrix is turned into a phylogeny, assuming that species with more similar DNA have more recent common ancestors (Figure 15.12).

---

[3] This is a key assumption. When looking at cladistic techniques earlier in the chapter, I pointed out that simple phenetic similarity (or phenetic distance) between species is not thought to reveal phylogenetic relations. Rates of phenetic evolution are so erratic that we need to break down phenetic similarity, to find the component due to shared derived characters. Chapter 16 will make much the same point. However, if molecular evolution is divergent and has a fairly constant rate, molecular distances can be used and cladistic analysis is unnecessary. In more advanced work, the molecular clock may not be a crucial assumption. If molecules, or lineages, with weird rates of evolution can be identified, they can be either corrected for or removed from the analysis.

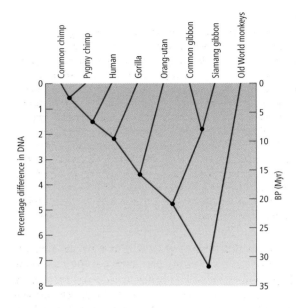

Figure 15.12
Phylogenetic relations of hominoids, as revealed by DNA hybridization. This result contains the evidence that the DNA of humans is 98.5% similar to that of chimpanzees. We meet another example of a classic distance method in Section 15.13 below. Redrawn, by permission of the publisher, from Sibley & Ahlquist (1987).

Distance methods revised the great ape phylogeny

Figure 15.12 mainly illustrates a distance method in action, but it has three particular features that are worth noting. One is that the DNA of humans and chimpanzees is 98.5% identical: DNA hybridization is the main evidence for this frequently encountered observation.[4] Secondly, humans and chimpanzees seem to have a more recent common ancestor than either has with gorillas. Thirdly, the human lineage banches from our nearest ape relatives a little over 5 million years ago. The second and third features matter for a controversy that we shall look at later in Section 15.13.

In practice, distance methods rarely use the simple fraction of sites that are identical between the DNA of two species, as the raw measurements of distance have first to be corrected for a problem known as "multiple hits." This problem comes up in some form or other in all molecular phylogenetic methods and we look at it next.

15.9.3 *Molecular evidence may need to be adjusted for the problem of multiple hits*

Multiple hits refers to the following problem. Imagine two species just after they have split from a common ancestor. Our 100-nucleotide stretch of DNA will probably be identical to them so the molecular distance between them is zero (Figure 15.13, at time zero). After a while, the nucleotide may change at one site in one of the species. Maybe it was initially T, and changed to C in one of the species. The molecular distance is now 1%. A while later a second change occurs, and then a third, and so on. The molecular

---

[4]   Britton (2002) have recently revised the figure down to about 95%, taking account of insertions and deletions. However, the lower figure does not alter the inferred time of human origin, because the distances between all species pairs are likely to be subject to similar adjustments.

Figure 15.13

As two species evolve apart over time, their DNA becomes increasingly different. Initially, each evolutionary change increases the difference between the two species and the line goes up. After a while, a second change may occur at a site where a change has already taken place; the second change then does not increase the difference between the two species. The line starts to level off. Eventually the two species are "saturated" with change and evolution has no average effect on the difference between them. The line is now flat. The line may flatten off at a 75% difference because there are four bases, but the exact figure may not be 75% for various reasons. See Figure 15.16 for examples. The regions I, II, and III correspond to regions where the molecular inference of phylogeny is (I) relatively easy, (II) possible but requires correction for multiple hits, and (III) impossible.

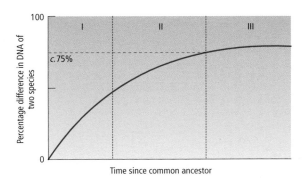

distance between the species increases over time. The molecular distance increases because each successive change is likely to be at a different site in the 100-nucleotide stretch. After a while, a second change may occur at a site where a change has already occurred. Maybe the species with C evolves to have G. This evolutionary change will not increase the molecular distance between the species. When the first change happened, and one species had T and the other species had C, that produced a 1% difference. If the T or C now changes, the difference is still 1%. Thus, beyond a certain level, the molecular distance between the two species flattens off even though they continue to evolve apart. The later changes do not add the distance — there are multiple hits at the same site.

The molecular distance between species is likely to level off at something like 75% (Figure 15.13) because DNA has four bases. Suppose the nucleotide is C at a site in one species. We then look at the equivalent site in a very different species — a species that is evolutionarily so distant that the site has changed many times and is effectively randomized. If we look at two random sequences of DNA, the chance of identity between the sequences at one site is approximately 25%. If the nucleotide at the site is C in one sequence, it could be C, G, T, or A in the other, chosen at random. Thus distance levels off at about 75% (and identity at 25%) for very different species. (These figures assume the base frequencies are equal. If C or G are more frequent than A or T in the species, molecular distance will level off at a figure below 75%.)

Molecular distances can be corrected for multiple hits. We use a model of sequence evolution (Box 15.1). The simplest model assumes that the chance that any nucleotide will change (p) is the same. We can estimate the value of p from the sequence data for the species. We then use an appropriate statistical model (such as the Poisson distribution) to calculate how many changes underlie the observed sequence data. The calculation might, for instance, show that in a 100-nucleotide stretch of DNA in two species, 30 sites have not changed, 30 have changed once, 20 have changed twice, 10 have changed three times, six have changed four times, and four have changed five times. We then add up the total number of changes: $(30 \times 1) + (20 \times 2) + (10 \times 3) + (6 \times 4) + (4 \times 5) = 144$. This is the corrected number of evolutionary events. Compare it with the 70 sites that differ between the two sequences: we have corrected a raw number of

More than one substitution may underlie one base difference between two species

Multiple hits can be corrected for

Box 15.1
Models of Sequence Evolution

A DNA sequence is made up of four kinds of nucleotide. Evolution consists of changes among the four nucleotide states. In the simplest model of evolution, we assume that the chance of any change, from one nucleotide to another, is the same, and has probability p. (p could be defined as the chance that a nucleotide at a site will change from one kind of nucleotide to another kind, per million years, in a population. In practice p is usually an instantaneous rate, rather than a rate per million years, but that does not matter here.) Figure B15.1 shows the evolutionary possibilities.

An A, for example, can change to a C, G, or T. In all, there are 12 kinds of change. The simplest model assumes that the chance of all 12 is the same, p. This model is a "one-parameter" model, called the Jukes–Cantor model after its originators. If two species have the same nucleotide at a site, it could be that the nucleotide has not changed (chance $1 - 3p$). Or it could have changed and then changed back (A → C → A, for instance), which has chance p^2. (The probabilities would need to be multiplied by an amount of time if they have been evolving apart for something other than 1 million years.) If the two species have different nucleotides (such as A in one species and C in the other) at a site, there could have been one change (chance p) or two (for instance A → G → C), with chance p^2. We can think through all the possibilities, and calculate the total probabilities that a site will be identical, or different, in the two species, when we sum over all the ways that a site can end up identical, or different.

The one-parameter Jukes–Cantor model is the simplest. In practice, the chance of transitions differs from the chance of transversions. This leads to the "two-parameter" model, first discussed by Kimura. We assume that the four transitions in Figure B15.1 have one chance, p_1, and the eight transversions have some other chance, p_2. More complex models allow for the possibility that some transitions are more likely than other transitions. Figure B15.1 has 12 arrows, and a complex model could have 12 parameters, one for each kind of nucleotide change. Models for maximum likelihood (see Box 15.2) usually also take account of differences in the rate of evolution between different sites.

For any given model of sequence evolution, we can use the sequence data to estimate the value of p (or of p_1 and p_2). Several statistical procedures are used, which can be found in an advanced text. The estimated value of p can then be used for various

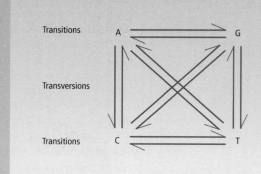

Figure B15.1
Possible kinds of evolutionary change between the four kinds of nucleotide.

purposes, such as correcting for multiple hits, or the calculation of maximum likelihood.

Inferences that employ models of sequence evolution are more or less accurate, depending on how good the model is and how well the parameters are estimated. For instance, if transition and transversion frequencies differ, then the use of the one-parameter Jukes–Cantor model would give misleading results, and might lead to a faulty phylogenetic inference. Also, the parameters (such as p) are estimated from sequence data, using a statistical model such as the Poisson or gamma distribution. The quality of the estimate depends on how good the data are — whether the stretch of sequence is long enough, for instance — and on whether the correct statistical model has been picked. Controversies in molecular phylogenetics can turn on details of these statistical models. In general, a trade-off exists between the quantity of data needed to estimate the parameters, and the accuracy of the model that can be used. A model with two parameters should be better than a model with one parameter, but requires more sequence data to estimate the parameters.

Further reading: Swofford *et al.* (1996), Page & Holmes (1998), Graur & Li (2000).

differences of about 70 to an inferred number of events of 144. The increase is due to the unobservability of multiple hits. (The numbers here are for illustration only. A real example would be more complex, and the numbers could look very different from those here.)

Figure 15.13 divides into three regions. For small amounts of change, the observed molecular distances accurately reflect the amount of evolution and no correction for multiple hits is needed. In the second region, we should correct for multiple hits. The corrected molecular distances are the figures to use in phylogenetic inference. Finally, in the third region, evolution has effectively randomized the sequences and, once the line has gone flat, we cannot recover the real amount of evolutionary change making correction for multiple hits impossible. Phylogenetic inference is impossible for sequences that have evolved this far apart. (The process by which changes occur at an increasing fraction of the sites in the sequences of two species as they evolve apart over time is referred to as *saturation*. When practically all the sites have changed, we are in region III of Figure 15.13 and the two sequences are referred to as "saturated," and are no longer any use for phylogenetic inference.)

The art of molecular phylogenetics consists in finding molecules that have evolved the right distance apart. For all techniques of molecular phylogenetics, inference is relatively easy in region I, becoming more difficult as we move through region II, and is impossible in region III. Section 15.10 looks at some examples to illustrate the point.

When two species are very different, phylogenetic inference is impossible

15.9.4 *A second class of phylogenetic techniques uses the principle of parsimony*

Phylogenies can be inferred on the assumption that change is rare

In phylogenetic inference, parsimony refers to the principle that the phylogeny requiring the fewest evolutionary changes is the best estimate of the true phylogeny. In a simplified case, we proceed as follows (Figure 15.14). First, write out all the possible unrooted trees for the species. Then count the smallest number of evolutionary events implied by each unrooted tree, given the observed data. The best estimate of the true phylogeny is the one that produces the lowest count.

How can the parsimony principle be justified? Why is a phylogeny requiring less evolutionary events a more plausible inference than one requiring more? The parsimony principle is reasonable because evolutionary change is improbable. Suppose we know that a modern species and one of its ancestors both have the same character state (Figure 15.15). Parsimony suggests that all the intermediate stages in the continuous lineage between ancestor and modern species possessed that same character state. As we have seen, an indefinitely large number of changes — indeed an infinite number — could logically have occurred between ancestor and descendant. However, a change followed by a reversal of that change is unlikely. Each change requires a gene (or set of genes) to arise by mutation and then to be substituted, either by drift if the change is neutral or by selection; both these processes are improbable. It is much more likely that the same character would have been continuously passed on, in much the same form, from ancestor to descendant by simple inheritance. We know that this is plausible because it happens every time a parent produces an offspring — the parental characters are passed on.

Figure 15.14

Phylogenetic inference by parsimony. (a) The inference uses observations such as DNA sequence data, shown here for five sites. (b) We then count the minimum number of evolutionary changes implied by the sequence data for all possible phylogenies (or unrooted trees, to be exact). The three possible unrooted trees for four species are shown. The marks within each branch indicate the location of an evolutionary change. For instance, the top row shows where the changes must be for the first two sites (AA in species 1 and 2, TT in species 3 and 4). In the left-hand tree, two changes is the minimum that can produce this pattern, and the two changes must be in the internal branch. We finally sum the number of changes for all the trees, and the tree requiring the fewest changes (seven, in this case) is inferred to be correct. The fifth site is ignored in the counting in (b) because it is the same in all species and does not help us to infer the phylogeny. Sites like this, which are equally compatible with all possible trees, are called uninformative. Sites (such as 1–4) that require different numbers of events in different trees are called informative.

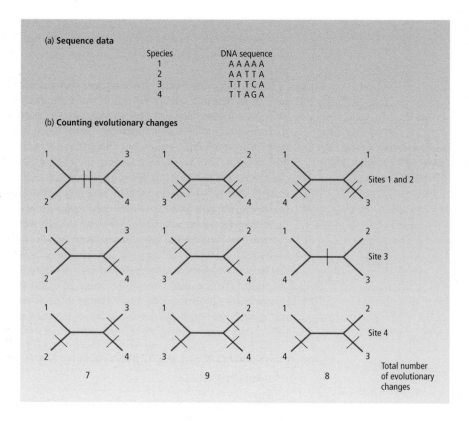

(a) **Sequence data**

Species	DNA sequence
1	A A A A A
2	A A T T A
3	T T T C A
4	T T A G A

(b) **Counting evolutionary changes**

Sites 1 and 2

Site 3

Site 4

Total number of evolutionary changes

7   9   8

For the characters shared between humans and chimps, the argument is particularly powerful. Chimps and humans share whole complex organ systems like hearts and lungs, eyes, brains, and spinal cords. The initial evolution of each of these characters required improbable mutations, and natural selection operating over millions of generations. It is evolutionarily improbable to the point of near impossibility that the same changes would have evolved independently in the two lineages after their common ancestor. By contrast, there is nothing improbable about postulating that the characters could have been passed on in passive inheritance from the common ancestor of chimps and humans to the modern descendants.

For some characters other than the complex morphological characters shared between humans and chimps, the argument is less powerful. At the other extreme, if we find one nucleotide, at a particular site in the DNA, shared between two species, there is a 25% probability that it could be shared by chance and the principle of parsimony does not strongly suggest that the nucleotide has not changed through all the evolutionary intermediates between the two.

The argument is more powerful in some cases than others. But evolutionary change in all characters is improbable to some extent, as compared with simple inheritance, and the principle of parsimony therefore has a sound evolutionary justification. In

Figure 15.15

The same character is found in both a descendant species and one of its ancestors. It is more likely (a) that the character has remained constant and has been passed on by inheritance than (b) that it has changed and reverted to its original state a number of times between the ancestor and descendant.

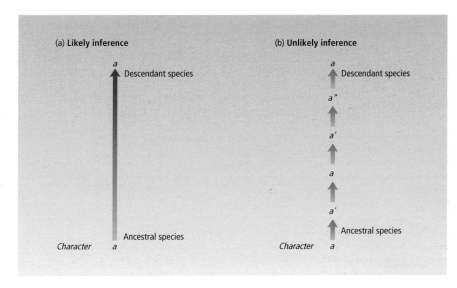

conclusion, it is more likely that a character will be shared by common descent than by independent, convergent evolution. For any set of species, a phylogeny requiring less evolutionary change is more plausible than one requiring more.

15.9.5 *A third class of phylogenetic techniques uses the principle of maximum likelihood*

The final technique we shall look at uses a statistical framework called maximum likelihood. The detailed calculations, when fully spelled out, are quite laborious, even for a simple case. (Box 15.2 works through the calculations for one nucleotide site in a four-species tree.) The basic procedure is to calculate (using a model of sequence evolution) the probability of observing the sequence data for a set of species, for all possible phylogenies. The most likely phylogeny is the one that has the highest probability of having produced the observed sequences.

The probability of a tree can be computed, given data and a model of evolution

Maximum likelihood is a computationally more demanding technique than parsimony. The method not only has to work through all possible phylogenies (just as parsimony has to), but also has to make detailed estimates and calculations for all the phylogenies. Maximum likelihood was little used until recently because it could only be implemented with small numbers of species. The advantage of maximum likelihood is that it can readily exploit information about rates of evolution. In the simple model used in Box 15.2, the chance of any evolutionary change was the same, p. But the same procedure can be used with more complicated models that have several parameters to describe evolution. Phylogenetic analysis with maximum likelihood can also use other information: the rate of evolution may vary between species, or between genes, or over time. Maximum likelihood is a very broad framework. It also has some other

Box 15.2
Phylogenetic Inference by Maximum Likelihood

Real sequence data consist of nucleotides at a long series of sites. In the calculations of maximum likelihood, each nucleotide site is subject to much the same calculation and we can look at any one site to see what the calculations are. Suppose we have one site and four species (called 1, 2, 3, and 4) and their nucleotides are:

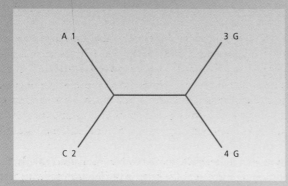

We now need a model of evolutionary change. The simplest is the model shown in Box 15.1, in which the chance of changing from one nucleotide to another is p. We can write out a matrix, with the chance of changing from one state to another (per time unit):

		Final state			
		A	C	G	T
Initial state	A	$1-3p$	p	p	p
	C	p	$1-3p$	p	p
	G	p	p	$1-3p$	p
	T	p	p	p	$1-3p$

If the nucleotide is A, for instance, it has a chance $1-3p$ of staying A and p each of changing into C, G, and T. Suppose that each branch is one time unit long. We now calculate the probability

of observing the data for all possible states of the internal nodes. We could start with:

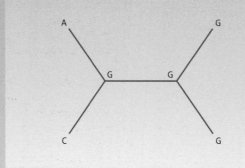

That is, we assume both internal nodes have G. The total chance of this is $p^2 + (1-3p)^3$. In two of the branches there has been a change (chance $1-3p$). We calculate the same sort of probability for all 16 possible combinations of the two nucleotides at the two internal nodes. That gives us the total probability of observing the data at this one site, given the model of evolution. Probabilities of this sort tend to be very small and they are usually converted to natural logarithms to make the numbers more manageable (so $p^2 + (1-3p)^3$ can be written as $\ln p + 3 \ln (1-3p)$.

In practice, we may have nucleotide data for 100 sites. The same sort of calculation is performed for every site, to find the total likelihood for the tree. We then need to do the same calculation for all the other possible unrooted trees. The best estimate of the true tree is taken to be the one with the highest probability (or maximum likelihood) of being observed. With data such as we used for parsimony in Figure 15.14, the result would usually be the same with maximum likelihood. The trees that require more evolutionary events will also be less probable, provided the value of p in the model of evolutionary change is low.

Further reading: Swofford *et al*. (1996), Page & Holmes (1998), Graur & Li (2000).

advantages, for instance it gives an exact probability for each unrooted tree, and this makes quantitative comparisons between trees straightforward. We can say that one tree is so many percent more probable than another. Quantitative comparisons of this kind are not so easy with the technique of parsimony.

15.9.6 Distance, parsimony, and maximum likelihood methods are all used, but their popularity has changed over time

Distance methods, parsimony, and maximum likelihood, in that order, require increasing amounts of data, and increasing computer power, in order for them to be used. Partly for this reason, the historic trend in phylogenetic research has been from using distance methods, which were used in the pioneering years of the late 1960s through to the early 1980s, to increasing use of parsimony, from the late 1970s to the 1990s, to an increasing use of maximum likelihood through the 1990s and into the twenty-first century. Maximum likelihood is likely now the most widely used method of molecular phylogenetics.

However, many biologists still use, and defend the use of, parsimony and distance methods. Some biologists think that molecules evolve in a basically clock-like manner, meaning that distance methods will usually give the right answer, and the sophistications of parsimony and maximum likelihood are unnecessary. But if some lineages evolve faster than others, distance methods misbehave — for much the same reason that simple phenetic similarity gives the wrong answer when comparing birds, crocodiles, and lizards (see Figure 15.6). Parsimony and maximum likelihood are less likely to go wrong.

Variable evolutionary rates upset distance methods

Parsimony has a particularly close relation with the methods of cladistics. The cladistic methods we looked at in the first part of this chapter are logically almost identical to the principle of parsimony. Parsimony counts evolutionary events, and each event generates a new derived character state. The use of homologies rather than homoplasies, and of derived rather than ancestral homologies, correspond to the principle of parsimony. Methods such as outgroup comparsion (Section 15.6.1) are simple applications of parsimony. It is therefore no coincidence that the use of parsimony in phylogenetic inference, and of cladistics in systematics, rose hand-in-hand from about 1980 onwards. (Chapter 16 considers cladistic systematics further.)

Cladists prefer parsimony

The sheer quantity of DNA data that are available now, along with increased computer power, makes maximum likelihood (arguably) the most powerful method in modern biology. However, the use of maximum likelihood is still seriously limited by the power of computers (for reasons we return to in Section 15.11.2).

15.10 Molecular phylogenetics in action

15.10.1 Different molecules evolve at different rates and molecular evidence can be tuned to solve particular phylogenetic problems

Different proteins, and stretches of DNA, evolve at different rates (Table 7.1, p. 161, and Table 7.6, p. 177), and they can be used like clocks with hands that revolve at different rates. If you use a rapidly evolving molecule for an ancient group, the molecule will have "turned over" many times during the phylogeny, and once multiple changes at the same site become common the phylogenetic information in the sequence similarity is lost — a stopwatch with only a seconds hand would be no use in comparing professors' lecture times. Likewise, slowly evolving molecules are useless for fine phylogenetic resolution because they will not have changed enough.

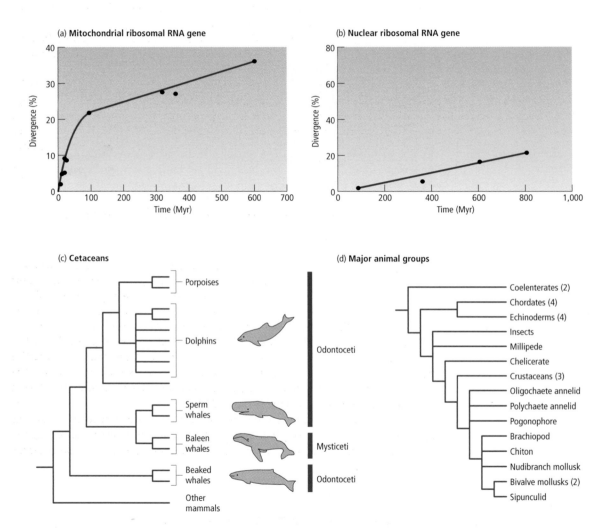

Figure 15.16

Matching the molecule to the phylogenetic problem. The ribosomal RNA genes in the mitochondria (a) evolve more rapidly than those in the nucleus (b). The different points are for species pairs, for which the date of their common ancestor can be estimated from fossils. The graphs tail off (at about 33% divergence) because of multiple substitutions at a site.

(c) Phylogeny of dolphins and whales, using mitochondrial rRNA genes; the deepest root is about 35 million years ago. (d) Relations of major animal groups, as revealed by nuclear rRNA genes; the deepest root is probably over 600 million years ago. Redrawn, by permission of the publishers; (a and b) from Mindell & Honeycutt (1990), (c) from Milinkovitch *et al.* (1993), and (d) from Lake (1990).

Ribosomal RNA genes are widely used in molecular phylogenetics

Ribosomal RNA genes are particularly valuable in phylogenetic reconstruction because they are found in almost all species: they are present in both mitochondrial and nuclear DNA. The mitochondrial genes evolve more rapidly than the nuclear (Figure 15.16a and b), and mitochondrial rRNA genes are useful for resolving phylogenetic problems in the 10–100 million year range, whereas the slowly evolving nuclear rRNA genes are useful in the hundreds of millions of years range.

Thus when Milinkovitch *et al.* (1993) wished to resolve the phylogeny of dolphins and whales, which the fossil record suggests to have originated less than 35–40 million years ago, the mitochondrial rRNA genes were appropriate (Figure 15.16c).

By contrast, Figure 15.16d shows the results of a study by Lake (1990) on the major groups in the animal kingdom. These groups originated about 1,000 million years ago (Section 18.4, p. 535) and the nuclear rRNA genes were the appropriate molecule for the problem. Some of the branch patterns in Lake's result have since been challenged, but the main point here is that slowly evolving molecules are needed to infer phylogenetic relations of this degree of antiquity.

15.10.2 *Molecular phylogenies can now be produced rapidly, and are used in medical research*

The origin of HIV has been dated

Human populations are recurrently infected with new, or apparently new, diseases. Many of the diseases are caused by viruses. Molecular phylogenetics has become, in the past decade, a key part of the medical research program to identify each new disease and where it came from. For example, HIV emerged as a mysterious new disease in the early 1980s. Since then, many copies of the virus, and other related viruses, have been sequenced. Figure 15.17 shows a phylogeny of HIV, which strongly suggests that HIV-1 entered human populations, perhaps more than once, from chimpanzees and HIV-2 came from sooty mangabeys. The molecular clock can be used to estimate the date when the virus moved between species, and Korber *et al.* (2000) estimate that HIV-1 moved from chimpanzees to humans in the 1930s.

The viruses that cause new diseases are now almost routinely sequenced, and the sequence run through a phylogeny program. We can identify the source, and something of the nature, of each new disease virus within months, or even weeks, after the disease breaks out.

15.11 Several problems have been encountered in molecular phylogenetics

Molecular phylogenetics is now among the most, perhaps *the* most, active areas of research in evolutionary biology. A number of problems have come up in this research program. None of them are insuperable, and in this section we shall look at five of the main problems and how they are being dealt with.

Figure 15.17
Tree for human immunodeficiency virus (HIV) and other related viruses (SIV) that infect other primate species. The tree was constructed using 38 amino acid sequences of the *pol* gene, by a phylogenetic method called neighbor joining. (Tree courtesy of Dr D. Robertson. See Holmes (2000a) for a similar tree, and further discussion.)

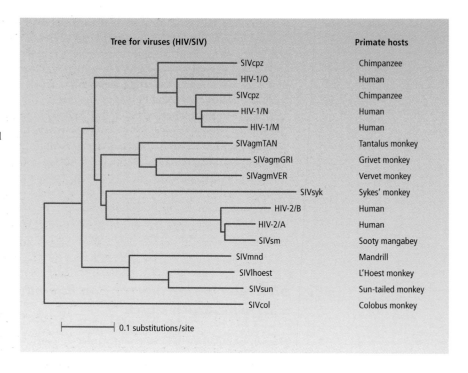

Tree for viruses (HIV/SIV) — Primate hosts

SIVcpz	Chimpanzee
HIV-1/O	Human
SIVcpz	Chimpanzee
HIV-1/N	Human
HIV-1/M	Human
SIVagmTAN	Tantalus monkey
SIVagmGRI	Grivet monkey
SIVagmVER	Vervet monkey
SIVsyk	Sykes' monkey
HIV-2/B	Human
HIV-2/A	Human
SIVsm	Sooty mangabey
SIVmnd	Mandrill
SIVlhoest	L'Hoest monkey
SIVsun	Sun-tailed monkey
SIVcol	Colobus monkey

0.1 substitutions/site

15.11.1 Molecular sequences can be difficult to align

When we compare a DNA sequence from two species, and count how many nucleotides have changed, we need to be sure that each site in one species corresponds to the same site in the other species. The two sequences need to be correctly aligned. Alignment is not simply a matter of putting the two sequences next to each other. With normal length sequences of more than 100 nucleotides, regions will usually have been deleted during evolution in some species and added to others, such that the sequences of the different species do not simply align, with nucleotide number 39 of species 1 corresponding to nucleotide 39 of the other species. There are ways of dealing with the problem, but they can sometimes go wrong. (See the references in the further reading section at the end of the chapter.)

15.11.2 The number of possible trees may be too large for them all to be analyzed

In Section 15.9.4 and 15.9.5, we saw that it is necessary to search through all the possible trees in order to find the most likely, or the most parsimonious, tree. The problem is that the number of possible trees may be impossibly large. With four species, three bifurcating unrooted trees are possible meaning it is not difficult to count the number of events implied by them all. For five species, however, 15 trees are possible. The general formula for the number of possible unrooted bifurcating trees for s species is:

$$\text{Number of possible unrooted trees} = \prod_{i=3}^{s} (2i - 5)$$

The number of possible trees can be astronomically large

The Π term means "product": we multiply (that is, take the product of) all the possible terms in the parentheses. For three species, $s = 3$ and there is only one term to take the product of (from $i = 3$ up to s, which is also 3); the parenthetic term for $i = 3$ is $6 - 5 = 1$, and the number of possible trees is therefore one. For $s = 4$, we have to multiply that 1 by the parenthetic term for $i = 4$, which is 3; $3 \times 1 = 3$, the number of unrooted trees for four species. For $s = 5$, the product is $5 \times 3 \times 1 = 15$, and so on. The number of possible trees increases explosively as the number of species goes up. For 50 species, there are about 3×10^{76} possible unrooted trees, and for the 30 million species that may be alive on Earth today the number is about $10^{300,000,000}$. No computer can search through that quantity of trees and about 25 or so species is the practical upper limit.

Students of molecular phylogenies distinguish between "algorithms" and "optimality criteria." Maximum likelihood and parsimony are examples of optimality criteria, which say that the best tree is the one requiring the least evolutionary change. An optimality criterion is a criterion that all the possible phylogenies can be compared against, and the best estimate of the phylogeny is the one that is closest to the criterion.[5] Optimality criteria run into the problem of limited computer search capacity, because all the trees have to be compared with the criterion. If the number of species is too big for all the possible trees to be searched, the search instead has to be done by means of an "algorithm." An algorithm is a rule about how to search from one tree to the next, and to assess which of the two trees is better. It will eventually find a tree that is better than any of the alternatives it compares it with, but it searches through only a limited number of trees to reach that end.

Algorithms are used to search a subsample of trees

Here is an analogy. Suppose you are in San Francisco and giving someone instructions on how to find Los Angeles. An optimality criterion would be to say "find the city with the largest population in the USA." The unfortunate person who receives this direction has to visit every city in the country, and measure their population sizes, in order to be sure he or she has found that destination. (We assume they have no other source of information.) An algorithm would be something like "face south and, keeping the Pacific Ocean on your right-hand side, move forwards until you arrive at a city with more than a million inhabitants." Now only a small proportion of the USA has to be searched, and the conclusion will be satisfactory so long as no other cities exist that meet the criterion between the starting and finishing points.

The particular algorithms used in phylogenetic research have constantly improved in recent years, and we shall not enter into details here. What does matter is that

---

[5] We could say, formally, that the optimality criterion of parsimony is zero evolutionary change: the tree of all the possible trees that comes closest to having zero change is the best. Notice that is not the same as saying we expect any tree to *have* zero change: we know that evolution has happened. It is a formal logical criterion, not a theory of reality.

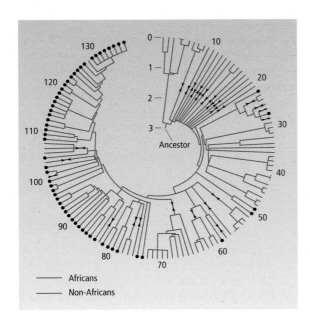

Figure 15.18
Phylogenetic relations within *Homo sapiens*, as revealed by mitochondrial DNA. Each of the 135 tips is a mitochondrial DNA type; the 135 types came from 189 individual human beings. The phylogeny suggests that humans originated in Africa and there have been successive colonizations from that source. The phylogeny is based on sequences of the control region within the mitochondrion, which evolves 4–5 times as fast as the average for the whole mitochondrion. The 135 types have the following ethnic sources: Western Pygmies (1, 2, 37–48), Eastern Pygmies (4–6, 30–2, 65–73), !Kung (7–22); African Americans (3, 27, 33, 35, 36, 59, 63, 100), Yorubans (24–6, 29, 51, 57, 60, 63, 77, 78, 103, 106, 107), Australian (49), Herero (34, 52–6, 105, 127), Asians (23, 28, 58, 74, 75, 84–8, 90–3, 95, 98, 112, 113, 121–4, 126, 128), Papua New Guineans (50, 79–82, 97, 108–10, 125, 129–35), Hadza (61, 62, 64, 83), Naron (76), and Europeans (89, 94, 96, 99, 101, 102, 104, 111, 114–20). The computational procedures for calculating the most parsimonious tree for 135 units are imperfect and the tree shown is only one possibility among many. The tree was inferred using PAUP and rooted using the chimpanzee. Arrows indicate branches where changes are inferred to occur. Reprinted, by permission, from Vigilant *et al.* (1991). © 1991 American Association for the Advancement of Science.

Algorithms may be fooled by local optima . . .

algorithms are vulnerable to becoming trapped on "local optima" when they search through the possible trees in a particular way. A local optimum is a tree that seems to be the best possible, by comparison with the other trees that the algorithm investigates, but is actually less parsimonious than other trees in a very different part of the space of possible trees. One practical response to the problem is to run the algorithm several times on a set of sequences, starting each run at a different starting point in the "tree space." If all the runs converge on the same answer, that strongly suggests it is the most parsimonious tree. If they give conflicting results, however, it may suggest the evidence is inadequate in some way.

A classic study of humans using mitochondrial DNA illustrates the problem (Vigilant *et al.* 1991). Figure 15.18 is a branching diagram for 135 human mitochondrial types. (A mitochondrial type is a particular mitochondrial sequence. Mitochondria were sequenced from 189 individual humans, and because the study had 189 humans and 135 mitochondrial types, each tip of the phylogeny represents one, or a few, individual human beings.) The number of possible phylogenies with 135 tips is astronomic; they cannot all be searched. The result in Figure 15.18 is the output after one run of a parsimony algorithm, and it has a number of interesting properties. One is that the deepest branch is African; it has African mitochondrial types to one side and a mix of African and non-African mitochondrial types on the other, implying that the root of the tree was an individual who lived in Africa. This is indeed part of the evidence that modern humans have an African ancestry (though the main evidence comes from fossils).

Another interesting result is that the mitochondrial types do not fall into the groupings that might have been expected. Look, for instance, at the Yorubans. The caption reveals which numbers in the picture are Yoruban mitochondrial types, and they are scattered through the phylogeny, even though all Yorubans live in Nigeria; likewise, Papua New Guineans do not form a discrete group. This might be because our naive expectations are incorrect — but it is more likely to be because the tree is unreliable. The tree is a "local optimum." It appears to be the most parsimonious tree, because it has only been compared with trees that are similar to itself and not with very different trees. When the program was rerun, starting in different regions of the tree space, many more trees were found that were more parsimonious than the one in Figure 15.18. Some had African deep roots, and others did not, and the different trees showed all sorts of groupings of human populations (Templeton 1993).

> ... as happened in one study of human phylogeny

In summary, when the number of species (or other taxa) at the tips of the phylogeny is large, the number of possible phylogenies may be too high for all of them to be searched. The algorithms that are used to search among the trees are often reliable, but not infallible. The main danger is that an algorithm will become stuck on a local optimum — a tree that on local comparisons seems to be the best estimate of the true tree, but is not in fact the best estimate among all the possible trees.

15.11.3 Species in a phylogeny may have diverged too little or too much

We saw above (Sections 15.9.3 and 15.10.1) how we need a molecule that has evolved an appropriate amount for the phylogeny under analysis. Molecular phylogenetics can run into difficulties if the molecules have not yet evolved far enough apart between the species, or if they have evolved apart too much and all the sites are "saturated" with change. In terms of Figure 15.13, the amount of change should not be so small that the data are all near the origin, and it should not be so large that the data are in the "leveled off" part of the graph (region III).

> A set of species may have too little ...

Vigilant *et al.*'s (1991) data illustrate the problem of too little evolutionary change. Figure 15.18 has 135 tips, but only 119 changes that can distinguish between alternative trees. The relations between human populations are better resolved by more rapidly evolving parts of our DNA (Cavalli-Sforza 2000).

> ... or too much change for phylogenetic inference

The opposite problem, of too much change, arises in rapidly evolving life forms such as RNA viruses. It is probably impossible to recover the phylogeny of different kinds of RNA viruses, such as HIV, influenza virus, and polio virus. We can find the phylogeny of different strains of HIV, or of influenza virus, but the relations between these major types are more uncertain (Holmes *et al.* 1996). Likewise, the phylogeny of life forms that had common ancestors in the deep past are difficult to recover. No molecules evolve slow enough to reveal the 3,000 million-year-old relations between the three major domains of life — Archaea, Bacteria, and Eukarya. In this case there is a further problem of horizontal gene transfer. Genes seem to move relatively readily between bacteria, and even between archaeans and bacteria. The Archaea, Bacteria, and Eukarya may not have a normal tree-like phylogeny. Some bacterial genes may be closer to archaeans, and other bacterial genes may be closer to eukaryotes. The true phylogeny would then be an anastomizing network rather than a branching tree (Figure 15.19).

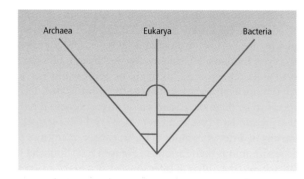

Figure 15.19
Horizontal gene transfer between lineages means that no unique phylogenetic tree exists. Some genes in eukaryotes have a more recent common ancestor with genes in Bacteria than in Archaea; other genes in eukaryotes have a more recent common ancestor with genes in Archaea than in Bacteria. Biologists disagree about the extent of gene transfer between the main domains of life, and about how clear-cut a tree exists for these three domains.

15.11.4 *Different lineages may evolve at different rates*

Figure 15.20
Long branch attraction. (a) The lineages leading to species 3 and 4 have evolved rapidly. The marks through those two lineages indicate a large number of evolutionary changes, such that all sites are saturated with multiple hits. The sequences in species 3 and 4 will be approximately 25% similar by chance (see Figure 15.13). The lineages leading to species 1 and 2 have changed little. In cases such as this, many methods of phylogenetic inference are liable in conclude (b) that species 3 and 4 are more closely related than in (c) the true tree. Most of the similarity between species 1 and 2 is ancestral and ignored in phylogenetic inference: if we exclude the ancestral G states, species 1 and 2 have zero similarity. Species 3 and 4 show 25% similarity. Application of parsimony (see Figure 15.14), for instance, will show that (b) is more parsimonious than (c).

Molecular phylogenetics is most reliable for molecules that evolve at a fairly constant rate, in the manner of a molecular clock. Phylogenetic inference becomes more difficult if some lineages evolve fast, and others evolve slowly. The statistical methods then become confused by two, related problems. One we met in the lizard–bird–crocodile case (see Figure 15.6): lineages that retain many ancestral homologies may be put together in the phylogeny, even though they are unrelated. This is mainly a problem for distance methods that do not distinguish ancestral from derived similarities. Parsimony and maximum likelihood should not be confused by ancestral similarity. However, they can suffer from the second problem, called *long branch attraction* (Figure 15.20). Two long branches will be 25% similar on average, and by chance could be more than 25% similar. They may be more similar than shorter branches, and are then put together in the phylogeny. The problem can be dealt with by discarding species in which evolution has been exceptionally rapid, or by analyzing new species that "break up" the long branches (Hillis 1996).

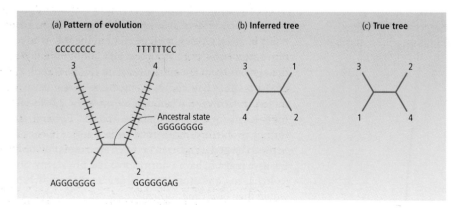

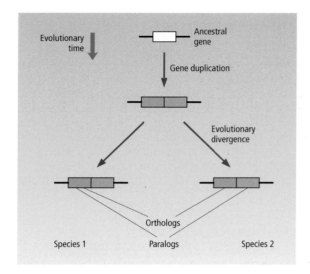

Figure 15.21

Orthologs and paralogs are two kinds of homology between genes. A gene here has duplicated in the past. In species 1 and 2, the genes that are descended from the same copy of the duplicated genes are orthologs; the descendants of different copies of the duplicates are paralogs. If the evolutonary divergence has taken place within one species, then the terms could be applied to different forms within one species, rather than two species as illustrated here.

15.11.5 *Paralogous genes may be confused with orthologous genes*

In evolution, there has been extensive gene duplication. Our genomes contain several closely related versions of genes such as the globins or the immunoglobulins. The set of closely related genes make up a gene family; some gene families consist of a linked cluster of genes, while other gene families are scattered among the chromosomes. Each gene family arose in evolution by a series of gene duplications. When we compare gene families between species, the term homology is too crude. We need to distinguish between *orthologs* — two copies of the same gene at the same locus within a set of duplicates — and *paralogs* — two genes at different loci produced by a duplication (Figure 15.21).

Orthology and paralogy are distinct forms of homology

The problem for molecular phylogenetics is that it is easy to confuse orthologs with paralogs. Phylogenetic inference ought to be based on orthologous genes, but genes are sometimes lost during evolution, and we may be deceived into comparing paralogous genes. Figure 15.22 shows how this can produce mistakes.

Gene trees can differ from species trees

Evolutionary biologists describe this problem by saying that the gene tree differs from the species tree. The gene tree (also called a gene genealogy) shows the evolutionary history of the genes in a gene family. The branching events can be either gene duplications or speciation events.[6] The species tree is the phylogeny in the sense of the present chapter. The branching events correspond to speciation in the past. The "phylogeny" in the lower half of Figure 15.22 accurately describes the history of the genes: the common ancestor of the paralogs is more distant than the common ancestor of the orthologs in species 1 and 2. The trouble is that the history of these genes is not

---

[6] Or the establishment of an intraspecific polymorphism. The comparison in Figure 15.22 would then be between two morphs within a species rather than "species 1" and "species 2."

Figure 15.22

Mistaken phylogenetic inference because of comparison between paralogous genes. Different copies of the gene family have been lost in different lineages. The genes remaining in species 1 and 2 are orthologs and more similar than either is to the paralogous gene in species 3. In species 1–3 we do not know that the genes are a mix of orthologs and paralogs. A similar problem arises if the duplicate genes have not been lost, but happen not to have been sequenced; the mistake is then due to absence of data rather than gene loss. (It is assumed that the two copies of the duplicated gene tend to evolve apart over time, whereas the orthologs remain more constant in different species.)

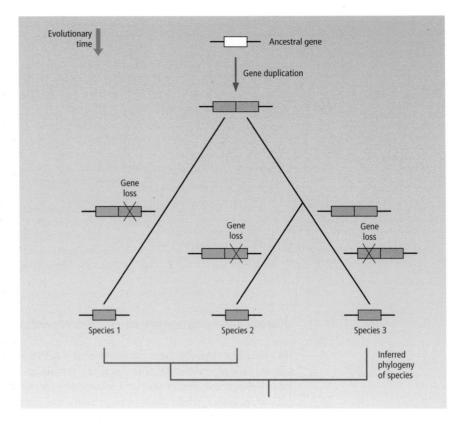

the same as the history of the species. In phylogenetic inference we use genes trees to infer species trees. In many, perhaps most, cases the method is reliable; but not in all, as Figure 15.22 illustrates.

15.11.6 *Conclusion: problems in molecular phylogenetics*

This section has concentrated on the problems of molecular phylogenetics. However, this does not mean that molecular phylogenetics is a weak, or more uncertain than average, research program. Indeed, it is because molecular phylogenetics is such a flourishing research program that biologists are so interested in where it can go wrong. Once the problem areas have been identified, we can do further research into how to fix or avoid the problems — or not be deceived by deceptive results. The problems we have looked at here are all tolerable problems that can cause local, temporary difficulties within molecular phylogenetics. But they are not insidious general problems that undermine the whole enterprise.

15.12 Paralogous genes can be used to root unrooted trees

The root of a tree . . .

The position of the root in an unrooted tree can be inferred by any of the cladistic methods for determining character polarity (Section 15.6). Outgroup comparison is the most widely used, but cannot be used in every case — we may be unsure which species to use as an outgroup, or different outgroups may give different answers, or evidence may be lacking as to the character states (or molecular sequences) of the outgroup. In one case, the deep root of life, no outgroup exists. If we want to find the root of the tree for the three domains of life (Archaea, Bacteria, Eukarya) — where all cellular life is grouped — we cannot do it by outgroup comparison. (Viruses cannot be used as an outgroup because they evolve too fast, and anyhow have probably evolved recently — they do not belong to a deeper branch, below the common ancestor of the three domains.)

. . . can be inferred by paralog rooting

Molecular phylogenetics has added a new method to root trees (and therefore to find character polarities). Its beauty is that it works internally, within the unrooted tree itself; it does not require us to find any external data, such as an outgroup. The method is called *paralog rooting*.

The method works as follows (Figure 15.23). We need a gene that has duplicated before the origin of the taxon we are studying. We then construct the unrooted gene tree for all the copies of the gene. For a duplicated gene in four species, there are eight tips to the unrooted gene tree (Figure 15.23b). Both genes in the duplicated pair have evolved through the same tree, and the gene tree is likely to have a "mirror image" shape. We can infer, from this tree alone, that the root lies in the long branch that connects the two mirror-image subtrees. We now know where the root is in the species tree (Figure 15.23c).

The method has been used to root the angiosperms

Paralog rooting was first applied to the problem of the deep root of all life (that is, the Archaea–Bacteria–Eukarya tree). But that problem is hard to solve because of saturation. The common ancestor of all cellular life lived 3,500–4,000 million years ago. The molecular differences between Archaea, Bacteria, and Eukarya lie well into the difficult region II or impossible region III of Figure 15.13. We can look at a more successful application of paralog rooting, to the phylogeny of angiosperms (Mathews & Donoghue 1999). Angiosperms are the group better known as flowering plants. Outgroup comparison tends to produce ambiguous results for angiosperms. Figure 15.24 shows Mathews and Donoghue's result using paralog rooting. The rooted tree has *Amborella* (one species, living in New Caledonia) forming a branch by itself from the root. The next deepest branch has the water lilies. Careful inspection of the two mirror images shows a few small mismatches, but no more than would be expected from the uncertainties of phylogenetic inference. In all, the two subtrees have impressively similar branching orders. Paralog rooting has given a major new insight into angiosperm phylogeny, and the method can be applied wherever molecular sequences are available for duplicate genes (with appropriate amounts of divergence) in a group of species.

Figure 15.23
Paralog rooting. (a) A gene duplicates into two copies, at different loci. The species then evolves into four descendants, each with copies of the two genes. (b) We use the molecular sequences of the eight genes to infer the unrooted gene tree. It has been drawn to show the logic of the method, with two sets of four genes arranged as mirror images. A more conventional tree would have the eight genes written down the page. Given this tree, we infer that the root is in the long branch connecting the two mirror-image subtrees. (c) Thus, if we have the unrooted species tree, we can use the pattern in (b) to infer where the root is. The answer is correct: it matches the real pattern of evolution in (a).

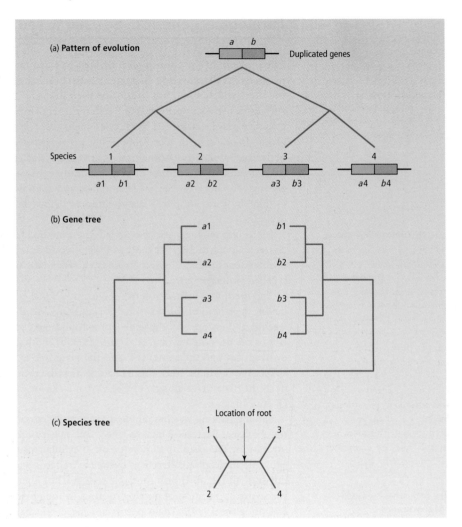

15.13 Molecular evidence successfully challenged paleontological evidence in the analysis of human phylogenetic relations

It is always interesting when two independent lines of evidence, from very different fields, are applied to the same question. This sections looks at a conflict between fossil and molecular evidence concerning the time of origin of the human evolutionary lineage.

The fossil ape *Ramapithecus* . . .

"*Ramapithecus*" (which is now classified in the genus *Sivapithecus*) is a group of fossil apes that lived about 9–12 million years ago. Until the late 1960s, almost all

Figure 15.24
The rooted angiosperm tree, inferred by paralog rooting. Species of the flowering plants are written down the center. The subtree on the left is based on sequences of the phytochrome *A* gene and the subtree on the right is based on sequences of the phytochrome *C* gene. The two genes are paralogs. The two subtrees are connected by a long branch at the bottom of the figure. The tree is arranged in this way, with two mirror-image subtrees, for reasons explained in Figure 15.23. The root is inferred to be in the long branch at the bottom. The correspondence between the subtrees for the two genes is impressive: nodes that are the same in the two subtrees are indicated by capital letters. Unlabeled nodes do not exactly correspond in the two. The tree was inferred by parsimony. Slightly modified, by permission of the publishers, from Mathews & Donoghue (1999).

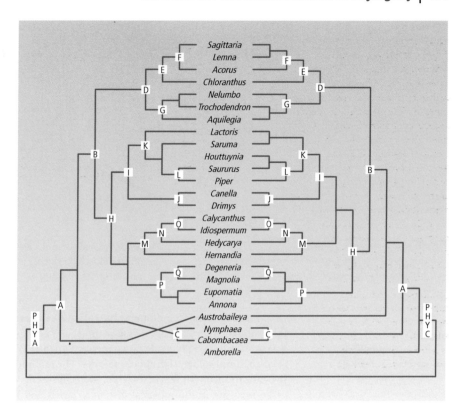

. . . used to be linked to humans by several characters . . .

. . . implying an ancient origin for humans

paleoanthropologists thought that *Ramapithecus* was a hominin: that is, it was more closely related to *Homo* than to chimpanzees and gorillas (Figure 15.25a). (Hominoids (formally superfamily Hominoidea) are the group of all great apes, including humans; hominins (formally subfamily Homininae) are the narrower group of *Homo* and the australopithecines.) *Ramapithecus* and *Homo* apparently shared a number of derived characters. For example, *Homo* has a rounded, "parabolic" dental arcade, whereas chimps have a more pointed dental arcade. The dental arcade of *Ramapithecus* was initially thought to be shaped more like *Homo*. Secondly, *Ramapithecus*'s canine teeth were thought to be relatively diminished compared with its other teeth, as in *Homo* but unlike chimps (in which the canines, especially in males, are large). Thirdly, *Homo* and *Ramapithecus* were thought to share, as a derived condition, a thickened layer of tooth enamel, unlike the thinner layer in other apes (and which was thought to be the condition in the ancestors of the Homininae).

This morphological and paleontological argument for a relation between *Homo* and *Ramapithecus* has a classic form: a set of character states are shown to be shared uniquely by these two species, and the characters are derived within the larger group of Hominoidea. The corollary was that the human lineage must have split from the great apes at least 12 million years ago, because *Ramapithecus* is nearer to us than to the great apes.

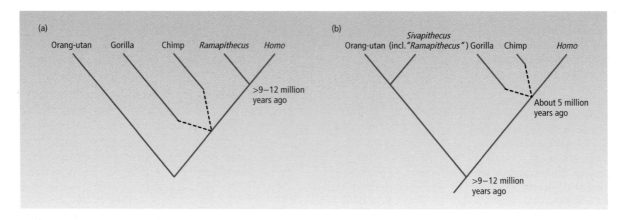

Figure 15.25

Relations of *Homo*, other great apes, and *Ramapithecus*, according to (a) original paleontological and morphological evidence, and (b) molecular (and revised paleontological and morphological) evidence. (The dashed lines imply uncertainty in the order of the human–chimpanzee–gorilla split.)

In the early 1960s, Goodman (1963) first demonstrated the molecular similarity of humans and other great apes; but the molecular argument for a recent human–ape split was most influentially made in a paper by Sarich & Wilson (1967). Sarich and Wilson used an immunological distance measure. The method is similar in philosophy to DNA hybridization, but differs in the exact molecule used.

Molecular evidence suggested a more recent origin for humans

To measure immunological distance, Sarich and Wilson first made an antiserum against human albumin by injecting human albumin into rabbits (albumin is a common protein which circulates in the blood). They then measured how much that antiserum cross-reacted with the albumin of other species, such as chimps, gorillas, and gibbons. The antiserum recognizes the albumins of closely related species, because they are similar to human albumin; but it does not recognize them quite as efficiently as it does human albumin. The degree of cross-reactivity gives a measure of the immunological distance (ID) between a pair of species. ID increases among phylogenetically more distant relatives, and the relative rate test (Box 7.2, p. 166) suggests that ID increases at a constant rate through time; immunological distance is a sort of molecular clock. The clock can be calibrated using the fossil record for some of the studied species, and the ID can then be used to estimate the divergence time for other pairs of species.

The results of this method suggest that *Homo* and the other great apes have too short an ID to fit with a pre-*Ramapithecus* divergence: Sarich and Wilson suggested humans and chimps diverged only about 5 million years ago. Subsequent molecular work has supported them. The DNA hybridization results that we looked at earlier suggest a similar, if perhaps slightly older, figure (see Figure 15.12), and other molecules suggest a figure of 3.75–4 million years. The corollary is that if *Homo* diverged from chimps and gorillas 5 million years ago, it cannot be more closely related to *Ramapithecus* than to the living great apes. The phylogeny must be more like Figure 15.25b.

So the molecular and fossil evidence disagreed. A controversy began, in which both the molecular and morphological evidence was challenged (often by experts in the other field). The controversy has now been settled (with a few dissenters) in favor of the original molecular evidence. The morphological characters previously believed to show a relation between *Homo* and *Ramapithecus* succumbed to reanalysis. The dental arcade of *Ramapithecus* had been wrongly reconstructed (originally by combining parts from different specimens). The reduced canine teeth may be because the fossil *Ramapithecus* specimens were female. Martin (1985) finally removed the last important character — thickened enamel — by reinterpreting it as an ancestral character. Moreover, when *Ramapithecus* was compared with another fossil (*Sivapithecus*) that was generally accepted to be a close relative of the orang-utan, and with the orang-utan itself, it was found to show clear similarities to them. The specimens formerly classified as *Ramapithecus* are now usually included in the genus *Sivapithecus*, which in turn is thought to be a close relative of the ancestors of modern orang-utans (Figure 15.25b).

The fossil characters were reinterpreted

In summary (simplifying things a little), molecular evidence helped to inspire a reanalysis of the fossil evidence for human origins — with the result that a figure of about 5 million years, and at any rate in the 4–8 million year range, is now widely accepted for the time of origin of the hominin lineage.

15.14 Unrooted trees can be inferred from other kinds of evidence, such as chromosomal inversions in Hawaiian fruitflies

Many other individual techniques are useful for inferring the phylogeny of particular taxonomic groups. We can look, for instance, at a particularly powerful technique that has been used with fruitflies. For some reason, an extraordinarily large number of species of fruitflies (*Drosophila*) live in the Hawaiian archipelago. There are probably about 3,000 drosophilid species in the world, and about 800 of them appear to be confined to this archipelago. The phylogeny of one subgroup of the Hawaiian fruitflies is better known than that of any other equivalently large group of living creatures. It was worked out, by Carson and his colleagues (see Carson 1983), from chromosomal banding patterns. Chromosome bands are clearly visible in fruitflies (Section 4.5, p. 82).

Fruitflies have radiated in Hawaii

Their phylogeny is inferred from chromosomal inversions

The banding patterns differ between species, and it soon becomes obvious that regions of the chromosomes have been inverted during evolution: a segment of genes within a chromosome has been inverted as a whole. The important event, for phylogenetic inference, is when a second inversion takes place across the end of an earlier inversion (Figure 15.26). When this happens, we can infer with near certainty that the unrooted tree is $1 \leftrightarrow 2 \leftrightarrow 3$, not $1 \leftrightarrow 3 \leftrightarrow 2$. If species 1 had evolved directly into species 3, and then 3 into 2, the two inversions in Figure 15.26 would be needed for the evolution of species 3; then, to go to species 2, the exact same two breaks (one at each end) of the second inversion would have to happen again in reverse — which is much less probable than evolution in the order $1 \leftrightarrow 2 \leftrightarrow 3$. As more species are added, with more overlapping inversions, the improbability of most alternative trees multiply to the point of practical impossibility.

Figure 15.26
Overlapping inversions, in different species, can be used to infer their phylogenetic relations, in the form of an unrooted tree. With this pattern of inversions, the tree must be $1 \leftrightarrow 2 \leftrightarrow 3$, and not $1 \leftrightarrow 3 \leftrightarrow 2$ or $3 \leftrightarrow 1 \leftrightarrow 2$.

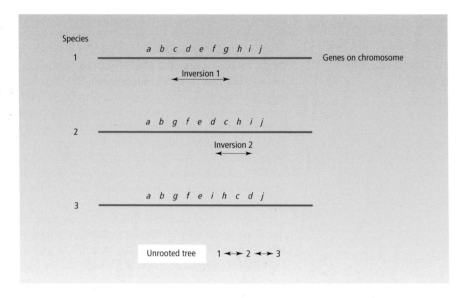

To apply the technique, we first have to work out the chromosomal banding patterns of the group of species. Then one species is picked, more or less arbitrarily, as the "standard" against which the other species are compared. Starting with the species that have a chromosomal banding pattern most like the standard, we gradually work outwards through the tree until all have been included. Carson concentrated on the "picture-wing" group of Hawaiian drosophilids. Figure 15.27 is a phylogeny, based on 214 inversions, of 103 of the 110 or so known species in this group. It is a marvellous piece of work. We can get some idea of how certain the inference is from the fact that *none* of the 214 inversions contradict the phylogeny; the characters all agree.

Figure 15.27 is an unrooted tree. The root can be located by two independent lines of evidence. One is to look outside the archipelago for the nearest outgroup and see what banding pattern it has. The fruitflies that are thought to be the nearest outgroup of the picture-wing group live in South America and are most similar to *Drosophila primaeva* (species number 1) and *D. attigua* (species number 2) among the fruitflies in Figure 15.27. These species are therefore probably closest to the root of the tree. The inference is supported by the geological history of the archipelago. Kauai is the oldest island and Hawaii the youngest, so the ancestor of the group would probably have colonized Kauai. If that ancestral species still survives, it is probably still on Kauai, because almost every Hawaiian drosophilid is confined to a single island; for example, *D. primaeva* and *D. attigua* live on Kauai. Indeed, much of the phylogenetic history of the picture-wing group consists of populations of species on the older islands moving to the younger islands, where they form new species by allopatric speciation. There are no examples of species on older islands that are derived from a species on a younger island. Thus the youngest island, Hawaii, has the most recently evolved species and the oldest islands in the west have the most ancient species (Figure 17.6, p. 504).

Different inversions show no conflict

The tree can be rooted

Figure 15.27
Phylogeny of 103 species of Hawaiian fruitflies (*Drosophila*) of the picture-wing group. The unrooted tree was inferred by patterns of chromosomal inversions. The tree can be rooted by geochronology of the islands, and comparison with closely related fruitflies in South America. Some details of the tree are inferred by biogeography rather than inversion patterns. Species shown as ancestors may actually be descendants of the (now lost) ancestor. The phylogenies shown above correspond to the islands below. Redrawn, by permission of the publisher, from Ridley (1986).

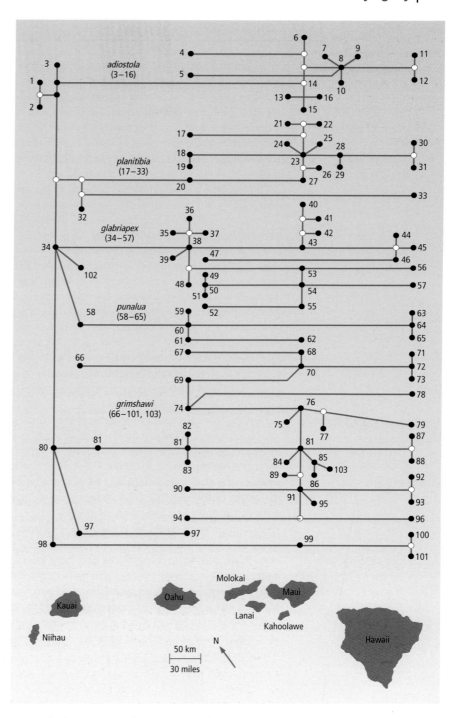

15.15 Conclusion

Molecular evidence has revived phylogenetics

Phylogenetic research has been transformed in two ways in recent years, one being purely scientific. Biologists have been trying to infer the tree of life since Darwin's time. Much progress was made with morphological evidence from living and fossil species, but some problems were insoluble with this kind of evidence alone and the pace of scientific progress had slowed by the 1960s and 1970s. Since then, increasing quantities of molecular evidence have become available. The next generation of biologists can hope to know what biologists of all past generations have wanted to know: the complete tree of life.

The second transformation has come from the same source — the huge quantity of molecular data. Phylogenetics has grown into a kind of applied evolutionary biology, and is used to solve forensic and medical problems. In all, phylogenetics has moved from being an unfashionable, slightly dusty topic to become one of the two or three hottest go-go areas of evolutionary biology.

Phylogenetic inference draws on all kinds of evidence, from molecular sequences, to chromosomal inversions, to morphology in modern and fossil forms. With morphological evolution, the commonest (though not universal) course of research is to distinguish homoplasies from homologies, and then infer the polarity of the homologous characters. The conflict between the characters shared among the species may be reduced (ideally to zero) during this analysis. A residue of reliable derived homologies may be left, and can be used to infer the (rooted) tree. With molecules, individual characters are analyzed less and a statistical method such as "distance," parsimony, or maximum likelihood is used to infer the unrooted tree. The location of the root can then be inferred by other evidence.

Some phylogenies are frustratingly uncertain . . .

Phylogenetic inference can sometimes seem to be an exceptionally uncertain, shaky kind of science. In a full discussion of an unsolved and controversial problem, an endless series of pieces of evidence may seem to support first one phylogeny, and then another. One student (of Professor C.F.A. Pantin of Cambridge University, England), when confronted with a classically recalcitrant problem in phylogenetic inference — the origin of the chordates — summed it up as "paleontology is mute, comparative anatomy meaningless, and embryology lies."

. . . others are satisfyingly solid

That impression could be misleading. Discussion (as well as research) naturally focuses on unsolved problems — and the unsolved problems tend to be the difficult ones. Many phylogenetic problems have been solved, so one can have reasonable certainty in a case such as the human–chimp–amoeba example, while reserving opinion in more slippery cases such as the relations between Eukarya, Archaea, and Bacteria. In addition, in the phylogeny of the picture-winged fruitflies of Hawaii, deduced from 214 conflict-free, multiply overlapping chromosomal inversions, the phylogenetic inference has a level of certainty that compares favorably with most of the facts known in the natural sciences.

Summary

1 Phylogenies show the ancestral relations between species. For each species, a phylogeny shows which other species (or group of species) it shares its most recent common ancestor with.

2 Phylogenetic relations are inferred using the shared characters of species. The characters can be morphological, in living and fossil species, or molecular.

3 When different characters imply the same phylogeny, phylogenetic inference is easy; when they do not, other methods are needed to unravel the disagreement.

4 Theoretical arguments suggest that some kinds of shared characters indicate phylogenetic relations reliably, whereas others do not. Homoplasies and ancestral homologies do not reliably indicate phylogenetic groups. Derived homologies do. Phylogenetic inference should be based on derived homologies.

5 Techniques exist to distinguish homologies from homoplasies, and ancestral homologies from derived homologies.

6 Character polarities can be inferred, with varying degrees of certainty, by outgroup comparison, the fossil record, and other criteria, including paralog rooting.

7 For molecular characters, the kinds of character analysis used with morphology are often inapplicable.

Phylogenies are usually inferred by statistical techniques. The three main classes of techniques are distance methods, parsimony, and maximum likelihood.

8 Distance methods group species according to their molecular similarity. With parsimony, the best estimate of the tree for a group of species is the tree that requires the fewest evolutionary changes. With maximum likelihood, the best estimate of the tree is the one that is most probable, given a model of sequence evolution.

9 Molecular evidence is often used to infer the phylogeny in the form of an unrooted tree. An unrooted tree specifies the branching relations among species, but not the direction of evolution.

10 Molecular phylogenetics is used in medical research to identify the source of emerging diseases.

11 Molecular phylogenetic inference encounters problems with sequence alignment, the large number of possible trees, multiple hits, inadequate data, unequal rates of evolution, and confusions between paralogous and orthologous genes.

12 The unrooted tree of the Hawaiian fruitflies is the most firmly established phylogeny of any large group of species. It has been reconstructed using chromosomal inversions.

Further reading

Felsenstein (2003) is an authoritative book about phylogenetic inference. However, for most of the further reading it makes sense to distinguish between cladistic references, in which rooted trees are inferred using character polarities, often with morphological evidence, and molecular references, in which unrooted trees are inferred by some statistical method.

Cladistics. For the cladistic method, see Wiley *et al*. (1991) and Kitching *et al*. (1998); references can be traced from these sources. Kemp (1999) concentrates on fossils, but introduce cladistics generally. Hennig (1966) is the classic reference. Sober (1989) is a philosophical discussion of most of the main methods, and he has also edited an anthology (Sober 1994) that reprints a number of relevant papers. Another philosophical question, not covered in this chapter, is whether phylogenetic reconstruction revolves a circular argument: see Hull (1967). Wagner (2000) is about the concept of a "character."

Homology has been much discussed recently, mainly because of the astonishing discoveries in "evo-devo." We look at that topic in Chapter 20 in this text; see the further reading there. Meanwhile, see many of the chapters in Bock & Cardew (1999). Fitch (2000) discusses molecular meanings of homology. Moore & Willmer (1997) discuss homology as part of a review of convergence in invertebrates.

For methods of determining character polarity, see Meier (1997) on the ontogenetic criterion (not discussed in this text), and its performance relative to outgroup comparison. Fox *et al.* (1999) discuss a further way fossil evidence can be used in phylogenetic inference.

Molecular phylogenetics. Page & Holmes (1998) and Graur & Li (2000) are introductory texts. Hall (2001) is a practical text, aimed at molecular biologists. Li (1997) and Nei & Kumar (2000) are more advanced. An authoritative general source is the book edited by Hillis *et al.* (1996). The chapter by Swofford *et al.* (1996) is a fine introduction to the theory. Whelan *et al.* (2001) is a modern overview of methods. Steel & Penny (2000) also discuss models, and different methods. Huelsenbeck & Crandall (1997) review maximum likelihood and discuss three main methods (distance, parsimony, and maximum likelihood). A fourth, Bayesian inference, may be emerging: see Huelsenbeck *et al.* (2001). Mooers & Holmes (2000) look at how to deal with codon bias. Diamond (1991, chapter 1) is a popular essay that discusses DNA hybridization.

For phylogenetic studies of disease, for example on the west Nile virus, see Lanciotti *et al.* (1999) and the newspiece in *Science* November 19, 1999, pp. 1450–1. There is another newspiece in *Science* May 11, 2001, pp. 1090–3. Also, see some chapters in Harvey *et al.* (1996) as well as Hahn *et al.* (2000) and Holmes (2000a) on HIV, and Page & Holmes's (1998) text.

For computer programs, MacClade (Maddison & Maddison 2000) is the most friendly program, but is designed more for morphological than molecular characters. The most widely used programs in molecular phylogenetic research are PAUP (Swofford 2002) and PHYLIP (Felsenstein 1993). Felsenstein's web page also discusses many computer packages for phylogenetic analysis.

The problems in molecular inference are covered by the general texts. See also Doolittle (2000) and a newspiece in *Science* May 21, 1999, pp. 1305–7 on the case of the deep root of life. Horizontal gene transfer has been discussed for the human genome, see several issues of *Nature* and *Science* in mid 2001.

On long branch attraction, and the question of whether breaking up the branches by more extensive taxon sampling solves the problem, see Hillis (1996), an exchange among several authors in *Trends in Ecology and Evolution* (1997), vol. 12, pp. 357–8, and Rosenberg & Kumar (2001).

Paralog rooting was first applied to find the deep root of all life, but see Philippe & Forterre (1999) for problems (saturation, mainly) in this work, as well as references to it. These problems probably do not apply to the angiosperm example in the text. Zanis *et al.* (2002) provide further analysis of the angiosperm case. Chapter 18 contains further references on angiosperms. The confusion caused by paralog rooting is sometimes said to be compounded by rapid evolution following gene duplication, but this seems not to be the case (Hughes 1999).

Ting *et al.* (2000) avoid one confusion between gene and species trees in an interesting way. They use the "speciation" gene *Odysseus*, which we met in Section 14.12. It should be invulnerable to the lineage sorting problem, if it caused speciation, making it a reliable indicator of phylogeny.

Many authors discuss the relation between molecular and morphological (or fossil) inferences. Benton (2001) measures congruence between stratigraphic and phylogenetic estimates of taxonomic times of origin, and finds that the congruence is as good for ancient branchings as for more recent branchings. Kemp (1999) discusses the general topic, as well as methods for combining different classes of evidence. Novacek (2001) discusses combinations of evidence for the special case of mammals. For human evolution, Lewin (2003) is an introductory book, and Klein (1999) a more advanced text.

On Hawaiian fruitflies, see the special issue of *Trends in Ecology and Evolution* (1987), vol. 2, pp. 175–228, and the book edited by Wagner & Funk (1995). And on flies in particular, see Powell (1997), Carson (1990), and Kaneshiro (1988).

The remarks I make about historic trends in the use of the various inference techniques could usefully be supplemented by Edwards (1996) and Felsenstein (2001).

Chapter 18 in this text looks at some examples of phylogenetic problems; see the references there too. The general topic can be followed in journals such as *Systematic Biology* and *Molecular Biology and Evolution*, as well as *Trends in Ecology and Evolution*, *Trends in Genetics*, and *Bioessays*.

A further theme is how a knowledge of phylogenies can be put to use in evolutionary biology. This is illustrated at various points in this text, but see the book edited by Harvey *et al.* (1996) and the review by Pagel (1999) in general, and the special issue of *Paleobiology* (2001), vol. 27 (2), pp. 187–310 for fossils in particular.

Study and review questions

1 Here is an unrooted tree for five species: draw all the rooted trees that are compatible with it.

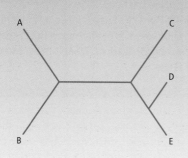

2 Here is a rooted tree: draw all the unrooted trees that are compatible with it.

3 Under what conditions (of evolutionary rates) are distance statistics reliable guides to phylogenetic relations?

4 What would you look at in some superficially similar structure, such as the tail of a dolphin and a salmon, to decide whether it was homologous or homoplasious?

5 Here are the character states of species 1–6.

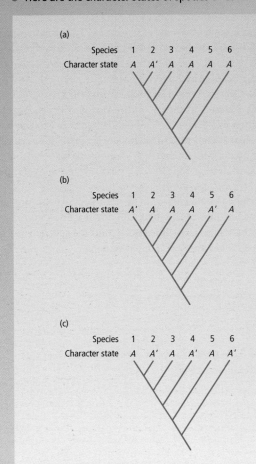

(a)

Species	1	2	3	4	5	6
Character state	A	A'	A	A	A	A

(b)

Species	1	2	3	4	5	6
Character state	A'	A	A	A	A'	A

(c)

Species	1	2	3	4	5	6
Character state	A	A'	A	A'	A	A'

Which of A and A' is ancestral, and which derived, in the group of species 1 + 2? Assess the relative certainty of the inference in the three cases.

6 Compare the attributes of molecular and morphological evidence that make character analysis, or statistical inference from unanalyzed characters, more or less applicable to evidence of each kind.

7 What is meant by the problem of multiple hits?

8 The simplest model of sequence evolution assumes that the chance that any nucleotide changes to any other nucleotide is the same, *p*. Here is a simple unrooted tree.

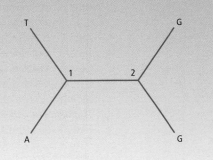

(a) What is the probability of the tree (in terms of *p*), if both internal nodes (labeled 1 and 2) are G? (b) If the nucleotide states of the internal nodes (labeled 1 and 2) are unknown, how many possible nucleotide states at the internal nodes need to be considered in calculating the probability of the tree?

9 Here are the orders of genes (or some other markers) along the chromosomes of three species: (1) adebcfg; (2) abcdefg; and (3) abedcfg. What is the unrooted tree of the three species?

16

Classification and Evolution

Biological classification is concerned with distinguishing and describing living and fossil species, and with arranging those species into a hierarchical, multilevel classification. The theory of evolution has a strong influence on classificatory procedures. Chapter 13 was about species; this chapter is about classification above the species level. We begin by looking at the two principles — phenetic and phylogenetic — that have been used to classify species hierarchically into groups (such as genera, families, and higher level categories), and see how the three main schools of biological classification put them to use. We then look at the conditions in which the two principles give the same, or differing, classifications of a set of species. The main question of the chapter is which (if any) of the two principles is the better justified. The answer comes from an argument that phylogenetic classification at its best is objective, whereas phenetic classification always suffers from subjectivity. We then look at some consequences of the strict use of phylogenetic relations to classify species into groups. We finish by considering why real evolution has resulted in a tree-like diverging pattern of relations among species.

16.1 Biologists classify species into a hierarchy of groups

Biologists have so far described approximately 1.75 million species of living plants and animals, and perhaps a further 0.25 million extinct fossil species. Estimates vary for the number of species that exist but have not yet been described: there may be between 10 and 100 million of them. Describing a species is a formalized activity, in which the taxonomist has to compare specimens from the new species and other, similar species, and then explain how the new species can be distinguished; the description also has to be published. Describing species is the most important task of taxonomists, but it has no particular connection with evolutionary biology.

The evolutionary interest of classification begins at the next stage. Biologists do not think of their million or so described species simply as a long list, beginning with the aardvark, working through buttercup, honeybee, and starfish, to end with zebra. Since Linnaeus, species have been arranged in a hierarchy; Figure 3.5 (p. 49) used the wolf as an example. Species are grouped in genera: the gray wolf species *Canis lupus* and the golden jackal *Canis aureus*, for example, are grouped in the genus *Canis*; genera are grouped into families: the genus containing dogs and wolves combines with several other genera, such as the fox genus *Vulpes*, to make up the family Canidae; several families combine to make up an order (Carnivora, in this example), several orders to make a class (Mammalia), classes to make a phylum (Chordata), and phyla to make a kingdom (Animalia).

Each species, therefore, is a member of a genus, a family, an order, and so on. The problem of biological classification above the species level is how to group the species into these higher categories. The problem has both a practical and a theoretical side. Any number of practical problems can arise in deciding which genus to put a species into and what level particular groups should have (genus or family?). But before these questions, there is the logically prior question of what procedures should be used, and what sort of hierarchy we should be trying to classify the species into.

If we take a million species and seek to arrange them into a classification, the arrangement could be made in a large number of ways. A classification does not even have to be hierarchical. Chemists, for example, classify elements by the periodic table, which is not hierarchical. Why biological classification should be hierarchical is an interesting question in itself (Section 16.8). However, we begin by assuming that classification is hierarchical, and ask what the exact form of the hierarchy should be. This chapter is about theoretical questions — of the relation between evolutionary trees and biological classification — rather than practical questions of how to classify species at the museum workbench.

Biological species are classified hierarchically . . .

. . . but what kind of hierarchy?

And why a hierarchy at all?

16.2 There are phenetic and phylogenetic principles of classification

In biology, two main methods are used to classify species into groups: the *phenetic* and the *phylogenetic* methods. (Some would prefer to substitute "phenotypic" for "phenetic" throughout this chapter.) The phenetic method groups species according to their observable phenetic attributes: if two species look more like each other than either does to any other species, they will usually be grouped together in a phenetic classification. The full classification consists of a hierarchy of levels, such that the members of different groups at higher and higher levels look decreasingly similar to one another. A wolf and a dog (same genus) look phenetically more similar than do a wolf and a dolphin (same class).

In formal classification, phenetic similarity has to be measured. Almost any observable attributes of organisms can be used for this purpose. Fossil vertebrates can be classified phenetically by the shape of their bones; modern species of fruitflies by the pattern of their wing venation; and birds by the shapes of their beaks or the color pattern of their feathers. Species can be grouped according to the number, shape, or banding pattern of their chromosomes, by the immunological similarity of their proteins, or by any other measurable phenotypic property.

Phenetic classification is non-evolutionary

Nothing needs to be known about evolution in order for species to be classified phenetically. The species are grouped by their similarity with respect to observable attributes alone, and the same principle can be applied to any sets of objects, non-living or living, whether or not they were produced by an evolutionary process. It could be applied to languages, furniture, clouds, songs, and styles of art and literature, as well as biological species.

Phylogenetic classification represents evolutionary relationships

The phylogenetic principle, however, is evolutionary. Only entities that have evolutionary relations can be classified phylogenetically. The clouds in the sky, for instance, cannot be classified phylogenetically (almost every cloud is formed independently, by physical processes — though a few clouds may be formed by the division of ancestral clouds, and those could be classified phylogenetically). The phylogenetic principle classifies species according to how recently they share a common ancestor. Two species that share a more recent common ancestor will be put in a group at a lower level than two species sharing a more distant common ancestor. As the common ancestor of two species becomes more and more distant, they are grouped further and further apart in the classification. In the end, all species are contained in the all-inclusive phylogenetic category — the set of all living things — which contains all the descendants of the most distant common ancestor of life.

In most real cases in biology, the phylogenetic and phenetic principles give the same classificatory groupings. If we consider how to classify a butterfly, a beetle, and a rhinoceros, the butterfly and beetle are more closely related both phenetically and phylogenetically (Figure 16.1a). The beetle and butterfly both look phenetically more alike and share a more recent common ancestor with each other than either does with the rhinoceros.

In other cases, the principles can disagree, for two main reasons. One reason is evolutionary convergence. Adult barnacles superficially look rather like limpets. If we were to classify an adult barnacle, limpet, and lobster phenetically we might well put the barnacle and limpet together even though the lobster and barnacle share a more recent common ancestor and are grouped together phylogenetically (Figure 16.1b). The other reason is illustrated by groups like reptiles. The phenetic and phylogenetic classifications of the reptilian groups differ because some descendants (such as birds) of the common ancestor of the group have evolved rapidly. The rapidly evolving groups

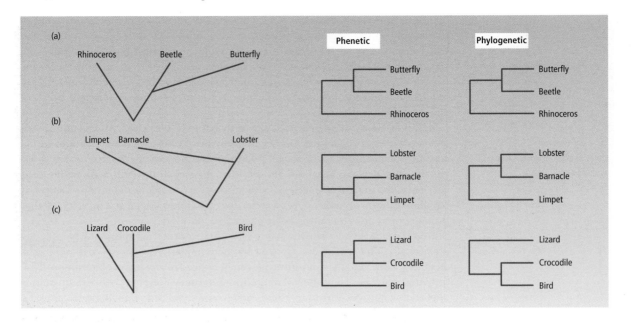

Figure 16.1
The phenetic and phylogenetic principles of classification may (a) agree or (b–c) disagree.

have left behind a rump of quite distantly related groups that resemble one another phenetically (Figure 16.1c). We discuss these two problematic cases further later on, but the illustrations in Figure 16.1 are enough to introduce the three main schools of classification.

16.3 There are phenetic, cladistic, and evolutionary schools of classification

We distinguish phenetic, phylogenetic, and evolutionary taxonomy

The phenetic and phylogenetic principles are the two fundamental types of biological classification, but three schools of thought exist about how classification should be carried out. The chapter will discuss these three schools, and Table 16.1 summarizes their main features.

The most influential school of phenetic classification is (or was) *numerical taxonomy*. It was particularly defended by Sneath & Sokal (1973). The terms phenetics, numerical phenetics, and numerical taxonomy are used almost interchangeably in modern biology.

Phylogenetic classification has been defended by the German entomologist Hennig (1966) and his followers; Hennig called it *phylogenetic systematics*, but *cladism* is now the commoner term. We saw how cladistic techniques are used to infer phylogeny in Chapter 15.

Table 16.1
Phenetic, cladistic, and evolutionary classifications can be distinguished by the characters they use to define groups, and the kinds of group they recognize.

	Groups recognized			Characters used	Homologies	
Classification	Monophyletic	Paraphyletic	Polyphyletic	Homoplasies	Ancestral	Derived
Phenetic	Yes	Yes	Yes	Yes	Yes	Yes
Cladistic	Yes	No	No	No	No	Yes
Evolutionary	Yes	Yes	No	No	Yes	Yes

The third school to be discussed uses a synthesis, or mixture, of phenetic and phylogenetic methods and is often called *evolutionary taxonomy*. In the reptilian example (Figure 16.1c) evolutionary taxonomy prefers the phenetic classification; in the barnacle example (Figure 16.1b), the phylogenetic. This school's best known advocates include Mayr (1981), Simpson (1961b), and Dobzhansky (1970).

16.4 A method is needed to judge the merit of a school of classification

We distinguish objective from subjective classificatory principles

How should we decide which school of classification, if any, is the best? To do so, we need a criterion to judge them against, and many biologists use the *objectivity* criterion for this purpose. An objective classification is one that represents a real, unambiguous property of nature. Objective classification can be contrasted with subjective classification, in which the classification represents some property arbitrarily chosen by the taxonomist.

For instance, I might arbitrarily choose to classify species into one group if I discovered them on a Monday or Tuesday and another group if I discovered them between Wednesday and Friday. The classification would then be subjective because I should have no method of justifying the choice — except my personal whim or convenience. If challenged about why I did not instead have one group for all days beginning with the letter "T" and another group for all other days, I should have no principled argument to defend my classification. The underlying classificatory principle — time of discovery — is ambiguous because it could be applied in a number of equally valid ways that would give differing classifications. It is also unreal because there is no inherent property shared by the organisms discovered on Monday and Tuesday and not shared by those discovered on Tuesday and Thursday. The objectivity test, therefore, is to ask whether a classificatory system has some compelling justification, external to the method it uses and the practitioners who practice it, for classifying in the way it does.

Objective classifications are preferable to subjective ones. If classification is objective, then different, rational people, working independently, should be able to agree that it is the way to classify. The results should be relatively stable and repeatable.

We can now look at how well the three classificatory schools meet the objectivity criterion. To do so, we need to look in more detail at how the three schools actually operate.

16.5 Phenetic classification uses distance measures and cluster statistics

The modern forms of phenetic classification are numerical and multivariate, and they were developed in reaction to the uncertainties and imprecision of evolutionary classification. Evolutionary classification, whether of the pure cladistic kind or the mixed evolutionary taxonomy of Mayr and others, requires a knowledge of phylogeny. Chapter 15 described how phylogenies can be inferred. Here, all we need to know is that, although the phylogenetic relations between species can often be inferred, the inferences are sometimes uncertain. Phylogenetic knowledge is subject to change, as improved evidence comes in, and a classification of a group based on its phylogeny is liable to be unstable — not because the phylogeny itself is unstable but because our knowledge of it is. For many groups of living things, hardly anything is known about phylogeny, and a "phylogenetic" classification of such a group will inevitably be poorly supported by evidence. Numerical phenetics aimed to avoid all the evolutionary uncertainty by classifying only by phenetic relations, and by using quantitative techniques to measure them. The classification would follow automatically, and therefore (it was thought) objectively, from the phenetic measurements. Let us consider the methods in some more detail, and see how well these aims can be achieved.

The simplest kind of phenetic classification is defined by only one or two characters. We might classify the vertebrates, for example, by the number of their legs, to form groups with 0, 2, or 4 legs. The trouble with this procedure is that it is likely to be subjective in the same way as classifying species by their order of discovery. Different individual characters show different distributions among species and therefore tend to produce different classifications. Consider the birds and some reptilian groups, such as crocodiles, lizards, and turtles. (We met this example in Chapter 15, and the conflicting character distributions are illustrated in Figure 15.2, p. 427. Figure 16.1 in this chapter also partly illustrates this example.)

Crocodiles are more similar to lizards and turtles than to birds if we look at their external surfaces, number of legs, and cold-blooded physiology. But crocodiles are more similar to lizards and turtles than to birds if we look at the anatomy of their skulls. The characters conflict. This is a universal problem, not just a peculiar problem in this example. A taxonomist working with one sample of characters will often produce a different classification from another taxonomist working with a different set of characters. As long as we stay with the principle of classifying according to a small number

Phylogenetic inference can be uncertain

Phenetic characters can be quantified

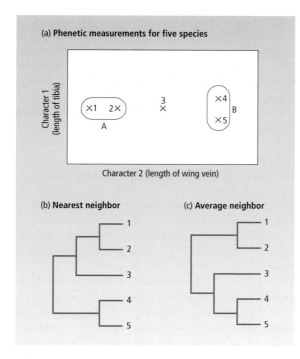

(a) **Phenetic measurements for five species**

Character 1 (length of tibia)

Character 2 (length of wing vein)

(b) **Nearest neighbor**

(c) **Average neighbor**

Figure 16.2
(a) The phenetic similarity between species can be expressed graphically. Suppose five species have been measured for two characters, for instance the length of a wing vein and the length of the tibia. The x-axis is the measurement for each species of the length of a wing vein, and the y-axis for the length of the tibia. The distance between two species on the graph is the phenetic distance between them. (Notice that the distance on the graph is different from the measure of mean character distance in Table 16.1.) (b) The phenetic classification by the nearest "nearest neighbor" technique puts species 3 with the group (cluster A) that has the nearest individual neighboring species (species 2). (c) But if it uses the nearest "average neighbor" technique it puts species 3 with the group (cluster B) that has the nearest average for all its species. Species 4–5 have a nearer average distance. (The average is simply the average of the distances of species 1–2 and of species 4–5 to species 3.)

Similarity can be measured for multiple phenetic characters

of phenetic characters, there is no way to decide which of the many classifications is the best.

The next step is to define the classification not by a few characters but by many. This became possible in the 1950s and 1960s as the statistical and computational apparatus became available for aggregating large numbers of phenetic measures into one grand measure of phenetic similarity. The aim of the numerical phenetic school was to measure so many characters that the idiosyncrasies of particular samples would disappear. The resulting classification groups the units according to their whole phenotype.

How do we aggregate a large number of measures into a single combined measure of phenetic similarity? Several methods exist, and we can illustrate one of them in a graph. We start with the simple case of two characters, though the extension to further dimensions is easy. Suppose that we wish to classify a group of fly species, and we have measured two characters, such as the length of a certain wing vein, and the length of the tibia of the hindleg. The average for each species can be represented as a point (Figure 16.2). For any pair of species, the average difference in their wing vein lengths is the distance between their points on the x-axis, and the difference between the lengths of their tibiae is the distance on the y-axis. If we used either character by itself, the classifications would differ; species 1 and 3 for instance have identical tibial lengths, but different wing vein measurements. The aggregate difference for the two characters can be measured simply by the *distance* between the two species in the two-dimensional space. The species are then classified by putting each with the species, or group of species, that it has the shortest distance to.

If we measured a third character, such as pulse interval between the sounds in the courtship song, it could be drawn as a third dimension into the paper; now each species would be represented by a point in the three-dimensional space. The aggregate distance between the species could be measured as before by the distance between the species' points. We could likewise measure dozens of characters and measure the distance between the species by the appropriate line through hyperspace. Numerical taxonomists recommend measuring as many characters as possible — even hundreds — and classifying according to the aggregate similarity for all of them. The more characters that are measured, the more likely it is that peculiar individual characters will be averaged out, and the classification will be better founded.

Is a numerical phenetic classification objective or subjective? Objective classifications, remember, must represent some unambiguous property of nature. The phenetic classification itself represents the measure of aggregate morphological similarity for large numbers of characters. The question, then, is whether there is some property of nature, some hierarchy of "real" phenetic similarity, that the measurements of aggregate morphological similarity may reasonably be said to be representing. (This topic is discussed in a related way in Section 13.5, p. 363.)

Different cluster statistics can give different hierarchies

We can start by looking more closely at the statistical methods used in numerical taxonomy. In Figure 16.2 there were five species with two characters. To form Figure 16.2a, we grouped each species with its phenetically nearest neighbor. Two clusters — of species 1–2 and 4–5 — immediately formed. But to which of these clusters should we join species 3? The nearest species is 2. If we join species 3 to the cluster with the nearest neighbor we put it with cluster A (the nearest neighbor to species 3 in cluster A is species 2, whereas in cluster B it is species 4 and 5 equally) (Figure 16.2b). However, if we had calculated the average distance of each cluster as a whole, the answer is the opposite (Figure 16.2c). The geometry of Figure 16.2 is such that cluster B rather than cluster A has the nearer average neighbor to species 3.

The nearest neighbor and average neighbor methods are both examples of cluster statistics. They are not the only ones, but they are enough to make a point of principle. We have here, within the phenetic philosophy, managed to produce two different classifications. If the numerical phenetic claim to repeatability and stability is to be upheld, it must have some way of deciding which of the two is the correct phenetic classification. To do so, it would need some higher criterion to fall back on. The problem is it does not have one. The higher criterion would presumably be "the" hierarchy of aggregate morphological similarity, but that hierarchy does not exist in nature independently of the statistics that measure it. And — as Figure 16.2 shows — different statistics produce different hierarchies.

Phenetic classification is inherently ambiguous

So there is an essential degree of subjectivity in the phenetic philosophy. If its classifications are to be consistent, it must pick on one statistic, such as the average neighbor statistic, and stick to it. Classification would then be repeatable, but at a price. The consistency does not follow from the phenetic system itself; it is imposed by the taxonomist — subjectively. In practice, numerical taxonomists have never been able to agree on which statistic to use, and this is one reason why the school has lost much of its influence since its origin in the early 1960s.

Moreover, the choice of cluster statistic is not the only subjective choice in phenetic classification. The measurement of distance poses an analogous problem. The measure

Distance can be measured in more than one way

used in Figure 16.2 is *Euclidean distance*: the straight line between two points; in two dimensions it is measured by Pythagoras' theorem. But other distance measures exist, such as *mean character distance* (MCD). MCD is the average distance between the groups for all characters measured. Thus, in two dimensions, if species 1 and 2 differ by x units in character A and y units in character B, then MCD = $(x + y)/2$ and Euclidean distance = $\sqrt{(x^2 + y^2)}$. The different measures of distance can give different hierarchies and the pheneticist is again faced with a subjective choice of which to use.

Phenetic classification, therefore, even in its modern numerical form, is not object-ive. It can produce classifications, but classifications that lack a deep philosophical justification. Let us see how the introduction of evolution into classification can help with the problem.

16.6 Phylogenetic classification uses inferred phylogenetic relations

16.6.1 *Hennig's cladism classifies species by their phylogenetic branching relations*

Phylogenetic classifications group species solely according to recency of common ancestry. When a species splits during evolution it will usually form two descendant species, called *sister species*, and in a cladistic classification sister species are classified together. The branching hierarchy of ancestral relations is a unique hierarchy, extend-ing back to the beginning, and including all, of life. The phylogenetic hierarchy is easy to convert into a classification (Figure 16.3). (I say it is "easy," but in Section 16.6.3 below we look at problems that arise in moving from a phylogenetic hierarchy to a Linnaean classificatory hierarchy.)

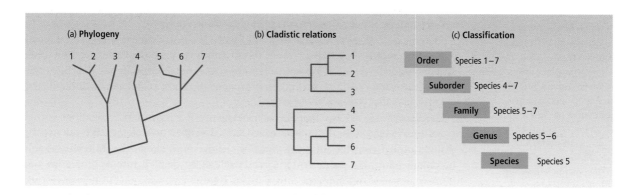

Figure 16.3

The phylogenetic (cladistic) classification of a group is simply related to their phylogenetic tree. (a) The evolutionary history of seven species. (b) Their cladistic classification. (c) The formal Linnaean classification, for species 5 as an example. This particular classification is an example only; it could be, depending on the detail in a particular case, that different Linnaean levels should be used.

Figure 16.4

Different kinds of character and taxonomic group. Homologies are characters shared between species that were present in the common ancestor. They can be derived or ancestral. (a) Shared derived homologies are found in all the descendants of the common ancestor, and are distributed in monophyletic groups. (b) Shared ancestral homologies are found in some but not all the descendants of the common ancestor, and are distributed in paraphyletic groups. (c) Homoplasies are characters shared between species that were not present in the common ancestor, and fall into polyphyletic groups. See Table 16.1 for the way the different characters are used by the different schools of classification. The crucial difference between paraphyletic and polyphyletic groups, in terms of the shaded gray zone, is at the bottom and not at the top of the tree: the paraphyletic group contains the ancestor whereas the polyphyletic group does not. The pattern at the top, in which the polyphyletic group seems to miss out a species between the included species whereas the paraphyletic group seems to contain a set of contiguous species, is an accident of the way the pictures are drawn — by revolving appropriate nodes it would be possible to make the species in the polyphyletic group contiguous or to intrude a gap in the paraphyletic group.

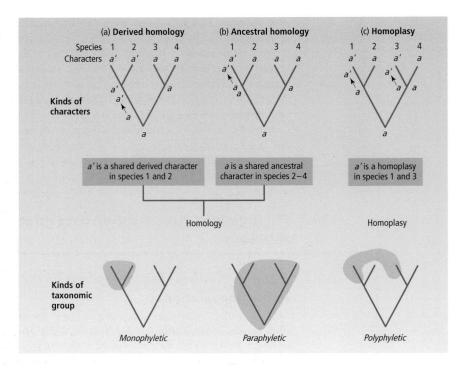

Cladistic classification has the advantage of objectivity. The phylogenetic hierarchy exists independently of the methods we use to discover it, and is unique and unambiguous in form. When different techniques for inferring phylogenetic relations disagree, there is always the external reference point to appeal to. When we cannot work out the phylogeny of some group or other, we do at least know that a solution exists to aim at. With the phenetic system there is no such external solution. There is no single natural phenetic hierarchy analogous to the phylogenetic hierarchy.

When a pair of species, like 5 and 6 in Figure 16.3b, are classified together cladistically, that means they share a more recent common ancestor than with any other species. Cladistic relations are fundamentally ancestral relations. In practice, the inference of ancestral relations (that is, the phylogeny in Figure 16.3a) can be difficult, and cladistic classification can be uncertain.

In Chapter 15, we saw that the main evidence for phylogenetic relations comes from a particular kind of character called shared derived homologies. We can use the distinction between the different kinds of characters to clarify the different schools of classification (see Table 16.1). Characters can be divided into homoplasies and homologies, and homologies into derived homologies and ancestral homologies. (The distinctions are reillustrated in Figure 16.4.) Only derived homologies indicate phylogenetic relations, and a cladistic classification is based on derived homologous characters, not on ancestral homologies or on homoplasies. Numerical phenetic classification groups species using as many characters as possible, and averaging over them regardless of their evolutionary meaning. Phenetic classification uses all three kinds of characters. Later in the chapter we shall discuss a third school of classification, called evolutionary classification, that uses homologies (both ancestral and derived) but rejects homoplasies.

16.6.2 Cladists distinguish monophyletic, paraphyletic, and polyphyletic groups

The groups of cladistic classifications are *monophyletic* in the sense that they contain all the descendants of a common ancestor: the group has a common ancestor unique to itself (Figure 16.4a). Cladism rejects *paraphyletic* and *polyphyletic* groups. A paraphyletic group contains some, but not all, of the descendants from a common ancestor (Figure 16.4b). The members that are included are the forms that have changed little from the ancestral state and the excluded species are those that have changed more. A paraphyletic group therefore contains the rump of conservative descendants from an ancestral species. Polyphyletic groups are formed when two lineages convergently evolve similar character states (Figure 16.4c). The key difference between paraphyletic and polyphyletic groups is that paraphyletic groups contain their common ancestor, whereas polyphyletic groups do not.

As Figure 16.4 shows, the different kinds of group are defined by the different kinds of characters. Derived homologies fall into monophyletic groups; ancestral homologies fall into both monophyletic and paraphyletic groups; and homoplasies fall into polyphyletic groups. Paraphyletic groups are defined when some descendants of a common ancestor retain their ancestral characters, while others evolve new derived character states. The difference between paraphyletic and polyphyletic groups therefore arises because ancestral characters would have been present in a group's common ancestor, but convergent characters would not.

Cladistic classifications only include monophyletic groups because only they have the unambiguous hierarchical arrangement of the phylogenetic tree. Only monophyletic groups are formed in the cladistic conversion of a phylogenetic tree into a classification (Figure 16.3). Monophyletic groups are defined unambiguously by their branching relations: they contain all the branches below a given ancestor, and nothing has to be said (or indeed be known) about the phenetic evolution of species within each branch. But in order to define paraphyletic and polyphyletic groups, we do need to know about phenetic similarity. We have to decide which species to include and which to exclude, and this is done by including in paraphyletic or polyphyletic groups only phenetically similar species. Because of the subjectivity of measures of phenetic similarity, the tree of life cannot be unambiguously divided into paraphyletic or polyphyletic groups. (Figure 16.7 below illustrates this point.)

Taxonomists had known about evolutionary convergence and polyphyletic groups for a long time before cladism, but paraphyletic groups proved a more insidious problem. Hennig was the first to recognize them clearly, and his work was not widely known until it was translated into English in 1966.

We can see how paraphyletic groups tend to crop up by means of an earlier example: reptiles (see Figure 16.1c). The phylogenetic relations of the main tetrapod groups are probably those of Figure 16.5. In traditional classifications, mammals, birds, and

Cladism admits only monophyletic groups

Figure 16.5
Phylogeny of the main
vertebrate groups. Reptiles are a
paraphyletic group made up of
turtles, lizards, snakes, and
crocodiles in this figure.

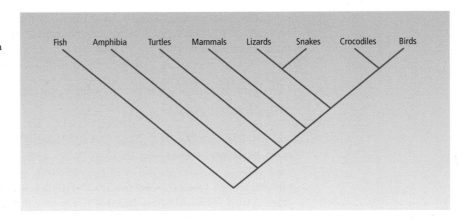

Reptiles are a paraphyletic group

reptiles are given equal taxonomic rank as classes. What has happened is that two groups, mammals and birds, have independently undergone relatively rapid phenetic evolution and have come to look very different from reptiles. The different reptilian lineages have changed more slowly and have been left looking more like each other than like birds or mammals. Crocodiles and lizards, for instance have cold blood, scales, four legs, and walk with a reptilian gait; birds are hot blooded, have feathers, two legs and two wings, and they fly. Yet crocodiles share a more recent common ancestor with birds than with lizards. The characters of crocodiles and lizards (scales, and so on) are ancestral for the group as a whole; and the paraphyletic group was formed because ancestral characters were used for definition.

Paraphyletic groups become a danger whenever one or more subgroups have evolved relatively quickly and left their former relatives behind. If the cladistic philosophy is accepted, paraphyletic groups have to be ruled out. They are defined phenetically (by ancestral characters) and their recognition is inevitably subjective. The class Reptilia, therefore, is disbanded in cladistic classification. One subgroup of the former Reptilia, called Archosauria, includes the crocodiles and birds (and dinosaurs). Another subgroup, called Lepidosauria, contains the lizards, snakes, and probably turtles.

Fish are another paraphyletic group

The cladistic classification of the tetrapods can seem odd. The Reptilia were recognized in almost every formal classification before cladism; but cladism rules them out. There are many other examples of paraphyletic groups in non-cladistic classifications: fish are one of them. The tetrapods evolved from one particular group of fish, the lobe-finned fish (Section 18.6.1, p. 540). If we consider the relations of any tetrapod (such as a cow), any lobe-finned fish (such as a lungfish), and any ray-finned fish (such as a salmon), the cow and the lungfish share a more recent common ancestor than do the lungfish and the salmon. The category "fish" (containing the lungfish and salmon, but excluding the cow) does not exist in a cladistic classification.

Some cladists have been rather fanatical in their insistence on ruling out fish and reptiles. In a way, what we do in practice in these cases does not matter much. Their paraphyletic status is well known, and can do little damage; it is not worth getting worked up about. In other, less well known, cases, it is more important to avoid paraphyletic groups. If a classification contains an unspecified mixture of monophyletic and paraphyletic groups, the evolutionary information in the classification becomes muddled. The beauty of a purely phylogenetic classification is that there can be no doubt what the branching relations of the classificatory groups are (Figure 16.3). But if taxonomists define some relations phenetically and others phylogenetically, it is no longer possible to say what any particular relation means. The branching relations are obscured and lost.

16.6.3 A knowledge of phylogeny does not simply tell us the rank levels in Linnaean classification

Phylogenetic classification faces several problems

The main advantage of phylogenetic classification is theoretical. It can run into many problems in practice. One is ignorance. We do not know the phylogenetic relations of many living creatures, and cannot classify them phylogenetically. Another problem is instability. Scientific knowledge can change when new evidence comes in, and our knowledge of phylogeny is no exception. When our knowledge of a phylogeny changes, a change will also be required in the classification of that group. A third problem has arisen recently, as the number of levels in the known phylogenetic hierarchy has increased beyond the ability of the Linnaean hierarchy to represent it. Chapters 15 and 18 look at our knowledge of phylogeny. Here we concentrate on the other two problems. (Other problems can arise when evolution is non-hierarchical, for instance because of hybrid speciation (Section 14.7, p. 405) or horizontal gene transfer (Figure 15.19, p. 456).)

We saw in Figure 16.3 how a phylogeny of a group of species can be converted into a Linnaean classification. However, the phylogenetic hierarchy only determines the pattern of groups within groups, not the ranks of the groups. For example, if we know that humans and chimpanzees share a more recent common ancestor with each other than either does with gorillas, then we know that humans and chimpanzees should be grouped together within the larger group of great apes. But the phylogenetic knowledge alone does not tell us whether the group of humans and chimpanzees should be ranked as a genus, subfamily, family, or something else.

A full phylogeny has many more levels than the Linnaean hierarchy

The Linnaean hierarchy has perhaps seven main levels (kingdom, phylum, class, order, family, genus, species); these can be multiplied by adding in super-, sub-, and infra-levels (e.g., superfamily, suborder, and so on) and other extra levels, such as the tribe (between genus and family). But even an expanded Linnaean hierarchy could not have much more than about 25 levels. This raises a problem of how to combine levels in the phylogenetic hierarchy to accommodate it in the Linnaean system. We cannot naively say that each successive branching point (or node) in the phylogeny can have its own Linnaean rank. There are just too many nodes.

For instance, Figure 15.27 (p. 465) showed the phylogeny of the Hawaiian picture-wing fruitflies. A cladistic classification of one part of it — such as the *Drosophila adiostola* species group — would need us to invent five new levels in the Linnaean hierarchy between the levels of genus and species (Figure 16.6). Historically, the problem has not been acute because we have been relatively ignorant of phylogenetic relations and the number of described species has not been too large. But we should now be planning for a complete phylogenetic knowledge of about 10–100 million species.

Figure 16.6
Possible classification of the *Drosophila adiostola* group of species of fruitflies. (Taken from the top branch in the phylogeny of Hawaiian *Drosophila* in Figure 15.27, p. 465.) These 14 species are only a small part of the Hawaiian *Drosophila*, which in turn are only a (large) part of the worldwide *Drosophila* fauna. And yet what we know about their phylogeny would require at least five new levels between the genus and species level.

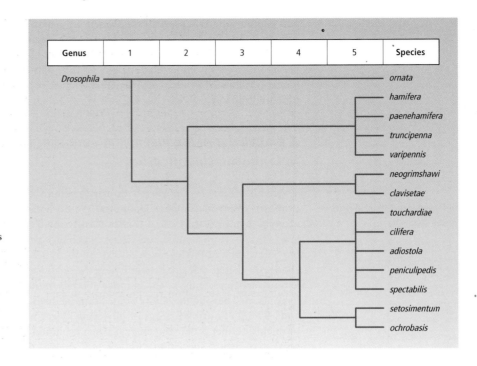

How can we fit the phylogenetic hierarchy into the Linnaean hierarchy? Hennig (1981) suggested using a numerical scheme in addition to the Linnaean terms. Alternatively, a number of levels of the phylogeny can be collapsed into one Linnaean level, to provide convenient and memorable categories. The classification is then phylogenetic and cladistic, but not all the phylogenetic knowledge is represented in the classification.

Another possibility is being considered by systematists at present: we might simply abandon the Linnaean ranks of genus, family, order, and so on. Taxonomic groups could consist simply of unranked clades (that is, monophyletic groups consisting of all the descendants of a common ancestor). When a clade was discovered by phylogenetic research, the clade could be named but not given a rank. Birds (or Aves) would be a clade, and passerine birds would be another clade. They would both continue to be used as taxonomic terms, just as they are now, but would lack a rank such as class or order. The idea of naming clades but not giving them ranks is one component of a proposed taxonomic system called *phylocode*. The proponents of phylocode have also suggested other changes to Linnaean classificatory methods. The suggestions are controversial, and the topic of active discussion. It is too early to say whether phylocode will catch on.

The phylogenetic hierarchy of life in principle provides a good basis for the Linnaean classification of life. Most biologists now accept that classification should be phylogenetic. However, our knowledge of biodiversity and of phylogeny is expanding at a huge rate. Phylocode may or may not come to replace the Linnaean classification, but it is a

Phylocode would abandon Linnaean ranks . . .

. . . but may not catch on

reasonable response to the problem of how to represent our phylogenetic knowledge in a formal classification. While only tens or hundreds of thousands of species had been named, and phylogenetics was a backwater of biological research, Hennig's cladistic principles posed little challenge to the Linnaean system of ranks. Now we know about millions of species, and molecular systematics is a major research program. Some adjustments to the Linnaean system may yet be needed to accommodate our expanding phylogenetic knowledge, but meanwhile, it remains useful to classify biological species into the Linnaean higher taxa. But decisions about which levels of the Linnaean hierarchy to apply to which nodes in a phylogeny are arbitrary and subjective.

16.7 Evolutionary classification is a synthesis of phenetic and phylogenetic principles

Evolutionary classification admits paraphyletic and monophyletic groups

Finally, we should look at *evolutionary classification*. Evolutionary classification incorporates both phenetic and phylogenetic elements, and both paraphyletic and monophyletic groups. In terms of the kinds of taxonomic groups (see Figure 16.4), evolutionary classification recognizes paraphyletic and monophyletic but not polyphyletic groups. In terms of the kinds of characters used to infer phylogeny (see Figure 16.3), it forms groups by homologies rather than homoplasies, but does not distinguish ancestral from derived homologies. Referring back to Figure 16.1, evolutionary classification picks the phenetic classification in the reptilian case (Figure 16.1c) but the phylogenetic where there is convergence (Figure 16.1b).

How can evolutionary taxonomy be justified? The evolutionary school predates both numerical phenetics and cladism and the main original discussions of the school therefore did not meet the phenetic and cladistic arguments head on. Moreover, no complete modern evolutionary taxonomic defense against numerical and cladistic taxonomy exists. As such, the school differs from the other two, which were conceived partly in opposition to evolutionary taxonomy and made their objections to it clear. Nevertheless, a case can be made.

Evolutionary taxonomists disagree with phenetic classification for much the same reasons we discussed above, though they express the argument differently. They criticized phenetic systems for being *idealistic*, that is for supposing that a phenetic classification represents some "ideal" phenetic relationship between species. The ideal relationship would be some "idea" or "plan" in nature. An example is the pre-Darwinian theory that classifications represent the thoughts of God. An idealist would then aim to classify species according to an idea that exists in the mind of God. That idea, arguably, manifests itself in (and is inferrable from) living nature. It is difficult for a modern scientist to make much sense of these old arguments, but "divine" taxonomy at least provides a concrete example of what idealism means. Other versions of idealism supposed that it was possible to deduce the existence of fundamental forms or plans of nature from a purely scientific analysis of species' morphology.[1]

---

[1] Section 13.5, p. 363, about "typological thinking," is concerned with much the same point. Idealism is an example of typological thinking.

Evolutionary taxonomists criticized phenetic classification

Notice that idealism could in principle solve the problem of subjectivity in phenetic classification. The plan of nature would provide an objective, external reference point for the phenetic classification to aim at. The only snag is that no plan of nature, in the idealist sense, exists. That is why modern numerical phenetics dropped the idealist philosophy of earlier phenetic classifications.

Section 16.5 argued that, in the absence of any natural hierarchy of phenetic similarity, phenetic classification lapses into subjectivity. The idealist error is the other side of the same coin. Idealists believe that a phenetic hierarchy exists "out there," but offer no good reason for their belief. When evolutionary taxonomists criticized phenetic classification for committing the error of idealism, they meant that no real phenetic hierarchy exists in nature, and a system which assumes such a hierarchy will be fundamentally subjective. Phenetic classifications try to group species according to a relationship — the ideal morphological system — that evolution does not produce. Evolution does not produce one particular privileged phenetic hierarchy that is more real than all other phenetic hierarchies.

Phenetic idealism can be avoided if taxonomists represent evolution, not phenetic similarity. But what does "evolution" mean? Evolutionary taxonomists exclude polyphyletic, but not paraphyletic, groups from classification. To see why, we must look at what evolutionary taxonomists say about cladism. Evolutionary taxonomists criticize cladism for its unnecessary puritanism. Cladism, as we have seen, leads to what at first sight can appear bizarre conclusions, such as the destruction of the Reptilia. It does so because of its distinction between paraphyletic and monophyletic groups. If both kinds of group are allowed, most of the long-recognized groups such as reptiles and fish can be retained.

That is how evolutionary classification sees itself. But cladists, and pheneticists, see it rather differently. According to a cladist, the argument that evolutionary taxonomists accept against the pheneticist's polyphyletic groups works just as well against paraphyletic groups. If you accept paraphyletic groups, you must accept polyphyletic ones too, or be inconsistent. Paraphyletic groups are defined phenetically, just like polyphyletic groups. If some of the descendants of a common ancestor are to be excluded from a group, it has to be decided how many such descendants are to be left out, and the decision is phenetic and arbitrary (Figure 16.7); paraphyletic groups are not formed by phylogenetic relations — as Figure 16.4 illustrated. The standard phenetic problem then re-enters: according to some measures of phenetic similarity, one paraphyletic group will seem appropriate; according to another, another will. The choice between them is subjective — or idealist. When paraphyletic groups are admitted, the argument against phenetic classification is lost. If phenetic criteria can be used in the case of paraphyletic groups, why not for polyphyletic ones? Paraphyletic groups presuppose idealism just as much as polyphyletic groups.

Classification with paraphyletic groups is ambiguous

Evolutionary taxonomy mixes phenetic and phylogenetic methods, but in a consistent and principled manner. It defines groups by homologies and excludes homoplasies; it therefore does not recognize polyphyletic groups. It allows phenetic groups such as reptiles and fish, but classifications in biology have a strong practical purpose and there is a case not to disband these long-established groups if a convincing reason can be found for keeping them. However, it is questionable whether the evolutionary taxonomist's reason is convincing enough.

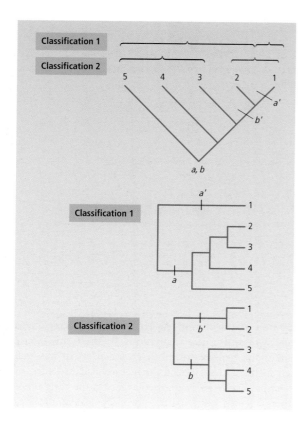

Figure 16.7

Paraphyletic groups contain some but not all of the descendants of a common ancestor. When they are defined, a decision has to be taken about which descendants to exclude. The decision is phenetic. *a*, *a′*, *b*, and *b′* are character states; *a′* and *b′* are derived from *a* and *b*, respectively. In the figure, the question is whether to exclude species 1 (and define the paraphyletic group of 2–5 by ancestral character *a*) or species 1–2 (and define the group 3–5 by ancestral character *b*). Notice there is no guarantee that shared ancestral characters can define such paraphyletic groups as 2–5 and 3–5, because *a* or *b* could in principle have undergone any number of changes within the branches leading to species 2–5 and 3–5.

16.8 The principle of divergence explains why phylogeny is hierarchical

Evolution produces a hierarchical tree of life

All three schools of classification — phenetic, cladistic, and evolutionary — aim at hierarchical classification. In the case of cladism, that is unsurprising. The phylogenetic tree is a hierarchy and phylogenetic classification will be hierarchical too. It is less obvious whether a phenetic classification has to be hierarchical. Nature presents us with an infinity of phenetic patterns. Some indeed are nested hierarchies, but others are overlapping hierarchies or non-hierarchical networks. If we aim at a phenetic classification, we have no strong reason to classify hierarchically. Biological classifications are hierarchical because evolution has produced a tree-like, diverging, hierarchical pattern of similarities among living things. Indeed, the hierarchical nature of biological classifications has since 1859 been part of the evidence for evolution (Section 3.9, p. 61). Evolutionary descent produces, in Darwin's words a pattern of "groups within groups."

But why should evolution proceed in this form? The question has an important place in the history of Darwin's thinking. He thought up natural selection in the late 1830s, as

Figure 16.8
The divergent pattern of evolution. From Darwin (1859).

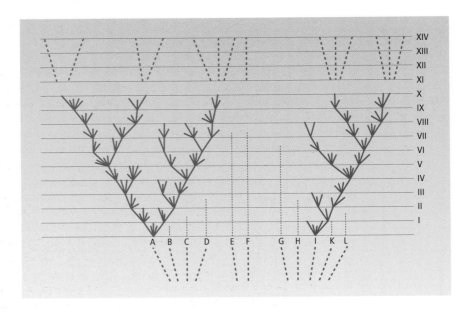

According to Darwin's principle of divergence . . .

a natural explanation for adaptation and evolution. As environments change, and competing species change, species will evolve new adaptations. By itself, this theory does not account for the tree-like, divergent course of evolution. Darwin was well aware that evolution had steered such a course, indeed the hierarchical structure, of groups within groups in classification, had been established as fact in the early nineteenth century — by (among others) Geoffroy St Hilaire's morphological, and Milne-Edwards's embryologic, work. So striking a fact had to fitted into the theory. Darwin recalled in his autobiography that in the early (1844) version of his theory:

> I overlooked one problem of great importance, and it is astonishing to me how I could have overlooked it and its solution. This problem is the tendency in organic beings descended from the same stock to diverge in character as they become modified. That they have diverged greatly is obvious from the manner in which species of all kinds can be classed under genera, genera under families, families under suborders and so forth. . . . The solution occurred to me long after I had come to Down. The solution, I believe, is that the modified offspring of all dominant and increasing forms tend to become adapted to many and diversified places in the economy of nature.

. . . competition pushes species apart

This was Darwin's *principle of divergence* (Figure 16.8). Why should it be that species apparently push one another apart in evolution? Darwin suggested that it mainly resulted from the relative strengths of competition for resources from more closely related individuals on the one hand, and from more distantly related individuals on the other. An individual of a species will compete strongly against other members of its own species, fairly strongly against members of other species in its own genus, and then more weakly against members of more distantly related groups. There is little or no competition at the taxonomic extremes, between an average plant and an average animal, for example.

Competition is strongest within a species. Each individual will, in many cases, encounter more members of its own species than of other species. Also, the members of its own species are more similar to it, exploiting more similar resources. One way to avoid competition is to become different from the competitors; hence there will be a force pushing similar competing types apart in evolution. Competition between similar individuals will make for the evolution of new adaptations in each that reduce the intensity of competition; divergence will thus result. Character displacement (Section 13.6, p. 366) provides evidence that competition can cause divergence between closely related species. Evolutionary divergence is not inevitable, however. It depends on the contingencies of competition in particular cases. Provided that, on average, more similar individuals compete more closely, divergence is likely to result.

Other factors are probably at work too

Other factors probably also contribute to causing divergence. For instance, speciation is often allopatric and each pair of sister species becomes increasingly isolated over time by the Dobzhansky–Muller process (Section 14.4, p. 389). While the members of two populations interbreed, natural selection favors genetic changes that are advantageous in both populations. The two populations are kept relatively similar. Once the two populations have evolved apart and no longer interbreed, no force constrains the genetic changes that occur in one population also to be favorable in members of the other population. Incompatible genetic changes accumulate in the two populations (or species, as they are by this stage). The two gene pools have evolutionarily "escaped" from each another and are free to diverge further. Our modern genetic understanding of speciation has added to Darwin's explanation for divergence.

16.9 Conclusion

Most biologists now accept that cladism is theoretically the best justified system of classification. It has a deep justification that phenetic and partly phenetic systems lack. Cladism is objective, and objective classifications are preferable to subjective ones. But despite cladism's theoretical advantages, it can run into practical problems. The uncertainties of phylogenetic inference make cladistic classifications liable to frequent revision. True cladists are not very worried about that, and remark that all healthy theories are modified as new facts come in.

Cladistic classification has become popular only recently. Evolutionary classification was the orthodox school from the "modern synthesis" of the 1930s (or even from Darwin's time in the 1860s) until about 20 years ago. Numerical phenetics enjoyed a surge of support from the late 1950s to the early 1970s. Now, however, both schools have succumbed to cladistic criticism and have relatively few supporters. At all events, the most important point is to understand the arguments that have been used for and against each school: these arguments are of permanent importance in evolutionary biology.

Summary

1 There are two main principles, and three main schools, of biological classification: phenetic and phylogenetic principles, and phenetic, cladistic, and evolutionary schools. The schools differ in how (if at all) they represent evolution in classification.

2 Phenetic classification ignores evolutionary relations and classifies species by their similarity in appearance; cladism ignores phenetic relations and classifies species by their recency of common ancestry; evolutionary taxonomy includes both phenetic and phylogenetic relationships.

3 Phylogenetic inference is uncertain and phenetic classification has the advantage that it is not subject to revision when new phylogenetic discoveries are made.

4 Phenetic classification is ambiguous because there is more than one way of measuring phenetic similarity and the different measures can disagree.

5 Cladism is unambiguous because there is only one phylogenetic tree of all living things.

6 The cladistic philosophy specifies the pattern of groups within groups in a classification, but not the ranking of those groups. The large number of nodes in a fully resolved phylogeny can be difficult to represent in a Linnaean classification.

7 Evolutionary taxonomy avoids some of the extraordinary properties of cladism. But it suffers from the ambiguity of phenetic taxonomy, and its argument for excluding one kind of phenetic relation (convergence) works equally well against the kind of phenetic relation (differential divergence) that it includes.

8 Living things show a diverging, tree-like pattern of relationships. Darwin explained this by his "principle of divergence": that competition is stronger between more similar forms, forcing them to evolve apart.

Further reading

Schuh (2000) is a recent book about biological systematics. I have previously discussed many of the points discussed in this chapter, at an introductory level and at greater length, in Ridley (1986). Sneath & Sokal (1973) is the standard work on numerical taxonomy; Sokal (1966) is a clear introduction. Hennig (1966) is the classic work on cladism — but he is not an easy read! Wiley *et al.* (1991) and Kitching *et al.* (1998) are two texts on cladism. Mayr (1976, 1981, and Mayr & Ashlock 1991), Dobzhansky (1970), and Simpson (1961b) are key works by key evolutionary taxonomists. Mayr & Diamond (2001) classify the birds of Melanesia according to traditional "Mayrian" principles.

Bryant & Cantino (2002) review criticisms of phylogenetic nomenclature. *Science* March 23, 2001 contains a newspiece about phylocode. Benton (2000b) defends the traditional (cladistic) practice. Ereshefsky (2001) is more critical. References to the primary sources can be traced through these sources. This controversial topic can be followed in occasional papers in the journals *Systematic Biology* and *Cladistics*. See also the website www.ohio.edu/phylocode.

On the principle of divergence, Darwin (1859) is the obvious source. Bolnick (2001) did an experiment showing how evolutionary divergence, and adaptation to exploit new resources, is more likely in the presence of competition.

The subject has been the topic of some interesting historic and philosophical work. Ritvo (1997) is a general, mainly historic book about classification. Hull (1988) gives an excellent history of the phenetic and cladistic movements to illustrate his evolutionary philosophy of science. Beatty (1994) uses the controversy among schools of systematics to discuss a broader philosophical question: why does evolution have so many controversies about relative frequencies, or relative significance, in contrast with the Newtonian paradigm of physics, and some other areas of biology, in which questions are asked that have a single answer. Sober (1994) contains some other philosophical chapters.

Study and review questions

1 Match the kinds of taxonomic groups to the schools that allow them:

Groups: polyphyletic, monophyletic, paraphyletic.
Schools: evolutionary, cladistic, phenetic.

2 Give (i) a phenetic, (ii) an evolutionary, and (iii) a cladistic classification of the cow, the lungfish, and the salmon.

3 Here are measurements of two characters in three species:

	Species 1	Species 2	Species 3
Character *a*	2	5	2
Character *b*	2	2	6

Calculate the differences for each character and then calculate (i) the Euclidean distance, and (ii) the mean character distance, for the three species pairs. Write the distances in the following matrix above the diagonal for calculation (i) and below it for calculation (ii).

	Species 1	Species 2	Species 3
Species 1			
Species 2			
Species 3			

4 Here are the pairwise distances (either mean character distance or Euclidean) among five species:

Species	2	3	4	5
1	1	2.31	4.28	5.27
2		2.31	4.28	5.27
3			2	3
4				1

(a) What is the average distance from species 3 to species 1 and 2? (b) What is the average distance from species 3 to species 4 and 5? (c) Which species is the nearest neighbor of species 3? (d) What is the phenetic classification of the five species? (e) What do the answers to (a–d) suggest about the objectivity of numerical phenetic taxonomic classification?

5 Supporters of phenetic classification have sometimes replied to the criticism that it can be ambiguous by saying that if enough extra characters are measured, the ambiguity will be resolved. (Thus in Figure 16.4, if further characters were measured in the species, the average neighbor and nearest neighbor statistics would come to agree.) Is this a good reply?

6 (a) Why are biological species classified hierarchically whereas chemical elements are classified non-hierarchically in the periodic table? (b) Why does evolution usually have a divergent branching pattern, in the form of a tree, rather than some other pattern (in the form, perhaps, of a row of telegraph poles, or of erratic zigzags)?

17 Evolutionary Biogeography

Biogeography is the science that seeks to explain the distribution of species, and higher taxa, on the surface of the Earth. The chapter begins by describing the elemental facts that are to be explained — the kinds of distribution. We then move on to the explanatory processes. We begin with short-term processes, such as species ecology and movement in relation to climate, that explain distributions at the species level. We move on to larger scale processes, such as adaptive radiations on island archipelagos, and then to grander biogeographic patterns and the longer term processes, particularly plate tectonics, that produce them. We see how to study the relation between the phylogenetic history of a species and the geological history of the area it has occupied. We finish by looking at another evolutionary phenomenon resulting from plate tectonics: encounters between faunas when previously separated areas are tectonically brought together. The classic example is the Great American Interchange.

Figure 17.1

The natural distribution of three species of toucans in the genus *Ramphastos* in South America: *R. vitellinus* and *R. culminatus* have endemic distributions, whereas *R. ariel*'s distribution is disjunct. There is an extensive hybrid zone between the species. Modified, by permission of the publisher, from Haffer (1974).

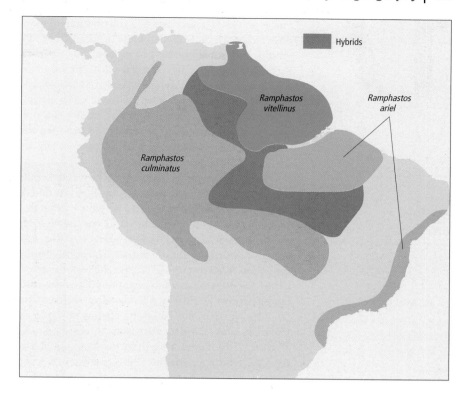

17.1 Species have defined geographic distributions

Geographic distributions of species may be endemic, cosmopolitan, or disjunct

The geographic distributions of species can be of a number of types. Consider Figure 17.1, which shows the distribution of three species of toucans in the genus *Ramphastos*, in South America. Two of the species, *R. vitellinus* and *R. culminatus*, have *endemic* distributions: they are limited to a particular area. Endemic distributions can be more or less widespread, and the extreme case of species that are found on all continents of the globe are called *cosmopolitan*. The pigeon, for example, is found on all continents except Antarctica; on a strict definition, the pigeon might not be allowed to be cosmopolitan, but the term is usually intended less strictly — and the pigeon is called a cosmopolitan species. Other species, like *R. ariel* in Figure 17.1, are not confined to a single area, but are distributed in more than one region with a gap between them: these are called *disjunct* distributions.

Maps like the one for species in Figure 17.1 can be drawn for a taxonomic group at any Linnaean level: just as species have geographic distributions, so too do genera, families, and orders. Biogeography aims to explain the distributions of the higher taxa too, in addition to those of species, and different explanatory processes are often appropriate at different levels. Short-term movements of individuals influence the distributions of populations and species, whereas slower acting geological processes may control the biogeography of higher taxa. The distributions of the higher taxonomic levels are, obviously, more widespread than those of species, but some taxonomically isolated higher groups, with small numbers of species (they are usually examples of living fossils, see Section 21.5, p. 606), have localized distributions. For example, the tuatara *Sphenodon punctatus* is the only surviving species of a whole order of reptiles (or almost the only survivor — there may be more than one surviving species of *Sphenodon*). Of about 20 orders of reptiles, 16 are completely extinct and only four have living survivors. Of those four, three contain the turtles and tortoises, lizards and snakes, and crocodiles, respectively. The fourth has only *Sphenodon*, which is now confined to some rocky islands off New Zealand.

When the biogeographers of the nineteenth century looked at the distributions of large numbers of species on the globe, they saw that different species often lived in the same broad areas. They suggested that there are large-scale faunal regions on the Earth. The first map of these faunal regions was drawn for birds by the British ornithologist Philip Lutley Sclater (1829–1913), and Alfred Russel Wallace soon generalized Sclater's regions to other groups of animals. The Earth was thus divided up into six main biogeographic regions (Figure 17.2a). The regions are mainly defined by the distribution of birds and mammals, and might not have been recognized if other groups had been used. Botanists, for instance, tend to draw different lines on the map: they usually combine the Nearctic and Palearctic regions into one larger region called the Boreal or Holarctic, and recognize a separate floral region, called the Cape, in Southern Africa (Figure 17.2b). Figure 17.2, therefore, does not illustrate a set of hard and fast facts as the regions are approximate. The regional terms — like Nearctic and Neotropical — are often used in biogeographic discussion.

The division into regions was made according to the degree of similarity between the lists of the species living in the various places. Biogeographic similarity can be quantified by various *indexes of similarity*. One of the simplest indexes is Simpson's index. If N_1 is the number of taxa in the area with the smaller number of taxa, and N_2 is the number of taxa in the other area, and C is the number of taxa in common between the two regions, then Simpson's index of similarity between the two areas is:

$$C/N_1$$

Table 17.1 gives the faunal similarities, for mammalian species, between several regions (they are expressed as percentages: that is, $(C/N_1) \times 100$). The indexes in the table show some of the justification for the division of the Earth into faunal regions as in Figure 17.2. For example, the faunas of Australia and New Guinea are 93% similar, whereas those of New Guinea and the Philippines are only 64% similar. The Philippines are more similar to Africa than to New Guinea. This Indonesian discontinuity, which can be seen in Figure 17.2a, is known as *Wallace's line*. It was not properly understood until the discovery of plate tectonics.

Major faunal . . .

. . . and floral . . .

. . . regions of the globe are recognized, . . .

. . . and are quantitatively described

Figure 17.2

(a) The six main faunal regions of the world, based on the distribution of animals, and particularly of birds and mammals (see Table 17.1). The discontinuity between the Australian and Oriental regions is called Wallace's line. (b) The six main floral regions of the world, based on the distribution of angiosperms (flowering plants). Redrawn, by permission of the publisher, from Cox & Moore (2000).

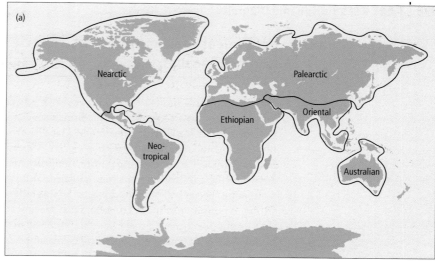

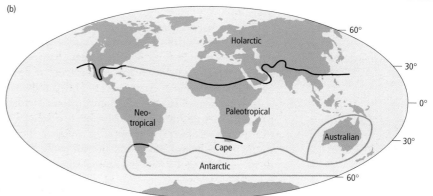

Table 17.1

Indexes of similarity for the mammalian species of various regions. Data from Flessa *et al.* (1979).

	North America	West Indies	South America	Africa	Madagascar	Eurasia	South East Asian islands	Philippines	New Guinea	Australia
North America										
West Indies	67									
South America	81	73								
Africa	31	27	25							
Madagascar	38	27	35	65						
Eurasia	48	27	36	80	69					
South East Asian islands	37	20	32	82	63	92				
Philippines	40	20	32	88	50	96	100			
New Guinea	36	21	36	64	50	64	79	64		
Australia	22	20	22	67	38	50	61	50	93	

17.2 Ecological characteristics of a species limit its geographic distribution

Species occupy particular ecological niches . . .

The distributional limits of a species are set by its ecological attributes. One way of understanding how ecological factors limit a species' distribution is in terms of a distinction, first made by Hutchinson and MacArthur in the 1950s, between the *fundamental niche* and the *realized niche* of a species. A species will be able to tolerate a certain range of physical factors — temperature, humidity, and so on — and could in theory live anywhere these tolerance limits were satisfied. This is its fundamental niche. However, competing species will often occupy part of this range and the competition may be too strong to permit both species to exist. Each species' realized niche will then be smaller than its physiology would make possible: each will occupy a smaller range than it could in the absence of competition. Much ecological research has been carried out to discover the factors — whether physical or biological — that act to limit particular species' distributions.

In some cases, a species' distribution is limited ecologically, for example the species could not live outside its existing range because a competing species is present elsewhere. In other cases, a historic rather than ecological explanation is needed. The species may be ecologically capable of living at a place, but it is absent because it has never arrived — that is, it has never migrated and established itself.

. . . but history can also be influential

In what sense are ecological and historic factors alternatives? If we consider the particular distributional limit of a species, we can ask whether it lies at the limit of the species' ecological tolerance, or whether the species could ecologically survive on the other side of the border but for some historic reason is not. It can therefore be meaningful to test between ecological and historic explanations. In most real cases, however, a complete account of a species' distribution needs both ecological and historic knowledge. A species cannot live outside its ecological tolerance range; its biogeography therefore cannot contradict its ecology. However, within its ecological tolerances, historic factors may have determined where it is living and where it is not. The two factors will then not be opposed, and the sensible method of analysis is to work out how ecology and history have combined to produce the species' distribution.

17.3 Geographic distributions are influenced by dispersal

A species' range will be changed if members of the species move in space, a process called *dispersal*. Individual animals and plants move, actively and passively, through space both in order to seek out unoccupied areas and in response to environmental change. (Plants move passively, at the seed stage.) When the climate cools, the ranges of species in the northern hemisphere move southwards, and tropical forests fragment into smaller forest patches. It would also be possible for the range of a species to change, when the climate changed, without the movement of individuals. Those in the colder regions (for example) might die off, and the range would shrink and move on average to the south. In practice, though, individuals would move southwards as well and extend the species range as they did so. If a species originated in one area and subsequently dispersed to fill out its existing distribution, the place where it originated is called its *center of origin*.

Various dispersal routes might have been followed in the biogeographic history of a species. Simpson distinguished dispersal by means of *corridors*, *filter bridges*, and *sweepstakes*. Two places are joined by a corridor if they are part of the same land mass — Georgia and Texas, for example. Animals can move easily along a corridor and any two places joined by a corridor will have a high degree of faunal similarity. A filter bridge is a more selective connection between two places, and only some kinds of animals will manage to pass over it. For instance, when the Bering Strait was above water, mammals moved from North America to Asia and vice versa, but no South American mammals moved to Asia and no Asian species moved to South America. The reason is presumably that the land bridges at Alaska and Panama were so far apart, so narrow, and so different in ecology that no species managed to disperse across them. Finally, sweepstakes

routes are hazardous or accidental dispersal mechanisms by which animals move from place to place. The standard examples are island hopping and natural rafts. Many land vertebrates live in the Caribbean Islands, and (if their biogeography is correctly explained by dispersal) they might have moved from one island to other, perhaps being carried on a log or some other sort of raft.

There is good evidence for the power of dispersal. In 1883, for example, a volcanic eruption covered the small Indonesian island of Krakatau with ash and killed all the plants and animals. Biologists then recorded the recolonization of the island, particularly for birds and plants. The recolonization was astonishingly rapid. Fifty years later, the island was already recovered with tropical forest, which supported 271 plant species and 31 bird species. Invertebrate animals, such as insects, had come too, though their numbers were less closely monitored. The immigrants mainly came from the neighboring islands of Java (25 miles (40 km) away) and Sumatra (50 miles (80 km) away); the birds would have dispersed by active flight and the plants would have been carried as seeds. Dispersal, therefore, in the right circumstances can have a clear effect on the ranges of species.

17.4 Geographic distributions are influenced by climate, such as in the ice ages

The current geological age is called the Quaternary, which began 2.5 million years ago (see Section 18.2, p. 525, for geological time). During the Quaternary, the climate has mainly been cooler than in the preceding Tertiary, and the temperature has cycled up and down. Many of the cooler times have been glacial periods, and the warmer times interglacials. These climatic changes have happened recently enough for the fossil record in some cases to be revealingly complete. When the weather turns cool, the ranges of animal and plant species tend to contract and move south (in the northern hemisphere). At any one site, the local ecology changes to one characteristic of the cooler climate. A change from a temperate to a tundra-type ecosystem, for example,

Figure 17.3
Changing American geographic distribution of beech (*Fagus*) and hemlock (*Tsuga*) as the polar ice cap retreated after the most recent ice age.

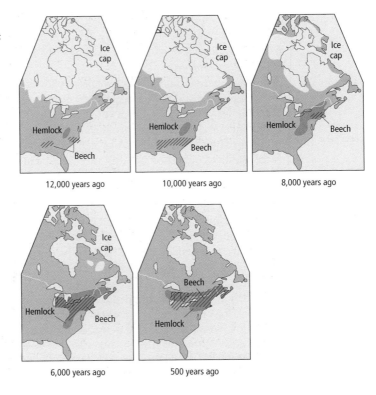

The ice age caused changes in geographic distributions

has been well documented from pollen records in the northern temperate zone through recent ice ages.

The change can also be seen in the distribution of single species (Figure 17.3). The most recent ice age ended about 10,000 years ago. Figure 17.3 shows how the geographic distributions of hemlock and beech trees moved north through the USA as the temperature warmed up and the ice cap retreated. The same movements, southwards and northwards with the advance and retreat of the ice caps, has been shown in many species. Indeed the climate of a locality in past times in the fossil record can be inferred from which species were present. As the interglacials and glacials come and go, many species do not evolve (or do not observably evolve) — they simply move north and south.

The ice age movements of species have had evolutionary consequences. In Europe, many species survived the cold by retreating to the southern extremes of the continent. Species such as bears and hedgehogs retreated into Spain, Italy, and the Balkans during the peak of the ice age. These local populations, surviving adverse conditions, are called *refuges* or *refugia*. The small populations in the different refuges would have evolved genetic differences, either by selection or drift. The populations in Spain, Italy, and the Balkans diverged. Then when the ice cap retreated north, all three populations expanded north too. This has had two detectable consequences.

One is that, within a species, approximately three genetic types can be distinguished (Figure 17.4a). If we look at the molecular phylogeny of European hedgehogs, we see three relatively distinct clades in east, central, and west Europe. The three are descended from the Balkan, Italian, and Spanish refugial populations, respectively (Figure 17.4b). Secondly, many different European species form hybrid zones in similar places. (A hybrid zone — see Section 14.9.1, p. 409 — is a region where two distinct forms of a species meet up and interbreed.) The reason is that many species formed refuges in a similar set of places, and expanded northwards at the same time.

In one species after another, west European and east European populations expanded northwards and met up in a north–south line through central Europe (Figure 17.4c). A *suture zone* is an area where many species form hybrid zones, and Figure 17.4c illustrates the suture zones of Europe. On the interpretation given here, suture zones have a historic explanation. An alternative explanation would be environmental: that suture zones form at the sites of major environmental discontinuities. But for European suture zones, the historic explanation is widely accepted. Analogous suture zones seem to exist in North America, such as the one in northern Florida (Remington 1968; Hewitt 2000).

The genetic changes in the fragmented refugial populations were not probably enough to produce full speciation. The European hedgehogs, for instance, are currently divided into two species. However, the molecular clock suggests the two split 3 million years ago or more, rather than 20,000 years ago as we would expect if they speciated in the most recent ice age. It was once suggested, following the ideas of Haffer (1969), that the latest ice age was a time when many modern species pairs evolved. Haffer suggested that the fragmentation of ranges accelerated the process of allopatric speciation, creating what was called a "speciation pump" that contributed to modern biodiversity.

Haffer stimulated research, but the results of that research have not supported his ideas. Evidence from molecular clocks, for instance, suggests that the speciation events that produced many modern species are too old to fit Haffer's hypothesis, nor do speciation rates seem to go up during ice ages. However, the periods of glacial refuges may have been accelerated times of genetic divergence between populations within some species. Although the latest ice age did not produce a burst of speciation, it may have helped to finish off speciation between populations that had already diverged, or started the divergence between populations that could lead to speciation in the future.

Refuges are not only formed during ice ages. The same principle is at work, in an inverted form, in species that now have local distributions but were more widely distributed in past climatic conditions. The Nevadan deserts contain the vestiges of former large lake systems, and the desert pupfish (*Cyprinodon*) occupies some of the scattered remaining waterholes (Brown 1971). The 20 or so isolated populations of

---

Many species show similar genetic patterns in space . . .

. . . creating suture zones

Molecular dates suggest speciation preceded the ice age

Figure 17.4
(a) Distribution of the main genetic clades of hedgehogs (*Erinaceus* species) in Europe. They are currently classified into two species, but the two probably hybridize.

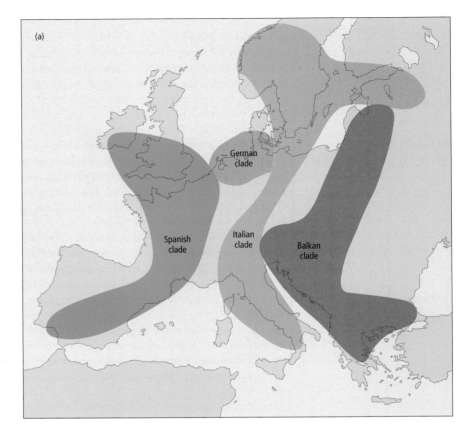

(a)

German clade

Spanish clade

Italian clade

Balkan clade

these remarkable fish have diverged into a number of (perhaps four) species, and when the next pluvial period brings water to the desert, they may expand from their inter-pluvial refuges to encounter one another in a process that is analogous to the expansion of European hedgehogs after the last ice age.

17.5 Local adaptive radiations occur on island archipelagos

Adaptive radiation means that an ancestral species evolves into a number of descendant species, each with distinct ecological adaptations. A single speciation event often occurs as two species, with different ecological adaptations, evolve from a single ancestral species (Section 14.3, p. 383). A local adaptive radiation occurs when several such speciation events occur in a local area. As we shall see in Chapter 23, adaptive radiation can be studied on a global scale, if the adaptive radiation of a taxon persists for a long enough time. But here we shall be looking at smaller scale adaptive radiations — those that are only a slight extension of the speciation process we looked at in Chapter 14.

Figure 17.4
(b) The three major clades originated by migration from ice age refugia in Spain, Italy, and the Balkans (the relation of the fourth, German, clade to these three is uncertain). The arrows indicate postglacial migration. (c) Many species have migrated from similar glacial refugia, and form hybrid zones where the different clades meet in rather similar areas of Europe. Areas where several species form hybrid zones, such as in central Europe, are called suture zones. Slightly modified, by permission of the publisher, from Hewitt (1999).

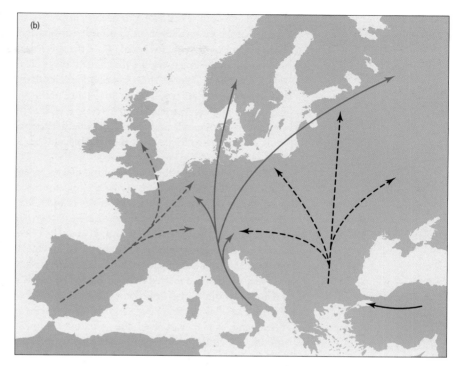

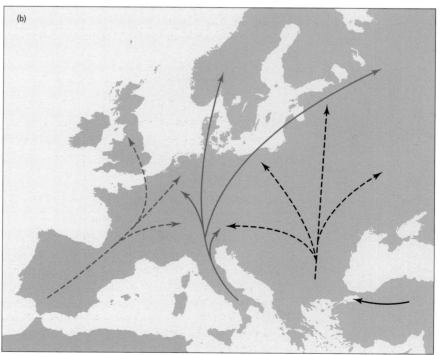

Lizards have radiated in the Caribbean . . .

The lizards of the genus *Anolis* in the Caribbean islands, are a well studied example. Species of *Anolis* have evolved to occupy various ecological niches, and have adaptations appropriate to their ways of life. Some species live on twigs, others in the canopy, and others in the grass. Species that live on twigs have long tails and short legs; species that live in the grass have short tails; and species that live on low tree trunks have long legs. *Anolis* are found on all the major islands of the Greater Antilles, and occupy a similar range of habitats on each island. The species that live on twigs (for example) all look similar, with long tails and short legs, whether they are the species from Cuba, Haiti, Jamaica, or Puerto Rico. The other ecological types also show similarities among the islands.

We can ask whether the twig-dwelling species on any one island shares a more recent common ancestor with other twig-dwelling species on the other islands, or with the ecologically different *Anolis* on the same island. That is, did the twig-dwelling ecological type evolve only once, and spread to all the islands? Or did each ecological type evolve independently on each island? Losos *et al.* (1998) answered the question by constructing a molecular phylogeny of the species. They found that for the most part each ecological type of lizard had evolved independently on each island (Figure 17.5).

. . . with a similar range of ecological forms on each island

Thus, each island tended to be colonized by one lizard population, which then radiated into a common set of ecological types on every island. Some exceptions exist in Figure 17.5. For instance, two sets of Cuban species are found in different parts of the phylogeny, as if they evolved after separate colonizations. But for the most part, the species are clustered by island rather than ecological type. The similarity between species in characters such as tail length is homoplasious rather than homologous (Section 15.3, p. 427). The force that drives the radiation is probably ecological competition. The adaptive radiation of the Caribbean *Anolis* lizards would then be a miniature example of Darwin's "principle of divergence" (Section 16.8, p. 487).

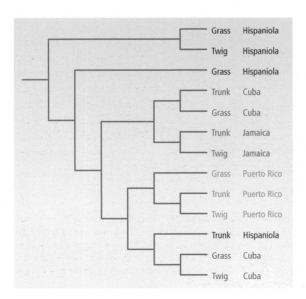

Figure 17.5
Phylogenetic relations of different ecological types of species of lizards (*Anolis*) on four Caribbean islands. The full results are for six ecological types, of which only three are shown here (grass, twig, and trunk). The other three, however, illustrate much the same pattern. Each ecological type tends to have evolved independently on different islands: the phylogeny groups species by island more than by ecological type. The phylogeny was produced by parsimony, using mitochondrial DNA sequences from the lizards. Modified from Losos *et al.* (1998).

Fruitflies and lizards show a
contrasting pattern

The fruitflies of the Hawaiian archipelago provide an instructive contrast. The phylogeny of this group is known from chromosomal inversions (Figure 15.27, p. 465). The phylogeny in Figure 15.27 is drawn on a map of the Hawaii archipelago. Inspection of it shows that, although many of the species evolved from ancestors on the same island, a large number of speciation events occurred following dispersal between islands. For example, at the bottom right of Figure 15.27, species 99 on Maui appears to have given rise to species 100 and 101 on Hawaii.[1] The reason for the difference between the Hawaiian fruitflies and the Caribbean lizards is unclear. It could be because the Hawaiian fruitflies on each island are not a regular set of ecological types, unlike the lizards, and did not evolve because of ecological competition.

Younger species inhabit younger
islands

A further hypothesis can be tested with the Hawaiian frutiflies, because in this case we also know something about time. The islands were geologically formed successively, perhaps as the tectonic plate moved from east to west over a volcanic "hot spot" that threw up one island after another. The oldest islands are in the west; the youngest island is Hawaii in the east. The fruitflies may have radiated as colonizing fruitflies gave rise to new species after the younger islands emerged from the ocean waves. If so, we can predict a tendency for descendant species to be on younger islands, and ancestral species on older islands. Figure 17.6 shows the inferred colonization events, which mainly conform to the prediction. The tarweeds, a group of plants, show the same pattern. The pattern has also been shown elsewhere. On the Galápagos Islands, for instance, the youngest species tend to be on the youngest islands, and to have evolved from ancestors on the older islands.

17.6 Species of large geographic areas tend to be more closely related to other local species than to ecologically similar species elsewhere in the globe

In the Caribbean *Anolis* lizards, a twig-dwelling species on (for instance) Cuba is more closely related to a Cuban grass-dwelling *Anolis* than to a Haitian twig-dwelling species (even though the two twig-dwelling species look more alike). A similar principle can be seen at work on a much larger geographic scale. For instance, Mediterranean-type ecosystems can be found at five places around the globe: the Mediterranean itself, California, Chile, South Africa, and Western Australia. In all five places, plants have evolved with a similar set of adaptations to local conditions. Mediterranean plants can resist drought and fire, but not frost. Many of the plants are shrubs, and are hard and spiny. Animals, too, have evolved distinct Mediterranean types.

---

[1] The phylogenetic test in Figure 17.5 can also be compared with the test of sympatric speciation in African lake cichlids (Figure 14.4, p. 414). In the Caribbean lizards, as in the African cichlids, the closest relatives of a species are usually found on the same island (analogous to a lake). However, Figure 17.5 is probably weaker evidence for sympatric speciation than is Figure 14.4. Lizards may be more likely to form local populations within an island, and speciate allopatrically, whereas the fish roam more widely through the lakes — but it is also possible that the cichlids speciate by microallopatric speciation.

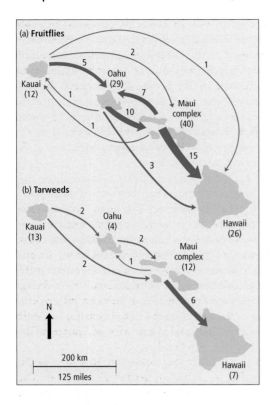

Figure 17.6
The dispersal events suggested by the phylogeny of the Hawaiian picture-wing group of fruitflies (*Drosophila*). The phylogeny is shown in Figure 15.27 (p. 465), but the numbers used here are not exactly those implied by the previous diagram because this figure is more recent. The numbers in the arrows are the inferred number of dispersal events; parenthetic numbers by the island names are the number of endemic species living on that island. (b) A comparable figure for the tarweed plant. The geological history of the archipelago, in which the islands have been successively formed toward the east, has imposed the same biogeographic histories on the two groups. Redrawn, by permission of the publisher, from Carr *et al.* (1989).

Mediterranean-type ecosystems have evolved convergently

The similarity of the plants in the five Mediterranean regions is due to convergent evolution. The shrubs of the Mediterranean itself are unrelated to the shrubs of California or Chile. The plants in the European Mediterranean are related to other European species; they have evolved from local ancestors. It is not the case that a Mediterranean set of species have evolved once and spread to all five regions.

The Mediterranean ecosystems illustrate a general point that Darwin discussed in *On the Origin of Species* (1859), and used as evidence for evolution. The species within any large geographic area tend to be more closely related to each other than to ecologically more similar species elsewhere in the globe. In Australia and South America (particularly before the Great American Interchange, which we shall look at below in Section 17.9), the main mammals have been marsupials. Both marsupials and eutherian mammal groups have evolved a saber-toothed "tiger," for example (Figure 15.4, p. 429), but the eutherian is a real cat (in the taxonomic sense) whereas the South American equivalent was a marsupial. The pattern makes sense, Darwin argued, if species evolved from other species in the same general area. That is, new mammal species in Australia were more often descended from other Australian mammals than from, say, North American mammals. If a species such as the saber-toothed tiger had been specially created, we might expect it to be much the same everywhere. There is

Biogeographic evidence supports evolution

no reason why the saber-tooth should be created in Australia with arbitrary similarities to marsupials but in North America with arbitrary similarities to eutherians. (The argument can be recognized as a geographic special case of the general argument for evolution from homology — Section 3.8, p. 55.) We could now add results such as those for the Caribbean lizards and Hawaiian fruitflies to strengthen Darwin's case.

17.7 Geographic distributions are influenced by vicariance events, some of which are caused by plate tectonic movements

A second factor influencing geographic distributions is *plate tectonics* (informally known as continental drift). The continents have moved over the surface of the globe through geological time. The positions of the main continents since the Permian have been reconstructed in some detail (Figure 17.7), and these maps immediately suggest the reason for many biogeographic observations. For example, when we looked at the faunal regions of the world (see Figure 17.2), we saw the difference between the faunas of the northern and southern Indonesian Islands known as Wallace's line. It turns out, as can be more or less seen in Figure 17.7, that the two regions have separate tectonic histories and have only recently come into close contact. The patterns of faunal similarity are therefore what we should expect, given plate tectonics.

Plate tectonics causes vicariance events

Let us look at one of the main modern research programs that studies the relation between biogeography and plate tectonics. It is called *vicariance biogeography*. The drifting apart of tectonic plates is the sort of event that could cause speciation (Section 14.2, p. 382). If the splitting of the land and of the species on it coincide, it results in two or more species occupying complementary parts of a formerly continuous area that was occupied by their common ancestor. This is an example of a *vicariance event*. (Vicariance means a splitting in the range of a taxon.) In theory, tectonic movements are just one process that could split a species' range; others could include mountain building or the formation of a river. According to the theory of vicariance biogeography, the distributions of taxonomic groups are determined by splits (or vicariance events) in the ranges of ancestral species.

Vicariance and dispersal are two biogeographic processes

We can contrast this idea with another, that distributions are determined more by dispersal. Before plate tectonics was known about, or at any rate accepted, the main process believed to alter biogeographic distributions was dispersal. Taxonomic groups were thought to originate in one confined area, called the center of origin, and then descendant populations dispersed away from it. Thus the geographic history of a group could have been either a series of splits within formerly larger ancestral ranges, or a series of dispersal events, or some mixture of the two (Figure 17.8). The events hypothesized by the dispersal and vicariance theories took place in the past, but they are not beyond study. Vicariance biogeographers test their idea by two methods.

One is to see whether the pattern of splitting in one group matches the geological history of the region where it lives. The first major piece of vicariance biogeographic research, by Brundin in 1966, was of this kind. He studied the Antarctic chironomid

Figure 17.7

Plate tectonics (or, informally, continental drift). (a) The movements of the continents during the past 200 million years. (b) The positions of the main tectonic plates today.

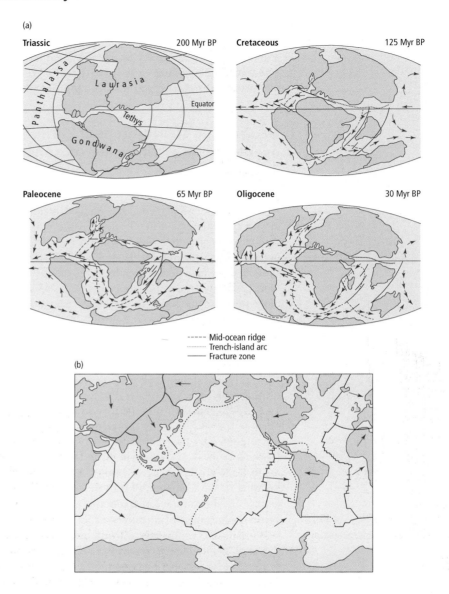

midges. These midges are distributed around the southern hemisphere (Figure 17.9). Brundin reconstructed their phylogeny by standard morphological techniques (Chapter 15) and then used the species' modern biogeographic distributions to draw a combined picture of their phylogeny and biogeography, called an *area cladogram* (Figure 17.10). If the successive splits in the phylogeny were driven by successive break-ups of the land, the phylogeny would imply a definable sequence of tectonic events.

To begin with, the common ancestor of the modern forms would have occupied a large area made up of all their modern distributional zones — which implies the

Vicariance biogeography predicts
. . .

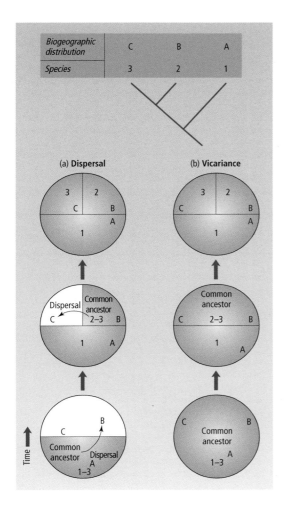

Figure 17.8
Dispersal and range splitting can be alternative hypotheses to explain the biogeography of a group. (a) An ancestral species with its center of dispersal in area A dispersed first to area B and a descendant from there then dispersed to area C. (b) An ancestor occupying area A + B + C had its range split first into A and B + C and then the descendant in B + C had its range split.

. . . the tectonic history of the Earth will match the phylogeny of species

existence, sometime in the past, of a southern supercontinent, Gondwanaland. Gondwanaland would then, Brundin's analysis predicts, have split in the following order. First, South Africa split from a combination of Australia, New Zealand, and South America; then New Zealand split from South America and Australia; and finally Australia split from South America. This prediction can be tested against the geological evidence, which was only accumulating during and after Brundin's work. The geology turned out to fit Brundin's prediction (see Figure 17.7a, but more detailed maps are needed for a strong test). It could also be tested by the molecular clock, but this is yet to be done.

Brundin's test concerns a single taxon. A second test is to compare the relation between phylogeny and biogeography in many taxa. As continents — or, in general, the habitats occupied by species — move in a particular pattern through time, all the groups of living things living in an area will be affected in a similar manner. If members of each group tend to speciate when their ranges are fragmented, they should all show similar

Figure 17.9
The biogeography of chironomid midges in the southern hemisphere. Reprinted, by permission of the publisher, from Brundin (1988).

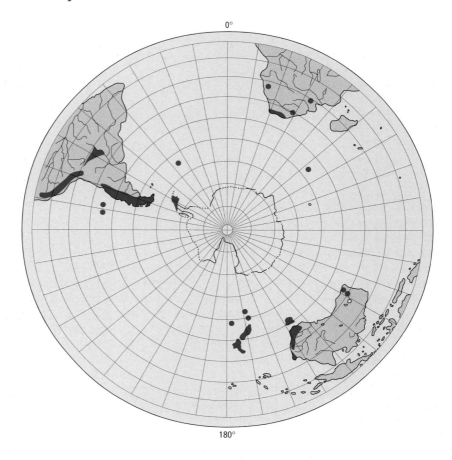

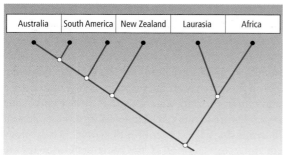

Figure 17.10
An area cladogram of chironomid midges. The diagram shows the phylogenetic relations between the midges from different areas: the midges of Australia, for example, are phylogenetically more closely related to those of South America than those of South Africa. (To see the location of Laurasia, see Figure 17.7.) Redrawn, by permission of the publisher, from Brundin (1988).

relations between phylogeny and biogeography: their area cladograms should match. This prediction can be tested (Figure 17.11).

Figure 17.11a shows the phylogeny and biogeography of three species in a hypothetical taxon 1. Now we can look at another taxon inhabiting the same region and see whether its area cladogram is the same. In technical language, vicariance biogeography

Figure 17.11
Testing vicariance biogeography by comparing the area cladograms of four taxa. (a) The phylogeny and biogeography of four species. The species are symbolized by numbers (1, 2, 3, 4) and the places where the species live by letters (A, B, C). (b) Inferred vicariant history of the distributions. (c) Taxa 2–3 have distributions that are congruent with taxon 1, but taxon 4 is incongruent. (d) Either there were dispersal events in the history of taxon 4, or its phylogeny is wrong. Species 15 may have been wrongly classified, for example, because it has evolved rapidly (the group of species 12–14 in (c) is then an example of a paraphyletic group). The suggested history with migration is only one of a number of possibilities that are compatible with a range split in the order A + B + C → A + B/C → A/B/C.

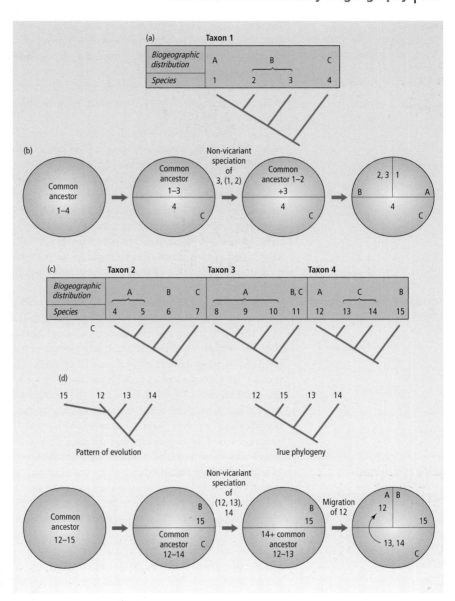

predicts their area cladograms will be *congruent*. Congruence is a term that can be applied to any sort of branching diagram (in phylogeny or biogeography). If two branching diagrams are congruent, the order of branching in the two do not contradict each other. The two diagrams do not have to be identical, because one place, or taxon, might be missing from one of the branching diagrams; but the order of branching in the entities that are present in both must be the same. In the figure, the area cladograms for taxa 2 and 3 are congruent, and that of taxon 4 is incongruent, with taxon 1. If the land area A + B + C had first split into A + B and C and then into A and B, the phylogeny of taxa 1–3 would all fit with it and the phylogeny can be understood as a series of vicariant events. Taxon 4 does not fit. If its common ancestor occupied the whole area, its first split suggests that the land first divided not into A + B and C, but into A + C and B. The congruence of taxa 1–3 conform with the ideas of vicariance biogeography, but taxon 4 does not.

Before these methods were developed — indeed before plate tectonics was widely accepted — the Venezuelan biogeographer Léon Croizat had established that different taxonomic groups often show correlated distributions. Croizat called them "generalized tracks," the distribution of any one species being its "track." He argued that if different species independently dispersed from centers of origin, they would not end up with correlated distributions. Correlated distributions are more likely to result from common vicariance events, such as plate tectonics, that split the ranges of several taxa in the same way. Modern vicariance biogeography adds to Croizat's ideas in two ways. One is that we now know more details about plate tectonics. The other is the importance of using a realistic phylogeny when testing whether different taxa have congruent distributions.

The test of vicariance biogeography requires cladistic classification

The analyses in Figures 17.10 and 17.11 are only possible for taxa that are monophyletic in the cladistic sense (Figure 16.4, p. 480). If a set of phylogenetic groups have been classified into a mixture of mono-, para-, and polyphyletic groups, then even if they have experienced the same sequence of range subdivisions, their area cladograms need not be congruent. Look at Figure 17.11 again. The area cladograms of taxon 4 and taxon 1 are incongruent. If taxon 4 has the phylogeny of Figure 17.11b — that is, if the groups there are monophyletic — then the incongruency between taxa 1 and 4 implies there must have been some dispersal events in the past (Figure 17.11d). But if the classification of taxon 4 was paraphyletic or polyphyletic, the theory of vicariance biogeography no longer predicts that the area cladograms of the taxa will be congruent. There is no reason to expect different paraphyletic or polyphyletic groups to have congruent biogeographic patterns with one another, or with monophyletic ones. It is therefore essential for vicariance biogeography that taxa are classified cladistically, to reflect the order of phylogenetic branching. If the classifications contain a mixture of phenetic and cladistic taxa, any general biogeographic study is liable to become meaningless.

Evidence from marsupials illustrates vicariance . . .

Let us now turn to an example from part of a larger study by Patterson (1981). His starting point was a probable area cladogram for the marsupials (Figure 17.12a). Recent marsupials live in Australia and New Guinea, and South and North America (where they are represented by the opossum *Didelphis*). Fossil marsupials can also be found in Europe, making five areas in the complete area cladogram for marsupials.

Now, marsupials have evolved on the same globe as all other species. If the modern distributions of vertebrates result from a history of range splitting, they should all share much the same geological history and their area cladograms should therefore all be more or less congruent. What do the area cladograms for the other vertebrates look like? Figure 17.12b reveals, for five other vertebrate groups, that the vicariance prediction is upheld: their area cladograms are congruent. The result could in theory be because all the taxa have dispersed in the same order and the same direction, but that

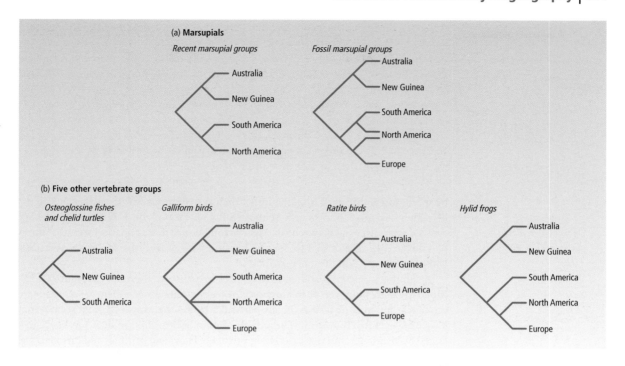

Figure 17.12
(a) Area cladograms of recent and fossil marsupials. (b) Area cladograms of five other taxa with congruent biogeographic distributions. Redrawn, by permission of the publisher, from Patterson (1981).

would require an unlikely series of coincidences. It is more likely that the common pattern is simply due to a shared history of range splits by tectonic events.

. . . other taxa illustrate dispersal and vicariance

Dispersal has probably had some influence on the history of the taxa in Patterson's study. The osteoglossine fish are found in South East Asia as well as Australia, New Guinea, and South America (Figure 17.12b). As it happens, none of the other four taxa are represented in South East Asia. Three explanations for this result are possible. One is that all six taxa used to live in South East Asia and that five of them have since gone extinct there. A second is that all six were originally absent from Asia and osteoglossine fish (in the form of *Scleropages*) arrived there by dispersal. The fossil record could in principle be used to show that a taxon once lived in Asia but is now extinct. But in the absence of any such evidence, Patterson reasoned that it is more likely that only one group (the osteoglossines) dispersed to Asia than that five groups went extinct there. Finally, it could be that osteoglossines originally had a broader distribution than the other five vertebrate groups, and were ancestrally present in South East Asia. The vicariance of the osteoglossines would then have taken place within a larger range. Vicariance biogeography has been successful in finding a number of area cladograms that are mainly consistent between different taxa and also consistent with tectonic history.

Some biogeographic distributions make sense in a history of range splitting (or vicariance). Others do not. We saw above how *Anolis* lizards in the Caribbean evolved

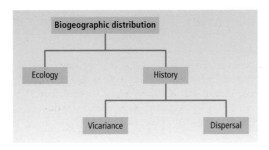

Figure 17.13
The relationship between different explanatory dichotomies.

by speciation within each island, but that their distributions within each island do not fit with vicariance. Also, many of the frutifly species in the Hawaiian archipelago evolved after dispersal between islands. We can be sure that speciation was not simply by splits in the range of a larger species, because the younger islands did not even exist while the flies were on the older islands.

In summary, dispersal and vicariance form two historic alternatives (Figure 17.13). Much the same point can be made about the distinction between them as was made for ecological and historic factors. In any particular case, either dispersal or vicariance may have been exclusively at work. The area cladogram of Brundin's midges was likely generated by vicariance, but the area cladogram of the Hawaiian fruitflies and tarweeds was likely generated by dispersal among an emerging archipelago of volcanic islands. The two processes can also operate together. The challenge is to work out the relative contributions of the two.

17.8 The Great American Interchange

The processes of plate tectonics and dispersal have both contributed to the events that take place when two previously separate faunas come into contact. These events are called biotic interchanges, and several are known from the history of life. The most famous is the Great American Interchange. Its deep geological cause is probably connected with the tectonic processes that have been raising the Andean mountains for the past 15 million years or so. The rate of this mountain building has varied from time to time, but during a period between 4.5 and 2.5 million years ago it intensified. At the same time — maybe 3 million years ago — the modern Isthmus of Panama rose out of the sea and the South and North American continents were reconnected. The connection had dramatic repercussions for the fauna, most noticeably the mammalian fauna, of the southern continent.

North and South America had been connected before, over 50 million years earlier. They may have had similar mammalian inhabitants, but the Cretaceous mammals of South American are too poorly known to be sure. Then, likely in the late Paleocene, the two halves of the American continent drifted apart. At that time, the modern orders of mammals — the groups such as horses, dogs, and cats that are still the dominant land vertebrates — evolved in North America, Africa, and Europe; however, South America

The formation of the Isthmus of Panama . . .

. . . led to an encounter between North . . .

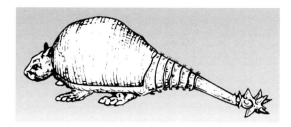

Figure 17.14
A reconstruction of *Doedicurus*, a Pleistocene glyptodont. The glyptodonts were a strange group of armored South American mammals related to armadillos. Reprinted, by permission of the publisher, from Simpson (1980). © 1983 Scientific American Books.

shows no sign of possessing these forms. It instead evolved its own distinctive mammalian fauna.

The South American mammals of the Paleocene and Eocene fall into three groups: marsupials, xenarthrans (armadillos, sloths, anteaters), and ungulates. Armadillos, tree sloths, and opossums still survive in South American forests, but they formerly lived along with many other curious, and now extinct, forms. There were marsupial saber-tooth carnivores (Figure 15.4, p. 429), ground sloths (the group from which the giant ground sloth *Megatherium* of the Pleistocene evolved), and the most heavily armored mammals that ever lived — the glyptodonts (Figure 17.14), which were first described from Darwin's collections made during the *Beagle* voyage.

New arrivals came in from the outside, on rare occasions, from the early Oligocene on. They probably immigrated by waif dispersal, hopping from island to island before there was a continuous land bridge between the continents. Rodents are a major group which first appeared in the Oligocene. It is so uncertain where they came from that experts still dispute whether the South American rodents are more closely related to African or North American species (though the latter is the more widely favored source). The South American rodents, like the other mammalian groups in that land, also in turn evolved peculiar South American forms, including one called *Telicomys gigantissimus* (in the Pleistocene) that is the biggest rodent ever to have lived and was almost as large as a rhinoceros.

In the late Miocene, about 8–9 million years ago, further small additions to the fauna arrived. These were the procyonids (racoons and allies) who came from North America, and the cricetid rodents. These too almost certainly entered by waif dispersal. It is possible that North and South America had tectonically wandered closer together at that time, but the connection can have been neither close nor lasting, because it was another 6 million years before the South America mammal fauna encountered the full range of the outside world's mammalian types.

Then, about 3 million years ago, the Bolivar trough finally disappeared and the modern Panamanian land bridge formed. The vegetation on both sides of the bridge was probably savannah, not the tropical rainforest of modern times. Mammals adapted to the similar vegetation of the two sides could move freely both ways, and it was now that the mustelids (skunks), canids (dogs), felids (cats), equids (horses), ursids (bears), and camels invaded South America from the north, while the dasypodids (armadillos), didelphids (opossums), callithricids (marmosets), and edentate anteaters moved rather less dramatically in the opposite direction — in both cases accompanied by many other less well known forms. This extraordinary clash and exchange of faunas is known

. . . and South American mammals

Some groups had migrated before . . .

. . . but the main event came after 3 million years ago

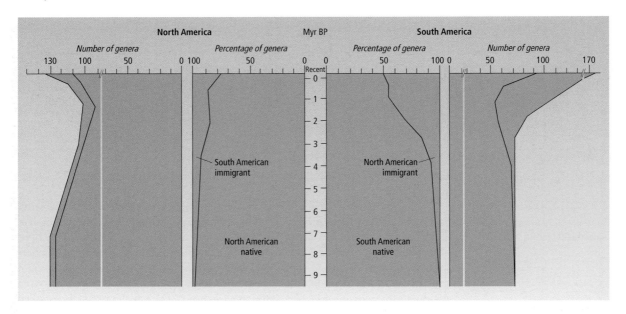

Figure 17.15
Numbers (and percentages) of genera of land mammals in the last 9 million years in North and in South America. Immigrant and native genera are distinguished in both places. Note the wave of immigration after about 3 million years ago. Redrawn, by permission of the publisher, from Marshall *et al.* (1982). © 1982 American Association of the Advancement of Science.

as the Great American Interchange, and popular biology still portrays it as a competitive rout of the South American mammals by the superior northern forms. There is some truth in that idea; but the increasing quantity of fossil evidence is allowing a more detailed reconstruction of the events.

A study by Marshall and colleagues (1982) has examined the time course of the Interchange in detail. They counted the number of mammalian genera in South and North America at successive times and divided the genera according to where they originally evolved. They then divided the immigrant genera into primary immigrants (genera that evolved in the south and immigrated to the north or vice versa) and secondary immigrants (genera descended from primary immigrants). They argued the primary invasions were roughly equal in both directions, and that the takeover of the south by northern mammals was partly a result of two other factors: weight of numbers and different rates of speciation after arrival.

Similar fractions of species moved in both directions

Figure 17.15 shows the numbers of genera, expressed both as absolute numbers and as proportions, of mammals in North and South America. On both sides, after 2.5–3 million years ago, an increasing proportion of the mammal genera were immigrants (or descendants of immigrants) from outside. At present, about 50% of South American genera are descended from originally North American mammals. The proportion of southern mammals in the north is much lower, at about 20%. The numbers become more revealing when we break them down further (Table 17.2).

We can begin by counting the total number of genera before the Interchange;

Table 17.2

Pattern of faunal exchange between North and South America in different time periods. The table gives the total numbers of genera of South or North American origin in each region (these are the numbers plotted in Figure 17.15), and breaks down the immigrant genera according to whether they were "primary" (that genus itself immigrated) or "secondary" (that genus descended from a primary immigrant genus, e.g., a secondary immigrant in North America evolved in there but came from a genus that itself evolved in South America). The total of the immigrant genera in the bottom two rows equals the number of alien genera in the "number of genera" row above. Note: (i) the similar proportions of primary immigrant genera moving in each direction, and (ii) the much greater numbers of secondary immigrants in South America than in the north. Modified from Marshall *et al.* (1982).

Time period (Myr BP):	South America						North America			
	9–5	5–3	3–2	2–1	1–0.3	0.3–Recent	9.5–4.5	4.5–2	2–0.7	0.7–Recent
Duration (Myr)	4	2	1	1	0.7	0.3	5	2.5	1.3	0.7
Number of genera										
North American	1	4	10	29	49	61	128	99	90	102
South American	72	68	62	55	58	59	3	8	11	12
Total	73	72	72	84	107	120	131	107	101	114
Number of immigrant genera										
Primary	1	1	2	10	18	20	2	6	8	9
Secondary	0	3	8	19	31	41	1	2	3	3

North America contained more species to begin with

the total number is higher in the north, perhaps because of the continent's greater area. (It is an important principle of island biogeography that a larger area supports a larger number of species.) Marshall *et al.* (1982) then counted the numbers of North American mammals moving south, and vice versa, and expressed them both as proportions of the total pool. They found that the proportions are about the same. Approximately 10% of available North American mammals invaded the south. (For instance, Table 17.2 shows that North America had about 100 endemic mammalian genera. Around 10 of them migrated south, with the number of primary immigrants going up from one to 10 between 3 and 1 million years ago.) Likewise, about 10% of South American mammals moved north. (In Table 17.2, there were 60 or so South American mammalian genera 3 million years ago. The number of South American genera in the north increased by about six between 4.5 and 1 million years ago.) The greater absolute numbers moving south is mainly due to the larger number of mammals in the north to begin with.

Patterns of speciation differed

The pattern of primary immigration is thus similar in both directions. Something like 10% of the genera from each side successfully invaded the other. But when we look at the subsequent proliferation of the immigrants, the pattern diverged markedly (Table 17.2). By the Recent period, a total of 12 (the nine in Table 17.2 is the number alive — three others had arrived and then gone extinct) immigrant southern mammal genera had produced only three new genera, while the 21 immigrant northern mammalian genera in the south produced 49 genera. In the Recent period the trend has

Table 17.3

Relative brain sizes (expressed as an encephalization quotient, EQ (Section 22.6, p. 632), which increases with increasing brain size) of North and of South American ungulates in the Cenozoic. From Jerison (1973).

Time (Myr BP)	Ungulate brain size (EQ)	
	South America	North America
65–22	0.44	0.38
	n = 9	n = 22
22–2	0.47	0.63
	n = 11	n = 13

continued. The North American mammals showed their superiority, therefore, not in the original invasion, but in their relative success afterwards.

Why did the North American mammals prove superior? The increase in number of originally North American genera can be seen across a wide range of mammal types, which suggests they had some general advantage. There are several ideas why. One is that the North American mammals had lived a more competitive life, in a larger continent with more species, than the isolated southern mammals. The "arms race" of competition had moved further in the north. The idea can be illustrated by Jerison's (1973) study of brain size (Section 22.6, p. 632).

In North American mammals, brain sizes, relative to body size, increased with time in both predators and prey in the past 65 million years. Jerison's interpretation is that brain sizes increased as predators and prey grew increasingly intelligent, in an escalating improvement of offensive and defensive behavior, and the pattern of brain evolution fits his interpretation (Figure 22.11, p. 633). However, in South American mammals, no such increase seems to have happened (Table 17.3). Arguably, then, when the North American mammals invaded the south they had been prepared by 50 million years or so of more demanding competition. They possessed advanced armaments, probably not only in intelligence, that enabled them to overrun the southern mammals.

Alternatively, as Marshall and his coauthors suggest, the North American mammals may have enjoyed some advantage in the environmental change of the past 3 million years. The Andean upthrust sheltered the Americas from the Pacific, creating a rain shadow east of the mountains. In South America, dryer pampas or even semidesert replaced the moist savannah and forest. Quite why such a change should benefit the North American mammals at the expense of the South American forms is unclear; but such a change would be likely to benefit one of the two groups more than the other. It was a large change, and was therefore probably influential in the faunal replacements of the time.

Several hypotheses exist to explain the greater proliferation of North American mammals

The Great American Interchange is one of the most dramatic case studies in historic biogeography. The mammalian faunas of North and South America have only been connected, by a narrow isthmus, for less than 3 million years. Yet 50% of the mammalian genera in the south are now of northern origin, and such wonderful animals as that rhinoceros-sized rodent, the giant ground sloth, and the saber-toothed borhyaenid were somehow involved in the general destruction of species during the Interchange. That the events of the Interchange were at least partly due to competition is very plausible, but to demonstrate it is a harder task.

17.9 Conclusion

Evolutionary biogeographers have been particularly interested in the historic processes that have shaped the geographic distributions of species — though they by no means rule out the well documented influence of modern ecology. They have mainly studied two sorts of historic process: movement and range splitting. Species undoubtedly do move by dispersal, and when a new corridor appears on Earth allowing a new encounter of faunas, it can precipitate dramatic evolutionary events. The Great American Interchange is a famous example. It is not easy to disentangle the exact causes that were at work, but the data allow a plausible inference that the faunal changes were substantially influenced by both the weight of numbers and competition.

Biogeography is one area of evolutionary biology that is particularly benefitting from the expansion of molecular phylogentic research. New molecular markers can be used to study the phylogeny of populations within a species, and of groups of related species, in relation to space. We have seen how the history of European species such as hedgehogs is written into the geographic distribution of their major clades, each of which is descended from a different ice age refuge. We also saw how the adaptive radiation of Caribbean lizards has been studied with molecular phylogenetic techniques. The combination of biogeography and phylogeny, now often called phylogeography, builds on older cladistic methods that were developed in the 1960s, 1970s, and 1980s, to study vicariance biogeography. Now, with the rise of molecular systematics, phylogenetic biogeography has flowered into a thriving and revealing research program.

Summary

1 Species, and higher taxa, have geographic distributions, and biogeographers aim to describe and explain them.

2 The similarity of the flora or fauna of two regions can be measured by indexes of similarity. The world can be divided up into six main faunal regions, based on the distributions of bird and mammal species. Other taxa, such as plants, form slightly different regional divisions.

3 The distributions of species are influenced by historic accidents of where species happened to be at certain times, and by their ecological tolerances.

4 The ranges of species may be altered by dispersal (when a species moves in space) and by plate tectonics (when movement of the land subdivides the ranges of species). The splitting of a species range is called vicariance.

5 When climates cooled in the most recent ice age, the ranges of species in the northern hemisphere moved to the south. In Europe, many species formed glacial refuges in Spain, Italy, and the Balkans. After the ice age they expanded north, resulting in a distrbution with a three-clade intraspecific phylogeny, and in suture zones, where several species form hybrid zones.

6 Some taxa have undergone local adaptive radiations on island archipelagos. The course of the radiation can be studied by molecular phylogenetic techniques.

7 The species of an area tend to be more closely related to other species in the same area than to ecologically more similar species elsewhere in the globe. Darwin used this observation to argue his case for evolution.

8 An area cladogram shows the geographic areas occupied by a group of phylogenetically related set of taxa.

9 Vicariance biogeography suggests that geographic distributions are determined mainly by splits in the ranges of ancestral species, not by dispersal. It predicts that the area cladogram of a taxon should match the geological history of the area, and the area cladograms of different taxa in an area should have compatible (congruent) area cladograms.

10 In the encounter between the North and South American faunas when the Isthmus of Panama formed 3 million years ago, similar proportions of mammals initially moved in both directions, but the immigrant North American mammals in the south proliferated at a greater rate.

Further reading

Cox & Moore (2000) is an introductory textbook, and Brown & Lomolino (1998) is more comprehensive. Avise (1999) is a text on phylogeography, and Hare (2001) looks at recent advances in phylogeography.

Simpson (1983) explains how the great faunal regions of the world were discovered, as well as the importance of movement, and how to measure faunal (and floral) similarity. Brown *et al.* (1996) review ecological ranges. For ecological influences, see an ecology text, such as as Ricklefs & Miller (2000). On niche concepts, see the entries by Griesemer and Colwell in Keller & Lloyd (1992). On Krakatau, see Thornton (1996) and the narrative in Wilson (1992). Van Oosterzee (1997) is a book about Wallace's line.

On ice age biogeography see Pielou (1991), and the general references above, which include a chapter in Cox & Moore (2000). See also Davies & Shaw (2001) on range changes. Hewitt (2000) reviews European refugia, and their modern genetic consequences. For their non-effect on modern species pairs, see also Da Silva & Patton (1998) and Klicka & Zink (1999), and Moritz *et al.*'s (2000) review. The idea that ice age refugia produced modern species diversity was classically suggested by Haffer (1969) for Amazonian birds, but skepticism prevails because of the time to common ancestors of modern species pairs (see above) and the pollen evidence suggesting the Amazonian forests did not form refugia. See Willis & Whitaker (2000) and Smith *et al.* (1997), who also draw conservation morals.

Losos (2001) describes the radiation of Caribbean lizards. Schluter (2000) contains more on ecologically powered local adaptive radiations. Losos & Schluter (2000) look at the additional topic of species–area relations on islands, and how ecological and evolutionary causes of the relationship combine. For the Darwinian argument, in which the relations between species within and between areas suggests evolution, see Darwin (1859) and Jones' (1999) update. Davis & Richardson (1995) contains more on Mediterranean ecosystems. Eldredge (1998) analyzes Darwin's argument and gives further evidence.

Molecular clocks are now increasingly used to study historic biogeography. For example, see Richardson *et al.* (2001) on the origin (in the past 8 million years or so) of one Mediterranean flora — the South African Cape — and Pellmyr *et al.* (1998) on the timing of species introductions into North America.

On vicariance biogeography see Brundin (1988), Wiley (1988), Humphries & Parenti (1999), and general texts. Sereno (1999) discusses the special case of dinosaurs. For Croizat's biogeography, see Croizat *et al.* (1974). On Hawaii generally, see Wagner & Funk (1995) and the special issue of *Trends in Ecology and Evolution* (1987), vol. 2, pp. 175–228.

Vermeij (1991) is a general study of biotic interchanges, as is an issue of *Paleobiology* (1991), vol. 17, pp. 201–324, which contains a paper by Webb on the Great American Interchange. Stehli & Webb (1985) is a book about the Great American Interchange; Jackson *et al.* (1996) contains more recent material. Simpson (1980) describes the South American mammals, and see also the chapters in Goldblatt (1993) for South American biogeography generally. On brain size difference, see Jerison (1973) and a popular essay by Gould (1977b, chapter 23). Part of another of Gould's (1983, chapter 27) popular essays is about the Interchange.

Study and review questions

1 Review the geographic terms Boreal, Nearctic, Palearctic, Holarctic, and Neotropical.

2 Calculate the indexes of similarity between areas 1 and 2:

Number of species in area 1	Number of species in area 2	Number of species common to areas 1 and 2	Index of similarity
10	15	5	
15	10	5	
10	10	5	
5	15	5	

3 Turn to the phylogeny of Figure 15.27 (p. 465). How many dispersal events does it imply from younger to older, and from older to younger, islands? (The ancestral species are numbers 1 and 2 at the top left; the oldest island is at the left, the youngest at the right. The species can be treated as four "columns" inhabiting the islands of Kauai, Oahu, Maui, and Hawaii, respectively: the small deviations to the right and left on the page are biogeographically insignificant.) You could draw the dispersal events on the map at the bottom. What relevance does the answer have for vicariance and dispersalist theories of area cladograms?

4 Using the same phylogeny (Figure 15.27), draw an area cladogram in the form of Figure 17.10 for fruitfly species 1–15.

5 Here are the geographic areas occupied by the species of two taxa:

Area	A	B	C	D
Species in taxon 1	1	2	3	4
Species in taxon 2	5	6	7	8

Here are three (rooted) phylogenies for the two taxa. Which pairs of area cladograms are congruent, and which not?

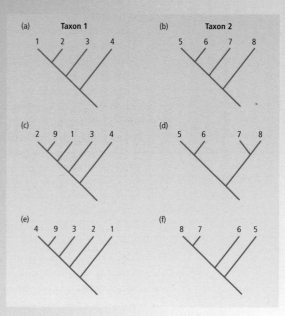

6 What are the main hypotheses to explain the proliferation of North American mammals in South America after the formation of the Isthmus of Panama?

Part five

Macroevolution

Part 5 is about macroevolution. Macroevolutionary changes are large: the kinds of events that can be studied in the fossil record, such as the origin of new organs, or body plans, or of new higher taxa (that is, taxa above the species level). These large-scale changes can be distinguished from "microevolution," which refers to changes in gene frequencies within a population. The conventional dividing line between macro- and microevolution is at speciation, so that events below that level are microevolution and those above it are macroevolution.

As said in the Preface to this book, the distinction between micro- and macroevolution has traditionally been not only between the timescales of the events but also between the methods used. Microevolution has been studied with genetic techniques, and has used observation and experiment on the timescale of human lifetimes. Macroevolution has been studied with fossil evidence, comparative morphology, and phylogenetic inference. However, modern biology has seen a breaking down of the methodological distinction as genetic techniques are being used to study large-scale, macroevolutionary questions. It is always interesting when two completely independent methods can be used to study the same question. We shall see a series of such cases in Part 5, as molecular and fossil evidence have been used to study the time of evolutionary events and the significance of mass extinctions.

Chapter 18 is a short history of life, from the origin of life to the origin of modern humans. The chapter begins with an introduction to paleontology (paleontology is the science that studies fossils). The history of life will lead us into an abstract question: is macroevolution really microevolution extrapolated over a long timescale or does macroevolution take place by different, though not incompatible, mechanisms from microevolution? This general question will recur at several points in the chapters of Part 5. Chapters 19 and 20 are about two emerging subdisciplines of evolutionary biology: evolutionary genomics and "evo-devo." Evolutionary genomics uses sequence data from whole or partial genomes to reconstruct the evolution of genomes. In a way, it is the DNA equivalent of the morphological history we looked at in Chapter 18. Evo-devo is concerned with how developmental processes change in evolution, and can cause changes in morphology.

In Chapter 21, we move on to the study of evolutionary rates. We see how they are measured and consider one controversy — about the relative rates of evolution during, and between, speciation events — in detail. Chapter 22 is about coevolution, in which the evolution of one species is directed by evolutionary changes in the other species that make up its environment.

The final chapter (Chapter 23) discusses the history of biological diversity — the number of forms of life on Earth. Diversity can be measured by the number of species, and this is controled by the relative rates of speciation and of extinction. One special topic is the importance of mass extinctions. We look at such questions as whether mass extinctions are real or artifactual, and whether they have shaped, or had little influence on, the history of biological diversity.

18 The History of Life

This chapter gives a brief history of life, from its origin to the present. We have two main sources of evidence for historic reconstruction: the molecular clock, which we looked at in Chapters 7 and 15, and the fossil record. We begin here by looking at how fossils are formed and how their age is estimated. This section of the chapter has, for paleontology, a similar purpose as did Chapter 2 for genetics. We then move on to the historic narrative. The chapter looks at the origin of life, of cells, and of multicellularity, at the Cambrian explosion, the colonization of land by plants and animals, the evolution of mammals from reptilian ancestors, and human evolution. These historic events are all examples of evolution on the large scale, or macroevolution. The chapter ends with a conceptual section on the possible relations between micro- and macroevolution, using the historic case studies as examples.

18.1 Fossils are remains of organisms from the past and are preserved in sedimentary rocks

A fossil is any trace of past life. The most obvious fossils are body parts, such as shells, bones, and teeth; but fossils also include remains of the activity of living things, such as burrows or footprints (called trace fossils), and of the organic chemicals they form (chemical fossils). For any organism to leave a fossil requires a series of events, each of which is unlikely. We can consider these events for the case of hard parts, though analogous points also apply for trace and chemical fossils.

Fossilization of soft parts, . . .

When an organism dies, its soft parts are usually either eaten by scavengers or decay by microbial action. For this reason, organisms that consist mainly of soft parts (such as worms and plants) are less likely to leave fossils than are organisms that have hard parts. Some fossils of soft parts do exist, but they were either deposited in exceptional circumstances or preserve exceptionally abundant life forms. Fossil plants often take the form of "compression fossils" in which the soft parts of a plant have been squeezed flat. Coal, for instance, contains huge numbers of compressed fossil ferns. However, even in plants, the majority of fossils are of hard parts, such as resistant spores or seeds.

. . . and even hard parts, is rare

Although an organism's hard parts stand the best chance of fossilization, even they are usually destroyed rather than fossilized. Hard parts may be crushed by rocks, stones, or wave action, or broken up by scavengers. If the hard parts survive, the next stage in fossilization is for them to be buried in sediment at the bottom of a water column — only sedimentary rocks contain fossils. (Geologists distinguish three main rock types: igneous rocks, often formed by volcanic action; sedimentary rocks, formed from sediments; and metamorphic rocks, formed deep in the Earth's crust by the metamorphosis of other rock types — when sedimentary rocks undergo metamorphosis, any fossils are lost.)

Animals that normally live within sediments are more likely to be buried in sediment before being destroyed. These animals are therefore more likely to leave fossils than are species that live elsewhere. Likewise, species that live on the surface of the sediment (i.e., on the sea bottom) are more likely to be fossilized than are species that swim in the water column. Terrestrial species are least likely of all to be fossilized. The further a

Fossils are preserved in sediments

species lives from sediments, the less likely it is to be fossilized. For most of the delicate kinds of animals that live on the sea bottom, such as feather stars and worms, practically the only way they may come to leave fossils is by "catastrophic" burial, such as a slide of sediment from shallower water into the depths that carries with it and buries some soft-skeletoned animals. Feather stars, for instance, are known to decay into nothing within 48 hours of death on the sea bottom; they therefore have to be buried rapidly to have any chance of fossilization.

Once an organism's remains have been buried in the sediment, they can potentially remain there for an indefinitely long period of time. As new sediment piles on top of older sediment, the lower sediments are compacted — the water is squeezed out and the sedimentary particles are forced closer together. The fossil hard parts may be destroyed or deformed in the process. As the sediments compact, they are gradually turned into sedimentary rock. They may subsequently be moved up, down, or around the globe by tectonic movement, and can be re-exposed in a terrestrial area. Any fossils they contain can then be picked up, or dug up, on land (a fossil is, etymologically, anything that is dug up). Sediments may also be lost by tectonic subduction and geological metamorphosis.

Any particular sedimentary rock will be made up of sediments that were deposited at a certain time, or through a certain range of times, in the geological past. Any fossils in it will be from organisms that lived at the time the sediments were deposited. It is possible to draw a geological map of an area showing the ages of the rocks that are either exposed at the surface or are near the surface but concealed beneath the topsoil. Plate 10 (between pp. 68 and 69) is a geological map of North America. Maps of this kind, at varying levels of detail, have been produced for many areas of the globe. A geological map is a first guide to where it may be possible to find fossils of particular ages. Dinosaurs, for example, lived in the Mesozoic. We can read directly from the geological map of the USA that the orange and red regions, for instance in Texas and New Mexico, and up through Colorado, Wyoming, and Montana, are appropriate regions to hunt for fossil dinosaurs. Abundant dinosaur remains can indeed be found at some sites in those regions. The pattern of rock types on a geological map can be understood in terms of the theory of plate tectonics.

Over geological time, the original hard parts of an organism will be transformed while lying in the sedimentary rock. Minerals from the surrounding rock slowly impregnate the bones or shell of the fossil, changing its chemical composition. Calcareous skeletons also change chemically. There are two forms of carbonate: aragonite and calcite. Aragonite is less stable and becomes rare in older fossils, and calcite may be replaced in some fossils by silica or pyrite. In extreme cases, the calcite may be dissolved away completely, and the space filled in by other material: the fossil then acts as a mold, or cast, for the new material. The remains still reveal the shape of the organism's hard parts.

Fossilization is an improbable eventuality. It is more likely for some kinds of species than others, and for some parts of an organism than others. After burial in sediment, the fossils slowly transform through time, but the transformed remains, if they are preserved, can still tell us (after expert interpretation) much about the original living form.

Geological maps show the locations of rocks of various ages

18.2 Geological time is conventionally divided into a series of eras, periods, and epochs

18.2.1 *Successive geological ages were first recognized by characteristic fossil faunas*

Figure 18.1 shows the main time divisions of the geological history of Earth during the past 550 million years. Earlier divisions are recognized too, but the past 550 million years are paleontologically the most important because fossils are much less common before this time.

The time divisions in Figure 18.1 were recognized by nineteenth century geologists on the basis of characteristic fossil faunas. The times of transition between two eras are times of transition between different characteristic fossil faunas: the fossils of the

Era	Period	Epoch	Myr BP (approx)
Cenozoic	Quaternary	Recent	0.01
		Pleistocene	1.8
	Neogene / Tertiary	Pliocene	5.3
		Miocene	24
	Paleogene	Oligocene	34
		Eocene	55
		Paleocene	65
Mesozoic	Cretaceous		144
	Jurassic		206
	Triassic		251
Paleozoic	Permian		290
	Carboniferous	Pennsylvanian	323
		Mississippian	354
	Devonian		417
	Silurian		443
	Ordovician		490
	Cambrian		543

Figure 18.1
The geological timescale.

Permian, for instance, characteristically differ from those of the Triassic and at the Permo-Triassic boundary there is a relatively sudden transition between the fossil faunas. In the nineteenth century it was not known whether the transition times corresponded to mass extinctions and sudden replacements, or to long gaps in the fossil record while a slower replacement was proceeding. It is now known they were mass extinctions in short periods. To demonstrate this, techniques to establish absolute times were necessary. For the smaller divisions of geological time, such as epochs, there is no exact agreement on the absolute dates and more than one geological timescale exists.

18.2.2 Geological time is measured in both absolute and relative terms

Geologists date events in the past both by relative and absolute techniques. An absolute time is a date expressed in years (or millions of years); whereas a relative time is a time relative to some other known event. The times on Figure 18.1 are absolute times and were established from the radioactive decay of elements. The exact method is explained in Box 18.1.

Box 18.1
Radioactive Decay and the Dates of Geological History

Radioisotopes of chemical elements decay over time. For example, the isotope of rubidium ^{87}Rb decays into an isotope of strontium, ^{87}Sr. The decay is very slow and has a half-life of about 48.6 billion years; that is, half of an initial sample of ^{87}Rb will have decayed into ^{87}Sr in 48.6 billion years (about 10 times the age of Earth).

Radioactive decay proceeds at an exponentially constant rate. Exponential decay means that a constant proportion of the initial material decays in each time unit. For example, suppose we start with 10 units and one-tenth of them decay per time interval; in the first time interval 1 unit will decay, and we shall have 9 units left. In the second time interval, a proportion equal to one-tenth of the remaining nine units (i.e., 0.9 units) will decay; and we shall be left with 8.1 units. In the third time interval, a further tenth of the 8.1 units will decay, leaving 7.29 units (8.1 − 0.81) at the beginning of the fourth time interval, and so on. In radioactive decay, the proportion of the isotope that decays each year is called the decay constant (l), and for $^{87}Rb/^{87}Sr$ the decay constant is 1.42×10^{-11} per year. Therefore, whatever the amount of ^{87}Rb that is present at any time, a proportion equal to 1.42×10^{-11} of it will decay into ^{87}Sr in the next year.

To estimate the age of a rock by the radioisotope technique, we need to be able to make two measurements and validate one assumption. The two measurements are the isotope composition of the rock now and when it was formed. The proportions of ^{87}Rb and ^{87}Sr are obviously measureable now. The composition of the rock was originally fixed when it crystallized as an igneous rock from liquid magma, and the ratios of ^{87}Rb and ^{87}Sr in modern magma can be measured: the ratio is a good estimate of the

isotope ratio when the rock first formed. The isotope ratio will slowly change from the original ratio as ^{87}Rb radioactively decays into ^{87}Sr. In order to estimate the age of the rock from the change in the isotope ratio, we must assume that all the change in the ratio is due to radioactive decay. For the case of $^{87}Rb/^{87}Sr$, the assumption is probably valid. Neither isotope seeps into, or leaks out of, the rock, and the ratios are therefore solely determined by time and radioactive decay. For some other radioisotopes, the assumption is less well met. Uranium, for example, can be oxidized into a mobile form and move among rocks (though the problem in this case can be dealt with by combining two uranium decay schemes, such that the time is inferred from the ratio of two lead isotopes and the concentration of uranium does not matter).

Table B18.1 lists the main radioisotopes used in geochronology. The decay of ^{40}K, for example, is a geochronologically useful decay scheme. When a volcano erupts the heat volatalizes all the ^{40}Ar out of the volcanic lava and ash, but not the ^{40}K. When the volcanic dust cools, therefore, it contains (of ^{40}K and ^{40}Ar) only ^{40}K. The ^{40}K then decays into both ^{40}Ca and ^{40}Ar. In practice, there is so much ^{40}Ca in the rock from other sources that it is not convenient to use it for dating purposes, but all the ^{40}Ar in the rock will have been produced by the decay of ^{40}K. The decay is so slow that it is not practical to use it for rocks less than about 100,000 years old; but there are other radioisotopes for shorter times. The decay of ^{14}C into nitrogen, for example, has a half-life of only 5,730 years.

The exact age of a rock is calculated as follows. Take $^{87}Rb/^{87}Sr$ as an example. Let N_0 be the number of ^{87}Rb atoms in the sample of rock when it was formed, and N be the number today. Then

Table B18.1
Radioactive decay systems used in geochronology.

Radioactive isotope	Decay constant ($\times 10^{-11}$/year)	Half-life (years)	Radiogenic isotope
^{14}C	1.2×10^7	5.73×10^3	^{14}N
^{40}K	$5.81 + 47.2$	1.3×10^9	$^{40}Ar + ^{40}Ca*$
^{87}Rb	1.42	4.86×10^{10}	^{87}Sr
^{147}Sm	0.654	1.06×10^{11}	^{143}Nd
^{232}Th	4.95	1.39×10^{10}	^{208}Pb
^{235}U	98.485	7×10^8	^{207}Pb
^{238}U	15.5125	4.4×10^9	^{206}Pb

* ^{40}K decays into both ^{40}Ar and ^{40}Ca, with the two decay constants given; the half-life is for the sum of the two.

$N = N_0 e^{-\lambda t}$, where λ is the decay constant and t is the age of the rock. Take logs and $t = (1/\lambda)\ln N_0/N$. It is practically easier to measure the quantity of the isotope generated by decay. Therefore, let N_R equal the number of ^{87}Sr atoms generated by radioactive decay up to any time. Because each ^{87}Sr atom has been generated by the decay of one ^{87}Rb atom, $N_R = N_0 - N$. This can be substituted into the formula for time:

$$t = \frac{1}{\lambda} \ln \frac{N + N_R}{N}$$

For example, if 3% of the original ^{87}Rb in a rock has decayed into ^{87}Sr, then the age of the rock is calculated as:

$$t = \frac{1}{1.42 \times 10^{-11}} \ln(100/97) = 2.08 \times 10^9 \text{ years}$$

Or about 2 billion years.

(The decay constant and half-life of a radioisotope are simply related. When half the original ^{87}Rb has decayed into ^{87}Sr, the number of ^{87}Sr atoms formed must equal the number of ^{87}Rb atoms that have decayed. $N = N_R$. Substitute that into the formula for time, and $t_{1/2} = \ln 2/\lambda = 0.693/\lambda$.)

In summary, if we know the isotope ratio in the rock when it was formed and in a modern sample, and if we can reasonably assume that the change in the ratio between then and now was only caused by radioactive decay, we can estimate the absolute age of the rock.

Fossils are dated by radioisotopes, . . .

In practice, the dating of fossils by the radioisotope method usually requires a combination of absolute and relative dating. The reason is that the radioisotope method can only be used for rocks that contain radioisotopes. Some fossils contain ^{14}C, because carbon is found in living material, and these can be dated directly if they are not too old. Some other radioisotopes are found in corals or shells. However, most of the radioisotopes used to find geological dates are not found in fossils. These isotopes (such as ^{87}Rb or ^{40}K) are only found in igneous rocks. In order to date a deposit of fossils, we need to find some associated igneous rocks that we can infer to have been deposited at about the same times as the fossils.

For instance, if some fossiliferous sediments are laid down on top of an igneous rock, we can infer that the fossils are no older than the date of the igneous rock (on the principle that younger rocks lie on top of older ones). If an igneous rock has been intruded into a sedimentary rock, we can infer that the sediments are older than the igneous rock (because igneous rocks only intrude into existing sedimentary rocks). In the best case, a fossiliferous sediment will lie on top of one, older set of igneous rocks, and have another younger igneous rock intruded in it. Then the fossils can be dated to a time between the age of the two igneous rocks.

. . . by the relative positions of rocks, . . .

The inference of the age of fossils from that of surrounding igneous rocks is an example of relative time measurement. If we know the relative date of a rock, or fossil, it means we know its date relative to that of another rock, or fossil: we have a statement of the form "rock A was laid down before/at the same time as/after rock B." Some of the procedures for finding relative times are as follows. At any one site, more recent sediments are deposited on top of older sediments. Fossils lower down a sedimentary column are therefore likely to be older (sometimes a large geologic convulsion, such as a volcanic explosion, may turn a sedimentary column upside down, but it is obvious when this has happened). The date of any one fossil deposit relative to those at different sites can also usually be estimated. It is done by comparing the fossil composition of the site, for some common fossils such as ammonites or foraminifers, with a standard reference collection. For these reference fossils, the fossils deposited at one place and time will be much the same as those being deposited at another place. They show that

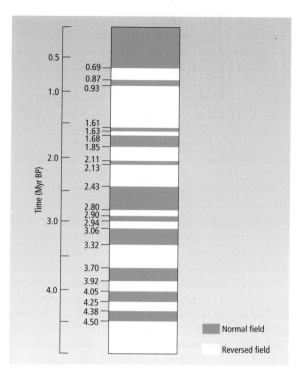

Figure 18.2

Polarity reversals of the Earth's magnetic field in the past 4.5 million years. The picture may not be complete: it is possible that further, shorter events have still to be discovered.

two sites had the same relative date. This kind of study is called correlation, and the paleontologist is said to be "correlating" the two sites.

. . . and by magnetic time zones

Magnetic time zones supply a similar principle. The magnetic field of the Earth has reversed its polarity at intervals through geological history. When the magnetic field is as it is now (compasses point north) it is called normal; when it has the opposite polarity it is called reversed. The history of Earth is an alternating sequence of normal and reversed time zones. (The reason for the reversals is not known for sure — though there are plenty of hypotheses.) Polarity switches have been commoner at some times than others; Figure 18.2 gives an idea of their frequency in recent times. All the rocks in the globe at any one time have the same polarity, and the polarity at the time rocks were formed can be detected. The polarities can the be used in fine-scale time resolution. If two rocks are known to have been formed at similar times, but we are not sure whether they are exact or only near contemporaries, magnetic polarities can provide the answer. If the rocks have different polarities they cannot have been exact contemporaries.

18.3 The history of life: the Precambrian

18.3.1 *The origin of life*

Most research on the origin of life is not on fossils, but consists of laboratory research on the kinds of chemical reactions that may also have taken place on Earth 4 billion years ago. Many of the molecular building blocks of life (such as amino acids, sugars, and nucleotides) can be synthesized from a solution of simpler molecules, of the sort that probably existed in the prebiotic seas, if an electric discharge or ultraviolet radiation is passed through it. Once the molecular building blocks exist, the next crucial step is the origin of a simple replicating molecule.

Although we do not know what the earliest ancestral replicating molecule was, several lines of evidence suggest that RNA preceded DNA. For instance, single-stranded RNA is simpler than DNA, which is always double-stranded. DNA needs enzymes to "unzip" the two strands in order to read or replicate the nucleotide information. DNA always takes on the structure of a double helix. RNA, by contrast, can interact directly with its environment. It can be read or replicated directly. Also, RNA can take on many different structures, depending on its nucleotide sequence. In some of those structural forms, RNA will act as an enzyme (or "ribozyme"), catalyzing biochemical reactions. RNA molecules are known that act as RNA polymerases, catalyzing the replication of RNA. However, no one has yet discovered an autocatalytic RNA that could catalyze its own replication. Such a self-replicating molecule would be one of the simplest imaginable living systems. Some other small lines of evidence also suggest that RNA preceded DNA, such as "prebiotic soup" experiments that have more readily yielded the nucleotide U than T.

The (hypothetical) early stage of life, when it used RNA as the hereditary molecule, is called the "RNA world." Life came to use DNA later in history. One reason for the transition from RNA to DNA may have been that RNA-based life was limited by the relatively high mutation rate of RNA. (This reasoning is similar to the argument in Section 12.2.1, p. 320, about the evolution of sex.) Asexual life forms cannot exist with a total deleterious mutation rate of more than about one.

Modern RNA viruses such as HIV have a mutation rate of about 10^4 per nucleotide. This limits their coding capacity to about 10^4 nucleotides, or about 10 genes. More complex life forms could not evolve until the mutation rate reduced. The evolution of DNA would have reduced, or led to a reduction of, the mutation rate.

The fossil record tells us little about the origin of life, because those events were on a molecular scale. However, the record does tell us something about timing, and leads us to the next stage. The Earth itself is about 4.5 billion years old. For the first few hundred million years, Earth was bombarded by huge asteroids that vaporized any oceans. Temperatures were too high to allow life. Life probably could not have originated before about 4 billion years ago.

The oldest known rocks are at a site at Isua, Greenland, and are 3.8 billion years old. These rocks contain chemical traces that may or may not be chemical fossils of life forms (van Zuilen *et al.* 2002). Chemical evidence of this kind is inevitably uncertain, because it could have been produced by a non-biological process. Some biologists and geologists tentatively accept it as evidence of life, but few place strong trust in it. The rocks have undergone too much metamorphosis to have any chance of retaining fossil cells — if cells existed at that time. Fossil evidence of cells comes from various sites in the period 3–3.5 billion years ago. The earliest fossil cells were until recently thought to come from 3.5 billion-year-old rocks from the Apex Chert in Western Australia (Schopf 1993). However, Brasier *et al.* (2002) have argued that the alleged fossils in these rocks are artifacts and not fossils. Other evidence for fossil cells exists from the 3–3.5 billion-

Early life may have used RNA for inheritance

Fossil cells exist earlier than 3 billion years ago

Figure 18.3

Around 2–3 billion years ago, prokaryotic life had evolved to live in a variety of habitats, using a variety of metabolisms. The illustration shows hydrothermal systems around structures called komatiite shields, chemotrophs in mid-ocean ridges, and various life forms in coastal, lacustrine, and oceanic waters. Redrawn, by permission of the publisher, from Nisbet (2000).

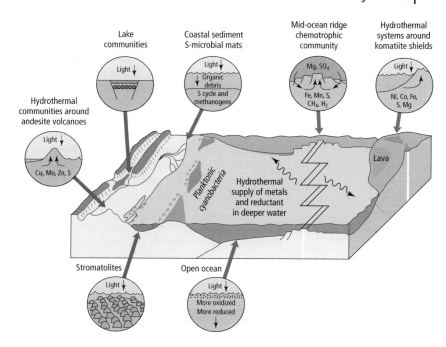

year period (for instance, Knoll & Baghoorn 1977, and Schopf 1999 reviews the evidence). Cells had therefore likely evolved by 3.5 billion years ago, or soon afterwards.

18.3.2 ***The origin of cells***

Early cellular life was prokaryotic

Fossils of cellular prokaryotic life have been found at several sites, aged between 3.5 and 2 billion years ago. They often exist in the form of stromatolites. Stromatolites are layered structures that are formed when cells grow on the sea surface, and sediments are deposited among or above the cells. The cells then grow up to the light, leaving a mineralized layer below them. As the process is repeated over time, a stromatolite builds up, consisting of many mineralized layers. Stromatolites are still formed at certain sites in the world today, but are rarer than in the past. One reason may be that grazers now consume any cell mats as they form. In the past, grazers did not exist and stromatolites could accumulate. Around 2–3 billion years ago, microbial prokaryotic life seems to have existed in several environments, and to have evolved several metabolic processes (Figure 18.3).

Thus, cellular life was flourishing by 2–3 billion years ago. However, the origin of cells was probably not an evolutionarily inevitable step. Unadorned replicating molecular systems could have persisted, the molecules being replicated as their component building blocks bonded to them and formed copies, or near copies, of the whole. For the system to become more complex, it needs enzymes and metabolic systems that enable it to harvest resources more powerfully, or to exploit the resources better by converting them into the molecular units needed for replication. This step is difficult for at least two reasons. One is mutational error, which becomes increasingly damaging as the replicating molecule increases in size. The other is that any advantageous innovation — for example, one that can produce useful molecules — will share its produce with all the other competing replicating molecules in the locality. A "selfish" replicating molecule, that used resources manufactured by others but did not itself manufacture them, would have a selective advantage over other replicating molecules that both manufactured and used resources (Section 11.2, p. 294).

This second difficulty was probably overcome by the evolution of cells. If the replicating molecules are enclosed within cells, the products of their metabolism are confined to the cell that produced them and are not available for any selfish replicating molecules outside. Another advantage of cell membranes is that metabolic enzymes can be arranged spatially; a chain of metabolic reactions can then operate in an efficient sequence. Thus the first cells were probably little more than replicating molecules either surrounded by, or arranged within, membranes. Modern prokaryotic cells are complex versions of this form of life.

The deepest classificatory division of cellular life is a three-way divide into archaeans, bacteria, and eukaryotes. Archaea and bacteria are both prokaryotes, and both existed on Earth 2–3 billion years ago. The other kind of cell, the eukaryotic cell, evolved after the prokaryotes. The time of origin of eukaryotes is uncertain. The oldest figure is about 2.7 billion years ago. Brocks *et al.* (1999) found chemical fossils of certain fats that are characteristic of eukaryotic metabolism in 2.7 billion-year-old Australian rocks. This may mean that eukaryotes had evolved by then. Or it may mean that the fats are not a good signature of eukaryotic life. After all, new prokaryotes, with an ever-expanding range of metabolic skills, are being discovered every year. Thus, the chemical fossils are not convincing evidence of eukaryotic origins, but they raise the possibility that eukaryotes had already evolved 2.7 billion years ago.

The earliest fossil cells that have been proposed to be eukaryotic were found in an abandoned mine in Michigan and are described by Han & Runnegar (1992). The fossils are corkscrew shaped and closely resemble later algae (algae are eukaryotes). The main criterion for distinguishing eukaryotic cells from prokaryotic cells in fossils is cell size. Eukaryotic cells are typically larger than prokaryotic cells, though there is some overlap in their size ranges. The Michigan corkscrews are huge — about 0.5 in (1 cm). If that is a single cell, it is surely eukaryotic. Unfortunately, the fossilization process did not preserve any signs of cell membranes. A skeptic can still suggest that the Michigan corkscrews are really multicellular, in which case they could be prokaryotic. If we move on to about 1.8 billion years ago, many fossil cells exist that are generally accepted to be eukaryotic.

Molecular clock studies suggest the eukaryotes originated in the 2.2–1.8 billion years ago range. Thus the body fossil and molecular evidence agree, but chemical fossil evidence hints at an earlier date. Around 2 billion years ago is a generally quoted, if uncertain, date for the origin of the eukaryotes.

The origin of the eukaryote cell could have been spread over many hundreds of millions of years. Modern eukaryotes differ from prokaryotes in a long list of features. The formally defining difference between eukaryotes and prokaryotes is the presence or absence of a nucleus. Eukaryotes also possess organelles, including mitochondria and (in plants) chloroplasts. Eukaryotes have a special process of cell division called

Cells had advantages over non-cellular life

Eukaryotic cells evolved maybe 2 million years ago . . .

mitosis, in which an apparatus of mobile spindles is formed and pulls the duplicated chromosomes apart. Eukaryotes also have meiosis. There are many other differences of structure between the two kinds of cell (Figure 2.1, p. 22).

Mitochondria and chloroplasts almost certainly originated by symbiosis (Section 10.4.3, p. 265). Their symbiotic origin was first suggested by the morphological similarity of the organelles to bacteria. The theory has since been strongly supported by molecular evidence. The genes in the mitochondria of a eukaryotic cell are more similar to genes in free-living bacteria than to comparable genes in the nucleus of the cell that the mitochondria is living in (Gray *et al.* 1999).

The evolution of the nucleus and then mitosis and meiosis were probably separate events, perhaps before — perhaps after — the origin of the organelles. The other differences between eukaryotes and prokaryotes could have evolved at other times. The origin of the eukaryotic cell would then have been a multistage process, extending over a long time period.

An important event associated with the origin of eukaryotes is the evolution of photosynthesis, or of photosynthesis on a mass scale. Photosynthesis itself probably originated earlier — indeed Schopf's 3.5 billion-year-old possible microbes may have been photosynthetic — but around the time that eukaryotic cells were evolving there was also an increase in the quantity of oxygen, suggesting that photosynthesis was becoming much more important. Atmospheric oxygen would not have increased immediately following the evolution of photosynthesis. The first oxygen would have been absorbed by rocks, which became oxidized (indeed the oxidized form of iron-containing rocks is the main way that photosynthesis is inferred in the geological record). Oxygen would have accumulated in the atmosphere only after the rocks had absorbed all the oxygen they could.

A little over 2 billion years ago, atmospheric oxygen concentration probably spurted up. The most likely reason is that photosynthesizing organisms had become more abundant and were pouring out oxygen as a by-product. Also, the chloroplast-containing cells of eukaryotes were more efficient photosynthesizers than the prior prokaryotes, and this is why the oxygen concentration increased at about the time when eukaryotes were evolving. Whatever the reason, when oxygen was first released in large amounts it was probably a poison to most existing forms of life, because they had evolved in environments with little oxygen; there may have been an ecological disaster. Subsequent forms of life have mainly descended from species that evolved to tolerate, and then make use of, this chemical novelty. Aerobic respiration, using mitochondria, may have become advantageous around this time.

18.3.3   *The origin of multicellular life*

"Multicellular" life is used to refer not simply to the presence of more than one cell in an organism, but to more than one kind of cell — that is, to cell differentiation. Life forms with more than one kind of cell have at least a rudimentary development. They develop from a single-celled zygote to an adult with specialized cell types. The origin of development is an important step in the evolution of life. Life forms consisting of rows or mats made up of many identical cells had existed early in life. Schopf's (1993) paper

on 3.5 billion-year-old possible fossils describes multicelled filaments of this sort. But multicellular life, in the sense of life with work development and cell differentiation, evolved much later. With only minor exceptions, all life forms with cell differentiation are eukaryotic. The molecular clock suggests that multicellular life originated about 1.5 billion years ago. This is somewhat, but not much, before the oldest multicellular fossils. Currently the earliest such fossils are algae from about 1.2 billion years ago (Butterfield 2000).

Multicellular life originated over 1 billion years ago

The earliest definite fossils of multicellular animals (Metazoa) come from the Ediacaran deposits in Australia. These, and similar deposits elsewhere in the world, date to the period from 670 to 550 million years ago. The Ediacaran fossils are of soft-bodied aquatic animals such as jellyfish and worms (Figure 18.4). Well preserved fossils of both multicellular aquatic animals and aquatic plants are also found in China from this time (Xiao *et al.* 1998). The Ediacaran fossils decline in abundance about 550 million years ago. The decline has been attributed to a mass extinction, but likely reflects changes in the conditions of fossil preservation; Ediacara-type fossils continue to exist in the Cambrian (Jensen *et al.* 1998). However, for the main fossil record of animal life, we need to move forward from the Precambrian.

18.4 The Cambrian explosion

The fossil record of multicellular plants and animals does not really take off until the Cambrian, which began about 540 million years ago. Indeed the main time periods of the fossils record begin with the Cambrian (Figure 18.1). Until the 1940s, no pre-Cambrian fossils were known and in Darwin's time it was assumed that they did not exist. Even though we now know they do, the picture is one of sudden proliferation, rather than a sudden beginning, of fossil life a little over 500 million years ago.

A good fossil record starts in the Cambrian

Figure 18.5 illustrates the Cambrian explosion. It shows, for all nine animal phyla for which we have a fossil record, when the earliest fossils date from. The majority of them date back to the early Cambrian, or some time near it. A superficial reading of the evidence could be dramatized as follows. Life has been evolving for about 4,000 million years and is today grouped into a series of major phyla — chordates, mollusks, arthropods, and so on. We might expect that these phyla originated at a relatively steady rate, yet it appears that they almost all arose within less than 40 million years of each other (Figure 18.5), or within a period of less than 1% of the history of life.

However, the molecular clock suggests a radically different view. If we measure the molecular distance between the major animal groups, and calibrate the clock, we find that the major groups diverged from a common ancestor more like 1,200 million years ago. Several molecular studies have been made, of which Wray *et al.* (1996) has been particularly influential. They inferred that the Metazoan Bilateria share a common

Figure 18.4 (*opposite*)
Some Ediacaran fossil animals. (a) *Charniodiscus arboreus*, an attached cnidarian; (b) *Cyclomedusa radiata*, a jellyfish; (c) *Spriggina*, a worm (its name honors R.C. Sprigg, Assistant Government Geologist of South Australia, who discovered the Ediacaran fauna in 1946); (d) *Dickinsonia costata*, another worm; and (e) *Tribrachidium heraldicum*, a possible proechinoderm. (a and b) reprinted, by permission of the publisher, from Glaessner & Wade (1996) and Wade (1972); (c–e) courtesy of M.F. Glaessner.

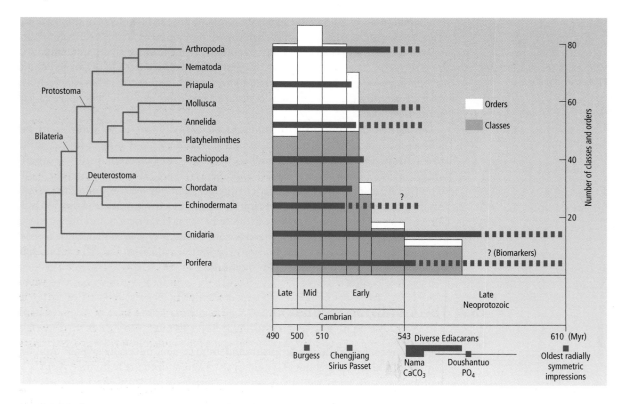

Figure 18.5

The fossil record of most major animal groups begins in the Cambrian. The horizontal lines to the right of each taxon show the times when the group is represented in the fossil record for the Cambrian and late Precambrian. The thick, continuous lines show the "crown group" fossils — fossils descended from the same common ancestor as modern members of the taxon. The broken lines show the "stem group" fossils — fossils that are members of the taxon but that branched off before the last common ancestor of the modern members of the taxon. Notice that the first fossils of all taxa except the Cnidaria and Porifera (and Nematoda) date in or near the early Cambrian. The histograms show the same event in terms of the numbers of orders and classes (the numbers are on the *y*-axis). Again, note the rapid increase in the early Cambrian. The phylogenetic relations of the taxa are shown to the left. Dates are along the bottom, along with times of major fossil deposits. Doushantuo PO_4 and Nama $CaCO_3$ refer to sites where the fossils are phosphates and calcium carbonates. The animal groups shown here are the main groups for which a reasonably good fossil record exists. Other animal groups also exist but are smaller, or have less certain fossil records. Modified, with permission of the publishers, from Knoll & Carroll (1999).

ancestor at about 1,200 million years ago. The common ancestor of all animals would then have lived earlier still. (The Bilateria include all the animal groups in Figure 18.5 except sponges and Cnidaria.)

How can we reconcile the fossil and molecular dates? Either (or both) could be wrong in some way. However, many biologists suspect that they are both correct. The molecular evidence tells us the date of the common ancestor, whereas the fossil evidence tells us when each animal group arose in its modern form. There could have been a period before the fossils were deposited when ancestors of each group existed but

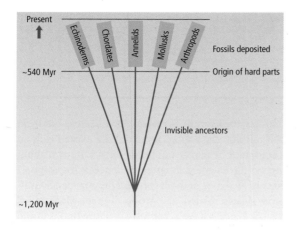

Figure 18.6

Reconciling fossil and molecular evidence on the timing of animal evolution. The main animal groups could have diverged around 1,200 million years ago, as the molecular evidence suggests, but been invisible in the fossil record until the origin of hard parts around 540 million years ago. The fossil invisibility of animals between 1,200 and 540 million years ago could be due to the animals being rare and soft. The time of proliferation in the fossil record and the time or origin of a group would then not be the same.

were too rare or fragile, or in the wrong place, to leave fossils (Figure 18.6). In Cooper & Fortey's (1998) image, a "phylogenetic fuse" preceded the Cambrian explosion.

Why should there have been a long period — 500 million years or more — when ancestors of the modern animal phyla existed but no fossils were deposited? The Cambrian explosion is a fossil event, and probably marks the time of origin of hard parts. Animals with hard skeletons or shells leave fossils much more often than animals with only soft parts. But if hard parts originated about 540 million years ago, that raises the question of why hard parts suddenly became advantageous in so many groups at about the same time.

Predators in general is one hypothesis and visually hunting predators in particular is a second hypothesis. After predators evolved, hard parts became advantageous for defensive reasons. Another factor is that oxygen levels may have increased towards the end of the Precambrian. This may have been caused by an increase in plant — that is, phytoplankton — productivity (Knoll & Carroll 1999). The higher plant production (if it occurred then) would have supported a greater mass and diversity of animals. The greater quantity of potential prey could have created an opportunity leading to the evolution of predators.

The "snowball Earth" hypothesis suggests a further environmental factor that could have been at work. For at least some of the period before the Cambrian, the Earth may have been almost completely covered with ice and glaciers. Life would have been rare, confined to areas near hot springs or ocean vents, or small localities where enough ice had melted to allow sunlight through and photosynthesis to occur. This would help explain the paucity of fossils before the Cambrian. Moreover, the ancestors of arthropods, mollusks, and chordates would then have been tiny creatures, small enough to have been supported by the very limited ecological productivity.

The Cambrian explosion is the subject of intense research at present. Biologists and paleontologists are studying just how suddenly the fossil event was: maybe it was less explosive than Figure 18.5 shows. Others are studying the molecular evidence, with further molecules and new calibration procedures. If, as seems likely, some major evolutionary event did occur around 540 million years ago, the big question is what caused it? The hypotheses at present are looking at external environmental

Several hypotheses exist to explain the Cambrian explosion

change, or internal biological innovations, or some mix of the two. But there is no consensus yet.

18.5 Evolution of land plants

The land was first colonized by microbes. Prokaryotic fossil cells have been found from terrestrial environments over 1,000 million years ago. Terrestrial life existed only in microbial form until the land was colonized by plants and fungi. Little is known about early terrestrial fungi, but for plants we have widely accepted evidence of fossil spores from 475 million years ago, and further evidence of fossil spores from maybe 550 million years ago. Land plants are most closely related to a group of green algae called charophyceans (Figure 18.7). Early fossil spores have a "tetrad" structure that is found in some modern bryophytes, and most early plant fossils seem to be related to the phylogenetic branches leading to modern bryophytes and pteridophytes.

The main events in the early evolution of land plants were the evolution of a resistant spore stage, then the evolution of vascular tissue, followed by roots and then leaves. Bryophyta such as moss have spores but lack vascular tissue. Pteridophyta such as ferns have vascular tissue. Vascular tissue, and particularly vascular tissue that is built of tracheid cells with lignin-containing cell walls, enables a plant to support itself on land.

The earliest fossils of whole plants, as distinct from spores, date back to around 430 million years ago. *Clarksonia* is one of the commonest fossils of this time. These early land plants lacked roots and leaves, and simply had branching stems. We can infer that they photosynthesized through their stems, because stomata are visible in the fossil stems. Fossils with leaves appear 390–350 million years ago.

The evolution of leaves coincides with a dramatic fall of about 90% in atmospheric carbon dioxide concentration. One hypothesis connects these events. The initial evolution of land plants may have removed carbon dioxide from the atmosphere, not only by their relatively minor photosynthetic activity through the stem but, more importantly,

Plants evolved adaptations for land life . . .

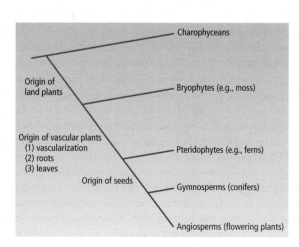

Figure 18.7
Phylogeny of the main groups of plants. Land plants are most closely related to a group of algae called charophyceans. The main events in plant evolution are the evolution of land adaptations, then seeds, and then flowers.

Figure 18.8

The rise of the angiosperms. Angiosperms (flowering plants) have gradually expanded in diversity since the Cretaceous. Gymnosperms (conifers, cycads, and ginkgos) have declined, as have pteridophytes. For some species it is not known which group they belong to, and they are called "incertae sedis." Redrawn, by permission of the publisher, from Niklas (1986).

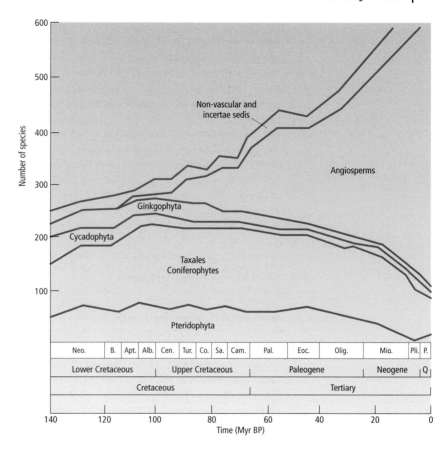

... then seeds ...

... and flowers

through the evolution of roots. Roots enhance weathering, and weathering removes large amounts of carbon dioxide from the atmosphere. This reduction in carbon dioxide set up a force of selection in favor of leaves — and the evolution of leaves, and more powerful photosynthesis, further reduced the carbon dioxide levels.

The seed plants make up the main modern land plant groups. Fossil seeds exist in the Carboniferous, when the coal deposits were formed. However, seed plants were a minor group at that time. Coal is mainly formed from fossil pteridophytes. The two groups of seed plants — gymnosperms (conifers) and angiosperms (flowering plants) — proliferated later (Figure 18.8).

Angiosperms first clearly appear in the fossil record in the early Cretaceous, about 125 million years ago (Sun *et al.* 2002). Darwin once remarked that the origin of the angiosperms was an "abominable mystery." Modern molecular phylogenetics has helped work out the relations within the angiosperms, and between angiosperms and gymnosperms (see, for example, Figure 15.24, p. 461). But molecular studies of angiosperm phylogeny have created a new abominable mystery in their turn. If we use the molecular clock to estimate the time of angiosperm origins we find a figure of perhaps 200–250 million years ago, well before the earliest fossils. As always with a "molecules versus fossils" controversy, the difference in dates could reflect incompleteness of the fossil record, inaccuracy of the molecular clock, or a delay between the origin and the proliferation of the taxon. However, the controversy about the time of angiosperm origin is currently unresolved.

The proliferation of the angiosperms in the Cretaceous and Tertiary is often explained in terms of coevolution with insect pollinators. We look at that hypothesis in Section 22.3.4 (p. 622). Later on, around 60 million years ago, the fossil record shows the origin and global proliferation of grasses. The proliferation of grasses has been explained by coevolution with mammals. Mammals proliferated at the same time, and included forms with specialized teeth for grazing. Grass is well adapted to thrive where mammalian grazers are present, because grass regrows from the base rather than the tip of its stem. The spread of grass may in turn have helped to set the stage for the future evolution of humans. Human evolution has often, if uncertainly, been associated with a shift from arboreal to savannah grassland habitats.

Some mammals coevolved with grass

18.6 Vertebrate evolution

18.6.1 *Colonization of the land*

The earliest vertebrate fossils are fish and date back to Cambrian, or even (in some recently described fossils from China) late Precambrian, times. Fish proliferated in the Ordovician fossil record, but we can pick up the story where we began with plants: the move on to land. The fossil evidence points to the late Devonian, around 360 million years ago, as the time when terrestrial vertebrates originated.

The terrestrial plants probably prepared the way. Terrestrial plants proliferated during the Devonian, at the water's edge. The presence of plants, and their roots growing down into the water, and the arthropod life associated with them, combined to create a new habitat at the water's edge. Fish would have evolved to exploit the resources there. The fossil record documents in excellent detail the evolutionary transition from fish to terrestrial amphibians (Figure 18.9). The amphibians were the first of the tetrapod groups to evolve. (The tetrapods are the group of four-legged vertebrates: amphibians, reptiles, birds and mammals. "Tetrapods" and "terrestrial vertebrates" refer to roughly the same group of animals.) We can notice a few features of the story.

There is a good fossil record from fish to amphibians

Modern fish (or, to be more exact, bony fish) divide into two main groups: ray-finned fish and lobe-finned fish. Most fish are ray-finned, but modern tetrapods are descended from lobe-finned fish ancestors. Modern lungfish and the coelocanth are lobe-finned fish. Within the lobe-finned fish, the lungfish, rather than the coelocanth, is thought to be the closest relative of tetrapods. Morphological evidence had been ambiguous, and in the 1980s an authoritative cladistic analysis suggested that coelocanths were closer to tetrapods than were lungfish (Rosen *et al.* 1981). However, molecular evidence in the 1990s pointed to the opposite conclusion. For now, the molecular evidence is generally accepted.

Between the lungfish and amphibians, a series of fossil forms range from the completely fish-like *Eusthenopteron*, through aquatic (*Acanthostega*), and partly terrestrial

Figure 18.9
The origin of tetrapods. A good series of fossil forms exist, connecting fish and the earliest tetrapods. Redrawn, by permission of the publisher, from Zimmer (1998).

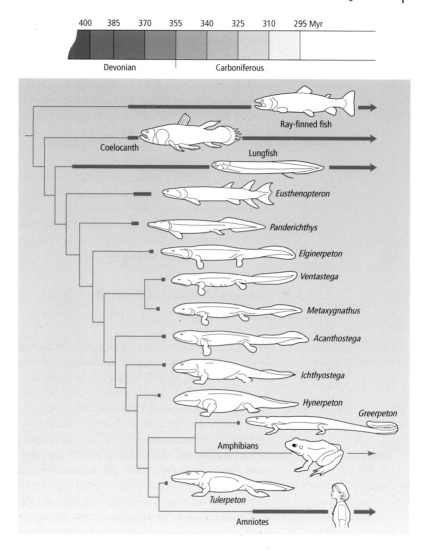

Tetrapody preceded life on land

(*Ichthyostega*) tetrapods, to amphibians. The fossil evidence showing the gradual transition is noteworthy in itself, because few major evolutionary transitions are so well documented. The evidence also has some important details. One is that the tetrapod condition seems to have evolved first in fully aquatic vertebrates. *Acanthostega* had four good legs, homologous with the four limbs of a cat and a lizard, but it also had gills and a tail for swimming. The fossil evidence therefore suggests that the tretrapod limb originally evolved as a paddle, for swimming. Its subsequent use for walking on land is an instance of preadaptation (Section 10.4.2, p. 264).

In modern tetrapods, the foot always has five digits (or if the number differs from five, it can be seen to be derived from a five-digit condition). However, the tetrapods of the Devonian include forms with different numbers of digits, such as seven or nine.

Modern tetrapods presumably just happen to be derived from five-digit ancestors, and have retained that condition.

The next big step in terrestrial vertebrate evolution was the origin of the amniotic egg: reptiles, birds, and mammals are amniotes, and members of these groups, unlike most amphibians, do not return to water for the early stages of the life cycle. The origin of egg types cannot be traced directly in the fossil record; however, at the origin of the reptiles there were changes in skeletal morphology as well as egg type and there is good evidence for the former. The reptiles probably evolved in the Carboniferous. The small lizard-like creature called *Hylonomus*, from fossil deposits in Nova Scotia, is an early reptile.

After the origin of the reptiles the two main events in vertebrate evolution were the origin of bird flight and the origin of mammals. We shall not look into bird evolution here (Section 10.4.2, p. 264, discusses how feathers are another example of preadaptation), but we shall look at the origin of mammals. It is the best documented of any of the major transitions in evolution, being even better documented in the fossil record than the origin of tetrapods.

18.6.2 Mammals evolved from the reptiles in a long series of small changes

The mammals are a distinct group of vertebrates in many respects: (i) they have warm blood and a constant body temperature, and the high metabolic rate and homeostatic mechanisms that go with it; (ii) they have a characteristic mode of locomotion, or gait, in which the body is held upright with the legs underneath (in contrast to the "sprawling" gait of reptiles, such as lizards, in which the legs stick out sideways); (iii) they have large brains; (iv) their method of reproduction, including lactation, is also distinctive; and (v) the active metabolism of mammals demands efficient feeding, so mammals have powerful jaws and a set of relatively durable teeth, differentiated into a number of tooth types. Therefore, when the mammals evolved from the reptiles, there had to be changes on a large scale in many characters. How did this transition take place?

Not all of the distinctively mammalian characters are preserved in the fossil record. The earliest mammalian fossils, such as *Megazostrodon* (Figure 18.10, at the top), date back to the late Triassic, about 200 million years ago. Whether *Megazostrodon* was viviparous and lactated is not known directly. But we can see that that it had a mammalian jaw, gait, and tooth structure, and can therefore infer that it probably also had warm-blooded physiology. The origin of the mammals can be traced back before 200 million years ago, through a series of reptilian groups informally called the *mammal-like reptiles* and formally called the Synapsida. They evolved over an approximately 100 million-year period from the Pennsylvanian to the end of the Triassic, when the first true mammals appeared. Some synapsids persisted into the Jurassic, but

Figure 18.10 (*opposite*)
The evolutionary radiation of the mammal-like reptiles. There were three main phases: (a) pelycosaurs (sphenacodontids and ophiacodontids in this picture), (b) therapsids, and (c) cynodonts. Within each phase there were many smaller evolutionary lineages. Some fossil forms are illustrated. Note again the evolution of the more powerful and precision-action mammalian jaw, and the change from a sprawling gait in *Dimetrodon* to an upright gait in *Probelesodon*. Redrawn, by permission of the publisher, from Kemp (1999).

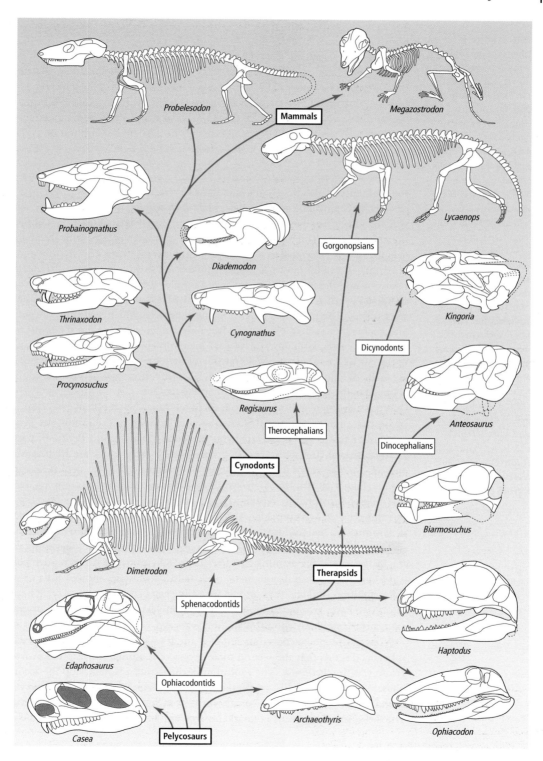

Probelesodon

Mammals

Megazostrodon

Probainognathus

Diademodon

Gorgonopsians

Lycaenops

Thrinaxodon

Cynognathus

Kingoria

Procynosuchus

Dicynodonts

Regisaurus

Therocephalians

Anteosaurus

Cynodonts

Dinocephalians

Dimetrodon

Sphenacodontids

Biarmosuchus

Therapsids

Edaphosaurus

Haptodus

Ophiacodontids

Casea

Archaeothyris

Pelycosaurs

Ophiacodon

Figure 18.11

Articulation of the jaw in mammal-like reptiles. (a) In the early form *Biarmosuchus*, the jaw muscle is at the basal articulation. In the evolution of mammals, the muscles move forward, and by *Probainognathus* (c), an advanced mammal-like reptile, the masseter has split into two. The superficial masseter joins a characteristic region of the upper jaw, and the presence of the advanced jaw condition can be recognized from the jaw alone (see Figure 18.10, where these three forms are again illustrated but without the muscles drawn in). *Thrinaxodon* (b) was probably like *Probainognathus*. The positions of muscles in fossils can be inferred from the bone shapes, which imply the attachment sites for muscles. Note also the increasing complexity and serial differentiation of the teeth. For the different gaits of reptiles and mammals, see Figure 15.2 (p. 427). From Carroll (1988). © 1988 WH Freeman & Company, with permission.

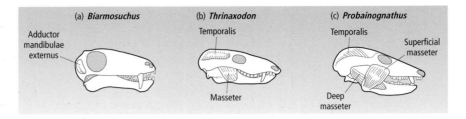

Stages in mammal evolution include the pelycosaurs . . .

. . . therapsids . . .

by then the dinosaurs had proliferated. No other terrestrial tetrapods thrived until after the dinosaurs went extinct at the end of the Cretaceous.

The characters that can most clearly be reconstructed in fossils are those concerned with locomotion and feeding, because these are simply related to the form of preserved bones and teeth. The reptilian jaw contrasts in many respects with the mammalian (Figure 18.11). Mammalian teeth have a complex, multicusped structure and are differentiated down the jaw into canines, molars, and so on, whereas reptilian teeth form a relatively undifferentiated row and have a simpler structure. The top and bottom of the reptilian jaw articulates (that is, hinges) at the back, where it has muscles that simply snap it shut. The mammalian jaw has cheek muscles which surround the cheek teeth and enable the jaw to close more powerfully and accurately than the reptile's. As the point of jaw articulation moved forward in mammalian evolution, the bones at the rear of the jaw were evolutionarily liberated, and went on to evolve into the ear bones — but we shall not follow that fascinating story. Here we concentrate on how the jaw, and gait, changed in the evolution of the mammal-like reptiles.

We can distinguish three main phases in mammal-like reptilian evolution. The first phase corresponds to one of the two major divisions of the group, the *pelycosaurs*. Pelycosaur fossils are preserved from the Pennsylvanian and Permian, particularly from southwest USA where rocks of this age are located (see Plate 10, between pp. 68 and 69). *Archaeothyris* lived there about 300 million years ago, and is an early pelycosaur (Figure 18.10). It was a lizard-like animal, about 20 in (50 cm) long. Its distinctive difference from other reptile groups is an opening in the bones behind the eye. The opening is called a temporal fenestra, and in the living animal a muscle passed through it. The muscle acted to close the jaw, and the opening up of the temporal fenestrae is the first sign of the more powerful jaw mechanism of the mammals. (The temporal fenestra, by the way, is the defining character of the Synapsida.) A better known pelycosaur was *Dimetrodon*, with its enigmatic back-sails. Pelycosaurs showed little or no tooth differentiation, and had the reptilian sprawling gait (Figure 15.2, p. 427). They evolved into three main groups during their 50 million-year history, and most of them went extinct quite suddenly about 260 million years ago.

A few of the sphenacodontids survived and it was from an unknown line within the sphenacodontids that the second main group of mammal-like reptiles evolved. This group was the *therapsids*. The evolution of the therapsids makes up the second main phase of the mammal-like reptiles, in the Permian and Triassic. Therapsid fossils are found in many regions of the world, with a site in South Africa having the best deposits. The therapsids underwent a remarkably similar pattern of evolution to the pelycosaurs but their temporal fenestrae are generally larger and more mammal-like than pelycosaurs, their teeth in some cases show more serial differentiation, and later forms had evolved a secondary palate. A secondary palate enables an animal to eat and breathe at the same time and is a sign of a more active, perhaps warm-blooded, way of life (Section 10.7.5, p. 284).

... and cynodonts

One subgroup of therapsids, the *cynodonts*, are of particular importance in tracing the origin of mammals, and they make up the third phase of mammal-like reptilian evolution. The jaws of cynodonts resemble modern mammal jaws more closely and their teeth are multicusped and differentiated down the jaw. Some cynodonts show a particularly interesting intermediate stage in jaw evolution. Recall that the reptilian jaw articulates in a different place from the mammalian jaw, a change associated with the evolution both of more precise chewing and of hearing in mammals. Some cynodonts seem to have had a double jaw articulation; their jaws articulated in both the mammalian and the reptilian positions. This suggests one way in which evolution can proceed from one structure to another without a non-functional intermediate stage: the structure evolved from state A to state A + B, then A was lost, giving state B alone. The jaw was a functional structure throughout. The cynodonts complete the story of the mammal-like reptiles, because it was from a line of cynodonts that the ancestors of the modern mammals evolved. The identity of the exact cynodont line from which modern mammals descended is uncertain, but *Probainognathus* (Figure 18.10) is close to it.

A further series of fossils connect the last mammal-like reptiles with modern mammals. Living mammals are divided into three groups: Prototheria (including the echidna), Metatheria, and Eutheria. Metatheria and Eutheria are also known as marsupials and placentals, respectively. The earliest known eutherian fossils are from the Yixian formation in China and date to the early Cretaceous (Ji *et al.* 2002). The three main modern mammal groups probably diverged in the Jurassic. The Eutheria in turn diverged into several main orders (that is, groups such as the primates, carnivores, proboscideans, and rodents). The timing of this divergence is controversial (Section 23.7.3, p. 671), but the primates probably originated in the Cretaceous even though their fossil record only takes off in the Tertiary. The next event we look at here is the origin of humans within the primate order.

18.7 Human evolution

18.7.1 *Four main classes of change occurred during hominin evolution*

Humans are primates, and our ancestors from about 60 (or more) million years ago to about 5–10 million years ago were tree-dwelling primates. Some of the trends we see in human evolution began in these ancestors. Primates have, compared with other mammals, relatively flat faces and large brains. Their flat faces provide their two eyes with a large overlap in their visual fields, giving good stereoscopic vision. Stereoscopic vision improves perception of depth, and is advantageous in leaping between branches.

Arboreal primates also have their thumbs and big toes (hallux) relatively separate from their other four digits. This allows them to grip branches. All primates have relatively opposable thumbs, as compared with other mammals, but a fully opposable thumb is (with minor exceptions) confined to the great apes — orang-utans, gorillas, chimpanzees, and humans. A fully opposable thumb means that you can touch the front tip of all four digits with your thumb. We can do it, but a cat or dog (for example) cannot.

In human evolution, the opposable hallux has been lost as our feet evolved for bipedality. The opposable thumb has been retained and modified. In our ancestors, it enabled a "power grip" used in gripping branches. We can still do the power grip, but changes in the hand bones allow us to use a "precision grip" not seen in other species. We use the precision grip in handling fine tools.

The big changes in human evolution may have occurred after our ancestors moved from forests to more savannah-type habitats. In fossils, the big changes can be understood in three categories. A fourth category concerns changes in social behavior, which is less easily studied in fossils.

1. *Brain enlargement.* Modern chimpanzees have brains of about 350–400 cm^3, and our ape ancestors 5 million years ago probably had brains of about the same size. Modern human brains are about 1,350 cm^3 in size.

2. *Changes to the jaw and teeth.* Chimpanzees, and our ape ancestors, are more prognathic than us, with their jaws sticking out more from their faces. During human evolution, the jaw shrunk back into the face, giving us flat faces. The jaw of our ape ancestors, and chimpanzees, has a semicircular shape (if viewed from above or below). In us, the semicircle has been pushed back to a shape more like a rectangle. Our teeth also got smaller, particularly the canine teeth, and our molars have evolved into grinding millstones.

3. *Bipedality.* The evolution of upright locomotion, on two legs, has resulted in changes throughout our bodies. Adaptations for bipedality are particularly clear in the anatomy of fossil feet and leg bones; but they can also be seen in the back vertebrae, the length of our arms, and the position of our skulls on our backbones.

Changes in these three categories are not independent. For instance, the themes of brain enlargement and bipedality combine in the evolution of human birth. Birth is relatively uncomplicated in chimpanzees, but our big brains and pelvic size, which is constrained by bipedality, have made birth more problematic in human beings. Brain size and gestation length are correlated across primates as a whole, and simple extrapolation from the general primate relationship suggests that humans might be expected to be born after 18 rather than 9 months. It may be that we are born relatively early because birth at a later fetal age would be impossible; 9 months is the latest possible point before the brain grows too big. Human babies are relatively undeveloped, compared with newborn chimps, in their motor development and other brain skills. Thus, newborn humans are relatively dependent on their mothers, and the intensity of parental care in humans is part of the other main trend of human evolution, which is not easily visible in fossils.

4. *Changes in social and cultural behavior.* The main way we differ from other apes is in our social and cultural lives. This development can be followed only indirectly in fossils. Sexual dimorphism, for example, is probably related to the breeding system. In apes other than humans, males weigh about twice as much as females on average

in gorillas and orang-utans, and about 1.35 times as much in chimpanzees. In humans sexual dimorphism is reduced; males on average weight about 1.2 times as much as females. Sexual dimorphism may have reduced in our ancestors when we evolved reproductive pair bonds — prolonged pairing is found in most human societies but not in other great apes. The cultural state of a society can be followed in tools and other artifacts associated with fossils. The main innovation underlying modern human culture is language. The origin of language is hard to study, with very indirect clues coming from jaw and throat anatomy and from the symbolic richness of artifacts associated with fossils.

. . . and sexual dimorphism all occurred in human evolution

18.7.2 *Fossil records show something of our ancestors for the past 4 million years*

We saw in Section 15.13 (p. 460) how the hominin lineage probably originated about 5 million years ago.[1] Currently, the earliest fossils that are generally accepted to be members of the hominin lineage are about 4.4 million years old and are classified in two species, *Australopithecus anamensis* and *A. afarensis*. Earlier fossils are known that may be hominins — but the fossils are fragmentary and could as easily be more closely related to other apes. *A. afarensis* is much the best known early australopithecine, because it includes the exceptionally complete specimen known as "Lucy." Lucy's skeleton, together with trace fossils of footprints, tell us that *A. afarensis* was bipedal. However, in other respects *A. afarensis* retained the ancestral conditions. Its brain size, relative to body size, was similar to that of a chimpanzee, and its jaws retain the ancestral shape. Australopithecines generally had evolved closer to the modern human condition in their method of locomotion than in their jaws and brains. Australopithecines are sometimes informally described as being like humans below the neck and like apes above the neck. Also, several dimorphism had not reduced in *A. afarensis*: males weighed about 1.5 times as much as females.

The fossil "Lucy" was bipedal

Species such as *A. afarensis* may have been direct ancestors of modern humans. Alternatively, they may have been relatives of our ancestors but not on the line leading to us. The fossil record is too incomplete for us to know which is true. In a figure such as Figure 18.12, the main fossil species are drawn as if some were ancestral to others, but it is more accurate to treat the figure as a simplification. For many points, we do not need to know whether one species is a direct ancestor of ours or not; for instance, we can conclude from the fossil evidence that the first main event in human evolution was bipedality. This is likely to be true whether *A. afarensis* is our ancestor or a relative of our ancestors.

Fossils may be relatives of our ancestors, not ancestors

[1] The "hominin" lineage is the one containing species more closely related to us than any other living species. Some use the word hominid — it depends on whether humans are a family (Hominidae) or subfamily (Homininae). The molecular evidence about our ape relations suggests to many that we should have a subfamily. Hence a recent trend toward the use of the term hominin, rather than the longer established hominid.

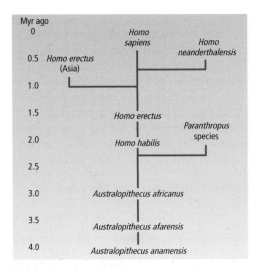

Figure 18.12

Hominin fossil species. The fossil record from East and South Africa shows the evolution of modern humans in many stages from ape-like ancestors. The picture is simplified: some fossil species, or possible species, have been omitted; the phylogenetic relations among the species that are shown is in most cases uncertain; and the species shown as an ancestor–descendant sequence up the center of the figure may well, in some cases, not be accurate. The descendant species might be descended from a close relative of the species shown as its ancestor. Many of the species names are controversial, as discussed further in the text for *Homo erectus, H. sapiens*, and *H. neanderthalensis*. *Paranthropus* is often included in the genus *Australopithecus*.

Fossils of *A. afarensis* have been found from about 4 to 3 million years ago. That brings us near to the time, about 3–2.5 million years ago, when the other fossil australopithecines are distinguished: *A. africanus* and *A. garhi*. *A. africanus* is much the better known and was the first australopithecine to be described, in 1924. It is somewhat more human-like than Lucy, and has often been taken for the next stage in human evolution. However, *A. africanus* fossils all come from South Africa, whereas Lucy and the later fossils that are thought to be closest to the human line are all from East Africa — particularly sites in Kenya and Ethiopia. Asfaw *et al.* (1999) described a new 2.5 million-year-old fossil from Ethiopia, and named it *A. garhi*. *A. garhi* may be closer to the human line than the South African *A. africanus*.

All the australopithecines mentioned so far are "gracile," with relatively light bones and jaws. Around 2.5–2 million years ago in both East and South Africa, the australopithecines diverged into gracile and robus forms. Two species, *Paranthropus robustus* and *P. boisei*, arose with much more powerful jaws, skulls, and cheek bones, who were able to eat tougher food than the gracile forms. Paleoanthropologists generally accept that *Homo* is more closely related to the gracile than the robust australopithecines. The robust species went extinct and have no descendants today.

By 2 or more million years ago, fossils classified as *Homo* start to be found. *Homo habilis* may date back as far as 2.5 million years ago. These fossils are closer to the modern human condition above as well as below the neck. *H. habilis* fossils are associated with stone tools, and the brain is larger — perhaps around 600–750 cm^3. Its jaws and teeth are reduced, though the jaw shape remains somewhat prognathic as compared with ours. Its sexual dimorphism was similar to modern humans, with males about 1.2 times heavier than females on average. *H. erectus* was the first hominid to move out of Africa and colonized Asia at least 1.5 million years ago, and Europe at an uncertain date. In Europe, *H. erectus* evolved into the Neanderthals — fossil humans that are found in Europe from about 200,000 to 40,000 years ago. (Some experts classify the African

Australopithecines included gracile and robust forms

Figure 18.13

Two hypotheses about the evolution of anatomically modern humans. (a) The "multiregional" hypothesis suggests that anatomically modern humans evolved independently, by parallel evolution, in Africa, Europe, and Asia, from populations that ultimately originated in Africa and emigrated from Africa perhaps about 1.8 million years ago. (b) The "out of Africa" hypothesis suggests that anatomically modern humans evolved uniquely in Africa some time between 500,000 amd 100,000 years ago, emigrated from there, and replaced the indigenous humans in Asia and Europe. The first emigration from Africa around 1.8 million years ago is not controversial. The two hypotheses differ over the second emigration. The timing of the first colonization of Europe is uncertain, as is whether it was from Africa or Asia; hence the dashed lines. Intermediates between hypotheses (a) and (b) are possible.

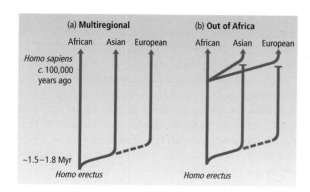

specimens as *H. ergaster* and reserve *H. erectus* for the Asian and perhaps European forms. Others use *H. erectus* for the African specimens as well.)

Paleoanthropologists refer to the kind of human beings that live today — *Homo sapiens* — as "anatomically modern humans." Anatomically modern humans differ from the fossils discussed so far in a series of details in skull anatomy. Our brains are a different shape from *H. erectus*, and our faces are flatter. When and where did anatomically modern humans originate? For the past 15 years, this question has been the topic of a debate between two hypotheses (Figure 18.13). By 500,000 years ago, human populations descended from *H. erectus* were established in Asia (and Australia) and Europe as well as Africa. The taxonomic names of these regional forms are not agreed on. Some taxonomists refer to all the regional forms as "archaic *Homo sapiens*"; some classify different regional forms as different species (*H. neanderthalensis* in Europe, for example); and some call them subspecies of *H. erectus*. The classificatory differences partly reflect the problem of squeezing continuous evolutionary change into discrete Linnaean categories and partly reflect the different theories in Figure 18.13.

Anatomically modern humans were fully established in Africa, Europe, and Asia by 30,000–40,000 years ago. The earliest definite fossils of anatomically modern humans are African, and over 100,000 years old. (Some earlier, also African, fossils may well be anatomically modern humans too.) Some paleoanthropologists argue that anatomically modern humans evolved independently in Asia, Europe, and Africa; this is the "multiregional" hypothesis (Figure 18.13a). Others argue that anatomically modern humans originated only in Africa, and then emigrated to Asia and Europe, replacing the indigenous people, with little or no interbreeding. Such is the "out of Africa" hypothesis (Figure 18.13b). (We can now see how the different names for regional forms make sense. On the multiregional hypothesis, it is appropriate to classify the 500,000-year-old populations as subspecies of *H. sapiens*, or to call them archaic *H. sapiens*. It is more appropriate to classify them as subspecies of *H. erectus* on the out of Africa hypothesis.)

Genetic evidence has tended to favor the out of Africa hypothesis. Two examples are discussed elsewhere. In Box 13.2 (p. 365) we saw that humans show little geographic variation, at a genetic level, as compared with the geographic races of other species.

Two hypotheses have been proposed for the origin of anatomically modern humans

Genetic evidence counts against the multiregional theory for Europe . . .

This suggests that modern humans share a recent common ancestor. Figure 15.18 (p. 454) showed mitochondrial DNA evidence for a recent, probably African, common ancestor of modern humans. Other pieces of evidence have been added. Ancient DNA has been extracted from Neanderthal fossil bones. Its sequence lies outside the range of modern humans. On the multiregional hypothesis, Neanderthal DNA should fit in the human phylogeny with modern European populations, inside the phylogeny of all modern humans. In fact it does not, suggesting that Neanderthals were completely replaced and made no genetic contribution to modern European populations.

The archeological and fossil evidence in Europe also fits with the out of Africa hypothesis. Anatomically modern humans appear suddenly in Europe, in the form called Cro-Magnon man, around 40,000 years ago. Neanderthals went extinct at that time. The artistic and symbolic artifacts associated with Cro-Magnons were far more elaborate than those connected with Neanderthals. The cave paintings of southern Europe, for instance, were created by early Cro-Magnons.

. . . but the Asian picture is more ambiguous

In Asia, the evidence is less clear-cut. No ancient DNA has been successfully obtained from Asian fossil humans. (Indeed none may ever be, because the fossils have to be preserved at low temperatures to maintain their DNA, and none of the Asian fossil sites have been cold for the past 100,000 years or more.) Some fossil evidence may fit with a continuous evolution of *H. erectus* into anatomically modern humans in Asia. However, it is controversial and many experts favor the out of Africa hypothesis for Asian populations too.

In summary, we have a fairly continuous fossil record of human evolution from about 4 million years ago to the present. It shows how at least most, and perhaps all, human evolution took place in Africa. The first major changes were locomotory: bipedality had evolved over 3 million years ago. Changes in brain size and prognathism came later. Brain size probably spurted up in early *Homo* around 2 million years ago. But our brains and jaws did not reach their final size and shape until anatomically modern humans originated — perhaps a little over 100,000 years ago.

18.8 Macroevolution may or may not be an extrapolated form of microevolution

The distinction between microevolution and macroevolution is the distinction between evolution on the small scale and evolution on the large scale. Microevolution refers to the topics we looked at in Part 2 of this book. It refers to changes in gene frequencies within populations, under the influence of natural selection and random drift. Macroevolution refers to the topics we are looking at in Part 5 of the book. It refers to the origin of higher taxa, such as the evolution of mammal-like reptiles into mammals, fish into tetrapods, and green algae into vascular plants. It also refers to long-term evolutionary trends, which we look at in Chapters 21 and 22, and to diversification, extinction, and replacements of higher taxa, which we look at in Chapter 23.

Microevolution and macroevolution can be thought of as vague terms, like "small" and "large," and as the ends of a continuum from evolution on the smallest scale to the largest scale. However, some biologists have argued that micro- and macroevolution proceed by distinct processes. Then the terms are not arbitrary: microevolution would refer to evolutionary phenomena driven by one set of processes and macroevolution to the evolutionary phenomena driven by a different set of processes.

We can ask, for any macroevolutionary phenomenon, whether it can be explained by microevolutionary processes that persist for a long time. That is, we can ask whether macroevolution is due to "extrapolated" microevolution. In this chapter we have looked at two major transitions in detail: the origin of mammals and the origin of humans. Some might question whether the past 4 million years of human evolution really amount to a macroevolutionary event. The origin of mammals, however, is unambiguously an example of macroevolution.

Two important points can be made about the origin of mammals. First, the changes from reptilian to mammalian characters evolved in gradual stages. Sidor & Hopson (1998) looked at the rate of evolution in mammal-like reptiles quantitatively. They measured the number of character changes per time unit and found not only that mammals evolved in many stages but also that the rate of morphological evolution was approximately constant over the 100 million-year period. The second is that the large-scale differences between mammals and reptiles concern adaptations. The mammals have a high-energy, high metabolic rate kind of physiology, with locomotory adaptations for rapid movements (upright rather than sprawling gait) and adaptations for powerful and efficient feeding (the mammalian teeth and jaw articulation). These are surely adaptive changes, which would have been brought about by natural selection.

The general evolutionary model suggested by the mammal-like reptiles, therefore, is one of cumulative action of natural selection over a long (100 million year) period. The accumulation of many small-scale changes resulted in the large-scale change from reptile to mammal. The theory of mammalian origins is therefore an extrapolative theory. A similar conclusion could be drawn about the origin of humans, and of terrestrial plants and vertebrates. In these example, macroevolution proceeds by the same process — natural selection and adaptive improvement — as has been observed within species and at speciation; but the process is operating over a much longer period. The extrapolative model is not the only model for the evolution of major groups, but it is the most important one and the only one that can be illustrated with detailed fossil evidence.

Figure 18.14 illustrates, as a theoretical alternative, how the origin of higher taxa might not be extrapolative. Higher taxa might originate when some rare process of large-scale change came into operation. Then the macroevolutionary event would not be extrapolatable from the normal processes of microevolution. No evidence exists for the process of Figure 18.14b, and it is unlikely in theory (Section 10.5.1, p. 266). However, the origins of many higher taxa have been little studied, and some biologists argue that some higher taxa may have originated by exceptional, revolutionary processes.

The two views in Figure 18.14 are only two of the possible relations between microevolution and macroevolution. Macroevolution might also be unextrapolatable from microevolution, not because their driving processes differ, but because the species that evolve into higher taxa are a non-random subset in some way. For example, Jablonski & Bottjer (1990) argued that major evolutionary break-throughs more often

Macroevolution may be due to extrapolated microevolution

The origin of mammals is an example

Figure 18.14
The origin of higher taxa could theoretically be by (a) extrapolated microevolution over long time periods, or (b) a distinct process that does not operate in microevolution. The two ideas are here illustrated for the evolution of mammals from reptiles. (a) and (b) are not the only two possible relations between microevolution and macroevolution. See also Figure 1.7 (p. 14) for another way of imagining (b).

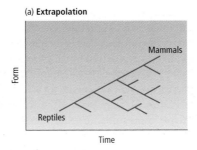

(a) **Extrapolation**

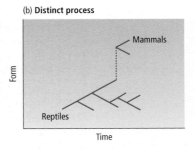

(b) **Distinct process**

But some features of macroevolution are not predictable from microevolution

occur in taxa living at the poles than in taxa living at the equator. Kemp (1999) argued that, in the origin of mammals, it was usually a small carnivorous taxa that gave rise to the next major radiation. Thus, at each stage (Figure 18.10) a variety of forms evolved — large and small herbivores, small carnivores, and others. We might expect that sometimes the next major radiation would begin with a large herbivore, and sometimes with a small carnivore. But in fact small carnivores are disproportionately represented.

If Kemp, Jablonski, and Bottjer, are right, macroevolution is not simply extrapolated microevolution. At any one time, natural selection will be favoring a variety of adaptations in different lineages — tropical adaptations in tropical species, polar adaptations in polar species. Something about the polar adaptation makes them more likely to contribute to macroevolutionary change. That something, whatever it is, cannot be seen simply by studying microevolution.

The theory of macroevolution in Figure 18.14b is controversial. If correct, it would challenge some deep tenets of neo-Darwinism. But the general idea that macroevolution is not simply predictable from microevolution need not be controversial. Kemp, Jablonski, and Bottjer's arguments are orthodox enough. In the other chapters of Part 5 we shall look at several macroevolutionary phenomena, and reflect on their conceptual relation with microevolution. In some cases, macroevolution will likely be extrapolatable from microevolution; in other cases macroevolution will likely not be extrapolatable from microevolution. In this chapter, we have seen that the origin of higher taxa can at least be mainly understood as the evolution of adaptation by natural selection over the long term.

Summary

1 Fossils are formed when the remains of an organism are preserved in the sediment deposited at the bottom of the water column; the sediment may then form a sedimentary rock by compaction over time. If that sedimentary rock is later exposed at the surface of the Earth, the fossils can be removed from it.

2 The history of the Earth is divided into a series of time stages. Most fossils are from organisms that lived in the past 600 million years. The 600 million-year period is divided into three eras (Paleozoic, Mesozoic, and Cenozoic); the eras in turn are divided into successively into periods and epochs.

3 Rock ages can be measured absolutely using their radioisotopic composition, and relatively by correlating their fossil content with other rocks elsewhere. Magnetic time zones also provide useful chronological evidence.

4 Life probably originated about 4 billion years ago. The oldest, chemical, evidence of life is from 3.8 billion years ago. The oldest (currently controversial) body fossils, in the form of cells, are 3.5 billion years old.

5 Eukaryotic cells evolved around 2 billion years ago in a series of events including symbiosis. The atmospheric oxygen concentration increased at a similar time.

6 Almost all multicellular life forms with cell differentiation are eukaryotic. Eukaryotic life forms with more than one cell type probably evolved about 1.5 years ago; the oldest fossils are algae from about 1.2 billion years ago.

7 Animal life underwent an apparently explosive radiation in the fossil record in the Cambrian, about 540 million years ago. Molecular evidence suggests that the common ancestor of the major animal groups lived about 1,200 million years ago. The fossil and molecular dates contradict each other on some interpretations, but can be reconciled.

8 The earliest fossil evidence of terrestrial plant life consists of spores from about 475, or even over 500, million years ago. Fossils from 420 million years ago show branching plant structures that lack roots and leaves. Leaves appear about 40 million years later, and their evolution may have followed a photosynthetically caused reduction in atmospheric carbon dioxide.

9 Terrestrial vertebrates evolved from fish in the Devonian. The tetrapod limb probably first evolved for paddling under water, not for walking on land.

10 The evolution of mammals from reptiles is an example of adaptive evolution and the fossil record reveals that it proceeded in a series of stages, through various groups of mammal-like reptiles.

11 Human evolution can be studied in the fossil record for the past 4 million years. The main observable changes are in bipedality, a reduction of jaws and teeth, and an increase in brain size. Other important changes were in culture and social behavior, including language, which are less easily studied in fossils.

12 Large-scale evolution, or macroevolution, may be caused by small-scale evolution, or microevolution, extended over a long period. The origin of mammals from reptiles is a possible example.

13 In other cases, macroevolution may not simply be due to microevolution over an extended period. For instance, the species in which evolutionary breakthroughs occur may be a non-random sample of all the species existing at the time.

Further reading

Texts on fossils include Clarkson (1998) for invertebrates, Carroll (1988, 1997) and Benton (2000a) for vertebrates, and Kemp (1999) for evolutionary ideas. Briggs & Crowther (2001) and Singer (1999) are encyclopedic introductions to paleobiology and paleontology, respectively. Fortey (2002) is a popular introduction. Martin (2000) is about taphonomy. McPhee (1998) is a literary book on the geological history of North America.

A further topic, not considered in this chapter, is the completeness or adequacy of the fossil record. See the general books above, also Donovan & Paul (1998) which is a multiauthor volume on it. Benton (2001) is a more recent approach. We also look at this topic in relation to the time of mammal origins in Chapter 23. A special issue of *Paleobiology* (2001), vol. 27, pp. 187–310 considers how phylogenies are being used in paleobiological research.

On the history of life, I have structured the narrative in this chapter taxonomically. Maynard Smith & Szathmáry (1995, 1999) provide a more conceptual narrative — they look at the main transitions in the way heredity occurs. From the origin of life to about the Cambrian explosion their narrative is structured much like the chapter here, but from then on the structure differs. The later book is written for a broader audience. *Nature* February 22, 2001 (vol. 409, pp. 1083–109) contains an "insight" section with several relevant reviews — on early life and habitats, extremophiles, and the rise of morphological complexity in animals.

On the origin of life, recent books include Wills & Bada (2000) and Fenchel (2002): Wills & Bada is more "popular," Fenchel more "professional." The relevant chapters in Maynard Smith & Szathmáry (1995, 1999) are good introductory summaries. *Science* March 15, 2002, pp. 2006–7 contains a journalistic profile of Wächtershäuser, who has influential alternative ideas about the prebiotic soup. Joyce (2002) is a recent review of the RNA world.

Schopf (1999) is a popular book on Precambrian fossils and research about them. *Nature* June 20, 2002, pp. 782–4 and December 5, 2002, pp. 476–8, and *Science* March 8, 2002, pp. 1812–13 and May 24, 2002, pp. 1384–5, contain news articles on Brasier *et al.*'s (2002) critique of Schopf's (1993) interpretation of the Apex Chert fossils or artifacts.

For the Cambrian explosion and origin of animals, Knoll & Carroll (1999) is an authoritative overview. Budd & Jensen (2000) look critically at the fossil evidence for an explosion. These two references also make a good "compare and contrast" on the influence of oxygen concentration. Hoffman & Schrag (2000) is an introduction to "snowball Earth" by two of its originators, and Runnegar (2000) provides a brief overview. Another recent hypothesis is that Precambrian life was limited by low concentrations of inorganic nutrients in the ocean (Anbar & Knoll 2002). The key molecular dating paper is by Wray *et al.* (1996). Cooper & Fortey (1998) look at molecular and fossil evidence, and how they can be reconciled. Gould (1989) and Conway Morris (1998) are two books on Cambrian life. Raff (1996) surveys views about the Ediacaran fauna. Nielsen (2001) is a book about the evolutionary relations between the main animal groups.

Ahlberg (2001) is about early vertebrates. On the colonization of land, see the books by Zimmer (1998) and Clack (2002) for vertebrates, and Kenrick & Crane (1997) for plants. See Shear (1991) in general. Kenrick (2001) gives a brief overview of ideas about the relation between roots, leaves, and carbon dioxide. Heckman *et al.* (2001) suggest, from a molecular clock analysis, that fungi and perhaps plants colonized the land earlier (maybe 700 million years ago for plants) than the fossil figures discussed in this text. Willis & McElwain (2002) is an introductory text on plant evolution, particularly on fossil evidence. Dilcher (2000) identifies the main themes in plant evolution. See Kellogg (2000) on grasses.

On the mammal-like reptiles see Kemp (1999), Hotton *et al.* (1986), Bramble & Jenkins (1989), Benton (2000a), and Carroll (1988, 1997). See Wellnhofer (1990) on *Archaeopteryx*. Flynn & Wyss (2002) describe Mesozoic fossils from Madagascar that help us understand the reptilian ancestry of mammals and dinosaurs. Novacek (1992) reviews mammalian phylogeny.

On human evolution, Klein (1999) is an authoritative text, Lewin (2003) a more introductory text, and Ehrlich (2000) a more personal book that nevertheless works as an introduction and text. *Science* February 15, 2002, pp. 1214–25 has a news focus on human evolution. *Science* April 23, 1999, pp. 572–3 has a newspiece on *Australopithecus garhi*. Asfaw *et al.* (2002) provide evidence for the global use of the species name *Homo erectus*.

Levinton (2001) discusses micro- and macroevolution, and refers to the literature. The topic can also be followed, more from the microevolutionists' perspectives, through several papers in the special issue of *Genetica* (2001), vols 112–113, which was also published as a separate book (Hendry & Kinnison 2001).

Many of Gould's popular essays, anthologized in Gould (1977b, 1980, 1983, 1985, 1991, 1993, 1996, 1998, 2000, 2002a), are paleobiological.

Study and review questions

1 Review (a) the events leading to fossilization, and (b) the divisions of geological time.

2 What is the age of a fossil that contains the following ratios of ^{14}C to ^{14}N? Assume all ^{14}N has been formed by the decay of ^{14}C.

	$^{14}C : ^{14}N$	Age
(a)	1 : 1	
(b)	2 : 1	
(c)	1 : 2	

3 What features of RNA make it likely to have preceded DNA as the hereditary molecule?

4 For a number of evolutionary events, including the radiation of animals and the origin of angiosperms, dates estimated from fossils seem to contradict dates from the molecular clock. How can the difference between the two estimates be reconciled?

5 What is the (possible) relation between the amount of atmospheric carbon dioxide and the evolution of terrestrial plants, roots, and leaves?

6 What features of the mammal-like reptile fossil record suggest that the mammals evolved by an extended period of microevolution?

7 What are the main changes (or classes of changes) in human evolution?

19 Evolutionary Genomics

The complete genome sequence for a species is rich in evolutionary information. This chapter looks at how genome sequences are being used to study the evolutionary history of genomes. We look at how the history of the human gene set can be inferred, by comparing the human genome with the genomes of other species. We then see how genomes expand and contract during evolution, by duplications, deletions, and gene transfers. The timing of duplication events can be inferred, and used to test whether major evolutionary events are associated with increases in gene number. We look at the history of the human sex chromosomes. We finish by looking at the evolution of non-coding DNA. Certain families of non-coding DNA seem to have proliferated at different times in the ancestry of humans and of mice.

19.1 Our expanding knowledge of genome sequences is making it possible to ask, and answer, questions about the evolution of genomes

Advances in any area of biology usually lead, in time, to a deeper understanding of evolution. From the 1960s to the 1990s, techniques to work out the amino acid sequences of proteins and the nucleotide sequence of genes were devised, perfected, and then industrialized. The resulting gush of data has allowed biologists to look again at the tree of life, as we have seen in Chapters 15 and 18. (It also led to the neutral theory of molecular evolution, as we saw in Chapter 7.) This chapter and the next will look at two new areas of evolutionary biology that have grown up along with advances in molecular genetics. In this chapter, we look at evolutionary genomics, which has grown out of whole-genome sequencing. In the next chapter we look at "evo-devo," which exploits our ability to identify the individual genes that control development.

The genome of an organism is its complete set of DNA. The genomic sequences, and partical sequences, of organisms from several species can be used to study how genomes change during evolution. *Evolutionary genomics* is concerned with any question that can be asked about the evolution of genomes. To introduce the subject, here are some examples of questions that can be asked about genome evolution.

1. How, and why, has the total size of the genomes changed?
2. Why is the DNA of some species longer than in other species?

We can break down the total genome into coding and non-coding parts, and ask questions such as:

3. Do some species have more coding DNA (that is, more or larger genes) than other species? And if so, why?
4. Why do some species have more non-coding (and perhaps "junk") DNA than others?
5. How do the different parts of genomes change in size during evolution?

The DNA is arranged in chromosomes, and we can ask about chromosome evolution:

6. How does the chromosomal arrangement of genes change during evolution? Do genes stay on the same chromosomes over evolutionary time, and in the same order, or do genes move about within and between chromosomes?

We can also try to relate genome evolution to other evolutionary events:

7. When have genomic changes occurred?
8. What genomic changes are associated with major evolutionary events, such as the origin of animals or the origin of vertebrates? Were these major events produced by changes in the number of genes, or by changes in the sequences of a constant number of genes?

Some of these questions could have been asked before the era of DNA sequencing, but the growth of DNA sequence evidence has stimulated evolutionary genomic research. The sequence data itself has also enabled many new kinds of tests. Research is at a preliminary stage, however, because research is limited to genomes that have been sequenced, or mainly sequenced. That has limited recent research to humans, the mouse (in part), the worm (*Caenorhabditis elegans*), the fruitfly (*Drosophila melanogaster*), and the weed (*Arabidopsis*) among multicellular eukaryotes, together with several prokaryotes.

Evolutionary genomics aims to answer questions about the evolution of genomes

In this chapter, I have mainly picked examples that use the human genome, not least because the human genome has been so intensively studied, since its partial publication in 2001. However, the field of evolutionary genomics is not limited to understanding the human genome, and aims to understand genome evolution in all life.

19.2 The human genome documents the history of the human gene set since early life

We can begin by looking at the part of the human genome that codes for genes. The two papers published in February 2001 (Celera 2001; International Human Genome Sequencing Consortium 2001) suggested that the human genome contains about 30,000 genes. (And the figure has been little changed in subsequent research.) The data can be refined further by concentrating on genes that code for proteins. Some genes code for RNA molecules, such as ribosomal RNA, that are not translated into protein, and these genes are excluded from the following analysis. We are concentrating on the proteome — the full set of proteins in an organism. Because most genes code for proteins, results for the proteome will be similar to results for the genome.

Figure 19.1 shows the percentage of human protein-coding genes that are homologous with genes in a range of other organisms. Humans share 21% of their genes with all cellular life forms. These are the "housekeeping" genes of each cell, the genes that regulate basic cellular machinery. The oldest fossil cells are 3,000–3,500 million years old (Section 18.3.1, p. 530). At least some, maybe all, the cellular housekeeping genes had evolved by then. Most housekeeping genes evolve slowly and have been copied with little change for billions of years. Our DNA has probably been copied 10–100 times a year on average since our bacterial ancestors. If a basic cellular gene such as a histone gene existed 3.5 billion years ago, then it would have been copied about 10^{11} times through a line of ancestors leading to each of us, with little change. The cellular housekeeping genes reverberate with "deep time" in all our DNA molecules.

Another 32% of our genes are homologous with genes in all eukaryotes, but not with bacteria. Those too are cellular housekeeping genes, reflecting the greater complexity of cellular metabolism in eukaryotes. The next stage for which we can make an inference is the origin of animals. About 24% of our genes are shared with other animals, but not with single-celled eukaryotes or prokaryotes. These "animal" genes include the genes such as *Hox* genes that control development. We look at these genes further in Chapter 20. Another 22% of our genes are shared only with vertebrates. These genes are 500 or more million years old. They include genes that operate in the immune system and the nervous system. Human beings, for example, have about 100 genes coding for the immune system, compared with more like 10 in the worm and the fly. The number of genes concerned with the nervous system has also expanded in the vertebrates, perhaps associated with the relatively complex vertebrate brain. Only 1%, or less, of human genes are "unique"' to humans, having no homologs with other vertebrates. (I put "unique" in quotes because the main other vertebrate for which we have data is the mouse. The 1% of genes we do not share with mice could well be shared with closer relatives, such as monkeys; the genes would then not be unique to us.)

Some human genes date back to early cellular life . . .

. . . and other genes were added later

Figure 19.1

The history of the human gene set. The analysis is based on protein-coding genes only. The figure shows the percentage of human genes that are shared with other taxa. The fraction of human genes that originated on each branch is shown in the tree, and the total percent similarity of genes is shown across the top. Thus, 32% of modern human genes originated after eukaryotes diverged from prokaryotes (but before animals diverged from the rest of the eukaryotes), and 53% of human genes are shared with all other eukaryotes. The total at the top is approximately equal to the sum of the percentages in the tree up to that branch (for example, 53% at the top equals 21 + 32% in the tree). Many modern human genes originated by duplication of ancestral genes. Thus, two genes in humans may be similar to one gene in yeast. If that gene is absent from bacteria, both the human genes are shown as "originating" in the branch between prokaryotes and eukaryotes even though one of the copies arose by duplication later. Alternative splicing is ignored. Non-protein-coding genes and non-coding DNA are excluded. From data in International Human Genome Sequencing Consortium (2001).

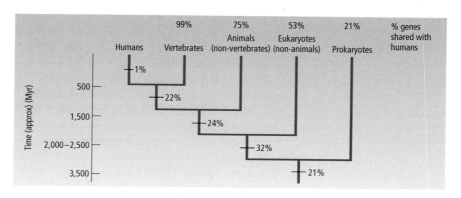

The history of our gene set, as described here, is very incomplete and uncertain. It is incomplete because it is mainly based on comparisons with a small number of other species, such as the mouse, fruitfly, worm, yeast, and *Escherichia coli*. The history will become better known as the genomes of more species are sequenced, and we can compare human DNA with DNA from a greater range of relatives. The history is also uncertain for several reasons. The methods used to recognize genes in raw DNA sequence data are subject to error. Genes may have been overlooked or wrongly compared. Also, the analysis in Figure 19.1 takes no account of "alternative splicing" (Section 2.2, p. 24). More than one protein can be read from a single gene, but the analysis considered only one protein per gene.

Thirdly, homology was inferred from relative sequence similarity. A human protein was taken to be a homolog of a protein in another species, if the human protein was much more similar to that other protein than to randomly picked proteins. For conserved, slowly evolving genes this criterion should give accurate, if approximate, results. But for genes that have evolved steadily over time, the results may be misleading. For instance, humans share a more recent common ancestor with all eukaryotes then with all prokaryotes. A gene in a prokaryote has been evolving away from us for longer, and may be less likely to be recognized as a homolog of a human gene, than a gene in a eukaryote. For this reason, the fraction of genes shown as homologous in Figure 19.1 could be biased by the technique used to recognize homologies. However, improved criteria of homology can be developed, together with other improvements in methods and data. We should then be able to flesh out the currently skeletal history of human DNA.

19.3 The history of duplications can be inferred in a genomic sequence

Genomes, as a whole or in part, change size during evolution by means of duplications and deletions (Section 2.5, p. 30). A duplication or deletion will initially be rare in the population; it may arise as a unique mutation. Its frequency may then increase by natural selection or random drift. Once a duplication or deletion has spread through the population, the genome size of the species will have gone up or down. The genome sequence of a modern species can be used to infer when duplications and deletions have occurred in the past, and we can ask what evolutionary events are associated with changes in the genome site.

For example, the genome sequence of the weed *Arabidopsis thaliana* was published in December 2000. *Arabidopsis* is a common small weed, and is the main flowering plant of genomic research. Vision *et al.* (2000) analyzed the *Arabidopsis* genome and found a large number of blocks of genes that looked like duplicates: that is, one stretch of DNA looked like a duplicate of another stretch of DNA in the same genome. The duplications are not for single genes, or the whole genome, but for intermediate lengths of DNA and contain several genes. Vision *et al.* concentrated on 103 blocks, each containing seven or more genes. For each pair of duplicate blocks, they compared the sequence of the two (paralogous) copies of each gene. They then used a molecular clock inference to estimate the time when the duplication occurred.

Most of the duplications seem to date back to three periods, at 100, 140, and 170 million years ago. Some of the duplications are older, perhaps 200 million years old. A few of them are more recent, being about 50 million years old. The modern *Arabidopsis* genome seems to reflect 3–5 major duplication periods in the past. These duplications mainly occurred in the Mesozoic geological era, 100–200 million years ago, which was a crucial time for the origin of the angiosperms. In Section 18.5 (p. 538), we saw that the time of origin of the angiosperms is uncertain. However, the molecular clock suggests that the dicotyledons split from monocotyledons about 180–210 million years ago. If so, the earliest (200 million-year-old) duplications now visible in the *Arabidopsis* genome may have originated at this time. In that case, one of the major breakthroughs of angiosperm evolution — the origin of dicotyledons — would have been associated with a round of increased gene numbers.

In summary, Vision *et al.* (2000) used the genome sequence of *Arabidposis* to identify duplicated regions of DNA and the time when the duplications evolved. They suggested some associations between these duplications and evolutionary events. The association of the 200 million-year-old duplications and the origin of dicotyledons is just one example. Their suggestions, however, are tentative and require more research. The idea that duplications may be associated with the origin of a major group is an example of the kind of hypothesis we can test, even if the test is at present rudimentary. (Box 22.1, p. 624, will look at another association between gene duplication and an evolutionary event — the origin of alcohol metabolism and the origin of fruit.)

We can now turn to another major evolutionary event that has been hypothesized to be associated with gene duplications: the origin of the vertebrates. Ohno (1970) used early measurements of the weight of DNA in several vertebrate and invertebrate species to argue that the whole genome had duplicated twice near the origin of the vertebrates. The vertebrate genomes were about four times the size of the invertebrate genomes in Ohno's sample. This is now sometimes called the "2R" hypothesis, named after the two rounds of genome doubling. If gene numbers did increase fourfold at that time, the extra genes might be part of the explanation for the origin of vertebrates. Ohno's hypothesis was relatively neglected until biologists noticed in the late 1980s that vertebrates contain four sets of *Hox* genes, compared with a single set in the fruitfly. The quadrupled *Hox* gene set seemed to fit Ohno's hypothesis.

The total number of genes in vertebrates, however, is not four times the number

Molecular clocks suggest duplications occurred at certain times in angiosperm evolution

Vertebrate origins may be associated with genome duplication

Figure 19.2

Testing the 2R hypothesis by gene tree shape. The 2R hypothesis suggests that two rounds of whole genome duplication occurred near the origin of the vertebrates. A, B, C, and D are four related copies of a gene, and they originated at gene duplications (numbered 1, 2, and 3). (a) If the 2R hypothesis is right, genes that have four related copies in a vertebrate genome should have a symmetric gene tree (1 and 2 correspond to the two whole-genome duplications). (b) In principle, these four genes could have many other gene tree shapes. For instance, if a gene initially duplicates and then only one of the copies duplicates again, and in turn only one of the new copies duplicates again. The result is four related genes, but not by two whole-genome duplication events.

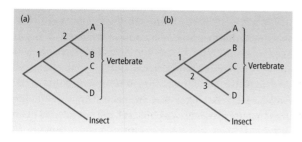

in invertebrates. Humans are thought to have about 30,000 genes, against 13,000 in fruitflies and 19,000 in the worm *Caenorhabditis elegans*. These gene numbers alone, however, do not refute Ohno's hypothesis, because some genes could have been lost after the two rounds of duplications. Gene numbers might have initially increased from 15,000 to 60,000, and then almost half the new genes may have been lost between then and modern humans. A stronger test of Ohno's 2R hypothesis can be made using the shape of gene trees (Hughes 1999; Martin 1999; Section 15.11.5, p. 457, defines gene trees) (Figure 19.2).

The hypothesis can be tested by the shape of gene trees

The test uses any gene that appears to contain four paralogs in vertebrate genomes, but only a single copy in invertebrate genomes. If Ohno's hypothesis is correct, these genes will all have originated in the same two rounds of duplication at the origin of the vertebrates and the resulting gene tree of the four will be symmetric (Figure 19.2a). In fact, in the majority of four-paralog sets analyzed by Hughes and Martin, the gene trees were not symmetric (Figure 19.2b). The evidence does not support the 2R hypothesis, which is one of the main reasons why many biologists currently doubt whether the origin of vertebrates was also the occasion of two great rounds of gene doubling. Vertebrates do contain more genes than invertebrates, but the extra genes probably evolved in a series of separate events in different gene families rather than in one or two big polyploidizations. Supporters of the 2R hypothesis still exist, however. They note that the tests performed so far are preliminary, and use only a small number of genes. They think there may be life in the old hypothesis yet, and it will certainly continue to be tested, as new evidence or methods become available.

The research program that we have looked at in this section is concerned to test whether new taxonomic groups evolve by means of gene duplications. In general, it is interesting to test what genomic events underlie events in morphological evolution. A new group might evolve by gene duplication, or sequence evolution within a fixed number of genes, or a mix of the two factors. Genome sequences can be used to find out.

19.4 Genome size can shrink by gene loss

Some bacteria live inside the cells of other species, either as parasites or intracellular symbionts. All such bacteria that have been appropriately studied share a common feature: massive gene loss and genome reduction. For example, the bacterium *Buchnera* lives symbiotically in the cells of aphids. *Buchnera* is descended from the group of enteric bacteria that includes *Escherichia coli*. *E. coli* has over 4,000 genes, but the common ancestor of the enteric bacteria probably had slightly less than 3,000 genes. *Buchnera* has only 590 genes: it has lost about 80% of its ancestor's genome. A gene loss originates as a deletion mutation, which may then spread by drift or selection. Many of the deletions in intracellular bacteria could have spread by drift. The environment inside the host cell contains many of the nutrients and defense systems that a bacterial cell needs. The resources are provided by the host, and natural selection on some of the genes in intracellular bacteria will be relaxed. Genes that are needed in a free-living bacterium to provide the resources that are present in the host cell are not needed in an intracellular bacterium. Alternatively, the gene loss may be positively advantageous. In general, a cell with less DNA can reproduce faster. Natural selection may favor gene reduction for this reason. (Box 19.1 discusses a medically interesting example. Yet another dramatic example of gene loss in an intracellular bacterium is provided by mitochondria. We look at mitochondria in the next section.)

Stretches of DNA are lost by deletion events in all species at a certain rate. Some of the non-coding DNA, for instance, is deleted from time to time, perhaps because copying it is burdensome. Differences between species in the rate of deletion of non-coding DNA may help to explain why some species have smaller genomes than others. Crickets in the genus *Laupala* have a genome size over 10 times larger than the fruitfly (*Drosophila*). Petrov et al. (2000) estimated the evolutionary rate of deletions in non-coding DNA from the two taxa. They found that DNA is deleted 40 times faster from fruitflies over evolutionary time than from crickets. Part of the explanation must be that when non-coding DNA arises in fruitflies it is more likely to be deleted. Natural selection discriminates more against fruitflies with non-coding DNA than against crickets with non-coding DNA. Why this should be is a question for the future. But in

Marginal note: Superfluous DNA can be deleted . . .

Marginal note: . . . but the efficiency of deletions varies

Box 19.1
Genome Reduction in Human Pathogens

Parasites tend to have reduced genomes, and intracellular parasites have particularly reduced genomes. *Shigella* is closely related to *Escherichia coli*. Indeed *Shigella* strains appear to evolve repeatedly and convergently from *E. coli* ancestors. Therefore, *Shigella* may not be a proper taxonomic term and we should refer to "shigella" strains within the species *E. coli* (Pupo et al. 2000).

Escherichia coli is a normally benign inhabitant of our guts, but certain strains of *Shigella* cause dysentery. During the evolution of *Shigella*, certain genes have been lost. For instance, the strains of *Shigella* that cause dysentery lack a gene (called *ompT*) that is present in benign strains of *E. coli*. The gene *ompT* can be experimentally introduced into *Shigella*, and has the effect of reducing the rate at which *Shigella* spreads between host cells. The experimental results suggests that natural selection positively favored the loss of *ompT* in the origin of these *Shigella* strains. The loss of *ompT* was advantageous not simply because it made the DNA more economic, but because the gene somehow increased the efficiency of cell infection. A knowledge of genome evolution in these bacteria provides useful clues for understanding their pathogenicity.

Further reading: Ochman & Moran (2001).

any case, we need to understand the rates of gene gain and loss in order to understand the sizes of the genomes (and of different parts of the genomes) in different species.

19.5 Symbiotic mergers, and horizontal gene transfer, between species influence genome evolution

Following symbiosis . . .

Most genome size increases in evolutionary history have occurred by duplications, of all or part of the genome. But duplications are not the only mechanism. Two species may also combine their genomes into one (or almost one), in a particularly intimate symbiosis that is rather like a business merger. In the history of human DNA, only one such event is known: the symbiosis between two bacteria that led to the eukaryotic cell containing a mitochondrion (Sections 10.4.3, p. 265, and 18.3.2, p. 533). The event probably took place 2,000–2,500 million years ago. The genome sizes of the two bacteria concerned is unknown. However, modern bacteria have a range of gene numbers, from less than 1,000 to over 6,000, with an approximate average of about 2,500 genes. The newly merged cell might have had two DNA molecules, each containing about 2,500 genes.

Since that time, one of the DNA molecules has expanded and evolved into the nuclear DNA while the other has shrunk and evolved into the mitochondrial DNA. All modern animals have mitochondria of about the same size, containing 13 protein-coding genes and 24 RNA-coding genes. The mitochondria of plants and microbes show a greater range of genome sizes, some being larger and others smaller than in animals; but even the largest mitochondrial genomes have only 100–200 genes. The reduction in gene numbers has mainly been by gene loss, in the same manner as in other bacterial intracellular symbionts (see Section 19.4). The genes were unnecessary after the symbiosis and were lost. But some mitochondrial genes were transferred to the nucleus. The nuclear DNA of modern human beings contains genes descended from both of the original eukaryotic merger partners. The process of gene transfer from mitochondria to nucleus is difficult to study in animals, because the mitochondrial genome is relatively constant. However, in plants, genes seem to be transferred more frequently and some revealing research has been done.

. . . genes have transferred to the nucleus

For example, in many plants the gene coding for ribosomal protein S14 (*rps 14*) is in the mitochondrion (Kubo *et al.* 1999). But in rice the *rps 14* gene in the mitochondrial genome is dysfunctional (it is a pseudogene). Instead the *rps 14* gene is found in the nucleus. The gene is found in an interesting place. It is inside an intron, which in turn is inside the gene for mitochondrial succinate dehydrogenase (*sdhb*). *sdhb* is an earlier mitochondrial gene that moved to the nucleus and codes for a protein that operates in the mitochondrion. One problem faced by a mitochondrial gene that is accidentally copied into the nucleus is that special targeting signals are needed for a protein to enter the mitochondrion. The mitchondrial gene must somehow acquire the targeting signal sequences if it is to work from the nucleus. The *rps 14* gene, by entering the *sdhb* gene, neatly solved this problem. Ribosomal protein S14 and succinate dehydrogenase are generated by alternative splicing (Section 2.2, p. 24) from the compound gene.

The *rps 14* story illustrates how genes transfer from the mitochondrion to the nucleus. The "targeting signal" problem is one of several detailed problems that have to

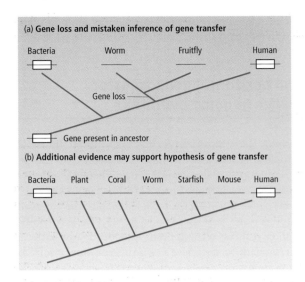

(a) Gene loss and mistaken inference of gene transfer

Bacteria   Worm   Fruitfly   Human

Gene loss

Gene present in ancestor

(b) Additional evidence may support hypothesis of gene transfer

Bacteria Plant Coral Worm Starfish Mouse Human

Figure 19.3

Extensive evidence of the phylogenetic distribution of a gene is needed to test hypotheses of gene transfer. (a) Certain "bacterial" genes are known in the human genome, and have been hypothesized to have originated by recent transfer from bacteria to humans. However, the DNA of few species have been sequenced. The observations can equally be explained by gene loss in the lineage leading to the worm and the fruitfly. (b) If we knew that the gene was lacking in many of the branches between bacteria and humans, gene loss would be less plausible, and the hypothesis of gene transfer better supported. Currently we lack the information in (b), and the origin of the "bacterial" genes in the human genome remains uncertain.

be overcome in a successful transfer. Transfers of mitochondrial genes to the nucleus have contributed to the evolutionary expansion of the nuclear genome, and evolutionary reduction of the mitochondrial genome. Gene transfer is a further mechanism, in addition to duplication and deletion, by which genomes can change in size.

Genomes also evolve by horizontal gene transfer. Horizontal gene transfer (also called lateral gene transfer) occurs when a gene is copied from the genome of one species into that of another species. It is rare event, but sequencing projects have shown it occurs at a non-trivial frequency over evolutionary time. It is probably most frequent in bacteria. Genes are even known to have transferred between Archaea and bacteria. Genes probably also occasionally transfer from bacteria into multicellular eukaryotes, but at present it is difficult to be sure that any apparently bacterial genes in a plant or animal genome are examples of horizontal gene transfer.

The reason for the current uncertainty is that we need a phylogeny with evidence for several species before we can identify examples of horizontal gene transfer. While we have evidence for only a few species, we cannot rule out the alternative hypothesis of gene loss (Figure 19.3). To see the problem, consider the human genome. The human genome contains about 100 genes that resemble bacterial genes but are not found in other animals. One interpretation is that these 100 genes have recently been transferred from bacteria to ancestral humans. The problem is that we do not have much evidence for other animals. The only complete sequences are for "the worm" and "the fly." The genes might have been present in the common ancestor of all life, and have been lost in the branch leading to worms and flies. We only need hypothesize one loss event to explain the facts. What is needed is evidence from more species, and from species that do not fall in the worm–fly branch (Figure 19.3b). If the genes were present in humans and bacteria, and absent from corals, flies, starfish, and mice, we could conclude more confidently that genes had recently moved horizontally from bacteria to us. As that information is not yet available, it is currently difficult to show that a human gene originated by horizontal gene transfer. Gene loss is at least as plausible an interpretation.

Genes can transfer between species . . .

. . . but phylogenetic evidence is needed

But horizontal gene transfer from bacteria into animals and plants probably does happen. As genome sequence evidence accumulates, we should be able to identify particular examples.

19.6 The X/Y sex chromosomes provide an example of evolutionary genomic research at the chromosomal level

Biologists are starting to use genome sequences to study the evolution of chromosomes. The evolution of chromosome numbers and size are long-standing research topics. Genome sequences enable biologists also to infer how genes have moved between chromosomes during evolution, and how the structure of chromosomes has changed over time. One illustrative example comes from the evolution of sex chromsomes.

All mammals have X/Y chromosomal sex determination. The reptilian ancestors of mammals probably did not have X/Y sex determination. The mammalian system likely originated 300 or more million years ago in a mammal-like reptile, when a "male-determining" gene arose on one chromosome. That chromosome has evolved into the modern Y chromosome.

The X and Y chromosomes are peculiar in that they do not recombine, except for small regions at the tips. Genes are not exchanged between the X and Y chromosomes. The genes on the X and on the Y chromosomes evolve apart over time, unlike the genes on the autosomes. If a superior version of a gene evolves on any chromosome except a sex chromosome, it has a good chance of spreading through all copies of its chromosome in the population. Natural selection can increase its frequency to 100%. Every individual will then have a chromosomal pair with the same gene on each chromosome. If a superior version of a gene arises on an X (or on a Y) chromosome, natural selection can only increase its frequency until it is present in every X (or Y) chromosome in the population.

At an early stage, when the chromosome pair that have now evolved into the X and Y were a normal chromosome pair, the genes did not on average differ between the two chromosomes in an individual. Then, as gene exchange between the X and Y chromosomes shut down, the genes on the X evolutionarily diverged from the genes on the Y. The amount of divergence between the genes on the X and Y chromsomes now depends on the time since gene exchange came to a stop.

Lahn & Page (1999) used the genetic difference between genes on the X and Y chromosomes to reconstruct the time course of chromosomal evolution. They found sequence evidence for 19 pairs of genes (pairs in which one version of the gene is on the X and the other on the Y), and looked at the genetic difference for each pair. The 19 gene pairs fell into four discrete categories, rather than showing a continuous range of differences (Figure 19.4). Moreover, the four categories of genes fell into four bands down the X chromosome.

Lahn and Page's interpretation is that gene exchange shut down in four discrete stages. The gene pair with the greatest difference belong to the chromosomal region

X and Y chromosomes evolve apart . . .

. . . with genes on the two showing four degrees of difference . . .

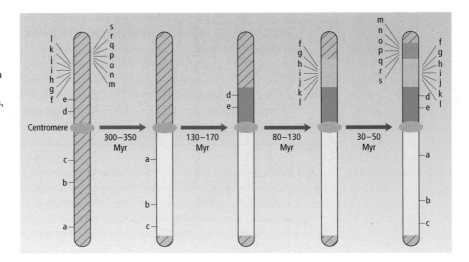

Figure 19.4
Evolution of the human X chromosome, according to Lahn & Page (1999). In the modern X chromosome, shown on the right, 19 genes (labelled a–s) fall into four age categories, according to the amount of divergence between equivalent genes on the X and Y chromosomes. Recombination between the sex chromosomes may have been shut down in four stages by four inversion events. A molecular clock inference gives the dates shown for the four inversions. The hatched regions show where recombination between the X and Y chromosomes occurred at each stage. Recombination still takes place between the tips of the two chromosomes. Free recombination occurs between the two X chromosomes in females. The Y chromosome is not illustrated.

. . . allowing a richly predictive theory of the history of the sex chromosomes

where gene exchange shut down first. A molecular clock inference suggests that gene exchange ceased for the genes 300–350 million years ago — about the time when the mammalian sex chromosomal system may have originated. Gene exchange was then shut off successively for the next three chunks of the chromosome. The most recent of the four bands has genes that have been diverging for 30–50 million years (Figure 19.4).

What events caused the shut down in gene exchange? Lahn and Page suggest there were four chromosomal inversions. Chromosomal inversions prevent recombination within the inverted region (Figure 19.5). We might predict, therefore, that the Y chromosome should have the same four bands as the X, but with the genes in an inverted order within each band. However, the genes are not arranged in the same four bands on the Y chromosome. This may be because the genes have moved about since the inversion events, or because some other mechanism than the inversions hypothesized by Lahn and Page was at work. At present, human beings are the only species for which enough sequence information exists to allow this kind of analysis. But Lahn and Page's hypothesis of a four-stage shut down in gene exchange between the X and Y chromosomes is rich in predictions about the sex chromosomes of other mammals. As genomic sequences accumulate, a stronger test should become possible.

The cessation of gene exchange between the X and Y chromosomes, whether it occurred in four stages or not, may explain another fact about genome evolution: the evolutionary shrinkage of the Y chromosome. The reason why the Y chromosome has become smaller over time is likely because recombination is advantageous (Sections 12.1–12.3, pp. 314–27). The Y chromosome now almost entirely lacks the advantages of sex, and the genetic information on it has decayed. The X chromosome has not shrunk in the same way, because recombination persists as normal (in females) between the X chromosomes.

In summary, Lahn & Page (1999) have used the genomic sequence information for human sex chromosomes to infer the evolutionary history of gene exchange. This led them to hypothesize four stages of gene rearrangement by inversion. They were also

Figure 19.5
(a) A chromosomal inversion has a set of genes inverted. The letters represent genes along the chromosomes.
(b) Recombination in a heterozygote for a chromosomal inversion can produce chromosomes that lack some genes and have others in double doses. These forms are probably selected against.

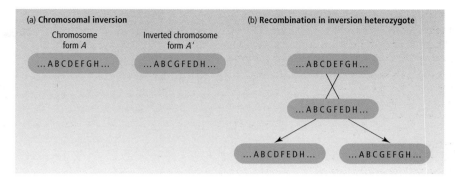

able to date the four events. Their hypothesis may or may not hold up as further research is done, but is a good example of the kind of inferences that genomic data are making possible for chromosomal evolution.

19.7 Genome sequences can be used to study the history of non-coding DNA

The "coding" part of human DNA — the part coding for the genes that regulate, build, and defend our bodies — makes up less than 5% of our genome. The rest is "non-coding" DNA. Non-coding DNA may be useless "junk" DNA that has no function in the body or it may have some structural or regulatory function. Here we can look at how the human genome sequence has been used to infer the evolutionary history of one large class of non-coding DNA, that derived from transposable elements (Section 2.5, p. 29).

Much of our DNA orginated from transposable elements

About 45% of the human genome is derived from transposable elements. These stretches of DNA are of four main kinds: short interspersed elements (SINEs), long interspersed elements (LINEs), long terminal repeat (LTR) retrotransposons, and DNA transposons. The first three kinds are "jumping genes" that are copied via an RNA intermediate. The most important SINE in our DNA is a sequence called *Alu*. The *Alu* unit sequence is somewhat less than 300 nucleotides long, and our DNA contains over a million copies of it: rather over 10% of human DNA consists of the *Alu* sequence. A LINE called LINE1 makes up even more of our DNA — about 17%.

We can take any two copies of a sequence such as *Alu* in our DNA, count the number of differences between them, and use a molecular clock to estimate how long ago the duplicative "jumping" event took place to give rise to them. The International Human Genome Sequencing Consortium (2001) performed an analysis of this kind for all the identified transposon-derived DNA in the human genome. Three patterns can be noticed. One is that the different kinds of transposable element have been more or less active at different times in human history. Table 19.1 gives the percentage of the human genome that consists of each of the three kinds of transposon, dating to particular times of origin. We can see that the *Alu* element, for example, had a burst of proliferation

Table 19.1
Time of origin of repetitive DNA in (a) humans, and (b) the mouse. The entries in the table show the percentage of the genome that belongs to the specified class of repetitive DNA and originated at the specified time. Not all dates and repetitive DNA classes are shown so the percentages do not add up to all the repetitive DNA in the genome. (a) In humans, note the burst of *Alu* origination 25–75 million years ago, and the relative quiescence of repetitive DNA origination in the past 25 million years. (b) In mice, neither feature is seen.

(a) Human.

Time (Myr)	Class of repetitive DNA		
	SINE (*Alu*)	LINE (LINE1)	LTR
0–25	0.5	0.5	0
25–50	4.5	2	0.25
50–75	3.5	1.5	1.5
75–100	1	2.5	2

(b) Mouse.

Time (Myr)	Class of repetitive DNA		
	SINE (*Alu*)	LINE (LINE1)	LTR
0–25	2	2.5	2
25–50	2	3	2
50–75	3	2.5	2
75–100	2	2.5	2

between 75 and 25 million years ago. About 10% of our DNA consists of *Alu* sequences, and 80% of it originated in that 50 million-year period.

A second pattern is that all transposable elements seem to have become quiet in the past 25 million years. Before 25 million years ago, new transposon-derived DNA was added to our DNA at a relatively steady rate (though different kinds of transposon were contributing more or less at different times). But little new DNA has been added in the recent evolutionary past.

Both these patterns are provisional, because the International Sequencing Consortium used a preliminary method of analysis. I said that we could infer the date of the common ancestor for any two copies of *Alu* (for example) by counting the differences and applying a molecular clock. The International Sequencing Consortium compared each sequence to a consensus sequence — counting the differences between each *Alu* and a consensus *Alu*. Ultimately, biologists will aim to reconstruct a gene tree for the *Alu* sequences in our DNA, and estimate the number of sequence changes in the branches of the tree. An analysis of that sort will give a more reliable history of our non-coding DNA than the figures in Table 19.1.

Transposable elements may have become quiescent in recent human history

However, the third pattern in the results suggests that the first two patterns are not simple artifacts of the method. The same analysis was performed on the available genome sequence for the mouse. The mouse showed neither pattern — neither a burst of *Alu* activity 25–75 million years ago, nor a slow down in all transposable element activity in the past 25 million years. Thus the methods alone do not automatically generate these patterns; but it is too early to say that the patterns are real. If the patterns are confirmed in more rigorous analyses, they will require explanation. Why, for example, has transposable element activity slowed down in the past 25 million years of human evolution?

Mice do not show the same pattern

The main point here is that by identifying particular kinds of repeat sequence, counting the differences between them, and applying a molecular clock, we can infer the history of non-coding DNA. The particular patterns found in a preliminary analysis are interesting, but they may not hold up in future research. In any case, some patterns will emerge in time from the sequence data. When they do, biologists will be after an explanation for them.

19.8 Conclusion

At one level, evolutionary genomics is not so new. Biologists have been studying the 3,500 million-year history of our bodies for a century and a half, and there must clearly be an equivalent history of our DNA as well. Biologists could have seen that the questions of evolutionary genomics could be asked. Indeed, some modern research has grown out of earlier ideas, such as Ohno's "2R" hypothesis. What has changed is our ability to answer the questions. We have more evidence, and many new techniques that have been invented to test hypotheses with that evidence.

The examples of research that we have looked at in this chapter illustrate a new science in that the research could not have been done much, if at all, before the year 2000. The evidence would have been lacking. The history of the human gene set can only be inferred when we know most of the DNA sequence for several species. Investigations of the timing of duplications, deletions, and gene transfers, require sequence data. The history of the human sex chromosomes could not be reconstructed until we had sequences for the coding parts of the X and Y chromosomes. The results we have seen in this chapter are more provisional than the results in much of the rest of this book. However, evolutionary genomics is worth looking at as much for its promise as for its achievements so far — interesting thought the initial results are. Evolutionary genomics is likely to be one of the fastest growing areas of evolutionary biology in the coming years.

Summary

1 Genome sequencing and development genetics are two areas of molecular genetics that are adding to our understanding of evolution, in particular macroevolution.

2 The protein-coding genes of the human genome can be compared with genes in prokaryotes, unicellular eukaryotics, invertebrate animals, and vertebrates. About 20% of our genes are shared with all life; a further 32% originated in single-celled eukaryotes; 24% were evolved before the origin of animals; and 22% near the origin of vertebrates. The human genome can be used to study the history of human DNA.

3 A genome sequence contains duplicated regions, and (using a molecular clock) the history of duplications has been inferred in the flower plant *Arabidopsis*. The evolutionary rate of gene duplication and loss in *Arabidopsis* is about as high as the rate of nucleotide substitution.

4 The history of duplication can be studied in the shape of the gene tree for paralogous genes in a modern genome. The evidence does not support the hypothesis that the genome was duplicated twice near the origin of vertebrates.

5 The genomes of intracellular parasites and symbionts tend to shrink, by gene loss, over evolutionary time.

6 The genomes of eukaryotes contain a compound gene set, descended from the symbiotic merger event that led to the evolution of the eukarotic cell.

7 The mammalian sex chromosomes seem to have evolved in four (dateable) stages, perhaps corresponding to four inversions that prevented gene exchange.

8 The history of non-coding DNA can be inferred using a molecular clock on regions of the genome that derive from transposition. Transposable elements may have become exceptionally immobile in the past 25 million years of human evolution.

Further reading

On evolutionary genomics in general, molecular evolution texts such as Page & Holmes (1998) and Graur & Li (2000) contain much material, as does Hughes (1999). The special issues of *Nature* and *Science* about the genome of particular species are informative. Bennetzen (2002) discusses another topic, the rice genome.

King & Wilson (1975) provide a classic view on regulatory genes — discussed in the evo-devo section of this chapter.

Ohno's (1970) "2R" hypothesis is the subject of a newspiece in *Science* December 21, 2001, pp. 2458–60. Lynch & Conery (2000) estimate the rate of duplications, and find it is about the same as the rate of base substitutions. A further topic is whether one gene in a newly duplicated pair experiences relaxed selection and evolves fast to a new function. This is a further example of the "valley crossing" theory of evolution. Hughes (1999) provides evidence against, but see also Lynch & Conery (2000).

Ochman & Moran (2001) review gene loss, in parasites and elsewhere. *Nature* May 23, 2002, pp. 374–6 contains a newspiece on genome size. For gene loss and gene transfers following symbiotic mergers, see Blanchard & Lynch (2000), and Martin *et al.* (1998) for chloroplasts. I discuss gene transfers and mergers in Ridley (2001).

On the general topic of chromososmes, and particularly sex chromosomes, see O'Brien *et al.* (1998) and O'Brien & Stanyon (1999). For the three phases of chromosome evolution in human history, see Burt *et al.* (1999). In addition to the work of Lahn & Page (1999) that we looked at in the text, several other hypotheses have been posed specifically about sex chromosomal genomics. Ohno (1970) put forward his hypothesis of the evolutionary conservation of the X chromosome (in mammals): that genes will move on and off the X chromosome less than for autosomes, because of the peculiar gene regulatory difficulties. Lahn & Page (1999) might lead to a fourfold complication of Ohno's hypothesis. Other classic material is covered in White (1973).

In the evolution of repetitive DNA, a further topic is "concerted evolution." Elder & Turner (1995) is a review.

Study and review questions

1 (a) How can we estimate the date when a gene duplicated? (b) What evolutionary events are thought to be associated with major rounds of gene duplication?

2 Some genes are present in the genomes of bacteria and of humans, but not in the genomes of worms and fruitflies. What are the two main hypotheses to explain this observation, and how could they be tested between?

3 In the human genome, pairs of genes on the X and Y chromosomes fall into four regions down the chromosome, according to the sequence similarity of the genes in the pair. The same is not true for pairs of genes on pairs of autosomes. Why is there this difference between sex chromosomes and autosomes?

20 Evolutionary Developmental Biology

Evolutionary developmental biology, now often known as "evo-devo," is the study of the relation between evolution and development. The relation between evolution and development has been the subject of research for many years, and the chapter begins by looking at some classic ideas. However, the subject has been transformed in recent years as the genes that control development have begun to be identified. This chapter looks at how changes in these developmental genes, such as changes in their spatial or temporal expression in the embryo, are associated with changes in adult morphology. The origin of a set of genes controlling development may have opened up new and more flexible ways in which evolution could occur: life may have become more "evolvable."

20.1 Changes in development, and the genes controlling development, underlie morphological evolution

Morphological structures, such as heads, legs, and tails, are produced in each individual organism by development. The organism begins life as a single cell. The organism grows by cell division, and the various cell types (bone cells, skin cells, and so on) are produced by differentiation within dividing cell lines. When one species evolves into another, with a changed morphological form, the developmental process must have changed too. If the descendant species has longer legs, it is because the developmental process that produces legs has been accelerated, or extended over time. Evolutionary changes in development, and developmental genetics, are the mechanism of all (or almost all) evolutionary change in morphology. We need to understand developmental evolution in order to understand morphological evolution. The same need not be said of molecular or chromosomal evolution: we do not need to study development in order to study molecular and chromosomal evolution. Some other kinds of evolution, such as behavioral evolution, can also have a developmental basis. But this chapter concentrates on the developmental basis of morphological evolution.

Morphological evolution is driven by developmental evolution

Biologists have recognized since the nineteenth century that development is the key to understanding morphological evolution. In the past 10–15 years, a new field of research has grown up. Many genes that control development have now been identified, and molecular techniques can be used to study how those genes have changed between species. The new field is often called by the informal term "evo-devo." In this chapter we shall look briefly at some older theories about developmental change and morphological evolution. We then look in more detail at some examples of modern "evo-devo" research. The ancient and modern research is imperfectly integrated because modern genetics has not yet identified the genes that underlie the structures and organs that were studied in earlier work. However, we can see how modern ideas can be used in an abstract way to explain earlier observations. The aim of all the research, from the nineteenth century to today, is to use a knowledge of development to explain how morphological evolution proceeds.

20.2 The theory of recapitulation is a classic idea (largely discredited) about the relation between development and evolution

Recapitulation is a bold and influential idea that is particularly associated with Ernst Haeckel (Section 1.3.3, p. 12) though many other biologists also supported it in the nineteenth and early twentieth centuries.

According to the theory of recapitulation, the stages of an organism's development correspond to the species' phylogenetic history: in a phrase, "ontogeny recapitulates phylogeny." Each stage in development corresponds to (that is, "recapitulates") an ancestral stage in the evolutionary history of the species. The transitory appearance of structures resembling gill slits in the development of humans, and other mammals, is a

Figure 20.1
Recapitulation, illustrated by fish tails. (a) The development of a modern teleost, the flatfish *Pleuronectes*, passes through (starting at the top) a diphycercal stage, to a stage in which the upper lobe of the tail is larger (heterocercal), to the adult, which has a tail with equal-sized lobes (homocercal). (b) Adult forms in order of evolution of tail form, from top to bottom: lungfish (diphycercal), sturgeon (heterocercal), and salmon (homocercal). Reprinted, by permission of the publisher, from Gould (1977a).

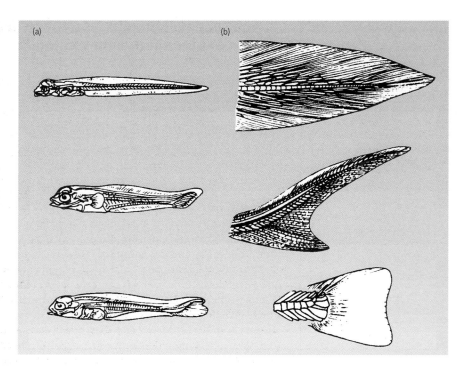

Recapitulation results from evolution by terminal addition

striking example. Mammals evolved from an ancestral fish stage and their embryonic gill slits recapitulate the piscine ancestry.

Another example, often quoted in the nineteenth century, is seen in the tail shapes of fish (Figure 20.1). During the development of an individual, evolutionarily advanced fish species, such as the flatfish *Pleuronectes*, the tail has a diphycercal stage in the larva. It then develops through a heterocercal stage, to the homocercal form of the adult. However, not all fish have homocercal tails in the adult. Indeed fish species can be found with all three kinds of tail in the adult. The lungfish, sturgeon, and salmon in Figure 20.1b are examples. The lungfish is thought to most resemble an early fish, the sturgeon to be a later stage, and the salmon to be the most recently evolved form. Thus evolution has proceeded by adding on successive new stages to the end of development. We can symbolize the diphycercal, heterocercal, and homocercal tails by A, B, and C, respectively. The development of the early fish advanced to stage A and then stopped. Then, in evolution, a new stage was added on to the end: the development of the fish at the second stage was A → B. The final type of development was A → B → C. Gould (1977a) named this mode of evolution *terminal addition* (Figure 20.2a).

When evolution proceeds by terminal addition, recapitulation is the result. An individual at the final evolutionary stage in Figure 20.2a grows up through stages A, B, and C, recapitulating the evolutionary history of the ancestral adult forms. However, evolution does not always proceed by terminal additions. We can distinguish two kinds of exception. One is that new, or modified, characters can be intruded at earlier

Figure 20.2

(a) Evolution by terminal addition. The stages in an individual's development are symbolized by alphabetic letters. (1), (2), and (3) up the page represent three successive evolutionary stages. With terminal addition, new stages are added only to the end of the life cycle. (b) Evolution by non-terminal addition. A new evolutionary stage has been added in early development, not on to the end of the life cycle in the adult.

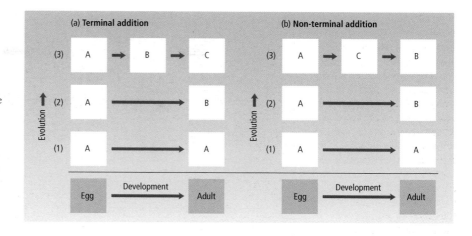

developmental stages (Figure 20.2b). Many specialized larval forms are not recapitulated ancestral stages (for example, the zoea of crabs, the Müller's larva of echinoderms, and the caterpillar of Lepidoptera). They probably evolved by modification of the larva, rather than by adding on a new stage in the adult.

The time of maturity may shift to an earlier developmental stage . . .

The second kind of exception arises when the members of a species evolve to reproduce at an earlier developmental stage. We need to distinguish the rate of reproductive development from the rate of somatic development. (The somatic cells make up all the cells in the body except the reproductive cells.) Somatic development proceeds through a series of stages, from egg to adult. If the organism becomes reproductively mature at an earlier stage, then its development will not fully recapitulate its ancestry. Its ancestral adult form has been lost. Reproduction in what was ancestrally a juvenile form is called *pedomorphosis*. Pedomorphosis can arise in two ways (Figure 20.3). One is *neoteny*, where somatic development slows down in absolute time, while reproduction development proceeds at the same rate. The other is *progenesis*, where reproductive development accelerates while somatic development proceeds at a constant rate.

. . . and axolotls are examples

Among modern species, the classic example of neoteny is the Mexican axolotl, *Ambystoma mexicanum*. The axolotl is an aquatic salamander. (Actually, we should say "axolotls" because there are a number of types, and Shaffer's (1984) fine-scale genetic work has shown that the kind of larval reproduction described below has evolved many times, independently, even within what appears to be one species.) Most salamanders have an aquatic larval stage that breathes through gills; the larva later emerges from the water as a metamorphosed terrestrial adult form, with lungs instead of gills. The Mexican axolotl, however, remains in the water all its life and retains its external gills for respiration. It reproduces while it has this juvenile morphology. However, a Mexican axolotl can be made to grow up into a conventional adult salamander by a simple treatment (it can be done, for instance, by injection of thyroid extract). This strongly suggests that the timing of reproduction has moved earlier in development during the axolotl's evolution. Otherwise there would be no reason for it to possess all the unexpressed adaptive information of the terrestrial adult.

Figure 20.3
Pedomorphosis, in which a descendant species reproduces at a morphological stage that was juvenile in its ancestors, can be caused by (a) progenesis, in which reproduction is earlier in absolute time, or (b) neoteny, in which reproduction is at the same age but somatic development has slowed down.

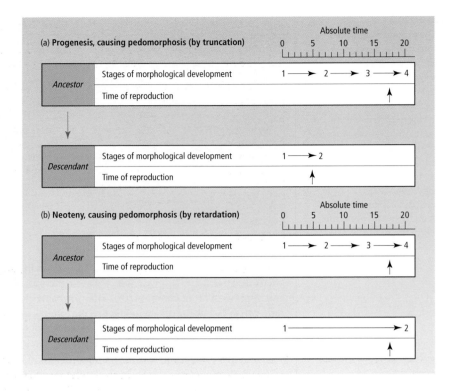

Pedomorphosis can evolve in two ways

So the Mexican axolotl is pedomorphic — but is it neotenous or progenetic? Its age of breeding (and the body size at which it breeds) is not abnormally early (or small) for a salamander. Its time of reproduction has therefore probably stayed roughly constant, while somatic development has slowed down. The axolotl is an example of neoteny. Humans have also been argued to be neotenous. As adults, we are morphologically similar to the juvenile forms of great apes. This pedomorphosis, if it is real (and there is a serious argument that it is not), would be neotenous rather than progenetic because our age of breeding has not shifted earlier relative to other apes. Our age of first breeding is actually later than other apes. Our somatic development has not simply slowed down while reproductive development has stayed the same. What might have happened was that our somatic development slowed down even more than our reproductive development.

In summary, Haeckel and others initially suggested that evolution almost always proceeds in one mode. Changes are made only in the adult, and new stages are added on to the end of the existing developmental sequence. Through the 1920s, biologists come to accept a broader view. Evolution does often proceed by terminal addition, and recapitulation results. But other developmental stages can also be modified, and the timing of reproductive and somatic development may be altered in any way — some of which result in recapitulation, and others which result in pedomorphosis (Table 20.1).

The changes that we have been considering in the relative rate of somatic and reproductive development are one example of an important general concept: *heterochrony*.

Table 20.1

Categories of heterochrony. In modern work, the term pedomorphosis is sometimes substituted for recapitulation. From Gould (1977a). © 1977 President and Fellows of Harvard College.

Developmental timing		Name of evolutionary result	Morphological process
Somatic features	**Reproductive organs**		
Accelerated	Unchanged	Acceleration	Recapitulation (by acceleration)
Unchanged	Accelerated	Progenesis	Pedomorphosis (by truncation)
Retarded	Unchanged	Neoteny	Pedomorphosis (by retardation)
Unchanged	Retarded	Hypermorphosis	Recapitulation (by prolongation)

Developmental change can be by heterochrony

Heterochrony refers to all cases in which the timing or rate of one developmental process in the body changes during evolution relative to the rate of another developmental process. In progenesis, neoteny, and so on (Table 20.1) the rate of reproductive development is sped up or slowed down relative to the rate of somatic development.

Heterochrony is a more general concept, however. It also refers to changes in the development of one somatic cell line relative to another. Consider, for example, a *D'Arcy Thompson transformation* (Figure 20.4). D'Arcy Thompson (1942) found that related species superficially looking very different could in some cases be represented as simple Cartesian transformations of one another. We met the most thoroughly worked out modern example in an earlier chapter (Raup's analysis of snail shell shapes: Figure 10.9, p. 278). With some simplification, the axes on the fish grids in Figure 20.4 or the snails of Figure 10.10 can be thought of as growth gradients. The evolutionary change between the species would then have been produced by a genetic change in the rates of growth in different parts of the fish's body.

Apparently complex change may have a simple basis

One general point is that evolutionary changes between species may be simpler than we might at first think. If we looked at, for example, *Scarus* and *Pomacanthus* without the grids of Figure 20.4 we might think that an evolutionary change from one into the other would be at least moderately complicated. The interest of D'Arcy Thompson's diagrams is then to show that shape changes could have been produced by simple regulatory changes in growth gradients. The more specific point here is that changes in the growth gradients of different parts of the body are further examples of heterochrony. Evolutionary changes in morphology are often produced by changes in the relative rates of different developmental processes: that is by heterochrony. Heterochrony also explains evolutionary changes in allometry, which we looked at in Section 10.7.3, (p. 279).

Figure 20.4

A D'Arcy Thompson transformational diagram. The shapes of two species of fish have been plotted on Cartesian grids. *Argyropelecus olfersi* could have evolved from *Sternoptyx diaphana* by changes in growth patterns corresponding to the distortions of the axes, or the direction of evolution could have been in the other direction, or they could have evolved from a common ancestral species. Likewise for *Scarus* and *Pomacanthus*. Reprinted, by permission of the publisher, from Thompson (1942).

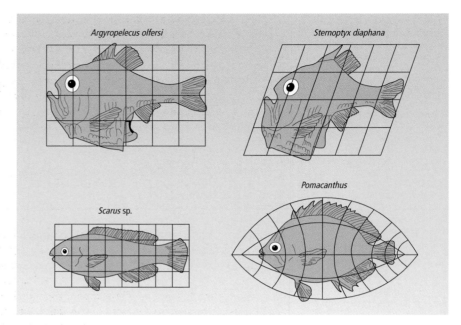

20.3 Humans may have evolved from ancestral apes by changes in regulatory genes

Biologists distinguish between regulatory genes and structural genes. Structural genes code for enzymes, building block proteins, and transport and defensive proteins. Regulatory genes code for molecules that regulate the expression of other genes (whether structural or regulatory). The distinction is imperfect, but can be used to make a point about evolution.

Britton and Davidson wrote some influential early papers in which they suggested that reorganizations within the genome's regulatory gene pathways could cause important evolutionary changes (for instance, Britton & Davidson 1971). King & Wilson (1975) then applied this general perspective to a striking example: human evolution. King and Wilson used several techniques to infer that the DNA of humans and chimpanzees is almost identical. Later work has supported their conclusion that only about 1.5% of nucleotide sites differ between human and chimpanzee DNA.[1] And yet, to our eyes, humans and chimpanzees are phenotypically very different. Human bodies have been redesigned for upright walking, human jaws have become shorter and weaker,

Humans and chimps are more similar genetically . . .

---

[1] See Figure 15.12 (p. 422). As also noted near that figure in Chapter 15, Britton (2002) has recently revised the percent similarity between human and chimpanzee DNA to more like 95%, after allowing for insertions and deletions. About 1.5% of nucleotide sites show substitutions, and another 3.5% of sites differ because of insertions or deletions. However, King and Wilson's essential argument is unaltered.

. . . than might be expected from morphology

and human brains have expanded, and we have acquired the use of language. In human evolution, a large phenotypic change appears to have been produced by a small genetic change. King and Wilson hypothesized that most of the genetic changes of human evolution were in regulatory genes. A small change in gene regulation might achieve a large phenotypic effect. We shall not know what genetic changes occurred in human evolution until we have (and understand) the genome sequences for chimpanzees and some other apes, as well as for human beings. But King and Wilson's hypothesis remains a popular idea about human evolution.

20.4 Many genes that regulate development have been identified recently

Figure 20.5
There are two main classes of developmentally influential genes. (a) Transcription factors (TF) that bind enhancers, which can switch genes on or off. The state of the enhancer determines whether RNA polymerase binds the promotor. The binding of RNA polymerase to the promotor is the first step in the transcription of a gene. A stretch of DNA may exist between the enhancer and promotor. (b) Signaling proteins. A signaling pathway in the cell may lead from a receptor molecule in the cell membrane, ultimately to a transcription factor which can be active or inactive. When the transcription factor is activated, it can switch a gene on by the process shown in (a). Many proteins may be able to interact with a receptor protein in the control of cellular metabolism: all such molecules are (provided they are proteins) examples of signaling proteins. Also, receptor proteins may be bound by molecules other than those conventionally classified as hormones. From Carroll *et al.* (2001).

A long list of genes that operate during development is now known, and the list is rapidly expanding. The genes fall into two main categories: genes that code for transcription factors and genes that code for signaling proteins (Figure 20.5). Transcription factors are molecules that bind *enhancers*. An enhancer is a stretch of DNA that can switch on a specific gene. Signaling proteins function in the cell's control pathways for switching specific genes on and off. For instance, a receptor protein in the cell membrane might change shape when bound by a hormone. The shape change might trigger further molecular changes in the cell, ultimately leading to the release of a transcription factor that switches on a specific gene. The protein in the cell membrane, or any other problem in the chain of reactions, would be an example of a signaling protein. Almost all the genes discussed in this chapter are transcription factors. The *Hox* genes, for

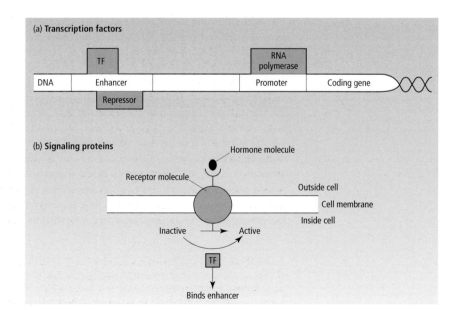

example, as well as such genes in fruitflies as *distal-less*, *eyeless*, and *engrailed* all code for transcription factors. However, other developmental genes, such as the genes in fruitflies called *hedgehog*, *notch*, and *wingless*, are signaling proteins, and most of the points of principle that we look at for transcription factors would also apply for signaling proteins.

The genes that regulate development are best understood in two species, the mouse and the fruitfly. However, geneticists have looked for the same genes in other species and their findings have led to an important generalization. All animals seem to use much the same set of genes to control development. For example, the *Hox* genes were first studied in fruitflies. After the genes were cloned it was possible to look for them in other species too, and they were duly found in every other animal taxon. The *Hox* genes have similar functions in all animals. They act as region-specific selector genes. The basic map coordinates of the early embryo are set out by another set of genes. Then, during development, specific sets of genes are switched on to cause the correct structures to develop in each region of the body. The genes for building a head have to be switched on at the top of the body, for example. Different *Hox* genes are expressed in different body regions, and act to switch on other genes that code for appropriate structures. The *Hox* genes mediate between the basic body map information and the genes that code for the structures in each body region.

The finding that all animals use much the same set of developmental genes might not have been predicted. The main groups of animals — the Protostoma and Deuterostoma (Figure 18.5, p. 536) — were initially defined by basic differences in how the animals develop. In the protostomes, cleavage in the egg is spiral; in the deuterostomes it is radial. In protostomes the embryonic structure called the blastopore develops into the mouth; in deuterostomes the blastopore develops into the anus. And so on. It might have been expected that these deep differences in development would reflect different genes regulating development. But in fact the same set of genes is at work in both taxa. The genes that regulate development presumably evolved once, when animals with development first originated, and has been conserved ever since.[2]

20.5 Modern developmental genetic discoveries have challenged and clarified the meaning of homology

The eyes of insects and the eyes of vertebrates were, until the early 1990s, considered to be a standard example of "analogous" structures. They perform the same function but have utterly different internal structures, suggesting that they evolved independently from a common ancestor that lacked eyes. Then the laboratory of Walter Gehring in Switzerland began to research genes that are crucial for eye development in fruitflies and mice. One gene, *ey*, was known to be needed in fruitflies; another gene, *Pax6*, was

The *Hox* genes function in the development of all animals

---

[2] The remarks here about "all animals" apply most clearly to triploblastic Bilateria: that is, to all animals except sponges, Cnidaria (corals, jellyfish, and sea anemones), and ctenophores (Figure 18.5, p. 536). The developmental genetics of sponges, Cnidaria, and Ctenophores are more uncertain.

A similar gene works in eye development in both mice and flies

needed in mice. The sequences of the two genes turned out to be similar, suggesting that they are really the same (that is, homologous) gene. The *ey* gene could be shown to cause eye development in fruitflies, because if the gene is switched on in inappropriate parts of the body, such as a leg, it induces the development of an "ectopic" eye.[3] Then genetic tricks were used to introduce the fruitfly *ey* gene into mice. These mice grew up with fly-type compound eyes. It seems that the same gene is used in both mice and fruitflies to cause eye development. If the insect and vertebrate eyes have evolved independently, we would hardly expect them to have hit on the same gene to act as the master gene of eye development.

Two interpretations are possible. One is that the common ancestor of fruitflies and mice had eyes. The structure of insect and vertebrate eyes are still so different that they probably evolved independently, but perhaps from a common ancestor that had, rather than lacked, eyes. The eye in that common ancestor might have been a much simpler structure (Section 10.3, p. 261), but there would be an element of homology between the insect and vertebrate eyes. The evolution of eyes in the two taxa would have been easier if they already possessed the developmental genetic machinery for specifying something about eye development.

The homology may be more, or less, specific

Alternatively the homology may be more abstract: *ey/Pax6*, or the ancestral gene from which they evolved, might have specified some activity only in a particular location in the body (the top front of the head). Then the use of the same gene in mice and fruitflies would reflect only the fact that the two animals grow eyes in a similar body region. The common ancestor of mice and fruitflies had a head, and would have had genes to work in the regions of the head. It would be less remarkable if mice and fruitflies have homologous genes for controlling development in a particular region of the head, than if they have homologous genes for developing eyes. At some level, homology must exist between mice and fruitfly eyes; the question is whether the homology is at the level of eyes, or head regions.

In general, structures that are not homologous at one level will be homologous at another, more abstract level. Ultimately, this reflects the fact that all life on Earth traces back to a common ancestor near the origin of life. Consider the wings of birds and bats. As wings, they are not homologous. They evolved independently from a common ancestor that lacked wings. But as forelimbs, they are homologous. Bird wings and bat wings are modified forelimbs, descended from a common ancestor that possessed forelimbs.

Since the Gehring lab's work on eyes, several other structures that had been thought to be analogous rather than homologous in insects and vertebrate have been found to have common genetic control. Some of these structures may turn out to be homologous in a specific sense, others only in an abstract sense. We shall not know which until the actions of the genes concerned are better understood. Meanwhile modern molecular techniques have added a new, genetic layer to our understanding of homology to add to the classic criteria we met in Chapter 15.

---

[3] An ectopic structure is one in the wrong place. An ectopic pregnancy, for instance, means that gestation is occurring somewhere other than the womb — the most common kind of ectopic pregnancies are in the Fallopian tubes.

20.6 The *Hox* gene complex has expanded at two points in the evolution of animals

Are changes in the developmental genes associated with major evolutionary changes in the history of life? The *Hox* genes are the most hopeful gene set for answering this question at present. More is known for the *Hox* genes about which genes are present in which animal taxa than is known for any of the other genes associated with development. We mainly know about the number of *Hox* genes in different taxa, and can therefore look at when in animal evolution the numbers of *Hox* genes changed. (The work is similar to the work we looked at in Section 19.3, p. 559, about how to test whether major evolutionary events are associated with duplications of genes.)

Figure 20.6 shows the *Hox* genes of 12 animal groups. It shows that the *Hox* gene complex clearly expanded at two points in the phylogeny. One is near the origin of the triploblastic Bilateria (see Figure 18.5, p. 536, for this taxon). Cnidaria have radial symmetry and only two cell layers. They are simpler than the other animal groups in the figure, which have three-cell layers and bilateral symmetry. Only two *Hox* genes have been found in Cnidaria, against a common set of at least seven *Hox* genes in Bilateria. Probably the number of *Hox* genes went up by about five some time near the origin of the Bilateria.

A second major expansion occurred near the origin of the vertebrates. Invertebrates have a single set of up to 13 *Hox* genes. This set is also found, in a single copy, in the closest relative of the vertebrates, the lancelet *Amphioxus* ("cephalochordates" in Figure 20.6). Vertebrates, including humans, have four copies of the 13-gene set. The *Hox* gene set was increased fourfold, perhaps in a series of duplications, during the origin of vertebrates. Some biologists have explained the fourfold increase in the *Hox* genes by Ohno's hypothesis that the genome as a whole was duplicated twice near the origin of the vertebrates. Ohno's hypothesis is not well supported (Section 19.3, p. 561), but even if the genome as a whole was not tetraploidized, the *Hox* gene set itself was. So also were some other sets of genes that operate in development. This increase in gene numbers may have contributed to the evolution of vertebrates.

Vertebrates are arguably more complex life forms than invertebrate animals, for one thing they have more cell types. Also, many biologists think that the anatomic complexity of vertebrates is greater than for invertebrates. Complexity is difficult to measure objectively, but if vertebrates are more complex than invertebrates, the increase in the number of *Hox* genes may be part of the explanation. Once life forms had evolved with extra *Hox* genes they may have become able to evolve, in the future, increased complexity. Figure 20.6 also hints at some other periods of *Hox* gene change. For instance, the number of *Hox* genes concerned with the posterior end of the body seems to have expanded in the origin of the deuterostomes (echinoderms plus chordates at the top of the figure; see also Figure 18.5, p. 536).

The accuracy of inferences about when *Hox* gene numbers changed depends on the accuracy of the phylogeny. For example, in the phylogeny of Figure 20.6, *Hox* gene numbers appear to have decreased in the nematodes (represented by the worm *Caenorhabditis elegans*). This may be correct. However, the position of the nematodes in a group with the arthropods is based on recent molecular evidence from a small

Figure 20.6

History of the *Hox* genes. Modern taxa contain many homologous *Hox* genes, and the distribution of the genes can be used to infer the time when new genes originated, and of a possible tetraploidization near the origin of vertebrates (compare Figure 19.2, p. 561). From Carroll *et al.* (2001).

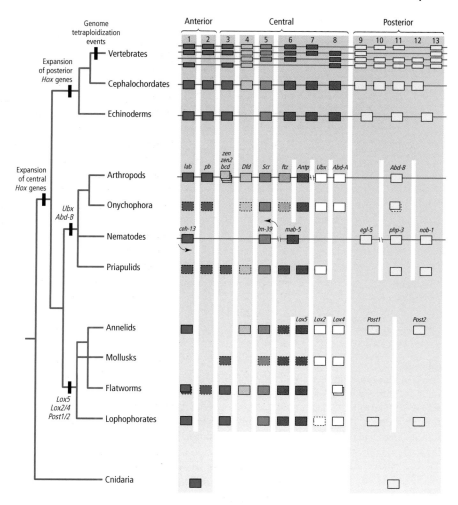

number of genes. Traditionally nematodes belonged to a branch nearer the base of the tree, between the Cnidaria and the rest of the Bilateria. Then we should not infer that they have lost genes, but that they are an intermediate stage in the early increase from two to seven *Hox* genes. The inferences for these early events are uncertain, and in any case we require a well substantiated phylogeny before we can draw confident conclusions.

20.7 Changes in the embryonic expression of genes are associated with evolutionary changes in morphology

The vertebrae that make up the spine, or backbone, of a mouse differ from head to tail. For instance, the cervical vertebrae in the mouse's neck differ in form from the thoracic

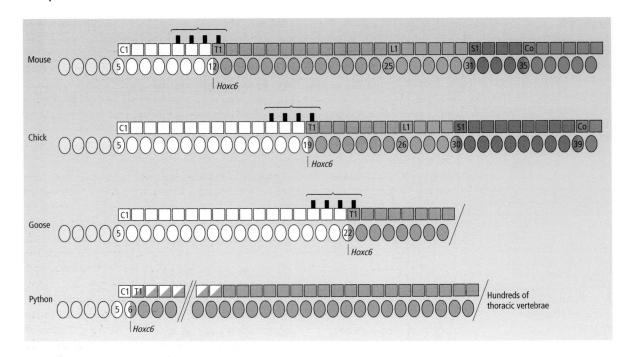

Figure 20.7

Change in gene expression associated with morphological evolution. The form of the vertebrae varies down the spine, with cervical vertebrae (C) in the neck and thoracic vertebrae (T) down the back. The vertebrae change from cervical to thoracic at different positions down the spine in the mouse, the chicken, the goose, and the python. The boundary of *Hoxc6* expression corresponds to the position where the vertebral form changes from cervical to thoracic. A change in the spatial expression of *hoxc6* could have contributed to the evolutionary change in the form of the backbone. Co, coccyx; L, lumbar; S, sacral. Modified from Carroll *et al.* (2001).

Changes in spine morphology . . .

. . . are associated with changes in a *Hox* gene's spatial development

vertebrae down the mouse's back. The cervical and thoracic vertebrae also differ in other vertebrate animals, such as chicken and geese. Geese and chickens have more neck vertebrae than mice do, and the division between cervical and thoracic vertebrae occurs further down the spine. The difference between species appears early in the embryo. The position of the boundary between cervical and thoracic vertebrae is further down the developing goose embryo than in a mouse embryo.

The boundary in the embryo between developing cervical and thoracic vertebrae is associated with the anterior boundary of expression of the *Hoxc6* gene (Figure 20.7). The *Hoxc6* gene is probably part of the control system that switches on the development of thoracic, rather than cervical, vertebrae. Thus, an evolutionary change in the morphology of the spine was probably partly produced, at a genetic level, by a change in the spatial expression of the *Hoxc6* gene in the embryo. Vertebrates develop in an anterior–posterior direction, with the head being specified first. A delay in switching on *hox6c* could cause the cervical–thoracic boundary to be shifted to the posterior, down the spine.

Changes in the timing of *Hox* gene expression can also contribute to morphological evolution. The five-digit limb of tetrapods, for example, has evolved from a fin in fish.

Hox genes are expressed in two phases during the development of fish fins. These phases might, for instance, help to cause an outward growth of bones to form the fin. In tetrapods, the *Hox* genes are also expressed in a third, later phase during limb development. The third phase is associated with the further growth outwards of the limb bones, to form the limb and hands. Thus, part of the mechanism by which fins may have evolved into limbs may have been for certain *Hox* genes to be switched on for a third time in the developing limb. Earlier in the chapter we met the concept of heterochrony (Section 20.2), which was based on classic morphological research. Here we can see a genetic example, in which a change in the timing of a developmental genetic process leads to evolutionary change in morphology.

Morphological evolution may be caused by a change in which genes a *Hox* gene interacts with. For example, insects differ from some other arthropods in lacking legs on their abdomens. An insect has legs on its thorax and not its abdomen, but myriapods and many crustaceans have abdominal legs. During evolution, leg development came to be switched off in the embryonic insect abdomen. The genetic mechanism, simplified, is that the *Hox* genes *ultrabithorax* (*Ubx*) and *Abd-A* are expressed down the abdomen of insects, crustaceans, and myriapods. They are regional controllers of development. In insects, *Ubx* and *Abd-A* repress the gene *distal-less* (*Dll*); *Dll* is the gene that directs leg development. In myriapods and crustaceans, *Ubx* and *Abd-A* do not repress *Dll*.

Two hypotheses can explain events such as the loss of limbs from the insect abdomen. One is a change in a transcription factor such as *Ubx*. In the evolution of insects, *Ubx* may have changed such that it became able to repress the genes, such as *Dll*, controlling limb development. The other hypothesis is that the enhancer of *Dll* may have changed during insect evolution. The enhancer may have ceased to bind *Ubx*. Alternatively, the enhancer may have continued to bind *Ubx*, but has changed its interaction with it such that *Ubx* now switches off limb development in the abdomen rather than switching it on. Some evidence supports the first hypothesis (Levine 2002). Crustacean *Ubx* is unable to repress *Dll* in fruitflies. That result suggests that *Ubx* itself has changed between crustaceans and insects. If *Ubx* were unchanged, crustacean *Ubx* should have the same effect in fruitflies as normal fruitfly *Ubx*.

In summary, we have seen three developmental mechanisms that are thought to have contributed to evolutionary changes in morphology. One is the change in the spatial expression of genes. A second is the change in which genes are switched on or off by transcription factors that have not themselves changed; this is achieved by changes in enhancers. A third is the change in transcription factors, such that they change their interactions with enhancers.

Changes in arthropod limbs are associated with . . .

. . . changes in Hox gene interactions

20.8 Evolution of genetic switches enables evolutionary innovation, making the system more "evolvable"

The examples in the previous section illustrate how evolutionary changes in gene regulatory networks can underlie morphological evolution. In the *hoxc6* example, in which the number of cervical vertebrae changed between mice and geese, the change concerned the regulatory relations between the *hoxc6* gene and some higher control gene. The anterior–posterior coordinates of the animal are probably given by a chemical gradient down the body. These chemicals may bind the enhancer of *hoxc6*, switching it off at some chemical concentrations and on at other concentrations. The *hoxc6* gene is then switched on in a certain region of the body. Morphological change can be produced if the enhancer of *hoxc6* changes such that it is switched on and off at somewhat different concentrations of the chemicals that specify the anterior–posterior axis. In the example of insect abdominal legs, the change was in which other genes were regulated by *Ubx* and *Abd-A*.

Whether the changes in these examples came about by the exact genetic mechanisms suggested here is not important. Several kinds of change in an enhancer, or the molecules that interact positively and negatively with an enhancer, could produce the same general outcome. What does matter, and is of broad interest, is that morphology can be altered by adding or subtracting switches that control existing genes. If a gene can cause, or help to cause, a leg to develop, then new legs can be added to (or old legs subtracted from) the body by switching the gene on or off. The gene may gain, or lose, an enhancer that binds to a transcription factor produced by one of the embryo's regional-specifier genes.

A gene may modify its function by sequence evolution . . .

It is instructive to compare evolutionary change produced by gain or loss of regulatory elements with change produced by sequence change in the gene itself. We have seen many examples in this book of changes in the sequence of a gene. The sequence of a globin gene may change, for example, such that the oxygen-binding attributes of the hemoglobin molecule are altered. This is an obvious way for a molecule to change its function, and much functional change has likely been produced by sequence changes.

The importance of genetic switches may be more in the evolutionary addition of new functions. Brakefield *et al.* (1996) and Keys *et al.* (1999) describe how a five-gene regulatory circuit has come to control the development of "eyespots" on the wings of butterflies. The gene circuit is able to produce borders, or boundaries, and is used in all insects to produce a certain boundary in the structure of the wing. Most insect wings do not have eyespots but some butterfly wings do. The eyespot has a distinct circular shape, with a boundary at the edge. Eyespots probably evolved when this "boundary-producing" gene circuit came to be expressed in a new gene network. In a butterfly eyespot, the boundary-producing genes are controlled by certain spatial-specifier genes within the wing, and they in turn control certain pigment-producing genes. Thus, a pre-existing set of genes came to be expressed in a new circumstance, probably by changes in the enhancers of the genes concerned. The boundary-producing gene circuit had gained a new function.

. . . but add new functions by evolution of its regulatory relations

When a gene adds an enhancer, which switches it on in a new circumstance, it can gain a new function without compromising its existing function. If a molecule, or morphological organ, changes to add a new function, it will usually perform its existing function less well. If a mouth is used for both eating and breathing, it is likely to do each less well than if it did one alone (see Section 10.7.5, p. 284, on trade-offs). A molecule can add a new function by changes in its internal sequence, although this evolutionary process is inherently difficult. However, the molecule is also likely to perform its old function less well as it adds its new function. The difficulty is avoided if the new function is added by a change in gene regulation. The existing, unchanged gene comes to be switched on in new circumstances and the old function need not be compromised at all.

Switching systems may have made life more evolvable

Enhancers, and their associated gene-regulatory relations, have not always existed in the history of life. They evolved in order to improve the precision with which genes were switched on and off. These improvements probably became more important as genomes evolved to be larger, and as life forms (that is, animals and plants) originated with development from egg to differentiated adult. But once genetic switches had originated, they arguably had the effect of making some kinds of evolutionary change easier. It became easier for genes to add new functions. Thus, a greater variety of animals and plants may have been able to evolve. Genetic switches did not evolve in order to promote biodiversity; but they may have done so, as a consequence.

The term evolvability has been used to refer to how probable, or "easy," it is that a species, or life form in general, will evolve into something new. Some species may be inherently more "evolvable" — more likely to evolve innovations and evolve into new, different species. Many suggestions have been made about factors that promote evolvability. Genetic switches are one example. Maybe, after the origin of genetic switches, life became more evolvable than it was before.

20.9 Conclusion

We can finish with some general reflections that apply to both this and the previous chapter. The two chapters have not had space for a full survey of either evolutionary genomics or evo-devo. Instead they have looked at a sample of examples, which are mainly intended to illustrate the promise — and the interest — of the two fields. However, they also illustrate one other general point. Traditionally in evolutionary biology, genetics provided the main methods and materials for studying microevolution. Evolutionary genomics and evo-devo are two ways in which genetics is now being used to answer macroevolutionary questions.

Genetics is increasingly used to study macroevolution

Evolutionary genomics, as we saw in Chapter 19, looks at questions that biologists had paid little attention to previously. The data that have made evolutionary genomics possible hardly existed before about the year 2000. In the case of evo-devo, biologists have always realized that morphological evolution must be driven by changes in development. They had concepts, such as heterochrony, for thinking about the development basis of evolution. The modern developmental genetic work provides a new way of thinking about these long-established problems. The modern work is more concrete than the earlier work, because it builds on a knowledge of individual genes and the developmental processes that they influence.

Maynard Smith & Szathmáry (1995, 1999) have identified a small number — 10 or so — of what they call the "major transitions" in evolution. These are events such as the origin of life, of chromosomes, of cells, of eukaryotic cells, of multicellular life, of the development of sexual reproduction, and of Mendelian inheritance. They are the big breakthroughs that made much of future evolution possible. The major transitions are all changes in the way inheritance occurs, and in the relation of genotype and phenotype. Understanding the major transitions is largely a matter of understanding evolutionary genomics and evo-devo. The advance of these two subjects should give us some insights into the grandest questions of macroevolution.

Summary

1 Morphological change in evolution usually occurs by changes in developmental processes. The identification of genes that influence development is a major area of modern biology, and its methods can be applied to study the relations of development and evolution, a field known as "evo-devo."

2 Heterochrony refers to evolutionary changes in the relative timing and rate of different developmental processes. For instance, the time of reproduction may shift relative to somatic development. Also, shape changes can result from changes in growth gradients, and D'Arcy Thompson's transformational diagrams can be interpreted in terms of heterochrony.

3 Regulatory genes influence the expression of other genes, and evolutionary change can result from changed regulatory relations among genes as well as changes in the sequence of genes.

4 Structures, such as the eyes of insects and vertebrates, that had been thought to be non-homologous, have been found to be developmentally controlled by the same gene. Insect and vertebrate eyes may share an element of homology, but it is uncertain what the level of the homology is.

5 The number of *Hox* genes increased from perhaps two to seven near the origin of the triploblastic Bilateria, and quadrupled from 13 to 52 near the origin of vertebrates. *Hox* genes control spatial differentiation within the body during development, and increases in the number of *Hox* genes may be associated with increases in developmental complexity.

6 Changes in the expression of developmental genes are likely achieved by gains, losses, and changes in the regulatory elements (particularly enhancers) of those genes.

7 Some forms of life may be more evolvable than others: that is, be more likely to undergo innovative evolutionary change. The origin of genetic switches may have made life more evolvable.

Further reading

General developmental biology texts, such as Gilbert (2000) and Wolpert (2002) contain chapters on evolution, as well as developmental biology background. Wilkins (2001), Carroll *et al.* (2001), and Hall (1998) are texts more specifically on evo-devo. *Proceedings of the National Academy of Sciences* (2000), vol. 97 (9), pp. 4424–540 contains the proceedings from a conference on evo-devo. Gerhart & Kirschner (1997) is a stimulating book, more about the evolution of cells, but containing much relevant material for this chapter. Meyerowitz (2002) gives an evo-devo comparison of plants and animals.

Gould's (1977a) book discusses the history of recapitulatory ideas and modern work on heterochrony. Gould (2002b) contains further material. Raff (1996) is a more recent general book, and Levinton's (2001) even broader book also covers the topic. Both Gould and Raff are good on heterochrony, but see also the review article by Klingenberg (1998), the web-page on heterochrony (and on D-Arcy Thompson's transformations) by Horder in www.els.org, and the think-piece by Smith (2001).

Britton & Davidson (1971) is an early work discussing gene regulation and evolution. See also the introductory article by A.C. Wilson (1985), the recent book by Davidson (2001), as well as the general references and some further references below.

Gehring & Ikeo (1999) is a recent paper on the *Pax6* gene and eye homology, and refers to the original papers in the early 1990s. Many authors have discussed what this and similar genetic findings reveals about homology. See Dickinson (1995), Abouheif *et al.* (1997), McGhee (2000), and Mindell & Meyer (2001).

On the origin of *Hox* genes see also the material on duplications in the genomics section of this chapter. Slack *et al.* (1993) discuss a further topic — the "phylotypic stage." They suggest: (i) that all animals are more similar at a certain developmental stage than earlier or later in development; (ii) the stage of maximum similarity is the stage at which *Hox* genes are expressed; and (iii) animals can be taxonomically defined by the possession of the phylotypic stage.

Carroll *et al.* (2001) give references for the examples in which gene expression in development is associated with morphological evolution. On butterfly spots, see also the general review by McMillan *et al.* (2002) and the particular contributions of Beldade *et al.* (2002a, 2002b), the second paper particularly connects with another classic theme, that of developmental constraints on evolution — discussed in this text in Chapter 10.

The general point about switches and evolvability is implicitly discussed in Carroll *et al.* (2001) and more explicitly in Ptashne & Gann (1998). The general concept of evolvability was introduced by Dawkins (1989b). It is also discussed in Gerhart & Kirschner (1997) and Kirschner & Gerhart (1998). Another, related finding concerns heat shock protein 90, which "canalizes" (Section 10.7.3, p. 276) development in animals and plants. The breakdown of canalization by *hsp90* increases the range of genetic variation in a population; *hsp90* could therefore normally reduce evolvability by decreasing variation but could increase evolvability in stressful times. Pigliucci (2002) introduces the topic and refers to the primary sources. Chapter 9 of this text has further material on canalization.

Study and review questions

1 If a descendant species, in its reproductive (adult) form, morphologically resembles a juvenile ancestral stage, what (a) is the descriptive term for this morphological pattern, and (b) are two possible heterochronic processes that could produce it?

2 The eyes of vertebrates and the compound eyes of insects have utterly different structures, and almost certainly evolved independently. And yet a related gene seems to control the development of eyes in both the mouse and fruitfly. How can we reconcile these two observations?

3 (a) What is meant by "evolvability"? (b) How can the evolution of gene regulatory circuits influence the evolvability of a life form?

21 Rates of Evolution

Evolutionary rates can be measured quantitatively for a character within a lineage, and we begin by seeing how this is conventionally done. We then look at a large compilation of over 500 such measurements and ask whether fossil evolutionary rates fit in with the theory of population genetics. Punctuated equilibrium is an influential modern idea about evolutionary rates in fossils and we discuss the theory, how to test it, and the evidence for and against it. We finish with two other measures of evolutionary rates: rates of change in arbitrarily coded character states — which we illustrate by a classic study of a living fossil (the lungfish) — and taxonomic rates, which are obtained from survivorship curves for fossil taxa.

21.1 Rates of evolution can be expressed in "darwins," as illustrated by a study of horse evolution

The six to eight modern species of the horse family (Equidae) are the modern descendants of a well known evolutionary lineage in the fossil record. The record extends back through forms such as *Merychippus* and *Mesohippus* to *Hyracotherium*, which lived 55 million years ago and was once called *Eohippus*. Horses have characteristic teeth, adapted to grind up plant material, and fossilized teeth provide the main evidence that has been used to trace the history of horses. Early members of the lineage were smaller on average than later forms, as Figure 21.1a illustrates. The Eocene ancestors of modern horses were about the size of a dog, and the smallest was the size of a cat. Their teeth were smaller too and had different shapes from modern horses. The ancestor–descendant relations of the equid species are known reasonably well. The rate of evolutionary change in the teeth can therefore be estimated by direct measurement, in fossils from successive times within a lineage.

Horse teeth are an example of how to measure evolutionary rates . . .

Horse teeth are classic subject matter in the study of evolutionary rates, and the most comprehensive modern work on them is by MacFadden (1992). He measured four properties of 408 tooth specimens, from 26 inferred ancestor–descendant pairs of species (Figure 21.1b–d). The measure of rate used by MacFadden, and many other paleontologists, was first suggested by Haldane (1949b). Suppose that a character has been measured at two times, t_1 and t_2; t_1 and t_2 are expressed as times before the present, in millions of years. t_1 might be 15.2 million years ago and t_2 14.2 million years ago (t_2 is the more recent sample and has a shorter time to the present). The time interval between the two samples can be written as $\Delta t = t_1 - t_2$, which is 1 million years if $t_1 = 15.2$ and $t_2 = 14.2$. The average value of the character is defined as x_1 in the earlier sample and x_2 in the later sample; we then take natural logarithms of x_1 and x_2 (the natural logarithm is the log to base e where $e \approx 2.718$, and it is symbolized by log or ln). The evolutionary rate (r) then is:

$$r = \frac{\ln x_2 - \ln x_1}{\Delta t}$$

. . . in "darwins"

The rate is positive if the character is evolutionarily increasing and negative if it is decreasing, but for many purposes the absolute rate of change, independent of the sign, is what matters. Haldane defined a "darwin" as a unit to measure evolutionary rates; 1 darwin is a change in the character by a factor of e ($e \approx 2.718$) in 1 million years. The formula above for r gives the rate in darwins provided that the time interval is in millions of years. If, for example, $x_1 = 1$, $x_2 = 2.718$, and $\Delta t = 10$ million years then $r = 0.1$ darwins.

The reason for transforming the measurements logarithmically is to remove spurious scaling effects. If logarithms were not taken, the rate of evolution of a character would appear to speed up when it became larger even if its proportional rate of change remained constant. With logarithmically transformed measurements, rates of change can be compared between species of very different size, such as mice and elephants.

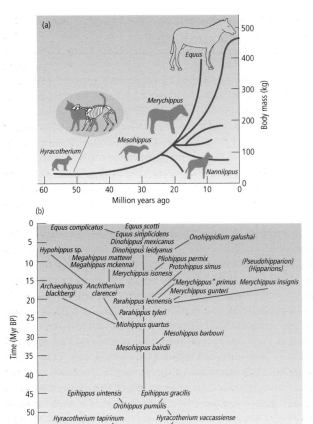

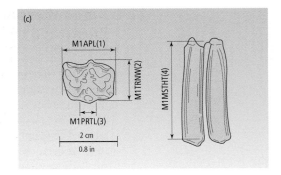

Species pair	Δt (Myr)	M1APL(1) (d)	M1TRNW(2) (d)	M1PRTL(3) (d)	M1MSTHT(4) (d)
Equus simplicidens-Equus complicatus	2.0	0.000	−0.014	0.115	0.054
Parahippus leonensis-Merychippus primus	3.0	−0.009	−0.026	−0.012	0.154
Parahippus leonensis-Protohippus simus	6.0	0.051	0.034	0.073	0.247
Miohippus quartis-Anchitherium clarencei	7.0	0.065	0.057	0.049	0.046
Mesohippus bairdii-Mesohippus barbouri	1.0	0.157	0.146	0.136	0.050
Equus simplicidens-Equus scotti	2.0	0.040	0.005	0.162	0.097
Dinohippus mexicanus-Equus simplicidens	2.0	0.064	0.081	0.088	0.142
Dinohippus leidyanus-Dinohippus mexicanus	2.0	−0.004	−0.026	0.074	−0.054
Dinohippus leidyanus-Onohippidium galushai	2.0	0.033	0.027	0.030	−0.074
Merychippus isonesis-Pliohippus permex	2.5	0.072	0.101	0.185	0.180
Megahippus mckennai-Megahippus matthewi	2.0	0.036	0.022	0.064	0.219
Anchitherium clarencei-Megahippus mckennai	4.0	0.083	0.096	0.094	0.109
Anchitherium clarencei-Hyphippus large sp.	8.0	0.062	0.072	0.077	0.077
Parahippus leonensis-Merychippus insignis	3.0	0.053	0.011	0.038	0.228
Parahippus leonensis-Merychippus isonesis	3.0	0.030	0.014	0.028	0.339
Parahippus leonensis-Merychippus gunteri	2.0	−0.023	−0.107	−0.116	0.172
Parahippus tyleri-Parahippus leonensis	2.0	−0.062	−0.039	−0.179	0.088
Miohippus quartus-Archaeohippus blackbergi	7.0	−0.017	−0.018	−0.029	−0.021
Miohippus quartus-Parahippus tyleri	5.0	0.067	0.052	0.066	0.091
Mesohippus bairdii-Miohippus quartus	6.0	0.022	0.018	0.012	0.022
Epihippus gracilis-Mesohippus bairdii	14.0	0.023	0.037	0.032	0.036
Orohippus pumulis-Epihippus uintensis	4.5	0.047	0.052	0.042	−0.007
Orohippus pumulis-Epihippus gracilis	4.5	0.026	−0.010	0.006	0.020
Hyracotherium vaccassiense-Oronippus pumulis	2.5	0.011	0.029	0.012	0.068
Hyrocotherium angustidens-Hyracotherium vaccassiense	3.0	0.010	0.026	−0.020	0.000
Hyracotherium angustidens-Hyracotherium tapirinium	3.0	0.101	0.106	0.091	0.096
Mean species pair evolutionary rate*		0.045	0.047	0.0690	0.104

Equus simplicidens-Equus complicatus

Figure 21.1

(a) Modern horses are descended from a group containing a number of lineages that have increased in average body size through the past 50 million years. The inset shows the smallest known species of *Hyracotherium*, *H. sandrae* (from the early Eocene of Wyoming) silhouetted against a cat for size comparison. (100 lb ≈ 45 kg.) (b) The phylogenetic relations of fossil and modern horse species. (c) First molar tooth in crown and side view, showing four aspects that were measured. (d) Evolutionary rates for the four measures, in 26 inferred ancestor–descendant species pairs, expressed in darwins (d). It is not important to study the numbers in detail! They are meant only to illustrate the results that come out of a study of evolutionary rates. Redrawn, by permission of the publisher, from MacFadden (1992).

However, the use of natural logarithms can be puzzling for people who think intuitively about changes in terms of percentages rather than logarithms. For them, a change of 10% is meaningful, a change of 0.1 natural logarithmic units less so. Fortunately, natural logarithms behave much like percentage changes for short time intervals. Suppose, for instance, a lineage is evolving at 1 darwin. For times up to about 1,000 years, the percent change will be approximately constant per year for all 1,000 years. That is, after 1,000 years the lineage will have changed by close on 0.1%, and by about one-thousandth of that amount (that is, 0.0001%) every year up to then. For longer time intervals things are not so simple. If the lineage continued to change by the same increment (0.0001%) per year for up to 1 million years, it would have increased by 100%. But a lineage that evolves at 1 darwin will in fact increase by 272% in 1 million years. Therefore, the familiar units of "percent change" give reasonable results even with logarithmic units such as darwins, but only over short time intervals. (The best way to familiarize yourself with the meaning of darwins is to calculate a few: see the study questions at the end of this chapter.)

The 26 ancestor–descendant species pairs and four dental characters measured by MacFadden produced $4 \times 26 = 104$ estimates of evolutionary rates (Figure 21.1d). The different tooth characters show different patterns, with height (M1MSTHT in the figure), for instance, evolving rapidly between *Parahippus* and *Merychippus*, while the other characters were evolving at normal rates. But the detailed pattern of the numbers is not important here,[1] though the approximate absolute values of the rates are worth bearing in mind.

Horse teeth show a representative range of values

The values in Figure 21.1 are mainly about 0.05–0.1 darwins, or about a 15–30% change per million years. They are mainly positive, indicating that the lineage was on average increasing in size. There are, however, negative values, as the horses in the lineage evolutionarily shrunk as well as expanded. The values in Figure 21.1 are averages for a lineage connecting an ancestral–descendant species pair, and do not imply that evolution had a constant rate throughout that time. An average is not a constant, and the rates for short periods may have been very different from the long-term average. However, as average figures, the values in Figure 21.1 are fairly typical for the fossil record, being neither exceptionally fast or slow. We shall see in a minute (Table 21.1 below) that the average figure for a large set of evolutionary rates in vertebrates is about 0.08, and rapidly evolving vertebrates show rates of more like 1–10 darwins over short periods. Simpson (1953), who did more than anyone to stimulate the study of fossil evolutionary rates, noticed that rates vary between taxa, characters, and times, and he invented the terms bradytelic, horotelic, and tachytelic, to describe slow, typical, and rapid evolution; horse evolution as such is horotelic.

21.1.1 How do population genetic, and fossil, evolutionary rates compare?

Rates of evolution in the fossil record have been measured for many characters, in many species, at many different geological times. A compilation by Gingerich (1983, 2001) included 521 different estimates, of which 409 were for the fossil record. The estimates vary between 0 and 39 darwins in fossil lineages. The main problem of evolutionary rates is to understand why they differ between times and taxa in the way they do.

Before we come to that problem, we can ask a more general question. Are the rates of change seen in the fossil record consistent with the mechanisms of evolutionary change

---

[1] The patterns mainly make sense in terms of the grinding functions of the teeth and the diets eaten by individual horse species. Diets in turn were influenced by changes in vegetation, particularly the spread of grass, and in climate. See Section 18.5 (p. 540).

Table 21.1

Gingerich's summary of evolutionary rates. The summary is large but not complete, and is based on 521 different measurements. Gingerich divided the measurements into four classes. The importance of the column for time intervals will become apparent in Section 21.2.

Domain	Sample size	Evolutionary rate (darwins)		Time interval	
		Range	Geometric mean	Range	Geometric mean
I Selection experiments	8	12,000–200,000	58,700	1.5–10 yr	3.7 yr
II Colonization	104	0–79,700	370	70–300 yr	170 yr
III Post-Pleistocene Mammalia	46	0.11–32.0	3.7	1,000–10,000 yr	8,200 yr
IV Fossil Invertebrata and Vertebrata	363	0–26.2	0.08	8,000 yr–350 Myr	3.8 Myr
Fossil Invertebrata alone	135	0–3.7	0.07	0.3–350 Myr	7.9 Myr
Fossil Vertebrata alone	228	0–26.2	0.08	8,000 yr–98 Myr	1.6 Myr
I to IV combined	521	0–200,000	0.73	1.5 yr–350 Myr	0.2 Myr

studied by population geneticists? Population genetics identifies two main mechanisms of evolution, natural selection and random drift, though drift is arguably unimportant in morphological evolution (Section 7.3, p. 165). For changes like those of tooth size in the history of horses, we cannot confirm directly that selection was the cause. That would require us to show that the character was inherited (that is, larger toothed horses gave rise to larger toothed offspring than average). We should also have to show that larger horses produced more offspring than average. That kind of study is usually impossible with fossils.

However, we can at at least find out whether the results of research in the two areas are consistent. We can first ask whether there is any contradiction between the rates of evolution observed in population genetics work, such as artificial selection experiments, and those observed in fossils. If, for example, the fossil rates are significantly higher, it would suggest that selection alone cannot be the only cause of evolution. Some other more rapid factor would be needed. In fact, it turns out, the rates of evolution in artificial selection experiments are far higher than those measured in fossils. Evolution under artificial selection has proceeded about five orders of magnitude faster than in the fossil record (Table 21.1). We can conclude that the known mechanisms of population genetics can comfortably accommodate the fossil observations.

Strictly speaking, this does not confirm that the fossil changes were driven by selection and (perhaps) drift. However, it does show that the observations are consistent. For this reason, and because no other mechanisms of evolution are known, no one seriously doubts that the microevolutionary processes of Chapters 4–9, 14 and 15 — even if operating indirectly (tooth sizes might increase because of selection for larger body size, for instance) — ultimately underlie the observed rates of evolution over geological time periods. We have no reason to think that some additional but unknown mechanisms of evolution were at work.

Rates of evolution in fossils are usually slower than in artificial selection experiments

Figure 21.2

Possible phylogeny of Darwin's finches, according to Lack. The dashed lines indicate uncertainty. Other phylogenies have been suggested too. Was there time enough for the evolution of 14 species, by selection within a population, since the Galápagos were colonized by the ancestral finch maybe 570,000 years ago? Redrawn, by permission of the publisher, from Lack (1947).

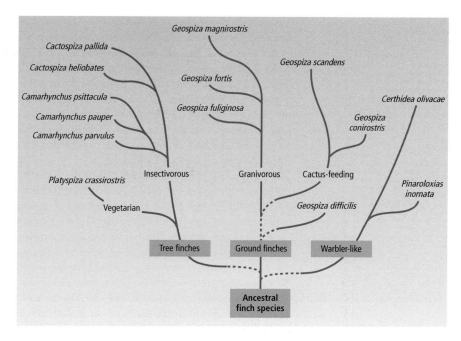

21.1.2 Rates of evolution observed in the short term can explain speciation over longer time periods in Darwin's finches

The same point — that rates of evolution over different time periods are consistent — can be made by another argument. We saw (Section 9.1, p. 223) how natural selection operates on the beaks of Darwin's finches. The evidence there was for natural selection within a species. It demonstrated that individuals with larger beaks are favored when the seeds and fruit that they eat are large, whereas smaller beaks are favored when the food size is smaller. In 1976–77 (and subsequently), the Grants measured the strength of selection on the finches' beaks, and its evolutionary results (Figure 9.9, p. 241). Today, 14 different finch species occupy the Galápagos (Figure 21.2). The species mainly differ in their beak and body proportions. What we can do is calculate whether the kind of selection observed in the short term would be enough to account for the origin of all the finches in the Galápagos in the time available.

The Grants observed rates of change in finches . . .

How long would it take for the process studied by the Grants in 1976–77 to convert one species of finch into another? During the 1977 drought, the beak size of *Geospiza fortis* on the island of Daphne Major increased by 4%. *G. magnirostris* is a close relative of *G. fortis*; the two species differ mainly in body and beak proportions and they coexist on many islands of the Galápagos. From the average difference in beak size between *G. fortis* and *G. magnirostris* on Daphne Major, Grant (1986) estimated that 23 bouts of evolution of the 1977 type would be enough to turn *G. fortis* into *G. magnirostris*. On other islands *G. fortis* is larger than on Daphne Major and, using one of the larger *G. fortis* populations as the starting point, only 12–15 such events would be needed.

. . . and those rates are more than adequate to explain the radiation on the Galápagos

And how much time is available? The Galápagos archipelago is made up of volcanic islands. The current islands probably first emerged from the ocean about 4 million years ago, and all the islands had appeared by 1–0.5 million years ago (see Sequeira *et al.* 2000 for differing views on the dates). The common ancestor of Darwin's finches has been estimated to have arrived from South America about 570,000 years ago, so the radiation of the 14 finch species has occurred in about 0.5 million years. Using either the low estimate of 12–15 events (where an "event" is a bout of evolution such as occurred in 1977 on Daphne Major) or the higher estimate of 25 events, we can see that even if only one such event took place per century, the evolutionary divergence between the two species could still have been accomplished well within the time available. In fact, the real evolutionary transition probably did not happen that way. The 1982–83 El Niño event reversed the evolution of 1977 and there was probably not a steady transition from one population into another, but frequent reversals of accumulated small changes. In any case, it is unlikely that *G. fortis* simply changed into *G. magnirostris* (or vice versa); they probably both diverged from another common ancestor.

The rough calculation is not intended to represent the exact history of the birds. Instead, it illustrates how we can extrapolate from the rate of evolution observed over a few years within a species to explain the diversification of the finches from a single common ancestor about 570,000 years ago to the present 14 species. If the extrapolation is correct, the reason for the speciation in the finches was the same process as has been observed in the present — natural selection for changes in beak shape, which were probably in turn due to changes in food types through time and between islands. Although the finches have speciated rapidly, no peculiar mechanism of evolution is needed to account for it. Arguments of this general kind are common in the theory of evolution. We met a similar argument in Section 18.6.2 (p. 542), where natural selection over long periods was used to explain the major evolutionary transition from the mammals to the reptiles.

21.2 Why do evolutionary rates vary?

Paleobiologists have studied a number of generalizations about evolutionary rates. For example, it has been suggested that species usually change more rapidly during, rather than between, speciation events; that structurally more complex forms evolve faster than simpler forms; and that some taxonomic groups evolve more rapidly than others, that mammals, for instance, evolve faster than mollusks (this is an old idea — it was one of Lyell's favorite generalizations). We shall examine the first of these issues in more detail later. But before doing so, let us return to Gingerich's (1983) compilation of evolutionary rates and consider a general point about their study.

Observed rates depend on the measurement interval . . .

Gingerich observed, in his compilation of evolutionary rates, an inverse relation between the rate and the time interval over which it was measured. The observed cases of rapid evolution have tended to be for shorter intervals than the cases of slower evolution (Figure 21.3). The relation is unlikely to be due to any strong force in the evolutionary process itself. Nothing in evolutionary theory constrains rapid evolution, at these speeds, to take place in short intervals and slower evolution in longer intervals. At

Figure 21.3
The relation between the estimated logarithmically transformed evolutionary rate and the time interval used, for the 521 studies summarized in Table 21.1. The relation is negative. For the meaning of samples I, II, III, and IV see Table 21.1. The digits higher than 1 on the graph mean that number of cases fell on that spot (x for numbers higher than 9). From Gingerich (1983). © 1983 American Association for the Advancement of Science.

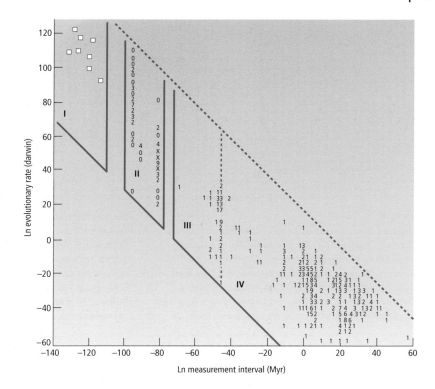

the molecular level, by way of comparison, the rates of evolution seem to be fairly constant over all time periods (e.g., Figure 7.3, p. 165).

Gingerich's basic observation can now be supplemented. Hendry & Kinnison (1999) reviewed 20 studies of microevolutionary change within a species (that is, studies like that by the Grants on the Galápagos finch — Section 9.1, p. 223). The timescales and measured rates of evolution fall between, and partly overlap with, categories I and II in Figure 21.3. The results also show a negative relation between measurement interval (from 2 to 125 years) and the observed rate of evolution.

Hendry and Kinnison point out several factors that can produce a negative relation as in Figure 21.3 and their data. The most important factor can be explained in terms of the Darwin's finch example. However, we need to concentrate on a slightly different feature. In the previous section, we considered the rate of evolution during a single burst of evolutionary change. Now we need to turn to the fluctuating selection pressure over several consecutive years (Section 9.1, p. 223).

. . . which may be due to fluctuations in evolutionary direction

The finches' beaks evolved to be larger in times of food shortage and smaller in times of abundance, and the food supply fluctuated through time according to the weather, particularly the periodic El Niño disturbance. Imagine measuring the rate of evolution both within one of these cycles and over the cycle as a whole (Figure 21.4a). If the direction of evolution fluctuates, the rate of evolution measured over a short interval is inevitable higher than the rate measured over a longer time interval because the short-term changes cancel out. The pattern in Figure 21.4a is simplified, giving zero net

(a)

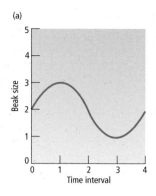

(b)

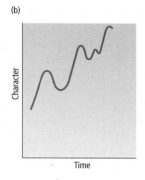

Measurement interval		Rate of evolution
1 unit	Within 1 or 2 or 3 or 4	1
2 units	1 + 2 2 + 3 3 + 4	0 ⎫ 1 ⎬ av. 1/3 0 ⎭
3 units	1 + 2 + 3 2 + 3 + 4	1/3 1/3
4 units	1 + 2 + 3 + 4	0

Figure 21.4

The inverse relation between measured evolutionary rate and time interval (Figure 21.3) will be found if the direction of evolution fluctuates through time. (a) Simplified cycle of evolutionary change. The rate of change measured for short time units is higher than for the cycle as a whole, there is no net change over the cycle, and the rate of evolution is zero. The numbers under "measurement interval" in the table refer to the time intervals in the x-axis of the graph (the arbitrary beak size units can be thought of as logarithmic, to make the rates properly comparable with the formula for calculating rates in Section 21.1). (b) With a more realistic pattern of evolution, the inverse relation between rate and measurement interval will still be found to some extent if there are any fluctuations in the direction of change.

change over a cycle, but if there are any fluctuations in evolutionary direction (Figure 21.4b) it will be true that a rate measured over a shorter interval will be higher. Likely almost all evolutionary lineages show some reversals in the direction of change, and the pattern illustrated by Darwin's finches may be quite common. This would explain the general relation in Figure 21.3.

Other factors may be contributing. For instance, the cases of rapid evolution over short time intervals are for artificial selection experiments (dataset I) and natural ecological colonizations (dataset II); it may be that these are extraordinary events and have higher than average selection intensities. (Alternatively, however, it might be argued that the rates are high only because the measurement interval is short enough to catch evolution in its unidirectional phase, and not because the intensity of selection is peculiar. Opinions differ about how representative the selection intensities in datasets I and II are of those in the lineages making up datasets III and IV.)

This interpretation, if it is correct, matters for some kinds of generalizations about evolutionary rates, but not others. It does not invalidate the measurements themselves. In the 14 million years between *Epihippus gracilis* and *Mesohippus bairdii*, horse teeth evolved at a rate of 0.023–0.037 darwins (Figure 21.1), and that is that. All questions about individual measurements, and comparisons between them, remain valid. It is for the more general patterns that Gingerich's result should make us suspicious. The generalization that mammals evolve faster than mollusks, for example, is reflected in Gingerich's data. He found that vertebrates as a whole tended to evolve faster than invertebrates (compare the mean rates of evolution for the two groups in Table 21.1). While it remains true that in the samples measured vertebrates did evolve 1.14 times as fast as invertebrates, this might mainly be due to the shorter time intervals for the

Other factors may contribute too

Some trends . . .

. . . disappear after the
measurement intervals are
corrected for

vertebrate than the invertebrate measurements (compare the mean time intervals for vertebrates and invertebrates in Table 21.1). When Gingerich corrected for the difference in intervals (by extrapolation from Figure 21.3) he deduced that invertebrates actually evolve faster than vertebrates. That particular correction may or may not be appropriate, but it is advisable to look at the time intervals when comparing the evolutionary rates of different lineages.

Rates of evolution can still be compared between different taxa, or between different kinds of taxa, despite the problem of time intervals. However, the problem does need to be taken into account. We shall concentrate on a question that should not be much influenced by the difficulties implied by Gingerich's result. It is also outstandingly the most lively modern controversy about evolutionary rates: the theory of punctuated equilibrium.

21.3 The theory of punctuated equilibrium applies the theory of allopatric speciation to predict the pattern of change in the fossil record

Eldredge & Gould (1972), in a famous essay, argued that paleontologists had misinterpreted neo-Darwinism. The fossil record had posed an apparent problem for Darwin because it does not show smooth evolutionary transitions. A common pattern is for a species to appear suddenly, to persist for a period, and then to go extinct. A related species may then arise, but with little sign of any transitional forms between the putative ancestor and descendant. Since Darwin, many paleontologists have explained this pattern by the incompleteness of the fossil record. If evolution was really gradual, but most of the record has been lost, the result would be the jerky pattern that we observe.

Phyletic gradualism . . .

Eldredge and Gould distinguished two extreme hypotheses about the pattern of evolution (Figure 21.5). They named one of them *phyletic gradualism*, which states that evolution has a fairly constant rate, that new species arise by the gradual transformation of ancestral species, and that the rate of evolution during the origin of a new species is much like that at any other time (Figure 21.5b).

. . . and punctuated equilibrium are
contrasting theories

They contrasted phyletic gradualism with their own preferred hypothesis, *punctuated equilibrium* (Figure 21.5a). They used the standard theory of speciation — allopatric speciation, which we looked at in Chapter 14 — to argue that the fossil record should show a pattern different from phyletic gradualism. If new species arise allopatrically and in small isolated populations then the fossil record at any one site may not reveal the speciation event. If the site preserves the record of the ancestral species, the descendant species will be evolving elsewhere. The newly evolving species will not be preserved at the same site as its ancestor. The new species will only leave fossils at the same site as its ancestor if it reinvades the same area. Reinvasion could happen if the descendant either was outcompeting its ancestor or was sufficiently different and could coexist ecologically. Either way, the new species would be fully formed by the time it turned up as fossils in the same place as its ancestor. The transitional forms would be unrecorded not because of the incompleteness of the fossil record at that site but because the interesting evolution took place elsewhere. The reason why the transitional

Figure 21.5

The crucial difference between punctuated equilibrium and phyletic gradualism concerns the observed rate of evolutionary change at, and between, splitting events. (a) Punctuated equilibrium. (b) Phyletic gradualism. (c) The theory of punctuated equilibrium also predicts that evolution will not occur except at times of speciation. Rapid change without splitting contradicts the theory.

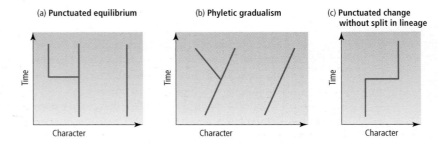

(a) **Punctuated equilibrium**   (b) **Phyletic gradualism**   (c) **Punctuated change without split in lineage**

Time — Character

Punctuated equilibrium exists in more, and less, orthodox versions

forms are absent is again the incompleteness of the fossil record — but not in the same way as the theory of phyletic gradualism suggested.

Between speciation events, species may have a low rate of evolutionary change — a condition Eldredge and Gould called *stasis*. In theory, the absence of evolutionary change in a species can be explained by stabilizing selection (Section 4.4, p. 76) or constraint (Section 10.7, p. 272). Constraint means that the species does not change because it lacks genetic variation, or lacks expressed genetic variation.[2] As we saw in Section 10.7.3 (p. 280), the evidence does not suggest that species stay constant because they lack genetic variation. Stabilizing selection, by contrast, is a well documented fact and is highly plausible in theory. Stabilizing selection, therefore, is the most likely (if not universally accepted) explanation for stasis in the fossil record. The theory of punctuated equilibrium holds that stasis is the normal condition within a species. Stasis breaks down only when speciation occurs. Evolutionary change is concentrated in speciation events. Any observation of change without speciation (Figure 21.5c) would contradict the theory of punctuated equilibrium.

In the account here, the theory of punctuated equilibrium is relatively "orthodox." Eldredge and Gould took the (or a) standard theory of speciation, and pointed out that is implies that fossils will usually show sudden, rather than smooth, change. However, the theory of punctuated equilibrium has stimulated much controversy, as Gould (2002b) documents. There are two main reasons. One is that punctuated equilibrium was sometimes said to challenge the fundamental "gradualism" of Darwin's theory of evolution. Box 21.1 distinguishes two meanings of the word "gradual." When the two meanings are distinguished, this line of controversy is defused.

The second source of controversy is that the theory of punctuated equilibrium has also drawn on and been associated with much less widely accepted ideas about speciation. The theory of punctuated equilibrium has been actively developed for about 30 years and exists in many different versions. In particular, "valley crossing" theories of speciation (in terms of Section 14.4.4, p. 394) have often been used to predict punctuated equilibrium. Speciation requires valley crossing if two species have different adaptations, and the intermediate forms between them have lower fitness. The two

---

[2] Genetic variation may be present, but not expressed, if it is concealed by developmental canalization (Sections 9.9, p. 242, and 10.7.3, p. 276). One version of punctuated equilbrium suggests that canalization creates a developmental constraint. Evolution is only possible in revolutionary circumstances, such as in a stressed subpopulation at the edge of a species' main range. See the further reading section of Chapter 20 for a reason why normally concealed genetic variation could be expressed in these conditions.

Box 21.1
Two Meanings of Gradualism

In the theory of evolution, the words "gradual" and "gradualism" have both been used in two distinct senses. One refers to the rate of evolution, and means that evolution has a fairly constant rate. This is the meaning in the term "phyletic gradualism." If evolution proceeds in the mode of phyletic gradualism, it has a constant rate; if it proceeds in the mode of punctuated equilibrium, it is slow within a species, and faster as new species evolve.

A second meaning refers to the evolution of adaptations, particularly complex adaptations such as the vertebrate eye. We saw in Section 10.3 (p. 259) that complex adaptations evolve via many intermediate stages. They do not arise suddenly, fully formed. It is a deep requirement of Darwinian theory that adaptations evolve gradually, in many stages. However, it is not a requirement at all of Darwinian theory that evolution should have a constant rate.

Darwin, in the *Origin of Species* (1859) and elsewhere, repeatedly stressed that evolution is slow and gradual. Gould has concluded, accordingly, that Darwin was a phyletic gradualist, and that the theory of punctuated equilibrium contradicts both Darwin's own ideas and also those of neo-Darwinism. By contrast, Dawkins argued that Darwin meant something crucially different by gradual evolution. Darwin did not make his remarks about gradualism particularly in the context of evolutionary rates at and between speciation events. When he did discuss that subject, he said things that sound quite like punctuated equilibrium, such as:

Many species once formed never undergo any further change . . . and the periods during which species have undergone modification, though long as measured by years, have probably been short in comparison with the periods during which they retained the same form. (Darwin 1859)

Darwin's theory and all subsequent versions of Darwinism, are strongly gradualist about the evolution of adaptation. But they are not gradualist about the rate of evolution. The only deep requirement that Darwinian theory has about evolutionary rates is that fossils should not evolve faster than the fastest rates seen in selection experiments, using normal genetic variation. If fossils evolved faster than that, it would suggest macromutations or some such factor were contributing to fossil evolution. That really would challenge neo-Darwinism. However, even the fastest rates of fossil evolution are slower than the rates seen in genetic experiments (Section 21.1.1). What appears to be fast on a geological timescale is slow — almost too slow to study genetically — on a genetic timescale.

We saw in Section 21.1 that the main neo-Darwinian authority on rates of evolution, Simpson, suggested that evolution shows a range of rates, from slow to fast. Neither Darwin, nor Simpson, argued that evolution has a constant rate. Thus, the theory of punctuated equilibrium is interesting to test, but if it turns out right and phyletic gradualism turns out wrong, no damage will have been done to any deep Darwinian principle of gradualism. Adaptations will still have to evolve in many small stages.

Further reading: Dawkins (1986), Gould (2002b).

species occupy different peaks on· an adaptive topography (Section 8.12, p. 214). Simple natural selection cannot then drive evolution from one species to the other. Some special circumstances, or evolutionary processes, will be required, and evolution may proceed by a rapid "peak shift."

When Eldredge and Gould first published their theory in the 1970s, valley crossing theories of speciation were more popular than they are now. As we saw (Section 14.4.4, p. 394), the evidence and theoretical trends have moved against valley crossing theories of speciation. Thus, punctuated equilibrium has been controversial because it has been associated with a controversial set of theories about speciation. Punctuated equilibrium has even been associated with the very unorthodox idea that evolution proceeds by macromutations (Section 10.5, p. 266). However, punctuated equilibrium does not depend on any of these valley crossing theories. Punctuated equilibrium can be derived, as we saw, from the well substantiated allopatric theory of speciation. Fossils can rarely be used to test between theories about the mechanism of speciation. The

But it does not require unorthodox ideas

methods discussed in Chapter 14 have been used for that kind of research. Instead we can concentrate here on the empirical question of what pattern of evolution is observed during speciation. Does the fossil record show new species evolving suddenly, or gradually with many intermediate stages?

21.4 What is the evidence for punctuated equilibrium and for phyletic gradualism?

21.4.1 A satisfactory test requires a complete stratigraphic record and biometrical evidence

In the fossil record, one species is often observed to be suddenly replaced by another. New species are rarely observed to evolve smoothly from their ancestors. However, these observations do not strongly count in favor of the theory of punctuated equilibrium. The fossil record is incomplete and the punctuated pattern will therefore appear in most fossil samples whether the underlying evolutionary pattern is gradual or punctuated. Two crucial conditions must be met by any test of the ideas. One is that the stratigraphic sequence should be relatively complete (that is, sediments should have been laid down fairly continuously). The other is that the evidence should be biometrical, not taxonomic.

Apparent punctuations may be due to . . .

Taxonomic evidence alone is inconclusive, because taxonomic categories such as species are discrete entities. The forms in a lineage will necessarily jump from being members of species A at one point to being members of species B at another point whether evolution in the lineage is sudden or gradual. Taxonomists, quite rightly, include a range of forms within a single species, and the observation that a single species persists for a certain amount of time tells us nothing about whether its morphology is changing gradually, or is constant in form. Thus, we need measurements of the forms in a population over time, to see whether the average changes suddenly or gradually.

. . . taxonomic artifacts . . .

It also helps to know whether any changes in a population are genetic. In some species, individuals can grow up with distinct forms, depending on the environmental conditions in which they develop. These changes in development are called "ecophenotypic switches." The phenotype switches from one form to another, depending on the environment; these switches are not genetic, evolutionary events. The theory of punctuated equilibrium is an evolutionary theory, and needs to be tested with evolutionary data. With fossils, we cannot be sure that any observed change in a population is not an ecophenotypic switch. However, we can at least avoid evidence in which the morphological change looks like the kind of change that can be induced in modern species by changes in the environmental conditions. Fryer *et al.* (1985) discuss ecophenotypic switches in snails and how they may have contaminated some research on punctuated equilibrium.

. . . or to ecophenotypic switches

The question is not simply whether "either" punctuated equilibrium "or" phyletic gradualism is right. The two theories represent extreme points in two continuous dimensions: the pattern of evolution itself in any one lineage, and the relative frequency

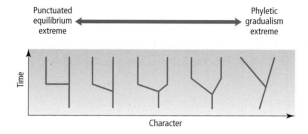

Punctuated
equilibrium
extreme

Phyletic
gradualism
extreme

Time

Character

Figure 21.6
Punctuated equilibrium and phyletic gradualism are extremes of
a continuum. Even here, the theories are simplified, for instance
phyletic gradualism may not proceed in a straight line, but may
contain reversals, as in Figure 21.4 (see Sheldon 1996).

of the patterns in different lineages, can occupy any point between these extremes.
Thus, according to the theory of punctuated equilibrium, the majority (perhaps more
than 90%) of evolutionary lineages should show a punctuated pattern whereas a
phyletic gradualist might claim the opposite. Nature itself could lie anywhere between
the two. Likewise there are plenty of patterns between the punctuated and gradual types
of change (Figure 21.6) and nature could show any of them. Punctuated equilibrium
and phyletic gradualism are not the only alternatives to be tested between. Research
aims to find out what the frequencies of the different patterns are.

Moreover, the two extreme positions are not schools of thought that are being advocated by two opposed camps of evolutionary biologists. They are disembodied theories,
not positions that large numbers of people are committed to. Some individual paleobiologists do think that the majority of cases fit the punctuational pattern; but the same
cannot be said for phyletic gradualism. It is even possible that no phyletic gradualists
exist (for reasons explained in Box 21.1).

the evidence supports a pluralistic
view . . .

However, the question of what the relative frequencies are of sudden and gradual
evolution during speciation merits an answer in itself. Eldredge & Gould (1972) posed
this question, and they have stimulated a major research program in the past 25 years.
They have also inspired paleontologists to collect data to new standards. The number of
biometrical studies, using relatively complete stratigraphic sequences, is growing, but is
not large. A review by Erwin & Anstey (1995) found some examples of gradual evolution, some of punctuated equilibrium, and some with a mix of the two. Jackson &
Cheetham (1999) found that the majority (29 of 31 studies) of the evidence that they
surveyed fitted punctuated equilibrium. Here we look at only two examples, to illustrate
two of the patterns and the kinds of evidence that are available.

21.4.2 *Caribbean bryozoans from the Upper Miocene and Lower Pliocene show a punctuated equilibrial pattern of evolution*

Cheetham (1986) studied in detail the evolution of a group of the sessile aquatic
invertebrates called Bryozoa (also called Polyzoa). His study included species in the
genus *Metrarabdotus*. Some members of the genus are alive in the seas today and they
also have an extensive fossil record. He worked on fossils dug up in the Dominican
Republic, which are the remains of animals that once lived in the Caribbean seas. The
main samples of fossils in the study date from the Upper Miocene and Lower Pliocene
(8–3.5 Myr BP), but some other species extend the age range.

Figure 21.7
The punctuated equilibrial pattern of evolution in Caribbean species of the bryozoan *Metrarabdotus*. Each point is the average for a large sample of individuals. Note that most lineages do not change through time; new species appear suddenly, without intermediates; and the ancestral species often persists alongside its daughter species. Time goes up the page and the *x*-axis indicates phenetic distance. Redrawn, by permission of the publisher, from Cheetham (1986).

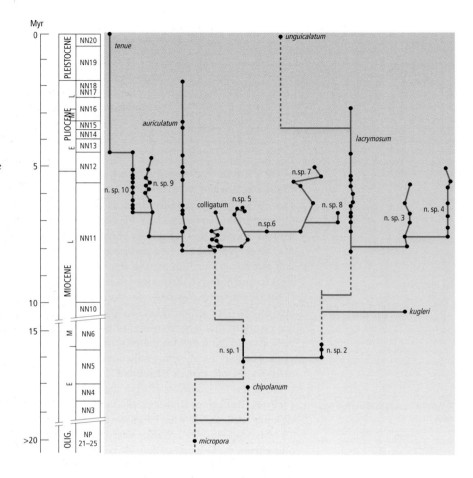

. . . as is illustrated by a thorough study of bryozoans . . .

Cheetham's was a thorough study. He measured up to 46 morphological characters per specimen in a total of about 1,000 specimens, from about 100 populations. The results show that these bryozoans mainly evolved in the punctuated equilibrial mode (Figure 21.7). Most of the species did not change in form over long periods of several million years, and most of the new species appeared suddenly without intermediate transitional populations. If there were intermediate forms, they lasted (on average) less than 160,000 years. Moreover, in a number of cases, the ancestral species persisted alongside its descendant species. Some species do occasionally appear to show short periods of gradual change within a lineage, but they are sufficiently rare that they could be interpreted as being due to sampling from a constant population. Cheetham also tested whether the forms classified as different species might just be ecophenotypic variants within one species. He raised members of a modern species in a range of environments. They all grew up much the same, and recognizably as members of one species. The punctuations, therefore, are likely to be real evolutionary events, not ecophenotypic switches.

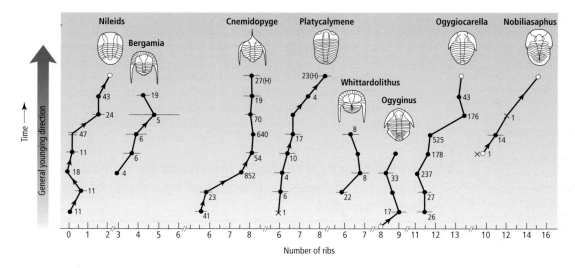

Figure 21.8
The gradual evolution in Sheldon's study of Ordovician Welsh trilobites. In eight lineages, the pattern of change is gradual rather than punctuated. Time goes up the page (total timespan 3 million years) and the biometrical variable (number of ribs) is along the bottom. The numbers besides the lines are the sample sizes. Redrawn, by permission of the publisher, from Sheldon (1987). © 1987 Macmillan Magazines Ltd.

21.4.3 Ordovician trilobites show gradual evolutionary change

. . . and a thorough study of trilobites

The extinct arthropod group of trilobites is classified by external morphological features such as the number of pygidial ribs (the pygidium is the tail region of a trilobite's body). Sheldon (1987) made a rigorous biometrical study of their evolution at a site in Wales. He measured the number of pygidial ribs in 3,458 specimens from eight generic lineages, taken from seven stratigraphic sections. The total time period spanned by the sections is about 3 million years.

In all eight genera, the average number of pygidial ribs increased through time, and in all eight the evolution was gradual (Figure 21.8); a population at any one time was usually intermediate between the samples before and after it. There is one possible artifactual explanation for the result, and Sheldon was able to rule it out. A gradual increase in the number of pygidial ribs would result if two populations, one with a higher number of ribs than the other, were mixed together, with successively later samples having increasing proportions of the high-rib-number population. Sheldon argued this was not the case because, with rare exceptions, his samples did not show bimodal frequency distributions, as they would if they contained a mixture of two distinct populations. These trilobites look like a good illustration of gradual evolution.

21.4.4 Conclusion

On the evidence so far we can conclude that both punctuated equilibrium and phyletic gradualism are real facts about fossil evolution. Some examples, like Cheetham's bryozoans, illustrate a punctuated equilibrial pattern of evolution; others, like Sheldon's trilobites, show a pattern of phyletic gradualism. Evolution has a range of rates, from sudden to smooth, in real examples of fossil speciation. Punctuated equilibrium may be somewhat commoner than phyletic gradualism, however (Erwin & Anstey 1995; Jackson & Cheetham 1999). A future research question will be to ask what conditions lead to more gradual evolution, and what conditions to punctuated evolution, but at present paleontologists are still answering the prior question of what the empirical rates of evolution are during, and between, speciation events.

21.5 Evolutionary rates can be measured for non-continuous character changes, as illustrated by a study of "living fossil" lungfish

The measurement of evolutionary rates in darwins is appropriate for metrical changes, such as a character evolving to be longer or shorter; but for larger changes, such as from a leg to a wing, this method ceases to be useful (Section 21.1). However, it is still possible to measure rates of evolution for larger changes. The last two sections of the chapter describe two methods. The first is a famous early study of evolutionary rates: Westoll's (1949) work on lungfish.

Lungfish (Dipnoi) form one of the four main divisions of fishes. They are an ancient group dating back over 300 million years, but only six modern species exist. The modern forms are examples of *living fossils*, species that have changed little from their fossil ancestors in the distant past. They should therefore show, at least recently, slow rates of evolution. Westoll investigated this question quantitatively. He distinguished 21 different skeletal characters of fossil Dipnoi. For each of the 21, he distinguished a number

Discrete character states . . .

of character states (like the character states discussed earlier for classification and phylogenetic inference). The 21 characters showed between three and eight different states. Character number 11, for example, was "degree of fusion of bones along the supraorbital canal." Westoll distinguished five different states, namely:

4. Irregular, more or less random fusions.
3. Tendency for fusions to be in twos, especially in some parts of the canal.
2. Still stronger tendency to fusion, rarely in threes or fours in specific sections.
1. Three or four elements (K–M) generally fuse, but there are numerous irregularities.
0. Three or four elements (K–L_2 or K–M) always fuse.

(Letters like K and L_2 refer to particular, identifiable bones.) The highest state (4) is the most primitive condition of the character, and 3, 2, 1, and 0 are successively later, more

. . . can be used to measure evolutionary rates

derived states. Westoll made an analogous list of states for all 21 characters. These character states are not the sort of metrical changes for which evolutionary rates can be measured in darwins (Section 21.1). The fusion of two bones into one is a discrete, not a continuous, evolutionary change.

For each fossil, Westoll calculated a total score, made up of its total for all 21 characters. The most advanced possible lungfish, with 21 characters in the most advanced state, would therefore have a total score of 0; the score for the most primitive possible lungfish, which had the highest scores for all 21, would have been 100. The rate of

(a) **Modernization of a lungfish character complex**

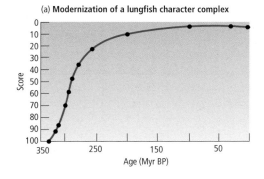

(b) **Rate of evolution in lungfish**

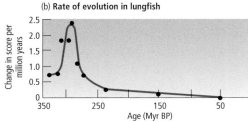

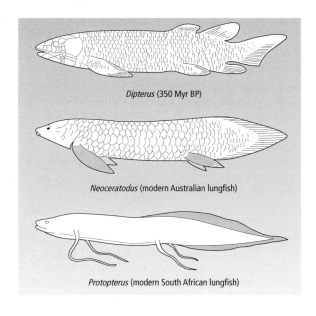

Dipterus (350 Myr BP)

Neoceratodus (modern Australian lungfish)

Protopterus (modern South African lungfish)

Figure 21.9

(a) Evolution in lungfish, shown as the total score for each fossil (the text explains the scoring). Score 0 is for the most primitive form, and 100 for the most advanced. The rate of evolution is the slope of the graph: when the graph is flat, evolution is not happening. (b) Rate of evolution. The graph is derived from (a) and shows the rate of change of the score through time. Lungfish have been living fossils since about 250 to 200 million years ago. Modified from Westoll (1949).

change in the score measures the rate of evolution of the group. The numbers assigned to the character states are arbitrary, but they can still be used to portray evolutionary rates. (By the way, the lungfish in Westoll's study are not a simple sequence of ancestors and descendants, in the way that the horses in MacFadden's work probably were. Westoll's rates are for evolution within the Dipnoi as a whole and are not rates of change down a single evolutionary lineage; the whole groups of Dipnoi would have contained many lineages.) Westoll's results are shown in Figure 21.9. Dipnoi, it reveals, have not always been "living fossils." Around 300 million years ago they were evolving rapidly, but since about 250 to 200 million years ago, their evolution has slowed right down. Their description as living fossils is accurate for the modern forms.

Lungfish have not always evolved slowly

The obvious biological question is why evolutionary change came almost to a stop in lungfish 200 million years ago. Lungfish are not the only examples of living fossils; other examples include the brachiopod *Lingula* and the horseshoe crab *Limulus* (Figure 21.10). The supreme examples of living fossils are the Cyanobacteria (sometimes called "blue-green algae") — 3 billion-year-old fossils look much like forms living today (Schopf 1994). There are many particular conjectures about why these groups have changed so little, but no general theory. The question is an instance of the general question of why there should be evolutionary stasis. Their stability may be due to stabilizing selection

Figure 21.10

The modern horseshoe "crab" (in fact a chelicerate, not a crustacean) *Limulus polyphemus*, which lives along the east coast of the USA, is a living fossil. It is morphologically very similar to forms that lived about 200 million years ago, and not all that different from Cambrian species. Redrawn, by permission of the publisher, from Newell (1959).

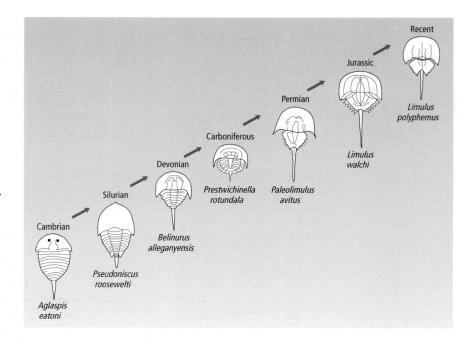

Character scores are arbitrary

or absence of genetic variation. Some living fossil species live in relatively isolated habitats, with no apparent competitors, and if their habitats have been stable there will have been no pressure for them to change. There is no evidence that living fossils have peculiar genetic systems that might prevent evolutionary change, for instance by constraining the degree of new genetic variation. The amount of genetic polymorphism in modern *Limulus polyphemus* it is not noticeably low.

But the point of the example here is methodological, not biological. It is to show how evolutionary rates can be studied quantitatively in characters whose evolutionary changes are not simply metrical. The characters can be divided into discrete states; the states assigned arbitrary scores; and the changes in those scores measured through time.

The quantification is mainly useful for purposes of illustration. Figure 21.9 neatly shows how rates of evolution have varied through time in a way that a table of raw character data could not. But the division of characters into states, and the assignment of scores to states, is arbitrary. The five states for supraorbital canal bone fusion could just as well have been scored 40, 16, 15, 14, 0 or 2, 8, 17, 39, 40 as 4, 3, 2, 1, 0. It would therefore be meaningless to compare exact numerical rates of change between characters, or between taxa. The scores are incommensurable. The approximate shape of a graph like Figure 21.10 could be compared with another such graph for another group; but there would be no point in asking why one group changed at, say, 2.1 units per million years and another at 1.3 units per million years. The scores are not intended for that kind of analysis. But as an illustration of how rates of change in lungfish have risen and fallen and declined to a virtual standstill, Westoll's analysis is a classic.

Figure 21.11

Taxonomic measurement of evolutionary rate. If a taxonomist has divided one group into two species, and another group into three species, in the same time interval, the latter group shows a 50% higher taxonomic rate of evolutionary change. The diagram illustrates only the logic of the argument. In real data, there would be gaps in the lineages, and the real pattern of evolution could have been smooth or jerky, with any number of branches in addition to the lineages shown here.

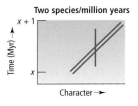

Two species/million years

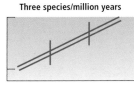

Three species/million years

21.6 Taxonomic data can be used to describe the rate of evolution of higher taxonomic groups

For a question such as "Do mammals evolve faster than bivalve mollusks?," a further kind of measurement of evolutionary rate can be used. The question could be tackled by either of the methods we have discussed so far. We could calculate the evolutionary rates for individual characters, measured either metrically or in discrete states, in mammals and bivalves, and then compare them, although there is a danger the comparisons might be meaningless. Any differences in evolutionary rate may reflect only the way we measure (for example) teeth in mammals and shell shape in bivalves, and nothing real. However, there may be some purposes — comparisons with rates of nucleotide change, perhaps — for which the measurements might be useful.

Another method is to use taxonomic evidence. When taxonomists divide a set of organisms into a number of species, they make their judgment according to the degree of phenetic differences among the forms. Their judgment will not usually be based on a single character, but on several characters, integrated in the taxonomist's mind into a single dimension of taxonomic similarity. Thus if there were two separate but comparable evolutionary lineages, and in the same time interval a taxonomist divided one lineage into two species and the other into three, it would suggest that the latter lineage had evolved at a higher rate (Figure 21.11). This comparison uses a *taxonomic rate* of evolution.

A taxonomic rate of evolution offers an abstract measure of how rapidly change is taking place in a group of species. The exact meaning of a taxonomic rate is less easy to specify than for an evolutionary rate of a single character and has a relatively imprecise meaning. It can be said in their defense that they summarize evidence from more than one character, and have greater generality. How reliable a taxonomic rate is depends on how reliable is the judgment of the taxonomist who divided the lineage up into species and genera.

Taxonomic rates of evolution are expressed in two main ways. One is the number of species or genera (or taxon) per million years. Table 21.2 gives some examples from the two taxa that we have discussed in this chapter: horses and lungfish. The rates for lungfish reillustrate how that group initially had a high rate of evolution, which then slowed down so much that they became living fossils.

The same data, but for a group made up of a larger number of lineages, can also be expressed as a *survivorship curve* (Figure 21.12). Survivorship curves are constructed by

The number of species per time unit is a rough measure of evolutionary rate

Figure 21.12

Survivorship curves for (a) bivalves mollusks and (b) carnivores (Mammalia). The curves express the numbers of genera surviving for different amounts of time. Note bivalve genera tend to last longer than carnivore genera, as is clear from the different scales of the x-axes in the two figures. The average duration of a bivalve genus is 78 million years, and 8.1 million years for a carnivore. Redrawn, by permission of the publisher, from Simpson (1953). © 1953 Columbia University Press.

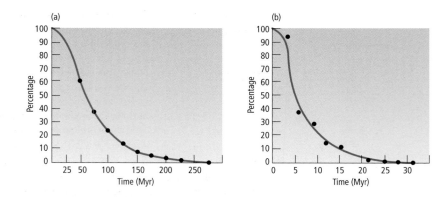

Table 21.2

Taxonomic rates of evolution in mammals and lungfish. Early in their evolution, lungfish evolved about as rapidly as mammals, but they have subsequently slowed down. *Hyracotherium–Equus* is the horse lineage discussed in Section 21.1. From Simpson (1953).

Group or line	Average duration of genus (Myr)
Hyracotherium–Equus	7.7
Lungfish	
Devonian	7
Permo-Carboniferous	34
Mesozoic	115

Taxonomic survivorship curves express evolutionary rates

taking a sample of a number (such as 100) of mammalian genera and measuring how long each one lasts in the fossil record. (Any taxonomic level within the mammals can be used; genera are only an example.) The survivorship curve plots the number of genera surviving for different times. Most of the genera are still surviving after a short time, but as time passes the members of the original sample drop out one by one. The slope of a survivorship curve measures the rate of evolution of the group. If the group is evolving fast, the survivorship curve falls rapidly, but is more drawn out for a slowly changing group. (Survivorship curves are more familiar for populations of individuals (Section 4.1, p. 72). An actuarial survivorship curve plots the survival of a sample of individuals through time, but the same type of graph can be plotted for species and other taxonomic groups.)

There is a similar problem in comparing taxonomic evolutionary rates between groups as there is for single characters. A bivalve taxonomist may be a good judge of a bivalve genus, and a mammal taxonomist of a mammal genus, but it is still difficult to know how to interpret any differences in the rates of turnover of the two sorts of genera. The differences in rate could just reflect some difference in the way the two taxonomists work. Within a group, survivorship curves can have revealing features (Section 22.7, p. 637). For now, however, we only need to know that taxonomic evidence can provide another sort of measurement of evolutionary rates, along with measurements of single characters.

21.7 Conclusion

This chapter has introduced three methods of measuring rates of evolution. For single characters that show continuous variation, we can measure the metrical rate of change and express it in darwins. For characters with discrete states, we can give each state an arbitrary score and measure the rate of change of the score. A cruder measure is provided by the duration of species in the fossil record, because taxonomists will recognize a higher turnover of species in groups that evolve rapidly. All three methods have their particular uses and applications. In this chapter we have seen how paleobiologists have used all three methods to study general evolutionary questions.

Summary

1 Evolutionary rates of fossil characters can be measured as simple rates of change through time; logarithms are often taken of the character measurements and the rate expressed in "darwins." Evolution in horse teeth is a classic example.

2 Rates of evolution measured in the fossil record are slower than those produced by artificial selection in the laboratory.

3 Evolutionary rates vary between different geological times, taxa, and types of taxa. The science of evolutionary rates is mainly concerned to explain the pattern of evolutionary rates.

4 Among published measurements of evolutionary rates, the rate and the time interval over which it was measured are inversely related: faster evolution is seen in shorter intervals. The reason is probably that the direction of evolution fluctuates through time.

5 Eldredge and Gould stimulated a controversy about evolutionary rates by their suggestion that rates have a strict pattern (called punctuated equilibrium) where evolution is fast at times of splitting (speciation) and comes to a halt between splits. The opposite pattern, in which evolution has a constant tempo, they called phyletic gradualism.

6 It is difficult to discover the pattern of evolutionary rates at, and between, speciation events because the fossil record is incomplete.

7 There is some evidence for punctuated equilibrium, such as Cheetham's study of Caribbean bryozoans from the Miocene and Pliocene, and some for phyletic gradualism, such as Sheldon's study of Ordovician trilobites in Wales. No general empirical conclusion is yet possible, though punctuated equilibrium is a well confirmed phenomenon.

8 Evolutionary processes and rates can be examined at all taxonomic levels from evolution within populations, through speciation, to the origin of the higher groups. Evolution may have characteristic mechanisms and rates at different levels, or the same set of rates and processes may operate equally at all levels.

9 For large changes, like from a limb into a wing, evolutionary rates cannot be measured as a continuous variable. The character can instead be divided into states. The evolutionary rate can then be studied in the rate of change between states. Westoll studied the evolution of lungfish by this method.

10 The number of species in a lineage per million years is a complex measure of evolutionary rate, called a taxonomic rate of evolution. Taxonomic rates can be expressed as survivorship curves.

Further reading

Simpson (1953) remains a good introduction to the study of evolutionary rates; Fenster & Sorhannus (1991) is a more recent review. MacFadden (1992) is an excellent book about evolution in fossil horses. Gingerich (2001) updates his work discussed in Sections 21.1.2 and 21.2. Hendry & Kinnison (1999) discuss some more sophisticated measures of evolutionary rate than the darwin; their compilation of measurements yields similar results to those of Gingerich's discussed in the text, but various adjustments need to be made before the numbers are directly comparable. See also some papers in Hendry & Kinnison (2001). See Grant (1986, 1991) on the finches.

The literature on punctuated equilibrium is now vast, but fortunately Gould (2002b) provides almost a one-stop shop for the theory, evidence, controversy, and broader implications, as well as references. Benton & Pearson (2001) is a short introduction and Levinton (2001) contains a critique. Dennett (1995, chapter 10) discusses the controversial relation with saltationism. Erwin & Anstey (1995) contains papers on fossil speciation. Several chapters in Jackson *et al.* (2001) also contain chapters on the topic. Jackson & Cheetham (1994) is a popular paper about the Caribbean bryozoans, and Jackson & Cheetham (1999) looks more broadly at the evidence from Neogene benthic fossils.

Study and review questions

1 A character (such as tooth size) has been measured in two populations, at two times (t_1 and t_2) (in Myr BP). x_1 and x_2 are the average sizes of the character (in size units) at the two times. Calculate the evolutionary rate in darwins. You may need a calculator that works out natural logarithms.

x_1	t_1	x_2	t_2	Rate
2	11	4	1	
2	11	20	1	
20	11	40	1	
20	6	40	1	

2 Gingerich (1983) plotted evolutionary rates against the time interval used to measure the rate for over 500 evolutionary lineages. (a) What did he find? (b) How would you interpret it?

3 Review the main predictions of the punctuated equilibrium and phyletic gradualist models of speciation in fossils.

4 Describe an evolutionary mechanism that could generate punctuated equilibrium. Does it imply that punctuated equilibrium contradicts orthodox Darwinism?

5 How can it be quantitatively shown whether a taxonomic group is a living fossil?

22 Coevolution

Coevolution happens when two or more species influence each other's evolution. It is often invoked to explain coadaptations between species, and we begin by considering whether coadaptation provides evidence of coevolution. Coevolution strictly speaking requires reciprocal influences between species, but there is a related phenomenon, called sequential evolution, in which changes in one species influence the other but not the reverse. The chapter then looks in turn at coevolution between flowering plants and insects, between parasites and hosts, at antagonistic coevolution in general, and the phenomenon of evolutionary escalation. Finally, we look at the "Red Queen" mode of coevolution. For plant–insect and parasite–host coevolution we look at cophylogenies — in which the phylogenetic trees of the two interacting taxa form mirror images. We consider the evolution of virulence in parasitic diseases, including diseases of humans. The Red Queen hypothesis suggests that species continually evolve to maintain a level of adaptation against competing species. Van Valen invented the hypothesis to explain a general result he discovered in the fossil record: the chance of extinction of the species in a taxonomic group is independent of the age of the species. The status of the Red Queen hypothesis is uncertain, not least because it is difficult to test.

Figure 22.1

A pair of complementary coadaptations in an ant and caterpillar. (a) The ant (*Formica fusca*) is tending a caterpillar of the lycaenid butterfly species *Glaucopsyche lygdamus*. The ant is drinking honeydew, secreted by the caterpillar from a special organ. (b) *Formica fusca* defending a caterpillar of *G. lygdamus* against a parasitic braconid wasp. The ant has seized the wasp in its mandibles. Bars indicate 1 mm. (Photos courtesy of Naomi Pierce.)

22.1 Coevolution can give rise to coadaptations between species

Figure 22.1a shows an ant (*Formica fusca*) feeding on the caterpillar of the lycaenid butterfly *Glaucopsyche lygdamus*. The ant is not eating the caterpillar; it is drinking "honeydew" from a special organ (Newcomer's organ), the sole purpose of which seems to be to provide food for ants. The reason why the caterpillars feed the ants has been the subject of several hypotheses. Pierce & Mead (1981) carried out an experiment which suggests that the caterpillars, at least in *G. lygdamus*, feed ants in return for protection from parasites.

The caterpillars are parasitized by braconid wasps and tachinid flies. Alone, they are almost defenseless against these lethal parasites; but the tending ants will fight off parasites from their caterpillars (Figure 22.1b). Pierce and Mead experimentally prevented ants from tending caterpillars. They then measured the rates of parasitism in the experimentally unprotected and in the normally protected (control) caterpillars. Their results show that ants reduce the rate of parasitism in *G. lygdamus* (Table 22.1). The ants and caterpillars are therefore closely adapted to each other; the ants gain food, and the caterpillars gain protection. They form a kind of interspecific *coadaptation*. (Here the term "coadaptation" refers to the mutual adaptation of two species; it has also been used to describe the mutual adaptation of genotypes (Section 8.2, p. 197) and of parts (Section 10.3, p. 260) within an organism.) Relationships like that between ants and lycaenids are called *mutualism*, many examples of which are known, and provide some of the most charming details in natural history.

How could the coadaptation between ant and lycaenid have evolved? The morphological structure and the behavior patterns of both ant and caterpillar appear to have

Interspecific coadaptation can be experimentally tested

Figure 22.2
Coevolution means that two separate lineages mutually influence each other's evolution. The two lineages tend to (a) change together, and (b) speciate together. Lineages 1 and 2 could be, for example, an ant lineage and a lycaenid butterfly lineage.

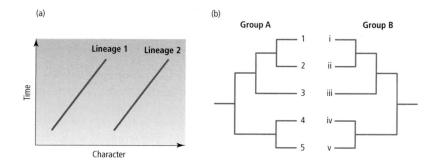

Interspecific relations can undergo coevolution

evolved in relation to each other. Natural selection probably favored mutually adapted changes in each species, after the ancestors of the two species had become associated. Changes in one species, such as to increase honeydew production, would favor changes in the other (to increase protection) as the caterpillars became more beneficial to the ants. This kind of reciprocal influence is what is meant by *coevolution*: each species exerts selection pressures on, and evolves in response to, the other species. The two lineages evolve together (Figure 22.2). In all examples of evolution, a species evolves in relation to changes in its environment. Coevolution refers to the special case of evolution in which the species' environment is itself evolving.

We look at several examples of coevolution in this chapter. In some, like the ants and caterpillars, coevolution promotes the mutual benefit of the coevolving lineages. In others, such as parasite–host coevolution, the process is antagonistic. An improvement in one party (such as improved host defense) is a deterioration in the environment of the other party (the parasites).

Table 22.1
Caterpillars of the lycaenid butterfly *Glaucopsyche lygdamus* are more likely to be parasitized if they are not tended by ants. Ants were experimentally excluded from some caterpillars, and the rate of parasitism on these and untreated control caterpillars was measured. The two sites are in Gunnis County, Colorado. The parasites were wasps and flies, and *n* is the sample size. Reprinted, by permission, from Pierce & Mead (1981). © 1981 American Association for the Advancement of Science.

Site	Caterpillars without ants		Caterpillars with ants	
	% parasitized	*n*	% parasitized	*n*
Gold Basin	42	38	18	57
Naked Hills	48	27	23	39

22.2 Coadaptation suggests, but is not conclusive evidence of, coevolution

Full evidence for coevolution is hard to obtain

Coadaptations, such as those between an ant and a caterpillar, likely arise by coevolution between the lineages leading to the two modern species. However, the observation of coadaptation between two species is not by itself enough to confirm that the two have coevolved together. Janzen (1980) pointed out that the two lineages could have been evolving independently, and at some stage the two forms just happened to be mutually adapted to each other. The ancestors of *Glaucopsyche lygdamus* might have evolved their Newcomer's organs for some other reason than feeding *Formica* and the ants might have evolved antiwasp behavior patterns for some other reason than defending caterpillars; when the two came together they were already coadapted. To demonstrate coevolution requires showing not only that two forms are coadapted now but also that their ancestors evolved together, exerting selective forces on each other.

That is a tall order. In practice, biologists tend to assume that interspecific coadaptations are due to a long history of coevolution unless a convincing alternative hypothesis can be put forward. Janzen's stricture is logically correct, but difficult to live up to in practical biology. Further evidence that a coadapted system arose by coevolution can come from comparison with related species. The relation between *G. lygdamus* and *Formica* is not unique. Lycaenids and ants have evolved a large number of relationships in different species and this suggests the two groups have been evolving together for some time.

22.3 Insect–plant coevolution

22.3.1 Coevolution between insects and plants may have driven the diversification of both taxa

Insect–food plant relations may result from biochemical evolution

In a paper that is perhaps the most influential modern discussion of coevolution, Ehrlich & Raven (1964) listed the food plants of the main butterfly taxa. Each family of butterflies feeds on a restricted range of plants, but these plants are in many cases not phylogenetically closely related. Ehrlich and Raven explained the diet patterns mainly in terms of plant biochemistry. Plants produce natural insecticides — chemicals like alkaloids that can poison herbivorous (phytophagous) insects. Insects, in the manner of pest species evolving resistance to artificial pesticides (Section 5.8, p. 115), may evolve resistance to these chemicals, for instance by means of detoxifying mechanisms. When a new detoxifying mechanism arises, it will open up a new array of food supplies, consisting of all those plants that produce the now harmless chemical. The insects can feed on them, and will diversify to exploit the resource. The result will be that each insect group can feed on a range of food plants, the range being set by the capabilities of the insect's detoxifying mechanisms. The range of food plants will form a biochemical group, but need not form a phylogenetic group because unrelated plants could use the same defensive chemicals. Ehrlich and Raven's pattern of butterfly–plant relationships could arise as a result.

In turn, natural selection on the plants favors the evolution of improved insecticides. Plant–insect coevolution should therefore consist of cycles, as plant groups are drawn into, and removed from, the diets of insect groups, and the insects evolutionarily "move" between plant types according to their biochemical abilities. The biochemical arms race between plants and insects should persistently favor new mechanisms on both sides, and might therefore have promoted the diversification of insects and angiosperms (Section 14.10.2, p. 412, and Figure 18.8, p. 539). In Ehrlich & Raven's (1964) words "the fantastic diversification of modern insects has developed in large measure as the result of a stepwise pattern of coevolutionary stages superimposed on the changing pattern of angiosperm variation."

Pollinator relationships have led to the evolution of specialized adaptations . . .

Coevolution between plant poisons and insect detoxification mechanisms is only one way in which insects and flowering plants may have influenced each other's evolution. Pollination is another example. A few gymnosperms are pollinated by insects, but insect pollination really took off with the evolution of flowers in angiosperms. Plants without flowers are mainly pollinated by abiotic mechanisms, such as the wind.

Once insect pollination had evolved, natural selection could favor increasingly specialized pollinator relations. In any one flower species, natural selection favors those flowers whose pollen is transported only to other flowers of the same species. If the insect flies to another flower species, the pollen is more likely to be wasted. A flower may put its nectar reward in a place that can only be reached by insects with a specialized organ, such as a long tongue. Only insects with long tongues can then obtain the reward — and those insects will be well rewarded because they have little competition from other insects. The insects with the specialized adaptation will probably next fly to another flower of the same type, because it will be well rewarded there too. The process can continue, as the plant places its nectar deeper and deeper, and the insects evolve longer and longer tongues. The final result could be something like the Madagascan orchid *Angraecum sesquipedale*, which puts its nectar in long spurs, up to 45 cm in length. Darwin knew of this species and predicted that a specialist pollinator would be discovered with a long tongue. Wasserthal (1997) recently confirmed that several hawkmoth species, with exceptionally long tongues, are able to obtain nectar from and pollinate this orchid (see the cover illustration of this book).[1]

. . . that may have promoted insect–angiosperm diversity

As natural selection favors specialized pollinator relationships, it will tend to increase the diversity of both plants and insects. Plants that are pollinated by a single insect species have an advantage, because less of their pollen is wasted. Insects that specialize on one plant species will make more efficient use of their specialized feeding adaptations. Other factors can also operate in the coevolution of plants and insects, but the two factors we have looked at here, diet and pollination, illustrate the general subject. The theoretical ideas have been tested in many ways, but one particularly active method at present is to use phylogenies.

---

[1] The evolutionary story may be more complex. Wasserthal (1997) found that pollinators with long tongues are less vulnerable to predation. They can feed on a flower without landing, and thereby avoid predation from spiders that sit on the flower and catch insects that land there. The hawkmoths could have evolved long tongues as an antipredator adaptation, perhaps while feeding on unrelated plants. The orchids may have evolved their long spurs to make use of hawkmoths that had already evolved long tonges pollinating other species. Wasserthal's argument illustrates how hard it is to distnguish coadaptation due to coevolution from coadaptation following unrelated histories in the two lineages.

Figure 22.3

Phylogenies of North American *Tetraopes* (beetles) and their food plants, milkweeds (*Asclepias*). The phylogenies are mainly mirror images, or cophylogenies. The two or three exceptions may be due to errors in phylogenetic inference or to host shifts (as discussed in Section 22.3.3). These beetles exploit their food plants both as larvae (which bore into the roots) and as adults (which eat the flowers and leaves). Redrawn, by permission of the publisher, from Farrell & Mitter (1994).

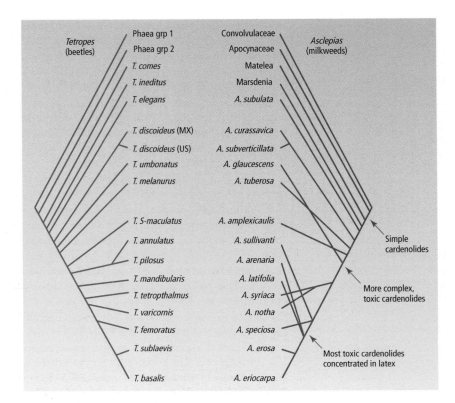

22.3.2 Two taxa may show mirror-image phylogenies, but coevolution is only one of several explanations for this pattern

Figure 22.3 shows on the left the phylogeny of 15 out of the 25 beetles in the genus *Tetraopes* that live in North America, together with two relatives. On the right is the phylogeny of their food plants, the milkweeds (*Asclepias*). Milkweeds contain poisons (cardenolides), but *Tetraopes* are not harmed by them. The beetles store the toxins in their own bodies, making them unattractive to birds and mammals. *Tetraopes* are brightly colored in orange and black. The striking feature of the two phylogenies in Figure 22.3 is that they are near mirror images. With two or three exceptions, the phylogeny of the beetles matches the phylogeny of the plants they eat. In technical language, the phylogenies of the two taxa are *cophylogenies*. The phylogenies of two taxa are cophylogenies if they have the same (or much the same) branching pattern. Statistical tests exist to determine whether two phylogenies are more similar than would be expected by chance.

Cophylogenies can arise because of . . .

. . . coevolution . . .

Cophylogenies can arise for at least three reasons. One is coevolution in the full sense of the word. The two taxa have exerted evolutionary influences on each other, and evolution leading to speciation in one taxa tends to cause speciation in the other taxon too. For instance, two subpopulations of one ancestral milkweed species might have

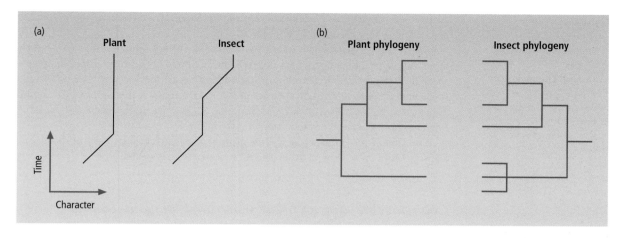

Figure 22.4

Sequential evolution means that change in one lineage selects for change in the other lineage, but not vice versa. Sequential evolution would apply to plants and insects if plant evolution influences insects, but insect evolution has little influence on plants. The pattern of change in (a) lineages, and (b) phylogenies differs from strict coevolution (compare with Figure 22.2). (a) Changes in plants coevolve with changes in insects, but changes (for some reason other than changes in plants) in insects do not cause changes in plants. (b) When plants speciate, so do insects, but when insects speciate it has no effect on plants.

become geographically separated from each other. Each subpopulation of milkweed might well have its own subpopulation of *Tetraopes*.

The two milkweed populations might diverge as they coevolved with their local insects. One population of milkweeds might evolve one set of cardenolides, while the other population evolved a different set of poisons. Each local beetle population would evolve detoxification mechanisms appropriate for the local milkweeds. Reproductive isolation would probably evolve as a by-product, by the classic process of allopatric speciation (Section 14.3, p. 383). After a while, the two forms of plant might meet up but be reproductively isolated because of the genetic differences that had built up between the two.

. . . with cospeciation . . .

. . . or sequential evolution . . .

Alternatively, cophylogenies can arise by *sequential evolution*. In sequential evolution, changes in one of the two taxa lead to changes in the other, but not the other way round. For instance, some biologists have argued that plants influence the evolution of insects, but insects have less effect on evolution in plants (Figure 22.4). This could be for a number of reasons. One is that many insects eat only one type of plant, whereas plants are eaten by many insects. When a plant changes, its insects will all have to change to keep up; but when one insect species changes, it alone will exert only a small selective pressure on its food plant. Sequential evolution may result in imperfect matches in the phylogenies of the two taxa (Figure 22.4b). In principle, the shape of the phylogenies can be used to distinguish sequential evolution from coevolution. However, cophylogenies are rarely perfectly matched, and many factors can influence the degree of match between the phylogenies of two taxa. It is statistically difficult to distinguish real coevolution from sequential evolution. The extent to which insect–plant evolution is sequential, or fully coevolutionary, is an active research topic that has reached no generally accepted conclusion.

. . . or cospeciation without coevolution

Finally, cophylogenies may arise if two taxa have no evolutionary influence on each other, but some independent factor leads to speciation in both. For example, allopatric speciation could occur in several non-interacting taxa if they occupy much the same range and something splits all their ranges. A river might cut through and divide the ranges of several species in an area, for instance. All the taxa might speciate at the same time, but not because of coevolution.

In Section 22.5.2 below we look some more at cophylogenies, for parasite–host relationships, and meet a fourth process that can lead to cophylogenies.

22.3.3 *Cophylogenies are not found when phytophagous insects undergo host shifts to exploit phylogenetically unrelated but chemically similar plants*

Insects and plants can coevolve without producing mirror-image phylogenies. Becerra (1997), for example, studied coevolution between plants in the genus *Bursera* and the specialist chrysomelid beetles of the genus *Blepharida* that feed on them. *Bursera* is a New World genus of the family Bursaraceae, a tropical family of trees and shrubs that is famous for its aromatic resins, which are used in perfume and incense. Frankincense and myrrh are two Old World relatives of *Bursera*. Figure 22.5a shows the phylogenies of some

A beetle group and its food plants do not form cophylogenies

associated *Bursera* and *Blepharida*. They are not cophylogenies. Becerra also analyzed the chemical defenses of *Bursera* and found that they fall into four main groups; but species with the four systems are found scattered around the phylogeny (Figure 22.5b).

The relations between the plant and its insects becomes clear when we look at the chemical defenses of the plants in relation to the beetle phylogeny (Figure 22.5c). Now the beetle phylogeny has a clear pattern, with four distinct regions, corresponding to the four chemical defenses of the plants. The caption explains further how the information in Figure 22.5c can be used to make sense of the initially confusing patterns in Figure 22.5a.

Probably what has happened is that during evolution a beetle species evolves a defense against a certain set of plant chemicals. The beetles can then colonize other plants that have similar chemical defenses. These colonizations, in which a beetle shifts from one host plant to another, are called *host shifts*. If chemically similar plants are not

The insects colonize plants according to biochemical similarity

phylogenetically closely related, the result is that the beetles evolutionarily jump around the plant phylogeny. Although *Bursera* and *Blepharida* are coevolving, they do not have cophylogenies. The pattern in Figure 22.5 illustrates Ehrlich & Raven's (1964) ideas discussed in Section 22.3.1 above.

In summary, the phylogenies of two interacting taxa, such as flowering plants and insects, can be used to study coevolution. A cophylogeny alone is not strong evidence of coevolution, because coevolution may or may not produce cophylogenies and coevolution is not the only factor that can cause cophylogenies. However, cophylogenies and the deviations from them can be used (in combination with other forms of evidence) to tease apart the coevolutionary forces in the history of two taxa.

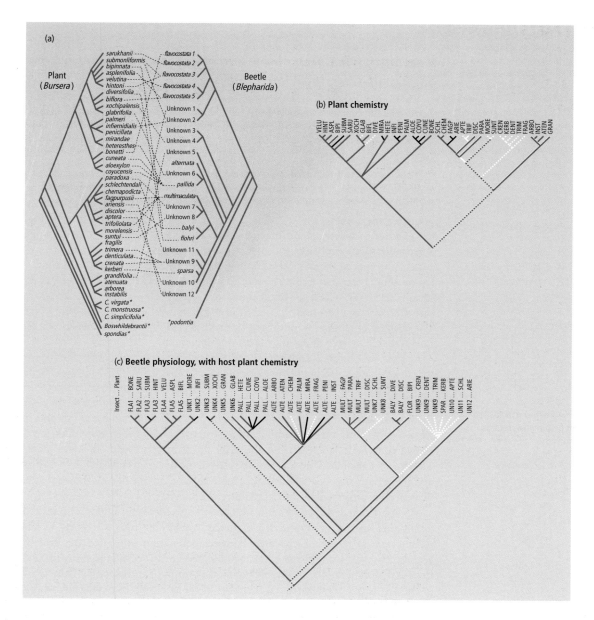

Figure 22.5

(a) Phylogenies of *Bursera*, a genus of plants (left), and of *Blepharida*, a genus of beetles (right) that feed on *Bursera*. The beetle species are mainly monophagous and each beetle species is written next to the plant species that it feeds on. The two phylogenies are not mirror images. (b) The phylogeny of the plants (*Bursera*) showing the distribution of four chemical defense systems. The four are phylogenetically scattered about, rather than falling into the four branches of the plant phylogeny. (c) Phylogeny of the beetles (*Blepharida*) with the chemical defenses of their host plants written on the phylogeny. The chemistry of the host plants forms neat phylogenetic groups on the beetle phylogeny. (Multiple branches are for polyphagous beetle species — those that feed on more than one plant species.) The three phylogenies can be read together as follows. Take a group of beetles in (c), such as FLA1–5 at the left. Their host plants are chemically similar. In (a) these five beetle species are at the top, and we can see that FLA1 and FLA2, for instance, feed on phylogenetically unrelated species of plants. Perhaps FLA2 evolved after a host shift in recent times. The host shift was possible because the two host plants *B. bonetti* and *B. sarukhanii* are chemically similar (as can be seen in (b)). Modified, by permission, from Becerra (1997). © 1997 American Association for the Advancement of Science.

22.3.4   *Coevolution between plants and insects may explain the grand pattern of diversification in the two taxa*

Terrestrial life on Earth today is dominated, among animals and plants, by insects and flowering plants. It is a plausible hypothesis that the two taxa have promoted each other's diversification, by coevolutionary mechanisms such as specialized pollinator relations. Here we can look at two tests of this hypothesis.

If the two taxa promoted each other's diversity, then they are predicted to have diversified at much the same time in the fossil record. Flowering plants diversified in the Cretaceous (Figure 18.8, p. 539 — though the molecular clock suggests an earlier origin for the angiosperms (Section 18.5, p. 538)). Insects were also diversifying, but the question is whether their diversification accelerated in the Cretaceous while the angiosperms were evolving.

One study suggests angiosperms did not promote insect diversification

Labandeira & Sepkoski (1993) counted the number of insect families through geological time, from 250 million years ago to the recent past. The number of families steadily increases, on a logarithmic scale, from about 100 in the Triassic, to 300 in the early Cretaceous, to around 400 in the early Tertiary, and 700 at the end of the Tertiary. The number of families went up as a (logarithmically) straight line, with no acceleration in the Cretaceous as the angiosperms proliferated.

At least superficially, Labandeira and Sepkoski's result contradicts the hypothesis that insects and flowering plants promoted each other's diversity. However, Grimaldi (1999) argued that diversification could have taken place within each insect family. Pollinating forms evolved within many insect groups independently, and the pollinating insects seem to have diversified at about the same time as the angiosperms in the fossil record. Thus, the total number of large insects groups, such as families, may have shown little or no increase, but the diversity of insects could still have increased. Therefore, Labandeira and Sepkoski's test may be inappropriate. But Grimaldi's argument has not yet been tested quantitatively, and no one has yet shown that insect diversity was promoted by the rise of flowering plants.

Another study found that biotic pollination promotes angiosperm diversity

A second kind of test looks at the numbers of species within angiosperm taxa today. Some angiosperms are pollinated biotically (usually by insects), others by abiotic means such as the wind or water. If insect pollination promoted the diversity of angiosperms, we would expect insect-pollinated groups of angiosperms to be more diverse than comparable abiotically pollinated groups of angiosperms. The exact form of the test is to compare related branches of the angiosperm phylogeny, where one branch has biotic and the other abiotic pollination (Figure 22.6). Dodd *et al.* (1999) identified 11 such comparisons, and the evidence as a whole strongly supported the hypothesis that biotic pollination is associated with increased angiosperm species diversity. Other tests of the same sort also suggest that insect, and particularly beetle, diversity is enhanced in groups that are associated with angiosperms rather than gymnosperms (Farrell 1998). Therefore, this second kind of test using phylogenetic comparisons within modern plants and insects, suggests that the two taxa have promoted each other's diversity.

Insects and flowering plants have likely influenced each other's evolution in many detailed ways, and research on particular insect–plant relationships seeks to understand

Figure 22.6

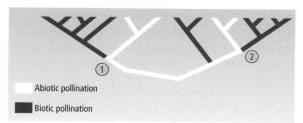

Abiotic pollination

Biotic pollination

Testing the effect of biotic as opposed to abiotic pollination on species diversity, using phylogenetically independent trials. The test starts with a phylogeny of angiosperms, distinguishing whether each species is pollinated by biotic or abiotic means. We then identify nodes in the tree where two sister branches have contrasting means of pollination, and count the species in the two branches. Each such node contributes one trial to the final test. (The nodes are statisitically independent. Statistical problems arise if we simply count the numbers of species with biotic and with abiotic pollination in the tree as a whole.) In this diagram, two nodes provide independent contrasts. In both nodes, the evidence supports the hypothesis that species diversity is higher in the branch with biotic pollination. The species in the middle branch are ignored in the test. Dodd *et al.* (1999) performed a test of this general form with real data, and found higher species diversity in branches with biotic pollination.

these influences. In this section we have looked at some grander comparisons. Grand comparisons are rarely conclusive, because so many factors can come into play. The hypothesis has been tested using evidence from fossils and from the shapes of modern phylogenies. The fossil test remains inconclusive, but the test with modern phylogenies is supportive. The evidence as a whole suggests that coevolution partly drove the insects and plants to cover the modern globe.

We have concentrated here on insect–plant relations, and on tests using phylogenetic methods. However, other methods are being used, and other taxa besides insects have evolutionary relations with plants. Box 22.1 looks at an example.

22.4 Coevolutionary relations will often be diffuse

The clearest examples of coevolution come from ecologically coupled pairs of species. In practice each species will experience, and exert, selective pressures on many other species. The evolution of a species will be an aggregate response to all its mutualists and competitors, and any evolutionary change may not be easy to explain in terms of any one competitor. The process is called *diffuse coevolution*. It undoubtedly operates in nature; indeed, it may be the main force shaping the evolution of communities of species. But it is difficult to study, and its importance is consequently controversial.

22.5 Parasite–host coevolution

Step-by-step coevolution is particularly likely to take place between parasites and their hosts. They can have specific and close relations, and it is easy to imagine how a change in a parasite, which improves its ability to penetrate its host, will reciprocally set up

Box 22.1
The Evolutionary Genomics of Fruit

Many plants produce large, nutritious fruits. Fruits probably evolved to attract birds and mammals, which disperse the plant seeds when they eat the fruit. Specialized seed dispersal relations with vertebrates are another coevolutionary mechanism by which angiosperm diversity may have been increased. However, once fruits had evolved, several other taxa evolved to exploit them, though not in ways that benefit plants. Fungi digest fruits, making them rot. Some insects, such as fruitflies, lay eggs on or in the fruit, and their larvae eat the fruit from within.

Fossil evidence of fruits is found from the early Tertiary, around 60 million years ago. The evolution of fruit manufacture in plants, and of fruit exploitation in vertebrates, insects, and fungi, required special genes coding for appropriate developmental and metabolic circuits. Yeast, for instance, have a special metabolic circuit for fruit digestion — the circuit that produces alcohol as a by-product and is the basis of beer, wine, and other drinks. The genes that code for this circuit have been identified in the yeast genome. The genes originated by duplications, and the time when that happened can be dated by the molecular clock (Section 19.3, p. 559) to about 80 million years ago. This date is somewhat earlier than the fossil date for the first fruit. Maybe the fossil date is too late, or the molecular clock is unreliable.

What about fruit-consuming insects? The Drosophilidae are a family of fruit exploiters. They originated about 65 million years ago, and probably evolved fruit-exploiting metabolism around that time. It would be interesting to date the duplications that generated fruitfly alcohol dehydrogenase (Sections 4.5, p. 83, and 7.8.1,

p. 180) and other fruit-related enzymes. Maybe they would date to around this time too. (Incidentally, fruitfly alcohol dehydrogenase is unrelated to the enzyme of the same name in humans.) Further evidence may also come from the genomes of angiosperms. If we could identify genes used in fruit manufacture, we could date them and see whether they also originated around this time.

Alternatively, we could do some evolutionarily inspired gene hunting. In the *Arabidopsis* genome, many duplications have been dated, but the functions of many of the duplicated genes remain unknown. Vision *et al.* (2000) published a picture of the time course of duplication events in the history of *Arabidopsis*. The picture has a minipeak around 75–80 million years ago. Maybe some of the genes in that minipeak will be related to fruit manufacture. (*Arabidopsis* itself does not produce fruits, but its genome could contain the ghosts or relatives of fruit genes.)

In conclusion, the origin of fruit led to evolutionary changes in several taxa — fungi, insects, and vertebrates. These changes occurred independently in each of the taxa. Genomic evidence can be used to date events in the past, once we have identified the genes concerned. We can predict that the evolutionary changes in all taxa should have occurred at much the same time, or at least that the changes in fungi and animals should follow the changes in plants. The evidence so far is incomplete, but tantalizing. It also illustrates how genomics is being integrated with existing paleontological and ecological methods in the study of coevolution.

Further reading: Ashburner (1998), Dilcher (2000), Benner *et al.* (2002).

---

selection for a change in the host. If the range of genetic variants in parasite and host is limited, coevolution can be cyclic (Section 12.2.3, p. 325); but if new mutants continually arise, the parasite and host may undergo unending coupled changes that may or may not be directional according to the type of mutations that arise. Coevolution in parasites and hosts is antagonistic, unlike the mutualistic coevolution of ants and caterpillars or of flowering plants and pollinators.

Parasite–host coevolution is antagonistic

Many biological properties of parasites and hosts have been attributed to coevolution. Here we concentrate on two. The first is parasitic virulence. In informal terms, virulence means how destructive the parasite is. In formal terms, *virulence* is expressed as the reduction in fitness of a parasitized host relative to an unparasitized host. A highly virulent parasite is one that kills its host quickly, reducing the host's fitness to zero. The virulence of a parasite is normally thought to be a side effect of the manner in which the parasite lives off its host. If, for instance, a parasite consumes a large proportion of its host's cells, it will be more likely to kill its host and therefore be more virulent than one that consumes less host cells. The second topic we look at is whether or not the phylogenies of parasites and their hosts have the same shape.

22.5.1   *Evolution of parasitic virulence*

Parasitic virulence and host resistance show evolutionary changes in Australian myxoma virus and rabbits

Myxomatosis reduced rabbit populations

The myxoma virus (which causes myxomatosis) in Australian rabbits provides the classic illustration that the virulence of a parasite can change evolutionarily. The rabbits in question belong to a species (*Oryctolagus cuniculus*) which is native to Europe but was introduced to Australia, where it thrived and became a pest. The natural host of the myxoma virus is another kind of rabbit, *Sylvilagus brasiliensis*, from South America, in which the virus probably has low virulence. In 1950 the virus was deliberately introduced into Australia in an attempt to control the pestiferous rabbits. It was, initially, a deadly success. It spreads (in Australia, at least) from rabbit to rabbit by means of mosquitoes and the large population of these biting insects enabled the myxoma virus to sweep through the southeast Australian rabbit population, and round the south coast as far as Perth in the west by 1953. Myxomatosis initially almost annihilated the rabbit population: it declined by 99% in some hard hit areas. The virus was introduced into France in 1952, and started to spread through Europe, and was surreptitiously introduced into the UK in 1953.

The myxoma virus was highly virulent when it first hit the Australian (and European) rabbit population; it killed 100% of infected hosts. Soon, however, the kill rate declined. This decline could result from any combination of increasing host resistance and decreasing viral virulence, and normally we should not know which was operating. But in this case, a carefully controlled set of experiments allowed the two factors to be teased apart.

The myxoma virus decreased its virulence

The decline in virulence of the myxoma virus was demonstrated by infecting standard rabbit strains in the laboratory with the viruses taken from the wild in successive years. Because the rabbit strain was controlled and constant, any decline in the kill rate must be due to a decline in virulence in the virus. Table 22.2 shows the results, in Australia and Europe. In both places the virus started off maximally virulent (killing 100% of infected rabbits), but there was then a rapid increase in the less virulent strains in the viral population — the less virulent strains kill a lower proportion of infected rabbits and take longer to kill them when they do. Meanwhile, the rabbits were also

Rabbits evolved resistance

evolving resistance. This could be shown by challenging wild rabbits through a series of times with standard strains of the virus; now the virus was held constant and any decline in kill rate must be due to changes in the rabbits. Table 22.3 shows the results of a series of such experiments through the 1960s and 1970s, in which resistance did indeed manifestly increase.

Therefore, both parasitic virulence and host resistance can evolve. Natural selection will clearly always favor increased resistance in hosts, but how will it operate on virulence in parasites?

Table 22.2
The myxoma virus has evolved lower virulence over time after its introduction into Australia, France, and Britain. Strains of the virus are classified into five virulence grades: I is the most virulent, V the least. The table shows the percentage occurance of the different strains in wild rabbits through time. Modified from Ross (1982), who compiled the results from a number of sources.

Country	Virulence grade					
	I	II	IIIA	IIIB	IV	V
Australia						
1950–51	100	0	0	0	0	0
1958–59	0	25	29	27	14	5
1963–64	0	0.3	26	34	31.3	8.3
France						
1953	100	0	0	0	0	0
1962	11	19.3	34.6	20.8	13.5	0.8
1968	2	4.1	14.4	20.7	58.8	4.3
Britain						
1953	100	0	0	0	0	0
1962–67	3	15.1	48.4	21.7	10.3	0.7
1968–70	0	0	78	22	0	0
1971–73	0	3.3	36.7	56.7	3.3	0
1974–76	1.3	23.3	55	11.8	8.6	0
1977–80	0	30.4	56.5	8.7	4.3	0

Table 22.3
Rabbits have evolved resistance over time after the introduction of the myxoma virus. These results are for wild rabbits (*Oryctolagus cuniculus*) caught at different times in two regions (Mallee and Gippsland) of Victoria, Australia; the rabbits were then challenged with a highly virulent standard laboratory strain (SLS) of the myxoma virus. The strain caused 100% mortality in unselected rabbits. From Fenner & Myers (1978) based on the data of Douglas *et al*.

	Mallee		Gippsland	
	Number tested	Mortality (%)	Number tested	Mortality (%)
Unselected rabbits		100		100
Selected rabbits				
1961–66	241	68	169	94
1967–71	119	66	55	90
1972–75	73	67	482	85

Natural selection can favor higher or lower virulence according to the transmission mode of the parasite, and other factors

One idea about how natural selection will work on virulence is that it will usually act to reduce it. Parasites depend on their hosts, and if they kill their hosts they will soon be dead too. For this reason, parasites (arguably) might evolve to keep their hosts alive. The objection to this argument, and the reason why it is almost universally rejected by evolutionary biologists, is that it is group selectionist (Section 11.2.5, p. 301). Although a parasite species has a long-term interest in not destroying the resource it lives off, natural selection on individual parasites will favor those parasites that reproduce themselves in the greatest numbers over those that restrain themselves in the interest of preserving their hosts. The short-term individual advantage of greater reproduction will usually outweigh any long-term group or species advantage of reproductive restraint.

The modern theory of virulence looks at other factors. One is the number of parasites that infect a host. If the host is infected by one parasite, all the parasitic individuals will be the offspring of the original colonizer and they will all be genetically related brothers and sisters. Kin selection (Section 11.2.4, p. 298) will then operate to reduce any selfish proliferation within the host. If the host has suffered multiple infections, by contrast, the parasites will be unrelated. Natural selection will favor individual parasites that can consume as much of the host as possible, as fast as possible, before any of the other parasites take advantage of the resource. Virulence will increase. If an individual restrains itself to preserve the host, other parasites will step in to take it over.

With multiple infections, evolution towards more virulent parasites can occur even within a single host, if the generation time of the parasites is short relative to the host's. There is abundant evidence that more virulent strains can evolve by competition between parasites within the host (Ebert 1998). In all, we can predict that diseases arising from single infections will have lower virulence than diseases arising from multiple infections.

A second factor is whether there is *vertical* or *horizontal transmission* of the parasites between hosts. In an external parasite, transmission may mean the movement of an adult parasite that has been living off one individual host on to another host. In internal parasites it typically means the movement of the offspring of parasites living inside one host on to another host. In vertical transmission a parasite transfers from its host to the offspring of that host; this can be done by a variety of mechanisms — by the mother's milk, or simply by jumping from host parent to host offspring when the two are near each other, or inside the gamete. In horizontal transmission, the parasite transfers between unrelated hosts, not particularly from parent to offspring, and this may be done through breathing, or by a vector such as a biting insect, or by copulation of one host with another. Some parasites are transmitted vertically, others horizontally: what consequence does this have for the evolution of virulence? A vertically transmitted parasite requires its host to reproduce to provide resources for itself or its immediate offspring, whereas horizontally transmitted parasites have no such requirement.

Consider the success of a more and a less virulent strain of parasite in the two cases. A vertically transmitted parasite experiences a trade-off between making more offspring and the success of those offspring. A parasite that reproduces more will be more virulent as it uses up more of the host; but it will reduce the host's reproduction. The

Virulence depends on . . .

. . . the relatedness between parasites within a host, . . .

. . . and whether transmission is vertical or horizontal

host's reproduction produces the resources (that is, host offspring) that the parasite's offspring exploit. This trade-off will place an upper limit on virulence. A horizontally transmitted parasite experiences no such trade-off: the success of its offspring is independent of the reproduction of its host. Virulence is therefore much less constrained.[2]

In nature, single infections and vertical transmission often occur together, and both factors may work side by side to reduce parasitic virulence. A comparative study, by Herre (1993), of 11 species of nematode worms that parasitize fig wasps in Panama illustrates the idea. The fig wasp life cycle is as follows. The adult fig wasp, who is carrying pollen from the fig from which she emerged, enters one of the structures that eventually ripen into a fig. There she pollinates the fig, lays her eggs, and dies. The eggs grow up and emerge within the fig; after emerging, they mate and the females pick up pollen and exit the fig in search of another to lay their eggs in. An important fact in the story is that fig wasp species vary in the number of fig wasps that enter a fig. In some, only one does, whereas in others several females may enter and lay their eggs in the same fig.

Nematode worms live off fig wasps, and in Panama there is a different species of nematode living off each species of fig wasp. The immature nematodes crawl on to a fig wasp after she emerges in the fig. "At some point [in Herre's words], the nematodes enter the body cavity of the wasp and begin to consume it." The nematodes emerge as adults from the body of the dead wasp and mate and lay their eggs in the same fig as the wasp did; the cycle can then repeat itself. Nematodes that live off fig wasps in which only one wasp enters a fig will tend to be vertically transmitted, and the nematodes on any one host will be genetically related. In contrast, the nematodes that live off fig wasps in which several females may enter the same fig can be horizontally transmitted — nematodes from a number of different parents may crawl on to the same wasp and they will be genetically unrelated. We can predict that the parasitic nematodes of fig wasps in which only one female typically enters a fig will have evolved lower virulence than those of fig wasps in which more than one female typically enters a fig. Herre's results are shown in Figure 22.7 and show the predicted relation. The virulence of the parasite appears to have been tuned by natural selection to the habits of the host.

The example here illustrates only one way in which natural selection works on virulence. In other cases virulence may not depend on the rate at which a parasite grows in, and uses up, its host. For other kinds of virulence, other theories may be needed. Even when virulence does depend on the parasite's growth rate, kin selection and vertical as opposed to horizontal transmission are just two of the evolutionary factors that have been hypothesized to influence it. Most of the other factors, however, have not been so well studied. We can often, if not always, expect lower virulence when the parasites on a host are genetically related and vertically transmitted than when they are unrelated and more horizontally transmitted.

The theory of the evolution of virulence is rich in implications for understanding human disease. Box 22.2 looks at an example.

Nematode parasites of fig wasps illustrate the theory

---

[2] Virulence has an upper limit, even with horizontal transmission. For instance, a parasite that is spread by biting insects requires its host to be attractive to those insects. The longer the host can stay alive, the longer it is available to be bitten. Indeed, the reduction in the virulence of myxoma virus that we looked at earlier may well have occurred as natural selection maximized the chance of transmission from one rabbit to the next by mosquitoes.

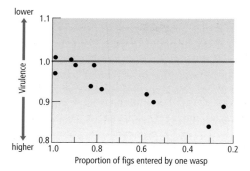

lower / higher — Virulence (y-axis)
1.1, 1.0, 0.9, 0.8

Proportion of figs entered by one wasp
1.0, 0.8, 0.6, 0.4, 0.2

Figure 22.7

Virulence is higher in nematode species that parasitize fig wasps in which more individual wasps lay their eggs in a fig. The results are for 11 species of fig wasps and the 11 species of nematodes that parasitize them (there is one nematode species per fig wasp species). Virulence is measured by the average number of offspring produced by parasitized relative to unparasitized wasps in each species; virulence is higher *down* the *y*-axis. A virulence of one means parasitized and unparasitized wasps leave the same number of offspring: these vertically transmitted parasites are so mild as practically to be commensals. The offspring leave a record inside the fig and can be counted accurately. The proportion of figs entered by one or more wasps can also be measured. Redrawn, by permission, from Herre (1993). © 1993 American Association for the Advancement of Science.

Box 22.2
Vaccines and the Virulence of Human Diseases

Vaccines have been developed against many parasitic diseases of humans and economically valuable non-human species. The parasites in turn may evolve resistance to the vaccine (for example, HIV — Section 3.2, p. 45). However, vaccination can have other effects on parasite evolution too. Gandon *et al.* (2001) modeled the effect of "imperfect vaccines" on the evolution of virulence. Imperfect vaccines are partially but not wholly effective vaccines, where the vaccine works against the parasites in some but not all infected members of the host populations. In practice, almost all vaccines are imperfect in this sense. Gandon *et al.* distinguished two cases.

1. *Vaccines that reduce parasitic growth inside the host.* Some vaccines work against parasites that have successfully infected a host, reducing the rate of growth and reproduction of the parasites. The more virulent strains of a parasite are those that grow and reproduce faster in the host. Virulent parasites "use up" the host more rapidly than less virulent parasites. We can assume that the

parasite has, in the absence of vaccination, some optimal level of virulence, such that it uses up its host at the best possible rate — depending on the number of infections, mode of transmission to new hosts, and other factors. Now a growth-inhibiting vaccine is applied. The effect will be to create a force of natural selection in favor of more virulent parasite strains. Suppose the best amount of time for a parasite to use a host up in is 10 days. More, and less, virulent strains of the parasite that use hosts up in 8 or 12 days are selected against. Suppose the vaccine cuts the parasite growth rate in half. A highly virulent parasitic strain, that would formerly have used up the host in only 5 days, will now be favored, and will use up a vaccinated host in 10 days. But in a non-vaccinated host it will show up its full virulence, and use up the host in 5 days. Growth-inhibiting vaccines tend to select for increased virulence.

2. *Vaccines that reduce the chance of infection by a parasite.* Other vaccines

may make it less likely a parasite will get into a vaccinated host. These vaccines tend to select for less virulent parasites. As we saw in the main text of this chapter, parasites that typically infect a host only once are less virulent than equivalent parasites that infect one host with many parasitic individuals. A vaccine that reduces the chance of infection will reduce the average number of parasites that infect each host individual, making it more likely that one parasite individual has a host individual to itself. Then the parasites evolve to become less virulent.

Parasite virulence can be influenced by other factors beside those in Gandon *et al.*'s model. However, the model illustrates two ways in which parasite evolution can be influenced, in humanly important ways, by vaccination programs. In general, when we interfere with nature, we set up new selection pressures, and evolutionary change is likely to occur in consequence. The theory of evolution enables use to work out what those consequences may be.

22.5.2 *Parasites and their hosts may have cophylogenies*

In Sections 22.3.1 and 22.3.2 above, we saw that the phylogenies of flowering plant taxa and the specialist insects that feed on them, or pollinate them, sometimes form mirror images, or cophylogenies. Host taxa and their specialist parasites can show the same pattern. Indeed, mirror-image phylogenies were first discovered in parasite–host relations. Cophylogenies between parasites and hosts are sometimes referred to as Fahrenholz's rule.

The cophylogeny of pocket gophers and lice . . .

Hafner *et al.*'s (1994) research on the rodent family Geomyidae (the pocket gophers) and their ectoparasitic lice (Mallophaga) is a good example. Hafner *et al.* sequenced a mitochondrial gene in 14 species of pocket gophers and their parasitic lice, and used the data to construct the phylogenies of the two groups (Figure 22.8a). The phylogenies are nearly mirror images, though there are some deviations. The deviations are probably due to host switching. For example, at the bottom of the figure, the parasitism of *Thomomys bottae* by *Geomydoecus actuosi* looks like host switching.

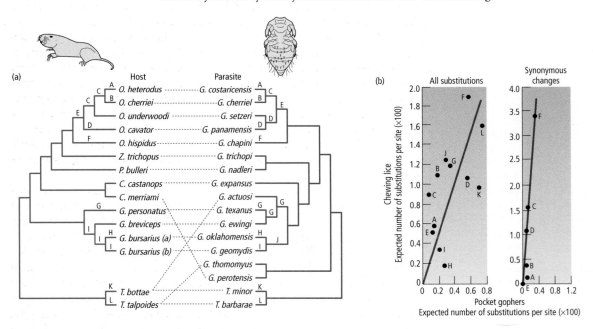

Figure 22.8

Mirror-image phylogenies in parasites and hosts. (a) The phylogenies of 14 species of pocket gophers (Geomyidae) and 17 of their mallophagan parasites. The phylogenies were reconstructed from the sequence of a mitochondrial gene (cytochrome oxidase subunit 1) using the parsimony principle. The phylogenies mainly form mirror images, but there are several cases of probable host switching. A pocket gopher (*Geomys bursarius*) and louse (*Geomydoecus geomydis*) are also illustrated. (b) Test of simultaneity of speciation in parasites and hosts. The estimated number of substitutions in various branches in the host phylogeny are plotted against the numbers in the mirror-image branches in the parasite phylogeny. Letters on the graph refer to lettered branches in (a). The clocks in the two taxa probably run at different rates because of differences in generation times. If the speciation events were really simultaneous, the points should fall on the line. The fit is better when only synonymous changes are counted. Redrawn, by permission, from Hafner *et al.* (1994). © 1994 American Association for the Advancement of Science.

. . . is probably due to cospeciation
. . .

The main mirror-image pattern between the pocket gopher and lice phylogenies is probably due to *cospeciation*. That is, a host species and its parasite species tend to split at the same time. For example, if we look at the top of the figure on both sides there is a split that produced the branches labeled E and F. The branch E on the host side probably represents an ancestral pocket gopher that was parasitized by the lousy ancestor of the four species in the branch E of the parasites. The ancestral gopher, and its ancestral louse, species then split twice. The events down the E → C → A + B branch (moving up the figure) look very like cospeciation.

Why should host and parasite speciate synchronously? Probably because the same circumstances favor speciation in both groups. For instance, the ranges of the two could be fragmented by some biogeographic factor, and the normal process of allopatric speciation occur in both parasites and hosts. (If any factor divides the range of the gophers, it will also divide the range of these lice, because the lice have limited independent powers of dispersal.) The same conditions drive speciation in both parasite and host and the result is cospeciation.

. . . as can be tested by molecular clocks

Figure 22.8b provides a stronger test of cospeciation. Hafner *et al.* used the estimated number of changes in each branch as a molecular clock to estimate the time when the branch originated. The clock runs faster in the lice, likely because their generation times are shorter than their hosts (Section 7.4, p. 169). If there was real cospeciation, the speciation events in host and parasite should have occurred simultaneously. Hafner *et al.* used two molecular clocks, one for all the nucleotide substitutions and the other for only the synonymous nucleotide changes. We know from Chapter 7 that synonymous changes are more likely to be neutral and therefore probably provide a more accurate clock. In both cases (Figure 22.8b) the points for the branch lengths cluster around the line for simultaneity, but the fit is better for the synonymous change clock. Figure 22.8b is good evidence that the host and parasite species tended to speciate at the same time.

The importance of the molecular clock test is shown in the next example (Figure 22.9). The phylogenies of primates and the primate lentiviruses are near mirror images. (The primate lentiviruses are the group that includes HIV. HIV and human beings are excluded from Figure 22.9 for technical reasons, but HIV-1 came from SIVcpz in chimpanzees and HIV-2 came from SIVsm in sooty mangabeys: see Section 15.10.2, p. 451.) In Figure 22.9, eight of the 11 splits are mirror images of each other in the two phylogenies, suggesting only three host switches. We might, naively, deduce that primate lentiviruses cospeciate with their primate hosts. However, a look at the timescales at the foot of Figure 22.9 suggests that deduction is false. The viruses evolve much faster than their hosts, and the split times for the viruses are only a few thousand years, against a few million years in the host monkeys. Unless the molecular clock is massively misleading, by three orders of magnitude, the cophylogenies here are not evidence of cospeciation.

Molecular clocks refute cospeciation in primates and lentiviruses

Why do the primate lentiviruses have a similar phylogeny to their hosts, despite the huge difference in splitting times? The answer is uncertain. One possibility is that viruses tend to switch between hosts that are phylogenetically closely related (Charleston & Robertson 2002). The immune systems of chimpanzees and humans are probably more similar than those of baboons and humans. A virus that is adapted to live in chimpanzees can probably more easily switch to exploit humans than a virus that is adapted

Figure 22.9
Phylogenies of primate hosts and primate lentiviruses (the group of viruses that includes HIV, though HIV and humans are not shown here). They are approximate, but not perfect, mirror images. The timescale of the phylogenies is shown, based on molecular clock inferences. Note the different timing of the splits in the two taxa: the cophylogenetic relations are not (if the molecular clock is reliable) due to cospeciation. (Figure courtesy of Dr D.L. Robertson.)

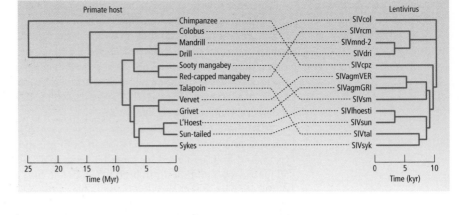

to exploit baboons. The splits in the lentivirus phylogeny then all represent host switches. The match between phylogenies would not be due to cospeciation but to phylogenetically constrained host switching. The influence of the host's physiology, and particularly its immune system, on host switching, could be analogous to the influence of plant chemistry on insect evolution (Section 22.3.3 above).

However, the reason for the cophylogenies in Figure 22.9 is an unsolved research problem. The main point of the example here is to show that cophylogenies alone are not complete evidence for cospeciation. Evidence about the timing of the splits is also needed, for instance from a molecular clock. For the gophers and lice, both phylogenetic and molecular clock evidence support cospeciation. For the primates and lentiviruses, the phylogenetic evidence is consistent with cospeciation but the molecular clock evidence counts strongly against it.

Some parasite phylogenies do not match their hosts

Finally, some taxa of parasites and their hosts do not show cophylogenies. For example, we looked at the phylogenies of pocket gophers and one group of parasitic lice. These showed cophylogenies because the lice have limited powers of dispersal independent of their hosts. Other taxa of lice that can move independently do not show mirror-image phylogenies with their hosts (Timm 1983).

In summary, we have looked at three possible relations between the phylogenies of parasites and hosts. One is that they have cophylogenies caused by cospeciation. A second is that they have cophylogenies, but for some reason other than cospeciation. A third is that they do not show cophylogenies. All three patterns can be found in different examples.

22.6 Coevolution can proceed in an "arms race"

Encephalization quotients . . .

A graph of brain size against body size for many vertebrate species reveals that larger vertebrates have larger brains (Figure 22.10). The brain size of a species can be expressed relative to this general trend, as an *encephalization quotient*. The encephalization quotient of a species is the ratio of its actual brain size to the brain size it would be expected to have given its body size and the general trend in Figure 22.10. If its actual brain size is

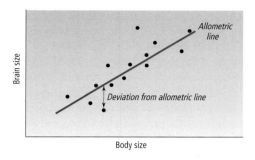

Figure 22.10
Relative brain size may be measured as relative encephalization by the deviation of a species' brain size from the allometric line for many species. Relative encephalization measures whether a species has a brain larger or smaller than would be expected for an animal with its body size. The species indicated in the figure has a relatively small brain, and an encephalization quotient below one.

. . . are used to measure intelligence

below the line, the encephalization quotient is less than one; if it is above the line, the encephalization quotient is more than one. Encephalization quotients are sometimes thought of as a crude measure of "intelligence" in a species. The more intelligent animals, in a loose sense, are those that deviate further above the line; they have greater relative encephalization.

Let us provisionally accept here that the encephalization quotient is an index of intelligence. We can then consider, from the classic work of Jerison (1973), a possible example of coevolution between prey and predator. (We met a related part of Jerison's general study when we discussed the Great American Interchange in Section 17.8, p. 512.)

In Cenozoic mammals, predators typically have relatively larger brains than their prey. This relation can be seen in Figure 22.11; but the same figure also shows another, more interesting fact. The relative brain sizes of both predators and prey have increased

Figure 22.11
The distribution of relative brain sizes for (a) ungulates (prey) and (b) carnivores (predators) through the Cenozoic. Brain size increased over time and at any one time carnivores had bigger brains than ungulates. Redrawn, by permission of the publisher, from Jerison (1973).

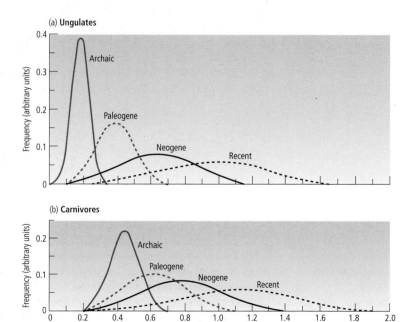

Carnivores and ungulates may have evolved in an "arms" race

through time. In order to estimate encephalization in a fossil, it is necessary to estimate body size, and the method used by Jerison has been criticized. The result is therefore uncertain, but Jerison's explanation is nevertheless interesting. He suggests that natural selection has favored higher intelligence both in the prey, to escape predators, and in predators, to catch prey. There has been a coevolutionary *arms race* between predator and prey, leading to ever larger brain sizes in both. The selective forces might well be truly coevolutionary, with each party exerting a reciprocal selective pressure on the other: as predators become cleverer at catching prey, the prey are selected to avoid them more intelligently, and vice versa. In this case, the evidence remains inconclusive. However, it is difficult to demonstrate coevolutionary relations in fossils and Jerison's work is well worth noticing as a rare example in which coevolution is a plausible explanation.

22.6.1 *Coevolutionary arms races can result in evolutionary escalation*

Evolution may be escalatory, . . .

Vermeij (1987, 1999) has applied the same argument as was used by Jerison, but much more generally. Vermeij suggests that predators and prey typically show an evolutionary pattern that he calls *escalation*. By escalation, he means that life has become more dangerous over evolutionary time: predators have evolved more powerful weapons and prey have evolved more powerful defenses against them. Vermeij distinguishes escalation from evolutionary progress. If evolution is progressive, organisms will become better adapted to their surroundings through evolutionary time; if it is escalatory, the improvement in predatory adaptations may be matched by improvements in prey defenses, and neither ends up any better off. The two concepts are easy to distinguish by a thought experiment. If evolution is progressive in predators (for example), then later predators would be better at catching their prey than were earlier predators. If, however, evolution is escalatory, later predators will be no better than their ancestors at catching their contemporary prey types. But if transported in a time machine and set loose on the prey hunted by ancestral predators, they should cut through them like a modern jet fighter in a dog-fight with an early biplane.

. . . as can be tested . . .

Vermeij and his followers have identified both biogeographic and paleontological evidence for escalation. Much of the evidence comes from mollusks in shallow-water marine environments — mollusks are abundant as fossils, and the nature of the shell itself can reveal how strongly adapted a species was, as a predator or a prey. More strongly defended shells have properties such as general thickening, or thickening concentrated around their apertures. Burrowing species, or those that cement themselves down, are better defended that those that lie loose on the bottom surface.

. . . more, or less, powerfully . . .

Some simple indicators of escalation can be misleading, but advanced research is needed to reveal the problems. Shell thickness, for instance, is usually a good indicator of defensive adaptation. But Dietl *et al.* (2002) point out that two species with equal shell thickness may differ in their degree of escalation if one grows faster than the other. A species in which the shell grows more rapidly would be more escalated than a species in which the shell grows more slowly. Dietl *et al.* estimated shell growth rates in fossils, using the shell's isotopic composition. However, information of this kind is rarely available, and inferences about the broad patterns of escalation are based on more

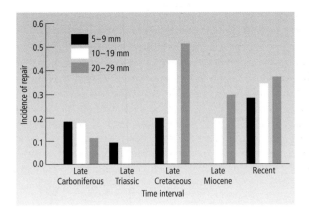

Figure 22.12

Incidence of repair at five successive time periods, for shells divided into three size classes. Note: (i) the incidence of repair is higher in more recent times; and (ii) larger shells show a relatively high incidence of repair compared with small shells in more recent times, as compared with earlier times. Large shells tend to be more resistant to breakage than small ones. One interpretation of the second trend is that more recent predators have become stronger, and therefore able to injure large-shelled animals. (0.4 in ≈ 10 mm.) Redrawn, by permission of the publisher, from Vermeij (1987).

limited evidence. Those broad patterns, as we shall see, have a wide scatter in the data. Dietl *et al.*'s point is worth keeping in mind, because problems in the data are likely to be one cause of the scatter.

Let us look at some of Vermeij's evidence. The frequency of shell repair is one indicator of predator–prey interactions in fossils. When a mollusk is non-lethally attacked, it repairs the damage to its shell and the repair pattern can be observed in the shell. Proportions of shells showing signs of repair have been measured in several fossil faunas, and the trend appears to be toward increasing amounts of repair over time (Figure 22.12). This Vermeij interprets as meaning that the prey have been suffering higher frequencies of predatory attacks over evolutionary time. (Logically it could also mean — though this is perhaps unlikely — that the predators have de-escalated from forms that destroyed their prey to forms that sometimes merely injured them!)

The escalation of molluskan prey defenses is also suggested by a trend in the proportion of different types of gastropod shells through time. The proportion of loosely attached forms, which are relatively poorly defended, has decreased through time relative to better defended types, such as burrowers and attached forms (Figure 22.13). Vermeij also found limited evidence that the better defended burrowers increased from being about 5–10% of genera in the late Carboniferous–late Triassic to about 37% in the late Cretaceous and 62–75% in modern formations. Internal thickening or narrowing of the aperture is another form of escalated defense and these types, too, have

. . . in fossil–predator prey relations . . .

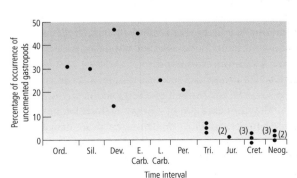

Figure 22.13

The incidence of sessile or sedentary uncemented gastropods through time. Note that the proportion decreases. Each point is for one fossil assemblage, except where marked. For each assemblage, Vermeij divided the gastropods into different types (burrowers, sessile, attached forms, etc.), adding up to 100%. This graph gives the proportion of gastropods that lie unattached on the bottom surface. The Neogene (Neog.) includes the Miocene and Pliocene (Figure 18.1, p. 526). Redrawn, by permission of the publisher, from Vermeij (1987).

Figure 22.14
(a) Total number of gastropod subfamilies through time.
(b) Proportion of subfamilies with members that have evolved internally thickened or narrowed apertures, a character that probably evolved as a defense against predators. Because the total number of subfamilies has increased, it is necessary (as here) to plot the *proportion*, not the total *number* of subfamilies, in order to show a trend. Dots marked with numbers represent that number of subfamilies. Redrawn, by permission of the publisher, from Vermeij (1987).

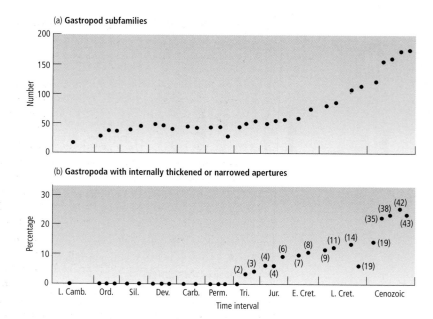

(a) Gastropod subfamilies

(b) Gastropoda with internally thickened or narrowed apertures

. . . and fossil plant–insect relations

proportionally increased through time (Figure 22.14). In all, more recent mollusks appear to be more strongly defended than were their earlier ancestors.

Some trends that look like escalation may be due to other factors. For example, Wilf & Labandeira (1999) counted the frequency of damage inferred to be caused by insects in fossil leaves. Paleocene leaves have lower levels of insect damage than Eocene leaves. In the samples, 29% of Paleocene leaves had insect damage, increasing to 36% in Eocene leaves. Plant–herbivore relations look more "escalated" in the Eocene. However, Wilf and Labandeira attribute the trend to warming temperatures. The Eocene was warmer than the Paleocene, and plants suffer more from herbivory when it is warmer. (Leaf damage is greater at the tropics than the poles today.) Although plant–herbivore relations came to be more dangerous in the Eocene, the reason may have been external climatic change rather than coevolutionary escalation between plants and insects.

Vermeij's evidence would not be persuasive enough to convince a skeptic. The evidence is not abundant; it is noisy; the patterns are not all consistent; alternative interpretations are sometimes possible; and the fossil record may be biased. However, in the absence of any particular argument that the trends are due to sampling biases, we can give the evidence the benefit of the doubt. It does then suggest that some escalation of predator–prey relations has occurred during evolution.

Escalation is a widely influential idea about macroevolution. It features again in the next section of this chapter. It also underlies many hypotheses about the Cambrian explosion (Section 18.4, p. 535). The diversification of animals with skeletons during the Cambrian has been attributed to the origin of predators, or more dangerously armed predators. This is an example of a hypothesis that invokes escalation. In the next chapter, we see how escalation may help us understand taxonomic replacements (Section 23.7.2, p. 670) and trends in species diversity (Section 23.8, p. 674).

22.7 The probability that a species will go extinct is approximately independent of how long it has existed

Now we are going to stay with the question of how influential coevolution has been in the history of life, but shift the scale of the evidence up to a more abstract, and general, level. We shall look at Van Valen's (1973) work. Van Valen inferred, from the shape of taxonomic survivorship curves, that macroevolution is shaped not only by coevolution, but by a particular mode of coevolution, called the "Red Queen" mode. The kind of escalatory coevolution we looked at in the previous section is (or can be) an example of Red Queen coevolution. But before we come to the coevolutionary interpretation, we must first look at the evidence from taxonomic survivorship curves.

We met taxonomic survivorship curves in Chapter 21 as a means of studying evolutionary rates (Section 21.6, p. 609). Here we use them to study extinction rates. To plot a taxonomic survivorship curve, take a higher taxonomic group, such as a family or order, measure the duration in the fossil record of its member species, and plot the number (or percent) of species that survive for each duration (Figure 21.12, p. 610). In 1973, Van Valen published a large study, based on measurements of the durations of 24,000 taxa, but with one crucial difference from earlier studies — he plotted the graphs after taking logarithms of the numbers surviving. He found that survivorship tends to be approximately linear on the log scale; that is, survivorship is log linear (Figure 22.15).

Van Valen's result gave a new interest to survivorship curves in evolutionary theory. With survivorship curves like Figure 21.12 on an arithmetic scale, it could be seen that different taxa evolved (taxonomically) at different rates, and it was possible to argue about why that should be so. But when it was noticed that the curves were linear on a log scale, more interesting questions could be asked. It has been questioned just how linear his results really are, and many of them are not for species, but for genera or even families within a higher taxon. Moreover, some of the extinctions will almost certainly have been pseudoextinctions, due to taxonomic division of a continuous lineage, rather than the true extinction of a lineage (Box 23.1, p. 647). However, Van Valen's results provide strongly suggestive evidence of logarithmic linearity in taxonomic survivorship curves. But what does the log linearity mean?

The log linearity of taxonomic survivorship curves means that species do not evolve to become any better (or worse) at avoiding extinction and that the chance a species will go extinct is independent of its age. Of the species that survived to a time t million years after their origins, a certain proportion go extinct by time $t + 1$ million years; then, of the survivors to time $t + 1$, the same proportion go extinct by time $t + 2$, and so on. Species decay at an exponential rate, with a constant proportion of the survivors going extinct in the next age unit. It could have been otherwise. If, for example, evolution is progressive, the probability of extinction might decrease with time, as the level of adaptation improves, so later species would last longer. For this reason, Van Valen's result is an important piece of evidence that evolution is not in general progressive.

Another theoretical possibility is that extinction rates might increase over time. For instance, if evolution is escalatory, the level of individual adaptation in a species relative to competitors may not change in any particular direction. However, all the species

Taxonomic survivorship curves are approximately log linear, . . .

. . . implying that evolution is not progressive

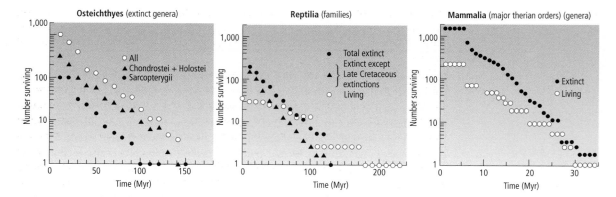

Figure 22.15

Three taxonomic survivorship curves that plot the number of genera (for mammals and Osteichthyes) or families (for reptiles) surviving for various durations in the fossil record.

Note that the lines, with this logarithmic *y*-axis, are approximately straight. Osteichthyes are bony fish and the two groups drawn on that graph are subgroups of bony fish. Redrawn, by permission of the publisher, from Van Valen (1973).

would be evolving to invest more heavily in armaments and defenses. Those armaments and defenses would probably develop via some trade-off with other adaptations (Section 10.7.5, p. 284). Heavily armed descendant species might then be more vulnerable to environmental stresses than their lightly armed ancestors. Extinction rates might increase over time. In fact that does not seem to happen. Van Valen's result suggest that no process leading to an increase or decrease in the chance of extinction generally operates.

22.8 Antagonistic coevolution can have various forms, including the Red Queen mode

Antagonistic coevolution can lead to . . .

Chapter 23 will look in detail at the factors causing extinctions. Here we concentrate on only one factor: antagonistic coevolution. As ecological competitors, or parasites and hosts, evolve against each other, if one competitor fails to evolve an adaptive improvement to keep up with its antagonists, it may go extinct. If a host species evolves a new kind of immunity, then if the parasite does not soon evolve a way of penetrating the defense, it will go extinct. What is the pattern of extinction rates likely to be for this process? Here is an analysis of the question, simplified from Stenseth & Maynard Smith (1984).

. . . the extinction of one or other party . . .

Two species (A and B) that are undergoing antagonistic coevolution will have a certain state of relative adaptation at any one time. One possibility is that one of the species, such as A, has superior adaptations. In this case, species B is heading for extinction unless it can soon evolve a better level of adaptation. Relations of this kind are unstable. They cannot exist for long, as the adaptively inferior species will go extinct.

Alternatively, species A and B may be at some kind of equilibrium. We can distinguish two kinds of equilibrium. One of them is static. The competing species have

. . . or a static equilibrium . . .

. . . or a dynamic equilibrium, known as the Red Queen

The Red Queen may explain the log-linear survivorship curves

But uncertainties remain

evolved to a set of optimal states and then simply stay there. This may be a common kind of coevolution. If there is one optimum form for an organism to have in order to compete with members of other species, its species will evolve to it and evolution will come to a stop.

The other kind of equilibrium is dynamic. This is the *Red Queen* equilibrium. Instead of evolving to an optimal state and then staying there, this result arises when adaptive improvement is always possible, and the species continually evolves to attain that improvement. Van Valen originally suggested that this mode of convolution would explain the log linear survivorship curves he had documented. His name for it — the Red Queen hypothesis — alludes to the Red Queen's remark in Lewis Carroll's *Alice Through the Looking Glass*: "here, you see, it takes all the running you can do, to keep in the same place." The analogy for running is coevolutionary change. In the Red Queen mode of coevolution, natural selection continually operates on each species to keep up with improvements made by competing species; each species' environment deteriorates as its competitors evolve new adaptations. This deterioration is the cause of extinctions in the model. On average, a group of competing species have balanced levels of adaptation, and they all lag behind their best possible states. At any one time, one species may experience some random run of bad reproductive luck, and go extinct. Coevolution will result in a log linear survivorship curve if the rate of environmental deterioration is roughly constant through time. If the species' competitive environments deteriorate in fits and starts, the survivorship curve will be non-linear.

Why should the rate of environmental deterioration be approximately constant? Van Valen reasoned that it would follow from the zero sum nature of competitive ecological interactions. The total resources available, he thought, will stay approximately constant. If one species adaptively improves, it will temporarily at least be able to take more of the resources and its population will expand. This increase will be experienced by its competitors as an equivalent decrease in the resources available to them. The selection pressure on them to improve will increase, by an amount proportional to the loss in resources caused by the competitor's improvement. They will then tend to improve their competitive abilities, and make up the ground lost to the competitor. The justification may be correct, but it is debatable. Resource levels, for example, have probably changed through evolutionary time and competitors may not always compete for a constant-sized pie. If resource levels increase, Van Valen's argument might predict that extinction rates would decrease, and the change in the resource level would cause a change in the extinction rate.

In summary, the Red Queen mode is not the only possible form of evolution among antagonistically coevolving species. Many species may in fact coevolve in Red Queen mode, but it is not an automatic theoretical consequence of antagonistic coevolution. An additional problem is that log linear survivorship curves like Figure 22.15 can arise by processes other than Red Queen coevolution. They can even arise if extinction rates vary in absolute time (McCune 1982).

Van Valen identified an important factual generalization in the log linearity of taxonomic survivorship curves. He also put forward a plausible explanation for it, in his Red Queen hypothesis. However, biologists remain uncertain both about how often extinction rates are constant, and about how good an explanation Red Queen coevolution provides for constant extinction rates. The Red Queen hypothesis continues to stimulate research, and it likely accounts for some fraction of macroevolution. Just how large a fraction that is remains to be seen.

22.9 Both biological and physical hypotheses should be tested on macroevolutionary observations

Coevolution occurs . . .

. . . but its relative contribution to macroevolution is unknown

Coevolution is one of several general processes that can account for evolution on a large scale — that is, macroevolution — as well as on a small scale. No one doubts its importance in microevolution, for instance in the evolution of mutualists or of parasites and their hosts. However, unanswered microevolutionary questions remain, such as the question of whether evolution in plants and insects is more often sequential evolution or fully reciprocal coevolution.

The contribution of coevolution to macroevolution is more controversial. Coevolution is not the only macroevolutionary force. Many macroevolutionary events are likely caused by changes in the physical environment — climatic change, or tectonic change, or asteroid impacts such as we look at in Chapter 23. Evolutionary biologists are interested in the relative contribution of physical and biological factors, and their interaction, in driving macroevolution.

One way to dramatize the issue is to ask a grand (and unanswered) question: if change in the physical environment ceased, do you think that evolution would soon come to a stop? If physical factors dominate evolution, it surely would, perhaps after a short period during which the species adjusted to the final permanent physical conditions. In the Red Queen view of evolution, the coevolutionary and biological relations between species might have a life of their own and evolution would carry on much as before after the cessation of physical environmental change. The question in its general form is too difficult for us to answer yet, but it puts under a spotlight many of the ideas that we need to examine to understand macroevolution.

Summary

1 Coevolution occurs when two or more lineages reciprocally influence each other's evolution. Coadaptation between species, such as in any example of mutualism, is probably but not necessarily the result of coevolution.

2 Insects and flowering plants have influenced each other's evolution. Adaptations concerned with phytophagy (animals feeding on plants) and pollination provide examples. The evolutionary relations of flowering plants and insects are sometimes fully coevolutionary. In other cases, evolution may be sequential, as plant evolution influences insect evolution but not vice versa.

3 Taxa of insects and flowering plants that interact with each other may show cophylogenies, or mirror-image phylogenies. Deviations from cophylogenies can be caused by host shifts. For instance, an insect species may colonize a new plant species that is chemically but not phylogenetically similar to the plant species it currently lives on.

4 Insects and flowering plants may have promoted each other's diversification, from the Cretaceous to the present. Biotic pollination, in particular, is associated with enhanced diversity in flowering plant taxa.

5 The level of virulence of parasites can evolutionarily decrease or increase. It can be understood in terms of the parasite–host relationship: two factors that influence it are kin selection and the mode of transmission of the parasite between hosts.

6 Some parasites speciate simultaneously with their hosts, in a process called cospeciation. Cospeciation is particulaly likely if the parasites have limited powers of dispersal independently of the host. Cospeciation is tested for by: (i) cophylogenies; and (ii) the molecular clock.

7 Coevolutionary "arms races" between predators and prey produce escalatory long-term evolutionary trends; they can be seen in the evolution of brain sizes in mammals and of armor and weapons in mollusks and their predators.

8 The extinction rates of species are independent of how long the species has existed for: a species does not become more likely to go extinct as time passes. Taxonomic survivorship curves are logarithmically linear.

9 Van Valen explained the log linearity of survivorship curves by his Red Queen hypothesis. It suggests that: (i) each species' environment deteriorates as competing species evolve new, superior adaptations; (ii) the competing species improve at a constant rate relative to each other; and (iii) the constant deterioration in the environment causes the chance of extinction of any one of them to be probabilistically constant.

Further reading

Thompson (1994) is a general book on coevolution. This chapter began by distinguishing mutualistic from antagonistic coevolution. A further point to note is that the two processes can be mixed within a single interaction, in different parts of a geographic range. Thompson & Cunningham (2002) describe a recent example. Bronstein (1994) reviews mutualism. Page (2002) is a multiauthor book on cophylogenies.

On plant–animal, and particularly plant–insect, coevolution, see the book by Schoonhoven *et al.* (1998) and the special issue of *Bioscience* (1992), vol. 42, pp. 12–57. Rausher (2001) reviews plant resistance to herbivores. On plant-pollinator evolution, see Waser (1998), Johnson & Steiner (2000), and the newspiece in *Science* October 4, 2002, pp. 45–6. Mant *et al.* (2002) is a further study of the topic by cophylogenies. They analyze the relation between orchids and their specialist wasp pollinators, finding a fair amount of congruence but some oddities in the branch lengths. The evolutionary forces at work are uncertain. Machado *et al.* (2001) describe another emerging case study in cophylogenies, between figs and fig wasps.

On the grand pattern of relations between plants and insects, see also Dilcher (2000), who puts the topic in the big picture of angiosperm evolution. Labandeira (1998) argues that pollinator evolution preceded angiosperm evolution, which would at least complicate (and might refute) the coevolutionary story.

On parasites and hosts, Clayton & Moore (1997) is a multiauthor book, concentrating on avian examples. Ebert (1998) reviews experimental work. See Ebert (1999) and his references for the evolution of virulence, and Ewald (1993). On the influence of number of infections, see also Chao *et al.* (2000). Moreover, increased rates of parasitism select for increased host resistance: see Green *et al.* (2000) for an experimental study. Fenner & Ratcliffe (1965) describe myxomatosis. Disease evolution is a related topic. See the special issues of *Science* May 11, 2001, pp. 1089–122, for human disease, and *Science* June 22, 2001, pp. 2269–89, for plant disease. For background on the primate–primate lentivirus example see also Hahn *et al.* (2000) and Holmes (2000a). Proctor & Owens (2000) describe bird-mite evolution, including research like that on gophers and mites in this chapter. Cophylogenies are also found in symbionts, such as bacteria in aphids (Clark *et al.* 2000).

Abrams' (2001) review on predator–prey evolution includes material on "arms races." Brodie & Brodie (1999) also discuss the topic, at a more introductory level. Gould (1977b, chapter 23) popularized Jerison's work. Bakker (1983) studied coevolution of the same ungulates and carnivores as Jerison, but looked at their morphological adaptations for running. Sereno (1999) mentions similar studies in dinosaurs.

Vermeij (1987, 1999) discusses evolutionary escalation, and he also has a chapter in Rose & Lauder (1996). See also several of the papers in a special issue of *Paleobiology* (1993), vol. 19, pp. 287–397. Dietl *et al.* (2000) provide evidence of escalation in oysters from the Cretaceous to the early Tertiary. The relation between escalation and extinction rate is controversial: see Vermeij (1987, 1999) and Dietl *et al.* (2000) and their references. Levinton (2001) also discusses (and provides references on) the topic, as well as Van Valen's law of constant extinction and the Red Queen hypothesis. Vrba (1993) integrates the Red Queen hypothesis with climatic change.

Study and review questions

1 What factors cause two interacting taxa to show: (a) cophylogenies, or (b) deviations from cophylogenies?

2 How have biologists tried to test whether insects and flowering plants have promoted each other's diversification over evolutionary time?

3 What two pieces of evidence are needed to show that two coevolving taxa show cospeciation?

4 It has been argued that the transmission mode of parasites influences the evolution of virulence. From this general idea, what rank order of virulence would you predict for otherwise similar parasites that are transmitted by the following means: (i) by the breathing of the host; (ii) by the water supply; (iii) by insect vectors; or (iv) by copulation of the host?

5 (a) Summarize the hypothesis of evolutionary escalation. (b) What kind of fossil evidence can be used to study it?

6 What evolutionary, or ecological, process might generate the Red Queen mode of coevolution?

23

Extinction and Radiation

This chapter looks at the two factors that govern diversity in the fossil record: extinction and radiation. Radiations increase, and extinctions decrease, the diversity of life. Interactions between the two processes explain much of the history of biological diversity. We begin the chapter by looking at the circumstances in which adaptive radiations occur. We then look at extinction rates since the Cambrian, and see that extinction rates have been elevated during certain mass extinctions. The causes of mass extinctions have been much studied, and we look at the evidence for an asteroid impact at the end of the Cretaceous. We look at statistical evidence that the causes of mass extinctions are not distinct from other extinctions — that extinction rates at all times fit the same power law. We also look at the possibility that most changes in observed extinction rates are artifacts, due to changes in the quality of the sedimentary record. We then turn to three topics that combine radiation and extinction. The first is species selection; the second is evolutionary replacement; and the third is the history of biological diversity on the global scale over geological time.

23.1 The number of species in a taxon increases during phases of adaptive radiation

The diversity of life through time reflects the rates of loss and gain of new life forms. The loss of species is by extinction, and the gain of species is by speciation. When the speciation rate exceeds the extinction rate for a taxon, its diversity increase. When the extinction rate exceeds the speciation rate, its diversity decreases. We can begin by looking at periods when the number of species in a taxon increases during *adaptive radiations*. An adaptive radiation (often just called a "radiation") means that a small number of ancestral species in one taxon diversifies into a larger number of descendant species, occupying a broader range of ecological niches. Adaptive radiations can occur at all taxonomic levels, and on all geographic scales. In a sense, the proliferation of life on Earth from the origin of life to the present is an adaptive radiation on the largest scale. However, adaptive radiations are particularly clear when they occur in a relatively small taxonomic group, in a confined geographic area.

We have seen several examples of local adaptive radiations earlier in this book. On a small scale, we saw in Section 17.5 (p. 502) how lizards have undergone radiations on Caribbean islands, with a similar set of ecological forms evolving independently many times as different islands have been colonized. Darwin's finches on the Galápagos are another example where a single ancestral species has evolved into 13–14 species with a range of ecological adaptations (Figure 21.2, p. 595). On a larger scale, the fruitflies have radiated on the Hawaiian islands into hundreds of species (Figure 15.27, p. 465). In the East African lakes, the cichlid fish have also evolved into hundreds of species (Section 13.3.3, p. 357). Many other examples exist. In Lake Baikal in Russia, aquatic invertebrates, and particularly crustaceans, have radiated into unknown hundreds of species. On Madagascar, there has been a radiation of lemurs — a taxon of primates that differs from the primates elsewhere on Earth.

Adaptive radiations can also be seen over wider geographic areas. Figure 23.1a shows an increase in the number of species of mammals in North America over the past 80 million years. The number increased rapidly after 65 million years ago. This is the early Tertiary radiation of mammals, which occurred after the extinction of the dinosaurs. The mammals in Figure 23.1a are eutherians, but the marsupial mammals also underwent a distinct radiation, across the land masses of Gondwanaland. The marsupial mammals in the south, and the eutherian mammals in the north, radiated into a similar set of ecological forms.

Radiations occur in a number of circumstances.

1. *Colonization of a new area where there are no competitors.* The radiation of fruitflies, lizards, and finches on island chains, and of cichlid fishes in African lakes, all occurred after ancestral species colonized the respective areas. Probably no competitors were present, because the islands had only recently emerged from the sea, or the lakes had only recently been flooded. The areas all contained unexploited resources, and the ancestral species radiated into a range of forms that could exploit those resources.

2. *Extinction of competitors.* The radiation of mammals followed the extinction of the dinosaurs. The dinosaur extinction vacated ecological space that was then occupied by mammals. We return to this topic in Section 23.7.3 later in the chapter.

Figure 23.1

(a) The number of families of mammals in North America increased abruptly in the Paleocene and Eocene, after which it has remained constant. (Alroy 1999 shows a similar pattern for an updated, taxonomically finer, dataset.) (b) The number of families of bivalves has increased steadily through time. Redrawn, by permission of the publisher, from Stanley (1979). © 1979 WH Freeman & Company.

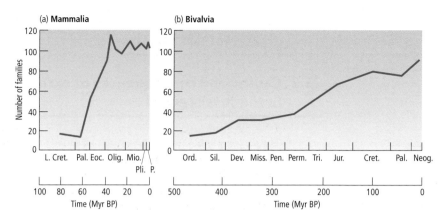

3. *Replacement of competitors.* One taxon may radiate if it is adaptively superior to its competitors. The superior taxon will take over from the inferior one, which will be driven extinct. The superior taxon may have become superior because of environmental change, or because it evolved a new, superior adaptation. We look at replacements in Section 23.7.

4. *Adaptive breakthroughs.* A taxon may evolve a new adaptation that allows it to out-compete another taxon (see point 3 above), or may allow it to exploit a previously unexploited resource. For instance, we saw in Sections 18.5 and 18.6 (pp. 538–42) how plants and animals colonized the land as they evolved a set of appropriate adaptations.

The colonization of land was somewhat analogous to the colonization of a new island chain (point 1 above). However, we can distinguish between colonization of an existing area, made possible by the evolution of a new adaptation, and colonization of a new area that requires no new adaptation. The colonization of land required new adaptations for support, respiration, and water retention. By contrast, the finch that first colonized the Galápagos probably did not have to evolve a new adaptation before it could colonize the island. After the colonization of land, plants and animals both underwent adaptive radiations. These radiations were made possible by an adaptive breakthrough.

The Cambrian explosion may have followed an adaptive breakthrough

The proliferation of animals with hard skeletons in the "Cambrian explosion" (Section 18.4, p. 535) is one of the most important adaptive radiations in the history of life. But the reason why it occurred remains uncertain. One hypothesis proposes that predators evolved escalated skills around this time, making hard skeletons advantageous. If so, the radiation may be an example of a replacement following changed conditions (factor 3 in the list above). The radiation of animals with hard parts may have occurred as they replaced soft-bodied predecessors.

Adaptive radiations can be understood in terms of Darwin's principle of divergence (Section 16.8, p. 487). Darwin was interested in why evolution usually shows a diverging, tree-like pattern. He explained the pattern by competition. More similar forms will compete more strongly than less similar forms, which tends to "push" species apart during evolution. Species diverge to escape competition. Darwin's principle of divergence likely needs to be slightly modified to incorporate the modern theory of allopatric speciation. However, his underlying reasoning is still part of the explanation why adaptive radiations occur when the conditions (numbered 1–4 above) are present.

23.2 Causes and consequences of extinctions can be studied in the fossil record

The discovery that species go extinct was made relatively recently in human history: it dates from the late eighteenth and early nineteenth centuries. Fossils had been known about long before that time, but when a fossil was found that differed from any known species, it could still have been alive in some unexplored region of the globe. As the global flora and fauna became better and better known through the eighteenth century, it became increasingly likely that some fossil forms were no longer alive. By the end of the century, several naturalists accepted that some marine invertebrate groups, such as the ammonites, were extinct.

Extinction is a relatively recent discovery

The best known taxa — the vertebrates, and mammals in particular — posed a special problem, however. Fossil bones are preserved as isolated, disarticulated fragments, and it is even more difficult to show that a single bone does not belong to any modern species, than it is for a complete specimen (such as a shell). The decisive work is usually credited to Cuvier. Cuvier reconstructed, with new standards of rigor, whole skeletons from bone fragments. It is easier to see whether the whole skeleton, rather than just the disarticulated bones, of a vertebrate belong to any living species. The most convincing cases of extinction were for gigantic forms like mastodons — it was hardly plausible that the explorers would have overlooked them. However, mistakes can still be made in recognizing extinctions, and Box 23.1 describes "pseudoextinctions."

Extinctions have two kinds of interest in evolutionary biology. One is the question of causality. Why do species go extinct? Some modern extinctions have been witnessed closely enough for the cause to be known with certainty. The enormous Steller's sea cow was discovered by a shipwrecked German naturalist called Georg Steller in 1742, but he was the only naturalist ever to see it alive. The animals were completely tame — Steller records how he could stroke them — and by 1769 they had been hunted to extinction. The extinction of modern species by analogous human means is all too easy to observe now, but for fossil species we do not have evidence as direct as sailors shooting sea cows. The quality of the evidence depends on how recent the fossils are, and for very recent fossils we can have quite convincing evidence about the cause of extinctions. The most recent ice age, for example, which was at its peak about 18,000 years ago, almost certainly caused many local extinctions. If a species disappears before the advancing ice cap and does not return, there is little doubt what the cause of the extinction was. The tulip tree and hemlock are only two of the species lost from the European flora at that time, though both survived in North America.

As we move further back in time, the causes of extinctions of particular species become more difficult to infer. We saw, in the well studied case of the Great American Interchange how uncertain the evidence is about the causes of the many extinctions, even though the Interchange took place only 2 million years ago and has left a good fossil record (Section 17.8, p. 512). However, the causes of extinctions can still be

Causes can sometimes be observed

Box 23.1
Pseudoextinction

Species (or higher taxa) may go extinct for two reasons. One is "real" extinction in the sense that the lineage has died out and left no descendants. For modern species, the meaning is unambiguous, but for fossils real extinction has to be distinguished from *pseudoextinction*. Pseudoextinction means that the taxon appears to go extinct, but only beause of an error or artifact in the evidence, and not because the underlying lineage really ceased to exist. We can distinguish three kinds of pseudoextinction, the first two of which are due to taxonomic artifacts (Figure B23.1).

1. *A continuously evolving lineage may change its taxonomic name.* As a lineage evolves, later forms may look sufficiently different from earlier ones that a taxonomist may classify them as different species, even though there is a continuous breeding lineage (Figure B23.1a). This may be because the species are classified phenetically (Section 13.2.3, p. 354), or it may be because the taxonomist only has a few specimens, some from early in the lineage and some from late in the lineage such that the continuous lineage is undetectable. Either way, this kind of taxonomic extinction is conceptually different from the literal death of a reproducing lineage. The taxonomic survivorship curves that we looked at in Sections 21.6, p. 609, and 22.7, p. 637, contain some (usually unknown) mix of real extinction and pseudoextinction.
2. *A higher taxon may cease to have any members if it is defined phenetically and only some divergent lineages persist.* A higher taxon, such as a family, can undergo pseudoextinction if the taxon is defined phenetically (Figure B23.1b). For instance, a paraphyletic group could go extinct even though some descendants of that group continue to exist. In this sense, the extinction of the dinosaurs was a pseudoextinction. Birds are lineal descendants of one dinosaur group, and birds continue to exist.
3. *Lazarus taxa.* A lineage may disappear temporarily from the fossil record, perhaps because appropriate sediments were not laid down for a while. Later it reappears. The first disappearance is a pseudoextinction, and may be misrecorded as a real extinction if the later reappearance is overlooked for some reason. (The term "Lazarus" taxa alludes to a man in the Christian Bible. Jesus Christ is there reported to have miraculously raised Lazarus from the dead.)

The topic of pseudoextinction is worth keeping in mind when considering theories to explain extinction and diversity patterns. Most of the theories we shall look at in this chapter make sense only for real extinctions of species lineages. For example, the theory that mass extinctions are caused by asteroid impacts makes sense if the mass extinctions are real but not if the evidence is largely composed of pseudoextinctions in the sense of Figure B23.1a and c. However, if the evidence for a mass extinction mainly consists of pseudoextinctions of higher taxa, in the sense of Figure B23.1b, that evidence could be compatible with an asteroid impact.

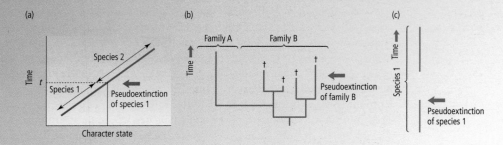

Figure B23.1
(a) Pseudoextinction within one lineage. If a continuous phylogenetic lineage is taxonomically subdivided, the earlier species will go "extinct" at the dividing line, even though the lineage persists just as it did before. The "extinction" of species 1 at time *t* is called pseudoextinction.
(b) Pseudoextinction of a higher taxon. Family A and B have been defined phenetically, and family B is a paraphyletic group (Section 16.6.2, p. 481). At time *t*, family B becomes extinct, even though some lineal descendants of the common ancestor of family B continue to exist.
(c) Lazarus taxa, with a fragmentary fossil record.

. . . but are usually statistically inferred

studied with fossil evidence. Instead of looking at particular species, we look at patterns among large numbers of species or higher taxa. The Red Queen hypothesis (Section 22.8, p. 638), provides an example of this general approach. The Red Queen hypothesis is a theory about the cause of extinctions. It suggests that species go extinct when they are outcompeted by other species that have made evolutionary advances. The validity of the hypothesis is uncertain, but what matters here is that it was deduced from a general pattern in fossil extinction — the log linearity of taxonomic survivorship curves. The Red Queen posits a biological cause for extinctions. In this chapter we shall be more concerned with non-biological causes, such as asteroid impacts and changes in the physical environment.

The second evolutionary interest of extinctions lies in their consequences. When a species goes extinct, it vacates ecological space that can be exploited by another species. The sudden extinction of an entire larger taxonomic group may vacate a larger space and permit a new adaptive radiation by a competing group (Figure 23.1). Radiations and extinctions can be related events, and we look at the relation between the two later in the chapter.

23.3 Mass extinctions

23.3.1 *The fossil record of extinction rates shows recurrent rounds of mass extinctions*

Sepkoski, in a series of papers from 1981 on, compiled, from the paleontological literature, the time distributions in the fossil record of all the families and genera of marine organisms. Sepkoski's compilation was not the first of its kind, but it is the most comprehensive and the most widely used. Figure 23.2 shows how extinction rates change over time in the fossil record.

A general survey of extinction rates . . .

We can notice two features of Figure 23.2. One is that the average extinction rate appears to decrease from the Cambrian to the present. The explanation for the decrease is not agreed on. The decrease may be an artifact of some kind — caused by changes over geological time in the quality of the sedimentary record or the degree to which taxa have been "split" by taxonomists. Or the decrease may be real. For instance, life may have initially colonized relatively "central" niches that became subject to intense competition. These niches may have a relatively rapid turnover of occupying species. Then, over time, life also colonized more marginal niches, where competition is less intense. The occupants of a marginal niche may stay there fairly permanently. These ideas are clearly vague and uncertain at present.

. . . shows a decreasing average rate . . .

The second remarkable feature in Figure 23.2 is the series of peak times when extinction rates appear to be exceptionally high. These peaks are called *mass extinctions*. The exact definition of a mass extinction is arbitrary, and different paleontologists recognize different numbers of mass extinctions in the history of life. The evidence for the Cambrian is sufficiently poor that we cannot say for sure whether extinction rates were exceptionally high at any time then. From the Ordovician onwards, the five largest extinction events were in the late Ordovician, the late Devonian, the end of the

. . . and up to five mass extinctions

Figure 23.2

The observed extinction rate for marine animals during the history of life, from the Cambrian to the present, expressed as percentages of genera going extinct per time unit (based on almost 29,000 genera). Note the general decline, and the series of peaks (for mass extinctions). Terrestrial life shows a similar pattern, though there is less evidence. Redrawn, by permission of the publisher, from Sepkoski (1996).

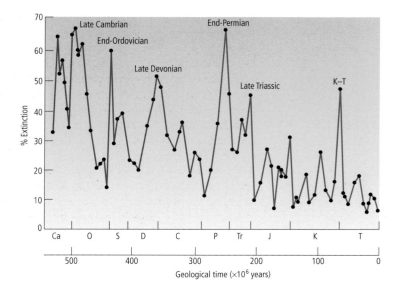

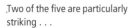

Two of the five are particularly striking . . .

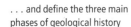

. . . and define the three main phases of geological history

Permian, the late Triassic, and the end of the Cretaceous. These are sometimes called the "big five."

Three of the "big five" are open to doubt. The three less certain mass extinctions are the ones in the late Ordovician, the late Devonian, and the late Triassic. The big five may therefore be reduced to a big four, big three, or even a big two, with the two most important mass extinctions being at the end of the Permian and the end of the Cretaceous. The end-Permian mass extinction is the biggest in history, with 80–96% (depending how the estimate is made) of species going extinct. In the Cretaceous–Tertiary mass extinction, at least half, and perhaps 60–75% of species went extinct.

The observations now explained by mass extinctions have been known about for a long time. The geologists of the nineteenth century who worked out the main eras of the Earth's history did so by looking for characteristic fossil faunas that lasted for a noticeable time (or rather, depth) in the sediments. Different characteristic faunas were recognized as different time periods. They recognized three large-scale faunal types and named them the Paleozoic, Mesozoic, and Cenozoic, with shorter term characteristic faunas within each of the three. Two major faunal transitions divide the three eras: the Permo-Triassic boundary, between the Paleozoic and Mesozoic, and the Cretaceous–Tertiary boundary, between the Mesozoic and Cenozoic. These two major faunal transitions correspond to the two main mass extinctions. Smaller transitions occurred between the stages within the main eras, and many of these correspond to smaller, but still elevated, peaks in Sepkoski's (1996) graph (Figure 23.2).

The extinction rate, in terms of number of species going extinct per million years, varies continuously through the history of life. Sometimes it is high, other times it is low, other times it is in between. No evidence exists for a distinct type of event that causes mass extinctions. The observed mass extinctions are just the times at the extreme of a continuum of extinction rates (Section 23.4 below). However, paleobiologists often study mass extinctions separately from the periods between, and for research

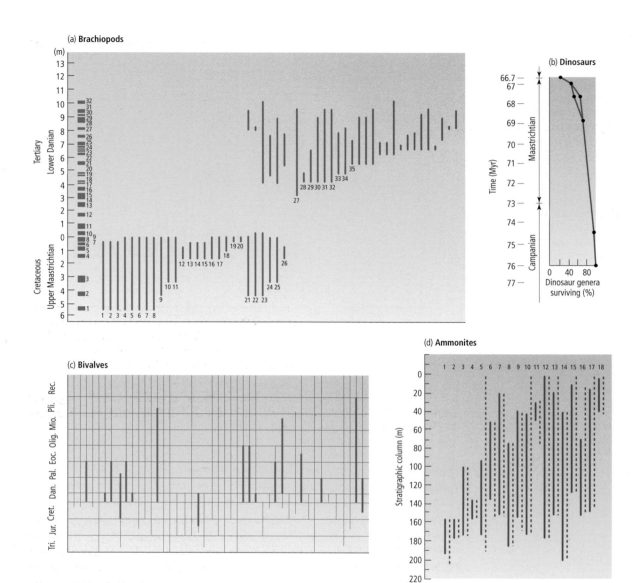

Figure 23.3

The mass extinction at the end of the Cretaceous affected all the main taxa, but the evidence about whether they all went extinct suddenly at the same time is controversial. Here are four examples. (a) Brachiopods from Nye Kløv, Denmark. These extinctions look synchronous. It has been argued that there was a sedimentary hiatus here; others disagree, but note the gap at the base of the Tertiary. (b) Dinosaurs from Hells Creek, Montana. These look gradual. It has been argued that the gradual extinctions, and persistence of dinosaurs into the Tertiary, is due to secondary reworking of the fossils, and the real extinction pattern is sudden and synchronous at the end of the Cretaceous (see Smit & van der Kaars 1984; Sheehan *et al.* 1991). (c) Bivalves from Stevens Klint, Denmark. These look synchronous. It has been argued that the sudden extinctions

are only apparent, being due to a gap in the sedimentary record, but most students of the site accept that the sedimentary record is continuous and that the extinctions are real. (d) Ammonites from the Zumaya section, northern Spain. The results are shown for two seasons of collecting. Note the improved evidence in the larger dataset (dashed lines) for synchronous extinctions at the Cretaceous–Tertiary boundary. The non-synchronous pattern in the earlier data (solid lines) might be due to incomplete evidence and the real pattern is synchronous. (100 ft ≈ 39.5 m.) Redrawn, by permission of the publishers: (a) from Surlyk & Johansen (1984), (b) from Sloan *et al.* (1986), (c) from Alvarez *et al.* (1984), and (d) from Ward (1990). (a–c) © 1984, 1986 American Association for the Advancement of Science.

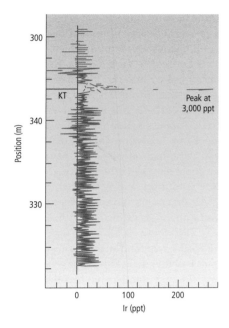

Figure 23.4

The iridium (Ir) concentration increases suddenly by 2–3 orders of magnitude at the Cretaceous–Tertiary (KT) boundary rocks at Gubbio, Italy. (100 ft ≈ 39.5 m.) Redrawn, by permission of the publisher, from Alvarez *et al.* (1990). © 1990 American Association for the Advancement of Science.

purposes it is convenient to distinguish between "mass extinctions" and "background extinctions." Background extinctions are the extinctions that are going on all the time, even when the extinction rate is not exceptionally high.

23.3.2 *The best studied mass extinction occurred at the Cretaceous–Tertiary boundary*

The mass extinction at the end of the Cretaceous has been found in all regions of the globe, and affected more or less every group of plants and animals (Figure 23.3). The fossil record of small, abundant, microfossil groups such as Foraminifera provides the best evidence for the fine-scale pattern of the extinction, but the demise of larger groups provides its drama. Some groups, such as the dinosaurs and ammonites, were driven finally extinct; most large groups were drastically reduced in diversity, though some odd groups, such as crocodiles, were not noticeably affected at all. The obvious question is: why did it happen?

In 1980, Alvarez *et al.* published an influential observation. They sampled the rocks of the Cretaceous–Tertiary boundary from Gubbio in Italy, and found exceptionally large concentrations of rare earth elements, particularly iridium (Figure 23.4). These elements also have high concentrations in extraterrestrial objects. Alvarez and his colleagues explained the biological mass extinction, and the geochemical *iridium anomaly*, by the collision of a large asteroid with the Earth. Since then, similar iridium anomalies have been found in Cretaceous–Tertiary boundary rocks at several other sites. Some geologists have argued that the iridium anomaly could have had a terrestrial cause, by volcanic eruptions; but asteroids are the most widely accepted explanation.

An iridium anomaly is associated with the KT boundary

The asteroid impact theory for the
Cretaceous–Tertiary mass
extinctions . . .

The exact means by which such an impact could have precipitated the mass extinction has been considered in detail by Alvarez, and other authors. Alvarez *et al.* originally suggested that the impact would have thrown up a global dust cloud, which would have blocked out sunlight for several years until it settled again. When Krakatoa erupted in 1883, it ejected an estimated 18 km³ of matter into the atmosphere, and this took 2.5 years to fall down again. The asteroid that hit the Earth at the end of the Cretaceous is estimated to have been 7.5–9 miles (12–15 km) in diameter. Such an asteroid, whose kinetic energy they described as "approximately equivalent to that of 108 megatons of TNT," would have produced an explosion about 1,000 times as large as the eruption of Krakatoa. The loss of sunlight alone would have been enough to cause the extinctions, but the impact could have had other destructive side effects too. Global warming, acid rain, extreme vulcanism, and perhaps an associated global fire, are some of the possibilities. An impact on the scale suggested by Alvarez *et al.* would have been capable of causing the mass extinction at the end of the Cretaceous.

. . . is supported by four lines of
evidence . . .

Since Alvarez *et al.*'s original publication, geologists have found an increasing quantity of evidence that supports his idea. The evidence is of four main kinds. The geochemical evidence, of which the iridium anomaly was the first example, has broadened in space, as similar anomalies have been found in Cretaceous–Tertiary boundary rocks at other sites, and in kind, as other chemical signatures of an asteroid impact have been detected. Secondly, we now have evidence of the impact crater itself. A geological structure (called the Chicxulub crater), buried beneath sediments off the Yucatan coast of Mexico, was the site of the impact. The structure is large enough, with a diameter of probably about 112 miles (180 km), and it dates to the Cretaceous–Tertiary boundary. The third kind of evidence is of physical structures that would have been generated by the impact. Rocks, such as shocked tektites and quartzes, which are suggestive of a high velocity collision, have been found from several Cretaceous–Tertiary sites, including Chicxulub. As the evidence has fallen into place, many geologists have come to accept that the mass extinction was caused by an asteroid. (But not all geologists, as we shall see in Section 23.5 later.)

A fourth kind of evidence comes from the pattern of extinctions in the fossil record. If Alvarez's theory is correct, the extinctions at the Cretaceous–Tertiary boundary should have been sudden, concentrated in a short interval of time, and not preceded by any decline through the Cretaceous; they should be synchronous in different taxa and geographic localities; and they should coincide with the iridium anomaly. This is a highly testable and stimulating set of predictions.

. . . including synchronous
extinctions

There are some problems in the evidence. You might think you could simply peer into the fossil record and observe whether extinctions were sudden or gradual, synchronous or spread out in time. In reality it is not so easy. How, for instance, do we observe the exact time of an extinction? The last appearance of a species in the fossil record will usually precede its final, true extinction (and a species certainly cannot appear *after* its true extinction). The species' population may decline before it finally disappears, which would reduce the chance of leaving fossils. Moreover, even if the population is constant, its chance of fossilization will still be much less than 100%. Species therefore appear to go extinct in the fossil record before they actually did. This "push backwards" is greater for forms that are less likely to leave fossils.

It can also be difficult to correlate events at different geographic localities, because absolute dates are often unavailable. The incompleteness of the fossil record also introduces uncertainty: a species may appear to go extinct suddenly at what is really a gap in the sedimentary record (look at Figure 23.3a and c for what, controversially, may be examples). For all these reasons, evidence from the fossil record is controversial when used to show either sudden or gradual, or synchronous or asynchronous, extinction patterns.

The fossil evidence can be tested

Despite these problems, the evidence can still be used (Figure 23.3). An increasing amount of evidence suggests that the mass extinction was sudden and synchronous. Let us look at one such study, by Ward (1990). He first collected ammonites from around the time of the Cretaceous–Tertiary boundary at a site in Spain, in 1986; further collections were made later and a larger study was possible in 1989. If the real extinctions were synchronous, the 1989 evidence should show more synchronous extinctions than the 1986 evidence; and vice versa if the real pattern was non-synchronous. The former was observed (Figure 23.3d). Ward's result tends to support the idea of an exactly synchronous extinction at the Cretaceous–Tertiary boundary; but it is not enough to convince a skeptic. It concerns only one taxon in one region, and the extinctions there could easily have been synchronous without the same being true of the rest of the world. Thus the supporters of Alvarez's theory accept evidence such as Figures 23.3a and c as showing synchroneity and attribute evidence like Figure 23.3b to the imperfections of the fossil record; critics argue the other way round.

In summary, there is good evidence for both suddenness and synchroneity of extinctions at the end of the Cretaceous, and the evidence appears to improve in more thorough fossil samples. However, the evidence is not complete enough to have persuaded everyone.

23.3.3 *Several factors can contribute to mass extinctions*

Asteroid impacts are one factor, among many, associated with mass extinctions

The Cretaceous–Tertiary mass extinction is only one of several mass extinctions, and asteroid impacts are only one of several factors hypothesized to cause mass extinctions. Figure 23.5 summarizes evidence for the main factors that have been hypothesized to cause mass extinctions. For asteroid impacts, the figure only shows evidence for impact craters. We see that the Cretaceous–Tertiary mass extinction is the only mass extinction to be associated with a large impact crater. Major craters exist for the Jurassic that are not associated with mass extinctions. Asteroid impacts therefore seem to be neither necessary nor sufficient to explain mass extinctions. Evidence from iridium anomalies tells the same story. Measurements of iridium have been made for other mass extinctions, some of which have small increases, but most do not. Small increases, of about one order of magnitude, may be better explained by terrestrial processes that concentrate iridium, rather than by an asteroid collision. The iridium spike at Gubbio is much larger, by three to four orders of magnitude (Figure 23.4). None of the five major mass extinctions except the Cretaceous–Tertiary one are generally accepted to have been caused by an asteroid impact.

Figure 23.5 also summarizes evidence for other factors that have been hypothesized to cause mass extinctions. These factors include: changes in sea level (and climate),

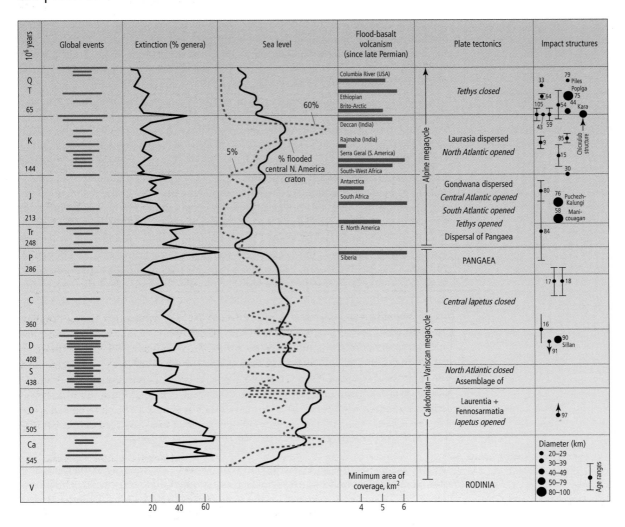

Figure 23.5

Summary of extinction events and the incidence of possible causes of these extinction events. Evidence is shown for: extinction rates (both as "global events" in which the bar width represents the magnitude of the extinction events, and as percentage rates, similar to Figure 23.2), sea levels, volcanic activity, plate tectonics, and asteroid impacts. The date of the end-Triassic volcanic rock has been updated from the source. (10 miles = 16 km.) Modified, by permission of the publisher, from Morrow *et al.* (1996).

high levels of volcanic eruptions, and changes in the shape of continents due to plate tectonic movements. Several periods of elevated extinction rates are associated with changes in sea level. Falls in sea level reduce the habitat available for marine life, driving species extinct. Changes in sea level will also be correlated with changes in climate. The combined influence of climate and sea level is widely thought to have contributed to some mass extinctions. However, sea level changes also occur at times when there is no mass extinction (Figure 23.5) and this factor is unlikely to be a general cause of all mass extinctions. Volcanic eruptions on a large scale may also cause mass extinctions. Three mass extinctions, including the two largest, are associated with large areas of rock that were deposited after volcanic eruptions.

Several factors may work interactively

The various factors are not mutually exclusive. A large asteroid impact could trigger volcanic activity or a change in climate — which could in turn affect sea level. The plate-tectonic pattern also influences sea level and climate, and tectonic activity influences vulcanism.

In summary, research on the causes of mass extinctions is considering various factors, including asteroid impacts, vulcanism, sea level changes, climate, and plate tectonics. Currently, it looks unlikely that any one factor acts as a general cause of all mass extinctions. The complex pattern of evidence in Figure 23.5 suggests that several factors may operate, in various combinations, to cause the observed pattern of extinctions. This impression will be strengthened in the next two sections.

23.4 Distributions of extinction rates may fit a power law

Are mass extinctions a distinct kind of event? We can distinguish two conceptual possibilities. One is that through the history of life the probability of extinction has been approximately constant, though the probability varies by chance. Sometimes, by chance, many taxa will go extinct in a time interval, at other times, by chance, few taxa will go extinct. The total distribution of extinction rates per unit time interval, for all history, will be continuous, ranging from low to high. The distribution, for instance, could look something like Figure 23.6a: a Poisson distribution. This distribution arises when the chance that any species goes extinct has, at all times, some small probability

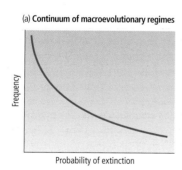

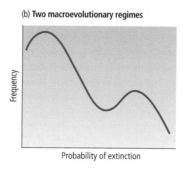

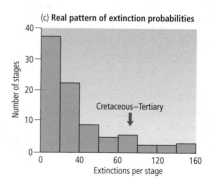

Figure 23.6
(a) If there is a continuum of macroevolutionary regimes, the frequency distribution of extinction probabilities in the fossil record will have a continuous Poisson distribution.
(b) Whereas if there are two macroevolutionary regimes, the frequency distribution of extinction probabilities in the fossil record should be bimodal. (c) The actual distribution for 2,316 marine animal families in the 79 generally recognized divisions of geological time since the Cambrian is continuous. The extinction intensity at the Cretaceous–Tertiary boundary is indicated for comparison. Redrawn, by permission of the publisher, from Raup (1986). © 1986 American Association for the Advancement of Science.

p per time unit. Then, many time intervals have a small number of species going extinct, and a few have many species going extinct. (Because if the chance that one goes extinct is *p*, the chance that two go extinct is p^2, the chance that three go extinct is p^3 and so on.) The different extinction rates observed at different times are simply due to chance effects.

Extinction rates could show a continuous or a two-peaked distribution

Alternatively, mass extinctions could be a distinct kind of event, with a distinct kind of cause from extinctions at other times. Then the extinction rates at times of mass extinction should be unpredictable, and distinct from the extinction rates at other times. For instance, suppose mass extinctions are caused by large asteroid impacts. Extinction rates would then occur at a certain rate between large asteroid impacts, and at a distinct, higher rate during and immediately after an impact. Unlike the random model, the frequency distribution will not be continuous. Mass extinctions will (or might) have a distinct peak (Figure 23.6b). When Alvarez's explanation for the Cretaceous–Tertiary mass extinction became widely accepted in the early and mid-1980s, some paleobiologists suggested that there are two *macroevolutionary regimes*. Evolution may alternate between "normal" periods with a "background" extinction rate, and mass extinctions. Extinctions would have different causes at the two times: asteroids (perhaps) for the mass extinctions and competition (perhaps) for the periods in between.

In fact, they look fairly continuous . . .

The distribution of extinction rates can be used to test between these ideas. Figure 23.6c shows one early study by Raup (1986). The extinction rates appear to fit the random, Poisson distribution. The rate during the Cretaceous–Tertiary mass extinction may be an exception, however, being above the curve. The generally good fit to the Poisson distribution supports the view that variations in extinction rate are mainly random, and that the history of life does not have two distinct macroevolutionary regimes.

More recently, the frequency distribution of extinction rates has been analyzed further to see whether it fits a "power law." A power law here refers to a family of mathematical equations that describe distributions like those in Figure 23.6. In particular, paleobiologists are interested in whether the distribution of extinction rates is "fractal," showing "self-similarity." "Self-similarity" and "fractal" are two ways of saying the same thing. A distribution shows self-similarity if its pattern on a large scale is an expanded version of its pattern on a small scale, that is the pattern is the same at all scales. For instance, the frequency distribution of extinction rates between 1 and 10 species per million years will show some pattern. If the pattern for extinction rates between 10 and 100 species per million years is the same, but multiplied up by a certain amount, then the whole distribution shows self-similarity.

. . . and may fit a power law

Solé *et al.* (1997) performed an analysis of this sort. They found that the distribution of extinction rates in a large compilation of fossil data appeared to be fractal — to show self-similarity. Extinction rate data are noisy, however, because of the many sources of error in the data. Tests of this sort are not all that strong.

If the distribution of extinction rates does show self-similarity, it is tempting to take the reasoning a step further. If extinction rates are fractal, the differences in extinction rates between different times are random and unpredictable. It would then be a mistake to ask what the "cause" or even "causes" of mass extinctions is, or are. Mass extinctions may not have a different cause from the periods when extinction rates are lower.

Mass extinctions may not have distinct causes

Consider, for instance, a simple model of extinctions that gives rise to fractal extinction rates. The species in an ecosystem have a certain degree of interdependence. Predators depend on prey; herbivores depend on food plants. If a food plant goes extinct, the herbivores that depend on it will go extinct too, and then the predators that depend on the herbivore. Thus if one species in an ecologically connected web of species goes extinct, it will take out a number of other species too. The number taken out depends on the number of interdependent species.

Over time, the degree of interconnectivity in an ecosystem may change. Sometimes many species depend on one another. At other times, ecological relations are more diffuse and few species strongly depend on one another. If one species accidentally goes extinct at a time of strong, extensive interconnectedness, many species will follow it to extinction. If one species accidentally goes extinct at a time of weak interconnectedness, few species will follow it. The same initial cause (the accidental loss of one species) can trigger a range of extinction rates, depending on the state of the ecosystem.

For this model to produce a fractal pattern of extinction rates, we would need to assume that the degree of interconnectedness in ecosystems evolves more or less at random, wandering up and down over time. Then if accidents happen at a steady rate, the resulting frequency distribution of extinction rates would be fractal in the way that appears to be observed.

The "ecosystem connectedness" model is not the only one that could explain the (tentative) observations. Another simple model could propose that almost all extinctions are caused by asteroid impacts. Asteroids vary in size, and small asteroid impacts probably cause fewer extinctions than large asteroid impacts. Then if the size distribution of asteroids fits a power law, the frequency distribution of the resulting extinctions will also fit a power law.[1]

More realistically, various causes of extinction such as asteroids and volcanic eruptions may interact with the condition of the ecosystem to determine the extinction rate. A more complex model could be produced. However, the point of the models here is to show that various factors could explain the observations. What the processes have in common is that they do not posit a distinct set of causes for mass extinctions as opposed to extinctions at other times. If extinction rates do fit a power law, we are led to think of causes for extinction rates that operate in much the same way over time. However, not all paleobiologists are agreed that extinction rates do fit a power law. There could be life yet in the search for a distinct set of causes for mass extinctions. This area of research, like several others in this chapter, will progress along with the quality of the fossil databases.

Several models could explain why extinction rates fit a power law . . .

. . . but the facts remain uncertain

23.5 Changes in the quality of the sedimentary record through time are associated with changes in the observed extinction rate

So far we have treated changes in extinction rates, and particularly the high extinction rates at times of mass extinction, as real. Factors such as asteroid impacts and

---

[1] In the next section we look at the possibility that fluctuations in the sedimentary record may explain changes in the observed extinction rate. If the processes determining sedimentation rates are fractal, this factor too could produce extinction rates that fit a power law.

vulcanism were invoked to explain real extinction patterns. However, at least since Lyell, in the mid-nineteenth century, some paleontologists have been skeptical about the observed changes in extinction rates. The apparently high extinction rates at the end of the major geological eras were known about in Lyell's time. But the high extinction rates could be artifacts, due to gaps in the fossil record, rather than real events. Darwin, for instance, wrote in the section on extinctions in *On the Origin of Species* (1859), "the old notion of all the inhabitants having been swept away by catastrophes at successive periods is very generally given up, even by those geologists, as Elie de Beaumont, Murchison, Barrande, &c., whose general views would naturally lead them to this conclusion. On the contrary, we have every reason to believe, from the study of the tertiary formations, that species and groups of species gradually disappear, one after another."

Darwin suspected mass extinctions were sedimentary artifacts

When Darwin wrote, absolute dates for rocks were not available. Absolute dates came in as the radioisotope method was developed in the twentieth century. Without these dates, a sudden transition such as from the Cretaceous to the Tertiary faunas could simply have reflected a prolonged gap in the fossil record. The end of the Cretaceous might have been 50 million years before the beginning of the Tertiary. Radioisotope dates ruled out that possibility. In fact the end of the Cretaceous runs straight into the beginning of the Tertiary, around 65 million years ago (Figure 18.1, p. 526). The mass extinctions, therefore, look real (Figure 23.2). Most paleontologists have come to accept that the history of life contains a number of catastrophic mass extinctions. This is an important respect in which the modern view of the history of life differs from Darwin's.

However, some of the changes in extinction rate observed in the fossil record could still be caused by changes in the sedimentary record. Box 23.2 looks at a comprehensive study by Peters & Foote (2002). The implications of their study remain undecided. A conservative conclusion would be that mass extinctions are real events, as is generally believed. But their work also allows a radical conclusion — that all the observed changes in extinction rates, including elevated extinction rates at "mass" extinctions, are sedimentary artifacts. Such a radical conclusion would take more work to establish, however. For now, paleobiologists will probably continue to study changes in extinction rates, but perhaps with more of an eye on artifacts in the data.

23.6 Species selection

23.6.1 *Characters that evolve within taxa may influence extinction and speciation rates, as illustrated by snails with planktonic and direct development*

What factors determine the patterns of speciation and radiation? The question has been studied in various ways and in this and the next section we shall concentrate on two ideas: one in which the attributes of the organisms may influence a taxon's probabilities of survival and speciation, and the other in which external ecological factors may show such an influence.

Box 23.2

Changes in Extinction Rates and Changes in the Sedimentary Record

Mass extinctions do not correspond to prolonged hiatuses in the fossil record — absolute dating of the rocks either side of the extinction events rule that possibility out. However, the quality of the sedimentary record could influence the observed extinction rates in other ways. The amount of sedimentary rock per geological time interval changes through time, due to changes in the amount of sedimentary rock originally laid down and the amount of it preserved up to now. Figure B23.2 illustrates how these changes

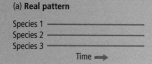

(a) **Real pattern**

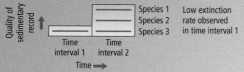

(b) **Quality of sedimentary record improves over time**

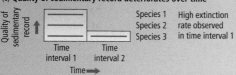

(c) **Quality of sedimentary record deteriorates over time**

Figure B23.2

Changes over time in the quality of the sedimentary record can influence the observed extinction rate. (a) Assume, for simplicity, that some species were continuously present through two successive geological time intervals. (b) If the first stage had a poor sedimentary record and the second stage a good sedimentary record, few species will have their last recorded representation in stage 1 and stage 1 will have an artifactually low extinction rate. (c) If the first stage has a good sedimentary record and the second stage a poor sedimentary record, many species will have their last representation in stage 1, and stage 1 will have an artifactually high extinction rate.

can influence the observed extinction rate. When a geological time interval with a good sedimentary record follows an interval with a poor sedimentary record, few species are likely to have their last representation in the earlier interval because many fossils are preserved in the later interval. The earlier interval then has an artifactually low observed extinction rate. The opposite result is observed when an interval with a poor record follows an interval with a good record. In this case, the earlier interval has an artifactually high observed extinction rate.

Peters & Foote (2002) used published data on the amount of exposed marine sedimentary rocks in the USA for 77 conventional time units from the Cambrian to the present (thus each unit averaged about 7 million years — the Cambrian began about 540 million years ago, Figure 18.1, p. 526). They also used Sepkoski's database for the time distributions of marine fossil genera. They constructed a model of the effect illustrated in Figure B23.2, and used it to predict the observed extinction rates given changes through time in the amount of sedimentary rock. The model had two versions, one in which the underlying real extinction rate was constant and another in which the real extinction rate steadily decreased from the Cambrian to the present.

Figure B23.3 illustrates the results for the two models. We can notice two things. One is the outstandingly good fit between the model and observations — outstandingly good, given the noise in the data from taxonomic and other sources of error (for instance, the sedimentary rock data are for the USA, but Sepkoski's database is global). Much of the variation in extinction rates is accounted for by variation in the amount of sedimentary rock. Secondly, some peaks in the observed extinction rate are explained by one version of the model but not the other. For instance, the Permian, Triassic, and Cretaceous extinctions at 240, 200, and 65 million years ago, respectively, are not explained by the model with decreasing extinction rates (Figure B23.3a) but are explained, or are better explained, by the model with constant extinction rates (Figure B23.3b). The reason is uncertain and requires further research. Meanwhile, we can fence-sittingly conclude that some mass extinctions may stand out from the extinction rates that would be predicted from the amounts of rock alone — or they may not.

Peters and Foote suggest two interpretations of their findings. One is more radical. Almost all changes in extinction rates, including the classic mass extinctions, may be artifactual — reflecting changes in the sedimentary record, not changes in the biological extinction rate. The search for causes of mass extinctions in such factors as

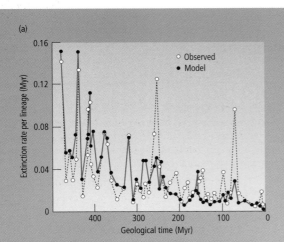

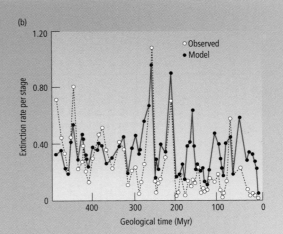

Figure B23.3

Changes in the quality of the sedimentary record alone can account for most of the observed changes in the extinction rate. The model used the basic idea of Figure B23.2 to predict the observed extinction rate, from the real changes in the sedimentary record between successive geological time intervals. (a) The observed and predicted ("model") extinction rates using absolute geological time in millions of years, and assuming that the extinction rate decreases steadily over time. (b) The observed and predicted ("model") extinction rates assuming all geological time intervals have the same length and that the extinction rate is constant. The "observed" graph in (a) is much the same as in Figure 23.2 although this one uses a slightly updated version of Sepkoski's dataset. The observed rates differ in (a) and (b) because of the different treatment of the geological time stages. From Peters & Foote (2002).

asteroid impacts could well be mistaken. The power laws of Section 23.4 would also need to be reinterpreted. Peters and Foote's work does not affect whether or not extinction rates fit a power law. However, the work suggests that any power law may arise not because of factors influencing biological extinction, but because of factors influencing how much sedimentary rock is deposited and then preserved and brought to the surface in successive geological time intervals. This radical interpretation returns to the skepticism of Lyell and Darwin, though with a new and exact model of how the quality of the fossil record influences observed extinction rates.

Their second interpretation is that some common factor may cause changes both in the sedimentary record and in the extinction rate. For instance, changes in the extinction rate may be explained by changes in the sea level. Extinction rates tend to go up when the sea level falls (see Figure 23.5), because (as noted above) the habitat available for many marine animals is reduced. But when the sea level goes down, the amount of sedimentary rock will also go down. Thus, the same factor could cause both an increase in real biological extinctions, and a change in the quality of the sedimentary record. Moreover, other extinction factors, such as climate, plate tectonics, and perhaps even vulcanism and asteroid impacts, may be associated with changes in sea level. Thus, the results of Peters and Foote do not rule out a role for the traditional causes of mass extinctions. However, their research raises the standard of evidence needed to demonstrate a real increase in extinction rates. A convincing demonstration of a mass extinction must take account of artifacts caused by the fluctuating sedimentary record.

Different mollusks grow up in different ways. In gastropod snails, planktonic and direct development are two of the main types of development. With planktonic development, the egg is released into the surface waters of the ocean and develops into a larval form which disperses among, and feeds on, the microscopic organisms (called "plankton") that float near the ocean surface. After a while, the larva settles and metamorphoses into an adult snail. With direct development, the eggs and young grow up near or (to begin with) inside the parental snail. Various ecological trends are known among modern forms, such as that planktonic development is commoner among shallow- than deep-water species, and commoner among tropical species than polar species. These results suggest that the mode of development in a species is an adaptation to the local ecological conditions.

The relation between larval type and speciation and extinction rates can be studied in fossil gastropods. Larval types in fossils are inferred by analogy with modern species. These kinds of inference were pioneered in the work of Thorson, and several criteria have now been used. Figure 23.7 shows one, which uses the size of regions in the larval shell. Modern species with planktonic development typically have small, yolk-poor eggs; in the larval shell, a region called prodissoconch I tends to be small and another region called prodissoconch II is larger (Figure 23.7a). Species with direct development have the reverse condition (Figure 23.7b). These morphological regions can be distinguished in fossil larval shells by scanning electron microscopy. We can reasonably assume that shell form is correlated with development type in the same way as modern forms.

Molluskan larval types can be inferred in fossils

Figure 23.7
Larval shell form is correlated with type of development in mollusks. The species in (a) and (b) are modern gastropods, and in (c) and (d) are late Cretaceous fossil bivalves. Note the relative sizes of the regions labeled PdI and PdII (prodissoconch I and II). (a) *Rissoa guerini*, which is known to have a planktonic larva (size bar = 50 μm); (b) *Barleeia rubra*, which is known to develop directly without a stage in the plankton (size bar = 50 μm); (c) *Uddenia texana*, which had small PdI and large PdII regions like (a) and is inferred to have had planktonic development (size bar = 20 μm); and (d) *Vetericardiella crenalirata*, which had large PdI and small PdII regions like (b) and is inferred to have had direct development (size bar = 20 μm). D, dissoconch. Reprinted, by permission of the publishers, from Jablonski & Lutz (1983).

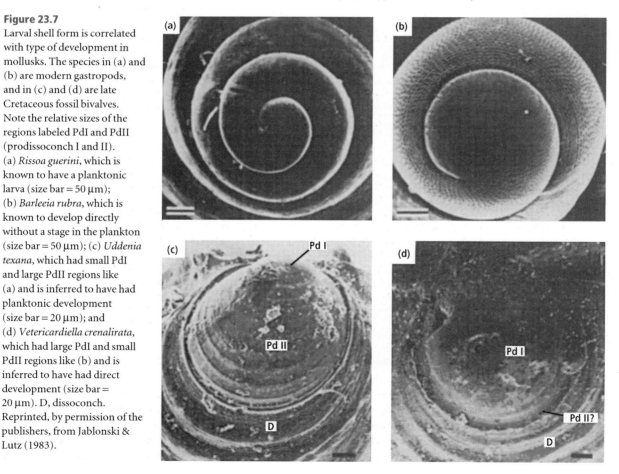

Figure 23.8

Duration in the fossil record and geographic ranges for late Cretaceous gastropods from North America. Species with planktonic larvae (a) last longer in the fossil record (i.e., they have lower extinction rates) and also have wider geographic ranges than species with direct development (b). The extinction rate is given as the chance that a species lineage will go extinct per million years. See Table 23.1 for related results. (500 miles ≈ 800 km.) Redrawn, by permission of the publisher, from Jablonski & Lutz (1983).

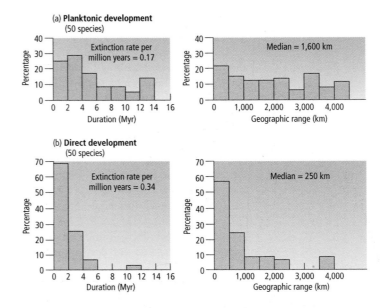

What is the relation between larval type and extinction rate? Several studies have found that species with planktonic larva have lower extinction rates (Figure 23.8). As the figure shows, they also have wider geographic ranges. This may be the reason for their lower extinction rate, because a species with a wider range is less vulnerable to local circumstances. Or it may mean merely that planktonic forms have a higher chance of being preserved as fossils than have directly developing forms, because their wider distribution increases the chance that in one site the conditions will be right for fossilization; the difference could then be just a bias in the fossil record.

Hansen (1978, 1983) looked at the relation between larval type and speciation rate. He predicted that snails with direct development will speciate more rapidly than species with planktonic larvae, because the species with non-planktonic development will be more likely to be geographically localized and isolated, which makes allopatric speciation easier. Planktonic development increases gene flow and makes allopatric speciation less probable. He used this idea to explain an observed trend in snails of the early Tertiary (Figure 23.9). The proportion of planktonically developing species declined through the Paleocene and Eocene. The trend was not being produced by the difference in extinction rates. The planktonically developing species, as usual, had lower extinction rates (Figure 23.9b), and that would tend to produce the opposite trend from that observed.

Two alternatives are left. Natural selection could have been favoring direct development within the majority of lineages. Hansen "suggested" this was not true (though he gave no evidence). The period was a time of global cooling, which might favor direct development, given the latitudinal trend mentioned earlier. Hansen said the decline in planktonically developing forms preceded the global cooling; concrete evidence, however, rather than a vague statement, would be needed to persuade a skeptic. The second alternative is that the increase was due to a higher speciation rate of the directly developing forms, simply because forms with lower dispersal rates are more likely to speciate.

Planktonic development is associated with low extinction rates . . .

. . . decreasing relative diversity . . .

. . . and (possibly) low speciation rates

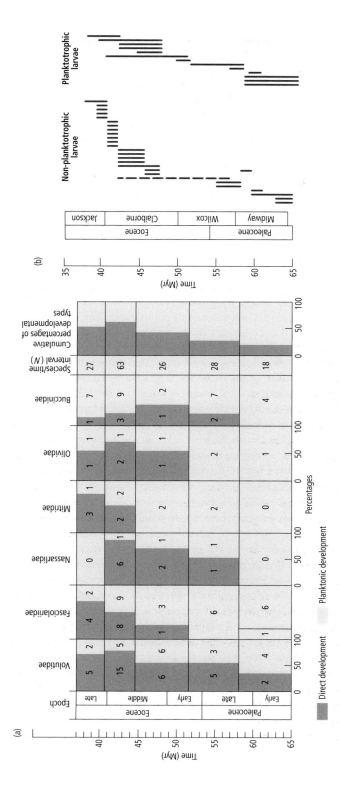

Figure 23.9

(a) The proportion of gastropod species with direct, rather than planktonic, development increases through the Paleocene and Eocene. The effect exists within several of the six families, and as an average for all six families. (b) Detailed observations for Volutidae, from the Gulf Coast of the USA. (Volutidae are furthest left of the six families in (a).) Note the proliferation of directly developing species. The extinction rate of species with planktonic development appears to be lower — which counters the trend toward more directly developing species. Small discrepancies between the numbers in (a) and the number of lineages in (b) may be due to extra data in the later paper (used to compile (a)). Redrawn, by permission of the publisher, from Hansen (1978, 1983). © 1983 American Association for the Advancement of Science.

Hansen's work has since been criticized

Since Hansen's work, Duda & Palumbi (1999) have cast doubt on one assumption of his argument. They show that, in a group of modern snails, species with planktonic development have repeatedly evolved from ancestral species with direct development. For the trend in Figure 23.9 to be driven by differences in speciation rate, it is important that the ancestor–descendant lines of species tend to retain the same mode of development. (In technical language, heritability is required at the level of species.) Duda and Palumbi's result for modern species suggests that the expanding group of species with direct development may not have been a clade with a constant mode of development. Species with direct development may have arisen from ancestors with planktonic development. Currently it is uncertain whether, as Hansen originally argued, the decline in planktonically developing forms in the early Tertiary occurred because they had a low speciation rate.

23.6.2 Differences in the persistence of ecological niches will influence macroevolutionary patterns

In the previous section, we considered the possibility that a character (larval type) might influence speciation, and extinction, rates. The influence, if real, is a straight consequence of the character itself: species in which there is direct development are more likely to split, in the process of allopatric speciation, than species in which there is planktonic development. A second factor that can influence speciation and extinction rates is the nature of the ecological niche occupied by species. Species that occupy niches that last longer will have lower extinction rates than species that occupy short-lived niches. Williams (1992) introduced this idea in terms of a concrete example — the three-spined stickleback (*Gasterosteus aculeatus*).

Sticklebacks that occupy long-lasting estuarine niches have low extinction rates

The three-spined stickleback is a fish with a widespread distribution in coastal waters in the northern hemisphere, on both sides of the North Atlantic and Pacific Oceans. From these coastal waters (it appears), many populations have separately colonized the local freshwater rivers and their tributaries inland. Some of these freshwater populations have been studied and they show various local adaptations to the rivers they occupy and have formed a complex set of local races or subspecies.

However, the populations that colonize the freshwater rivers are probably evolutionarily short lived. Ecological and geographic changes, for example, may be more frequent in these habitats. A river may dry up, or change its course or nature in such a way that the fish are driven extinct. The main coastal niche persists for longer. Thus when a new freshwater tributary opens up, it is usually colonized from the main coastal population rather than another freshwater population. The populations in the coastal niche have a low extinction rate, and probably a higher speciation rate. The populations in the freshwater tributaries have high extinction rates, and probably low speciation rates. The difference in extinction rates is not a straight consequence of the characters of the organisms. The coastal and freshwater populations have evolved different adaptations. The different adaptations are associated with, but do not directly cause, differences in the extinction rate. (By contrast, for instance, species with asexual reproduction go extinct at higher rates than species with sexual reproduction (Section 12.1.4, p. 318). The difference in extinction rate is partly a consequence of sexual and asexual reproduction.)

23.6.3 *When species selection operates, the factors that control macroevolution differ from the factors that control microevolution*

Figure 23.10

Species selection by differences among lineages in: (a) extinction rates, and (b) speciation rates. There is a trend over time toward more species with large body size. (a) Species with large body sizes have lower extiction rates (last longer) than species with smaller body size. Speciation is equally likely to produce a new species with a smaller or a larger body size than its ancestor; the speciation rate is also constant over time. (b) Species with larger body sizes have higher speciation rates than species with smaller body sizes. Speciation is equally likely to produce a descendant species with smaller or larger body size than its ancestor. Each species has the same longevity (extinction rate). In both cases, natural selection within the lineage does not favor individuals with a larger body size. This is shown by the "blown up" insets in the center: within a species, selection maybe stabilizing or inoperative. The insets are attached to (a) but they implictly apply to (b) too. (The figures have a punctuated pattern of evolution, but whether evolution is really gradual or punctuated is irrelevant in the theory of species selection.)

The trend toward increasing numbers of snail species with direct development is an example of what is sometimes called *species selection*. Species selection is a higher level analog of normal natural selection within a population. Species selection means, other things being equal, that those kinds of species that have lower extinction and higher speciation rates will tend to increase in frequency over evolutionary time.

The key question, for determining whether a trend is caused by species selection, is whether natural selection within a species is driving evolution in the direction of the trend. Consider a trend toward increasing body size (Figure 23.10). If natural selection within each species is stabilizing, but species in which body size is larger have lower extinction rates, then the trend to larger body size is driven by species selection. If natural selection within each species favors larger body size, then the trend is probably driven by conventional natural selection. The question is difficult to study. However, Alroy (1998) studied it for a trend to increasing body size in North American fossil land mammals. He found that the trend, on average, could be accounted for by increases within each lineage suggesting that species selection is at most a minor factor in this case.

Species selection should not be confused with group selection (Section 11.2.5, p. 303). Group selection aims to explain why individuals sacrifice themselves for the

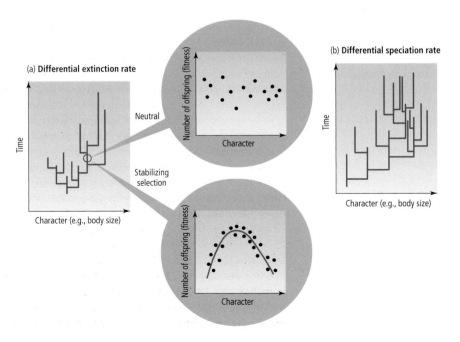

good of the group (or species) they belong to, and we have seen that it is difficult for adaptations of this sort to arise. In species selection, there is no question of individuals using a disadvantageous developmental mode in order to boost the speciation rate of their taxonomic groups. Direct and planktonic development are favored by natural selection in different taxonomic groups for good ecological reasons within each species: but they can then have different long-term consequences for radiation and extinction. We have no reason to suppose that what is favored by the short-term process of natural selection will always be the same as that which allows species to last a long time or split at a high rate. Natural selection may favor adaptations within some species that result in reduced long-term survival and adaptations that increase it in others.

Species selection is another example of a reason why macroevolution cannot simply be extrapolated from microevolution (Section 18.8, p. 550). Within a species natural selection favors one character in one species and another in a different species; but species selection over long periods may cause the species with one of the characters to proliferate, because of the character's consequences for speciation or extinction rates. This does not mean that the long-term process contradicts, or is incompatible with, the short-term process, only that we cannot understand the long-term evolutionary pattern by studying natural selection in the short-term alone and extrapolating it.

A similar conclusion can be drawn from the argument about niches. Again, macroevolution cannot simply be predicted from microevolution. A microevolutionary study would reveal how natural selection was favoring various characters in the stickleback populations, according to the aquatic environments they were occupying. The key to macroevolution is the persistency of the niches over time, and that is irrelevant to the short-term process of natural selection and to investigations of it. (Natural selection does not favor one adaptation over another because it allows the organisms to occupy a longer lasting niche.) Thus additional factors beside those studied in the short term come to matter when we try to understand evolutionary phenomena on the grand scale.

Taxa with certain attributes tend to proliferate over evolutionary time

23.6.4 *Forms of species selection may change during mass extinctions*

The relation between extinction rate and mode of development . . .

We saw in Section 23.6.1 that, in the normal times of the late Cretaceous before the mass extinction, the extinction rate was higher in species with direct than with planktonic development. Jablonski (1986) found similar relations for the other two variables: taxa that contained more species and that had broader geographic ranges had lower extinction rates than taxa with smaller ranges or with less species. He compared these results with those from the Cretaceous–Tertiary mass extinction and found that at that time two of the three correlations disappeared (Table 23.1). Species-rich taxa had the same chance of extinction as species-poor ones, and planktonic species had the same chance of extinction as directly developing ones. Only broad geographic range continued to be associated with a lower extinction rate. The extinction seems to have been so massive as to have taken out groups almost at random.

At any rate, the relations between the characters of a taxon and its extinction probability were significantly altered. In normal times, planktonically developing and species-rich taxa have lower probabilities of extinction than directly developing

Table 23.1

Survival of different kinds of snail taxa through the Cretaceous–Tertiary mass extinction and at other times (showing a "background" extinction pattern). Background extinction rates vary with snail type, whereas survival through the mass extinction may have been a matter of luck. (a) Relation between the chance of generic extinction and developmental mode. The background rate of extinction is lower for snails with planktonic, than with direct, development (this evidence is the same as in Figure 23.8), but in mass extinctions snail genera of the two types have the same chance of surviving. (b) Relation between the chance of generic extinction and number of species in the genus. The background extinction rate is lower for genera that are species rich (contain three or more species) than for genera that are species poor (contain one or two species). But in mass extinctions a species-rich genus had about the same chance of going extinct as a species-poor genus; in both cases about 40% of genera went extinct and about 60% survived. n is number of genera although the genera studied at the two times are not all the same. From Jablonski (1986).

(a) Extinction rate and mode of development.

Mode of development	Background extinctions		Mass extinction		
	n	Median geological longevity (Myr)	n	Genera surviving (%)	Genera extinct (%)
Planktonic development	50	6	28	60	40
Direct development	50	2	21	60	40

(b) Extinction rate and species richness of genus.

Species richness	Background extinctions		Mass extinction	
	n	Median geological longevity (Myr)	Genera surviving (n)	Genera extinct (n)
Species poor	145	32	31	38
Species rich	114	49	22	25

. . . changed during a mass extinction

and species-poor taxa. In contrast, in the Cretaceous–Tertiary mass extinction the difference disappeared. The conditions had altered and the form of species selection altered too.

In Jablonski's research, the extinction pattern through a mass extinction became less selective — perhaps because the extinction was so massive that almost all snail species succumbed regardless of their adaptations. But in other periods, or other taxa, the extinction patterns remained selective during mass extinctions. The form of the selectivity can even provide clues about the nature of the extinction event. For instance, a major (if not a mass) extinction occurred at the Oligocene–Eocene boundary (Figures 23.2 and 23.5). The extinction is widely thought to have been caused by global cooling. Evidence suggests that species with adaptations to warm temperatures were more likely to go extinct at that time than species adapted to cool temperatures. The selective pattern of the extinctions fits with the climatic explanation.

Table 23.2

Specialist insects were more likely to go extinct in the Cretaceous–Tertiary (KT) mass extinctions. Insects were divided into three categories, according to whether they had dietically specialist, generalist, or intermediate relations with plants. Specialists probably fed on only one plant species, generalists on many plant species. The evidence came from the type of damage found in fossil leaves. From Labandeira *et al.* (2002).

Diet type	Number before KT extinction	Number after KT extinction	Percent extinct
Generalist	12	12	0
Intermediate	16	10	37.5
Specialist	20	6	70

Fossil leaf evidence suggests specialist insects had higher extinction rates

In the Cretaceous–Tertiary mass extinction, Labandeira *et al.* (2002) found that ecologically specialist species were more vulnerable than ecologically generalist species. Specialist insects feed on only one plant species whereas generalists feed on several plant species. Labandeira *et al.* used evidence from damage in fossil leaves. Microanatomic study of leaf damage can suggest whether it was caused by insects. The different kinds of insect damage can be divided into three categories — those caused by generalist insects (which cause, for instance, damage to leaf margins or make holes in the leaf), those caused by specialist insects (which cause, for example, galls, or act as leaf-miners), and intermediate cases. Table 23.2 summarizes their results. None of the generalist species went extinct, but most of the specialist ones did. Specialist phytophagous insects were more vulnerable. The reason is probably that plant resources were reduced. Suppose that 70% of plant species disappeared. Then any insect specialist on those 70% of species would also go extinct. However, the generalists could survive by eating on the 30% of plant species that survived.

The main point of these examples is that species selection can be studied in mass extinctions, and that the form of species selection may change during mass extinctions from other times. However, we can also notice that the results provide an independent source of evidence that mass extinctions were real rather than artifactual events. Initially we saw that mass extinctions were inferred as peaks in the graph for extinction rates through time (see Figure 23.2). We then saw that this evidence was inconclusive, because changes in the sedimentary record could account for the observations (see Box 23.2). Here we have seen that the form of extinctions was non-random, and non-random in a pattern that fits with a real mass extinction. Sedimentary sampling (Figure B23.2) alone would not disproportionately take out specialist insects. However, a real mass extinction would be expected to remove specialists disproportionately. The argument in itself is not conclusive, but needs to be weighed in the balance. The evidence for mass extinctions does not solely come from the total extinction rate.

23.7 One higher taxon may replace another, because of chance, environmental change, or competitive replacement

23.7.1 *Taxonomic patterns through time can provide evidence about the cause of replacements*

After the dinosaurs went extinct at the end of the Cretaceous, the mammals radiated rapidly and filled the ecological niches for large land vertebrates formerly occupied by dinosaurs (see Figure 23.1). Earlier in the Cretaceous, the angiosperms had radiated, apparently at the expense of the gymnosperms, which simultaneously declined (Figure 18.8, p. 539). These are both examples of evolutionary *replacements*, in which one taxonomic group comes to occupy the ecological space formerly occupied by another taxonomic group.

Taxonomic replacements can be competitive or independent

Why should one higher taxon replace another higher taxon of ecologically similar species? Two theories can be tested. One (competitive displacement) says that the later group outcompeted the first, and drove it extinct. The other (independent replacement) says that the first group declined and went extinct for some reason unrelated to the presence of the second group, and the second group only radiated after the first had been cleared away. The pattern of change in diversity of the two groups provides the best evidence to test between the two theories (Figure 23.11). If the first group declines before the second expands, it suggests competition was not influential. If the first group declines in proportion to the increase in the second, it suggests competition; this pattern (Figure 23.11b) is sometimes called a *double-wedge* pattern.

In the case of independent replacement, we can distinguish two possibilities (Figure 23.11a). One is that the environment changed, and the earlier group went extinct due to poor adaptation to the new environmental conditions. The second is that a catastrophic mass extinction occurred, for instance following an asteroid impact, and one dominant taxonomic group went extinct while another taxonomic group had a few survivors. The reason why one group went completely extinct while the other survived might mainly be luck.

The test between competitive and independent replacement in Figure 23.11 is not foolproof. The double-wedge pattern characteristic of competitive replacement could

Figure 23.11
The exact pattern of replacement of one group by another suggests whether or not competition was at work. (a) If the initially dominant group declines before the second group expands, it suggests that the replacement was not caused by competitive displacement. The dominant group may decline either gradually or catastrophically. (b) If the dominant group declines as the other group gains at its expense, competition and relative adaptation are more likely to have influenced the replacement.

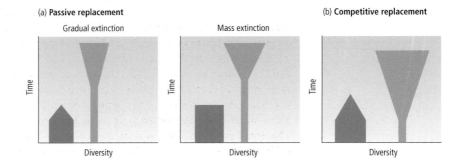

also arise without competition. For instance, environmental change alone might produce a simultaneous decline of one group and radiation of the other. However, the test still has some interest. If the earlier group does go extinct clearly before the rise of the later one, it is difficult to explain the replacement by competition. And if the takeover is correlated in time and the two groups are ecological analogous, competition is at least suggested. Other kinds of evidence of competition can also be brought in, as we shall see.

23.7.2 *Two bryozoan groups are a possible example of a competitive replacement*

One of the most plausible examples of competitive replacement concerns two groups of Bryozoa. (Perhaps I should say "least implausible" — any conclusion on the influence of competition in the past will be uncertain. It takes hard work to show that competition is at work in a modern ecosystem, and the evidence for fossils is much more limited.) Bryozoans are sessile, aquatic invertebrate animals that live attached to rocks or other surfaces. The two main taxa of bryozoans are called Cyclostomata and Cheilostomata. Between approximately 150 and 50 million years ago, the Cheilostomata steadily replaced the Cyclostomata (Figure 23.12). The pattern in the figure itself suggests competitive replacement: one taxon (Cheilostomata) rises as the other (Cyclostomata) falls.

One group of Bryozoa replaced another group

In this case, we also have other evidence of competition and of a competitive advantage for Cheilostomata. Bryozoans compete with one another by "overgrowth": one bryozoan grows over the top of another. The animal that does the growing over expands in size and can feed over a larger area. The overgrown animal is prevented from feeding, and is killed. Overgrowth can be seen in nature today, and is also seen in fossils (Figure 23.13). In the majority of cases, when members of the two main taxa are involved, a Cheilostome is overgrowing a Cyclostome. Cheilostomes therefore seem to have a competitive advantage over Cyclostomes and are more aggressive growers. This competitive superiority is probably part of the explanation for the taxonomic replacement over time.

We have direct evidence of competition

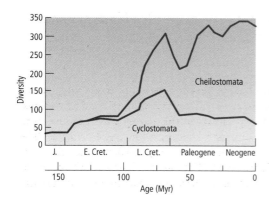

Figure 23.12
Diversity (measured as number of genera) of two bryozoan taxa through time. Cheilostomata have replaced Cyclostomata as the main group. From Sepkoski *et al.* (2000).

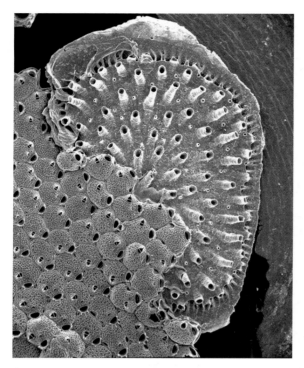

Figure 23.13
Competitive overgrowth in fossil bryozoans. The cheilostome bryozoan *Microporella ciliata* forms an approximate triangle at the lower left of the figure. The cyclostome bryozoan *Diplosolen obelium* forms an approximate crescent shape to the upper right. The cheilostome is overgrowing the cyclostome in the center of the photo, growing upwards and to the right. The cyclostome is also in a small way striking back at its aggressor, overgrowing the cheilostome leftwards at the top. The sign that the "victim" is stiking back provides evidence that the main overgrowth is not simply one live colony overgrowing a dead colony. The fossils come from the Adriatic Sea, near Rovinj, Croatia and date to the Recent. (Photo courtesy of F.M. McKinney.)

The replacement of Cyclostomata by Cheilostomata looks like an example of evolutionary escalation (Section 22.6.1, p. 234). However, McKinney (1995) has argued that the escalation was a one-off event rather than a continual process. At some point, Cheilostomata gained their evolutionary advantage, perhaps by their superior growing power. They then maintained that advantage. It was not the case that the two taxa evolved incremental escalatory adaptations against each other over 100 million years. Rather, the Cheilostomata became stronger 150 million years ago and gradually took over from the Cyclostomata, without any further escalatory evolution.

23.7.3 *Mammals and dinosaurs are a classic example of independent replacement, but recent molecular evidence has complicated the interpretation*

The replacement of dinosaurs by mammals is the classic example of an independent replacement. Dinosaurs were the main land vertebrates of the Jurassic and Cretaceous, while mammals took over in the Tertiary. However, this example has become more complicated recently, as molecular results have challenged the fossil evidence. The fossil record shows a rapid rise of mammals in the early Tertiary, after the end-Cretaceous mass extinction (see Figure 23.1). The earliest fossils of all the main groups (orders, to be taxonomically exact) of modern mammals come from the early

Mammals proliferated in the early Tertiary

Tertiary: this is when we find the earliest fossil Carnivora, ungulates (Perissodactyla and Cetartiodactyla), ant-eaters, elephants, and primates. The dinosaurs had gone extinct in the mass extinction that preceded the mammal radiation. The fossil evidence fits the pattern of Figure 23.11a exactly, implying independent, rather than competitive, replacement.

A Tertiary origin for the modern mammal orders is also supported by the fossil record for more ancestral mammals. Eutherian mammal fossils are known from the Cretaceous. Until recently (see below), the oldest clearly eutherian mammal fossils were from about 80 million years ago. These fossils do not belong in any of the modern mammal orders. They are classified as relatives of the moderns orders, connected by deep branches in the mammal tree. It would have taken some time for the modern orders of mammals to have evolved from ancestral eutherians. If the eutherians originated around 80 million years ago, it makes sense that the modern groups evolved around 55 million years ago. That leaves 20 million years for the evolutionary change from ancestral to modern eutherian forms. The modern orders could hardly have existed *before* 80 million years ago, if that was when the earliest ancestral eutherians lived.

However, when the molecular differences among the modern mammal orders are measured, and the rate of the molecular clock calibrated, the inferred time for the common ancestor of these groups is much older than the early Tertiary. The molecular date for the common ancestor is about 90–100 million years ago. The molecular evidence implies that the modern mammal orders — Carnivora, Primata, Proboscidea, and so on — already existed in the mid-Cretaceous. Indeed they already existed before the earliest known fossil of any eutherian mammal.

If the molecular date is correct, the mammal groups that now occupy the niches of dinosaurs coexisted with dinosaurs for the last 30 million years of the Cretaceous. This does not prove that mammals competed with dinosaurs, or that any such competition contributed to the demise of the dinosaurs. The Cretaceous mammals may have been small in size and ecologically different from their modern descendants, so they would not have competed with dinosaurs. Or they may have been numerically too rare to affect the dinosaurs. All that is uncertain. For now, the main point is that the molecular evidence has spoiled the clear Figure 23.11a-like pattern for dinosaurs and mammals, and weakened the case for an independent replacement.

The conflict between molecular and fossil dates for the modern mammal orders is another example of a common kind of conflict in modern evolutionary biology (see Section 15.13, p. 460, for hominins, and Section 18.4, p. 535, for the Cambrian explosion). The modern groups of birds — gulls, ducks, and passerines, for instance — are a further example. The fossil evidence suggests that they radiated after the Cretaceous extinction, in the early Tertiary, but the molecular clock suggests they originated far earlier. As with the hominin and Cambrian explosion examples, the conflict could be resolved in three ways. The molecular evidence may be wrong, the fossil evidence may be wrong, or the two may be reconciled.

For the mammal orders, the molecular evidence has been analyzed and reanalyzed, but has not been seriously challenged. Fossil research has been more revealing. One approach has been to estimate, statistically, how likely it is that the main mammal orders existed in the Cretaceous but left no fossils. The estimates are made using independent evidence about the completeness of the fossil record. For one lineage, we can

Molecular evidence suggests an earlier origin

The fossil–molecular conflict has inspired several lines of research

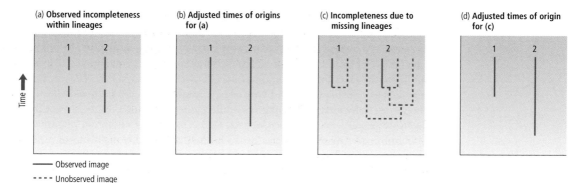

(a) Observed incompleteness within lineages

(b) Adjusted times of origins for (a)

(c) Incompleteness due to missing lineages

(d) Adjusted times of origin for (c)

—— Observed image

- - - - Unobserved image

Figure 23.14

Use of observed incompleteness in the fossil record to adjust estimated times of origin for a taxon. (a) Observed fossil records for two lineages that both appear to originate at the same time. (The x-axis has no meaning.) (b) Taxon 1 has a more incomplete record, suggesting that it had an earlier time of origin than taxon 2. (c) Taxon 1 and 2 are observed to originate at the same time in the fossil record, though taxon 1 has fewer missing lineages than taxon 2 (dashed lines indicate lineages unrepresented in the fossil record). (d) Taxon 2 probably had an earlier origin than taxon 1. The illustrated adjustments to time of origin in (b) and (d) are not quantitatively exact.

look at the time from its first to its last fossil appearance, During that time, there will often be gaps, for times when the lineage is not represented. We can use the extent of those gaps to estimate how much earlier the lineage really originated than its first appearance in the fossil record (Figure 23.14a and b). Foote *et al.* (1999) used this method, and concluded it was unlikely that the mammal orders had originated much before 55 million years ago.

Foote *et al.*'s (1999) method only used incompleteness within observed lineages. The fossil record is also incomplete in missing out whole lineages (Figure 23.14c). Tavaré *et al.* (2002) recalculated the chance that the primates originated in the mid-Cretaceous, taking account of both kinds of incompleteness — incompleteness within lineages, and missing whole lineages. Their adjusted fossil estimate for the origin of the Primata was 81 million years ago. This is well before the observed date of about 55 million years ago and not far off the molecular date (around 90 million years ago). Tavaré *et al.*'s (2002) adjusted date implies that primates existed for an extensive ancestral period that happens to be unrepresented in the fossil record.

The earlier date for the origin of the modern mammal orders has been supported by discoveries of earlier mammal fossils. Ji *et al.* (2002) reported a 125 million-year-old fossil eutherian from the Yixian formation in China. The fossil pushes back the eutherian record by about 50 million years. While the oldest eutherian was from about 80 million years ago, it was difficult to see how the modern mammal orders could have originated much before the Tertiary. Now, with the oldest eutherian at 125 million years ago, ample time exists for those orders to have originated by 80–90 million years ago.

The molecular evidence has led biologists to accept an earlier time for the origin of the modern mammal orders. However, many biologists suspect that mammals "lay

The fossil dates are being "corrected" back in time

low" for 40 million years until after the extinction of the dinosaurs. Only then did the mammals radiate into formerly dinosaurian niches. Thus, the fossil evidence may still be essentially correct. It shows when the main mammal orders proliferated, rather than when they originated. This reconciliation between the molecular and fossil evidence is similar to the case of the Cambrian explosion (Section 18.4, p. 535). The reconciliation is plausible, and popular, but by no means confirmed — after all, we know nothing about the Cretaceous representatives of the modern mammal orders, except that they probably existed.

In summary, many examples exist through the history of life when one taxon replaces another. One question we can ask about replacements is whether they were competitive or independent. The main class of evidence comes from the time course of the rise of one taxon, and the fall of the other (see Figure 23.11). Other evidence comes from direct competition, such as bryozoan overgrowth. One possible example of independent replacement, that of dinosaurs and mammals, has been reanalyzed recently and, although the revised picture does not show that the replacement was competitive, a formerly watertight case for independent replacement has sprung a leak. The role of the Cretaceous mass extinction in the rise of the mammals is less certain than it appeared to be 10 years ago, in the early to mid-1990s.

23.8 Species diversity may have increased logistically or exponentially since the Cambrian, or it may have increased little at all

Changes in the total diversity of life may fit . . .

The number of species alive on Earth today is uncertain, with estimates ranging from 10 to 100 million species. About 2 million species have been described. At some point in the past, fewer species must have existed than now (this almost has to be true, if we trace back far enough — even to the origin of life). But how have the number of species changed over time? Early attempts to answer this question were made by Simpson, Valentine, and others, but modern thinking about it begins with Sepkoski's database — the database for marine fossil animals we have met previously in this chapter.

Figure 23.15a illustrates Sepkoski's classic result. The y-axis is for numbers of families, but we can assume that numbers of species would show a similar pattern. Sepkoski distinguished three faunas: the Cambrian, Paleozoic, and Modern. The three differ in the kind of animals that were alive at the time. The data for the Cambrian is poor, and the important features of the graph are: (i) the rapid increase in diversity after the Cambrian, starting about 500 million years ago; (ii) the apparent "Paleozoic plateau" of constant diversity; (iii) the reduction in diversity at the Permian mass extinction; followed by (iv) the steadily increasing diversity since then (the end-Cretaceous mass extinction is only a blip in the steady rise).

. . . a logistic model . . .

The increase of diversity, followed by a plateau, in the Paleozoic can be explained by a *logistic* model. Logistic increase is the kind of increase seen by ecologists when new resources are colonized. Numbers increase exponentially to begin with, due to the absence of competition. Then, as competitors fill up the resource space, no new species can be added except following the extinction of an existing species.

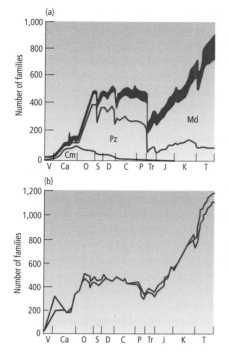

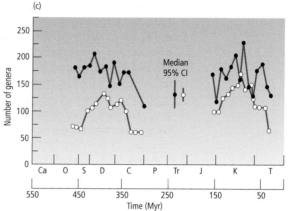

Figure 23.15

The history of global biological diversity. (a) Sepkoski originally found a logistic increase with a Paleozoic plateau, followed by a steady increase since the Permian mass extinction. He also recognized three distinct faunas: Cambrian (Cm), Paleozoic (Pz), and Modern (Md). (b) Benton compiled the data in a slightly different way, and found a pattern that was indistinguishable from a steady exponential increase. (c) A new compilation aims to correct for stratigraphic incompleteness, and preliminary findings suggest no increase in diversity between the Ordovician and Tertiary. Redrawn, by permission of the publisher, from Miller (1998) and Alroy *et al.* (2002). © 1998 American Association for the Advancement of Science.

The difference between the Paleozoic and Modern parts of Sepkoski's graph suggest that the Permian mass extinction had a major creative influence for life on Earth today. In the absence of the Permian mass extinction, maybe the Paleozoic fauna would have continued to dominate life on Earth, and diversity might still be stuck at the Paleozoic plateau. The "Modern" fauna seems to have been able to diversify more than the Paleozoic fauna.

. . . or an exponential model . . .

A second model — the *exponential* model —for the history of diversity was proposed by Benton (1997). He used a different data compilation, including land organisms. He suspected that Sepkoski's result, including the Paleozoic plateau, was peculiar to marine life. If all life is looked at together (Figure 23.15b), the full history of diversity could just as well be explained by a steady (if noisy) exponential increase. Two implications of Benton's model are as follows. First, species have been persistently dividing and subdividing ecological niches into finer and finer units — if there is a limit on total diversity, as the logistic model assumes, then that limit has not yet been reached.[2] Secondly, mass extinctions have been less important for the history of diversity. Diversity now would be much the same if no mass extinctions had occurred. Other models beside those by Sepkoski and Benton have also been proposed (see Miller 1998).

Which model is correct? The question has been difficult to answer, because of limitations in the data. Even Sepkoski's compilation has gaps and biases. Jackson & Johnson (2001), for instance, investigated Sepkoski's data for tropical Bryozoa. They found a huge underestimate of the diversity of this regional taxon for the Pliocene. The reason was simply that Europe and North America were no longer tropical in the Pliocene. Most paleontologists are European or North American, and most of them work locally. Thus, the high diversity of tropical bryozoan fauna has been much better studied for earlier times, when Europe and North America were tropical, than for the Pliocene, when they were temperate. If Sepkoski's data are adjusted for this bias, Jackson and Johnson suggest it fits better with the exponential model. However, the main point is that we need a database that has been more systematically corrected for bias. The two main kinds of bias come from the amount of rock preserved from different time intervals, and the amount of that preserved rock that has been studied by paleontologists. Geographic and taxonomic biases (as paleontologists study some regions, and some taxa, more than others) interact with these two main kinds of bias.

A new database is being compiled by a team of paleobiologists, and is being housed at the University of California at Santa Barbara. The new database aims to correct for biases, by appropriate statistical corrections to the raw data. The preliminary results are remarkable (Figure 23.15c). The estimated global species diversity seems to show little change from the Paleozoic to the Oligocene. If the result holds up as the database is developed, it will suggest that diversity has a tighter limit than had previously been suspected. Speciation and extinction may have been in balance for much of the past 500 million years, with mass extinctions having had little effect on global species diversity. That such a dramatically different picture can be produced now highlights the importance of statistical corrections to the data. The main difference between Figure 23.15a and 23.15c is due to the statistical corrections for bias.

In summary, we have seen how there is an active research program that aims to describe the history of global species diversity. The main problems lie in data compilation, and correction for biases and gaps. Paleobiologists can be found who support at least three models of history: a logistic increase to a Paleozoic plateau, followed by a steady increase; steady exponential increase; and a prolonged plateau. The influence of mass extinctions on diversity is uncertain. It is also uncertain whether modern life subdivides ecological niches more finely than earlier life did. These uncertainties exist not because the problems have been forgotten about, but because of a multipronged modern attack by research paleobiologists.

---

[2]   Ultimately, species diversity will be limited, among sexual creatures, by the time it takes to find a mate. As species split and split, the numbers of individuals per species will go down (because the total biomass on Earth is limited by the energy input from the sun). As the members of a species become rarer, and sparser, the energetic cost of finding a mate will ultimately become prohibitive. Hybridization, despite its genetic costs (Section 14.4, p. 389) will eventually be favored.

23.9 Conclusion: biologists and paleontologists have held a range of views about the importance of mass extinctions in the history of life

Biological diversity — the range of life forms on Earth — is clearly influenced by radiations and extinctions. The most controversial question in this subject has concerned the importance of mass extinctions. We can distinguish four historic phases in our understanding of mass extinctions. The four phases correspond to changing views about the reality, distinctiveness, and evolutionary significance of mass extinctions.

Early research established the main faunal transitions

The faunal transitions that we now recognize as mass extinctions were first discovered in the early nineteenth century as the main stages, and substages, of the fossil record were also established. This research forms Phase 1. The faunal transitions between the main stages of the fossil record were then often explained by rounds of catastrophic extinctions. Phase 2 begins with Lyell in the 1830s. Lyell doubted whether the observed faunal transitions were real catastrophes. He explained the faunal changes by changes in the environmental and sedimentary conditons. Darwin continued this line of thought. However, absolute geological dates later showed that the faunal transitions did not correspond to hiatuses in the fossil record. They seemed to be real mass extinctions.

The asteroid theory inspired a new set of ideas . . .

Phase 3 can conveniently be dated to about 1980. Strong evidence then suggested that the Cretaceous–Tertiary mass extinction was caused by an asteroid impact. This was one of several components in a "1980s" view that mass extictions were real, distinct events, and had a major influence in the history of life. We saw in this chapter how mass extinctions have been thought to clear space and permit radiations of new taxa, such as the modern mammal and bird orders in the Tertiary. Also, the Permian mass extinction seemed to precipitate a faunal transition, leading to a modern set of life forms that diversified more than the earlier Paleozoic fauna. Rare asteroid impacts are not the sort of event that life will evolve adaptations to survive. Which taxa do survive, and which go extinct, in the exceptional circumstances of mass extinctions, may be largely a matter of luck, and have little to do with the microevolutionary process of adaptation and natural selection. Thus, there could be distinct "macroevolutionary regimes" during, and between, mass extinctions. In this set of ideas, mass extinctions are the key to understanding much of evolutionary history.

. . . some of which are holding up better than others

Some of these Phase 3 ideas will probably endure. However, recent research has moved on to a Phase 4. Paleobiologists and molecular phylogeneticists have at least picked holes in, and sometimes seriously challenged, earlier work. Extinction rates appear to be "fractal." A similar set of causes are at work when extinction rates are low, high, or anywhere between. Random factors determine whether that set of causes results in high or low extinction rates. The search for the "cause" (or "causes") of mass extinctions may be mistaken; the same set of causes are at work all the time. It is even possible that mass extinctions are artifacts, though few experts advance that as more than a hypothetical possibility.

The molecular date for the origin of modern mammal orders could revise our understanding of the nature of mass extinctions, and of the influence of mass extinctions in evolutionary replacements. Almost all the modern orders of mammals (and birds) may have existed in the Cretaceous, and survived the extinction at the end of the Cretaceous. The fossil evidence overestimates the massiveness of the mass extinction, because no Cretaceous fossils of these taxa are known. The Cretaceous mass extinction may still have enabled the rise of the mammals, but the picture is less clear than before.

Finally, the influence of mass extinctions on global diversity is being challenged. In the 1980s, the Permian mass extinction was thought to be a key event, as diversity steadily increased after it, following a plateau before. The most recent, and statistically best adjusted, evidence suggests that diversity could have been constant. Maybe the diversity of life today would be much the same if the mass extinctions had never happened.

Much remains uncertain. The future lies with improvements in data collection, taxonomy, and statistical adjustment for biases in the data through time. As further results unfold, we shall see whether the Phase 3 or Phase 4 ideas hold up better — and what Phase 5 will bring in.

Summary

1 An adaptive radiation occurs when one or a small number of ancestral species evolves over time into a larger number of descendant species, occupying a range of ecological niches. Adaptive radiations occur following the colonization of a new area where competitors are absent; the extinction of competitors; and adaptive break-throughs.

2 The observed extinction rate varies through geological time. At certain moments, extinction rates have increased to a peak; these moments are called mass extinctions. From two to five major mass extinctions have been observed in the past. The two most important are at the end of the Permian and the end of the Cretaceous. Three others were at the end of the Ordovician, Devonian, and Triassic.

3 Alvarez *et al.*'s discovery of anomalously large concentrations of the rare earth element iridium at the Cretaceous–Tertiary boundary rocks at Gubbio, in Italy, suggested that the mass extinction may have been caused by the collision of an asteroid, about 7.5 miles (12 km) in diameter, with the Earth. Much evidence now supports their idea.

4 The impact theory of mass extinctions predicts that extinctions in different taxa should be sudden, synchronous, and global; the evidence for the Cretaceous–Tertiary extinction mainly fits the prediction, though other interpretations of the pattern are possible.

5 Asteroidal collisions, volcanic eruption, climatic cooling, sea level changes, and changes in habitat area caused by plate-tectonic movements are the five potentially general theories of mass extinction. The evidence does not suggest mass extinctions are generally caused by asteroidal collisions; the effects of climatic sea level changes need to be tested systematically; and the effect of plate tectonics is difficult to test at present.

6 The distribution of extinction rates may fit a power law. This would suggest that the same basic causal pattern is always at work, and random factors determine whether extinction rates are high, low, or in between. "Mass" extinctions may not have distinct causes as opposed to extinctions at other times (sometimes called "background" extinctions).

7 Almost all changes in extinction rates may be accounted for by changes in the amount of sedimentary rock per geological time interval. This would suggest either that many changes in extinction rates are artifactual, or that a common factor drives changes both in the real extinction rate and the amount of sedimentary rock.

8 If natural selection favors one form of a character in one species and another form in another, and if the different forms of the character cause different speciation or extinction rates, then there may be a trend toward more of the kind of species with higher speciation, or lower extinction, rates. The process is called species selection.

9 If the niches of some species last longer than others, those species will have lower extinction rates. If some niches are so positioned that new species can easily evolve from them, then the species occupying them will be more likely than average to give rise to new taxa.

10 Different kinds of species may suffer differentially in mass extinctions. The relation between the characters of taxa and their extinction rates changed between the late Cretaceous and the Cretaceous–Tertiary mass extinction.

11 Large-scale evolutionary replacements of one taxon by another occur by either competitive or independent replacement. The frequencies of the taxa through time provide a partial test between the two explanations.

12 The global number of species since the Cambrian may show a logistic increase up to the Permian followed by a steady increase; or a persistent exponential increase; or may have been constant. The different results depend mainly on differing statistical corrections to the observed number of fossil species through time.

13 Various views are possible about the importance of mass extinctions in the history of life. At one extreme, mass extinctions may be a distinct kind of event and a creative historic force, responsible for shaping many of the observed changes in the fossil record. At the other extreme, mass extinctions may differ little from extinctions at other times, or may even be artifacts, and have made little difference to the course of evolutionary history.

Further reading

Wilson (1992) is a book for a broad audience on biological diversity. It covers many themes not covered in this chapter, such as the number of extinctions that modern humans are causing. Gould's *Natural History* column (1977b, 1980, 1983, 1985, 1991, 1993, 1996, 1998, 2000, 2002a) included a number of essays on extinction, particularly mass extinction. Jablonski (1986, 2000) looks at the macroevolutionary significance of mass extinctions. See also Gould (2002b). Magurran & May (1999) contains research-level papers. Givnish & Sytsma (1997) is a multiauthor research-level book using molecules to study adaptive ratiation.

Sereno (1999) compares the radiation of dinosaurs at the end of the Triassic with the Tertiary radiation of mammals (see his fig. 1). The dinosaur radiation was much slower, perhaps reflecting the less catastrophic extinction at the end of the Triassic than the end of the Cretaceous — but comparative analysis of radiation rates is largely a research problem for the future. See also Chapter 13 on character displacement and Chapter 16 on divergence.

A further factor that causes extinction, perhaps particularly in plants, is that a species increasingly hybridizes with other species as it becomes rare. Its genes then become diluted out of existence. See Levin (2000) and his references.

For Sepkoski's work, Sepkoski (1992) is a standard source and cites earlier papers. But his database continued to be updated until his death in 1999, and Peters & Foote

(2002), for instance, used an unpublished update of "Sepkoski." Sepkoski's database has often been criticized for taxonomic errors. Adrain & Westrop (2000) made a study of the problem. They made an expert assessment of part of the trilobite database in Sepkoski. They found that as many as 70% of the entries in the database were inaccurate, but that the errors were random and did not introduce bias. See Pease (1992) on the trend toward declining extinction rates with time.

On mass extinctions, see Hallam & Wignall (1997), particularly their overview chapter. For the Cretaceous–Tertiary mass extinction, see the pair of papers by Alvarez & Asaro and by Courtillot in *Scientific American* (October 1990) for asteroidal and volcanic interpretations. Courtillot (1999) is a book on the topic. See Grieve (1990) on impact craters. The other mass extinctions, including the Permian mass extinctions, can be followed up through Hallam & Wignall (1997) and the general references on extinction. Benton (2003) is a popular book mainly about the Permo-Triassic mass extinction.

On fractal extinction statistics, Kirchner & Weil (1998) and Hewzulla *et al.* (1999) followed the topic up, respectively more differing from and more agreeing with Solé *et al.* (1997). On fractals in nature generally, see Bak (1996). On a related theme, Hubbell (2001) looks at the clade shapes expected on random models of speciation and extinction.

On replacements, Cooper & Fortey (1999) and Tavaré *et al.* (2002) give references for the molecular work. McKinney *et al.* (1998) is a related analysis for the bryozoan example. For dinosaurs and mammals, see Gould (1983, chapter 30) and Van Valen & Sloan (1977). Gould argues for independent replacement, Van Valen and Sloan (long before the molecular data!) for competitive replacement. Novacek (1992) reviews mammal fossils and phylogeny. A further classic case study is the replacement of mammal-like reptiles by dinosaurs after the Triassic mass extinction: it looks non-competitive (Sereno 1999). Levinton (2001) and Gould (1989) discuss the number of higher groups through geological time.

On species selection, Gould (2002b) is now a standard source for one side of the question. Further discussion is in, among many others: Williams (1966, 1992 — who prefers the term "clade selection"), Levinton (2001), and the exchange between Alroy and McShea in *Paleobiology* (2000), vol. 26, pp. 319–33. As to evidence, in the text of this chapter I concentrated on fossils. This is one of two methods. The other is to compare the number of modern species between different branches of a phylogeny with different character states. In Section 22.3.4, we looked at an example in the case of wind versus biotic pollination. Another example, that fits with a recurrent theme in this edition, is in Arnqvist *et al.* (2000). Polyandrous insect clades have a four times higher rate of speciation than sister monandrous clades. Sexual conflict, of the type we met in Section 12.4.7, is the prime suspect, and fits with ideas about speciation in Section 14.12.

On global species diversity, a further topic is how diversity recovers after mass extinctions. Miller (1998) includes discussion and references. Kirchner (2002) shows that the speciation rate can have an upper limit that delays recovery following massive extinctions. The trend to increasing diversity is one of a number of possible macrotrends in the history of life, a topic reviewed by McShea (1998).

Study and review questions

1 Why could it not be known until relatively recently in human history that a species had gone extinct?

2 (a) What is the distinction between a real and a pseudoextinction? (b) How does the distinction matter for theories of species selection and mass extinctions?

3 (a) What is the best evidence that the mass extinction at the Cretaceous–Tertiary boundary was caused by an asteroidal impact? (b) What predictions about the pattern of extinctions in the fossil record can be made if it was indeed caused by an asteroidal impact? What difficulties arise in testing them?

4 When in the history of life were the two, three, and five best documented mass extinctions?

5 How do we expect the observed extinction rate to change if: (a) a geological time interval with a good sedimentary record is followed by a geological time interval with a poor sedimentary record; and (b) the other way round, where a geological time interval with a poor sedimentary record is followed by a geological time interval with a good sedimentary record?

6 What is the relation between the developmental mode of snail taxa and the chance of extinction at the time of mass extinctions and at times of normal (or background) extinction rates? How do you explain the trends?

7 Why cannot the macroevolutionary pattern of extinction rates in different taxa be simply predicted from microevolution in the taxa?

8 In Chapter 11, we saw that natural selection does not generally favor adaptations at the group level, because group attributes are not heritable. So how is species selection possible? Do species, but not groups, show heritability? Or is heritability irrelevant in species selection?

9 How can we use the patttern of radiation and extinction over time in two taxa to test whether one of the taxa replaced the other by outcompeting it?

10 How it is it possible that much the same basic observations of the numbers of fossil species through time have been explained by such different models as logistic or exponential increase, or even constant numbers over time?

Glossary

Words in italics cross-refer to a separate entry for that word elsewhere in the glossary.

adaptation A feature of an organism enabling it to survive and reproduce in its natural environment better than if it lacked the feature.

adaptive topography A graph of the average *fitness* of a *population* in relation to the frequencies of *genotypes* in it. Peaks on the landscape correspond to genotypic frequencies at which the average fitness is high, valleys correspond to genotypic frequencies at which the average fitness is low. It is also called an adaptive landscape, or a fitness surface.

allele A variant of a single *gene*, inherited at a particular genetic *locus*; it is a particular sequence of *nucleotides*, coding for *messenger RNA*.

allometry The relation between the size of an organism and the size of any of its parts: for example, there is an allometric relation between brain size and body size, such that (in this case) animals with bigger bodies have bigger brains. Allometric relations can be studied during the growth of a single organism, between different organisms within a *species*, or between organisms in different species.

allopatric speciation Speciation via geographically separated *populations*.

allopatry Living in separate places. Compare *sympatry*.

amino acid A unit molecular building block of *proteins*. A protein is a chain of amino acids in a certain sequence. There are 20 main amino acids in the proteins of living things, and the properties of a protein are determined by its particular amino acid sequence.

amniotes The group comprising reptiles, birds, and mammals. They all develop through an embryo that is enclosed within a membrane called an amnion. The amnion surrounds the embryo with a watery substance, and is probably an *adaptation* for breeding on land.

analogy A term mainly not used in this edition of the text, but close in meaning to *homoplasy*. That is, a *character* shared by a set of *species* but not present in their common ancestor — a convergently evolved character. Some biologists distinguish between homoplasies and analogies. In Chapter 3, the term is used to contrast with pre-evolutionary *homology*. An analogy is

then a structure like a bird wing and an insect wing. It is similar for functional reasons and not deeply similar in structure. Compare *homology*.

anatomy (i) The structure itself of an organism, or one of its parts. (ii) The science that studies those structures.

ancestral homology *Homology* that evolved before the common ancestor of a set of *species*, and is present in other species outside that set of species. Compare *derived homology*.

area cladogram A branching diagram (or *phylogeny*) of a set of *species* (or other *taxa*) showing the geographic areas they occupy. According to the theory of vicariance biogeography, the branching diagram represents the history of range splits (probably driven by geological processes such as continental drift) in the ancestry of the species.

artificial selection Selective breeding, carried out by humans, to alter a *population*. The forms of most domesticated and agricultural *species* have been produced by artificial selection; it is also an important experimental technique for studying *evolution*.

asexual reproduction The production of offspring by virgin birth or by vegetative reproduction; that is, reproduction without sexual fertilization of eggs.

assortative mating Tendency of like to mate with like. It can be for a certain *genotype* (e.g., individuals with genotype *AA* tend to mate with other individuals of genotype *AA*) or *phenotype* (e.g., tall individuals mate with other tall individuals).

atomistic (as applied to theory of inheritance) Inheritance in which the entities controlling heredity are relatively distinct, permanent, and capable of independent action. *Mendelian inheritance* is an atomistic theory because, in it, inheritance is controlled by distinct *genes*.

autosome Any *chromosome* other than a *sex chromosome*.

base The *DNA* is a chain of *nucleotide* units, and each unit consists of a backbone made of a sugar and a phosphate group, with a nitrogenous base attached. The base in a unit is one of adenine (A), guanine (G), cytosine (C), or thymine (T). In *RNA*, uracil (U) is used instead of thymine. A and G belong to the chemical class called *purines*; C, T, and U are *pyrimidines*.

Batesian mimicry A kind of *mimicry* in which one non-poisonous *species* (the Batesian mimic) mimics another poisonous species.

Biogenetic law The name given by Haeckel to *recapitulation*.

biological species concept A concept of *species*, according to which a species is a set of organisms that can interbreed with each other. Compare *ecological species concept, phenetic species concept, recognition species concept*.

biometrics The quantitative study of the *characters* of organisms.

blending inheritance The historically influential, but factually erroneous, theory that organisms contain a blend of their parents' hereditary factors and pass that blend on to their offspring. Compare *Mendelian inheritance*. (And see Section 2.9, p. 37.)

character Any recognizable trait, feature, or property of an organism.

character displacement The increased difference between two closely related *species* where they live in the same geographic region (*sympatry*) as compared with where they live in different geographic regions (*allopatry*). Explained by the relative influences of intra- and interspecific competition in sympatry and allopatry.

chloroplast A structure (or *organelle*) found in some cells of plants that is involved in photosynthesis.

chromosomal inversion See *inversion*.

chromosome A structure in the cell *nucleus* that carries the *DNA*. At certain times in the cell cycle they are visible as string-like entities. Chromosomes consist of DNA with various *proteins*, particularly histones, bound to it.

clade A set of *species* descended from a common ancestral species. It is a synonym of *monophyletic group*.

cladism Phylogenetic *classification*. The members of a group in a cladistic classification share a more recent common ancestor with each other than with the members of any other group. A group at any level in the classificatory hierarchy, such as "family," is formed by combining a subgroup (at the next lowest level, perhaps the genus in this case) with that other subgroup it shares its most recent common ancestor with. Compare *evolutionary classification, phenetic classification*.

classification The arrangement of organisms into hierarchical groups. Modern biological classifications are *Linnaean* and classify organisms into *species*, genus, family, order, class, phylum, kingdom, and certain intermediate categorical levels. *Cladism, evolutionary classification*, and *phenetic classification* are three methods of classification.

cline A geographic gradient in the frequency of a *gene*, or in the average value of a *character*.

clock See *molecular clock*.

clone A set of genetically identical organisms asexually reproduced from one ancestral organism.

coadaptation The beneficial interaction between a number of: (i) *genes* at different loci within an organism; (ii) different parts of an organism; or (iii) organisms belonging to different *species*.

codon A triplet of *bases* (or *nucleotides*) in the *DNA* coding for one *amino acid*. The relation between codons and amino acids is given by the *genetic code*. The triplet of bases that is complementary to a codon is called an anticodon; conventionally, the triplet in the *messenger RNA* is called the codon and the triplet in the *transfer RNA* is called the anticodon.

coevolution *Evolution* in two or more *species* in which the evolutionary changes of each species influence the evolution of the other species.

comparative biology The study of patterns among more than one *species*.

comparative method The study of *adaptation* by comparing many *species*.

convergence The process by which a similar *character* evolves independently in two *species*. It is a synonym for *homoplasy*, i.e., an instance of a convergently evolved character is a similar character in two species that was not present in their common ancestor.

co-option Evolutionary change, or addition, of a function of a molecule or a part of an organism. For instance, a molecule with one function may, following *duplication*, gain a second, different function. Compare *preadaptation*.

Cope's rule An evolutionary increase in body size over geological time in a *lineage* of *populations*.

creationism See *separate creation*.

crossing-over The process during *meiosis* in which the *chromosomes* of a *diploid* pair exchange genetic material. It is visible in the light microscope. At a genetic level, it produces *recombination*.

cytoplasm The region of a *eukaryotic cell* outside the *nucleus*.

Darwinism Darwin's theory, that *species* originated by *evolution* from other species and that evolution is mainly driven by *natural selection*. Differs from *neo-Darwinism* mainly in that Darwin did not know about *Mendelian inheritance*.

derived homology *Homology* that first evolved in the common ancestor of a set of *species* and is unique to them. Compare *ancestral homology*.

diploid Having two sets of *genes* and two sets of *chromosomes* (one from the mother, one from the father). Many common *species*, including humans, are diploid. Compare *haploid, polyploid*.

directional selection *Selection* causing a consistent directional change in the form of a *population* through time, e.g., selection for larger body size.

disruptive selection *Selection* favoring forms that deviate in either direction from the *population* average. Selection favors forms that are larger or smaller than average, but works against the average forms between.

distance In *taxonomy*, this refers to the quantitatively measured difference between the phenetic appearance of two groups of individuals, such as *populations* or *species* (phenetic distance), or the difference in their *gene* frequencies (genetic distance).

DNA Deoxyribose nucleic acid, the molecule that controls inheritance.

dominance (genetic) An *allele* (*A*) is dominant if the *phenotype* of the *heterozygote Aa* is the same as the *homozygote AA*. The allele *a* does not influence the heterozygote's phenotype and is called *recessive*. An allele may be partly, rather than fully, dominant: then the heterozygous phenotype is nearer to, rather than identical with, the homozygote of the dominant allele.

drift Synonym of *genetic drift*.

duplication The occurrence of a second copy of a particular sequence of *DNA*. The duplicate sequence may appear next to the original, or be copied elsewhere into the *genome*. When the duplicated sequence is a *gene*, the event is called gene duplication. A distinction exists between the *mutation* that creates a duplication, and the evolutionary process that substitutes a duplicated form of a gene. The word is sometimes used to refer to the mutation, and sometimes to the combination of the mutation and its *substitution*.

ecological genetics Study of *evolution* in action in nature, by a combination of fieldwork and laboratory genetics.

ecological species concept A concept of *species*, according to which a species is a set of organisms adapted to a particular, discrete set of resources (or "*niche*") in the environment. Compare *biological species concept*, *phenetic species concept*, *recognition species concept*.

electrophoresis A method of distinguishing entities according to their motility in an electric field. In evolutionary biology, it has been mainly used to distinguish different forms of *proteins*. The electrophoretic motility of a molecule is influenced by its size and electric charge.

epistasis An interaction between the *genes* at two or more loci, such that the *phenotype* differs from what would be expected if the loci were expressed independently.

eukaryote Made up of *eukaryotic cells*. Almost all multicellular organisms are eukaryotic. Compare *prokaryote*.

eukaryotic cell A cell with a distinct *nucleus*.

eutherian (Eutheria) One of two or three major subdivisions of mammals. The other two are Prototheria (echidnas) and Metatheria (marsupials). Most familiar mammals (at least, outside Australia) are eutherians: cats, elephants, dolphins, monkeys, and rodents are all eutherians.

evo-devo The term used for research on the relation between individual development (from egg to adult) and *evolution*.

evolution Darwin defined it as "descent with modification." It is the change in a *lineage* of *populations* between generations.

evolutionary classification The method of *classification* using both *cladistic* and *phenetic* classificatory principles. To be exact, it permits *paraphyletic groups* (which are allowed in phe-

netic but not in cladistic classification) and *monophyletic groups* (which are allowed in both cladistic and phenetic classification) but excludes *polyphyletic groups* (which are banned from cladistic classification but permitted in phenetic classification).

exon The *nucleotide* sequences of some *genes* consist of parts that code for *amino acids*, and other parts interspersed among them that do not code for amino acids. The coding parts, which are translated, are called exons; the interspersed non-coding parts are called *introns*.

fitness The average number of offspring produced by individuals with a certain *genotype*, relative to the number produced by individuals with other genotypes. When genotypes differ in fitness because of their effects on survival, fitness can be measured as the ratio of a genotype's frequency among the adults divided by its frequency among individuals at birth.

fixation A *gene* has achieved fixation when its frequency has reached 100% in the *population*.

fixed (i) In *population genetics*, a *gene* is "fixed" when it has a frequency of 100%. (ii) In the theory of *separate creation*, *species* are described as "fixed" in the sense that they are believed not to change their form, or appearance, through time.

founder effect The loss of genetic variation when a new colony is formed by a very small number of individuals from a larger *population*.

frequency-dependent selection *Selection* in which the *fitness* of a *genotype* (or *phenotype*) depends on its frequency in the *population*.

gamete The *haploid* reproductive cells that combine at fertilization to form the *zygote*: sperm (or pollen) in the male and eggs in females.

gene Sequence of *nucleotides* coding for a *protein* (or, in some cases, part of a protein).

gene duplication See *duplication*.

gene family A set of related *genes* occupying various loci in the *DNA*, almost certainly formed by *duplication* of an ancestral gene, and having recognizably similar sequence. The globin gene family is an example.

gene flow The movement of *genes* into, or through, a *population* by interbreeding or by migration and interbreeding.

gene frequency The frequency in the *population* of a particular *gene* relative to other genes at its *locus*. Expressed as a proportion (between 0 and 1) or percentage (between 0% and 100%).

gene pool All the *genes* in a *population* at a particular time.

genetic code The code relating *nucleotide* triplets in the *messenger RNA* (or *DNA*) to *amino acids* in the proteins. It has been decoded (see Table 2.1, p. 26).

genetic distance See *distance*.

genetic drift Random changes in *gene* frequencies in a *population*.

genetic load A reduction in the average *fitness* of the members of a *population* because of the deleterious *genes*, or gene combinations, in the population. It has many particular forms, such as "mutational load," "segregational load," or "recombinational load."

genetic locus See *locus*.

genome The full set of *DNA* in a cell or organism.

genomics The study of *genomes*, particularly using *DNA* sequence data.

genotype The set of two *genes* at a *locus* possessed by an individual.

geographic isolation See *reproductive isolation*.

geographic speciation See *allopatric speciation*.

germ line See *germ plasm*.

germ plasm The reproductive cells in an organism that produce the *gametes*. All the cells in an organism can be divided into the *soma* (the cells that ultimately die) and the germ cells (that are perpetuated by reproduction). The cell line within the body that form the gametes is called the germ line.

group selection *Selection* operating between groups of individuals rather than between individuals. It would produce attributes beneficial to a group in competition with other groups, rather than attributes beneficial to individuals.

haploid The condition of having only one set of *genes* or chromsomes. In normally *diploid* organisms such as humans, only the *gametes* are haploid.

haplotype The set of *genes* at more than one *locus* inherited by an individual from one of its parents. It is the multilocus analog of an *allele*.

Hardy–Weinberg ratio The ratio of *genotype* frequencies that evolve when mating is random and neither *selection* nor *drift* are operating. For two *alleles* (A and a) with frequencies p and q, there are three genotypes AA, Aa, and aa; and the Hardy–Weinberg ratio for the three is $p^2 AA : 2pq Aa : q^2 aa$. It is the starting point for much of the theory of *population genetics*.

heritability Broadly, the proportion of variation (more strictly *variance*) in a phenotypic *character* in a *population* that is due to individual differences in *genotypes*. Narrowly, the proportion of variation (more strictly variance) in a phenotypic character in a population that is due to individual genetic differences that will be inherited in the offspring.

heterogametic Sex with two different *sex chromosomes* (males in mammals, because they are XY). Compare *homogametic*.

heterozygosity A measure of the amount of genetic diversity in a *population*. For a population in *Hardy–Weinberg equilibrium* it equals the proportion of individuals in a population that are *heterozygotes*.

heterozygote Individual having two different *alleles* at a genetic *locus*. Compare *homozygote*.

heterozygote advantage The condition in which the *fitness* of a *heterozygote* is higher than the fitness of either *homozygote*.

homeostasis (developmental) A self-regulating process in development, such that the organism grows up to have much the same form independently of the external influences it experiences while growing up.

homeotic mutation *Mutation* causing one structure of an organism to grow in the place appropriate to another. For example, in the mutation called "antennapedia" in the fruitfly, a foot grows in the antennal socket.

homogametic Sex with two of the same kind of *sex chromosomes* (females in mammals, because they are XX). Compare *heterogametic*.

homology A *character* shared by a set of *species* and present in their common ancestor. Compare *homoplasy*. (Some molecular biologists, when comparing two sequences, call the corresponding sites "homologous" if they have the same *nucleotide* — regardless of whether the similarity is evolutionarily shared from a common ancestor or convergent; they likewise talk about percent *homology* between the two sequences. Homology then simply means similarity. This usage is frowned on by many evolutionary biologists, but is established in much of the molecular literature.)

homoplasy A *character* shared between two *species* but not present in their common ancestor. Homoplasies can arise by *convergence* (powered by *natural selection*), or by reversion, or by *random drift* in *DNA* sequences. Compare *homology*.

homozygote An individual having two copies of the same *allele* at a genetic *locus*. It is also sometimes applied to larger genetic entities, such as a whole *chromosome*: a homozygote is then an individual having two copies of the same chromosome.

***Hox* genes** A group of *genes* important in development. They act as regional specifiers, and thus help determine which kind of cells differentiate in the various regions of a body.

hybrid The offspring of a cross between two *species*.

idealism A philosophical theory that there are fundamental non-material "ideas," "plans," or "forms" underlying the phenomena we observe in nature. It has been historically influential in *classification*.

inheritance of acquired characters Historically influential but factually erroneous theory that an individual inherits *characters* that its parents acquired during their lifetimes.

intron The *nucleotide* sequences of some *genes* consist of parts that code for *amino acids*, and other parts interspersed among them that do not code for amino acids. The interspersed non-coding parts, which are not translated, are called introns; the coding parts are called *exons*.

inversion An event (or the product of the event) in which a sequence of *nucleotides* in the *DNA* are reversed, or inverted. Sometimes inversions are visible in the structure of the *chromosomes*.

isolating mechanism Any mechanism, such as a difference between *species* in courtship behavior or breeding season, that results in *reproductive isolation* between the species.

isolation Synonym for *reproductive isolation*.

Lamarckian inheritance A historically misleading synonym for *inheritance of acquired characters*.

larva (and **larval stage**) Prereproductive stage of many animals; the term is used particularly when the immature stage has a different form from the adult.

lineage An ancestor–descendant sequence of: (i) *populations*; (ii) cells; or (iii) *genes*.

linkage disequilibrium The condition in which the *haplotype* frequencies in a *population* deviate from the values they would have if the *genes* at each *locus* were combined at random. (When there is no deviation, the population is said to be in linkage equilibrium.)

linked Refers to *genes* present on the same *chromosome*.

Linnaean classification Hierarchical method of naming classificatory groups, invented by the eighteenth century Swedish naturalist Carl von Linné, or Linnaeus. Each individual is assigned to a *species*, genus, family, order, class, phylum, and kingdom, and some intermediate classificatory levels. Species are referred to by a Linnaean binomial of its genus and species, such as *Magnolia grandiflora*. Universally used by educated persons.

locus The location in the *DNA* occupied by a particular *gene*.

macroevolution *Evolution* on the grand scale: the term refers to events above the *species* level. The origin of a new higher group, such as the vertebrates, would be an example of a macroevolutionary event.

macromutation The *mutation* of a large phenotypic effect; one that produces a *phenotype* well outside the range of variation previously existing in the *population*.

mean The average of a set of numbers. For example, the mean of 6, 4, and 8 is $(6 + 4 + 8)/3 = 6$.

meiosis A special kind of cell division that occurs during the reproduction of *diploid* organisms to produce the *gametes*. The double set of *genes* and *chromosomes* of the normal diploid cells is reduced during meiosis to a single *haploid* set. *Crossing-over* and therefore *recombination* occur during a phase of meiosis.

Mendelian inheritance The mode of inheritance of all *diploid species*, and therefore of nearly all multicellular organisms. Inheritance is controlled by *genes*, which are passed on to the offspring in the same form as they were inherited from the previous generation. At each *locus*, an individual has two genes, one inherited from its father and the other from its mother. The two genes are represented in equal proportions in its *gametes*.

messenger RNA (mRNA) The kind of *RNA* produced by *transcription* from the *DNA* and which acts as the message that is decoded to form *proteins*.

microevolution Evolutionary changes on the small scale, such as changes in *gene* frequencies within a *population*.

mimicry A case in which one *species* looks more or less similar to another species. See *Batesian mimicry*, *Müllerian mimicry*.

mitochondrion A kind of *organelle* in *eukaryotic cells*; mitochondria burn the digested products of food to produce energy. They contain *DNA* coding for some mitochondrial *proteins*.

mitosis Cell division. All cell division in multicellular organisms is by mitosis except for the special division called *meiosis* that generates the *gametes*.

modern synthesis Synthesis of *natural selection* and *Mendelian inheritance*. Also called *neo-Darwinism*.

molecular clock The theory that molecules evolve at an approximately constant rate. The difference between the form of a molecule in two *species* is then proportional to the time since the species diverged from a common ancestor, and molecules become of great value in the inference of *phylogeny*.

monophyletic group A set of *species* containing a common ancestor and all its descendants.

morphology The study of the form, shape, and structure of organisms.

Müllerian mimicry A kind of *mimicry* in which two poisonous *species* evolve to look like each other.

mutation When parental *DNA* is copied to form a new DNA molecule, it is normally copied exactly. A mutation is any change in the new DNA molecule from the parental DNA molecule. Mutations may alter single *bases*, or *nucleotides*, short stretches of bases, or parts of or whole *chromosomes*. Mutations can be detected both at the DNA level or the phenotypic level.

natural selection The process by which the forms of organisms in a *population* that are best adapted to the environment increase in frequency relative to less well adapted forms over a number of generations.

neo-Darwinism (i) Darwin's theory of *natural selection* plus *Mendelian inheritance*. (ii) The larger body of evolutionary thought that was inspired by the unification of natural selection and Mendelism. A synonym of *modern synthesis*.

neutral drift Near synonym of *genetic drift*.

neutral mutation *Mutation* with the same *fitness* as the other *allele* (or alleles) at its *locus*.

neutral theory (and neutralism) A theory that most *evolution* at the molecular level occurs by *neutral drift*.

niche The ecological role of a *species*: the set of resources it consumes, and habitats it occupies.

nucleotide A unit building block of *DNA* and *RNA*. A nucleotide consists of a sugar and phosphate backbone with a *base* attached.

nucleus The region of *eukaryotic cells* containing the *DNA*.

numerical taxonomy In general, any method of *taxonomy* using numerical measurements; in particular, it often refers to *phenetic classification* using large numbers of quantitatively measured *characters*.

organelle Any of a number of distinct small structures found in the *cytoplasm* (and so outside the *nucleus*) of eukaryotic cells, e.g., a *mitochondrion* or *chloroplast*.

orthogenesis The erroneous idea that *species* tend to evolve in a *fixed* direction because of some inherent force driving them to do so.

paleobiology The biological study of fossils.

paleontology The scientific study of fossils.

panmixis *Random mating* throughout a *population*.

parapatric speciation Speciation in which the new *species* forms from a *population* contiguous with the ancestral species' geographic range.

paraphyletic group A set of *species* containing an ancestral species together with some, but not all, of its descendants. The species included in the group are those that have continued to resemble the ancestor; the excluded species have evolved relatively rapidly and no longer resemble their ancestor.

parsimony The principle of phylogenetic reconstruction in which the *phylogeny* of a group of *species* is inferred to be the branching pattern requiring the smallest number of evolutionary changes.

parthenogenesis Reproduction by virgin birth, a form of *asexual reproduction*.

particulate (as a property of theory of inheritance) Synonym of *atomistic*.

peripheral isolate speciation A form of *allopatric speciation* in which the new *species* is formed from a small *population* isolated at the edge of the ancestral population's geographic range. Also called peripatric speciation.

phenetic classification A method of *classification* in which *species* are grouped together with other species that they most closely resemble phenotypically.

phenetic species concept A concept of *species*, according to which a species is a set of organisms that are phenetically similar to one another. Compare *biological species concept*, *ecological species concept*, *recognition species concept*.

phenotype The *characters* of an organism, whether due to the *genotype* or environment.

phylogeny "Tree of life": a branching diagram showing the ancestral relations among *species*, or other *taxa*. It shows, for each species, which other species it shares its most recent common ancestor with.

plan of nature A philosophical theory that nature is organized according to a plan. It has been influential in *classification*, and is a kind of *idealism*.

plankton The microscopic animals and plants that float in the water near the surface. In the top few feet of water, both in the sea and in freshwater, small plants can photosynthesize and there is abundant microscopic life. Many organisms that are sessile as adults disperse by means of a planktonic *larval stage*.

plasmid A genetic element that exists (or can exist) independently of the main *DNA* in the cell. In bacteria, plasmids can exist as small loops of DNA and be passed between cells independently.

Poisson distribution The frequency distribution for number of events per unit time, when the number of events is determined randomly and the probability of each event is low.

polymorphism A condition in which a *population* possesses more than one *allele* at a *locus*. Sometimes it is defined as the condition of having more than one allele with a frequency of over 5% in the population.

polyphyletic group The set of *species* descended from more than one common ancestor. The ultimate common ancestor of all the species in the group is not a member of the polyphyletic group.

polyploid An individual containing more than two sets of *genes* and *chromosomes*.

polytypic A *species* with many distinct forms (that is, the species does not simply show continuous, or normal, variation).

population A group of organisms, usually a group of sexual organisms that interbreed and share a *gene pool*.

population genetics The study of processes influencing *gene frequencies*.

postzygotic isolation *Reproductive isolation* in which a *zygote* is successfully formed but then either fails to develop or develops into a sterile adult. Donkeys and horses are postzygotically isolated from each other: a male donkey and a female horse can mate to produce a mule, but the mule is sterile.

prezygotic isolation *Reproductive isolation* in which the two *species* never reach the stage of successful mating, and thus no *zygote* is formed. Examples would be species with different breeding seasons or courtship displays, and which therefore never recognize each other as potential mates.

prokaryote Made up of *prokaryotic cells*. Bacteria and some other simple organisms are prokaryotic. In classificatory terms, the group of all prokaryotes is *paraphyletic*. Compare *eukaryote*.

prokaryotic cell A cell without a distinct *nucleus*.

protein A molecule made up of a sequence of *amino acids*. Many of the important molecules in a living thing are proteins; all enzymes, for example, are proteins.

pseudogene A sequence of *nucleotides* in the *DNA* that resembles a *gene* but is non-functional for some reason.

purine A kind of *base*; in *DNA*, adenine (A) and guanine (G) are purines.

pyrimidine A kind of *base*; in *DNA*, cytosine (C) and thymine (T), and in *RNA*, cytosine (C) and uracil (U) are pyrimidines.

quantitative character A *character* showing continuous variation in a *population*.

random drift Synonym of *genetic drift*.

random mating A mating pattern where the probability of mating with another individual of a particular *genotype* (or *phenotype*) equals the frequency of that genotype (or phenotype) in the *population*.

recapitulation Partly or wholly erroneous theory that an individual, during its development, passes through a series of stages corresponding to its successive evolutionary ancestors. An individual thus develops by "climbing up its family tree."

recessive An *allele* (A) is recessive if the *phenotype* of the *heterozygote Aa* is the same as the *homozygote* (*aa*) for the alternative allele *a* and different from the homozygote for the recessive (*AA*). The allele *a* controls the heterozygote's phenotype and is called *dominant*. An allele may be partly, rather than fully, recessive: then the heterozygous phenotype is nearer to, rather than identical with, the homozygote for the dominant allele.

recognition species concept A concept of *species*, according to which a species is a set of organisms that recognize one another as potential mates; they have a shared mate recognition system. Compare *biological species concept*, *ecological species concept*, *phenetic species concept*.

recombination An event, occurring by the *crossing-over* of *chromosomes* during *meiosis*, in which *DNA* is exchanged between a pair of chromosomes. Thus two *genes* that were previously unlinked, being on separate chromosomes, can become *linked* because of recombination, and vice versa. Linked genes may become unlinked.

refuge (plural sometimes **refugia**) The contracted biogeographic range of *species* during ice ages or other adverse times.

reinforcement An increase in *reproductive isolation* between incipient *species* by *natural selection*. Natural selection can only directly favor an increase in *prezygotic isolation*; reinforcement therefore amounts to selection for *assortative mating* between the incipiently speciating forms.

reproductive character displacement An increased *reproductive isolation* between two closely related *species* where they live in the same geographic region (*sympatry*) as compared with where they live in separate geographic regions. It involves a kind of *character displacement*, in which the character concerned influences reproductive isolation, not ecological competition.

reproductive isolation Two *populations*, or individuals of opposite sexes, are reproductively isolated from each other if they cannot together produce fertile offspring.

ribosomal RNA (rRNA) The kind of *RNA* that constitutes the *ribosomes* and provides the site for *translation*.

ribosome The site of *protein* synthesis (or *translation*) in the cell, mainly consisting of *ribosomal RNA*.

ring species A situation in which two reproductively isolated *populations* (see *reproductive isolation*), living in the same region, are connected by a geographic ring of populations that can interbreed.

RNA Ribonucleic acid. *Messenger RNA*, *ribosomal RNA*, and *transfer RNA* are its three main forms. They act as the intermediaries by which the hereditary code of *DNA* is converted into *proteins*. In some *viruses*, RNA is itself the hereditary molecule.

secondary Such expressions as "secondary contact" or "secondary *reinforcement*" mean that two *species*, or near species, have been geographically separated in the past and have met up again. The term usually alludes to the theory of *allopatric speciation* and implies that *sympatric speciation* is not at work.

selection A short-hand synonym of *natural selection*.

selectionism The theory that some class of evolutionary events, such as molecular or phenotypic changes, have mainly been caused by *natural selection*.

selective sweep An increase in homozygosity (that is, genetic uniformity) at neighboring *nucleotide* sites, when *natural selection* fixes a favored nucleotide variant. The increased homozygosity is due to hitch-hiking, because there is little *recombination* between neighboring nucleotide sites. It can be used to test for the recent action of selection in genomic sequences.

separate creation The theory that *species* have separate origins and never change after their origin. Most versions of the theory of separate creation are religiously inspired and suggest that the origin of species is by supernatural action.

sex chromosome A *chromosome* that influences sex determination. In mammals, including humans, the X and Y chromosomes are the sex chromosomes (females are XX, males XY). Compare *autosome*.

sexual selection The selection on mating behavior, either through competition among members of one sex (usually males) for access to members of the other sex or through choice by members of one sex (usually females) for certain members of the other sex. In sexual selection, individuals are favored by their *fitness* relative to other members of the same sex, whereas *natural selection* works on the fitness of a *genotype* relative to the whole *population*.

soma (and **somatic cells**) All the cells in the body except the reproductive cells (or *germ plasm*): that is, the skin, bone, blood, nerve cells and so on.

spacer region A sequence of *nucleotides* in the *DNA* between coding *genes*.

species An important classificatory category, which can be variously defined by the *biological species concept*, *ecological species concept*, *phenetic species concept*, and *recognition species concept*. The biological species concept, according to which a species is a set of interbreeding organisms, is the most widely used definition, at least by biologists who study vertebrates. A particular species is referred to by a *Linnaean* binomial, such as *Homo sapiens* for human beings.

stabilizing selection *Selection* tending to keep the form of a *population* constant: individuals with the *mean* value for a *character* have high *fitness*, those with extreme values have low fitness.

stepped cline A *cline* with a sudden change in *gene* (or *character*) frequency.

substitution The evolutionary replacement of one *allele* by another in a *population*.

sympatric speciation Speciation via *populations* with overlapping geographic ranges.

sympatry Living in the same geographic region. Compare *allopatry*.

systematics A near synonym of *taxonomy*.

taxon (plural **taxa**) Any named taxonomic group, such as the family Felidae, or genus *Homo*, or *species Homo sapiens*. A formally recognized group, as distinct from any other group (such as the group of herbivores or tree-climbers).

taxonomy The theory and practice of biological *classification*.

tetrapod A member of the group made up of amphibians, reptiles, birds, and mammals.

transcription The process by which *messenger RNA* is read off the *DNA* forming a *gene*.

transfer RNA (tRNA) The kind of *RNA* that brings the *amino acids* to the *ribosomes* to make *proteins*. There are 20 kinds of tRNA molecules, one for each of the 20 main amino acids. A tRNA molecule has an amino acid attached to it, and has the anticodon corresponding to that amino acid in another part of its structure. In protein synthesis, each *codon* in the *messenger RNA* combines with the appropriate tRNA's anticodon, and the amino acids are thus arranged in order to make the protein.

transformism The evolutionary theory of Lamarck, in which changes occur within a *lineage* of *populations*, but in which lineages to do not split and do not go extinct.

translation The process by which a *protein* is manufactured at a *ribosome*, using *messenger RNA* code and *transfer RNA* to supply the *amino acids*.

transition A *mutation* changing a *purine* into the other purine, or a *pyrimidine* into the other pyrimidine (i.e., changes from A to G or vice versa, and changes from C to T or vice versa).

transversion A *mutation* changing a *purine* into a *pyrimidine* or vice versa (i.e., changes from A or G to C or T, and changes from C or T to A or G).

typology (i) The definition of classificatory groups by phenetic similarity to a "type" specimen. A *species*, for example, might be defined as all individuals less than *x* phenetic units from the species' type. (ii) The theory that distinct "types" exist in nature, perhaps because they are part of some *plan of nature* (see also *idealism*). The type of the species is then the most important form of it, and variants around that type are "noise" or "mistakes." *Neo-Darwinism* opposes typology because in a *gene pool* no one variant is any more important than any other.

unequal crossing-over *Crossing-over* in which the two *chromosomes* do not exchange equal lengths of *DNA*; one receives more than the other.

variance A measure of how variable a set of numbers are. Technically, it is the sum of squared deviations from the *mean* divided by $n - 1$ (where n is the number of numbers in the sample). Thus to find the variance of the set of numbers, 4, 6, and 8, we first calculate the mean, which is 6; then sum the squared deviations from the mean $(4 - 6)^2 + (6 - 6)^2 + (8 - 6)^2$, which comes to 8; and divide by $n - 1$ (which is 2 here). The variance of the three numbers is $8/2 = 4$. The more variable a set of numbers are, the higher the variance. The variance of a set of identical numbers (such as 6, 6, and 6) is zero.

virus A kind of intracellular parasite that can only replicate inside a living cell. In its dispersal stage between host cells a virus consists only of nucleic acid that codes for a small number of *genes*, surrounded by a *protein* coat. (Less formally, on Medawar's definition, a virus is "a piece of bad news wrapped in a protein.")

vitamin A member of a chemically heterogeneous class of organic compounds that are essential, in small quantities, for life.

wild type The *genotype*, or *phenotype*, out of a set of genotypes, or phenotypes, of a *species* that is found in nature. The expression is mainly used in laboratory genetics, to distinguish rare mutant forms of a species from the lab stock of normal individuals.

wobble The ability of the third *base* in some anticodons of *transfer RNA* to bond with more than one kind of base in the complementary position in the *messenger RNA codon*.

zygote The cell formed by the fertilization of male and female *gametes*.

Answers to Study and Review Questions

Answers are given below for calculations, problems, and short answers; a reference to the relevant section(s) of the chapter is given for longer answers and definitions or explanations of technical terms; or a reference to further reading is given for topics not explicitly discussed in the text.

Chapter 1

1. See Section 1.1.
2. Adaptation.
3. The popular concept had evolution as progressive, with species ascending a one-dimensional line from lower forms to higher. Evolution in Darwin's theory is tree-like and branching, and no species is any "higher" than any other — forms are adapted only to the environments they live in.
4. Darwin's theory of natural selection and Mendel's theory of heredity.

Chapter 2

1. The terms are explained in the chapter.
2. (i) 100% AA; (ii) 1 AA : 1 Aa; (iii) 1 AA : 2 Aa : 1 aa; (iv) 100% AB/AB; and (v) 100% AB/AB.
3. As fractions: (i) 1/4 AB, 1/4 Ab, 1/4 aB, 1/4 ab; and (ii) $(1-r)/2\ AB, (1-r)/2\ ab, r/2\ Ab, r/2\ aB$. As ratios: (i) 1 AB : 1 Ab : 1 aB : 1 ab; and (ii) $(1-r)\ AB : (1-r)\ ab : r\ Ab : r\ aB$.

Chapter 3

1. Approximately the species level; the pigeon example might stretch it to genera, but higher categories do not evolve in human lifetimes.
2. (i) If you look at any one time and place, living things usually fall into distinct, recognizable groups that could be called "kinds." (ii) If you look over a range of space (if the "kinds" in question are species) the kinds break down; if you look through a range of times (if the "kinds" are species or any higher category) the kinds also break down. The differences between higher categories can also be broken down by studying the full range of diversity on Earth: you might think that plants and animals are less clear categories after studying the range of unicellular organisms.
3. (a), (b), and (d) are homologies; (c) is an analogy.

4. It is an accident in the sense that other codes with the same four letters could work equally well. It is frozen in the sense that changes in it are selected against. See Section 3.8.
5. This is intended more as a discussion topic: for the idea see Section 3.9.
6. Because a form would have existed in time before the series of fossils (of vertebrates from fish to mammals) that can be strongly argued to be its ancestors.

Chapter 4

1.

Age interval (days)	Number (as density/m²) surviving to day x	Proportion of original cohort surviving to day x	Proportion of original cohort dying during interval
0–250	100	100	0.9989
251–500	0.11	0.11	0.273
501–750	0.08	0.08	0.75
751–1,000	0.06	0.06	–

2. (a) See Section 4.2. (b) In technical terms, drift (on which see Chapter 6). The gene frequencies would change between generations because there is heritability (condition 2) and some individuals produce more offspring than others (condition 3). But if the differences in reproduction are not systematically associated with some character or other, the changes in gene frequency between generations will be random or directionless. (c) No evolution at all. If the character conferring higher than average fitness is not inherited by the individual's offspring, natural selection cannot increase its frequency in the population.
3. The requirements of inheritance and association between high reproductive success and some character have also to be met.
4. The mechanism has to: (i) perceive the change in environment; (ii) work out what the appropriate adaptation is to the new environment; (iii) alter the genes in the germ line in a manner to code for the new adaptation. (i) is possible; (ii) could vary from possible in a case such as simple camouflage to impossible in a case requiring a new complex adaptation, such as the adaptations for living on land of the

first terrestrial tetrapods; and (iii) would contradict what is known about genetics and it is difficult to see how it could be done. The mechanism would have to work backwards from the new phenotype (something like a long neck in a giraffe) to deduce the needed genetic changes, even though the phenotype was produced by multiple, interacting genetic and environmental effects.

5. (a) Directional selection (for smaller brains); (b) stabilizing selection; and (c) no selection.

6. (a) Here are two arguments. (i) If every pair produced two offspring, natural selection would favor new genetic variants that produced three, or more, offspring. After the more fecund form had spread through the population the average would still be two but the greater competition among individuals to survive would lead to variation in the success of the broods of different parents. (ii) Random accidents alone will guarantee that some individuals fail to breed; then for the average to be two, as it must be for any population that is reasonably stable in the long term, all successfully reproducing individuals will produce more than two offspring. (b) Ecologists discuss this in terms of r and K selection, or life history theory: in some environments there is little competition and selection favors producing large numbers of small offspring; whereas in others there is massive competition and selection favors producing fewer offspring and investing a large amount in each. Many other factors can also operate.

Chapter 5

1. Populations 1 and 5 are in Hardy–Weinberg equilibrium; populations 2–4 are not. As to why they are not, in population 4 it looks like AA is lethal, in 2 there may be a heterozygote advantage, and in 3 a heterozygote disadvantage. Population 2 could also be produced by disassortative mating, and 3 by assortative mating. In population 3 there could also be a Wahlund effect. All the deviations are so large that random sampling is unlikely to be the whole explanation.

2. (a) $p^2/(1 - sq^2)$; (b) $p/(1 - sq^2)$; and (c) $1 - sq^2$.

3. $(1/3)(3 - s) = 1 - (s/3)$.

4. If you do it in your head, $s \approx 0.1$. To be exact, $s = 0.095181429619$.

5. That the fitness differences are in survivorship (not fertility), and in particular in survivorship during the life stage investigated in the mark–recapture experiment. (By the way, mark–recapture experiments are also used by ecologists to estimate absolute survival rates: they require the additional assumptions that the animals do not become "trap shy" or "trap happy," and that the mark and release treatment does not reduce survival. These assumptions are not needed when estimating relative survival. However,

we do need the second-order assumption that these factors are the same for all genotypes.)

6. AA 1/2; aa 1/2. The gene frequencies are 0.5 and the Hardy–Weinberg ratio 1/4 : 1/2 : 1/4. The observed to expected ratios are 2/3 : 4/3 : 2/3, which when scaled to a maximum of 1 give the fitnesses 1/2 : 1 : 1/2.

7. (a) 0.5; and (b) 0.5625. You need equation 5.13; $t = 1$. For (a) it looks like this: $0.625 = 0.5 + (0.75 - 0.5)(1 - m)$. And for (b) it looks like this: $x = 0.5 + (0.625 - 0.5)(1 - 0.5)$.

8. (a) The aa genotype is likely to be fixed. If aa mate only among themselves and AA and Aa mate only with AA and Aa, whenever there is an $Aa \times Aa$ mating, some aa progeny are produced, who will subsequently only mate with other aa individuals. (b) Now AA mate only with AA, Aa with Aa, and aa with aa. The homozygous matings preserve their genotypes, but when Aa mate together they produce 1/4 aa and 1/4 AA progeny. In an extreme case, the population diverges into two species, one AA the other aa and the heterozygotes are lost. (c) (i) The dominant allele will be fixed; and (ii) the recessive allele will be fixed.

9. $$p' = \frac{p(1 - s)}{1 - p^2 s - 2pqs}$$

    The denominator can be variously rearranged.

10. $$p^* \approx \sqrt{\frac{m}{s}}$$

    The derivation starts with the equilibrium condition, $p^2 s = qm$. We then note that $q \approx 1$, and $p^2 s \approx m$; divide both sides by s and take square roots.

Chapter 6

1. Either 100% A or 100% a (there is an equal chance of each).

2. See (a) Section 6.1, and (b) Section 6.3.

3. (1) 0.5; (2) 0.5; (3) 0.375; and (4) 0. See Section 6.5.

4. (a) and (b) 10^{-8}. Population size cancels out in the formula for the rate of neutral evolution.

5. (a) $1/(2N)$; and (b) $(1 - (1/(2N)))$.

6. Both manipulations requires substituting $1 - H$ for f and then some canceling and multiplying though by -1 to make the sign positive.

Chapter 7

1. See Figure 7.1a and b.

2. The main observations suggesting neutral molecular evolution are not also seen in morphology. This was discussed for the constancy of evolutionary rates in Section 7.3. The other original observations (for absolute rates and heterozygosities,

and for the relation between rate and constraint) either have not been made, or such observations as there are do not suggest that the problems found in selective explanations for molecular evolution also apply to morphology.

3. (i) A high rate of evolution; (ii) high levels of polymorphism; (iii) a constant rate of evolution; and (iv) functionally more constrained changes have lower evolutionary rates.

4. (i) The molecular clock is not constant enough; (ii) generation time effects seem to differ between synonymous and non-synonymous substitutions; (iii) genetic variation is too similar between species with different population sizes, and heterozygosity is too low in species with high N; and (iv) [not discussed in the text, but for completeness] rates of evolution do not have a predicted relation with levels of genetic variation.

5. (a) The key variable in the neutral explanation is the chance that a mutation is neutral: it is arguably higher for regions with less functional constraints (Section 7.6.2). (b) The key variable in the selective explanation is the chance that a mutation has a small rather than large effect, and so may cause a fine-tuning improvement (Section 7.6.2).

6. No; the main evidence is from codon usage biases (Section 7.11.4).

7. (a) This is a fairly standard figure. Non-synonymous substitutions are rarer, probably because more of them are deleterious than synonymous substitutions. It could be that almost all evolution for both kinds of change is by neutral drift. (b) Either selection has positively favored amino acid changes, elevating the rate of non-synonymous evolution, or selection has been relaxed and non-synonymous changes that are normally disadvantageous are here neutral. (c) It looks like selection is driving amino acid changes in the protein coded for by this gene.

Chapter 8

1.

Population	Frequency of			Value of D
	A_1B_1	A_1	B_1	
1	7/16	1/2	1/2	+3/16
2	1/4	1/2	1/2	0
3	1/9	1/3	1/3	0
4	11/162	1/3	1/3	−7/162

Note that the haplotype frequency is found by the sum of homozygotes plus 1/2 the heterozygotes (as for a gene frequency, Section 5.1). If you have figures in the A_1B_1 frequency column for population 1 such as 11/16 or 14/16 you may have not divided the frequency of A_1B_1/A_1B_2 by 2.

2. Populations 1 and 4 may show fitness epistasis, in which, in population 1, A_1 has higher fitness in combination with B_1 than with B_2, and vice versa in poulation 2. Fitnesses are independent (maybe multiplicative or additive) in populations 2 and 3.

3. Populations 1 and 2 should equilibrate at haplotype frequencies of 1/4, 1/4, 1/4, and 1/4 for the four haplotypes; populations 3 and 4 should equilibrate at 1/9, 2/9, 2/9, and 4/9. Populations 2 and 3 are already at equilibrium and should not change through time; populations 1 and 4 will evolve toward the equilibrium frequencies at a rate determined by the recombination rate between the two loci.

4. Observed heterozygosities are arguably a little on the low side for the neutral theory (Section 7.6); the effect of a section at one locus is on average to reduce heterozygosities at linked loci, producing a net reduction in average heterozygosity through the genome.

5. See Figure 8.8b. The equilibrium is at the top of the hill.

Chapter 9

1. See Section 9.2, particularly Figures 9.3 and 9.4. In statistical theory, the argument is formalized as the central limit theorem.

2. +1. The answer is incomplete without the sign.

3. (a) $V_P = 300/8 = 37.5$; $V_A = 48/8 = 6$; and $h^2 = 6/37.5 = 6.16$. (b) +3: you add the additive effects inherited from each parent.

4. 106.

5. This is not explicitly discussed in the chapter, but see Section 13.x (p. 000). You might predict it will evolve toward a canalizing type relation, as in Figure 9.11b, because then an individual is most likely to have the optimal phenotype.

6. It will go through an intermediate phase with many recombinant genotypes produced by crossing-over between the three initial chromosomes; it should end up with only the chromosomal type that yields the optimal character by means of a homozygote: all +++−−−−.

Chapter 10

1. They cannot explain adaptation. There is no reason except chance why a new genetic variant should be in the direction of improved adaptation, and random chance change will not produce adaptation. If (as in the "Lamarckian" theory) the new genetic variants are in the direction of adaptation, it implies there is some adaptive mechanism behind the production of new variants. Natural selection is the only known theory that could explain such a mechanism.

2. See Figures 10.2 and 10.3.

3. Superficially, yes; but the adaptive information — all the metabolic processes of the cyanobacteria that evolved

photosynthesis—probably evolved in small steps and therefore nothing deep in Fisher's or Darwin's arguments is violated.

4. (a) Many small steps; and (b) some larger initial steps, followed by more small steps — the full distribution may be a negative exponential (Orr 1998).

5. No, it just turned out that way. Sometimes, by chance, an organ that works well in one function turns out to work well in another function after relatively little adjustment.

6. (a) (i) Natural selection, in the form of negative selection. The absent regions represent maladaptive forms which, when they arise as mutations, are selected out. (ii) Developmental constraint. Something about the way the organisms develop embryonically makes it impossible, or at least difficult, for these forms to arise. (b) Four kinds of evidence were mentioned in Section 10.7.3. The kind that most was said about was the use of artificial selection: if the character can be altered, its form is unlikely to be due to constraint.

Chapter 11

1. Many answers are possible, but the main examples in this chapter were: (a) adaptations for finding food; (b) eating as much food as possible to maximize reproductive rate; or cannibalism; or destructive fighting; or producing a 50 : 50 sex ratio in a polygynous species; (c) restraining reproduction to preserve the local food supply; and (d) segregation distortion in which the total fertility of the organism is reduced.

2. You might explain it in terms of two factors, the relative rates of extinction of altruistic and selfish groups and the rate of migration. Or you might reduce them to the one variable *m*, which is the average number of successful emigrants produced by a selfish group during the time the group exists (before it goes extinct). The fate of the model is then determined by whether *m* is greater or smaller than 1 (Section 11.2.5).

3. *b* can be estimated as the number of extra offspring produced by the nests with helpers: 2.2 − 1.24 ≈ 1. But that is produced by 1.7 helpers, giving *b* ≈ 1/1.7 ≈ 0.6. *c* can be estimated either as zero (if the helper has no other option) or as the number of offspring produced by an unhelped pair (if it could breed alone), in which case *c* = 1.24. With *r* = 1/2 it should help if it cannot breed alone but should breed if it can. This way of estimating *b* and *c* does have problems, however.

4. Kin selection applies to a family group, or more generally a group of kin (indeed it is not theoretically necessary that the kin live in groups, though they do have to be able to influence one another's fitness); group selection, at least in the pure sense, applies to groups of unrelated individuals. Kin selection is a plausible process, because the conditions

for an individual to produce more copies of a gene may be improved more by helping relatives than by breeding more itself. Group selection requires more awkward conditions (see question 2!).

5. The average individual is likely to be worse off, as Figure B11.1 illustrates. Competition between individuals reduces the efficiency of the group. Whole bodies would have the same problem if there were no mechanisms to suppress competition between genes, or cells, within a body.

6. (a) The whole genome; and (b) the chromosome.

Chapter 12

1. (a) 33%; and (b) 67%.

2. (a) Crudely, it has to be high; more exactly, a total deleterious mutation rate of more than one per organism per generation is needed. On its realism, see the end of Section 12.2.2: the evidence is inconclusive and neither rules it out or in. (b) Relation 1 in Figure 12.6. The *y*-axis is logarithmic. Relation 2 corresponds to independent fitness effects, in which sex (before the 50% cost) is indifferent. Relation 3 is the diminishing returns type of epistasis, in which sex is positively daft, even before the 50% cost. Again, you can argue reality either way: see the end of Section 12.2.2.

3. The material in the text (Section 12.2.3) would suggest looking at the relation between the frequency of sex and parasitism in taxa that can reproduce both ways; or looking into the genetics of host–parasite relations and measuring the frequency of resistance genes in hosts or penetration genes in parasites. Other answers would be possible too, going beyond the textual materials.

4. See Section 12.4.4: if the character were cheap to produce, males of all genetic qualities would evolve to produce it.

5. See Section 12.4.3: a female who did not choose extreme males would on average mate with a less extreme male than would other females in the population; she would produce less extreme than average sons; and they would grow up into a population in which most females prefer extreme males. Her sons would have low reproductive success and their mother's lack of preference would be selected against.

6. See Section 12.5.1: if more daughters than sons were produced by most members of the population, the fitness of a male would be higher than that of a female. Individuals who produced more sons than daughters would be favored by selection. A sex ratio of one is a stable point at which there is no advantage to producing more offspring of either sex.

7. (a) Positive, and (b) negative frequency-dependent selection.

8. (a) Yes, and (b) no. When different levels of selection conflict, adaptation cannot be perfect at all levels. You can find another example of a constraint on perfection in Holland & Rice's (1999) imposed monogamy experiment.

Chapter 13

1. Look at Section 13.2.
2. (i) 1, (ii) 2, and (iii) 1.
3. See Section 13.3, particularly Table 13.1.
4. G_{ST} is 0 for species 1, 0.5 for species 2, and 0 for species 3. Biological factors influencing G_{ST} include: the recency of origin of the species, the speed of evolution, how uniform the environment is through space, and the amount of gene flow between populations.
5. It can be argued both ways; see the second part of Section 13.7.2. If asexual species are discrete in the same way that sexual species seem to be, that suggests the force maintaining species as discrete clusters is ecological rather than interbreeding.
6. (i) Typological; (ii) population (or so I would argue); (iii) population; (iv) two schools of thought implicitly argue it each way (do you think we have a set number of real, distinct emotions?); (v) typological; and (vi) most would argue typological, I suspect, but Hull (1988) makes the opposite case, that scientific theories are like biological species.
7. See the work of Grant and Grant described in Section 13.7.3. Chapter 14 contains more material on the genetic theory of postzygotic isolation.
8. (i) (a), (b), and (c) (probably) yes; (ii) (a) yes, (b) and (c) no; and (iii) (a), (b), and (c) can be yes.

Chapter 14

1. Pleiotropy and hitch-hiking (Section 14.3.2), perhaps partly due to sexual selection (Section 14.11). Figure 14.3 shows an example from Darwin's finches.
2. (a) Postzygotic isolation can be expressed by the fitness reduction of hybrid offspring compared with offspring of crosses within a population (or within a near species). (b) The index we saw (Figure 14.2) was (number of matings to same type − number of matings to other type)/(total number of matings), which gives: (i) $I = 1$; (ii) $I = 0.5$; and (iii) $I = 0$.
3. It shows that the neighboring populations are more closely related: the northeast populations are more closely related to the southeast populations than to any other populations (such as southwest or northwest). It could have been that the populations evolved from formerly fragmented ranges, and expanded to the current distribution, but the phylogeny suggests a gradual evolution of the current songs in the current places. Also, the phylogeny shows that the gap in the range on the east side is probably only because there is a desert; there is an underlying continuity. The birds still evolved in a ring, with the northeast birds derived from the southeast birds.

4. The intermediate stages (heterozygotes) would be selected against.
5. (a) "When in the F_1 offspring of the two different animal races one sex is absent, rare, or sterile, that sex is the heterozygous one" (Section 14.4.6). (b) Males. (c) By postulating that some of the genes in the Dobzhansky–Muller theory are on the X chromosome, and recessive (Box 14.1).
6. Valley crossing means that evolution passes through a phase in which fitness goes down. (a) No, (b) no, and (c) yes.
7. Reproductive character displacement, or character displacement for prezygotic isolation. There are two main explanations. (i) Reinforcement. Females in allopatry have not been selected to discriminate against heterospecific males, because in evolutionary history the ancestors of the modern females have never met those males; females in sympatry are descended from females that have been exposed to both kinds of male. Females who mated with heterospecific males produced hybrid offspring of low fitness, so selection favored discrimination. (ii) Without reinforcement. There are various versions of the alternative explanation; the one most explained in the text (Section 16.8) is as follows. Different individuals of the two species in the past may have shown various degrees of isolation from the other species. In areas where they now coexist in sympatry, if reproductive isolation was low, the two species would fuse and probably now look more like one of the species (and so be classified as a member of it); if reproductive isolation was high, the two would coexist and remain distinct. Thus only where isolation was high do we now see the two species in sympatry. In areas where the species are now allopatric, whatever the reproductive isolation, they continue to exist. Thus the average isolation will be lower than for sympatry.
8. Theoretical reasons: the conditions required for reinforcement maybe too short lived. Empirical reasons: the evidence from artificial selection is poor, and the evidence from reproductive character displacment is open to alternative interpretations.
9. (a) "Secondary": divergent evolution in separate populations occurred in the past, followed by range expansion, and the two populations come into contact at what is now a hybrid zone. (b) "Primary": a stepped cline evolved within the population, which became large enough for the forms on either side of the step to be recognized as distinct taxonomic forms.
10. See Figure 14.14. On sympatric speciation, the closest relatives of a species should live in the same area; on allopatric speciation, the closest relatives should be in a different area.

Chapter 15

1.

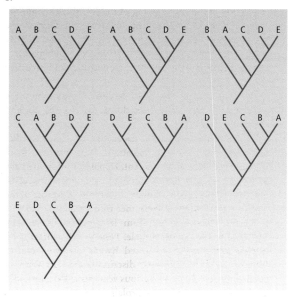

2.

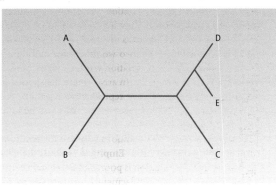

3. Evolutionary rates are approximately equal in all lineages.
4. See Section 15.4.
5. A is ancestral and A′ derived in all three, but the inference is most certain in (a) and least certain in (c). (If A is ancestral in the group of species 1 + 2, then the minimum number of events in (a), (b), and (c), respectively, are 2, 3, and 3; whereas if A′ were ancestral, the minimums would be 3, 4, and 4.)
6. See Section 15.8.

7. When more than one evolutionary change underlies an observed difference (or identity) between two sequences.
8. (a) $(1-3p)^3p^2$; (b) 16.
9. $2 \times 3 \times 1$.

Chapter 16

1. See Table 16.1.

   Evolutionary: paraphyletic, monophyletic.
   Cladistic: monophyletic.
   Phenetic: polyphyletic, paraphyletic, monophyletic.

2. (i) Cow (lungfish, salmon); (ii) cow (lungfish, salmon); (iii) (cow, lungfish), salmon. See the end of Section 16.3.
3. (i) The Euclidean distances are obtained by Pythagoras's theorem, and I picked the numbers to give a 3, 4, 5 triangle: the three species can be drawn on a graph with one character per axis. (ii) The mean character distance is the average of the distances for the two characters. See Section 16.5.

	Species 1	Species 2	Species 3
Species 1		3	4
Species 2	1.5		5
Species 3	2	3.5	

   The two distance measures imply contradictory groupings of the three species. This is a case where the classification chosen by the numeric phenetic method would be ambiguous, and therefore arguably subjective.

4. (a) 2.31; (b) 2.5; (c) species 1 and 2 equally; (d) with a nearest neighbor cluster statistic the grouping is (1,2)(3(4,5)); with a nearest average neighbor cluster statistic it is ((1,2)3) (4,5); and (e) see Section 16.5 for the moral.
5. A critic would reply that the same problem would resurface in another form. Maybe the average and nearest neighbor statistics in the case of Figure 14.5 could be made to agree by adding five further characters. However, those two are just two of many cluster statistics, and the result would almost certainly still be ambiguous with respect to some other cluster statistic. The ambiguity could only be removed if there were one non-ambiguous phenetic hierarchy in nature, and there is no reason to suppose such a hierarchy exists.
6. (a) The difference reflects evolutionary theory's scientific peculiarity as a historic theory. The hierarchy in a phylogenetic classification is historic, and is used for the reasons discussed in the chapter. The periodic table is non-historic and

non-hierarchical. Its structure represents two of the fundamental properties that determine the nature of an element, and the position of an element in the table can be used to predict what the element will be like. The position of an organism in a phylogenetic classification cannot be used to predict much about what the organism will be like. Much more could be said about the nature of different theories in science, and the way different theories imply different kinds of classifications. (b) See Section 16.8!

Chapter 17

1. See Section 17.1 and Figure 17.2.
2. Going down the column: 1/2, 1/2, 1/2, and 1.
3. My first three counts gave 26, 27, and 26 dispersal events from older to younger islands, and 13, 12, and 12 dispersals from younger to older islands, respectively! The correct answer is something close to these numbers, but you have the idea if your figures are in this region. Compare Figure 17.6. There is no reason why there should be so many more dispersal events from older to younger islands if speciation was created by splitting a larger range, whereas it makes sense if it was the result of dispersal because the older island would have been occupied first.
4.

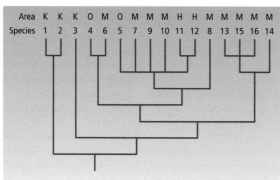

Some minor variants would also be possible, depending on how ancestral species like species 8 and 14 are represented.
5. (a)–(b), (a)–(c), and (e)–(f) are congruent. (a)–(d), (a)–(f), (c)–(d), (c)–(f), (e)–(b), and (e)–(d) are incongruent.
6. There are two main hypotheses. (i) The competitive superiority of North American mammals, perhaps due to a history of more intense competition, and reflected in their relative encephalization. (ii) Environmental change, such that the North American mammals were competitively superior in South American environments after the Interchange.

Chapter 18

1. See Section 18.1 and Figure 18.1.
2. Using the rounded figure of 1.2×10^{-4} for the decay constant:

	$^{14}C : ^{14}N$	Age
(a)	1 : 1	5,776
(b)	2 : 1	3,379
(c)	1 : 2	9,155

3. RNA can be single-stranded, allowing both metabolism and replication in one molecule. RNA can also have many shapes, enabling many reactions, including catalysis.
4. There are several possibilities. (i) The molecular evidence is wrong, for instance because of an error in calibration or non-constant rates of evolution. (ii) The fossil date is wrong, for instance because the record is incomplete or a fossil has been misdated. (iii) The estimates concern different events — the molecular clock gives a time of common ancestor, and the fossil evidence gives a time of proliferation.
5. See Section 18.5.
6. The mammals evolved in many stages, and the changes were in adaptive characters.
7. (i) Brain size, (ii) bipedality, and (iii) jaw reduction and associated changes in teeth. You might also mention changes in cultural, social, and linguistic behavior, and even changes in the thumb and big toe in the hand and foot.

Chapter 19

1. (a) By the molecular clock. (b) Several answers are possible, but the chapter noticed, for instance, the 2R hypothesis about the origin of vertebrates, and the possible association of gene duplications with the origin of dicotyledons.
2. (i) Gene transfer between bacteria and humans, or (ii) gene loss in a lineage leading to worms and fruitflies. They can be tested once we have an expanded knowledge of the phylogenetic distribution of the genes (see Figure 19.3).
3. Genes on the X and Y chromosomes do not recombine, and will have been diverging since recombination stopped. The four regions of gene similarity suggest that recombination was shut down in four stages, perhaps by inversions. Autosomal genes recombine and this prevents them from diverging.

Chapter 20

1. (a) Pedomorphosis; and (b) neoteny and progenesis (see Table 20.1).

2. The eyes of insects and vertebrates are at some level homologous, but not necessarily as eyes. The gene, for instance, could simply be a regional selector for a certain part of the head, that happens to have eyes in these two taxa. Alternatively, the common ancestor may have had eyes of some sort, developmentally controlled by the gene, but the structures we now observe as eyes in insects and vertebrates still built up independently in evolution.

3. (a) Evolvability is the chance that a species will undergo evolutionary change. The "evolutionary change" in the definition could refer to: (i) any genetic change, (ii) change in the form of speciation, or (iii) macroevolutionary, innovative change. (b) See Section 20.8: genetic switches enable genes to be recruited to act in new circumstances. Genes can acquire new functions without compromising their old function.

Chapter 21

1.

x_1	t_1	x_2	t_2	Rate
2	11	4	1	0.0693
2	11	20	1	0.2303
20	11	40	1	0.0693
20	6	40	1	0.1386

If you have answers like 0.2, 1.8, 2, and 4 you forgot to take logs. If you have minus numbers you have x_1 and x_2, or t_1 and t_2, the wrong way round.

2. (a) An inverse relation (see Figure 21.3). (b) One possibility is that long periods with rapid change and short periods with slow change have been excluded from the study, perhaps because the former would transform the character beyond commensurability and the latter seemed unworthy of notice.

3. See Section 21.5.

4. There are three possible answers. (i) Allopatric speciation, in which case punctuated equilibrium is orthodox. (ii) Speciation by valley crossing, in which case the theory is backing an unorthodox — some would say discredited — theory of speciation. (iii) Saltational macromutations, in which case the theory is unorthodox to the point of probably being erroneous.

5. The text contains two types of evidence: (i) rates of change in arbitrarily coded characters (see Figure 21.9), and (ii) taxonomic rates, in which the longevity of living fossil genera is longer than average (see Table 21.2).

Chapter 22

1. (a) Cospeciation and host shifts are more likely when hosts are phylogenetically closer. (b) Host shifts that are independent of phylogeny, for instance between hosts that are chemically similar but phylogenetically distant.

2. Biologists have seen whether the diversity of each taxon increases simultaneously in the fossil record. They have also made phylogenetically controlled comparisons between plants that do and do not interact with insects to see if the former have higher diversity. (And phylogenetically controlled comparisons between insects that do and do not interact with flowering plants to see if the former have higher diversity.)

3. Cophylogeny, and some evidence about the timing of the branches, for instance from molecular clocks.

4. In order of increasing virulence: (iv) < (i) < (ii) < (iii). (ii) and (iii) might be about the same, but this was not specifically discussed in the text. See Ewald (1993).

5. (a) See Section 22.6.1. Antagonistic biological interactions, such as between predator and prey, have evolved to become more dangerous over time: predators have become more dangerously armed, prey more powerfully defended. (b) The level of defensive adaptation in prey can be measured in such features as the thickness of molluskan shells and the habitats they occupy. Predatory adaptations have primarily been studied by the numbers of specialist as opposed to generalist predators: the presence of specialists suggests a more dangerous condition. It is important to test escalation by the proportion of species types through time, because there is more of everything in more recent fossil records. See Figures 22.12–22.14.

6. Antagonistic coevolution in general, and antagonistic coevolution with a dynamic equilibrium in particular. Van Valen suggested that total ecological resources may be constant through time, and the selective pressure on a species is proportional to the loss of resources it suffers due to lagging behind competing species.

Chapter 23

1. Because there was a lack of knowledge of the global distribution of species and (for large-bodied animals, whose geographic distributions were best known) the difficulty of assigning disarticulated fossil bone fragments to species (Section 23.1).

2. (a) In a real extinction all the members of a lineage die without leaving descendants; in a pseudoextinction the lineage continues to reproduce but its taxonomic name changes in mid-lineage, or the lineage persists but is temporarily unrepresented in the fossil record (Lazarus taxa). See Box 23.1. (b) Pseudoextinction of type (a) in Figure B23.1 snarls up

tests of both. If the extinctions of species with differing developmental modes in Figure 23.9, or in the test of synchroneity in Figure 23.3, were pseudoextinctions, the explanation for the trend, or synchronous pattern, would be something to do not with nature but with the habits of taxonomists. Lazarus-type pseudoextinction would also suggest the results are artifacts. Pseudoextinction of type (b) in Figure B23.1 may be less damaging. (Also, the test of the Red Queen hypothesis by survivorship curves in Chapter 22 may be little damaged by either of the taxonomic causes of pseudoextinction. The hypothesis might be recast in terms of rate of change rather than chance of extinction.)

3. (a) The iridium anomaly (Figure 23.4), perhaps combined with the dated Chicxulub crater. (b) The extinctions should be sudden and synchronous in all taxa, rather than gradual, and they should not in general be preceded by reductions in population size. For the difficulties of testing these predictions, look at Section 23.3.2.

4. At the end of the Cretaceous and Permian (for the top two extinctions); plus either the end Ordovician or the end Triassic (for the top three); plus the Devonian (for the top five).

5. See Figure B23.2. The observed extinction rate in the earlier interval will be (a) high, and (b) low.

6. At mass extinctions there is no relation. In background extinctions the taxa with planktonic development have an extinction rate that is half that of taxa with direct development. The background extinction difference can be explained either by bias in the fossil record (if planktonically developing species are more likely to be preserved) or by their being more likely to survive local difficulties by their dispersing larval stage.

7. Two possibilities we looked at are differences in speciation or extinction rates caused by differences in adaptations of different species, or by differential persistency of niches (see Section 23.6.2).

8. The heritability criterion is as relevant as ever here. In classic group selection problems, the character (such as altruism) is disadvantageous to individuals but advantageous to groups. Selfish individuals can invade groups. Once the group is infected by selfishness, it loses altruism. The character (altruism) is not inherited by groups for long. In classic cases of species selection, there is no question of a species being invaded by some alternative adaptation. Selection favors different adaptations in different species — direct development in some, planktonic development in others, for example. Those attributes are passed down from ancestral species to descendant species. Species selection is possible because there is no conflict between individual and species selection; heritability therefore is possible. Species selection is not a theory of the evolution of adaptation — only of the consequences of adaptations.

9. See Figure 23.11. A double wedge pattern suggests a competitive replacement. If one taxon goes extinct before the other radiates, it suggests non-competitive replacement.

10. Partly by different data compilations (with different taxonomic make ups) but mainly by different statistical corrections for biases in: (i) the amount of rock preserved from different times, and (ii) the amount of rock studied.

References

Abouheif, E., Akam, M., Dickinson, W.J. *et al.* (1997). Homology and developmental genes. *Trends in Genetics* **13**, 432–435.

Abrams, P.A. (2001). The evolution of predator–prey interactions: theory and evidence. *Annual Review of Ecology and Systematics* **31**, 79–105.

Adams, D.C. & Rohlf, F.J. (2000). Ecological character displacement in *Plethodon*: biomechanical differences found from a geometric morphological study. *Proceedings of the National Academy of Sciences USA* **97**, 4106–4111.

Adams, M.B. (ed.) (1994). *The Evolution of Theodosius Dobzhansky*. Princeton University Press, Princeton, NJ.

Adrain, J.M. & Westrop, S.R. (2000). An empirical assessment of taxic paleobiology. *Science* **289**, 100–112.

Ahlberg, P. (ed.) (2001). *Major Events in Early Vertebrate Evolution*. Taylor & Francis, London.

Allen, C., Bekoff, M. & Lauder, G. (eds) (1998). *Nature's Purposes*. MIT Press, Cambridge, MA.

Allison, A.C. (1954). Protection afforded by sickle-cell trait against subtertian malarial infection. *British Medical Journal* **1**, 290–294. (Reprinted in Ridley 1997.)

Alroy, J. (1998). Cope's rule and the dynamics of body mass evolution in North American fossil mammals. *Science* **280**, 731–734.

Alroy, J. (1999). The fossil record of North American mammals: evidence for a Paleocene evolutionary radiation. *Systematic Biology* **48**, 107–118.

Alroy, J., Marshall, C.R., Bambach, R.K. *et al.* (2002). Effects of sampling standardization on estimates of Phanerozoic marine diversification. *Proceedings of the National Academy of Sciences USA* **98**, 6261–6266.

Alvarez, L.W., Alvarez, W., Asaro, F. & Michel, H.V. (1980). Extraterrestrial cause for the Cretaceous–Tertiary extinction. *Science* **208**, 1095–1108.

Alvarez, W., Asaro, F. & Montanari, A. (1990). Iridium profile for 10 million years across the Cretaceous–Tertiary boundary at Gubbio (Italy). *Science* **250**, 1700–1702.

Alvarez, W., Kauffman, E.G., Surlyk, F., Alvarez, L.W., Asaro, F. & Michel, H.V. (1984). Impact theory of mass extinctions and the invertebrate fossil record. *Science* **223**, 1135–1141.

Anbar, A.D. & Knoll, A.H. (2002). Proterozoic ocean chemistry and evolution: a bioinorganic bridge. *Science* **297**, 1137–1142.

Andersson, D.I., Slechta, E.S. & Roth, J.R. (1998). Evidence that gene amplification underlies adaptive mutability of the bacterial *lac* operon. *Science* **282**, 1133–1135.

Andersson, M. (1994). *Sexual Selection*. Princeton University Press, Princeton, NJ.

Antolin, M.F. & Herbers, J.M. (2001). Evolution's struggle for existence in America's public schools. *Evolution* **55**, 2379–2388.

Antonovics, J. & van Tienderen, P.H. (1991). Ontoecogenophyloconstraints? The chaos of constraint terminology. *Trends in Ecology and Evolution* **6**, 166–168.

Arnold, M.L. (1997). *Natural Hybridization and Evolution*. Oxford University Press, New York.

Arnold, M.L. & Bennett, B.D. (1993). Natural hybridization in Louisiana irises: genetic variation and ecological determinants. In Harrison, R.G. (ed.) *Hybrid Zones and the Evolutionary Process*, pp. 115–139. Oxford University Press, New York.

Arnqvist, G., Edvardsson, M., Friberg, U. & Nilsson, T. (2000). Sexual conflict promotes speciation in insects. *Proceedings of the National Academy of Sciences USA* **97**, 10460–10464.

Asfaw, B., Gilbert, W.H., Beyene, Y. *et al.* (2002). Remains of *Homo erectus* from Bouri, Middle Awash, Ethiopia. *Nature* **416**, 317–320.

Asfaw, B., White, T., Lovejoy, O. *et al.* (1999). *Australopithecus garhi*: a new species of early hominid from Ethiopia. *Science* **284**, 629–635.

Ashburner, M. (1998). Speculations on the subject of alcohol dehydrogenase and its properties in *Drosophila* and other fruitflies. *Bioessays* **20**, 949–954.

Averof, M., Rokas, A., Wolfe, K.H. & Sharp, P.M. (2001). Evidence for a high frequency of simultaneous double-nucleotide substitutions. *Science* **287**, 1283–1286.

Avise, J.C. (1999). *Phylogeography*. Harvard University Press, Cambridge, MA.

Bak, P. (1996). *How Nature Works*. Springer, New York. (Paperback: Oxford University Press, Oxford.)

Bakker, R.T. (1983). The deer flees, the wolf pursues: incongruencies in predator–prey coevolution. In Futuyma, D.J. & Slatkin, M. (eds) *Coevolution*, pp. 350–382. Sinauer, Sunderland, MA.

Barbujani, G., Magagni, A., Minch, E. & Cavalli-Sforza, L.L. (1997). An apportionment of human DNA diversity. *Proceedings of the National Academy of Sciences USA* **94**, 4516–4519.

Barlow, C. (2000). *The Ghosts of Evolution*. Basic Books, New York.

Barraclough, T.G. & Nee, S. (2001). Phylogenetics and speciation. *Trends in Ecology and Evolution* **16**, 381–390.

Barraclough, T.G. & Vogler, A.P. (2001). Detecting the geographical pattern of speciation from species-level phylogenies. *American Naturalist* **155**, 419–434.

Barthélemy-Madaule, M. (1982). *Lamarck the Mythical Precursor*. MIT Press, Cambridge, MA.

Barton, N.H. & Charlesworth, B. (1998). Why sex and recombination? *Science* **281**, 1986–1990.

Barton, N.H. & Hewitt, G.M. (1985). Analysis of hybrid zones. *Annual Review of Ecology and Systematics* **16**, 113–148.

Barton, N.H. & Partridge, L. (2000). Limits to natural selection. *Bioessays* **22**, 1075–1084.

Beatty, J. (1992). Drift. In Keller, E.F. & Lloyd, E. (eds) *Keywords in Evolutionary Biology*. Harvard University Press, Cambridge, MA.

Beatty, J. (1994). Theoretical pluralism in biology, including systematics. In Grande, L. & Rieppel, O. (eds) *Interpreting the Hierarchy of Nature: from systematic patterns to evolutionary process theories*, pp. 33–60. Academic Press, New York.

Becerra, J.X. (1997) Insects on plants: macroevolutionary chemical trends in host use. *Science* **276**, 253–256.

Beldade, P., Brakefield, P.M. & Long, A.D. (2002a). Contribution of *Distal-less* to quantitative variation in butterfly eyespots. *Nature* **415**, 315–318.

Beldade, P., Koops, K. & Brakefield, P.M. (2002b). Developmental constraints and flexibility in morphological evolution. *Nature* **416**, 844–847.

Bell, G. (1997a). *The Basics of Selection*. Chapman & Hall, New York.

Bell, G. (1997b). *Selection: the mechanism of evolution*. Chapman & Hall, New York.

Benner, S.A., Caraco, M.D., Thomson, J.M. & Gaucher, E.A. (2002). Planetary biology — paleontological, geological, and molecular histories of life. *Science* 296, 864–868.

Bennetzen, J. (2002). Opening the door to comparative plant biology. *Science* 296, 60–63.

Benton, M.J. (1997). Models for the diversification of life. *Trends in Ecology and Evolution* 12, 490–495.

Benton, M.J. (2000a). *Vertebrate Paleontology*, 2nd edn. Blackwell Science, Boston, MA.

Benton, M.J. (2000b). Stems, nodes, crown clades, and rank-free lists: is Linnaeus dead? *Biological Reviews* 75, 633–648.

Benton, M.J. (2001). Finding the tree of life: matching phylogenetic trees to the fossil record through the 20th century. *Proceedings of the Royal Society of London B* 268, 2123–2130.

Benton, M.J. (2003). *When Life Nearly Died*. Thames & Hudson, London.

Benton, M.J. & Pearson, P.N. (2001). Speciation in the fossil record. *Trends in Ecology and Evolution* 16, 405–411.

Berlin, B. (1992). *Ethnobiological Classification: principles of categorization of plants and animals in traditional societies*. Princeton University Press, Princeton, NJ.

Berry, A. (ed.) (2002). *Infinite Tropics: an Alfred Russel Wallace anthology*. Verso, New York.

Bertram, B. (1978). *Pride of Lions*. Scribner, New York.

Blanchard, J.L. & Lynch, M. (2000). Organellar genes: why do they end up in the nucleus? *Trends in Genetics* 16, 315–320.

Blondel, J., Dias, P.C., Perret, P., Maistre, M. & Lambrecht, M.M. (1999). Selection-based biodiversity at a small spatial scale in a low-dispersing insular bird. *Science* 285, 1399–1402.

Bock, G.R. & Cardew, G. (eds) (1999). *Homology*. Novartis Foundation Symposium No. 222. John Wiley, New York.

Bodmer, W.F. & Cavalli-Sforza, L. (1976). *Genetics, Evolution, and Man*. W. H. Freeman, San Francisco.

Bolnick, D.I. (2001). Intraspecific competition favours niche width expansion in *Drosophila melanogaster*. *Nature* 410, 463–466.

Bowler, P.J. (1989). *Evolution: the history of an idea*, revised edn. University of California Press, Berkeley, CA.

Bowler, P.J. (1996). *Life's Splendid Drama*. University of Chicago Press, Chicago, IL.

Box, J.F. (1978). *R. A. Fisher. The Life of a Scientist*. John Wiley, New York.

Bradshaw, A.D. (1971). Plant evolution in extreme environments. In Creed, E.R. (ed.) *Ecological Genetics and Evolution*, pp. 20–50. Appleton-Century-Croft, New York.

Brakefield, P.M. (1987). Industrial melanism: do we have the answers? *Trends in Ecology and Evolution* 2, 117–122.

Brakefield, P.M., Gates, J., Keys, D. *et al.* (1996). The development, plasticity, and evolution of butterfly eyespot patterns. *Nature* 384, 236–242.

Bramble, D.M. & Jenkins, F.A. (1989). Structural and functional integration across the reptile–mammal boundary: the locomotor system. In Wake, D.B. & Roth, G. (eds) *Complex Organismal Functions*, pp. 133–146. Wiley, Chichester.

Brasier, M.D., Green, O.R., Jephcoat, A.P. *et al.* (2002). Questioning the evidence for Earth's oldest fossils. *Nature* 416, 76–81.

Breeuwer, J.A.J. & Werren, J.H. (1990). Microorganisms associated with chromosome destruction and reproductive isolation between two insect taxa. *Nature* 346, 558–560. (Reprinted in Ridley 1997.)

Briggs, D.E.G. & Crowther, P.R. (eds) (2001). *Paleobiology II*. Blackwell Science, Oxford.

Britton, R.J. (2002). Divergence between samples of chimpanzee and human DNA sequences is 5%, counting indels. *Proceedings of the National Academy of Sciences USA* **99**, 13633–13635.

Britton, R.J. & Davidson, E.H. (1971). Repetitive and non-repetitive DNA sequences and a speculation on the origins of evolutionary novelty. *Quarterly Review of Biology* **46**, 111–138.

Britton-Davidian, J., Catalan, J., Ramalhinho, J. da G. *et al.* (2000). Rapid chromosomal evolution in island mice. *Nature* **403**, 158.

Brocks, J.J., Logan, G.A., Buick, R. & Summons, R.E. (1999). Archean molecular fossils and the early rise of eukaryotes. *Science* **285**, 1033–1036.

Brodie, E.D. & Brodie, E.D. (1999). Predator–prey arms races. *Bioscience* **49**, 557–568.

Bronstein, J.L. (1994). Our current understanding of mutualism. *Quarterly Review of Biology* **69**, 31–51.

Brookfield, J.F.Y. (2001). The signature of selection. *Current Biology* **11**, R388–390.

Brown, J.H. (1971). The desert pupfish. *Scientific American* **225** (November), 104–110.

Brown, J.H. & Lomolino, M.V. (1998). *Biogeography*, 2nd edn. Sinauer, Sunderland, MA.

Brown, J.J., Stevens, G.C. & Kaufman, D.M. (1996). The geographic range: size, shape, boundaries, and internal structure. *Annual Review of Ecology and Systematics* **27**, 597–623.

Brown, W.L. & Wilson, E.O. (1958). Character displacement. *Systematic Zoology* **5**, 49–64.

Browne, J. (1995–2002). *Charles Darwin*, 2 vols. Knopf, New York and Jonathan Cape, London.

Brundin, L. (1988). Phylogenetic biogeography. In Myers, A.A. & Giller, P.S. (eds) *Analytical Biogeography*, pp. 343–369. Chapman & Hall, New York.

Bryant, H.N. & Cantino, P.D. (2002). A review of criticisms of phylogenetic nomenclature: is taxonomic freedom the fundamental issue? *Biological Reviews* **77**, 39–56.

Budd, G.E. & Jensen, S. (2000). A critical appraisal of the fossil record of the bilaterian phyla. *Biological Reviews* **75**, 253–295.

Bull, J.J. & Wichman, H.A. (2001). Applied evolution. *Annual Review of Ecology and Systematics* **32**, 183–217.

Burch, C.L. & Chao, L. (1999). Evolution by small steps and rugged landscapes in the RNA virus F6. *Genetics* **15**, 921–927.

Burkhardt, F. & Smith, S. (eds) (1985–). *The Correspondence of Charles Darwin*, vol. 1–. Cambridge University Press, Cambridge, UK.

Burkhardt, R.W. (1977). *The Spirit of System: Lamarck and evolutionary biology*. Harvard University Press, Cambridge, MA.

Burt, A. (2000). Sex, recombination, and the efficacy of selection — was Weismann right? *Evolution* **54**, 337–351.

Burt, D.W., Bruley, C., Dunn, I.C. *et al.* (1999). The dynamics of chromosome evolution in birds and mammals. *Nature* **402**, 411–413.

Buss, L.W. (1987). *The Evolution of Individuality*. Princeton University Press, Princeton, NJ.

Bustamente, C.D., Nielsen, R., Sawyer, S.A., Olsen, K.M., Purugganan, M.D. & Hartl, D.L. (2002). The cost of inbreeding in *Arabidopsis*. *Nature* **416**, 531–533.

Butlin, R. (2002). The costs and benfits of sex: new insights from old asexual lineages. *Nature Reviews Genetics* **3**, 311–317.

Butterfield, N.J. (2000). *Bangiomorpha pubescens* n. gen., n. sp.: implications for the evolution of sex, multicellularity, and the Mesoproterozoic/Neoproterozoic radiation of eukaryotes. *Paleobiology* **26**, 386–404.

Byers, J.A. (1997). *American Pronghorn: social adaptations and the ghost of predators past.* University of Chicago Press, Chicago, IL.

Cain, A.J. (1954). *Animal Species and their Evolution.* Hutchinson, London. (Reprinted 1993 by Princeton University Press, Princton, NJ.)

Cain, A.J. (1964). The perfection of animals. In: Carthy, J.D. & Duddington, C.L. (eds) *Viewpoints in Biology,* vol. 3, pp. 36–63. Butterworth, London. (Reprinted in *Biological Journal of the Linnean Society* **36** (1989), 3–29, and in Ridley 1997.)

Cairns, J., Overbaugh, J. & Miller, S. (1988). The origin of mutants. *Nature* **335**, 142–145.

Calvo, R.N. (1990). Inflorescence size and fruit distribution among individuals in three orchid species. *American Journal of Botany* **77**, 1378–1381.

Carr, G.D., Robichaux, R.H., Witter, M.S. & Kyhos, D.W. (1989). Adaptive radiation of the Hawaiian silversword alliance (Compositae — Madiinae): a comparison with Hawaiian picture-winged Drosophila. In Giddings, L.V., Kaneshiro, K.Y. & Anderson, W.W. (eds) *Genetics, Speciation, and the Founder Principle,* pp. 79–95. Oxford University Press, New York.

Carrington, M., Nelson, G.W., Martin, M.P., Kissner, T. *et al.* (1999). HLA and HIV-1: heterozygote advantage and B*35-Cw*04 disadvantage. *Science* **283**, 1748–1752.

Carroll, R.L. (1988). *Vertebrate Paleontology and Evolution.* W. H. Freeman, New York.

Carroll, R.L. (1997). *Pattern and Process of Vertebrate Evolution.* Cambridge University Press, New York.

Carroll, S.B., Grenier, J.K. & Weatherbee, S.D. (2001). *From DNA to Diversity: molecular genetics and the evolution of animal design.* Blackwell Science, Malden, MA.

Carson, H.L. (1983). Chromosomal sequences and inter-island colonizations in Hawaiian *Drosophila. Genetics* **103**, 465–482.

Carson, H.L. (1990). Evolutionary process as studied in population genetics: clues from phylogeny. *Oxford Surveys in Evolutionary Biology* **7**, 129–156.

Cavalli-Sforza, L.L. (2000). *Genes, Peoples, and Languages.* North Point Press, New York.

Celera (2001). The sequence of the human genome. *Science* **291**, 1304–1351.

Chao, L. & Carr, D.E. (1993). The molecular clock and the relationship between population size and generation time. *Evolution* **47**, 688–690.

Chao, L., Hanley, K.A., Burch, C.L., Dahlberg, C. & Turner, P.E. (2000). Kin selection and parasite evolution: higher and lower virulence with hard and soft selection. *Quarterly Review of Biology* **75**, 261–275.

Charleston, M. & Robertson, D.L. (2002). Preferential host switching by primate lentiviruses can account for phylogenetic similarity with the primate phylogeney. *Systematic Biology* **51**, 5–12.

Cheetham, A. (1986). Tempo of evolution in a Neogene bryozoan: rates of morphologic change within and across species boundaries. *Paleobiology* **12**, 199–202.

Civetta, A. & Clark, A.G. (2000). Correlated effects of sperm competition and postmating female mortality. *Proceedings of the National Academy of Sciences USA* **97**, 13162–13165 (with commentary by W. R. Rice, pp. 12953–12955).

Clack, J. (2002). *Gaining Ground.* Indiana University Press, Bloomington, IN.

Clark, M.A., Moran, N.A., Baumann, P. & Wernegreen, J.J. (2000). Cospeciation between bacteria endosymbionts (*Buchnera*) and a recent radiation of aphids (*Uroleucon*) and the pitfalls of testing for phylogenetic congruence. *Evolution* **54**, 417–525.

Clark, R.W. (1969). *JBS: the life and work of J. B. S. Haldane.* Coward-McCann, New York.

Clarke, C.A. & Sheppard, P.M. (1969). Further studies on the genetics of the mimetic butterfly *Papilio memnon. Philosophical Transactions of the Royal Society of London B* **263**, 35–70.

Clarke, C.A., Sheppard, P.M. & Thornton, I.W.B. (1968). The genetics of the mimetic butterfly *Papilio memnon*. *Philosophical Transactions of the Royal Society of London B* **254**, 37–89.

Clarkson, E.N.K. (1998). *Invertebrate Palaeontology and Evolution*, 4th edn. Chapman & Hall, New York.

Clayton, D.H. & Moore, J. (eds) (1997). *Host–Parasite Evolution: general principles and avian models*. Oxford University Press, New York.

Clayton, D.H. & Walther, B.A. (2001). Influence of host ecology and morphology on the diversity of Neotropical bird lice. *Oikos* **94**, 455–467.

Clutton-Brock, T.H. (ed.) (1988). *Reproductive Success*. University of Chicago Press, Chicago, IL.

Clutton-Brock, T.H. (2002). Breeding together: kin selection and mutualism in cooperative vertebrates. *Science* **296**, 69–72.

Cohen, F.M. (2001). Bacterial species and speciation. *Systematic Biology* **50**, 513–524.

Conover, D.O. & Munch, S.R. (2002). Sustaining fishery yields over evolutionary time scales. *Science* **297**, 94–96.

Conway Morris, S. (1998). *The Crucible of Creation*. Oxford University Press, New York.

Cook, A. (1975). Changes in the carrion/hooded crow hybrid zone and the possible importance of climate. *Bird Study* **22**, 165–168.

Cook, L.M. (2000). Changing views on melanic moths. *Biological Journal of the Linnean Society* **69**, 431–441.

Cook, L.M., Dennis, R.L.H. & Mani, G.S. (1999). Melanic morph frequency in the peppered moth in the Manchester area. *Proceedings of the Royal Society of London B* **266**, 293–297.

Cooper, A. & Fortey, R. (1998). Evolutionary explosions and the phylogenetic fuse. *Trends in Ecology and Evolution* **13**, 151–156.

Courtillot, V.E. (1999). *Evolutionary Catastrophes*. Cambridge University Press, New York.

Cox, C.B. & Moore, P.D. (2000). *Biogeography*, 6th edn. Blackwell Science, Boston, MA.

Coyne, J.A. (1985). The genetic basis of Haldane's rule. *Nature* **314**, 736–738.

Coyne, J.A. (1994). Ernst Mayr and the origin of species. *Evolution* **48**, 19–30.

Coyne, J.A. & Orr, H.A. (1989). Patterns of speciation in *Drosophila*. *Evolution* **43**, 362–381.

Coyne, J.A. & Orr, H.A. (1997). Patterns of speciation in *Drosophila* revisited. *Evolution* **51**, 295–303.

Coyne, J.A. & Orr, H.A. (1998). The evolutionary genetics of speciation. *Philosophical Transactions of the Royal Society of London B* **353**, 287–305. (Reprinted in Singh & Krimbas 2000.)

Coyne, J.A. & Orr, H.A. (2003). *Speciation*. Sinauer, Sunderland, MA.

Coyne, J.A., Barton, N.H. & Turelli, M. (1997). A critique of Sewall Wright's shifting balance theory of evolution. *Evolution* **51**, 643–671.

Coyne, J.A., Orr, H.A. & Futuyma, D.J. (1989). Do we need a new species concept? *Systematic Zoology* **37**, 190–200.

Creed, E.R., Lees, D.R. & Bulmer, M.G. (1980). Pre-adult viability differences of melanic *Biston betularia* (L.) (Lepidoptera). *Biological Journal of the Linnean Society* **13**, 25–62.

Crick, F.H.C. (1968). The origin of the genetic code. *Journal of Molecular Biology* **38**, 367–379. (Reprinted in Ridley 1997.)

Croizat, L., Nelson, G. & Rosen, D.E. (1974). Centers of origin and related concepts. *Systematic Zoology* **23**, 265–287.

Cronin, H. (1991). *The Ant and the Peacock*. Cambridge University Press, Cambridge, UK.

Cronin, T.M. & Schneider, C.E. (1990). Climatic influences on species: evidence from the fossil record. *Trends in Ecology and Evolution* 5, 275–279.

Crow, J.F. (1986). *Basic Concepts in Population, Quantitative, and Evolutionary Genetics.* W. H. Freeman, New York.

Crow, J.F. & Kimura, M. (1970). *An Introduction to Population Genetics Theory.* Harper & Row, New York.

Curtis, C.F., Cook, L.M. & Wood, R.J. (1978). Selection for and against insecticide resistance and possible methods of inhibiting the evolution of resistance in mosquitoes. *Ecological Entomology* 3, 273–287.

Cushing, D.H. (1975). *Marine Ecology and Fisheries.* Cambridge University Press, Cambridge, UK.

Cutler, D. (2000). Understanding the overdispersed molecular clock. *Genetics* **154**, 1403–1417.

Da Silva, M.N.F. & Patton, J.L. (1998). Molecular phylogeography and the evolution and conservation of Amazonian mammals. *Molecular Ecology* 7, 473–486.

Darwin, C.R. (1859). *On the Origin of Species.* John Murray, London. (There are many modern editions.)

Darwin, C.R. (1871). *The Descent of Man, and Selection in Relation to Sex.* John Murray, London. (Modern edition by Princeton University Press, Princeton, NJ.)

Darwin, C.R. (1872). *The Expression of the Emotions in Man and Animals.* John Murray, London. (Modern edition edited by P. Ekman, 1998, Harpercollins, London and Oxford University Press, New York.)

Davidson, E. (2001). *Genomic Regulatory Systems: development and evolution.* Academic Press, San Diego, CA.

Davies, M.B. & Shaw, R.G. (2001). Range shifts and adaptive responses to Quaternary climate change. *Science* **292**, 673–679.

Davis, G.W. & Richardson, D.M. (eds) (1995). *Mediterranean-type Ecosystems.* Springer-Verlag, New York.

Dawkins, R. (1982). *The Extended Phenotype.* W. H. Freeman, Oxford, UK. (Paperback edition by Oxford University Press, Oxford, UK.)

Dawkins, R. (1986). *The Blind Watchmaker.* W. W. Norton, New York and Longman, London.

Dawkins, R. (1989a). *The Selfish Gene,* 2nd edn. Oxford University Press, Oxford, UK.

Dawkins, R. (1989b). The evolution of evolvability. In C. Langton (ed.) *Artificial Life.* Addison Wesley, Sante Fe, NM.

Dawkins, R. (1996). *Climbing Mount Improbable.* W. W. Norton, New York and Viking Penguin, London.

de Beer, G.R. (1971). *Homology: an unsolved problem.* Oxford Biology Readers. Oxford University Press, Oxford. (reprinted in Ridley 1997.)

Dean, G. (1972). *The Porphyrias.* Pitman, London.

Dennett, D. (1995). *Darwin's Dangerous Idea.* Simon & Schuster, New York.

Diamond, J. (1990). Alone in a crowded universe. *Natural History* 1990 (June), 30–34.

Diamond, J. (1991). *The Third Chimpanzee.* Harpercollins, New York and Hutchinson Radius, London.

Dickinson, W.J. (1995). Molecules and morphology: where's the homology? *Trends in Genetics* 11, 119–121. (Reprinted in Ridley 1997.)

Dieckmann, U. & Doebeli, M. (1999). On the origin of species by sympatric speciation. *Nature* **400**, 354–357.

Dietl, G.P., Alexander, R.R. & Bien, W.F. (2000). Escalation in Late Cretaceous–early Paleocene oysters (Gryphaeidae) from the Atlantic Coastal Plain. *Paleobiology* **26**, 215–237.

Dietl, G.P., Kelley, P.H., Barrick, R. & Showers, W. (2002). Escalation and extinction selectivity: morphology versus isotopic reconstruction of bivalve metabolism. *Evolution* **56**, 284–291.

Dilcher, D. (2000). Toward a new synthesis: major evolutionary trends in the angiosperm fossil record. *Proceedings of the National Academy of Sciences USA* **97**, 7030–7036.

Dobzhansky, T. (1970). *Genetics of the Evolutionary Process*. Columbia University Press, New York.

Dobzhansky, T. (1973). Nothing in biology makes sense except in the light of evolution. *American Biology Teacher* **35**, 125–129. (Reprinted in Ridley 1997.)

Dobzhansky, T. & Pavlovsky, O. (1957). An experimental study of interaction between genetic drift and natural selection. *Evolution* **11**, 311–319.

Dodd, D.M.B. (1989). Reproductive isolation as a consequence of adaptive divergence in *Drosophila pseudoobscura*. *Evolution* **43**, 1308–1311.

Dodd, M.E., Silvertown, J. & Chase, M.W. (1999). Phylogenetic analyses of trait evolution and species diversity among angiosperm families. *Evolution* **53**, 732–744.

Donovan, S.K. & Paul, C.R.C. (eds) (1998). *The Adequacy of the Fossil Record*. John Wiley, New York.

Doolittle, W.F. (2000). Uprooting the tree of life. *Scientific American* **282** (February), 90–95.

Dowling, T.E. & Secor, C.L. (1997). The role of hybridization and introgression in the diversification of animals. *Annual Review of Ecology and Systematics* **28**, 593–619.

Duda, T.F. & Palumbi, S.R. (1999). Developmental shifts and species selection in gastropods. *Proceedings of the National Academy of Sciences USA* **96**, 10272–10277.

Dudley, J.W. & Lambert, R.J. (1992). Ninety generations of selection for oil and protein in maize. *Maydica* **37**, 96–119.

Duret, L. & Mouchiroud, D. (1999). Expresssion patterns and, surprisingly, gene length shape codon usage in *Caenorhabditis*, *Drosophila*, and *Arabidopsis*. *Proceedings of the National Academy of Sciences USA* **96**, 4482–4487.

Dybdahl, M.F. & Lively, C.M. (1998). Host–parasite coevolution: evidence for rare advantage and time-lagged selection in a natural population. *Evolution* **52**, 1057–1066.

Eanes, W.F. (1999). Analysis of selection on enzyme polymorphisms. *Annual Review of Ecology and Systematics* **30**, 301–326.

Ebert, D. (1998). Experimental evolution of parasites. *Science* **282**, 1432–1435.

Ebert, D. (1999). The evolution and expression of parasite virulence. In Stearns, S.C. (ed.) *Evolution in Health and Disease*, pp. 161–172. Oxford University Press, Oxford, UK.

Edwards, A.W.F. (1996). The origin and early development of the method of minimum evolution for the recognition of phylogenetic trees. *Systematic Biology* **46**, 79–91.

Ehrlich, P.R. (2000). *Human Natures*. Island Press, Washington, DC.

Ehrlich, P.R. & Raven, P.H. (1964). Butterflies and plants: a study in coevolution. *Evolution* **18**, 586–608.

Elder, J.F. & Turner, B.J. (1995). Concerted evolution of repetitive DNA sequences in Eukaryotes. *Quarterly Review of Biology* **70**, 297–320.

Eldredge, N. (1998). *The Pattern of Evolution*. W. H. Freeman, New York.

Eldredge, N. (2000). *The Triumph of Evolution: and the failure of creationism*. W. H. Freeman, New York.

Eldredge, N. & Gould, S.J. (1972). Punctuated equilibria: an alternative to phyletic gradualism. In Schopf, T.J.M. (ed.) *Models in Paleobiology*, pp. 82–115. Freeman, Cooper & Co., San Francisco.

Endler, J.A. (1977). *Geographic Variation, Speciation, and Clines*. Princeton University Press, Princeton, NJ.

Endler, J.A. (1986). *Natural Selection in the Wild*. Princeton University Press, Princeton, NJ.

Ereshefsky, M. (ed.) (1992). *The Units of Evolution: essays on the nature of species*. MIT Press, Cambridge, MA.

Ereshefsky, M. (2001). *The Poverty of the Linnaean Hierarchy*. Cambridge University Press, New York.

Erwin, D.H. & Anstey, R.L. (eds) (1995). *New Approaches to Speciation in the Fossil Record*. Columbia University Press, New York.

Ewald, P.W. (1993). The evolution of virulence. *Scientific American* **268** (April), 56–62.

Eyre-Walker, A. & Keightley, P. (1999). High genomic deleterious mutation rates in hominoids. *Nature* **397**, 344–347.

Falconer, D.S. & Mackay, T. (1996). *Introduction to Quantitative Genetics*, 4th edn. Longman, London.

Farrell, B.D. (1998). "Inordinate fondness" explained: why are there so many beetles? *Science* **281**, 555–559.

Farrell, B.D. & Mitter, C. (1994). Adaptive radiation in insects and plants: time and opportunity. *American Zoologist* **34**, 57–69.

Fay, J.C., Wyckoff, G.J. & Wu, C.-I. (2002). Testing the neutral theory of molecular evolution with genomic data from *Drosophila*. *Nature* **415**, 1024–1026.

Felsenstein, J. (1993). *PHYLIP (Phylogeny inference package). Version 3.5*. (Computer software package: see evolution.genetics.washington.edu/phylip/software.html.)

Felsenstein, J. (2001). The troubled growth of statistical phylogenies. *Systematic Biology* **50**, 465–467.

Felsenstein, J. (2003). *Inferring Phylogeny*. Sinauer, Sunderland, MA.

Fenchel, T. (2002). *Origin and Early Evolution of Life*. Oxford University Press, Oxford, UK.

Fenner, F. & Myers, K. (1978). Myxoma virus and myxomatosis in retrospect: the first quarter century of a new disease. In Kurstak, E. & Maramorosch, K. (eds) *Viruses and Environment*, pp. 539–570. Academic Press, New York.

Fenner, F. & Ratcliffe, R.N. (1965). *Myxomatosis*. Cambridge University Press, London.

Fenster, E.J. & Sorhannus, U. (1991). On the measurement of morphological rates of evolution: a review. *Evolutionary Biology* **25**, 375–410.

Fisher, R.A. (1918). The correlation between relatives under the supposition of Mendelian inheritance. *Transactions of the Royal Society of Edinburgh* **52**, 399–433.

Fisher, R.A. (1930). *The Genetical Theory of Natural Selection*. Oxford University Press, Oxford, UK. (2nd edn, 1958, published by Dover Books, New York. Variorum edition, 2000, by Oxford University Press, Oxford, UK.)

Fishman, L. & Willis, J.H. (2001) Evidence for Dobzhansky–Muller incompatibilities contributing to the sterility of hybrids between *Mimulus guttatus* and *M. masutus*. *Evolution* **55**, 1932–1942.

Fitch, W.M. (2000). Homology: a personal view of the problems. *Trends in Genetics* **16**, 227–231.

Flessa, K.W., Barnett, S.G., Cornue, D.B. *et al.* (1979). Geologic implications of the relationship between mammalian faunal similarity and geographic distance. *Geology* **7**, 15–18.

Flynn, J.J. & Wyss, A.R. (2002). Madagascar's Mesozoic secrets. *Scientific American* **286** (February), 54–63.

Foote, M., Hunter, J.P., Janis, C.M. & Sepkoski, J.J. (1999). Evolutionary and preservational constraints on origins of biologic groups: divergence times of eutherian mammals. *Science* **283**, 1310–1314.

Ford, E.B. (1975). *Ecological Genetics*, 3rd edn. Chapman & Hall, London.

Fortey, R. (2002). *Fossils: the key to the past*, 3rd edn. Natural History Museum, London.

Foster, P.L. (2000). Adaptive mutation: implications for evolution. *Bioessays* **22**, 1067–1074.

Fox, D.L., Fisher, D.C. & Leighton, L.R. (1999). Reconstructing phylogeny with and without temporal data. *Science* **284**, 1816–1819.

Fryer, G. (2001). On the age and origin of the species flock of haplochromine cichlid fishes of Lake Victoria. *Proceedings of the Royal Society of London B* **268**, 1147–1152.

Fryer, G., Greenwood, P.H. & Peake, J.F. (1985). The demonstration of speciation in fossil molluscs and living fishes. *Biological Journal of the Linnean Society of London* **26**, 325–336.

Fu, Y-X. & Li, W-H. (1999). Coalescing into the 21st century: an overview and prospects of coalescent theory. *Theoretical Population Biology* **56**, 1–20.

Futuyma, D.J. (1997). *Science on Trial*. Sinauer, Sunderland, MA.

Galis, F., van Alphen, J.J.M. & Metz, J.A.J. (2001). Why five fingers? Evolutionary constraints on digit numbers. *Trends in Ecology and Evolution* **16**, 637–646.

Gandon, S., Mackinnon, M.J., Nee, S. & Read, A.F. (2001). Imperfect vaccines and the evolution of pathogen virulence. *Nature* **414**, 751–756.

Gavrilets, S. & Boake, C.R.B. (1998). On the evolution of premating isolation after a founder event. *American Naturalist* **152**, 706–716.

Gehring, W.J. & Ikeo, K. (1999). *Pax6*: mastering eye morphogenesis and eye evolution. *Trends in Genetics* **15**, 371–377.

Gerhart, J. & Kirschner, M. (1997). *Cells, Embryos, and Evolution*. Blackwell Science, Boston, MA.

Ghiselin, M.T. (1997). *Metaphysics and the Origin of Species*. State University of New York Press, Albany, NY.

Gibbs, H.L. & Grant, P.R. (1987). Oscillating selection in Darwin's finches. *Nature* **327**, 511–513.

Gibson, G. & Wagner, G. (2000). Canalization in evolutionary genetics: a stabilizing theory? *Bioessays* **22**, 372–380.

Gigord, L.D.B., Macnair, M.R. & Smithson, A. (2001). Negative frequency-dependent selection maintains a dramatic flower color polymorphism in the rewardless orchid *Dactylrhiza sambucina* (L.) Soò. *Proceedings of the National Academy of Sciences USA* **98**, 6253–6255.

Gilbert, S.F. (2000). *Developmental Biology*, 6th edn. Sinauer, Sunderland, MA.

Gill, D.E. (1989). Fruiting failure, pollinator inefficiency, and speciation in orchids. In Otte, D. & Endler, J. (eds) *Speciation and its Consequences*, pp. 458–481. Sinauer, Sunderland, MA.

Gillespie, J.H. (1991). *The Causes of Molecular Evolution*. Oxford University Press, New York.

Gillespie, J.H. (1998). *Population Genetics: a concise guide*. Johns Hopkins University Press, Baltimore, MD.

Gillespie, J.H. (2001). Is the population size of a species relevant to its evolution? *Evolution* **56**, 284–291.

Gingerich, P.D. (1983). Rates of evolution: effects of time and temporal scaling. *Science* 222, 159–161.

Gingerich, P.D. (2001). Rates of evolution on the timescale of evolutionary processes. *Genetica* 112–113, 127–144. (Also in Hendry & Kinnison 2001.)

Gingerich, P.D., Smith, B.H. & Simons, E.L. (1990). Hind limbs of Eocene *Basilosaurus*: evidence of feet in whales. *Science* 249, 154–157.

Givnish, T.J. & Sytsma, K.J. (eds) (1997). *Molecular Evolution and Adaptive Radiation.* Cambridge University Press, New York.

Glaessner, M.F. & Wade, M.J. (1996). *Palaeontology* 9.

Goldblatt, P. (ed.) (1993). *Biological Relationships between Africa and South America.* Yale University Press, New Haven, CT.

Golding, G.B. & Dean, A.M. (1998). The structural basis of molecular adaptation. *Molecular Biology and Evolution* 15, 355–369.

Goldschmidt, R.B. (1940). *The Material Basis of Evolution.* Yale University Press, New Haven, CT.

Goodman, M. (1963). Man's place in the phylogeny of the primates as reflected in serum proteins. In Washburn, S.L. (ed.) *Classification and Human Evolution*, pp. 204–234. Aldine, Chicago, IL.

Gould, S.J. (1977a). *Ontogeny and Phylogeny.* Harvard University Press, Cambridge, MA.

Gould, S.J. (1977b). *Ever Since Darwin.* W. W. Norton, New York.

Gould, S.J. (1980). *The Panda's Thumb.* W. W. Norton, New York.

Gould, S.J. (1983). *Hen's Teeth and Horse's Toes.* W. W. Norton, New York.

Gould, S.J. (1985). *The Flamingo's Smile.* W. W. Norton, New York.

Gould, S.J. (1989). *Wonderful Life.* W. W. Norton, New York.

Gould, S.J. (1991). *Bully for Brontosaurus.* W. W. Norton, New York.

Gould, S.J. (1993). *Eight Little Piggies.* W. W. Norton, New York.

Gould, S.J. (1996). *Dinosaurs in a Haystack.* W. W. Norton, New York.

Gould, S.J. (1998). *Leonardo's Mountain of Clams and the Diet of Worms.* W. W. Norton, New York.

Gould, S.J. (2000). *The Lying Stones of Marrakech.* W. W. Norton, New York.

Gould, S.J. (2002a). *I have Landed.* W. W. Norton, New York.

Gould, S.J. (2002b). *The Structure of Evolutionary Theory.* Harvard University Press, Cambridge, MA.

Gould, S.J. & Johnston, R.F. (1972). Geographic variation. *Annual Review of Ecology and Systematics* 3, 457–498.

Gould, S.J. & Lewontin, R.C. (1979). The spandrels of San Marco and the panglossian paradigm: a critique of the adaptationist program. *Proceedings of the Royal Society of London B* 205, 581–598. (Reprinted in Ridley 1997.)

Grant, B. (1999). Fine-tuning the peppered moth paradigm. *Evolution* 53, 980–984.

Grant, B.S. & Wiseman, L.L. (2002). Recent history of melanism in North American moths. *Journal of Heredity* 93, 86–90.

Grant, P.R. (1986). *Ecology and Evolution of Darwin's Finches.* Princeton University Press, Princeton, NJ. (Reprinted 1999 with new afterword.)

Grant, P.R. (1991). Natural selection and Darwin's finches. *Scientific American* 265 (October), 83–87.

Grant, P.R. & Grant, B.R. (1995). Predicting microevolutionary responses to directional selection on heritable variation. *Evolution* 49, 241–251.

Grant, P.R. & Grant, B.R. (2000). Quantitiative genetic variation in populations of Darwin's finches. In Mousseau, T.A., Sinervo, B. & Endler, J. (eds) *Adaptive Genetic Variation in the Wild*, pp. 3–40. Oxford University Press, New York.

Grant, P.R. & Grant, B.R. (2002). Unpredictable evolution in a 30-year study of Darwin's finches. *Science* 296, 707–711.

Grant, V. (1981). *Plant Speciation*, 2nd edn. Columbia University Press, New York.

Graur, D. & Li, W-H. (2000). *Fundamentals of Molecular Evolution*, 2nd edn. Sinauer, Sunderland, MA.

Graveley, B.R. (2001). Alternative splicing: increasing diversity in the proteomic world. *Trends in Genetics* 17, 100–107.

Gray, M.W., Burger, G. & Lang, B.F. (1999). Mitochondrial evolution. *Science* 283, 1476–1482.

Green, D.M., Kraaijeveld, A.R. & Godfray, H.C.J. (2000). Evolutionary interaction between *Drosophila melanogaster* and its parasitoid *Asobara tabida*. *Heredity* 85, 450–458.

Greene, E., Lyon, B.E., Muehter, V.R., Ratcliffe, L., Oliver, S.J. & Boag, P.T. (2000). Disruptive sexual selection in a passerine bird. *Nature* 407, 1000–1003. (Also see news and views piece on p. 955 of the same issue.)

Grieve, R.A.F. (1990). Impact cratering on the Earth. *Scientific American* 262 (April), 66–73.

Griffiths, A.J.F., Miller, J.F., Suzuki, D.T., Lewontin, R.C. & Gelbart, W.M. (2000). *An Introduction to Genetic Analysis*, 7th edn. W. H. Freeman, New York.

Grimaldi, D. (1999). The co-radiations of pollinating insects and angiosperms in the Cretaceous. *Annals of the Missouri Botanic Garden* 86, 373–406.

Haffer, J. (1969). Speciation in Amazonian forest birds. *Science* 165, 131–137.

Haffer, J. (1974). Avian speciation in tropical South America. *Publications of the Nuttall Ornithological Club* 14, 1–390.

Hafner, M.S., Sudman, P.D., Villablanca, F.X., Spradling, T.A., Demastes, J.W. & Nadler, S.A. (1994). Disparate rates of molecular evolution in cospeciating hosts and parasites. *Science* 265, 1087–1090.

Hahn, B.H., Shaw, G.M., De Cock, K.M. & Sharp, P.M. (2000). AIDS as a zoonosis: scientific and public health implications. *Science* 287, 607–614.

Haldane, J.B.S. (1922). Sex ratio and unisexual sterility in animals. *Journal of Genetics* 12, 101–109.

Haldane, J.B.S. (1924). A mathematical theory of natural and artificial selection. Part I. *Transactions of the Cambridge Philosophical Society* 23, 19–41.

Haldane, J.B.S. (1932). *The Causes of Evolution*. Longman, London. (Reprinted 1966 by Cornell University Press, New York, and 1990 by Princeton University Press, Princeton, NJ.)

Haldane, J.B.S. (1949a). Disease and evolution. *La Ricercha Scientifica* 19 (suppl.), 69–76. (Reprinted in Ridley 1997.)

Haldane, J.B.S. (1949b). Suggestions as to quantitative measurement of rates of evolution. *Evolution* 3, 51–56.

Haldane, J.B.S. (1957). The cost of natural selection. *Journal of Genetics* 55, 511–524.

Hall, B.K. (1998). *Evolutionary Developmental Biology*, 2nd edn. Chapman & Hall, New York.

Hall, B.K. (2001). *Phylogenetic Trees made Easy: a how-to manual for molecular biologists*. Sinauer, Sunderland, MA.

Hallam, A. & Wignall, P.B. (eds) (1997). *Mass Extinctions and their Aftermath*. Oxford University Press, Oxford, UK.

Hamilton, W.D. (1996). *The Narrow Roads of Geneland*, vol. 1. Oxford University Press, Oxford, UK.

Hamilton, W.D. (2001). *The Narrow Roads of Geneland*, vol. 2. Oxford University Press, Oxford, UK.

Han, T.-M. & Runnegar, B. (1992). Megascopic eukaryotic algae from the 2.1-billion-year-old Negaunee iron-formation, Michigan. *Science* **257**, 232–235.

Hansen, T.A. (1978). Larval dispersal and species longevity in Lower Tertiary neogastropods. *Science* **199**, 885–887.

Hansen, T.A. (1983). Modes of larval development and rates of speciation in early Tertiary neogastropods. *Science* **220**, 501–502.

Hardison, R. (1999). The evolution of hemoglobin. *American Scientist* **87**, 126–137.

Hardy, I.C.W. (ed.) (2002). *Sex Ratios: concepts and research methods.* Cambridge University Press, Cambridge, UK.

Hare, M.P. (2001). Prospects for nuclear gene phylogeography. *Trends in Ecology and Evolution* **16**, 700–706.

Harrison, R.G. (ed.) (1993). *Hybrid Zones and the Evolutionary Process.* Oxford University Press, New York.

Harrison, R.G. (2001). Book review. *Nature* **411**, 635–636.

Hartl, D.L. (2000). *A Primer of Population Genetics*, 3rd edn. Sinauer, Sunderland, MA.

Hartl, D.L. & Clark, A.G. (1997). *Principles of Population Genetics*, 3rd edn. Sinauer, Sunderland, MA.

Harvey, P.H. & Pagel, M.D. (1991). *The Comparative Method in Evolutionary Biology.* Oxford University Press, Oxford, UK.

Harvey, P.H., Leigh Brown, A.J., Maynard Smith, J. & Nee, S. (eds) (1996). *New Uses for New Phylogenies.* Oxford University Press, New York.

Hayden, M. (1981). *Huntington's Chorea.* Springer-Verlag, Berlin.

Heckman, D.S., Geiser, D.M., Eidell, B.R. *et al.* (2001). Molecular evidence for the early colonization of land by fungi and plants. *Science* **293**, 1129–1133.

Hedrick, P.W. (2000). *Genetics of Populations*, 2nd edn. Jones & Bartlett, Boston, MA.

Hedrick, P.W., Klitz, W., Robinson, W.P., Kuhner, M.K. & Thomson, G. (1991). Evolutionary genetics of HLA. In Selander, R.K., Clark, A.G. & Whittam, T.S. (eds) *Evolution at the Molecular Level*, pp. 248–271. Sinauer, Sunderland, MA.

Hendry, A.P. & Kinnison, M.T. (1999). The pace of modern life: measuring rates of contemporary microevolution. *Evolution* **53**, 1637–1653.

Hendry, A.P. & Kinnison, M.T. (eds) (2001). *Microevolution: rates, pattern, process.* Kluwer Academic, Dordecht, Netherlands.

Hennig, W. (1966). *Phylogenetic Systematics.* University of Illinois Press, Urbana, IL.

Hennig, W. (1981). *Insect Phylogeny.* John Wiley, Chichester, UK.

Herre, E.A. (1993). Population structure and the evolution of virulence in nematode parasites of fig wasps. *Science* **259**, 1442–1446.

Hewison, A.J.M. & Gaillard, J-M. (1999). Successful sons or advantaged daughters? *Trends in Ecology and Evolution* **14**, 229–234.

Hewitt, G. (1999). Post-glacial re-colonization of European biota. *Biological Journal of the Linnean Society* **68**, 87–112.

Hewitt, G. (2000). The genetic legacy of the Quaternary ice ages. *Nature* **405**, 907–913.

Hewzulla, D., Boulter, M.C., Benton, M.J. & Halley, J.M. (1999). Evolutionary patterns from mass originations and mass extinctions. *Philosophical Transactions of the Royal Society of London B* **354**, 463–469.

Hey, J. (2001). *Genes, Categories, and Species*. Oxford University Press, New York.

Higashi, M., Takimoto, G. & Yamamura, N. (1999). Sympatric speciation by sexual selection. *Nature* 402, 523–526.

Higgie, M., Chenoweth, S. & Blows, M.W. (2000). Natural selection and the reinforcement of mate recognition. *Science* 290, 519–521.

Hillis, D.M. (1996). Inferring complex phylogenies. *Nature* 383, 130–131.

Hillis, D.M., Moritz, C. & Mable, B.K. (eds) (1996). *Molecular Systematics*, 2nd edn. Sinauer, Sunderland, MA.

Hoekstra, H.E., Hoekstra, J.M., Berrigan, D. *et al.* (2001). Strength and tempo of directional selection in the wild. *Proceedings of the National Academy of Sciences USA* 98, 9157–9160.

Hoffmann, A.A. (2000). Laboratory and field heritabilities: some lessons from *Drosophila*. In Mousseau, T.A., Sinervo, B. & Endler, J.A. (eds) *Adaptive Genetic Variation in the Wild*, pp. 200–218. Oxford University Press, New York.

Hoffman, P.F. & Schrag, D.P. (2000). Snowball Earth. *Scientific American* 282 (January), 68–75.

Holland, B. & Rice, W.R. (1999). Experimental removal of sexual selection reverses intersexual antagonistic coevolution and removes a reproductive load. *Proceedings of the National Academy of Sciences USA* 96, 5083–5088.

Holman, E.W. (1987). Recognizability of sexual and asexual species of rotifers. *Systematic Zoology* 36, 381–386.

Holmes, E.C. (2000a). On the origin and evolution of the human immunodeficiency virus (HIV). *Biological Reviews* 76, 239–254.

Holmes, E.C., Gould, E.A. & Zanotto, P.M. de A. (1996). An RNA virus tree of life? In Roberts, D.McL., Sharp, P., Alderson, G. & Collins, M. (eds) *Evolution of Microbial Life*, pp. 127–144. Cambridge University Press, Cambridge, UK.

Holmes, R. (2000b). Fatal flaw. *New Scientist* October 28, 34–37.

Hori, M. (1993). Frequency-dependent natural selection in the handedness of scale-eating cichlid fish. *Science* 260, 216–219.

Hosken, D.J., Garner, T.W.J. & Ward, P.I. (2001). Sexual conflict selects for male and female reproductive characters. *Current Biology* 11, 489–493.

Hostert, E. (1997). Reinforcement: a new perspective on an old controversy. *Evolution* 51, 697–702.

Hotton, N.H., MacLean, P.D., Roth, J.J. & Roth, E.C. (eds) (1986). *Ecology and Biology of Mammal-like Reptiles*. Smithsonian Institution, Washington, DC.

Howard, D.J. (1993). Reinforcement: origins, dynamics, and fate of an evolutionary hypothesis. In Harrison, R.G. (ed.) *Hybrid Zones and the Evolutionary Process*, pp. 46–69. Oxford University Press, New York.

Howard, D.J. (1999). Conspecific sperm and pollen precedence and speciation. *Annual Review of Ecology and Systematics* 30, 109–132.

Howard, D. & Berlocher, S. (eds) (1998). *Endless Forms: species and speciation*. Oxford University Press, New York.

Hubbell, S.P. (2001). *The Neutral Theory of Biodiversity and Biogeogaphy*. Princeton University Press, Princeton, NJ.

Hudson, R.R., Kreitman, M. & Aguadè, M. (1987) A test of neutral molecular evolution based on nucleotide data. *Genetics* 116, 153–159.

Huelsenbeck, J.P. & Crandall, K.A. (1997). Phylogeny estimation and hypothesis testing using maximum likelihood. *Annual Review of Ecology and Systematics* 28, 437–466.

Huelsenbeck, J.P., Ronquist, F., Nielsen, R. & Bollback, S. (2001). Bayesian inference of phylogeny and its impact on evolutionary biology. *Science* **294**, 2310–2314.

Huey, R.B., Gilchrist, G.W., Carlson, M.L., Berrigan, D. & Serra, L. (2000). Rapid evolution of a geographic cline in size in an introduced species. *Science* **287**, 308–309.

Hughes, A.L. (1999). *Adaptive Evolution of Genes and Genomes*. Oxford University Press, New York.

Hull, D.L. (1967). Certainty and circularity in evolutionary taxonomy. *Evolution* **21**, 174–189.

Hull, D.L. (1988). *Science as a Process*. University of Chicago Press, Chicago, IL.

Humphries, C.J. & Parenti, L.R. (1999). *Cladistic Biogeography*. Oxford University Press, Oxford, UK.

Hunt, H.R., Hoppert, C.A. & Rosen, S. (1955). Genetic factors in experimental rat caries. In Sognnaes, R.F. (ed.) *Advances in Experimental Caries Research*, pp. 66–81. American Association for the Advancement of Science, Washington, DC.

Huxley, J.S. (1932). *Problems of Relative Growth*. Methuen, London.

Huxley, J.S. (ed.) (1940). *The New Systematics*. Oxford University Press, Oxford, UK.

Huxley, J.S. (1942). *Evolution: the modern synthesis*. Allen & Unwin, London.

Huxley, J.S. (1970–73). *Memories*, 2 vols. Allen & Unwin, London.

International Human Genome Sequencing Consortium (2001). Initial sequencing and analysis of the human genome. *Nature* **409**, 860–921.

Irwin, D.E., Bensch, S. & Price, T.D. (2001a). Speciation in a ring. *Nature* **409**, 333–337.

Irwin, D.E., Irwin, J.H. & Price, T.D. (2001b). Ring species as bridges between macroevolution and microevolution. *Genetica* **112–113**, 223–243. (Also in Hendry & Kinnison 2001.)

Jablonski, D. (1986). Background and mass extinctions: the alternation of macroevolutionary regimes. *Science* **231**, 129–133.

Jablonski, D. (2000). Micro- and macroevolution: scale and hierarchy in evolutionary biology and paleobiology, *Paleobiology* **26** (suppl.), 15–52.

Jablonski, D. & Bottjer, D.J. (1990). The origin and diversification of major groups: environmental patterns and macroevolutionary lags. In Taylor, P.D. & Larwood, G.P. (eds) *Major Evolutionary Radiations*, pp. 17–57. Oxford University Press, Oxford, UK.

Jablonski, D. & Lutz, R.A. (1983). Larval ecology of marine benthic invertebrates: paleobiological implications. *Biological Reviews* **58**, 21–89.

Jackman, T.R. & Wake, D.B. (1994). Evolutionary and historical analysis of protein variation in the blotched forms of salamanders of the *Ensatina* complex (Amphibia: Plethodontidae). *Evolution* **48**, 876–897.

Jackson, J. & Cheetham, A. (1994). On the importance of doing nothing. *Natural History* June, 56–59.

Jackson, J.B.C. & Cheetham, A.H. (1999). Tempo and mode of speciation in the sea. *Trends in Ecology and Evolution* **14**, 72–77.

Jackson, J.B.C. & Johnson, K.G. (2001). Measuring past diversity. *Science* **293**, 2401–2404.

Jackson, J.B.C., Budd, A.F. & Coates, A.G. (eds) (1996). *Evolution and Environment in Tropical America*. University of Chicago Press, Chicago, IL.

Jackson, J.B.C., Lidgard, S. & McKinney, F.K. (eds) (2001). *Evolutionary Patterns: growth, form, and tempo in the fossil record*. University of Chicago Press, Chicago, IL.

Janzen, D.H. (1980). When is it coevolution? *Evolution* **34**, 611–612.

Janzen, D.H. & Martin, P.S. (1982). Neotropical anachronisms: the fruits the gomphotheres ate. *Science* **215**, 19–27.

Jeffreys, A.J., Royle, N.J., Wilson, V. & Wong, Z. (1988). Spontaneous mutation rate to new length alleles at tandem-repetitive hypervariable loci in human DNA. *Nature* **332**, 278–281.

Jensen, S., Gehling, J.G. & Droser, M.L. (1998). Ediacara-type fossils in Cambrian sediments. *Nature* **393**, 567–569.

Jepsen, G.L., Mayr, E. & Simpson, G.G. (eds) (1949). *Genetics, Paleontology, and Evolution*. Princeton University Press, Princeton, NJ.

Jerison, H.J. (1973). *Evolution of the Brain and Intelligence*. Academic Press, New York.

Ji, Q., Luo, Z-X., Yuan, C-X., Wible, J.R., Zhang, J-P. & Georgi, J.A. (2002). The earliest known eutherian mammal. *Nature* **416**, 816–822.

Johnson, N.A. (2002). Sixty years after "Isolating mechanisms, evolution, and temperature": Muller's legacy. *Genetics* **161**, 939–944.

Johnson, S.D. & Steiner, K.E. (2000). Generalization and specialization in plant-pollination systems. *Trends in Ecology and Evolution* **15**, 140–143.

Johnston, R.F. & Selander, R.K. (1971). Evolution in the house sparrow. II. Adaptive differentiation in North American populations. *Evolution* **25**, 1–28.

Jones, S. (1999). *Almost Like a Whale*. Transworld Publishers, London.

Joyce, G.F. (2002). The antiquity of RNA-based evolution. *Nature* **418**, 214–221.

Kaneshiro, K. (1988). Speciation in Hawaiian *Drosophila. Bioscience* **38**, 258–263.

Karn, M.N. & Penrose, L.S. (1951). Birth weight and gestation time in relation to maternal age, parity, and infant survival. *Annals of Eugenics* **16**, 147–164. (Extracted in Ridley 1997.)

Keightley, P.D. & Eyre-Walker, A. (1999). Terumi Mukai and the riddle of deleterious mutation rates. *Genetics* **153**, 515–523.

Keller, E.F. & Lloyd, E.A. (eds) (1992). *Keywords in Evolutionary Biology*. Harvard University Press, Cambridge, MA.

Keller, L. (ed.) (1999). *Levels of Selection in Evolution*. Princeton University Press, Princeton, NJ.

Keller, M.J. & Gerhardt, H.C. (2001). Polyploidy alters advertisement call structure in treefrogs. *Proceedings of the Royal Society of London B* **268**, 341–345.

Kellogg, E.A. (2000). The grasses: a case study in macroevolution. *Annual Review of Ecology and Systematics* **31**, 217–238.

Kemp, T.S. (1999). *Fossils and Evolution*. Oxford University Press, Oxford, UK.

Kenrick, P. (2001). Turning over a new leaf. *Nature* **410**, 309–310.

Kenrick, P. & Crane, P.R. (1997). The origin and early evolution of plants on land. *Nature* **389**, 33–39.

Kessler, S. (1966). Selection for and against ethological isolation between *Drosophila pseudoobscura* and *Drosophila persimilis. Evolution* **20**, 634–645.

Kettlewell, H.B.D. (1973). *The Evolution of Melanism*. Oxford University Press, Oxford, UK.

Keys, D.N., Lewis, D.L., Selegue, J.E. *et al.* (1999). Recruitment of a *hedgehog* regulatory circuit in butterfly eyespot evolution. *Science* **283**, 532–534.

Kimura, M. (1968). Evolutionary rate at the molecular level. *Nature* **217**, 624–626.

Kimura, M. (1983). *The Neutral Theory of Molecular Evolution*. Cambridge University Press, Cambridge, UK.

Kimura, M. (1991). Recent developments of the neutral theory viewed from the Wrightian tradition of theoretical population genetics. *Proceedings of the National Academy of Sciences USA* **88**, 5969–5973. (Reprinted in Ridley 1997.)

King, L. & Jukes, T. (1969). Non-darwinian evolution. *Science* **164**, 788–789.

King, M-C. & Wilson, A.C. (1975). Evolution at two levels: molecular similarities and biological differences between humans and chimpanzees. *Science* **188**, 107–116.

Kingman, J.F.C. (2000). Origins of the coalescent. *Genetics* **156**, 1461–1463.

Kingsolver, J.G., Hoekstra, H.E., Hoekstra, J.M. *et al.* (2001). The strength of phenotypic selection in natural populations. *American Naturalist* **157**, 245–261.

Kirchner, J.W. (2002). Evolutionary speed limits inferred from the fossil data. *Nature* **415**, 65–68.

Kirchner, J.W. & Weil, A. (1998). No fractals in fossil extinction statistics. *Nature* **395**, 337–338.

Kirschner, M. & Gerhart, J. (1998). Evolvability. *Proceedings of the National Academy of Sciences USA* **95**, 8420–8427.

Kitching, I.J., Forey, P.L., Humphries, C.J. & Williams, D.M. (1998). *Cladistics: the theory and practice of parsimony analysis*, 2nd edn. Oxford University Press, Oxford, UK.

Klein, R.G. (1999). *The Human Career*, 2nd edn. University of Chicago Press, Chicago, IL.

Klicka, J. & Zink, R.M. (1999). Pleistocene effects on North American songbird evolution. *Proceedings of the Royal Society of London B* **266**, 695–700.

Klingenberg, C.P. (1998). Heterochrony and allometry: the analysis of evolutionary change in ontogeny. *Biological Reviews* **73**, 79–123.

Knoll, A.H. & Baghoorn, E.S. (1977). Archaean microfossils and showing cell division from the Swaziland system of South Africa. *Science* **198**, 396–398.

Knoll, A.H. & Carroll, S.B. (1999). Early animal evolution: emerging views from comparative biology and geology. *Science* **284**, 2129–2137.

Koepfer, H.R. (1987). Selection for isolation between geographic forms of *Drosophila mojavensis*. I. Interactions between the selected forms. *Evolution* **41**, 37–48.

Kohn, M.H., Pelz, H-J. & Wayne, R.K. (2000). Natural selection mapping of the warfarin-resistance gene. *Proceedings of the National Academy of Sciences USA* **97**, 7911–7915.

Komdeur, J. (1996). Facultative sex ratio bias in the offspring of Seychelles warblers. *Proceedings of the Royal Society of London B* **263**, 661–666.

Kondrashov, A.S. (1988). Deleterious mutations and the evolution of sexual reproduction. *Nature* **336**, 435–440.

Kondrashov, A.S. & Kondrashov, F.A. (1999). Interactions among quantitative traits in the course of sympatric speciation. *Nature* **400**, 351–354.

Kondrashov, A.S. & Turelli, M. (1992). Deleterious mutations, apparent stabilizing selection, and the maintenance of quantitative variation. *Genetics* **132**, 603–618.

Korber, B., Muldoon, M., Theiler, J. *et al.* (2000). Timing the ancestor of the HIV-1 pandemic strains. *Science* **288**, 1789–1796.

Korol, A., Rashkovetsky, E., Iliadi, K. *et al.* (2000). Nonrandom mating in *Drosophila melanogaster* laboratory populations derived from closely adjacent ecologically contrasting slopes at "Evolution canyon". *Proceedings of the National Academy of Sciences USA* **97**, 12637–12642.

Kreitman, M. (1983). Nucleotide polymorphism at the alcohol dehydrogenase locus of *Drosophila melanogaster*. *Nature* **304**, 412–417.

Kreitman, M. & Antezana, M. (2000). The population and evolutionary genetics of codon bias. In Singh, R.S. & Krimbas, C. (eds) *Evolutionary Genetics*, pp. 82–101. Cambridge University Press, New York.

Kruckeberg, A.R. (1957). Variation in fertility of hybrids between isolated populations of the serpentine species *Streptanthus glandulosus* Hook. *Evolution* **11**, 185–211.

Kruuk, L.E.B., Merilä, J. & Sheldon, B.C. (2001). Phenotypic selection on a heritable size trait revisited. *American Naturalist* **158**, 557–571.

Kubo, N., Harada, K., Hirai, A. & Kadowaki, K-I. (1999). A single nuclear transcript encoding mitochondrial RSP14 and SDHB of rice is processed by alternative splicing. *Proceedings of the National Academy of Sciences USA* **96**, 9207–9211.

Kumar, S. & Subramanian, S. (2002). Mutation rates in mammalian genomes. *Proceedings of the National Academy of Sciences USA* **99**, 803–808.

Labandeira, C.C. (1998). How old is the flower and the fly? *Science* **280**, 57–59.

Labandeira, C.C. & Sepkoski, J.J. (1993). Insect diversity in the fossil record. *Science* **261**, 310–315.

Labandeira, C.C., Johnson, K.R. & Wilf, P. (2002). Impact of the terminal Cretaceous event on plant–insect associations. *Proceedings of the National Academy of Sciences USA* **99**, 2061–2066.

Lack, D. (1947). *Darwin's Finches*. Cambridge University Press, Cambridge, UK.

Lahn, B.T. & Page, D.C. (1999). Four evolutionary strata on the human X chromosome. *Science* **286**, 964–967.

Lake, J.A. (1990). Origin of the Metazoa. *Proceedings of the National Academy of Sciences USA* **87**, 763–766.

Lamarck, J-B. (1809). *Philosophie Zoologique*. Paris.

Lambert, D.M. & Spencer, H.E. (eds) (1994). *Speciation and the Recognition Concept: theory and application*. Johns Hopkins University Press, Baltimore, MD.

Lan, R. & Reeves, P.R. (2001). When does a clone deserve a name? *Trends in Microbiology* **9**, 419–424.

Lanciotti, P.S., Roehrig, J.T., Deubel, V. *et al.* (1999) Origin of the west Nile virus responsible for an outbreak of encephalitis in the northeastern United States. *Science* **286**, 2333–2337.

Land, M.F. & Nilsson, D-E. (2002). *Animal Eyes*. Oxford University Press, Oxford, UK.

Lande, R. (1976). The maintenance of genetic variability by mutation in a polygenic character with linked loci. *Genetical Research* **26**, 221–235.

Langley, C.H. (1977). Nonrandom associations between allozymes in natural poulations of *Drosophila melanogater*. In Christiansen, F.B. & Fenchel, T.M. (eds) *Measuring Selection in Natural Populations*, pp. 265–273. Springer-Verlag, Berlin.

Laporte, L.F. (2000). *George Gaylord Simpson: paleontologist and evolutionist*. Columbia University Press, New York.

Larson, E.J. (2003). *Trial and Error: the American controversy over creation and evolution*, 3rd edn. Oxford University Press, New York.

Law, R. (1991). Fishing in evolutionary waters. *New Scientist* March 2, 35–37.

Lederburg, J. (1999). J.B.S. Haldane (1949) on infectious disease and evolution. *Genetics* **153**, 1–3.

Lees, D.R. (1971). Industrial melanism: genetic adaptation of animals to air pollution. In Bishop, J.A. & Cook, L.M. (eds) *Genetic Consequencies of Man-made Change*, pp. 129–176. Academic Press, London.

Leigh, E.G. (1987). Ronald Fisher and the development of evolutionary theory. II. *Oxford Surveys in Evolutionary Biology* **4**, 214–223.

Leitch, I. & Bennett, M. (1997). Polyploidy in angiosperms. *Trends in Plant Sciences* **2**, 470–476.

Lenormand, T., Bourguet, D., Guillemaud, T. & Raymond, M. (1999). Tracking the evolution of insecticide resistance in the mosquito *Culex pipiens*. *Nature* **400**, 861–864.

Lens, L., van Dongen, S., Kark, S. & Matthysen, E. (2002). Fluctuating asymmetry as an indicator of fitness: can we bridge the gap between studies? *Biological Reviews* 77, 27–38.

Levene, H. (1953). Genetic equilibrium when more than one niche is available. *American Naturalist* 87, 331–333.

Leverich, W.J. & Levin, D.A. (1979). Age-specific survivorship and reproduction in *Phlox drummondii. American Naturalist* 113, 881–903.

Levin, B.R., Perrot, V. & Walker, N. (2000). Compensatory mutations, antibiotic resistance, and the population genetics of adaptive evoltuion in bacteria. *Genetics* 154, 985–997.

Levin, D.A. (2000). *The Origin, Expansion, and Demise of Plant Species*. Oxford University Press, New York.

Levine, M. (2002). How insects lose their limbs. *Nature* 415, 848–849.

Levinton, J. (2001). *Genetics, Paleontology, and Macroevolution*, 2nd edn. Cambridge University Press, Cambridge, UK.

Lewin, B. (2000). *Genes VII*. Oxford University Press, New York.

Lewin, R. (2003). *Principles of Human Evolution*. Blackwell Science, Malden, MA.

Lewontin, R.C. (1974). *The Genetic Basis of Evolutionary Change*. Columbia University Press, New York.

Lewontin, R.C. (1986). How important is population genetics for an understanding of evolution? *American Zoologist* 26, 811–820.

Lewontin, R.C. (2000). *The Triple Helix*. Harvard University Press, Cambridge, MA.

Lewontin, R.C., Moore, J.A., Provine, W.B. & Wallace, B. (eds) (1981). *Dobzhansky's Genetics of Natural Populatons I–XLIII*. Columbia University Press, New York.

Li, W-H. (1997). *Molecular Evolution*. Sinauer, Sunderland, MA.

Li, W-H., Tanimura, M. & Sharp, P.M. (1987). An evaluation of the molecular clock hypothesis using mammalian DNA sequences. *Journal of Molecular Evolution* 25, 330–342.

Lively, C.M. (1996). Host–parasites coevolution and sex. *Bioscience* 46, 107–114.

Lively, C.M. & Dybdahl, M.F. (2000). Parasite adaptation to locally common host genotypes. *Nature* 405, 679–681.

Losos, J.B. (2000). Ecological character displacement and the study of adaptation. *Proceedings of the National Academy of Sciences USA* 97, 5693–5695.

Losos, J.B. (2001). Evolution: a lizard's tale. *Scientific American* 284 (March), 56–61.

Losos, J.B. & Schluter, D. (2000). Analysis of an evolutionary species–area relationship. *Nature* 408, 847–850.

Losos, J.B., Jackman, T.R., Larson, A., de Queiroz, K. & Rodríguez-Schettino, L. (1998). Contingency and determinism in replicated adaptive radiations of island lizards. *Science* 279, 2115–2118.

Luria, S.E. & Delbruck, M. (1943). Mutations of bacteria from virus sensitivity to virus resistance. *Genetics* 28, 491–511.

Lyell, C. (1830–33). *Principles of Geology*, 3 vols. John Murray, London.

Lynch, M. & Conery, J.S. (2000). The evolutionary fate and consequences of duplicate genes. *Science* 290, 1151–1155.

Lynch, M. & Walsh, B. (1998). *Genetics and Analysis of Quantitative Traits*. Sinauer, Sunderland, MA.

Lynch, M., Blanchard, J., Houle, D. *et al.* (1999). Perspective: spontaneous deleterious mutation. *Evolution* 53, 645–663.

MacArthur, R.H. (1958). Population ecology of some warblers of northeastern coniferous forests. *Ecology* 39, 599–619.

MacFadden, B.J. (1992). *Fossil Horses. Systematics, paleobiology, and evolution of the family Equidae.* Cambridge University Press, New York.

Macgregor, H.C. (1991). Chromosomal heteromorphism in newts (*Triturus*) and its significance in relation to evolution and development. In Green, D.M. & Sessions, S.K. (eds) *Amphibian Cytogenetics and Evolution*, pp. 175–196. Academic Press, San Diego, CA.

Macgregor, H.C. & Horner, H.A. (1980). Heteromorphism for chromosome 1, a requirement for normal development in crested newts. *Chromosoma* 76, 111–122.

Machado, C.A., Jousselin, E., Kjellberg, F., Compton, S.G. & Herre, E.A. (2001). Phylogenetic relationships, historical biogeography, and character evolution of fig-pollinating wasps. *Proceedings of the Royal Society of London B* 268, 685–694.

Maddison, W.P. & Maddison, D.R. (2000). *MacClade, Version 4.* Sinauer Associates, Sunderland, MA.

Magurran, A.E. & May, R.M. (eds) (1999). *Evolution of Biological Diversity.* Oxford University Press, Oxford, UK.

Majerus, M.E.N. (1998) *Melanism: evolution in action.* Oxford Univesity Press, Oxford, UK.

Majerus, M.E.N. (2002). *Moths.* New Naturalist series. HarperCollins, London.

Mant, J.G., Schiestl, F.P., Peakall, R. & Weston, P.H. (2002). A phylogenetic study of pollinator conservatism among sexually deceptive orchids. *Evolution* 56, 888–898.

Mark Welch, D. & Meselsohn, M. (2000). Evidence for the evolution of bdelloid rotifers without sexual reproduction or genetic exchange. *Science* 288, 1211–1215.

Marshall, L.G., Webb, S.D., Sepkoski, J.J. & Raup, D.M. (1982). Mammalian evolution and the Great American Interchange. *Science* 215, 1351–1357.

Martin, A.P. (1999). Increasing genome complexity by gene duplication and the origin of the vertebrates. *American Naturalist* 154, 111–128.

Martin, L. (1985). Significance of enamel thickness in hominoid evolution. *Nature* 314, 260–263.

Martin, R.E. (2000). *Taphonomy.* Cambridge University Press, New York.

Martin, W., Stoebe, B., Goremykin, V., Hansmann, S., Hasegawa, M. & Kowallik, K.V. (1998). Gene transfer to the nucleus and the evolution of chloroplasts. *Nature* 393, 162–165.

Mather, K. (1943). Polygenic inheritance and natural selection. *Biological Reviews* 18, 32–64.

Mathews, S. & Donoghue, M.J. (1999). The root of angiosperm phylogeny inferred from duplicate phytochrome genes. *Science* 286, 947–950.

May, A.W. (1967). Fecundity of Atlantic cod. *Journal of the Fisheries Research Board of Canada* 24, 1531–1551.

Maynard Smith, J. (1976). Group selection. *Quarterly Review of Biology* 51, 277–283.

Maynard Smith, J. (1978). Optimization theory in evolution. *Annual Review of Ecology and Systematics* 9, 31–56.

Maynard Smith, J. (1986). *The Problems of Biology.* Oxford University Press, Oxford, UK.

Maynard Smith, J. (1987). How to model evolution. In: Dupré, J. (ed.) *The Latest on the Best*, pp. 119–31. MIT Press, Cambridge, MA.

Maynard Smith, J. (1998). *Evolutionary Genetics*, 2nd edn. Oxford University Press, Oxford, UK.

Maynard Smith, J. & Szathmáry, E. (1995). *The Major Transitions in Evolution.* W. H. Freeman/Spektrum, Oxford, UK and New York.

Maynard Smith, J. & Szathmáry, E. (1999). *Origins of Life.* Oxford University Press, Oxford, UK.

Maynard Smith, J., Burian, R., Kauffman, S. *et al.* (1985). Developmental constraints and evolution. *Quarterly Review of Biology* **60**, 265–287.

Maynard Smith, J., Smith, N.H., O'Rourke, M. & Spratt, B.G. (1993). How clonal are bacteria? *Proceedings of the National Academy of Sciences USA* **90**, 4384–4388.

Mayr, E. (1942). *Systematics and the Origin of Species.* Columbia University Press, New York. (Paperback reprint with new introduction, 1999, by Harvard University Press, Cambridge, MA.)

Mayr, E. (1963). *Animal Species and Evolution.* Harvard University Press, Cambridge, MA.

Mayr, E. (1976). *Evolution and the Diversity of Life.* Harvard University Press, Cambridge, MA.

Mayr, E. (1981). Biological classification: toward a synthesis of opposing methodologies. *Science* **214**, 510–516.

Mayr, E. (2001). *What Evolution Is.* Basic Books, New York and Weidenfeld & Nicolson, London.

Mayr, E. & Ashlock, P.D. (1991). *Principles of Systematic Zoology*, 2nd edn. McGraw-Hill, New York.

Mayr, E. & Diamond, J. (2001). *The Birds of Northern Melanesia.* Oxford University Press, New York.

Mayr, E. & Provine, W.B. (eds) (1980). *The Evolutionary Synthesis.* Harvard University Press, Cambridge, MA.

McCune, A.R. (1982). On the fallacy of constant extinction rates. *Evolution* **36**, 610–614.

McDonald, J.H. & Kreitman, M. (1991). Adaptive evolution at the *Adh* locus in *Drosophila. Nature* **351**, 652–654.

McGhee, J.D. (2000). Homologous tails? Or tales of homology? *Bioessays* **22**, 781–785.

McKenzie, J.A. (1996). *Ecological and Evolutionary Aspects of Insecticide Resistance.* Academic Press, San Diego, CA.

McKenzie, J.A. & Batterham, P. (1994). The genetic, molecular, and phenotypic consequences of selection for insecticide resistance. *Trends in Ecology and Evolution* **9**, 166–169.

McKenzie, J.A. & O'Farrell, K. (1993). Modification of developmental instability and fitness: malathion-resistance in the Australian sheep blowfly. *Genetica* **89**, 67–76.

McKinney, F.K. (1995). One hundred million years of competitive interactions between bryozoan clades: asymmetrical but not escalating. *Biological Journal of the Linnean Society* **56**, 465–481.

McKinney, F.K., Lidgard, S., Sepkoski, J.J. & Taylor, P.D. (1998). Decoupled temporal patterns of evolution and ecology in two post-Paleozoic clades. *Science* **281**, 807–809.

McMillan, W.O., Monteiro, A. & Kapan, D.D. (2002). Development and evolution on the wing. *Trends in Ecology and Evolution* **17**, 125–133.

McPhee, J. (1998). *Annals of the Former World.* Farrar, Straus & Giroux, New York.

McShea, D.W. (1998). Possible large-scale evolutionary trends in organismal evolution: eight "live hypotheses". *Annual Review of Ecology and Systematics* **29**, 293–318.

Meffert, L.M. (1999). How speciation experiments relate to conservation biology. *Bioscience* **49**, 701–711.

Meier, R. (1997). A test and review of the empirical performance of the ontogenetic criterion. *Systematic Biology* **46**, 699–721.

Messier, W. & Stewart, C-B. (1997). Episodic adaptive evolution of primate lysozymes. *Nature* **385**, 151–154.

Meyerowitz, E.M. (2002). Plants compared to animals: the broadest comparative study of development. *Science* **295**, 1482–1485.

Milinkovitch, M.C., Ortí, G. & Meyer, A. (1993). Revised phylogeny of whales suggested by mitochondrial ribosomal DNA sequences. *Nature* **361**, 346–348.

Miller, A.I. (1998). Biotic transitions in global marine diversity. *Science* **281**, 1157–1160.

Mindell, D.P. & Honeycutt, R.L. (1990). Ribosomal RNA in vertebrates and phylogenetic applications. *Annual Review of Ecology and Systematics* **21**, 541–566.

Mindell, D.P. & Meyer, A. (2001). Homology evolving. *Trends in Ecology and Evolution* **16**, 434–440.

Mitton, J. (1998). *Selection in Natural Populations*. Oxford University Press, New York.

Mivart, G.J. (1871). *The Genesis of Species*. Macmillan, London.

Møller, A.P. (1994). *Sexual Selection and the Barn Swallow*. Oxford University Press, Oxford, UK.

Mooers, A.Ø. & Holmes, E.C. (2000). The evolution of base composition and phylogenetic inference. *Trends in Ecology and Evolution* **15**, 365–369.

Moore, J. & Willmer, P. (1997). Convergent evolution in invertebrates. *Biological Reviews* **72**, 1–60.

Moore, J.A. (2002). *From Genesis to Genetics: the case of evolution and creationism*. University of California Press, San Francisco, CA.

Moritz, C., Patton, J.L., Schneider, C.J. & Smith, T.B. (2000). Diversification of rainforest faunas: an integrated molecular approach. *Annual Review of Ecology and Systematics* **31**, 533–563.

Morrow, J.R., Schindler, E. & Walliser, O.H. (1996). Phanerozoic development of selected global environmental features. In Walliser, O.H. (ed.) *Global Events and Event Stratigraphy in the Phanerozoic*, pp. 53–61. Springer-Verlag, Berlin.

Muir, G., Fleming, C.C. & Schlötterer, C. (2000). Species status of hybridizing oaks. *Nature* **405**, 1016.

Muir, W.M. (1995). Group selection for adaptation to multiple-hen cages: selection program and direct responses. *Poultry Science* **75**, 447–458.

Mukai, T., Chigusa, S.I., Mettler, L.E. & Crow, J.F. (1972). Mutation rate and dominance of genes affecting viability in *Drosophila melanogaster*. *Genetics* **72**, 335–355.

Muller, H.J. (1959). One hundred years without Darwinism are enough. *School Science and Mathematics* **49**, 314–318.

Mumme, R.L. (1992) Do helpers increase reproductive success: an experimental analysis in the Florida scrub jay. *Behavioral Ecology and Sociobiology* **31**, 319–328.

Murray, J. & Clarke, B. (1980). The genus *Partula* on Moorea: speciation in progress. *Proceedings of the Royal Society of London B* **211**, 83–117.

Nachman, M.W. & Crowell, S.L. (2000) Estimate of the mutation rate per nucleotide in humans. *Genetics* **156**, 297–304.

Nachman, M.W. & Searle, J.B. (1995). Why is the house mouse karyotype so variable? *Trends in Ecology and Evolution* **10**, 397–402.

Nei, M. & Kumar, S. (2000). *Molecular Evolution and Phylogenetics*. Oxford University Press, New York.

Nesse, R.M. & Williams, G.C. (1995). *Why We get Sick: the new science of Darwinian medicine*. Times Books, New York. (Also published as *Evolution and Healing*, 1995, by Weidenfeld & Nicolson, London.)

Nevo, E. (1988). Genetic diversity in nature. *Evolutionary Biology* **23**, 217–246.

Newell, N.D. (1959). The nature of the fossil record. *Proceedings of the American Philosophical Society* **103**, 264–285.

Nielsen, C. (2001). *Animal Evolution: interrelationships of the living phyla*, 2nd edn. Oxford University Press, Oxford, UK.

Nielsen, R. (2001). Statistical tests of selective neutrality in the age of genomics. *Heredity* **86**, 641–647.

Niklas, K.J. (1986). Large-scale changes in animal and plant terrestrial communities. In Raup, D.M. & Jablonski, D. (eds) *Patterns and Processes in the History of Life*, pp. 383–405. Dahlem Workshop. John Wiley, Chichester, UK.

Niklas, K.J. (1997). *The Evolutionary Biology of Plants*. University of Chicago Press, Chicago, IL.

Nilsson, D-E. & Pelger, S. (1994). A pessimistic estimate of the time required for an eye to evolve. *Proceedings of the Royal Society of London B* **256**, 53–58.

Nisbet, E. (2000). The realms of Archaean life. *Nature* **405**, 625–626.

Nitecki, M.H. (ed.) (1990). *Evolutionary Innovations*. University of Chicago Press, Chicago, IL.

Nixon, K.C. & Wheeler, Q.D. (1990). An amplification of the phylogenetic species concept. *Cladistics* **6**, 211–223.

Noor, M. (1999). Reinforcement and other consequences of sympatry. *Heredity* **83**, 503–508.

Noor, M., Grams, K.L., Bertucci, L.A. & Reiland, J. (2001). Chromosomal inversions and the reproductive isolation of species. *Proceedings of the National Academy of Sciences USA* **98**, 12084–12088.

Nordenskiöld, E. (1929). *The History of Biology*. Knopf, New York.

Nosil, P., Crespi, B.J. & Sandoval, C.P. (2002). Host–plant adaptation drives the parallel evolution of reproductive isolation. *Nature* **417**, 440–443.

Novacek, M.J. (1992). Mammalian phylogeny: shaking the tree. *Nature* **356**, 121–125.

Novacek, M.J. (2001). Mammalian phylogeny: genes and supertrees. *Current Biology* **11**, R573–575.

Novak, S.J., Soltis, D.E. & Soltis, P.S. (1991). Ownbey's tragopogons: 40 years later. *American Journal of Botany*, **78**, 1586–1600.

Numbers, R.L. (1992). *The Creationists: the evolution of scientific creationism*. Knopf, New York. (Paperback edition, 1993, by University of California Press, Berkeley, CA.)

Numbers, R.L. (1998). *Darwinism Comes to America*. Harvard University Press, Cambridge, MA.

Nurminsky, D.M., De Aguiar, D., Bustamante, C.D. & Hartl, D. (2001). Chromosomal effects of rapid gene evolution in *Drosophila melanogaster*. *Science* **291**, 128–130.

O'Brien, S.J. & Stanyon, R. (1999). Ancestral primate viewed. *Nature* **402**, 365–366.

O'Brien, S.J., Menotti-Raymond, M., Murphy, W.J. *et al.* (1998). The promise of comparative genomics in mammals. *Science* **286**, 358–463.

Ochman, H. & Moran, N.A. (2001). Genes lost and genes found: evolution of bacteria, pathogenesis and symbiosis. *Science* **292**, 1096–1098.

Ochman, H., Elwyn, S. & Moran, N. (1999). Calibrating bacterial evolution. *Proceedings of the National Academy of Sciences USA* **96**, 12638–12643.

Ochman, H., Jones, J.S. & Selander, R.K. (1983). Molecular area effects in *Cepaea*. *Proceedings of the National Academy of Sciences USA* **80**, 4189–4193.

Ohno, S. (1970). *Evolution by Gene Duplication*. Springer, New York.

Ohta, T. (1992). The nearly neutral theory of molecular evolution. *Annual Review of Ecology and Systematics* **23**, 263–286.

Ohta, T. (2000). Near-neutrality in evolution of genes and gene regulation. *PNAS* **99**, 16134–16137.

Ohta, T. & Gillespie, J.H. (1996). Development of neutral and nearly neutral theories. *Theoretical Population Biology* **49**, 128–142.

Orr, H.A. (1998). The population genetics of adaptation: the distribution of factors fixed during adaptive evolution. *Evolution* **52**, 935–949.

Orr, H.A. (2001). The genetics of species differences. *Trends in Ecology and Evolution* **16**, 343–350.

Orr, H.A. & Coyne, J.A. (1992). The genetics of adaptation: a reassessment. *American Naturalist* **140**, 725–742.

Orr, H.A. & Presgraves, D.C. (2000). Speciation by postzygotic isolation: forces, genes and molecules. *Bioessays* **22**, 1085–1094.

Orzack, S.H. & Sober, E. (1994). Optimality models and the test of adaptationism. *American Naturalist* **143**, 361–380.

Osawa, S. (1995). *Evolution of the Genetic Code*. Oxford University Press, New York.

Otte, D. & Endler, J.A. (eds) (1989). *Speciation and its Consequences*. Sinauer, Sunderland, MA.

Otto, S.P. & Lenormand, T. (2002). Resolving the paradox of sex and recombination. *Nature Reviews Genetics* **3**, 252–261.

Ownbey, M. (1950). Natural hybridization and amphiploidy in the genus *Tragopogon*. *American Journal of Botany* **27**, 487–499.

Page, R. (ed.) (2002). *Tangled Trees*. University of Chicago Press, Chicago, IL.

Page, R. & Holmes, E.C. (1998). *Molecular Evolution*. Blackwell Science, Oxford, UK.

Pagel, M. (1999). Inferring the historical patterns of biological evolution. *Nature* **401**, 877–884.

Pagel, M. (ed.) (2002). *Encyclopedia of Evolution*. Oxford University Press, New York.

Palumbi, S.R. (2001a). Humans as the world's greatest evolutionary force. *Science* **293**, 1786–1790.

Palumbi, S.R. (2001b). *The Evolution Explosion: how humans cause rapid evolutionary change*. W. W. Norton, New York.

Panhuis, T.M., Butlin, R., Zuk, M. & Tregenza, T. (2001) Sexual selection and speciation. *Trends in Ecology and Evolution* **16**, 364–372.

Parker, G.A. & Maynard Smith, J. (1990). Optimality theory in evolutionary biology. *Nature* **348**, 27–33.

Paterson, H.E.H. (1993). *Evolution and the Recognition Concept of Species: collected writings*. Johns Hopkins University Press, Baltimore, MD.

Patterson, C. (1981). Methods of paleobiogeography. In Nelson, G. & Rosen, D.E. (eds) *Vicariance Biogeography: a critique*, pp. 446–489. Columbia University Press, New York.

Pease, C. (1992). On the declining extinction and origination rates of fossil taxa. *Paleobiology* **18**, 89–92.

Pellmyr, O., Leebens-Mack, J. & Thompson, J.N. (1998). Herbivores and molecular clocks as tools in plant biogeography. *Biological Journal of the Linnean Society* **63**, 367–378.

Pennock, R.T. (2000). *Tower of Babel: the evidence against the new creationism*. MIT Press, Cambridge, MA.

Pennock, R.T. (ed.) (2001). *Intelligent Design Creationism and its Critics*. MIT Press, Cambridge, MA.

Penny, D., Foulds, L.R. & Hendy, M.D. (1982). Testing the theory of evolution by comparing the phylogenetic trees constructed from five different protein sequences. *Nature* **297**, 197–200.

Peters, S.E. & Foote, M. (2002). Determinants of extinction in the fossil record. *Nature* **416**, 420–424.

Petrov, D.A., Sangster, T.A., Johnston, J.S., Hartl, D.L. & Shaw, K.L. (2000). Evidence for DNA loss as a determinant of genome size. *Science* **287**, 1060–1062.

Philippe, H. & Forterre, P. (1999). The rooting of the universal tree of life is not reliable. *Journal of Molecular Evolution* **49**, 509–523.

Pielou, E.C. (1991). *After the Ice Age: the return of life to glaciated North America*. University of Chicago Press, Chicago, IL.

Pierce, N.E. & Mead, P.S. (1981). Parasitoids as selective agents in the symbiosis between lycaenid butterfly larvae and ants. *Science* **211**, 1185–1187.

Pigliucci, M. (2002). Buffer zone. *Nature* **417**, 598–599.

Pigliucci, M. & Kaplan, J. (2000). The rise and fall of Dr Pangloss: adaptationism and the spandrels paper 20 years later. *Trends in Ecology and Evolution* **15**, 66–70.

Podos, J. (2001). Correlated evolution of morphology and vocal signal structure in Darwin's finches. *Nature* **409**, 185–187.

Powell, J.R. (1997). *Progress and Prospects in Evolutionary Biology: the Drosophila model*. Oxford University Press, New York.

Primack, R.B. & Kang, H. (1989). Measuring fitness and natural selection in wild plant populations. *Annual Review of Ecology and Systematics* **20**, 367–396.

Proctor, H. & Owens, I. (2000). Mites and birds: diversity, parasites, and coevolution. *Trends in Ecology and Evolution* **15**, 358–364.

Provine, W.B. (1971). *The Origins of Theoretical Population Genetics*. University of Chicago Press, Chicago, IL. (Reprinted 2001 with afterword.)

Provine, W.B. (1986). *Sewall Wright and Evolutionary Biology*. University of Chicago Press, Chicago, IL.

Prum, R.D. & Brush, A.H. (2002). The evolutionary origin and diversification of feathers. *Quarterly Review of Biology* **77**, 261–295.

Przeworski, M., Hudson, R.R. & Di Rienzo, A. (2000). Adjusting the focus on human variation. *Trends in Genetics* **16**, 296–302.

Ptashne, M. & Gann, A. (1998). Imposing specificity by localization: mechanism and evolvability. *Current Biology* **8**, R812–822.

Pupo, G.M., Lan, R. & Reeves, P.R. (2000). Multiple independent origins of *Shigella* clones of *Escherichia coli* and convergent evolution of many of their characteristics. *Proceedings of the National Academy of Sciences USA* **97**, 10567–10572.

Raff, R.A. (1996). *The Shape of Life*. University of Chicago Press, Chicago, IL.

Ramsey, J. & Schemske, D.W. (1998). Pathways, mechanisms, and rates of polyploid formation in flowering plants. *Annual Review of Ecology and Systematics* **29**, 467–501.

Rand, D. (2000). Mitochondrial genomics flies high. *Trends in Ecology and Evolution* **16**, 2–4.

Rand, D.M. (2001). The units of selection on mitochondrial DNA. *Annual Review of Ecology and Systematics* **32**, 415–448.

Raup, D.M. (1966). Geometric analysis of shell coiling. *Journal of Paleontology* **40**, 1178–1190.

Raup, D.M. (1986). Biological extinction in earth history. *Science* **231**, 1528–1533.

Rausher, M.D. (2001). Co-evolution and plant resistance to natural enemies. *Nature* **411**, 857–864.

Reeve, H.K. & Sherman, P.W. (1993). Adaptation and the goals of evolutionary research. *Quarterly Review of Biology* **68**, 1–32.

Reich, D.E., Cargill, M., Bolk, S. *et al.* (2001). Linkage disequilibrium in the human genome. *Nature* **411**, 199–204.

Remington, C.L. (1968). Suture-zones of hybrid interaction between recently joined biotas. *Evolutionary Biology* **2**, 321–428.

Reznick, D.N., Shaw, F.H., Rodd, F.H. & Shaw, R.G. (1997). Evaluation of the rate of evolution in natural populations of guppies (*Poecilia reticulata*). *Science* **275**, 1934–1936.

Rice, W.R. (2002). Experimental tests of the adaptive significance of sexual reproduction. *Nature Reviews Genetics* 3, 241–251.

Rice, W.R. & Chippindale, A.K. (2001). Sexual recombination and the power of natural selection. *Science* 294, 555–559.

Rice, W.R. & Hostert, E.E. (1993). Laboratory experiments on speciation: what have we learned in 40 years? *Evolution* 47, 1637–1653. (Reprinted in Ridley 1997.)

Richardson, J.E., Weitz, F.M., Fay, M.F. *et al.* (2001). Rapid and recent origin of species richness in the Cape flora of South Africa. *Nature* 412, 181–183.

Ricker, W.E. (1981). Changes in the average size and average age of Pacific salmon. *Canadian Journal of Fisheries and Aquatic Sciences* 38, 1636–1656.

Ricklefs, R.E. & Miller, G.L. (2000). *Ecology*, 4th edn. W. H. Freeman, New York.

Ridley, M. (1986). *Evolution and Classification: the reformation of cladism.* Longman, London.

Ridley, M. (ed.) (1997). *Evolution.* Oxford Readers. Oxford University Press, New York.

Ridley, M. (2001). *The Cooperative Gene.* Free Press, New York. (Also published as *Mendel's Demon*, 2000, by Weidenfeld & Nicolson, London.)

Rieseberg, L.H. (1997). Hybrid origins of plant species. *Annual Review of Ecology and Systematics* 28, 359–389.

Rieseberg, L.H. (2001). Chromosomal rearrangements and speciation. *Trends in Ecology and Evolution* 16, 351–357.

Rieseberg, L.H. & Wendel, J.F. (1993). Introgression and its consequences in plants. In Harrison, R.G. (ed.) *Hybrid Zones and the Evolutionary Process*, pp. 70–109. Oxford University Press, New York.

Rieseberg, L.H., Sinervo, B., Linder, C.R., Ungerer, M.C. & Arias, D.M. (1996). Role of gene interactions in hybrid speciation: evidence from ancient and experimental hybrids. *Science* 272, 741–745. (Reprinted in Ridley 1997.)

Ritchie, M.G. & Phillips, S.D.F. (1998). The genetics of sexual isolation. In Howard, D. & Berlocher, S. (eds) *Endless Forms: species and speciation*, pp. 291–308. Oxford University Press, New York.

Ritvo, H. (1997). *The Platypus and the Mermaid and Other Figments of the Classifying Imagination.* Harvard University Press, Cambridge, MA.

Robson, G.C. & Richards, O.W. (1936). *The Variations of Animals in Nature.* Longman, London.

Roff, D.A. (1997). *Evolutionary Quantitative Genetics.* Chapman & Hall, New York.

Rose, M.R. & Lauder, G.V. (eds) (1996). *Adaptation.* Academic Press, San Diego, CA.

Rosen, D.E., Forey, P.L., Gardiner, B.C. & Patterson, C. (1981). Lungfishes, tetrapods, paleontology, and plesiomorphy. *Bulletin of the American Museum of Natural History* 167, 159–276.

Rosenberg, M.S. & Kumar, S. (2001). Incomplete taxon sampling is not a problem for phylogenetic inference. *Proceedings of the National Academy of Sciences USA* 98, 10751–10756.

Ross, J. (1982). Myxomatosis: the natural evolution of the disease. In Edwards, M.A. & McDonnell, U. (eds) *Animal Disease in Relation to Animal Conservation*, pp. 77–95. Symposia of the Zoological Society of London No. 50. Academic Press, London.

Rudwick, M.J.S. (1964). The inference of function from structure in fossils. *British Journal for the Philosophy of Science* 15, 27–40.

Rudwick, M.J.W. (1997). *Georges Cuvier, Fossil Bones, and Geological Catastrophes: new translations and interpretations of the primary texts.* University of Chicago Press, Chicago, IL.

Runnegar, B. (2000). Loophole for snowball Earth. *Nature* **405**, 403–404.

Saetre, G-P., Moum, T., Bures, S., Král, M., Adamjan, M. & Moreno, J. (1997). A sexually selected character displacement in flycatchers reinforces premating isolation. *Nature* **387**, 589–592.

Sarich, V. & Wilson, A.C. (1967). Immunological time scale for hominid evolution. *Science* **158**, 1200–1203.

Scharloo, W. (1987). Constraints in selective response. In Loeschcke, V. (ed.) *Genetic Constraints on Adaptive Evolution*, pp. 125–149. Springer-Verlag, Berlin.

Scharloo, W. (1991). Canalization: genetic and developmental aspects. *Annual Review of Ecology and Systematics* **22**, 65–94.

Schemske, D.W. & Bierzychudek, P. (2001). Evolution of flower color in the desert annual *Linanthus parryae*: Wright revisited. *Evolution* **55**, 1269–1282.

Schemske, D.W. & Bradshaw, H.D. (1999). Pollinator preferences and the evolution of floral traits in monkey flowers (*Mimulus*). *Proceedings of the National Academy of Sciences USA* **96**, 11910–11915. (Pop summary by Charlesworth, B. (2000) in *Current Biology* **10**, R68–70.)

Schiltuizen, M. (2001). *Frogs, Flies, and Dandelions: the making of a species.* Oxford University Press, Oxford, UK.

Schliekelman, P., Garner, C. & Slatkin, M. (2001). Natural selection and resistance to HIV. *Nature* **411**, 545.

Schliewen, U.K., Tautz, D. & Pääbo, S. (1994). Sympatric speciation suggested by monophyly of crater lake cichlids. *Nature* **368**, 629–632.

Schluter, D. (2000). *The Ecology of Adaptive Radiation.* Oxford University Press, Oxford, UK.

Schoonhoven, L.M., Jermy, T. & van Loon, J.J.A. (1998). *Insect–Plant Biology.* Chapman & Hall, London.

Schopf, J.W. (1993). Microfossils of the Early Archean Apex Chert: new evidence of the antiquity of life. *Science* **260**, 640–645.

Schopf, J.W. (1994). Disparate rates, differing fates: tempo and mode of evolution changed from the Precambrian to the Phanerozoic. *Proceedings of the National Academy of Sciences USA* **91**, 6735–6742. (Reprinted in Ridley 1997.)

Schopf, J.W. (1999). *Cradle of Life.* Princeton University Press, Princeton, NJ.

Schuh, R.T. (2000). *Biological Systematics: principles and application.* Comstock Publishing, Ithaca, NY.

Schuurman, R., Nijhuis, M., van Leeuwen, R. *et al.* (1995). Rapid changes in human immunodeficiency virus type I RNA load and appearance of drug-resistant virus populations in persons treated with lamivudine (3TC). *Journal of Infectious Diseases* **175**, 1411–1419.

Seehausen, O. & van Alphen, J.J.M. (1998). The effect of male coloration on female mate choice in closely related Lake Victoria cichlids (*Haplochromis nyererei* complex). *Behavioral Ecology and Sociobiology* **42**, 1–8.

Seehausen, O., van Alphen, J.M. & Witte, F. (1997). Cichlid fish diversity threatened by eutrophication that curbs sexual selection. *Science* **277**, 1808–1811.

Sepkoski, J.J. (1992). Ten years in the library: new data confirm paleontological patterns. *Paleobiology* **19**, 43–51.

Sepkoski, J.J. (1996). Patterns of Phanerozoic extinction: a perspective from global databases. In Walliser, O.H. (ed.) *Global Events and Event Stratigraphy in the Phanerozoic*, pp. 35–51. Springer-Verlag, Berlin.

Sepkoski, J.J., McKinney, F.M. & Lidgard, S. (2000). Competitive displacement among post-Paleozoic cyclostome and cheilostome bryozoans. *Paeobiology* **26**, 7–18.

Sequeira, A.S., Lanteri, A.A., Scataglini, M.A., Confalmieri, V.A. & Farrell, B.D. (2000). Are flightless *Galapaganus* weevils older than the Galápagos Islands they inhabit? *Heredity* 85, 20–29.

Sereno, P.C. (1999). The evolution of dinosaurs. *Science* 284, 2137–2147.

Servedio, M.R. (2001). Beyond reinforcement: the evolution of premating isolation by direct selection on preferences and postmating, prezygotic incompatibilities. *Evolution* 55, 1909–1920.

Shabalina, S.A., Ogurtsov, A.Y., Kondrashov, V.A. & Kondrashov, A.S. (2001). Selective constraint in intergenic regions of human and mouse genomes. *Trends in Genetics* 17, 373–376.

Shaffer, H.B. (1984). Evolution in a paedomorphic lineage. I. An electrophoretic analysis of the Mexican ambystomatid salamanders. *Evolution* 38, 1194–1216.

Sharp, P.M., Averof, M., Lloyd, A.T., Matassi, G. & Peden, F.J. (1995). DNA sequence evolution: the sounds of silence. *Philosophical Transactions of the Royal Society of London B* 349, 241–247.

Shear, W.A. (1991). The early development of terrestrial ecosystems. *Nature* 351, 283–289.

Sheehan, P.M., Fastovsky, D.E., Hoffmann, R.G., Berghaus, C.B. & Gabriel, D.L. (1991). Sudden extinction of the dinosaurs: latest Cretaceous, Upper Great Plains, USA. *Science* 254, 835–839.

Sheldon, P.R. (1987). Parallel gradualistic evolution of Ordovician trilobites. *Nature* 330, 561–563.

Sheldon, P.R. (1996). Plus ça change — a model for stasis and evolution in different environments. *Palaeogeography, Palaeoclimatology, Palaeoecology* 127, 209–227.

Sibley, C.G. & Ahlquist, J.E. (1987). DNA hybridization evidence of hominoid phylogeny: results from an expanded data set. *Journal of Molecular Evolution* 26, 99–121.

Sidor, C.A. & Hopson, J.A. (1998). Ghost lineages and "mammalness": assessing the temporal pattern of character acquisition in the Synapsida. *Paleobiology* 24, 254–273.

Silberglied, R.E., Ainello, A. & Windsor, D.M. (1980). Disruptive coloration in butterflies: lack of support in *Anartia fatima*. *Science* 209, 617–619.

Silva, J.C. & Kondrashov, A.S. (2002). Patterns in spontaneous mutation revealed by human–baboon sequence comparison. *Trends in Genetics* 18, 544–546.

Simmons, E.L. (1996). The evolutionary genetics of plant–pathogen systems. *Bioscience* 46, 136–145.

Simpson, G.G. (1944). *Tempo and Mode in Evolution*. Columbia University Press, New York.

Simpson, G.G. (1949). *The Meaning of Evolution*. Yale University Press, New Haven, CT.

Simpson, G.G. (1953). *The Major Features of Evolution*. Columbia University Press, New York.

Simpson, G.G. (1961a). One hundred years without Darwin are enough. *Teachers College Record* 60, 617–626. (Reprinted in Simpson, G.G. (1964). *This View of Life*. Harcourt, Brace & World, New York.)

Simpson, G.G. (1961b). *Principles of Animal Taxonomy*. Columbia University Press, New York.

Simpson, G.G. (1978). *Concession to the Improbable*. Yale University Press, New Haven, CTt.

Simpson, G.G. (1980). *Splendid Isolation*. Yale University Press, New Haven, CT.

Simpson, G.G. (1983). *Fossils and the History of Life*. Scientific American Library, New York.

Singer, R. (ed.) (1999). *Encycopedia of Paleontology*, 2 vols. Fitroy Dearborn, Chicago, IL.

Singh, R.S. & Krimbas, C. (eds) (2000). *Evolutionary Genetics*. Cambridge University Press, New York.

Slack, J., Holland, P.H.H. & Graham, C.F. (1993). The zootype and the phylotypic stage. *Nature* **361**, 490–493. (Reprinted in Ridley 1997.)

Sloan, R.E., Rigby, J.K., Van Valen, L.M. & Gabriel, D. (1986). Gradual dinosaur extinction and simultaneous ungulate radiation in the Hell Creek Formation. *Science* **232**, 629–633.

Smit, J. & van der Kaars, S. (1984). Terminal Cretaceous extinctions in the Hell Creek area, Montana: compatible with catastrophic extinction. *Science* **223**, 1177–1179.

Smith, K.K. (2001). Heterochrony revisited: the evolution of developmental sequences. *Biological Journal of the Linnean Society* **73**, 169–186.

Smith, N.G.C. & Eyre-Walker, A. (2002). Adaptive protein evolution in *Drosophila*. *Nature* **415**, 1022–1024.

Smith, T.B. & Girman, D.J. (2000). Reaching new adaptive peaks: evolution of alternative forms in an African finch. In Mousseau, T.A., Sinervo, B. & Endler, J. (eds) *Adaptive Genetic Variation in the Wild*, pp. 139–156. Oxford University Press, New York.

Smith, T.B., Wayne, R.K., Girman, D.J. & Bruford, M.W. (1997). A role for ecotones in generating rainforest biodiversity. *Science* **276**, 1855–1857.

Sneath, P.H.A. & Sokal, R.R. (1973). *Numerical Taxonomy*, 2nd edn. W. H. Freeman, New York.

Sniegowski, P.D., Gerrish, P.J., Johnson, T. & Shaver, A. (2000). The evolution of mutation rates: separating causes from effects. *Bioessays* **22**, 1067–1074.

Sober, E. (1989). *Reconstructing the Past*. MIT Press, Cambridge, MA.

Sober, E. (ed.) (1994). *Conceptual Issues in Evolutionary Biology*, 2nd edn. MIT Press, Cambridge, MA.

Sober, E. & Wilson, D.S. (1998). *Unto Others*. Harvard University Press, Cambridge, MA.

Sokal, R.R. (1966). Numerical taxonomy. *Scientific American* **215** (December), 106–116.

Solé, R.V., Manrubia, S.C., Benton, M. & Bak, P. (1997). Self-similarity of extinction statistics in the fossil record. *Nature* **388**, 764–767.

Soltis, D.E. & Soltis, P.S. (1999). Polyploidy: recurrent formation and genome evolution. *Trends in Ecology and Evolution* **14**, 348–352.

Sommer, S.S. (1995). Recent human germ-line mutation: inferences from patients with hemophilia B. *Trends in Genetics* **11**, 141–147.

Stanley, S.M. (1979). *Macroevolution*. W. H. Freeman, San Francisco.

Stebbins, G.L. (1950). *Plant Variation and Evolution*. Columbia University Press, New York.

Stebbins, R. (1994). Biology's four horsemen of the apocalypse [interview]. In *Life on the Edge*, pp. 228–239. Heyday Books, San Francisco.

Steel, M. & Penny, D. (2000). Parsimony, likelihood, and the role of models in molecular phylogenetics. *Molecular Biology and Evolution* **17**, 839–850.

Stehli, F.G. & Webb, S.D. (eds) (1985). *The Great American Biotic Interchange*. Plenum Press, New York.

Stenseth, N.C. & Maynard Smith, J. (1984). Coevolution in ecosystems: Red Queen evolution or stasis? *Evolution* **38**, 870–880.

Stiassny, M.L. & Meyer, A. (1999). Cichlids of the rift lakes. *Scientific American* **280** (February), 44–49.

Strickberger, M. (1990). *Evolution*. Jones & Bartlett, Boston, MA.

Sun, G., Ji, Q., Dilcher, D.L. *et al.* (2002). Archaefructaceae, a new basal angiosperm family. *Science* **296**, 899–904.

Surlyk, F. & Johansen, M.B. (1984). End-Cretaceous brachiopod extinctions in the chalk of Denmark. *Science* **223**, 1174–1177.

Swanson, W.J. & Vacquier, V.D. (1998). Concerted evolution in an egg receptor for a rapidly evolving abalone sperm protein. *Science* **281**, 710–712.

Swanson, W.J. & Vacquier, V.D. (2002). The rapid evolution of reproductive proteins. *Nature Reviews Genetics* **3**, 137–144.

Swofford, D.L. (2002). *PAUP: phylogenetic analysis using parsimony*. Sinauer, Sunderland, MA.

Swofford, D.L., Olsen, G.J. & Waddell, P. (1996). Phylogeny reconstruction. In Hillis, D.M., Moritz, C. & Mable, B.K. (eds) *Molecular Systematics*, 2nd edn, pp. 407–514. Sinauer, Sunderland, MA.

Tao, Y., Hartl, D.L. & Laurie, C.C. (2001). Sex ratio distortion associated with reproductive isolation in *Drosophila*. *Proceedings of the National Academy of Sciences USA* **98**, 13183–13188.

Taper, M.L. & Case, T.J. (1992). Coevolution among competitors. *Oxford Surveys in Evolutionary Biology* **8**, 63–109.

Tavaré, S., Marshall, C.R., Will, O., Soligo, C. & Martin, R.D. (2002). Using the fossil record to estimate the age of the last common ancestor of extant primates. *Nature* **416**, 726–729.

Taylor, C.E. (1986). Genetics and evolution of resistance to insecticides. *Biological Journal of the Linnean Society* **27**, 103–112.

Templeton, A.R. (1993). The "Eve" hypothesis: a genetic critique and reanalysis. *American Anthropologist* **95**, 51–72.

Templeton, A.R. (1996). Experimental evidence for the genetic-transilience model of speciation. *Evolution* **50**, 909–915.

Templeton, A.R. (1998). Species and speciation: geography, population structure, ecology, and gene trees. In Howard, D. & Berlocher, S. (eds) *Endless Forms: species and speciation*, pp. 32–43. Oxford University Press, New York.

Thompson, D'A.W. (1942). *On Growth and Form*, 2nd edn. Cambridge University Press, Cambridge, UK.

Thompson, J.N. (1994). *The Coevolutionary Process*. University of Chicago Press, Chicago, IL.

Thompson, J.N. & Cunningham, B.M. (2002). Geographic structure and dynamics of coevolutionary selection. *Nature* **417**, 735–738.

Thornton, I. (1996). *Krakatau: the destruction and reassembly of an island ecosystem*. Harvard University Press, Cambridge, MA.

Timm, R.M. (1983). Fahrenholz's rule and resource tracking: a study of host–parasite coevolution. In Nitecki, R.M. (ed.) *Coevolution*, pp. 225–265. University of Chicago Press, Chicago, IL.

Ting, C-T., Tsaur, S-C. & Wu, C-I. (2000). The phylogeny of closely related species as revealed by the speciation gene, *Odysseus*. *Proceedings of the National Academy of Sciences USA* **97**, 5313–5316.

Ting, C-T., Tsaur, S-C., Wu, M-L. & Wu, C-I. (1998). A rapidly evolving homeobox at the site of a hybrid sterility gene. *Science* **282**, 1501–1504.

Travis, J. (1989). The role of optimizing selection in natural populations. *Annual Review of Ecology and Systematics* **20**, 279–296.

Travisano, M. (2001). Towards a genetical theory of adaptation. *Current Biology* **11**, R440–442.

Trivers, R.L. & Willard, D.E. (1973). Natural selection of parental ability to vary the sex ratio of offspring. *Science* **179**, 90–92.

True, J.R., Weir, B.S. & Laurie, C.C. (1996). A genome-wide survey of hybrid incompatibility factors. *Genetics* **142**, 819–837.

Tudge, C. (1992). Last stand for Society snails. *New Scientist* **135**, July 11, 25–29.

Turelli, M., Barton, N.H. & Coyne, J.A. (2001a). Theory and speciation. *Trends in Ecology and Evolution* **16**, 330–342.

Turelli, M., Schemske, D.W. & Bierzychudek, P. (2001b). Stable two-allele polymorphisms maintained by fluctuating fitnesses and seed banks: protecting the blues. *Evolution* **55**, 1283–1298.

Turner, J.R.G. (1976). Muellerian mimicry: classical "beanbag" evolution, and the role of ecological islands in race formation. In Karlin, S. & Nevo, E. (eds) *Population Genetics and Ecology*, pp. 185–218. Academic Press, New York.

Turner, J.R.G. (1977). Butterfly mimicry: the genetical evolution of an adaptation. *Evolutionary Biology* **11**, 163–206.

Turner, J.R.G. (1984). Mimicry: the palatability spectrum and its consequences. In Vane-Wright, R.I. & Ackery, P.R. (eds) *The Biology of Butterflies*, pp. 141–161. Academic Press, London.

Turner, J.R.G. & Mallett, J. (1996). Did forest islands drive the diversity of warningly coloured butterflies? Biotic drift and the shifting balance. *Philosophical Transactions of the Royal Society of London B* **351**, 835–845.

Ulizzi, L. & Manzotti, C. (1988). Birth weight and natural selection: an example of selection relaxation in man. *Human Heredity* **38**, 129–135.

Ulizzi, L., Astolfi, P. & Zonta, L.A. (1998). Natural selection in industrialized countries: a study of three generations of Italian newborns. *Annals of Human Genetics* **62**, 47–53.

Ungerer, M.C., Baird, S.J.E., Pan, J. & Rieseberg, L.H. (1998). Rapid hybrid speciation in wild sunflowers. *Proceedings of the National Academy of Sciences USA* **95**, 11757–11762.

Van Oosterzee, P. (1997). *Where Worlds Collide: the Wallace Line*. Cornell University Press, Ithaca, NY.

Van Valen, L.M. (1973). A new evolutionary law. *Evolutionary Theory* **1**, 1–30.

Van Valen, L.M. (1976). Ecological species, multispecies, and oaks. *Taxon* **25**, 233–239.

Van Valen, L.M. & Sloan, R.E. (1977). Ecology and the extinction of the dinosaurs. *Evolutionary Theory* **2**, 37–64.

van Zuilen, M.A., Lepland, A. & Arrhenius, G. (2002). Reassessing the evidence for the earliest traces of life. *Nature* **418**, 627–630.

Veen, T., Borge, T., Griffiths, S.C. *et al.* (2001). Hybridization and adaptive mate choice in flycatchers. *Nature* **411**, 45–50.

Vermeij, G.J. (1987). *Evolution and Escalation*. Princeton University Press, Princeton, NJ.

Vermeij, G.J. (1991). When biotas meet: understanding biotic interchange. *Science* **253**, 1099–1104.

Vermeij, G.J. (1999). Inequality and the directionality of history. *American Naturalist* **153**, 243–253.

Via, S. (2001). Sympatric speciation in animals. *Trends in Ecology and Evolution* **16**, 381–390.

Vickery, R.K. (1978). Case studies in the evolution of species complexes in *Mimulus*. *Evolutionary Biology* **11**, 405–507.

Vigilant, L., Stoneking, M., Harpending, H., Hawkes, K. & Wilson, A.C. (1991). African populations and the evolution of human mitochondrial DNA. *Science* **25**, 1503–1507.

Vision, T.J., Brown, D.G. & Tanksley, S.D. (2000). The origins of genomic duplications in *Arabidopsis*. *Science* **290**, 2114–2116.

Vrba, E.S. (1993). Turnover-pulses, the Red Queen, and related topics. *American Journal of Science* **293A**, 418–452.

Wade, M.J. (1972). *Palaeontology* **15**.

Wade, M.J. (1976). Group selection among laboratory populations of *Tribolium*. *Proceedings of the National Academy of Sciences USA* **73**, 4604–4607.

Wade, M.J., Patterson, H., Chang, N. & Johnson, N.A. (1993). Postcopulatory, prezygotic isolation in flour beetles. *Heredity* **71**, 163–167.

Wade, M.J., Wintherm R.G., Agrawal, A.F. & Goodnight, C.J. (2001). Alternative definitions of epistasis: dependence and interaction. *Trends in Ecology and Evolution* **16**, 498–504.

Wagner, G.P. (ed.) (2000). *The Character Concept in Evolutionary Biology*. Academic Press, San Diego, CA.

Wagner, W.L. & Funk, V.A. (eds) (1995). *Patterns of Speciation and Biogeography of Hawaiian Biota*. Smithsonian Press, Washington.

Wake, D.B., Yanev, K.P. & Brown, C.W. (1986). Intraspecific sympatry in allozymes in a "ring species," the plethodontid salamander *Ensatina eschscholtzii*, in southern California. *Evolution* **40**, 866–868.

Wang, R-L., Stec, A., Hey, J., Lukens, L. & Doebly, J. (1999). The limits of selection during maize domestication. *Nature* **398**, 236–239.

Wang, W., Thornton, K., Berry, A. & Long, M. (2002). Nucleotide variation along *Drosophila melanogaster* fourth chromosome. *Science* **295**, 134–137.

Ward, P.D. (1990). The Cretaceous/Tertiary extinctions in the marine realm; a 1990 perspective. *Geological Society of America Special Papers* **247**, 425–432.

Waser, N.M. (1998). Pollination, angiosperm speciation, and the nature of species boundaries. *Oikos* **82**, 198–201.

Wasserman, M. & Koepfer, H.R. (1977). Character displacement for sexual isolation between *Drosophila mojavensis* and *Drosophila arizonensis*. *Evolution* **31**, 812–823.

Wasserthal, L.T. (1997). The pollinators of the Malagasy star orchids *Angraecum sesquipedale*, *A. sororium*, and *A. compactum* and the evolution of extremely long spurs by pollinator shifts. *Botanica Acta* **110**, 343–359.

Weaver, R.F. & Hedrick, P.W. (1997). *Genetics*, 3rd edn. Wm. C. Brown, Dubuque, IA.

Weiner, J. (1994). *The Beak of the Finch*. Knopf, New York and Cape, London.

Welch, A.M., Semlitsch, R.D. & Gerhardt, H.C. (1998). Call duration as an indicator of genetic quality in male gray tree frogs. *Science* **280**, 1928–1930.

Wellnhofer, P. (1990). *Archaeopteryx*. *Scientific American* **262** (May), 70–77.

Werren, J.H. (1997). Biology of *Wolbachia*. *Annual Review of Entomology* **42**, 587–609.

West, S.A., Herre, E.A. & Sheldon, B.C. (2000). The benefits of allocating sex. *Science* **290**, 288–290.

Westoll, T.S. (1949). On the evolution of the Dipnoi. In Jepsen, G.L., Mayr, E. & Simpson, G.G. (eds) *Genetics, Paleontology, and Evolution*, pp. 121–184. Princeton University Press, Princeton, NJ.

Whelan, S., Liò, P. & Goldman, N. (2001). Molecular phylogenetics: state of the art methods for looking into the past. *Trends in Genetics* **17**, 262–272.

White, M.J.D. (1973). *Animal Cytology and Evolution*, 3rd edn. Cambridge University Press, Cambridge, UK.

Whitham, T.G. & Slobodchikoff, C.N. (1981). Evolution by individuals, plant–herbivore interactions, and mosaics of genetic variability: the adaptive significance of somatic mutations in plants. *Oecologia* **49**, 287–292.

Wiley, E.O. (1988). Vicariance biogeography. *Annual Review of Ecology and Systematics* **19**, 513–542.

Wiley, E.O., Siegel-Causey, D., Brooks, D.R. & Funk, V.A. (1991). *The Compleat Cladist*. Museum of Natural History, University of Kansas, Lawrence, KS.

Wilf, P. & Labandeira, C.C. (1999). Response of plant–insect associations to Paleocene–Eocene warming. *Science* **284**, 2153–2156.

Wilkins, A.S. (2001). *The Evolution of Developmental Pathways*. Sinauer, Sunderland, MA.

Wilkinson, G.S. (1993). Artificial sexual selection alters allometry in the stalk-eyed fly *Cyrtodiopsis dalmanni* (Diptera: Diopsidae). *Genetical Research* **62**, 213–222.

Wilkinson, G.S., Presgraves, D.C. & Crymes, L. (1998). Male eye span in stalk-eyed flies indicates genetic quality by meiotic drive suppression. *Nature* **391**, 276–279.

Williams, G.C. (1966). *Adaptation and Natural Selection*. Princeton University Press, Princeton, NJ.

Williams, G.C. (1975). *Sex and Evolution*. Princeton University Press, Princeton, NJ.

Williams, G.C. (1992). *Natural Selection: domains, levels, and challenges*. Oxford University Press, New York.

Willis, K.J. & McElwain, J.C. (2002). *The Evolution of Plants*. Oxford University Press, Oxford, UK.

Willis, K.J. & Whittaker, R.J. (2000). The refugial debate. *Science* **287**, 1406–1407.

Wills, C. & Bada, J. (2000). *The Spark of Life*. Perseus Books, Cambridge, MA and Oxford University Press, Oxford, UK.

Wilson, A.C. (1985). The molecular basis of evolution. *Scientific American* **253** (October), 164–173.

Wilson, A.C., Carlson, S.S. & White, T.J. (1977). Biochemical evolution. *Annual Review of Biochemistry* **46**, 573–639.

Wilson, E.O. (1992). *The Diversity of Life*. Harvard University Press, Cambridge, MA.

Winsor, M.P. (2003). Non-essentialist methods in pre-Darwinian taxonomy. *Biology and Philosophy* **18**, 1–14

Wolf, J., Brodie, B. & Wade, M.J. (eds) (2000). *Epistasis and the Evolutionary Process*. Oxford University Press, New York.

Wolpert, L. (2002). *Principles of Development*, 2nd edn. Oxford University Press, Oxford, UK.

Woolfenden, G.E. & Fitzpatrick, J.W. (1990). Florida scrub jays: a synopsis after 18 years of study. In Stacey, P.B. & Koenig, W.D. (eds) *Cooperative Breeding in Birds*, pp. 240–266. Cambridge University Press, New York.

Wootton, J.C., Fang, X., Ferdig, M.T. *et al.* (2002). Genetic diversity and chloroquine selective sweeps in *Plasmodium falciparum*. *Nature* **418**, 320–323.

Wray, G.A., Levinton, J.S. & Shapiro, L.H. (1996). Molecular evidence for deep Precambrian divergences among Metazoan taxa. *Science* **274**, 568–573. (Also included in editorially simplified form in Ridley 1997.)

Wright, S. (1931). Evolution in Mendelian populations. *Genetics* **16**, 97–159.

Wright, S. (1932). The roles of mutation, inbreeding, crossbreeding, and selection in evolution. In *Proceedings of the VI International Congress of Genetics*, vol. 1, pp. 356–366. (Reprinted in Ridley 1997.)

Wright, S. (1968). *Evolution and Genetics of Populations*, vol. 1. University of Chicago Press, Chicago, IL.

Wright, S. (1969). *Evolution and Genetics of Populations*, vol. 2. University of Chicago Press, Chicago, IL.

Wright, S. (1977). *Evolution and Genetics of Populations*, vol. 3. University of Chicago Press, Chicago, IL.

Wright, S. (1978). *Evolution and Genetics of Populations*, vol. 4. University of Chicago Press, Chicago, IL.

Wright, S. (1986). *Evolution: selected papers.* (Provine, W.B., ed.). University of Chicago Press, Chicago, IL.

Wyckoff, G.J., Wang, W. & Wu, C-I. (2000). Rapid evolution of male reproductive genes in the descent of man. *Nature* **403**, 304–309.

Wynne-Edwards, V.C. (1962). *Animal Dispersion in Relation to Social Behaviour.* Oliver & Boyd, Edinburgh, UK.

Xiao, S., Zhang, Y. & Knoll, A.H. (1998). Three-dimensional preservation of algae and animal embryos in a Neoproterozoic phosphorite. *Nature* **391**, 553–558.

Zahavi, A. (1975). Mate selection — a selection for a handicap. *Journal of Theoretical Biology* **53**, 205–214.

Zanis, M.J., Soltis, D.E., Soltis, P.S., Mathews, S. & Donoghue, M.J. (2002). The root of the angiosperms revisited. *Proceedings of the National Academy of Sciences USA* **99**, 6848–6853.

Zimmer, C. (1998). *At the Water's Edge.* Free Press, New York.

Zimmer, C. (2001). *Evolution: the triumph of an idea.* Harpercollins, New York.

Index